Study and Solutions Guide for

PRECALCULUS FUNCTIONS AND GRAPHS: A GRAPHING APPROACH

and

PRECALCULUS WITH LIMITS: A GRAPHING APPROACH
SECOND EDITION

Larson/Hostetler/Edwards

Bruce H. Edwards
University of Florida
Gainesville, Florida

Dianna L. Zook
Indiana University
Purdue University at Fort Wayne, Indiana

HOUGHTON MIFFLIN COMPANY   Boston   New York

Sponsoring Editor: Christine B. Hoag
Senior Associate Editor: Maureen Brooks
Managing Editor: Catherine B. Cantin
Assistant Editor: Carolyn Johnson
Supervising Editor: Karen Carter
Associate Project Editor: Rachel D'Angelo Wimberly
Editorial Assistant: Caroline Lipscomb
Production Supervisor: Lisa Merrill
Art Supervisor: Gary Crespo
Marketing Manager: Ros Kane
Marketing Assistant: Kate Burden Thomas

Printed in the United States of America.

International Standard Book Number: 0–669–41729–7

3456789 - POO 01 00 99 98

TO THE STUDENT

The *Study and Solutions Guide* for *Precalculus Functions and Graphs: A Graphing Approach*, Second Edition, and *Precalculus with Limits: A Graphing Approach*, Second Edition, is a supplement to the text by Roland E. Larson, Robert P. Hostetler, and Bruce H. Edwards.

As mathematics instructors, we often have students come to us with questions about the assigned homework. When we ask to see their work, the reply often is "I didn't know where to start." The purpose of the *Study Guide* is to provide brief summaries of the topics covered in the text book and enough detailed solutions to problems so that you will be able to work the remaining exercises.

This *Study Guide* is the result of the efforts of Larson Texts, Inc.

If you have any corrections or suggestions for improving this *Study Guide*, we would appreciate hearing from you.

Good luck with your study of precalculus.

Bruce H. Edwards
358 Little Hall
University of Florida
Gainesville, FL 32611
(be@math.ufl.edu)

Dianna L. Zook
Indiana University
Purdue University
at Fort Wayne, Indiana 46805

CONTENTS

STUDY STRATEGIES

- Attend all classes and come prepared. Have your homework completed. Bring the text, paper, pen or pencil, and a calculator (scientific or graphing) to each class.

- Read the section in the text that is to be covered before class. Make notes about any questions that you have and, if they are not answered during the lecture, ask them at the appropriate time.

- Participate in class. As mentioned above, ask questions. Also, do not be afraid to answer questions.

- Take notes on all definitions, concepts, rules, formulas and examples. After class, read your notes and fill in any gaps, or make notations of any questions that you have.

- DO THE HOMEWORK!!! You learn mathematics by doing it yourself. Allow at least two hours outside of each class for homework. Do not fall behind.

- Seek help when needed. Visit your instructor during office hours and come prepared with specific questions; check with your school's tutoring service; find a study partner in class; check additional books in the library for more examples—just do something before the problem becomes insurmountable.

- Do not cram for exams. Each chapter in the text contains a chapter review and this study guide contains a practice test at the end of each chapter. (The answers are at the back of the study guide.) Work these problems a few days before the exam and review any areas of weakness.

CHAPTER P
Prerequisites

CHAPTER P
Prerequisites

Section P.1   The Cartesian Plane

- ■ You should be able to plot points.
- ■ You should know that the distance between (x_1, y_1) and (x_2, y_2) in the plane is

$$d = \sqrt{(x_2 - x_1)^2 + (y_2 - y_1)^2}.$$

- ■ You should know that the midpoint of the line segment joining (x_1, y_1) and (x_2, y_2) is

$$\left(\frac{x_1 + x_2}{2}, \frac{y_1 + y_2}{2} \right).$$

- ■ You should know the equation of a circle: $(x - h)^2 + (y - k)^2 = r^2$.

Solutions to Odd-Numbered Exercises

1.

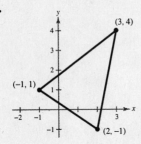

3.

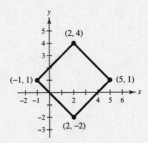

5. $A: (2, 6)$, $B: (-6, -2)$, $C: (4, -4)$, $D: (-3, 2)$

7. $A:(0, 5)$, $B:(-3, -6)$, $C:(1, -4.5)$, $D:(-4, 2)$

9. $(-3, 4)$

11. $(-5, -5)$

13. On the x-axis, $y = 0$.
On the y-axis, $x = 0$.

15. $x > 0 \implies x$ lies in Quadrant I or in Quadrant IV.

$y < 0 \implies y$ lies in Quadrant III or in Quadrant IV.

$x > 0$ and $y < 0 \implies (x, y)$ lies in Quadrant IV.

17. $x = -4 \implies x$ is negative $\implies x$ lies in Quadrant II or Quadrant III.

$y > 0 \implies y$ lies in Quadrant I or Quadrant II.

$x = -4$ and $y > 0 \implies (x, y)$ lies in Quadrant II.

19. $y < -5 \implies y$ is negative $\implies y$ lies in either Quadrant III or Quadrant IV.

21. Since $(x, -y)$ is in Quadrant II, we know that $x < 0$ and $-y > 0$. If $-y > 0$, then $y < 0$.

$x < 0 \implies x$ lies in Quadrant II or in Quadrant III.

$y < 0 \implies y$ lies in Quadrant III or in Quadrant IV.

$x < 0$ and $y < 0 \implies (x, y)$ lies in Quadrant III.

23. If $xy > 0$, then either x and y are both positive, or both negative. Hence, (x, y) lies in either Quadrant I or Quadrant III.

25.

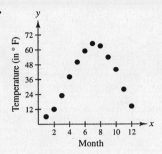

27. The x-coordinates are increased by 2, and the y-coordinates are increased by 5: $(0, 1)$, $(4, 2)$, $(1, 4)$.

29. $y = 2 - \frac{1}{2}x$

x	-2	-1	$-\frac{1}{2}$	0	$\frac{1}{2}$	1	2
y	3	$\frac{5}{2}$	$\frac{9}{4}$	2	$\frac{7}{4}$	$\frac{3}{2}$	1

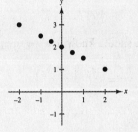

31. $\left(\dfrac{1200 - 60}{60}\right)(100) = 1900\%$

33. The highest price of milk is approximately \$13.70 per 100 lb. This occurred in 1990.

35. The minimum wage increased most rapidly in the 1970s.

37. The point $(65, 83)$ represents an entrance exam score of 65.

39. (a) The distance between $(0, 2)$ and $(4, 2)$ is 4.
    The distance between $(4, 2)$ and $(4, 5)$ is 3.
    The distance between $(0, 2)$ and $(4, 5)$ is
$$\sqrt{(4 - 0)^2 + (5 - 2)^2} = \sqrt{16 + 9} = \sqrt{25} = 5.$$

    (b) $4^2 + 3^2 = 16 + 9 = 25 = 5^2$

41. (a) The distance between $(-1, 1)$ and $(9, 1)$ is 10.
    The distance between $(9, 1)$ and $(9, 4)$ is 3.
    The distance between $(-1, 1)$ and $(9, 4)$ is
$$\sqrt{(9 - (-1))^2 + (4 - 1)^2} = \sqrt{100 + 9} = \sqrt{109}.$$

    (b) $10^2 + 3^2 = 109 = \left(\sqrt{109}\right)^2$

43. $d = |5 - (-3)| = 8$

45. $d = |2 - (-3)| = 5$

47. (a)

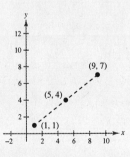

    (b) $d = \sqrt{(9 - 1)^2 + (7 - 1)^2}$
$$= \sqrt{64 + 36} = 10$$

    (c) $\left(\dfrac{9 + 1}{2}, \dfrac{7 + 1}{2}\right) = (5, 4)$

49. (a)

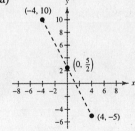

    (b) $d = \sqrt{(4 + 4)^2 + (-5 - 10)^2}$
$$= \sqrt{64 + 225} = 17$$

    (c) $\left(\dfrac{4 - 4}{2}, \dfrac{-5 + 10}{2}\right) = \left(0, \dfrac{5}{2}\right)$

51. (a)

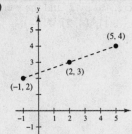

(b) $d = \sqrt{(5 + 1)^2 + (4 - 2)^2}$

   $= \sqrt{36 + 4} = 2\sqrt{10}$

(c) $\left(\dfrac{-1 + 5}{2}, \dfrac{2 + 4}{2}\right) = (2, 3)$

53. (a)

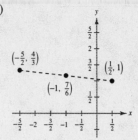

(b) $d = \sqrt{\left(\dfrac{1}{2} + \dfrac{5}{2}\right)^2 + \left(1 - \dfrac{4}{3}\right)^2}$

   $d = \sqrt{9 + \dfrac{1}{9}} = \dfrac{\sqrt{82}}{3}$

(c) $\left(\dfrac{-\frac{5}{2} + \frac{1}{2}}{2}, \dfrac{\frac{4}{3} + 1}{2}\right) = \left(-1, \dfrac{7}{6}\right)$

55. (a)

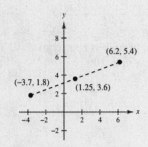

(b) $d = \sqrt{(6.2 + 3.7)^2 + (5.4 - 1.8)^2}$

   $= \sqrt{98.01 + 12.96}$

   $= \sqrt{110.97}$

(c) $\left(\dfrac{6.2 - 3.7}{2}, \dfrac{5.4 + 1.8}{2}\right) = (1.25, 3.6)$

57. (a)

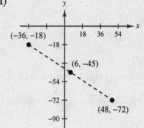

(b) $d = \sqrt{(48 + 36)^2 + (-72 + 18)^2}$

   $= \sqrt{7056 + 2916}$

   $= \sqrt{9972} = 6\sqrt{277}$

(c) $\left(\dfrac{-36 + 48}{2}, \dfrac{-18 - 72}{2}\right) = (6, -45)$

59. $\left(\dfrac{1991 + 1995}{2}, \dfrac{\$520,000 + \$740,000}{2}\right) = (1993, \$630,000)$

61. $d_1 = \sqrt{(4 - 2)^2 + (0 - 1)^2} = \sqrt{5}$

   $d_2 = \sqrt{(4 + 1)^2 + (0 + 5)^2} = \sqrt{50}$

   $d_3 = \sqrt{(2 + 1)^2 + (1 + 5)^2} = \sqrt{45}$

   $\left(\sqrt{5}\right)^2 + \left(\sqrt{45}\right)^2 = \left(\sqrt{50}\right)^2$

63. $d_1 = \sqrt{(0 - 1)^2 + (0 - 2)^2} = \sqrt{5}$

   $d_2 = \sqrt{(0 - 2)^2 + (0 - 1)^2} = \sqrt{5}$

   $d_3 = \sqrt{(3 - 1)^2 + (3 - 2)^2} = \sqrt{5}$

   $d_4 = \sqrt{(3 - 2)^2 + (3 - 1)^2} = \sqrt{5}$

   $d_1 = d_2 = d_3 = d_4$

65. $d_1 = \sqrt{(0 - 2)^2 + (9 - 5)^2} = \sqrt{4 + 16} = \sqrt{20} = 2\sqrt{5}$

   $d_2 = \sqrt{(-2 - 0)^2 + (0 - 9)^2} = \sqrt{4 + 81} = \sqrt{85}$

   $d_3 = \sqrt{(0 - (-2))^2 + (-4 - 0)^2} = \sqrt{4 + 16} = \sqrt{20} = 2\sqrt{5}$

   $d_4 = \sqrt{(0 - 2)^2 + (-4 - 5)^2} = \sqrt{4 + 81} = \sqrt{85}$

Opposite sides have equal lengths of $2\sqrt{5}$ and $\sqrt{85}$.

67. $(x - 0)^2 + (y - 0)^2 = 3^2$

   $x^2 + y^2 = 9$

69. $(x - 2)^2 + (y + 1)^2 = 4^2$

   $(x - 2)^2 + (y + 1)^2 = 16$

71. $(x + 1)^2 + (y - 2)^2 = r^2$

$(0 + 1)^2 + (0 - 2)^2 = r^2 \implies r^2 = 5$

$(x + 1)^2 + (y - 2)^2 = 5$

73. $r = \dfrac{1}{2}\sqrt{(6 - 0)^2 + (8 - 0)^2} = \dfrac{1}{2}\sqrt{100} = 5$

$\text{Center} = \left(\dfrac{0 + 6}{2}, \dfrac{0 + 8}{2}\right) = (3, 4)$

$(x - 3)^2 + (y - 4)^2 = 25$

75. Center: $(0, 0)$

Radius $= 2$

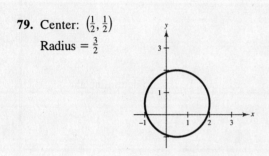

77. Center: $(1, -3)$

Radius $= 2$

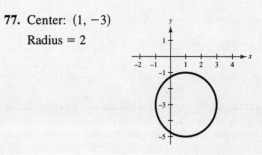

79. Center: $\left(\frac{1}{2}, \frac{1}{2}\right)$

Radius $= \frac{3}{2}$

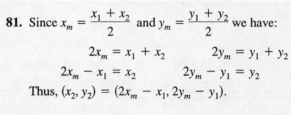

81. Since $x_m = \dfrac{x_1 + x_2}{2}$ and $y_m = \dfrac{y_1 + y_2}{2}$ we have:

$$2x_m = x_1 + x_2 \qquad 2y_m = y_1 + y_2$$
$$2x_m - x_1 = x_2 \qquad 2y_m - y_1 = y_2$$

Thus, $(x_2, y_2) = (2x_m - x_1, 2y_m - y_1)$.

83. The midpoint of the given line segment is $\left(\dfrac{x_1 + x_2}{2}, \dfrac{y_1 + y_2}{2}\right)$.

The midpoint between (x_1, y_1) and $\left(\dfrac{x_1 + x_2}{2}, \dfrac{y_1 + y_2}{2}\right)$ is

$$\left(\dfrac{x_1 + \dfrac{x_1 + x_2}{2}}{2}, \dfrac{y_1 + \dfrac{y_1 + y_2}{2}}{2}\right) = \left(\dfrac{3x_1 + x_2}{4}, \dfrac{3y_1 + y_2}{4}\right).$$

The midpoint between $\left(\dfrac{x_1 + x_2}{2}, \dfrac{y_1 + y_2}{2}\right)$ and (x_2, y_2) is

$$\left(\dfrac{\dfrac{x_1 + x_2}{2} + x_2}{2}, \dfrac{\dfrac{y_1 + y_2}{2} + y_2}{2}\right) = \left(\dfrac{x_1 + 3x_2}{4}, \dfrac{y_1 + 3y_2}{4}\right).$$

Thus, the three points are

$$\left(\dfrac{3x_1 + x_2}{4}, \dfrac{3y_1 + y_2}{4}\right), \left(\dfrac{x_1 + x_2}{2}, \dfrac{y_1 + y_2}{2}\right), \text{ and } \left(\dfrac{x_1 + 3x_2}{4}, \dfrac{y_1 + 3y_2}{4}\right).$$

85. $d = \sqrt{(45 - 10)^2 + (40 - 15)^2} = \sqrt{35^2 + 25^2} = \sqrt{1850}$

$\qquad = 5\sqrt{74} \approx 43$ yards

87.

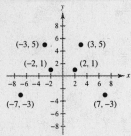

The points are reflected through the *y*-axis.

89. (a) It appears that the number of artists elected alternates between 7 and 8 per year in the 1990s. If this pattern continues, 7 or 8 would be elected in 1996.

(b) Since 1986 and 1987 were the first two years that artists were elected, there was a larger number of artists chosen

Section P.2   Graphs and Graphing Utilities

- ■ You should be able to use the point-plotting method of graphing.
- ■ You should be able to find *x*- and *y*-intercepts.
  - (a) To find the *x*-intercepts, let $y = 0$ and solve for *x*.
  - (b) To find the *y*-intercepts, let $x = 0$ and solve for *y*.
- ■ You should know how to graph an equation with a graphing utility. You should be able to determine an appropriate viewing rectangle.
- ■ You should be able to use the zoom and trace features of a graphing utility.

Solutions to Odd-Numbered Exercises

1. $y = \sqrt{x + 4}$

(a) $(0, 2)$: $2 \overset{?}{=} \sqrt{0 + 4}$

$\qquad 2 = 2$ ✓

Yes, the point *is* on the graph

(b) $(5, 3)$: $3 \overset{?}{=} \sqrt{5 + 4}$

$\qquad 3 = \sqrt{9}$ ✓

Yes, the point *is* on the graph.

3. $y = 4 - |x - 2|$

(a) $(1, 5)$: $5 \overset{?}{=} 4 - |1 - 2|$

$\qquad 5 \neq 4 - 1$

No, the point *is not* on the graph.

(b) $(6, 0)$: $0 \overset{?}{=} 4 - |6 - 2|$

$\qquad 0 = 4 - 4$ ✓

Yes, the point *is* on the graph.

5. $2x - y - 3 = 0$

(a) $(1, 2)$: $2(1) - (2) - 3 \overset{?}{=} 0$

$\qquad\qquad -3 \neq 0$

No, the point *is not* on the graph.

(b) $(1, -1)$: $2(1) - (-1) - 3 \overset{?}{=} 0$

$\qquad\qquad 2 + 1 - 3 = 0$ ✓

Yes, the point *is* on the graph.

7. $x^2 y - x^2 + 4y = 0$

(a) $\left(1, \frac{1}{5}\right)$: $(1)^2\left(\frac{1}{5}\right) - (1)^2 + 4\left(\frac{1}{5}\right) \overset{?}{=} 0$

$\qquad\qquad \frac{1}{5} - 1 + \frac{4}{5} = 0$ ✓

Yes, the point *is* on the graph

(b) $\left(2, \frac{1}{2}\right)$: $(2)^2\left(\frac{1}{2}\right) - (2)^2 + 4\left(\frac{1}{2}\right) \overset{?}{=} 0$

$\qquad\qquad 2 - 4 + 2 = 0$ ✓

Yes, the point *is* on the graph.

9. $y = -2x + 3$

x	-1	0	1	$\frac{3}{2}$	2
y	5	3	1	0	1

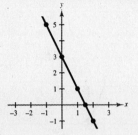

13. $y = \frac{1}{4}x - 3$

x	-2	-1	0	1	2
y	$-\frac{7}{2}$	$-\frac{13}{4}$	-3	$-\frac{11}{4}$	$-\frac{5}{2}$

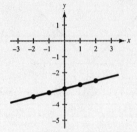

17. $y = -\frac{1}{4}x - 3$

x	-2	-1	0	1	2
y	$-\frac{5}{2}$	$-\frac{11}{4}$	-3	$-\frac{13}{4}$	$-\frac{7}{2}$

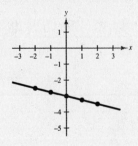
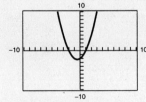

11. $y = x^2 - 2x$

x	-1	0	1	2	3
y	3	0	-1	0	3

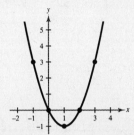

15. $y = \sqrt{x - 1}$

x	1	2	5	10	17
y	0	1	2	3	4

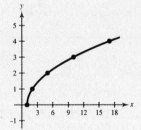

19.

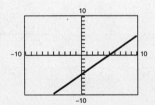

$y = x - 5$

Intercepts: $(5, 0), (0, -5)$

21.

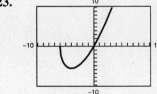

$y = x^2 + x - 2$

Intercepts:
$(1, 0), (-2, 0), (0, -2)$

23.

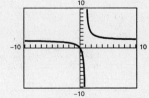

$y = x\sqrt{x + 6}$

Intercepts: $(0, 0), (-6, 0)$

25.

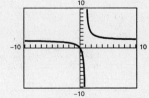

$y = \dfrac{2x}{x - 1}$

Intercept: $(0, 0)$

27. $y = 1 - x$ has intercepts $(1, 0)$ and $(0, 1)$. Matches graph (d).

29. $y = \sqrt{9 - x^2}$ has intercepts $(\pm 3, 0)$ and $(0, 3)$. Matches graph (f).

31. $y = x^3 - x + 1$ has a y-intercept of $(0, 1)$ and the points $(1, 1)$ and $(-2, -5)$ are on the graph. Matches (a).

33. $y = -3x + 2$

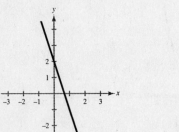

35. $y = 1 - x^2$

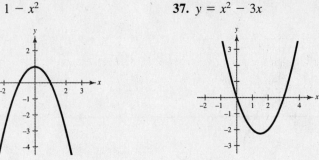

37. $y = x^2 - 3x$

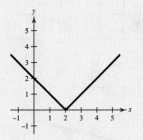

39. $y = x^3 + 2$

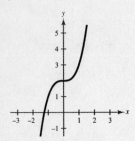

41. $y = \sqrt{x - 3}$

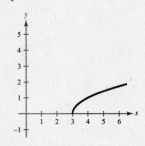

43. $y = |x - 2|$

45. $x = y^2 - 1$

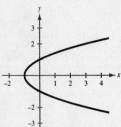

47. $y = 3 - \frac{1}{2}x$

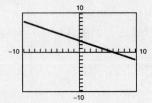

Intercepts: $(6, 0)$, $(0, 3)$

49. $y = x^2 - 4x + 3$

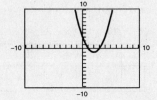

Intercepts: $(3, 0)$, $(1, 0)$, $(0, 3)$

51. $y = x(x - 2)^2$

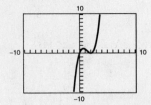

Intercepts: $(0, 0)$, $(2, 0)$

53. $y = \sqrt[3]{x}$

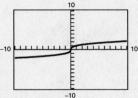

Intercepts: $(0, 0)$

55. $y = \frac{5}{2}x + 5$

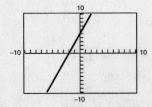

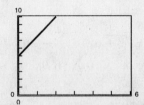

Both settings show the line and its intercept. The first setting is better.

57. $-x^2 + 10x - 5$

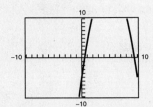 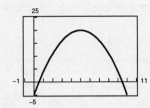

Both settings show the line and its intercept. The second setting is better.

59. $y = 4x^2 - 25$

Range/Window

Xmin = -5
Xmax = 5
Xscl = 1
Ymin = -30
Ymax = 10
Yscl = 5

61. $y = |x| + |x - 10|$

Range/Window

Xmin = -30
Xmax = 30
Xscl = 5
Ymin = -10
Ymax = 50
Yscl = 5

63. $y = 0.25x - 50$

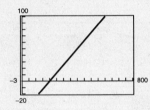 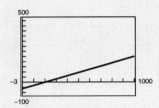

The first graph shows a sharper increase.

65. $x^2 + y^2 = 64$
$$y^2 = 64 - x^2$$
$$y = \pm\sqrt{64 - x^2}$$
Use: $y_1 = \sqrt{64 - x^2}$
$$y_2 = -\sqrt{64 - x^2}$$

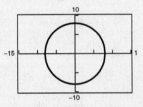

67. $x^2 + y^2 = 49$
$$y^2 = 49 - x^2$$
$$y = \pm\sqrt{49 - x^2}$$
Use: $y_1 = \sqrt{49 - x^2}$
$$y_2 = -\sqrt{49 - x^2}$$

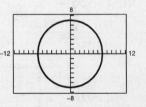

69. $y_1 = \frac{1}{4}(x^2 - 8)$
$$y_2 = \frac{1}{4}x^2 - 2$$

The graphs are identical.
The Distributive Property is illustrated.

71. $y_1 = \frac{1}{5}[10(x^2 - 1)]$
$$y_2 = 2(x^2 - 1)$$

The graphs are identical.
The Associative Property of Multiplication is illustrated.

73. $y = 225,000 - 20,000t, \ 0 \le t \le 8$

(a) Range/Window

Xmin = -1
Xmax = 9
Xscl = 1
Ymin = -60000
Ymax = 230000
Yscl = 10000

(b)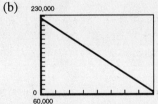

75. Most likely you would need to change the viewing window. For example, let $y_1 = x^2 + 12$. This graph would not show up on the standard window. Change the range/window to the following setting and try again.

Xmin = -5
Xmax = 5
Xscl = 1
Ymin = -5
Ymax = 40
Yscl = 5

77. (a)

Year	1920	1930	1940	1950	1960	1970	1980	1990
Life Expectancy	54.1	59.7	62.9	68.2	69.7	70.8	73.7	75.4
Model	52.8	58.7	63.3	66.9	69.9	72.4	74.6	76.4

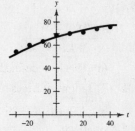

(b) When $t = 48$, $y \approx 77.7$ years.

(c) When $t = 50$, $y \approx 78.0$ years.

79. (a) $y = 0.086t + 0.872$, $0 \le t \le 4$

Year	1990	1991	1992	1993	1994
t	0	1	2	3	4
y	0.872	0.958	1.044	1.130	1.216

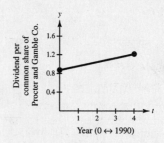

81. $y = \sqrt{5 - x}$

(a) $(2, y) \approx (2, 1.73)$

(b) $(x, 3) = (-4, 3)$

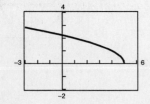

83. $y = x^5 - 5x$

(a) $(-0.5, y) \approx (-0.5, 2.47)$

(b) $(x, -4) = (1, -4)$ or $(x, -4) \approx (-1.65, -4)$

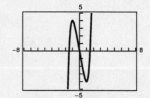

Section P.3   Lines in the Plane

You should know the following important facts about lines.

- The graph of $y = mx + b$ is a straight line. It is called a linear equation.
- The slope of the line through (x_1, y_1) and (x_2, y_2) is

    $$m = \frac{y_2 - y_1}{x_2 - x_1}.$$

- (a) If $m > 0$, the line rises from left to right.

    (b) If $m = 0$, the line is horizontal.

    (c) If $m < 0$, the line falls from left to right.

    (d) If m is undefined, the line is vertical.

- Equations of Lines

    (a) Slope-Intercept: $y = mx + b$

    (b) Point-Slope: $y - y_1 = m(x - x_1)$

    (c) Two-Point: $y - y_1 = \dfrac{y_2 - y_1}{x_2 - x_1}(x - x_1)$

    (d) General: $Ax + By + C = 0$

    (e) Vertical: $x = a$

    (f) Horizontal: $y = b$

- Given two distinct nonvertical lines

    $$L_1: y = m_1 x + b_1 \quad \text{and} \quad L_2: y = m_2 x + b_2$$

    (a) L_1 is parallel to L_2 if and only if $m_1 = m_2$ and $b_1 \neq b_2$.

    (b) L_1 is perpendicular to L_2 if and only if $m_1 = -1/m_2$.

Solutions to Odd-Numbered Exercises

1. (a) $m = \frac{2}{3}$. Since the slope is positive, the line rises. Matches L_2.

(b) m is undefined. The line is vertical. Matches L_3.

(c) $m = -2$. The line falls. Matches L_1.

3. Slope $= \dfrac{\text{rise}}{\text{run}} = \dfrac{8}{5}$

5. Slope $= \dfrac{\text{rise}}{\text{run}} = \dfrac{0}{1} = 0$

7. Slope $= \dfrac{\text{rise}}{\text{run}} = \dfrac{-8}{2} = -4$

9.

11.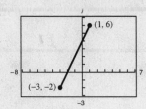

13.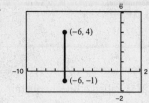

15.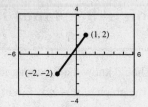

Slope $= \dfrac{6+2}{1+3} = 2$

Slope is undefined.

Slope $= \dfrac{2+2}{1+2} = \dfrac{4}{3}$

17. Since $m = 0$, y does not change. Three points are $(0, 1)$, $(3, 1)$, and $(-1, 1)$.

19. Since $m = 1$, y increases by 1 for every one unit increase in x. Three points are $(6, -5)$, $(7, -4)$, and $(8, -3)$.

21. Since m is undefined, x does not change. Three points are $(-8, 0)$, $(-8, 2)$, and $(-8, 3)$.

23. $m_{L_1} = \dfrac{9+1}{5-0} = 2$

$m_{L_2} = \dfrac{1-3}{4-0} = -\dfrac{1}{2} = -\dfrac{1}{m_{L_1}}$

L_1 and L_2 are perpendicular.

25. $m_{L_1} = \dfrac{0-6}{-6-3} = \dfrac{2}{3}$

$m_{L_2} = \dfrac{\frac{7}{3}+1}{5-0} = \dfrac{2}{3} = m_{L_1}$

L_1 and L_2 are parallel.

27. Yes, any pair of points on a line can be used to calculate the slope of the line. The rate of change remains the same on a line.

29. (a) $m = 135$. The sales are increasing 135 units per year.

(b) $m = 0$. There is no change is sales.

(c) $m = -40$. The sales are decreasing 40 units per year.

31. (a)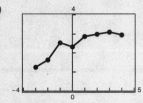

(b) Decreased most rapidly in 1989
Increased most rapidly in 1988.·

33. $y = 0.5x - 3$

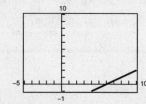

The second setting shows the x and y intercepts more clearly.

35. $5x - y + 3 = 0$

$\qquad y = 5x + 3$

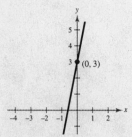

Slope: $m = 5$

y-intercept: $(0, 3)$

37. $5x - 2 = 0$

$\qquad x = \frac{2}{5}$

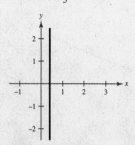

Slope: undefined

No y-intercept

39. $7x + 6y - 30 = 0$

$\qquad y = -\frac{7}{6}x + 5$

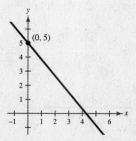

Slope: $m = -\frac{7}{6}$

y-intercept: $(0, 5)$

41. $y + 1 = \dfrac{5 + 1}{-5 - 5}(x - 5)$

$\qquad y = -\dfrac{3}{5}(x - 5) - 1$

$\qquad y = -\dfrac{3}{5}x + 2 \implies 3x + 5y - 10 = 0$

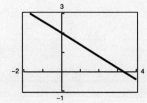

43. $y - \dfrac{1}{2} = \dfrac{\frac{5}{4} - \frac{1}{2}}{\frac{1}{2} - 2}(x - 2)$

$\qquad y = -\dfrac{1}{2}(x - 2) + \dfrac{1}{2}$

$\qquad y = -\dfrac{1}{2}x + \dfrac{3}{2} \implies x + 2y - 3 = 0$

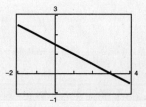

45. Since both points have $x = -8$, the slope is undefined.

$\quad x = -8 \implies x + 8 = 0$

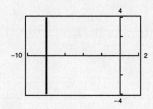

47. $y - 0.6 = \dfrac{-0.6 - 0.6}{-2 - 1}(x - 1)$

$\qquad y = 0.4(x - 1) + 0.6$

$\qquad y = 0.4x + 0.2 \implies 2x - 5y + 1 = 0$

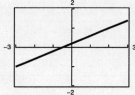

49. Slope $= \dfrac{\text{rise}}{\text{run}}$

$\quad -\dfrac{12}{100} = -\dfrac{2000}{y}$

$\quad -12y = -200,000$

$\qquad y = 16,666\frac{2}{3}$ feet ≈ 3.16 miles

51. $y + 2 = 3(x - 0)$

$\qquad y = 3x - 2 \implies 3x - y - 2 = 0$

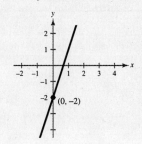

53. $y - 6 = -2(x + 3)$

$\qquad y = -2x \implies 2x + y = 0$

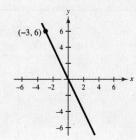

55. $y - 0 = -\frac{1}{3}(x - 4)$

$\qquad y = -\frac{1}{3}x + \frac{4}{3} \implies x + 3y - 4 = 0$

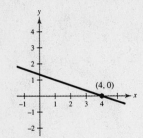

57. $\qquad x = 6$

$\qquad x - 6 = 0$

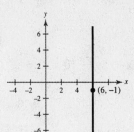

59. $y - \frac{5}{2} = \frac{4}{3}(x - 4) \implies 8x - 6y - 17 = 0$

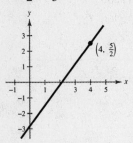

61. $\qquad \frac{x}{5} + \frac{y}{-3} = 1$

$\qquad -3x + 5y + 15 = 0$

$a = 5$ and $b = -3$ are the x- and y-intercepts.

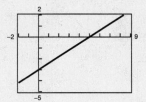

63. $\qquad \frac{x}{2} + \frac{y}{3} = 1$

$\qquad 3x + 2y - 6 = 0$

65. $4x - 2y = 3$

$\qquad y = 2x - \frac{3}{2}$

Slope: $m = 2$

(a) $y - 1 = 2(x - 2)$

$\qquad y = 2x - 3 \implies 2x - y - 3 = 0$

(b) $y - 1 = -\frac{1}{2}(x - 2)$

$\qquad y = -\frac{1}{2}x + 2 \implies x + 2y - 4 = 0$

67. $y = -3$

slope: $m = 0$

(a) $y = 0$

(b) $x = -1 \implies x + 1 = 0$

69. $L_1 : y = 2x \qquad L_2 : y = -2x \qquad L_3 : y = \frac{1}{2}x$

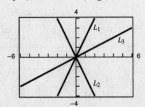

L_2 and L_3 are perpendicular.

71. L_1: $y = -\frac{1}{2}x$ L_2: $y = -\frac{1}{2}x + 3$ L_3: $y = 2x - 4$

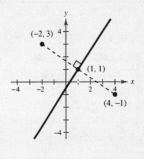

L_1 and L_2 are parallel.

L_3 is perpendicular to L_1 and L_2.

73. $(6, 2540)$, $m = 125$

$V - 2540 = 125(t - 6)$

$V - 2540 = 125t - 750$

$\quad\quad V = 125t + 1790$

75. $(6, 20400)$, $m = -2000$

$V - 20400 = -2000(t - 6)$

$V - 20400 = -2000t + 12000$

$\quad\quad V = -2000t + 32400$

77. The slope is $m = -10$. This represents the decrease in the amount of the loan each week. Matches graph (b).

79. The slope is $m = 0.25$. This represents the increase in travel cost for each mile driven. Matches graph (a).

81. Set the distance between $(4, -1)$ and (x, y) equal to the distance between $(-2, 3)$ and (x, y).

$$\sqrt{(x - 4)^2 + [y - (-1)]^2} = \sqrt{[x - (-2)]^2 + (y - 3)^2}$$

$$(x - 4)^2 + (y + 1)^2 = (x + 2)^2 + (y - 3)^2$$

$$x^2 - 8x + 16 + y^2 + 2y + 1 = x^2 + 4x + 4 + y^2 - 6y + 9$$

$$-8x + 2y + 17 = 4x - 6y + 13$$

$$0 = 12x - 8y - 4$$

$$0 = 4(3x - 2y - 1)$$

$$0 = 3x - 2y - 1$$

This line is the perpendicular bisector of the line segment connecting $(4, -1)$ and $(-2, 3)$.

83. Using the points $(0, 32)$ and $(100, 212)$, we have

$$m = \frac{212 - 32}{100 - 0} = \frac{180}{100} = \frac{9}{5}$$

$$F - 32 = \frac{9}{5}(C - 0)$$

$$F = \frac{9}{5}C + 32.$$

85. Using the points $(1994, 28{,}500)$ and $(1996, 32{,}900)$, we have:

$$m = \frac{32900 - 28500}{1996 - 1994} = \frac{4400}{2} = 2200$$

$$S - 28500 = 2200(t - 1994)$$

$$S = 2200t - 4{,}385{,}300$$

When $t = 2000$ we have $S = 2200(2000) - 4{,}360{,}500$ or $\$41{,}700$.

87. (a) Using the points $(0, 875)$ and $(5, 0)$, where the first coordinate represents the year t and the second coordinate represents the value V, we have

$$m = \frac{0 - 875}{5 - 0} = -175$$

$$V = -175t + 875, \ 0 \le t \le 5.$$

(b)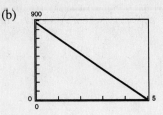

(c)

t	0	1	2	3	4	5
V	875	700	525	350	175	0

89. Sale price = List price − 15% of the list price

$$S = L - 0.15L$$

$$S = 0.85L$$

91. (a) $C = 36{,}500 + 5.25t + 11.50t$

$\quad = 16.75t + 36{,}500$

(b) $R = 27t$

(c) $P = R - C$

$\quad = 27t - (16.75t + 36{,}500)$

$\quad = 10.25t - 36{,}500$

(d) $\quad 0 = 10.25t - 36{,}500$

$\quad 36{,}500 = 10.25t$

$\quad t \approx 3561$ hours

93. (a) Using the regression capabilities of a graphing utility, you obtain $y = 71.08t - 10.29$.

(b)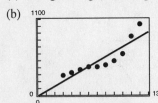

(c) For $t = 20$ (year 2000), $y \approx \$1{,}411{,}000$.

(d) The slope is the average increase per year.

95. (a) $x = 50 - \dfrac{p - 580}{15}$.   Note that when $p = 580$, $x = 50$ and when $p = 625$, $x = 47$.

(b)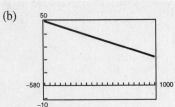

If $p = 655$, $x = 45$.

(c) If $p = 595$, $x = 49$.

Section P.4   Solving Equations Algebraically and Graphically

- You should know how to solve linear equations.
  $ax + b = 0$
- An identity is an equation whose solution consists of every real number in its domain.
- To solve an equation you can:
  - (a) Add or subtract the same quantity from both sides.
  - (b) Multiply or divide both sides by the same nonzero quantity.
- To solve an equation that can be simplified to a linear equation:
  - (a) Remove all symbols of grouping and all fractions.
  - (b) Combine like terms.
  - (c) Solve by algebra.
  - (d) Check the answer.
- A "solution" that does not satisfy the original equation is called an extraneous solution.
- You should be able to solve equations graphically.
- You should be able to solve a quadratic equation by factoring, if possible.
- You should be able to solve a quadratic equation of the form $u^2 = d$ by extracting square roots.
- You should be able to solve a quadratic equation by completing the square.
- You should know and be able to use the Quadratic Formula: For $ax^2 + bx + c = 0, a \neq 0$,

$$x = \frac{-b \pm \sqrt{b^2 - 4ac}}{2a}.$$

- You should be able to solve polynomials of higher degree by factoring.
- For equations involving radicals or fractional powers, raise both sides to the same power.
- For equations with fractions, multiply both sides by the least common denominator to clear the fractions.
- For equations involving absolute value, remember that the expression inside the absolute value can be positive or negative.
- Always check for extraneous solutions.

Solutions to Odd-Numbered Exercises

1. $5x - 3 = 3x + 5$

(a) $5(0) - 3 \stackrel{?}{=} 3(0) + 5$

   $-3 \neq 5$

   $x = 0$ *is not* a solution.

(b) $5(-5) - 3 \stackrel{?}{=} 3(-5) + 5$

   $-28 \neq -10$

   $x = -5$ *is not* a solution.

(c) $5(4) - 3 \stackrel{?}{=} 3(4) + 5$

   $17 = 17$

   $x = 4$ *is* a solution.

(d) $5(10) - 3 \stackrel{?}{=} 3(10) + 5$

   $47 \neq 35$

   $x = 10$ *is not* a solution.

3. $(x + 5)(x - 3) = 20$

    (a) $(3 + 5)(3 - 3) \overset{?}{=} 20$

                $0 \neq 20$

      $x = 3$ *is not* a solution.

    (c) $(0 + 5)(0 - 3) \overset{?}{=} 20$

              $-15 \neq 20$

      $x = 0$ *is not* a solution.

    (b) $(-2 + 5)(-2 - 3) \overset{?}{=} 20$

                 $-15 \neq 20$

      $x = -2$ *is not* a solution.

    (d) $(-7 + 5)(-7 - 3) \overset{?}{=} 20$

                 $20 = 20$

      $x = -7$ *is* a solution.

5. $2(x - 1) = 2x - 2$ is an *identity* by the Distributive Property. It is true for all real values of x.

7. $3 + \dfrac{1}{x + 1} = \dfrac{4x}{x + 1}$ is *conditional*. There are real values of x for which the equation is not true.

9. (a) Equivalent equations are derived from the substitution principle and simplification techniques. They have the same solution(s).

      $2x + 3 = 8$ and $2x = 5$ are equivalent equations.

    (b) Equivalent equations are produced by removing symbols of grouping, adding or subtracting the same quantity from both sides of the equation, multiplying or dividing both sides of the equation by the same nonzero quantity, or by interchanging the two sides of the equation.

11. (a) $3x = 15$

       $x = 5$

    (c) $s + 12 = 18$

       $s = 6$

    (b) $\frac{1}{2}t = 7$

       $t = 14$

    (d) $2u - 3 = 25$

       $u = 14$

13. $8x - 5 = 3x + 10$

      $5x = 15$

       $x = 3$

15. $2(x + 5) - 7 = 3(x - 2)$

    $2x + 10 - 7 = 3x - 6$

       $2x + 3 = 3x - 6$

         $-x = -9$

           $x = 9$

17. $6[x - (2x + 3)] = 8 - 5x$

      $6[-x - 3] = 8 - 5x$

      $-6x - 18 = 8 - 5x$

          $-x = 26$

           $x = -26$

19. $\dfrac{5x}{4} + \dfrac{1}{2} = x - \dfrac{1}{2}$

    $4\left(\dfrac{5x}{4}\right) + 4\left(\dfrac{1}{2}\right) = 4(x) - 4\left(\dfrac{1}{2}\right)$

        $5x + 2 = 4x - 2$

            $x = -4$

21. $\dfrac{3}{2}(z + 5) - \dfrac{1}{4}(z + 24) = 0$

    $4\left(\dfrac{3}{2}\right)(z + 5) - 4\left(\dfrac{1}{4}\right)(z + 24) = 4(0)$

        $6(z + 5) - (z + 24) = 0$

      $6z + 30 - z - 24 = 0$

              $5z = -6$

                $z = -\dfrac{6}{5}$

23. $0.25x + 0.75(10 - x) = 3$

    $4(0.25x) + 4(0.75)(10 - x) = 4(3)$

        $x + 3(10 - x) = 12$

        $x + 30 - 3x = 12$

            $-2x = -18$

              $x = 9$

25. $\dfrac{100 - 4u}{3} = \dfrac{5u + 6}{4} + 6$

    $12\left(\dfrac{100 - 4u}{3}\right) = 12\left(\dfrac{5u + 6}{4}\right) + 12(6)$

      $4(100 - 4u) = 3(5u + 6) + 72$

      $400 - 16u = 15u + 18 + 72$

           $-31u = -310$

              $u = 10$

27. $\dfrac{5x - 4}{5x + 4} = \dfrac{2}{3}$

$3(5x - 4) = 2(5x + 4)$

$15x - 12 = 10x + 8$

$5x = 20$

$x = 4$

29. $\dfrac{1}{x - 3} + \dfrac{1}{x + 3} = \dfrac{10}{x^2 - 9}$

$\dfrac{(x + 3) + (x - 3)}{x^2 - 9} = \dfrac{10}{x^2 - 9}$

$2x = 10$

$x = 5$

31. $\dfrac{x}{x + 4} + \dfrac{4}{x + 4} + 2 = 0$

$\dfrac{x + 4}{x + 4} + 2 = 0$

$1 + 2 = 0$

$3 = 0$

Contradiction : no solution

33. $\dfrac{7}{2x + 1} - \dfrac{8x}{2x - 1} = -4$

$7(2x - 1) - 8x(2x + 1) = -4(2x + 1)(2x - 1)$

$14x - 7 - 16x^2 - 8x = -16x^2 + 4$

$6x = 11$

$x = \dfrac{11}{6}$

35. $\dfrac{1}{x} + \dfrac{2}{x - 5} = 0$

$1(x - 5) + 2x = 0$

$3x - 5 = 0$

$3x = 5$

$x = \dfrac{5}{3}$

37. $\dfrac{3}{x(x - 3)} + \dfrac{4}{x} = \dfrac{1}{x - 3}$

$3 + 4(x - 3) = x$

$3 + 4x - 12 = x$

$3x = 9$

$x = 3$

A check reveals that $x = 3$ is an extraneous solution, so there is no solution.

39. $(x + 2)^2 + 5 = (x + 3)^2$

$x^2 + 4x + 4 + 5 = x^2 + 6x + 9$

$4x + 9 = 6x + 9$

$-2x = 0$

$x = 0$

41. $W_1 x = W_2(L - x)$

$50x = 75(10 - x)$

$50x = 750 - 75x$

$125x = 750$

$x = 6$ feet from the 50-pound child.

43. $y = x - 5$

Let $y = 0$: $0 = x - 5 \implies x = 5 \implies (5, 0)$ x-intercept

Let $x = 0$: $y = 0 - 5 \implies y = -5 \implies (0, -5)$ y-intercept

45. $y = x\sqrt{x + 2}$

Let $y = 0$: $0 = x\sqrt{x + 2} \implies x = 0, -2 \implies (0, 0), (-2, 0)$ x-intercepts

Let $x = 0$: $y = 0\sqrt{0 + 2} = 0 \implies (0, 0)$ y-intercept

47. $y = |x - 2| - 3$

Let $y = 0$: $|x - 2| - 3 = 0 \implies |x - 2| = 3 \implies x = -1, 5 \implies (-1, 0), (5, 0)$ x-intercepts

Let $x = 0$: $|0 - 2| - 3 = |-2| - 3 = 2 - 3 = -1 = y \implies (0, -1)$ y-intercept

49. $f(x) = 12 - 4x$

51. $f(x) = x^2 - 2.5x - 6$

53. $f(x) = \dfrac{x + 2}{3} - \dfrac{x - 1}{5} - 1$

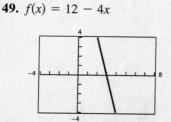

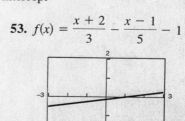

55.

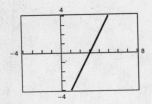

$(3, 0)$

$$2(x - 1) - 4 = 0$$
$$2x - 6 = 0$$
$$x = 3$$

57.

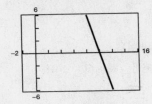

$(10, 0)$

$$y = 0 = 20 - (3x - 10) = 20 - 3x + 10$$
$$= 30 - 3x$$
$$3x = 30$$
$$x = 10$$

59. $\dfrac{3x}{2} + \dfrac{1}{4}(x - 2) = 10$

$$\dfrac{6x}{4} + \dfrac{x}{4} = 10 + \dfrac{1}{2}$$
$$\dfrac{7x}{4} = \dfrac{21}{2}$$
$$x = 6$$

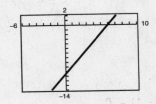

$$f(x) = \dfrac{3x}{2} + \dfrac{1}{4}(x - 2) - 10 = 0$$
$$x = 6.0$$

61. $3(x + 3) = 5(1 - x) - 1$

$$3x + 9 = 5 - 5x - 1$$
$$8x = -5$$
$$x = -\dfrac{5}{8}$$
$$f(x) = 3(x + 3) - 5(1 - x) + 1 = 0$$
$$x = -0.625$$

63. $y = 2 - x$

$y = 2x - 1$

$(x, y) = (1, 1)$

65. $(x, y) = (6, 4)$

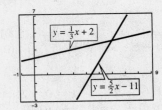

$y = \frac{1}{3}x + 2$

$y = \frac{5}{2}x - 11$

67. $2x + y = 6 \implies y = -2x + 6$

$-x + y = 0 \implies y = x$

$(2, 2)$

69. $y = 9 - 2x$

$y = x - 3$

$(4, 1)$

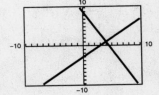

71. $y = 8$
$y = 3x^2 + 2x$
$(x, y) = (-2, 8),\ (1.333, 8)$

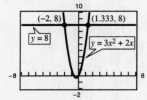

73. $y = 2x^2$
$y = x^4 - 2x^2$
$(x, y) = (0, 0),\ (2, 8),\ (-2, 8)$

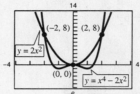

75. $x^2 - 2x - 1 = 0$
$x^2 - 2x = 1$
$x^2 - 2x + 1^2 = 1 + 1^2$
$(x - 1)^2 = 2$
$x - 1 = \pm\sqrt{2}$
$x = 1 \pm \sqrt{2}$

77. $(x + 3)^3 = 81$
$x + 3 = \pm 9$
$x + 3 = 9$ or $x + 3 = -9$
$x = 6$ or $x = -12$

79. $4x^4 - 18x^2 = 0$
$2x^2(2x^2 - 9) = 0$
$2x^2 = 0$
$x = 0$
$2x^2 - 9 = 0$
$x = \pm\dfrac{3\sqrt{2}}{2}$

81. $x^4 - 4x^2 + 3 = 0$
$(x^2 - 3)(x^2 - 1) = 0$
$(x + \sqrt{3})(x - \sqrt{3})(x + 1)(x - 1) = 0$
$x + \sqrt{3} = 0 \Rightarrow x = -\sqrt{3}$
$x - \sqrt{3} = 0 \Rightarrow x = \sqrt{3}$
$x + 1 = 0 \implies x = -1$
$x - 1 = 0 \implies x = 1$

83. $\dfrac{1}{t^2} + \dfrac{8}{t} + 15 = 0$
$1 + 8t + 15t^2 = 0$
$(1 + 3t)(1 + 5t) = 0$
$1 + 3t = 0 \implies t = -\dfrac{1}{3}$
$1 + 5t = 0 \implies t = -\dfrac{1}{5}$

85. $2x + 9\sqrt{x} - 5 = 0$
$(2\sqrt{x} - 1)(\sqrt{x} + 5) = 0$
$\sqrt{x} = \dfrac{1}{2} \Rightarrow x = \dfrac{1}{4}$
$(\sqrt{x} = -5$ is not a solution.$)$

87. $\sqrt{x - 10} - 4 = 0$
$\sqrt{x - 10} = 4$
$x - 10 = 16$
$x = 26$

89. $\sqrt[3]{2x + 5} + 3 = 0$
$\sqrt[3]{2x + 5} = -3$
$2x + 5 = -27$
$2x = -32$
$x = -16$

91. $\sqrt{x + 1} - 3x = 1$
$\sqrt{x + 1} = 3x + 1$
$x + 1 = 9x^2 + 6x + 1$
$0 = 9x^2 + 5x$
$0 = x(9x + 5)$
$x = 0$
$9x + 5 = 0 \implies x = -\dfrac{5}{9}$, extraneous

93. $\sqrt{x} - \sqrt{x - 5} = 1$
$\sqrt{x} = 1 + \sqrt{x - 5}$
$(\sqrt{x})^2 = (1 + \sqrt{x - 5})^2$
$x = 1 + 2\sqrt{x - 5} + x - 5$
$4 = 2\sqrt{x - 5}$
$2 = \sqrt{x - 5}$
$4 = x - 5$
$9 = x$

95. $(x - 5)^{2/3} = 16$

$$x - 5 = \pm 16^{3/2}$$
$$x - 5 = \pm 64$$
$$x = 69, \ -59$$

97. $\dfrac{20 - x}{x} = x$

$$20 - x = x^2$$
$$0 = x^2 + x - 20$$
$$0 = (x + 5)(x - 4)$$
$$x + 5 = 0 \implies x = -5$$
$$x - 4 = 0 \implies x = 4$$

99. $\dfrac{1}{x} - \dfrac{1}{x + 1} = 3$

$$x(x + 1)\dfrac{1}{x} - x(x + 1)\dfrac{1}{x + 1} = x(x + 1)(3)$$
$$x + 1 - x = 3x(x + 1)$$
$$1 = 3x^2 + 3x$$
$$0 = 3x^2 + 3x - 1; \ a = 3, \ b = 3, \ c = -1$$
$$x = \dfrac{-3 \pm \sqrt{(3)^2 - 4(3)(-1)}}{2(3)} = \dfrac{-3 \pm \sqrt{21}}{6}$$

101. $x = \dfrac{3}{x} + \dfrac{1}{2}$

$$(2x)(x) = (2x)\left(\dfrac{3}{x}\right) + (2x)\left(\dfrac{1}{2}\right)$$
$$2x^2 = 6 + x$$
$$2x^2 - x - 6 = 0$$
$$(2x + 3)(x - 2) = 0$$
$$2x + 3 = 0 \implies x = -\dfrac{3}{2}$$
$$x - 2 = 0 \implies x = 2$$

103. $\dfrac{4}{x + 1} - \dfrac{3}{x + 2} = 1$

$$4(x + 2) - 3(x + 1) = (x + 1)(x + 2), \ x \neq -2, \ -1$$
$$4x + 8 - 3x - 3 = x^2 + 3x + 2$$
$$x^2 + 2x - 3 = 0$$
$$(x - 1)(x + 3) = 0$$
$$x - 1 = 0 \implies x = 1$$
$$x + 3 = 0 \implies x = -3$$

105. $|2x - 1| = 5$

$$2x - 1 = 5 \implies x = 3$$
$$-(2x - 1) = 5 \implies x = -2$$

107. $|x| = x^2 + x - 3$

$$x = x^2 + x - 3 \qquad \text{OR} \qquad -x = x^2 + x - 3$$

$x^2 - 3 = 0$	$x^2 + 2x - 3 = 0$
$x = \pm\sqrt{3}$	$(x - 1)(x + 3) = 0$
	$x - 1 = 0 \implies x = 1$
	$x + 3 = 0 \implies x = -3$

Only $x = \sqrt{3}$, and $x = -3$ are solutions to the original equation. $x = -\sqrt{3}$ and $x = 1$ are extraneous. Note that the graph of $y = x^2 + x - 3 - |x|$ has two x-intercepts.

109. $y = x^3 - 2x^2 - 3x$

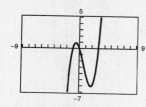

x-intercepts: $(-1, 0), (0, 0), (3, 0)$

$0 = x^3 - 2x^2 - 3x$
$0 = x(x + 1)(x - 3)$
$x = 0$
$x + 1 = 0 \implies x = -1$
$x - 3 = 0 \implies x = 3$

111. $y = x^4 - 10x^2 + 9$

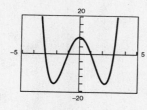

x-intercepts: $(\pm 1, 0), \ (\pm 3, 0)$

$0 = x^4 - 10x^2 + 9$
$0 = (x^2 - 1)(x^2 - 9)$
$0 = (x + 1)(x - 1)(x + 3)(x - 3)$
$x + 1 = 0 \implies x = -1$
$x - 1 = 0 \implies x = 1$
$x + 3 = 0 \implies x = -3$
$x - 3 = 0 \implies x = 3$

113. $y = \sqrt{11x - 30} - x$

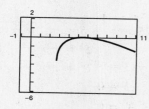

x-intercepts: $(5, 0), (6, 0)$

$0 = \sqrt{11x - 30} - x$
$x = \sqrt{11x - 30}$
$x^2 = 11x - 30$
$x^2 - 11x + 30 = 0$
$(x - 5)(x - 6) = 0$
$x - 5 = 0 \implies x = 5$
$x - 6 = 0 \implies x = 6$

115. $y = \sqrt{7x + 36} - \sqrt{5x + 16} - 2$

x-intercepts: $(0, 0), (4, 0)$

$$0 = \sqrt{7x + 36} - \sqrt{5x + 16} - 2$$
$$\sqrt{7x + 36} = 2 + \sqrt{5x + 16}$$
$$\left(\sqrt{7x + 36}\right)^2 = \left(2 + \sqrt{5x + 16}\right)^2$$
$$7x + 36 = 4 + 4\sqrt{5x + 16} + 5x + 16$$
$$7x + 36 = 5x + 20 + 4\sqrt{5x + 16}$$
$$2x + 16 = 4\sqrt{5x + 16}$$
$$x + 8 = 2\sqrt{5x + 16}$$
$$x^2 + 16x + 64 = 4(5x + 16)$$
$$x^2 + 16x + 64 = 20x + 64$$
$$x^2 - 4x = 0$$
$$x(x - 4) = 0$$
$$x = 0$$
$$x - 4 = 0 \implies x = 4$$

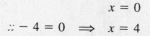

117. $y = \dfrac{1}{x} - \dfrac{4}{x-1} - 1$

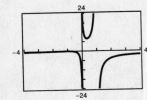

x-intercept: $(-1, 0)$

$$0 = \dfrac{1}{x} - \dfrac{4}{x-1} - 1$$
$$0 = (x-1) - 4x - x(x-1)$$
$$0 = x - 1 - 4x - x^2 + x$$
$$0 = -x^2 - 2x - 1$$
$$0 = x^2 + 2x + 1$$
$$0 = (x+1)^2$$
$$x + 1 = 0 \implies x = -1$$

119. $y = |x+1| - 2$

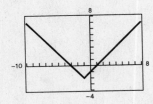

x-intercept: $(1, 0), (-3, 0)$

$$0 = |x+1| - 2$$
$$2 = |x+1|$$
$$x + 1 = 2 \quad \text{or} \quad -(x+1) = 2$$
$$x = 1 \quad \text{or} \quad -x - 1 = 2$$
$$-x = 3$$
$$x = -3$$

121. $A = \dfrac{1}{2}bh$

$$2A = bh$$
$$\dfrac{2A}{b} = h$$

123. $A = \dfrac{1}{2}(a+b)h$

$$2A = ah + bh$$
$$\dfrac{2A - ah}{h} = b$$

125. $S = \pi r \sqrt{r^2 + h^2}$

$$S^2 = \pi^2 r^2 (r^2 + h^2)$$
$$S^2 = \pi^2 r^4 + \pi^2 r^2 h^2$$
$$\dfrac{S^2 - \pi^2 r^4}{\pi^2 r^2} = h^2$$
$$h = \dfrac{\sqrt{S^2 - \pi^2 r^4}}{\pi r}$$

127. (a) $4x + 3y = 100$ (amount of fence)

$$y = \tfrac{1}{3}(100 - 4x)$$
$$\text{Area} = A(x) = (2x)y = (2x)\tfrac{1}{3}(100 - 4x)$$
$$= \tfrac{8}{3}x(25 - x).$$

(b)

x	y	Area
2	$\frac{92}{3}$	$\frac{368}{3}$
4	28	224
6	$\frac{76}{3}$	304
8	$\frac{68}{3}$	$\frac{1088}{3}$
10	20	400
12	$\frac{52}{3}$	416
14	$\frac{44}{3}$	$\frac{1232}{3}$

(c)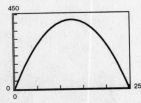

$x = 12.5$, $y = 16.67$ or $25 \times \frac{50}{3}$

(d) The graphs $y_1 = \frac{8}{3}x(25 - x)$ and $y_2 = 350$ intersect at $x = 7.5$ and $x = 17.5$. The dimensions are therefore $15 \times 23\frac{1}{3}$ or 35×10.

Approximate dimension for maximum area: $x = 12$, $y = \frac{52}{3}$ or $24 \times \frac{52}{3}$.

(e)
$$\tfrac{8}{3}x(25 - x) = 350$$
$$-8x^2 + 100x = 1050$$
$$4x^2 - 50x - 525 = 0$$
$$(2x - 35)(2x - 15) = 0$$
$$x = \tfrac{35}{2} = 17.5 \quad \text{or} \quad x = \tfrac{15}{2} = 7.5$$

The dimensions are therefore $15 \times 23\frac{1}{3}$ or 35×10.

Section P.5   Solving Inequalities Algebraically and Graphically

■ You should know the properties of inequalities.

(a) Transitive: $a < b$ and $b < c$ implies $a < c$.

(b) Addition: $a < b$ and $c < d$ implies $a + c < b + d$.

(c) Adding or Subtracting a Constant: $a \pm c < b \pm c$ if $a < b$.

(d) Multiplying or Dividing by a Constant: For $a < b$,

   1. If $c > 0$, then $ac < bc$ and $\dfrac{a}{c} < \dfrac{b}{c}$.

   2. If $c < 0$, then $ac > bc$ and $\dfrac{a}{c} > \dfrac{b}{c}$.

■ You should know that

$$|x| = \begin{cases} x & \text{if } x \geq 0 \\ -x & \text{if } x < 0 \end{cases}.$$

■ You should be able to solve absolute value inequalities.

(a) $|x| < a$ if and only if $-a < x < a$.

(b) $|x| > a$ if and only if $x < -a$ or $x > a$.

■ You should be able to solve polynomial inequalities.

(a) Find the critical numbers.

   1. Values that make the expression zero

   2. Values that make the expression undefined

(b) Test one value in each interval on the real number line resulting from the critical numbers.

(c) Determine the solution intervals.

■ You should be able to solve rational and other types of inequalities.

Solutions to Odd-Numbered Problems

1. $x < 3$

Matches (d).

3. $-3 < x \leq 4$

Matches (c).

5. (a) $x = 3$

$5(3) - 12 \overset{?}{>} 0$

$3 > 0$

Yes, $x = 3$ is a solution.

(b) $x = -3$

$5(-3) - 12 \overset{?}{>} 0$

$-27 \not> 0$

No, $x = -3$ is not a solution.

(c) $x = \frac{5}{2}$

$5\left(\frac{5}{2}\right) - 12 \overset{?}{>} 0$

$\frac{1}{2} > 0$

Yes, $x = \frac{5}{2}$ is a solution.

(d) $x = \frac{3}{2}$

$5\left(\frac{3}{2}\right) - 12 \overset{?}{>} 0$

$-\frac{9}{2} \not> 0$

No, $x = \frac{3}{2}$ is not a solution.

7. (a) $x = 13$

$$|13 - 10| \overset{?}{\geq} 3$$

$$3 \overset{?}{\geq} 3$$

Yes, $x = 13$ is a solution.

(b) $x = -1$

$$|-1 - 10| \overset{?}{\geq} 3$$

$$11 \geq 3$$

Yes, $x = -1$ is a solution.

(c) $x = 14$

$$|14 - 10| \overset{?}{\geq} 3$$

$$4 \geq 3$$

Yes, $x = 14$ is a solution.

(d) $x = 9$

$$|9 - 10| \overset{?}{\geq} 3$$

$$1 \not\geq 3$$

No, $x = 9$ is not a solution.

9.

$$-10x < 40$$

$$-\tfrac{1}{10}(-10) > -\tfrac{1}{10}(40)$$

$$x > -4$$

11. $4(x + 1) < 2x + 3$

$$4x + 4 < 2x + 3$$

$$2x < -1$$

$$x < -\tfrac{1}{2}$$

13. $1 < 2x + 3 < 9$

$$-2 < 2x < 6$$

$$-1 < x < 3$$

15. $-4 < \dfrac{2x - 3}{3} < 4$

$$-12 < 2x - 3 < 12$$

$$-9 < 2x < 15$$

$$-\dfrac{9}{2} < x < \dfrac{15}{2}$$

17. $-1 < -\dfrac{x}{3} < 1$

$$-1(-3) > x > (-3)$$

$$3 > x > -3$$

or $-3 < x < 3$

19. $6x > 12$

$$x > 2$$

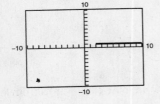

21. $5 - 2x \geq 1$

$$-2x \geq -4$$

$$x \leq 2$$

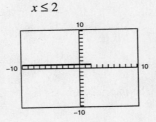

23. $0 \leq 2(x + 4) < 20$

$$0 \leq x + 4 < 10$$

$$-4 \leq x < 6$$

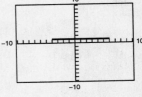

25. $y = 2x - 3$

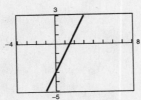

(a) $y \geq 1$

$$2x - 3 \geq 1$$

$$2x \geq 4$$

$$x \geq 2$$

(b) $y \leq 0$

$$2x - 3 \leq 0$$

$$2x \leq 3$$

$$x \leq \tfrac{3}{2}$$

27. $y = -\tfrac{1}{2}x + 2$

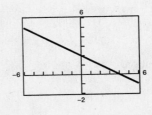

(a) $0 \leq y \leq 3$

$$0 \leq -\tfrac{1}{2}x + 2 \leq 3$$

$$-2 \leq -\tfrac{1}{2}x \leq 1$$

$$4 \geq x \geq -2$$

(b) $y \geq 0$

$$-\tfrac{1}{2}x + 2 \geq 0$$

$$-\tfrac{1}{2}x \geq -2$$

$$x \leq 4$$

29. $x - 5 \geq 0$

   $x \geq 5$

   $[5, \infty)$

31. $\left|\dfrac{x}{2}\right| > 3$

   $\dfrac{x}{2} < -3$ or $\dfrac{x}{2} > 3$

   $x < -6$ $x > 6$

33. $|x - 20| \leq 4$

   $-4 \leq x - 20 \leq 4$

   $16 \leq x \leq 24$

35. $|x - 20| \geq 4$

   $x - 20 \leq -4$ or $x - 20 \geq 4$

   $x \leq 16$ $x \geq 24$

37. $\left|\dfrac{x - 3}{2}\right| \geq 5$

   $\dfrac{x - 3}{2} \leq -5$ or $\dfrac{x - 3}{2} \geq 5$

   $x - 3 \leq -10$ $x - 3 \geq 10$

   $x \leq -7$ $x \geq 13$

39. $|x - 5| < 0$

   No solution. The absolute value of a number can never be less than zero.

41. $y = |x - 3|$

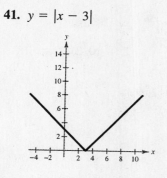

   (a) $y \leq 2$

   $|x - 3| \leq 2$

   $-2 \leq x - 3 \leq 2$

   $1 \leq x \leq 5$

   (b) $y \geq 4$

   $|x - 3| \geq 4$

   $x - 3 \leq -4$ or $x - 3 \geq 4$

   $x \leq -1$ $x \geq 7$

43. The midpoint of the interval $[-3, 3]$ is 0. The interval represents all real numbers x no more than 3 units from 0.

   $|x - 0| \leq 3$

   $|x| \leq 3$

45. The graph shows all real numbers at least 3 units from 7.

   $|x - 7| \geq 3$

47. All real numbers within 10 units of 12

   $|x - 12| \leq 10$

49. (a) and (b) $y = 0.067x - 5.638$

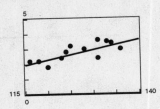

The plot looks linear.

(c) $\quad 3 \geq 0.067x - 5.638$

$\quad 8.638 \geq 0.067x$

$\quad\quad x \geq 129$

(d) IQ scores are not a good predictor of GPAs. Other factors include study habits, class attendance, and attitude.

51. $(x + 2)^2 < 25$

$\quad x^2 + 4x + 4 < 25$

$\quad x^2 + 4x - 21 < 0$

$\quad (x + 7)(x - 3) < 0$

Critical numbers: $x = -7, x = 3$

Test intervals: $(-\infty, -7), (-7, 3), (3, \infty)$

Test: Is $(x + 7)(x - 3) < 0$?

Solution set: $(-7, 3)$

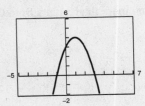

53. $x^2 + 4x + 4 \geq 9$

$\quad x^2 + 4x - 5 \geq 0$

$\quad (x + 5)(x - 1) \geq 0$

Critical numbers: $x = -5, x = 1$

Test intervals: $(-\infty, -5), (-5, 1), (1, \infty)$

Test: Is $(x + 5)(x - 1) \geq 0$?

Solution set: $(-\infty, -5] \cup [1, \infty)$

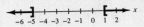

55. $x^2 + x < 6$

$\quad x^2 + x - 6 < 0$

$\quad (x + 3)(x - 2) < 0$

Critical numbers: $x = -3, x = 2$

Test intervals: $(-\infty, -3), (-3, 2), (2, \infty)$

Test: Is $(x + 3)(x - 2) < 0$?

Solution set: $(-3, 2)$

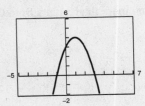

57. $x^3 - 4x \geq 0$

$\quad x(x + 2)(x - 2) \geq 0$

Critical number: $x = 0, x = \pm 2$

Test intervals: $(-\infty, -2), (-2, 0), (0, 2), (2, \infty)$

Test: Is $x(x + 2)(x - 2) \geq 0$?

Solution set: $[-2, 0] \cup [2, \infty)$

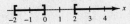

59. $y = -x^2 + 2x + 3$

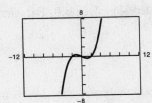

(a) $y \leq 0$ when $x \leq -1$ or $x \geq 3$.

(b) $y \geq 3$ when $0 \leq x \leq 2$.

61. $y = \frac{1}{8}x^3 - \frac{1}{2}x$

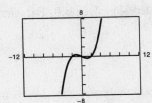

(a) $y \geq 0$ when $-2 \leq x \leq 0, 2 \leq x < \infty$.

(b) $y \leq 6$ when $x \leq 4$.

63. $\dfrac{1}{x} - x > 0$

$\dfrac{1 - x^2}{x} > 0$

Critical numbers: $x = 0, x = \pm 1$

Test intervals: $(-\infty - 1), (-1, 0), (0, 1), (1, \infty)$

Test: Is $\dfrac{1 - x^2}{x} > 0$?

Solution set: $(-\infty, -1) \cup (0, 1)$

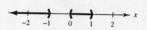

65. $\dfrac{x + 6}{x + 1} - 2 < 0$

$\dfrac{x + 6 - 2(x + 1)}{x + 1} < 0$

$\dfrac{4 - x}{x + 1} < 0$

Critical numbers: $x = -1, x = 4$

Test intervals: $(-\infty, -1), (-1, 4), (4, \infty)$

Test: Is $\dfrac{4 - x}{x + 1} < 0$?

Solution set: $(-\infty, -1) \cup (4, \infty)$

67. $y = \dfrac{3x}{x - 2}$

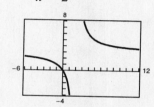

(a) $y \le 0$ when $0 \le x < 2$.

(b) $y \ge 6$ when $2 < x \le 4$.

69. $y = \dfrac{2x^2}{x^2 + 4}$

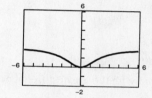

(a) $y \ge 1$ when $x \le -2$ or $x \ge 2$.

This can also be expressed as $|x| \ge 2$.

(b) $y \le 2$ for all real numbers x.

This can also be expressed as $-\infty < x < \infty$.

71. $4 - x^2 \ge 0$

$(2 + x)(2 - x) \ge 0$

Critical numbers: $x = \pm 2$

Test intervals: $(-\infty, -2), (-2, 2), (2, \infty)$

Test: Is $4 - x^2 \ge 0$?

Domain: $[-2, 2]$

73. $P = 5.9556 + 1.492t + 0.0056t^2, \ 0 \le t \le 40$

(a)

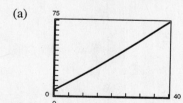

(b) $5.9556 + 1.492t + 0.0056t^2 \ge 25$

$0.0056t^2 + 1.492t - 19.0444 \ge 0$

By the Quadratic Formula, the critical numbers are $t \approx -278.63$ and $t \approx 12.21$. Only $t \approx 12.21$ makes sense, and this corresponds to the year 1962.

75. According to the graph, if the frequency is 600 vibrations per second, then the thickness is approximately 3.6 mm.

77. According to the graph, if the thickness is less than 3 mm, then the frequency is less than 500: $0 < v < 500$.

Section P.6   Exploring Data: Representing Data Graphically

- ■ You should know how to construct line plots.
- ■ You should know how to construct stem and leaf plots.
- ■ You should be able to draw and interpret histograms and frequency distributions.
- ■ You should be able to construct a line graph.

Solutions to Odd-Numbered Exercises

1. (a) The price 1.109 occurred with the greatest frequency (6).

    (b) The prices range from 0.999 to 1.189.

3. By scanning the data, we see that the largest number is 24 and the smallest number is 10. We construct the line plot on the interval [10, 24] as follows.

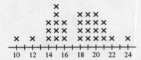

    The score of 15 occurred with the greatest frequency (5).

5. By scanning the data, we see that the largest number is 100 and the smallest number is 70. We construct the line plot on the interval [70, 100] as follows.

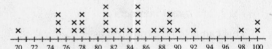

    The scores of 81 and 85 occurred with the greatest frequency (4).

7. Since the scores range from 70 to 100, the stems are 7, 8, 9 and 10.

Stems	Leaves
7	0 5 5 5 7 7 8 8 8
8	1 1 1 1 2 3 4 5 5 5 5 7 8 9 9 9
9	0 2 8
10	0 0

9. Since the expenditures range from 663 to 1800, the stems are 6, 7, . . . ,17, 18.

Stems	Leaves
6	63 76 92
7	62 64
8	04 18 34 34 41 48 50 63 64 65 85 98
9	03 13 15 16 19 19 32 40 43 64 64 74 74
10	03 24 24 29 45 49 56 61 96 98
11	23 24 31 38 55 97
12	52 65
13	36 51
14	
15	
16	
17	
18	00

11. The projected loss of 500 million in 1995 is 500% of the loss in 1991.

13.

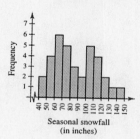

15.

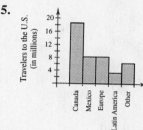

17. (a) The savings decreased from 9% to 4%. The decrease is 55.6%.

(b) No. The trend of low savings limits the amount of funds available for capital improvements and research in industries.

19.

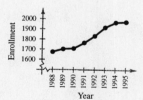

21.

You can conclude that the receipts are increasing in a steady rate.

❑ **Review Exercises for Chapter P**

Solutions to Even-Numbered Exercises

1.

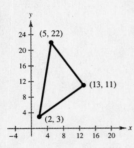

$$d_1 = \sqrt{(13-5)^2 + (11-22)^2} = \sqrt{8^2 + (-11)^2} = \sqrt{64+121} = \sqrt{185}$$

$$d_2 = \sqrt{(2-13)^2 + (3-11)^2} = \sqrt{(-11)^2 + (-8)^2} = \sqrt{121+64} = \sqrt{185}$$

$$d_3 = \sqrt{(2-5)^2 + (3-22)^2} = \sqrt{(-3)^2 + (-19)^2} = \sqrt{9+361} = \sqrt{370}$$

$$d_1^2 + d_2^2 = 185 + 185 = 370 = d_3^2$$

The points form a right triangle.

3. $x > 0$ and $y = -2$ is in Quadrant IV

5. (a)

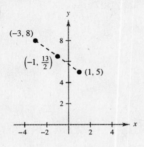

(b) $d = \sqrt{(1-(-3))^2 + (5-8)^2}$

$ = \sqrt{4^2 + (-3)^2}$

$ = \sqrt{16+9}$

$ = \sqrt{25}$

$ = 5$

(c) $\left(\dfrac{-3+1}{2}, \dfrac{8+5}{2}\right) = \left(-1, \dfrac{13}{2}\right)$

7. $y - 2x - 3 = 0$

$y = 2x + 3$

Line with x-intercept $\left(-\frac{3}{2}, 0\right)$ and
y-intercept $(0, 3)$

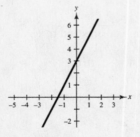

9.

11. $y + 2x^2 = 0$

$y = -2x^2$ is a parabola.

x	0	± 1	± 2
y	0	-2	-8

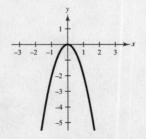

13. $y = \sqrt{25 - x^2}$

Domain: $-5 \le x \le 5$

x	0	± 3	± 4	± 5
y	5	4	3	0

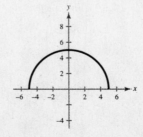

15. $y = \frac{1}{4}(x + 1)^3$

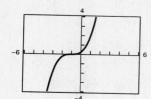

Intercepts: $(-1, 0)$, $\left(0, \frac{1}{4}\right)$

17. $y = \frac{1}{4}x^4 - 2x^2$

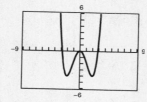

Intercepts: $(0, 0)$, $\left(\pm 2\sqrt{2}, 0\right)$

19. $y = x\sqrt{9 - x^2}$

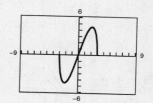

Intercepts: $(0, 0)$, $(\pm 3, 0)$

21. $y = |x - 4| - 4$

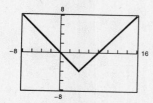

Intercepts: $(0, 0)$, $(8, 0)$

23. Center: $(3, -1)$
Radius: $\sqrt{9} = 3$

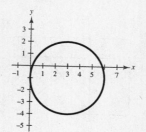

25. $y = 10x^3 - 21x^2$

Xmin = -2
Xmax = 3
Xscl = 1
Ymin = -20
Ymax = 15
Yscl = 5

27. $(-4.5, 6)$, $(2.1, 3)$

$$m = \frac{3 - 6}{2.1 - (-4.5)} = \frac{-3}{6.6} = -\frac{30}{66} = -\frac{5}{11}$$

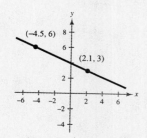

29. $m = \dfrac{5/2 - 1}{5 - 3/2} = \dfrac{3/2}{7/2} = \dfrac{3}{7}$

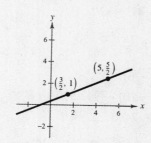

31. $(2 + 4, -1 + 1) = (6, 0)$

$\quad (6 + 4, 0 + 1) = (10, 1)$

$\quad (2 - 4, -1 - 1) = (-2, -2)$

33. $(-6 + 1, -5 - 2) = (-5, -7)$

$\quad (-6 + 2, -5 - 4) = (-4, -9)$

$\quad (-6 + 3, -5 - 6) = (-3, -11)$

35. $(0, 0)$, $(0, 10)$

$$m = \frac{10 - 0}{0 - 0} = \frac{10}{0} \text{ undefined.}$$
The line is vertical.

$x = 0$

37. $y - 1 = \dfrac{6 - 1}{14 - 2}(x - 2) = \dfrac{5}{12}(x - 2)$

$\qquad\quad = \dfrac{5}{12}(x - 2)$

$\qquad y = \dfrac{5}{12}x + \dfrac{1}{6}$ or $5x - 12y + 2 = 0$

39. $y - 0 = \dfrac{2 - 0}{6 + 1}(x + 1)$

$\qquad\quad = \dfrac{2}{7}(x + 1)$

$\qquad y = \dfrac{2}{7}x + \dfrac{2}{7}$ or $2x - 7y + 2 = 0$

41. $y - (-5) = \frac{3}{2}(x - 0)$

$\quad\quad y + 5 = \frac{3}{2}x$

$\quad\quad\quad y = \frac{3}{2}x - 5 \text{ or } 0 = 3x - 2y - 10$

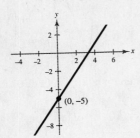

43. $\quad\quad y - 0 = -\frac{2}{3}(x - 3)$

$\quad\quad\quad 3y = -2(x - 3)$

$\quad\quad\quad 3y = -2x + 6$

$\quad 2x + 3y - 6 = 0$

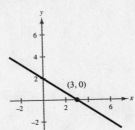

45. $5x - 4y = 8$

$\quad\quad y = \frac{5}{4}x - 2 \text{ and } m = \frac{5}{4}$

(a) Parallel slope: $m = \frac{5}{4}$

$\quad y - (-2) = \frac{5}{4}(x - 3)$

$\quad\quad 4y + 8 = 5x - 15$

$\quad\quad\quad 0 = 5x - 4y - 23$

(b) Perpendicular slope: $m = -\frac{4}{5}$

$\quad y - (-2) = -\frac{4}{5}(x - 3)$

$\quad\quad 5y + 10 = -4x + 12$

$\quad 4x + 5y - 2 = 0$

47. $(6, 12{,}500), \quad m = 850$

$y - 12{,}500 = 850(t - 6)$

$\quad\quad y = 850t - 5100 + 12{,}500$

$\quad\quad y = 850t + 7400$

49. The distance between $(-2, -5)$ and (x, y) equals the distance between $(6, 3)$ and (x, y).

$$\sqrt{(x + 2)^2 + (y + 5)^2} = \sqrt{(x - 6)^2 + (y - 3)^2}$$

$$(x + 2)^2 + (y + 5)^2 = (x - 6)^2 + (y - 3)^2$$

$$x^2 + 4x + 4 + y^2 + 10y + 25 = x^2 - 12x + 36 + y^2 - 6y + 9$$

$$4x + 10y + 29 = -12x - 6y + 45$$

$$16x + 16y - 16 = 0$$

$$x + y - 1 = 0$$

This line is the perpendicular bisector of the line segment joining the two points.

51. $\quad 6 - (x - 2)^2 = 2 + 4x - x^2$

$\quad 6 - (x^2 - 4x + 4) = 2 + 4x - x^2$

$\quad\quad 2 + 4x - x^2 = 2 + 4x - x^2$

$\quad\quad\quad\quad 0 = 0 \quad\quad$ Identity

All real numbers are solutions.

53. $4(x + 3) - 3 = 2(4 - 3x) - 4$

$\quad 4x + 12 - 3 = 8 - 6x - 4$

$\quad\quad 4x + 9 = -6x + 4$

$\quad\quad 10x = -5$

$\quad\quad\quad x = -\dfrac{1}{2}$

55. $3\left(1 - \dfrac{1}{5t}\right) = 0$

$1 - \dfrac{1}{5t} = 0$

$1 = \dfrac{1}{5t}$

$5t = 1$

$t = \dfrac{1}{5}$

57. $6x = 3x^2$

$0 = 3x^2 - 6x$

$0 = 3x(x - 2)$

$3x = 0 \implies x = 0$

$x - 2 = 0 \implies x = 2$

59. $(x + 4)^2 = 18$

$x + 4 = \pm\sqrt{18}$

$x = -4 \pm 3\sqrt{2}$

61. $x^2 - 12x + 30 = 0$

$x^2 - 12x = -30$

$x^2 - 12x + 36 = -30 + 36$

$(x - 6)^2 = 6$

$x - 6 = \pm\sqrt{6}$

$x = 6 \pm \sqrt{6}$

63. $5x^4 - 12x^3 = 0$

$x^3(5x - 12) = 0$

$x^3 = 0 \quad \text{or} \quad 5x - 12 = 0$

$x = 0 \quad \text{or} \qquad x = \dfrac{12}{5}$

65. $\dfrac{4}{(x - 4)^2} = 1$

$4 = (x - 4)^2$

$\pm 2 = x - 4$

$4 \pm 2 = x$

$x = 6 \quad \text{or} \quad x = 2$

67. $\sqrt{x + 4} = 3$

$\left(\sqrt{x + 4}\right)^2 = (3)^2$

$x + 4 = 9$

$x = 5$

69. $\sqrt{2x + 3} + \sqrt{x - 2} = 2$

$\left(\sqrt{2x + 3}\right)^2 = \left(2 - \sqrt{x - 2}\right)^2$

$2x + 3 = 4 - 4\sqrt{x - 2} + x - 2$

$x + 1 = -4\sqrt{x - 2}$

$(x + 1)^2 = \left(-4\sqrt{x - 2}\right)^2$

$x^2 + 2x + 1 = 16(x - 2)$

$x^2 - 14x + 33 = 0$

$(x - 3)(x - 11) = 0$

$x = 3$, extraneous   or $x = 11$, extraneous

No solution

71. $(x - 1)^{2/3} - 25 = 0$

$(x - 1)^{2/3} = 25$

$(x - 1)^2 = 25^3$

$x - 1 = \pm\sqrt{25^3}$

$x = 1 \pm 125$

$x = 126 \quad \text{or} \quad x = -124$

73. $|x - 5| = 10$

$x - 5 = -10 \quad \text{or} \quad x - 5 = 10$

$x = -5 \qquad\qquad x = 15$

75. $y = 4x^3 - 12x^2 + 8x$

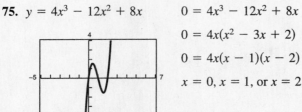

$0 = 4x^3 - 12x^2 + 8x$

$0 = 4x(x^2 - 3x + 2)$

$0 = 4x(x - 1)(x - 2)$

$x = 0, \, x = 1, \text{ or } x = 2$

x-intercepts: $(0, 0)$, $(1, 0)$, $(2, 0)$

77. $y = \dfrac{1}{x} + \dfrac{1}{x+1} - 2$

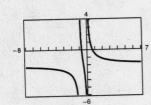

x-intercepts: $\left(\pm\dfrac{\sqrt{2}}{2}, 0\right)$

$0 = \dfrac{1}{x} + \dfrac{1}{x+1} - 2$

$2 = \dfrac{1}{x} + \dfrac{1}{x+1}$

$2x(x+1) = (x+1) + x$

$2x^2 + 2x = 2x + 1$

$2x^2 = 1$

$x^2 = \dfrac{1}{2}$

$x = \pm\sqrt{\dfrac{1}{2}} = \pm\dfrac{\sqrt{2}}{2}$

79.

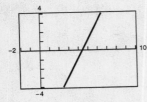

Intercept $\left(\dfrac{40}{9}, 0\right) \approx (4.44, 0)$

$\sqrt{x^2+1} + x - 9 = 0$

$\left(\sqrt{x^2+1}\right)^2 = (9-x)^2$

$x^2 + 1 = 81 - 18x + x^2$

$18x = 80$

$x = \dfrac{40}{9}$

81. $V = \dfrac{1}{3}\pi r^2 h$

$3V = \pi r^2 h$

$\dfrac{3V}{\pi h} = r^2$

$r = \sqrt{\dfrac{3V}{\pi h}}$

Since r represents the radius of a cone, r is positive only.

83. $L = \dfrac{k}{3\pi r^2 p}$

$3\pi r^2 p L = k$

$p = \dfrac{k}{3\pi r^2 L}$

85. $\dfrac{1}{2}(3-x) > \dfrac{1}{3}(2-3x)$

$3(3-x) > 2(2-3x)$

$9 - 3x > 4 - 6x$

$3x > -5$

$x > -\dfrac{5}{3}, \left(-\dfrac{5}{3}, \infty\right)$

87. $\dfrac{x-5}{3-x} < 0$

Critical numbers: $x = 5, x = 3$

Test intervals: $(-\infty, 3), (3, 5), (5, \infty)$

Test: Is $\dfrac{x-5}{3-x} < 0$?

Solution set: $(-\infty, 3) \cup (5, \infty)$

89. $|x - 2| < 1$

$-1 < x - 2 < 1$

$1 < x < 3$

which can be written as $(1, 3)$.

91. $\left|x - \dfrac{3}{2}\right| \geq \dfrac{3}{2}$

$x - \dfrac{3}{2} \leq -\dfrac{3}{2}$ or $x - \dfrac{3}{2} \geq \dfrac{3}{2}$

$x \leq 0$ or $x \geq 3$

which can be written as $(-\infty, 0] \cup [3, \infty)$.

93.

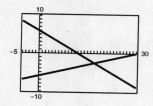

The graphs of $y_1 = \dfrac{x}{5} - 6$ and $y_2 = -\dfrac{x}{2} + 6$ cross at $x = \dfrac{120}{7}$. Hence, $\dfrac{x}{5} - 6 \leq -\dfrac{x}{2} + 6$ for $x \leq \dfrac{120}{7}$.

95. The graph of $y = (x - 4)|x|$ is above the x-axis for $x > 4$. Hence, $(x - 4)|x| > 0$ for $x > 4$.

97. *Model:* $\begin{pmatrix} \text{Distance of} \\ \text{outer track} \end{pmatrix} - \begin{pmatrix} \text{Distance of} \\ \text{inner track} \end{pmatrix} = \begin{pmatrix} \text{Distance between} \\ \text{starting positions} \end{pmatrix}$

Labels: Distance of outer track $= 2l + 2\pi r$

Distance of inner track $= 2l + 2\pi(r - 1)$

Distance between starting positions $= d$

Equation: $d = 2l + 2\pi r - [2l + 2\pi(r - 1)]$

$\quad = 2l + 2\pi r - 2l - 2\pi r + 2\pi$

$\quad = 2\pi$ meters

99.

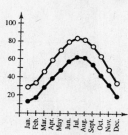

❏ Chapter Test Solutions for Chapter P

1.

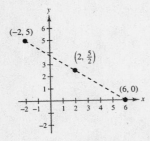

$$\text{Midpoint} = \left(\frac{-2 + 6}{2}, \frac{5 + 0}{2}\right) = \left(2, \frac{5}{2}\right)$$

$$\begin{aligned}
\text{Distance} &= \sqrt{(6 - (-2))^2 + (0 - 5)^2} \\
&= \sqrt{64 + 25} \\
&= \sqrt{89}
\end{aligned}$$

2.

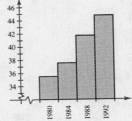

3. $y = 4 - \frac{3}{4}|x|$

y-axis symmetry

x-intercept: $\left(\pm\frac{16}{3}, 0\right)$

y-intercept: $(0, 4)$

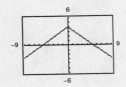

4. $y = 4 - (x - 2)^2$

Parabola; vertex: $(2, 4)$

No symmetry

x-intercepts: $(0, 0)$ and $(4, 0)$

$$0 = 4 - (x - 2)^2$$
$$(x - 2)^2 = 4$$
$$x - 2 = \pm 2$$
$$x = 2 \pm 2$$
$$x = 4 \ \text{ or } \ x = 0$$

y-intercept: $(0, 0)$

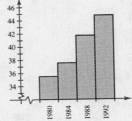

5. $y = x - x^3$

Origin symmetry

x-intercepts: $(0, 0), (1, 0), (-1, 0)$

$$0 = x - x^3$$
$$0 = x(1 + x)(1 - x)$$
$$x = 0, x = \pm 1$$

y-intercept: $(0, 0)$

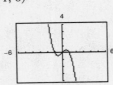

6. $y = \sqrt{3 - x}$

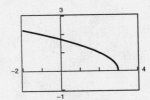

$x = 0 \implies y = \sqrt{3}$

$y = 0 \implies 0 = \sqrt{3 - x} \implies x = 3$

Intercepts: $\left(0, \sqrt{3}\right), (3, 0)$

7. $2x - 3y = 12$

$$-3y = -2x + 12$$
$$y = \frac{2}{3}x - 4$$

Intercepts: $(0, -4), (6, 0)$

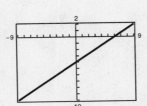

8. $(x - 3)^2 + y^2 = 9$

Circle radius: 3

Center: $(3, 0)$

Symmetry: x-axis

Intercepts: $(0, 0), (6, 0)$

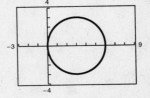

9. $y - (-1) = \frac{3}{2}(x - 3) = \frac{3}{2}x - \frac{9}{2}$

$$y = \frac{3}{2}x - \frac{11}{2}$$

Additional points: $(1, -4), \left(4, \frac{1}{2}\right), (5, 2)$

10. (a) slope $= \dfrac{3 - 0}{2 - (-4)} = \dfrac{3}{6} = \dfrac{1}{2}$

       line: $y - 0 = \dfrac{1}{2}(x - (-4)) = \dfrac{1}{2}(x + 4)$

$$y = \dfrac{1}{2}x + 2$$

$$2y - x = 4$$

(b) Slope of $5x + 2y = 3$ is $-\dfrac{5}{2}$.

$$y - 4 = \dfrac{2}{5}(x - 0)$$

$$y = \dfrac{2}{5}x + 4$$

$$5y - 2x = 20$$

11. $2x - 3(x - 4) = 5$

$2x - 3x + 12 = 5$

$-x = -7$

$x = 7$

12. $\dfrac{2}{t - 3} + \dfrac{2}{t - 2} = \dfrac{10}{t^2 - 5t + 6} = \dfrac{10}{(t - 3)(t - 2)}$

$2(t - 2) + 2(t - 3) = 10$

$4t - 10 = 10$

$t = 5$

13. $3y^2 + 6y + 2 = 0$

$$y = \dfrac{-6 \pm \sqrt{6^2 - 4(3)(2)}}{2(3)}$$

$$= \dfrac{-6 \pm \sqrt{12}}{6}$$

$$= -1 \pm \dfrac{1}{3}\sqrt{3}$$

14. $\sqrt{x + 10} = x - 2$

$x + 10 = (x - 2)^2 = x^2 - 4x + 4$

$x^2 - 5x - 6 = 0$

$(x - 6)(x + 1) = 0$

$x = 6$ ($x = -1$ is extraneous.)

15. $2\sqrt{x} - \sqrt{2x + 1} = 1$

$4x + (2x + 1) - 4\sqrt{x(2x + 1)} = 1$

$6x = 4\sqrt{2x^2 + x}$

$36x^2 = 16(2x^2 + x) = 32x^2 + 16x$

$4x^2 - 16x = 0$

$x(x - 4) = 0$

$x = 4$ ($x = 0$ is extraneous.)

16. $|3x - 1| = 7$

$3x - 1 = 7$ or $3x - 1 = -7$

$3x = 8$ $3x = -6$

$x = \dfrac{8}{3}$ $x = -2$

17. $-3 \le 2(x + 4) < 14$

$-3 \le 2x + 8 < 14$

$-11 \le 2x \quad\quad < 6$

$-\dfrac{11}{2} \le x \quad\quad < 3$

$\left[-\dfrac{11}{2}, 3\right)$

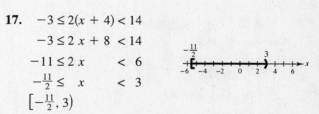

18. $\dfrac{2}{x} > \dfrac{5}{x + 6}$

$\dfrac{2}{x} - \dfrac{5}{x + 6} > 0$

$\dfrac{2x + 12 - 5x}{x(x + 6)} > 0$

$\dfrac{12 - 3x}{x(x + 6)} > 0$

Critical numbers: $-6, 0, 4$

Test intervals: $(-\infty, -6), (-6, 0), (0, 4), (4, \infty)$

Test: Is $\dfrac{3(4 - x)}{x(x + 6)} > 0$?

By testing these intervals, you obtain $(-\infty, -6)$ and $(0, 4)$.

19. Area circle $= \pi(40)^2 = \pi ab = \pi a(100 - a)$

$$40^2 = 100a - a^2$$

$a^2 - 100a + 1600 = 0$

$(a - 20)(a - 60) = 0 \Rightarrow a = 60$ ($a > b$)

and $b = 20$.

C H A P T E R 1
Functions and Their Graphs

CHAPTER 1
Functions and Their Graphs

Section 1.1   Functions

■ Given a set or an equation, you should be able to determine if it represents a function.

■ Given a function, you should be able to do the following.

(a) Find the domain.

(b) Evaluate it at specific values.

Solutions to Odd-Numbered Exercises

1. Yes, it does represent a function. Each domain value is matched with only one range value.

3. No, it does not represent a function. The domain values are each matched with three range values.

5. Yes, it does represent a function. Each input value is matched with only one output value.

7. No, it does not represent a function. The input values of 10 and 7 are each matched with two output values.

9. (a) Each element of A is matched with exactly one element of B, so it does represent a function.

(b) The element 1 in A is matched with two elements, -2 and 1 of B, so it does not represent a function.

(c) Each element of A is matched with exactly one element of B, so it does represent a function.

11. Each are functions. For each year there corresponds one and only one circulation.

13. $x^2 + y^2 = 4 \implies y = \pm\sqrt{4 - x^2}$

Thus, y *is not* a function of x. For instance, the values $y = 2$ and -2 both correspond to $x = 0$.

15. $x^2 + y = 4 \implies y = 4 - x^2$

Thus, y *is* a function of x.

17. $2x + 3y = 4 \implies y = \frac{1}{3}(4 - 2x)$

Thus, y *is* a function of x.

19. $y^2 = x^2 - 1 \implies y = \pm\sqrt{x^2 - 1}$

Thus, y *is not* a function of x. For instance, the values $y = \sqrt{3}$ and $-\sqrt{3}$ both correspond to $x = 2$.

21. $y = |4 - x|$

This is a function of x.

23. $f(s) = \dfrac{1}{s + 1}$

(a) $f(4) = \dfrac{1}{(4) + 1} = \dfrac{1}{5}$

(b) $f(0) = \dfrac{1}{(0) + 1} = 1$

(c) $f(4x) = \dfrac{1}{(4x) + 1} = \dfrac{1}{4x + 1}$

(d) $f(x + c) = \dfrac{1}{(x + c) + 1} = \dfrac{1}{x + c + 1}$

25. $f(x) = 2x - 3$

(a) $f(1) = 2(1) - 3 = -1$

(b) $f(-3) = 2(-3) - 3 = -9$

(c) $f(x - 1) = 2(x - 1) - 3 = 2x - 5$

27. $h(t) = t^2 - 2t$

(a) $h(2) = 2^2 - 2(2) = 0$

(b) $h(1.5) = (1.5)^2 - 2(1.5) = -0.75$

(c) $h(x + 2) = (x + 2)^2 - 2(x + 2) = x^2 + 2x$

29. $f(y) = 3 - \sqrt{y}$

(a) $f(4) = 3 - \sqrt{4} = 1$

(b) $f(0.25) = 3 - \sqrt{0.25} = 2.5$

(c) $f(4x^2) = 3 - \sqrt{4x^2} = 3 - 2|x|$

31. $q(x) = \dfrac{1}{x^2 - 9}$

(a) $q(0) = \dfrac{1}{0^2 - 9} = -\dfrac{1}{9}$

(b) $q(3) = \dfrac{1}{3^2 - 9}$ is undefined.

(c) $q(y + 3) = \dfrac{1}{(y + 3)^2 - 9} = \dfrac{1}{y^2 + 6y}$

33. $f(x) = \dfrac{|x|}{x}$

(a) $f(2) = \dfrac{|2|}{2} = 1$

(b) $f(-2) = \dfrac{|-2|}{-2} = -1$

(c) $f(x - 1) = \dfrac{|x - 1|}{x - 1}$

35. $f(x) = \begin{cases} 2x + 1, & x < 0 \\ 2x + 2, & x \geq 0 \end{cases}$

(a) $f(-1) = 2(-1) + 1 = -1$

(b) $f(0) = 2(0) + 2 = 2$

(c) $f(2) = 2(2) + 2 = 6$

37. $f(x) = x^2 - 3$

x	-2	-1	0	1	2
$f(x)$	1	-2	-3	-2	1

39. $h(t) = \frac{1}{2}|t + 3|$

t	-5	-4	-3	-2	-1
$h(t)$	1	$\frac{1}{2}$	0	$\frac{1}{2}$	1

41. $f(x) = \begin{cases} -\frac{1}{2}x + 4, & x \leq 0 \\ (x - 2)^2, & x > 0 \end{cases}$

x	-2	-1	0	1	2
$f(x)$	5	$\frac{9}{2}$	4	1	0

43. $15 - 3x = 0$

$3x = 15$

$x = 5$

45. $x^2 - 9 = 0$

$x^2 = 9$

$x = \pm 3$

47.
$$f(x) = g(x)$$
$$x^2 = x + 2$$
$$x^2 - x - 2 = 0$$
$$(x + 1)(x - 2) = 0$$
$$x = -1 \quad \text{or} \quad x = 2$$

49.
$$f(x) = g(x)$$
$$\sqrt{3x} + 1 = x + 1$$
$$\sqrt{3x} = x$$
$$3x = x^2$$
$$0 = x^2 - 3x$$
$$0 = x(x - 3)$$
$$x = 0 \quad \text{or} \quad x = 3$$

51. $f(x) = 5x^2 + 2x - 1$

Since $f(x)$ is a polynomial, the domain is all real numbers x.

53. $h(t) = \dfrac{4}{t}$

Domain: All real numbers except $t = 0$

55. $g(y) = \sqrt{y - 10}$

Domain: $y - 10 \geq 0$
$$y \geq 10$$

57. $f(x) = \sqrt[4]{1 - x^2}$

Domain: $1 - x^2 \geq 0$
$$-x^2 \geq -1$$
$$x^2 \leq 1$$
$$x^2 - 1 \leq 0$$
$$-1 \leq x \leq 1$$

59. $g(x) = \dfrac{1}{x} - \dfrac{1}{x + 2}$

Domain: All real numbers except
$x = 0, \ x = -2$.

61. $f(x) = x^2$

$$\{(-2, 4), (-1, 1), (0, 0), (1, 1), (2, 4)\}$$

63. $f(x) = \sqrt{x + 2}$

$$\{(-2, 0), (-1, 1), \left(0, \sqrt{2}\right), \left(1, \sqrt{3}\right), (2, 2)\}$$

65. The domain is the set of inputs of the function and the range is the set of corresponding outputs.

67. By plotting the points, we have a parabola, so $g(x) = cx^2$. Since $(-4, -32)$ is on the graph, we have $-32 = c(-4)^2 \implies c = -2$. Thus, $g(x) = -2x^2$.

69. Since the function is undefined at 0, we have $r(x) = \dfrac{c}{x}$. Since $(-8, -4)$ is on the graph, we have $-4 = \dfrac{c}{-8} \implies c = 32$. Thus, $r(x) = \dfrac{32}{x}$.

71.
$$f(x) = x^2 - x + 1$$
$$f(2 + h) = (2 + h)^2 - (2 + h) + 1$$
$$= 4 + 4h + h^2 - 2 - h + 1$$
$$= h^2 + 3h + 3$$
$$f(2) = (2)^2 - 2 + 1 = 3$$
$$f(2 + h) - f(2) = h^2 + 3h$$
$$\frac{f(2 + h) - f(2)}{h} = h + 3, \ h \neq 0$$

73. $f(x) = x^3$
$$f(x + c) = (x + c)^3 = x^3 + 3x^2c + 3xc^2 + c^3$$
$$\frac{f(x + c) - f(x)}{c} = \frac{(x^3 + 3x^2c + 3xc^2 + c^3) - x^3}{c}$$
$$= \frac{c(3x^2 + 3xc + c^2)}{c}$$
$$= 3x^2 + 3xc + c^2, \ c \neq 0$$

75. $g(x) = 3x - 1$
$$\frac{g(x) - g(3)}{x - 3} = \frac{(3x - 1) - 8}{x - 3} = \frac{3x - 9}{x - 3} = \frac{3(x - 3)}{x - 3} = 3, \ x \neq 3$$

77. $A = \pi r^2, \ C = 2\pi r$
$$r = \frac{C}{2\pi}$$
$$A = \pi \left(\frac{C}{2\pi} \right)^2 = \frac{C^2}{4\pi}$$

79. (a)

Height, x	Width	Volume, V
1	$24 - 2(1)$	$1[24 - 2(1)]^2 = 484$
2	$24 - 2(2)$	$2[24 - 2(2)]^2 = 800$
3	$24 - 2(3)$	$3[24 - 2(3)]^2 = 972$
4	$24 - 2(4)$	$4[24 - 2(4)]^2 = 1024$
5	$24 - 2(5)$	$5[24 - 2(5)]^2 = 980$
6	$24 - 2(6)$	$6[24 - 2(6)]^2 = 864$

The volume is maximum when $x = 4$.

(b)

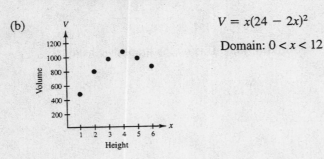

$V = x(24 - 2x)^2$

Domain: $0 < x < 12$

V is a function of x.

81. $A = \frac{1}{2}bh = \frac{1}{2}xy$

Since $(0, y)$, $(1, 2)$, and $(x, 0)$ all lie on the same line, the slopes between any pair are equal.

$$\frac{1-y}{2-0} = \frac{0-1}{x-2}$$

$$1 - y = -\frac{2}{x-2}$$

$$y = \frac{2}{x-2} + 1$$

$$y = \frac{x}{x-2}$$

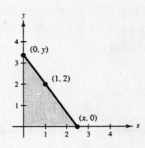

Therefore,

$$A = \frac{1}{2}x\left(\frac{x}{x-2}\right) = \frac{x^2}{2(x-2)}.$$

The domain of A includes x-values such that $2x^2/[2(x-2)] > 0$. Solving this inequality we find that the domain is $x > 2$.

83. $V = l \cdot w \cdot h = x \cdot y \cdot x = x^2 y$ where $4x + y = 108$.

Thus, $y = 108 - 4x$ and $V = x^2(108 - 4x) = 108x^2 - 4x^3$ where $0 < x < 27$.

85. (a) Cost = variable costs + fixed costs

$$C = 12.30x + 98,000$$

(b) Revenue = price per unit × number of units

$$R = 17.98x$$

(c) Profit = Revenue − Cost

$$P = 17.98x - (12.30x + 98,000)$$

$$P = 5.68x - 98,000$$

87. (a) $f(1992) = 28$

(b) $\dfrac{f(1994) - f(1991)}{1994 - 1991} = \dfrac{9 - 60}{3} = \dfrac{-51}{3} = -17$

This is the average decrease per year in the population.

(c)

t	1988	1989	1990	1991	1992	1993	1994	1995
N	9.0	19.8	43.9	53.6	30.9	16.7	10.2	6.9
actual number	10	16	50	60	28	15	9	8

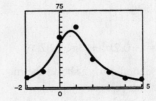

Section 1.2   Graphs of Functions

■   You should be able to determine the domain and range of a function from its graph.

■   You should be able to use the vertical line test for functions.

■   You should be able to determine when a function is constant, increasing, or decreasing.

■   You should be able to find relative maximum and minimum values of a function.

■   You should know that f is
   (a) Odd if $f(-x) = -f(x)$.
   (b) Even if $f(-x) = f(x)$.

Solutions to Odd-Numbered Exercises

1. $f(x) = \sqrt{x^2 - 1}$

Domain: $(-\infty, -1] \cup [1, \infty)$

Range: $[0, \infty)$

3. $h(x) = \sqrt{16 - x^2}$

Domain: $[-4, 4]$

Range: $[0, 4]$

5. $g(x) = 1 - x^2$

Domain: All real numbers

Range: $(-\infty, 1]$

7. $f(x) = |x + 3|$

Domain: All real numbers

Range: $[0, \infty)$

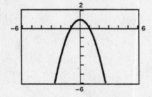

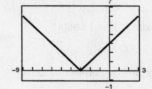

9. $y = \frac{1}{2}x^2$

A vertical line intersects the graph just once, so y is a function of x.

11. $x - y^2 = 1 \implies y = \pm\sqrt{x - 1}$

y is not a function of x. Graph
$y_1 = \sqrt{x - 1}$ and $y_2 = -\sqrt{x - 1}$.

13. $x^2 = 2xy - 1$

A vertical line intersects the graph just once, so y is a function of x. Solve for y and graph

$$y = \frac{x^2 + 1}{2x}.$$

15. $f(x) = -0.2x^2 + 3x + 32$

The second setting shows the most complete graph.

17. $f(x) = 4x^3 - x^4$

The first setting shows the most complete graph.

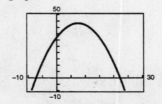

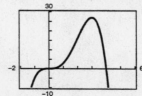

19. $f(x) = \frac{3}{2}x$

   (a) f is increasing on $(-\infty, \infty)$.

   (b) Since $f(-x) = -f(x)$, f is odd.

21. $f(x) = x^3 - 3x^2 + 2$

   (a) f is increasing on $(-\infty, 0)$ and $(2, \infty)$.

   f is decreasing on $(0, 2)$.

   (b) $f(-x) \neq -f(x)$

   $f(-x) \neq f(x)$

   f is neither odd nor even.

23. Yes, for every value of y there corresponds exactly one value of x.

25. $f(x) = 3x^4 - 6x^2$

   (a)

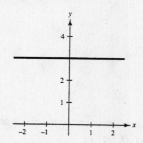

   (b) Increasing on $(-1, 0)$ and $(1, \infty)$

   Decreasing on $(-\infty, -1)$ and $(0, 1)$

   (c) Since $f(-x) = f(x)$, f is even.

27. $f(x) = x\sqrt{x + 3}$

   (a)

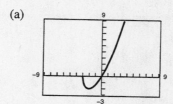

   (b) Increasing on $(-2, \infty)$

   Decreasing on $(-3, -2)$

   (c) $f(-x) \neq -f(x)$

   $f(-x) \neq f(x)$

   f is neither odd nor even.

29. $f(-x) = (-x)^6 - 2(-x)^2 + 3$

   $= x^6 - 2x^2 + 3$

   $= f(x)$

   f is even.

31. $g(-x) = (-x)^3 - 5(-x)$

   $= -x^3 + 5x$

   $= -g(x)$

   g is odd.

33. $f(-t) = (-t)^2 + 2(-t) - 3$

   $= t^2 - 2t - 3$

   $\neq f(t) \neq -f(t)$

   f is neither even nor odd.

35. $\left(-\frac{3}{2}, 4\right)$

   (a) If f is even, another point is $\left(\frac{3}{2}, 4\right)$.

   (b) If f is odd, another point is $\left(\frac{3}{2}, -4\right)$.

37. $f(x) = 3$, even

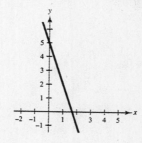

39. $f(x) = 5 - 3x$, neither even nor odd

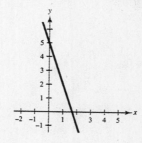

41. $f(x) = \sqrt{1 - x}$, neither even nor odd

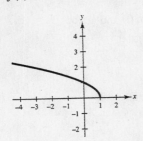

43. $g(t) = \sqrt[3]{t - 1}$, neither even nor odd

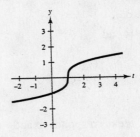

45. $f(x) = \begin{cases} x + 3, & x \leq 0 \\ 3, & 0 < x \leq 2, \\ 2x - 1, & x > 2 \end{cases}$

Neither even nor odd

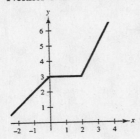

47. $f(x) = \begin{cases} 2x + 3, & x < 0 \\ 3 - x, & x \geq 0 \end{cases}$

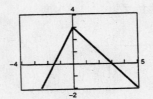

49. $f(x) = \begin{cases} x^2 + 5, & x \leq 1 \\ -x^2 + 4x + 3, & x > 1 \end{cases}$

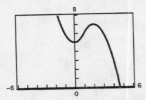

51. $f(x) = x^2 - 6x$

Relative minimum: $(3, -9)$

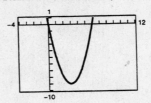

53. $y = 2x^3 + 3x^2 - 12x$

Relative minimum: $(1, -7)$

Relative maximum: $(-2, 20)$

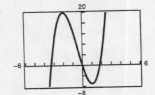

55. $h(x) = (x - 1)\sqrt{x}$

Relative minimum: $(0.33, -0.38)$

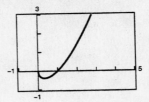

((0, 0) is not a relative maximum because it occurs at the endpoint of the domain $[0, \infty)$.)

57. (a) Let x and y be the length and width of the rectangle. The $100 = 2x + 2y$ or $y = 50 - x$. Thus, the area is

$$A = xy = x(50 - x).$$

(b)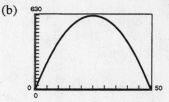

(c) The maximum area is 625 m² when $x = y = 25$ cm. That is, the rectangle is a square.

59. $s(x) = 2\left(\frac{1}{4}x - \left[\!\left[\frac{1}{4}x\right]\!\right]\right)$

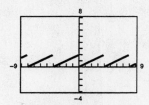

Domain: $(-\infty, \infty)$

Range: $[0, 2)$

Sawtooth pattern

61. (a) $C_2(t) = 0.65 - 0.4\left[\!\left[-(t - 1)\right]\!\right]$ is the appropriate model since the cost does not increase until after the next minute of conversation has started.

(b)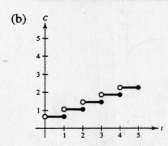

The cost of an 18-minute 45 second call is

$$C = 0.65 - 0.4[[-(18 - 1)]] = 0.65 + 0.4(18)$$
$$= \$7.85.$$

63. $f(x) = 4 - x \geq 0$

$\qquad 4 \geq x$

$(-\infty, 4]$

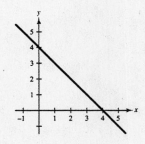

65. $f(x) = x^2 - 9 \geq 0$

$\qquad x^2 \geq 9$

$\qquad x \geq 3 \quad \text{or} \quad x \leq -3$

$\qquad [3, \infty) \quad \text{or} \quad (-\infty, -3]$

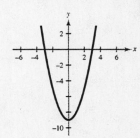

67. $f(x) = 1 - x^4 \geq 0$

$\qquad\qquad 1 \geq x^4$

$\qquad\quad -1 \leq x \leq 1$

$[-1, 1]$

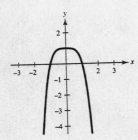

71. $h = \text{top} - \text{bottom}$

$\qquad = (-x^2 + 4x - 1) - 2$

$\qquad = -x^2 + 4x - 3$

75. $L = \text{right} - \text{left}$

$\qquad = \frac{1}{2}y^2 - 0$

$\qquad = \frac{1}{2}y^2$

69. $f(x) = x^2 + 1 \geq 0$

All x

$(-\infty, \infty)$

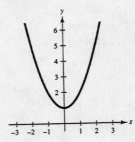

73. $h = \text{top} - \text{bottom}$

$\qquad = (4x - x^2) - 2x$

$\qquad = 2x - x^2$

77. $y = -87.49 + 16.28t - 4.82t^2 - 1.17t^3$

(a) Domain: $-4 \leq t \leq 3$

(b)

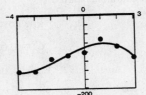

(c) Most accurate in 1992

Least accurate in 1990

(d) The balance would continue to decrease.

79. $f(x) = a_{2n+1}x^{2n+1} + a_{2n-1}x^{2n-1} + \cdots + a_3x^3 + a_1x$

$\quad f(-x) = a_{2n+1}(-x)^{2n+1} + a_{2n-1}(-x)^{2n-1} + \cdots + a_3(-x)^3 + a_1(-x)$

$\qquad = -a_{2n+1}x^{2n+1} - a_{2n-1}x^{2n-1} - \cdots - a_3x^3 - a_1x = -f(x)$

Therefore, $f(x)$ is odd.

Section 1.3   Shifting, Reflecting, and Stretching Graphs

■ You should know the graphs of the most commonly used functions in algebra, and be able to reproduce them on your graphing utility.

(a) Constant function: $f(x) = c$

(b) Identity function: $f(x) = x$

(c) Absolute value function: $f(x) = |x|$

(d) Square root function: $f(x) = \sqrt{x}$

(e) Squaring function: $f(x) = x^2$

(f) Cubing function: $f(x) = x^3$

■ You should know how the graph of a function is changed by vertical and horizontal shifts.

■ You should know how the graph of a function is changed by reflection.

■ You should know how the graph of a function is changed by nonrigid transformations, like stretches and shrinks.

■ You should know how the graph of a function is changed by a sequence of transformations.

Solutions to Odd-Numbered Exercises

1.
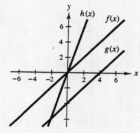

3.
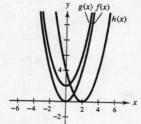

5.
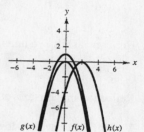

7.
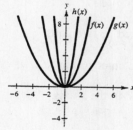

9.
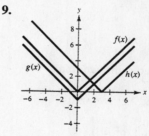

11. (a) $y = f(x) + 2$ (b) $y = -f(x)$ (c) $y = f(x - 2)$

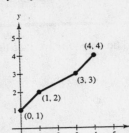   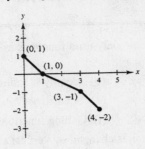   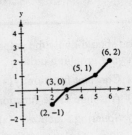

(d) $y = f(x + 3)$ (e) $y = f(2x)$ (f) $y = f(-x)$

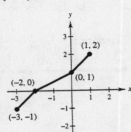   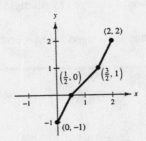   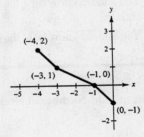

13. Horizontal shift two units to the right of $y = x^3$

$$y = (x - 2)^3$$

15. Reflection in the x-axis of $y = x^2$

$$y = -x^2$$

17. Reflection in the x-axis and a vertical shift one unit upward of $y = \sqrt{x}$

19. Vertical shift one unit downward of $y = x^2$
$$y = x^2 - 1$$

21. Reflection in the x-axis and a vertical shift one unit upward
$$y = 1 - x^3$$

23. Vertical shift three units upward of $y = x$

25. $y = \sqrt{x} + 2$ is $f(x)$ shifted up two units.

27. $y = \sqrt{x - 2}$ is $f(x)$ shifted right two units.

29. $y = \sqrt{2x}$ is a vertical stretch of $f(x)$ by $\sqrt{2}$.

31. $y = |x + 2|$ is $f(x)$ shifted left two units.

33. $y = -|x|$ is $f(x)$ reflected in the x-axis.

35. $y = \frac{1}{3}|x|$ is a vertical shrink of $f(x)$.

37. $f(x) = x^3 - 3x^2$

$g(x) = f(x + 2) = (x + 2)^3 - 3(x + 2)^2$ horizontal shift 2 units to the left

$h(x) = f\left(\frac{1}{2}x\right) = f\left(\frac{1}{2}x\right)^3 - 3\left(\frac{1}{2}x\right)^2$ stretch horizontally by factor of 2

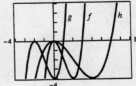

39. $f(x) = x^3 - 3x^2$

$g(x) = -\frac{1}{3}f(x) = -\frac{1}{3}(x^3 - 3x^2)$ reflection in the x-axis and vertical shrink

$h(x) = f(-x) = (-x)^3 - 3(-x)^2$ reflection in the y-axis

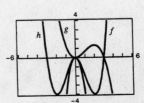

41. $g(x) = 4 - x^3$ is obtained from $f(x)$ by a reflection in the x-axis followed by a vertical shift upward of four units.

43. $h(x) = \frac{1}{4}(x + 2)^3$ is obtained from $f(x)$ by a left shift of two units and a vertical shrink by a factor of $\frac{1}{4}$.

45. $p(x) = \left(\frac{1}{3}x\right)^3 + 2$ is obtained from $f(x)$ by a horizontal stretch, followed by a vertical shift of two units upward.

47. The graph of g is obtained from that of f by first negating f, and then shifting vertically one unit upward: $g(x) = -x^3 + 3x^2 + 1$.

49. (a) $P(x) = 80 + 20x - 0.5x^2, 0 \le x \le 20$

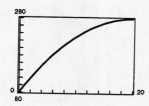

(b) $P(x)$ is shifted downward by a vertical shift of -2500.

$P(x) = -2420 + 20x - 0.5x^2, 0 \le x \le 20$

(c) $P(x)$ is changed by a *horizontal stretch*.

$$P(x) = 80 + 20\left(\frac{x}{100}\right) - 0.5\left(\frac{x}{100}\right)^2$$

$$= 80 + 0.2x - 0.00005x^2 \text{ or } 80 + \frac{1}{5}x - \frac{x^2}{20,000}$$

51. (a)

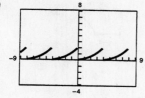

(b)

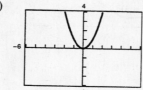

(c)

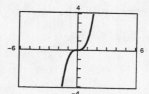

(d)

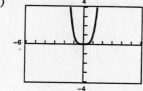

(e)

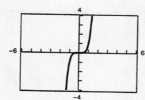

(f)

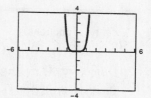

All the graphs pass through the origin. The graphs of the odd powers of x are symmetric to the origin and the graphs of the even powers are symmetric to the y-axis. As the powers increase, the graphs become flatter in the interval $-1 < x < 1$.

53.

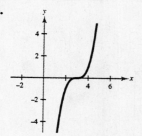

The graph of $y = (x - 3)^3$ is a horizontal shift of $f(x) = x^3$.

55. $f(x) = x^2(x - 6)^2$

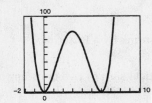

57. $f(x) = x^2(x - 6)^3$

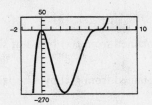

59. (a) For each time t there corresponds one and only one temperature T.

(b) $T(4) \approx 60°$, $T(15) \approx 72°$

(c) All the temperature changes would be one hour later.

(d) The temperature would be decreased by one degree.

61. $f(x - 4)$

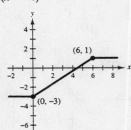

63. $f(x + 4)$

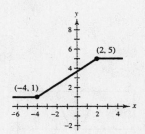

65. $2f(x)$

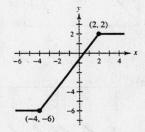

Section 1.4   Combinations of Functions

■ Given two functions, f and g, you should be able to form the following functions (if defined):

1. Sum: $(f + g)(x) = f(x) + g(x)$
2. Difference: $(f - g)(x) = f(x) - g(x)$
3. Product: $(fg)(x) = f(x)g(x)$
4. Quotient: $(f/g)(x) = f(x)/g(x), g(x) \neq 0$
5. Composition of f with g: $(f \circ g)(x) = f(g(x))$
6. Composition of g with f: $(g \circ f)(x) = g(f(x))$

Solutions to Odd-Numbered Exercises

1.

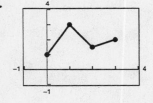

3.

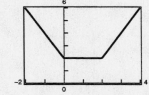

5. $f(x) = x + 1, g(x) = x - 1$

$(f + g)(x) = f(x) + g(x) = (x + 1) + (x - 1) = 2x$

$(f - g)(x) = f(x) - g(x) = (x + 1) - (x - 1) = 2$

$(fg)(x) = f(x) \cdot g(x) = (x + 1)(x - 1) = x^2 - 1$

$\left(\dfrac{f}{g}\right)(x) = \dfrac{f(x)}{g(x)} = \dfrac{x + 1}{x - 1}, \ x \neq 1$

7. $f(x) = x^2, g(x) = 1 - x$

$(f + g)(x) = f(x) + g(x) = x^2 + (1 - x) = x^2 - x + 1$

$(f - g)(x) = f(x) - g(x) = x^2 - (1 - x) = x^2 + x - 1$

$(fg)(x) = f(x) \cdot g(x) = x^2(1 - x) = x^2 - x^3$

$\left(\dfrac{f}{g}\right)(x) = \dfrac{f(x)}{g(x)} = \dfrac{x^2}{1 - x}, x \neq 1$

9. $f(x) = x^2 + 5, g(x) = \sqrt{1 - x}$

$(f + g)(x) = f(x) + g(x) = (x^2 + 5) + \sqrt{1 - x}$

$(f - g)(x) = f(x) - g(x) = (x^2 + 5) - \sqrt{1 - x}$

$(fg)(x) = f(x) \cdot g(x) = (x^2 + 5)\sqrt{1 - x}$

$\left(\dfrac{f}{g}\right)(x) = \dfrac{f(x)}{g(x)} = \dfrac{x^2 + 5}{\sqrt{1 - x}}, \ x < 1$

11. $f(x) = \dfrac{1}{x}, g(x) = \dfrac{1}{x^2}$

$(f + g)(x) = f(x) + g(x) = \dfrac{1}{x} + \dfrac{1}{x^2} = \dfrac{x + 1}{x^2}$

$(f - g)(x) = f(x) - g(x) = \dfrac{1}{x} - \dfrac{1}{x^2} = \dfrac{x - 1}{x^2}$

$(fg)(x) = f(x) \cdot g(x) = \dfrac{1}{x}\left(\dfrac{1}{x^2}\right) = \dfrac{1}{x^3}$

$\left(\dfrac{f}{g}\right)(x) = \dfrac{f(x)}{g(x)} = \dfrac{1/x}{1/x^2} = \dfrac{x^2}{x} = x, \ x \neq 0$

13. $(f + g)(3) = f(3) + g(3) = (3^2 + 1) + (3 - 4) = 9$

15. $(f - g)(0) = f(0) - g(0) = [0^2 + 1] - (0 - 4) = 5$

17. $(f - g)(2t) = f(2t) - g(2t) = [(2t)^2 + 1] - (2t - 4) = 4t^2 - 2t + 5$

19. $(fg)(4) = f(4)g(4) = (4^2 + 1)(4 - 4) = 0$

21. $\left(\dfrac{f}{g}\right)(5) = \dfrac{f(5)}{g(5)} = \dfrac{5^2 + 1}{5 - 4} = 26$

23. $\left(\dfrac{f}{g}\right)(-1) - g(3) = \dfrac{f(-1)}{g(-1)} - g(3)$

$$= \dfrac{(-1)^2 + 1}{-1 - 4} - (3 - 4)$$

$$= -\dfrac{2}{5} + 1 = \dfrac{3}{5}$$

25. $f(x) = \frac{1}{2}x,\ g(x) = x - 1,\ (f + g)(x) = \frac{3}{2}x - 1$

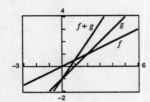

27. $f(x) = x^2,\ g(x) = -2x,\ (f + g)(x) = x^2 - 2x$

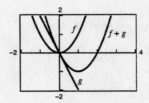

29. $f(x) = 3x,\ g(x) = -\dfrac{x^3}{10},\ (f + g)(x) = 3x - \dfrac{x^3}{10}$

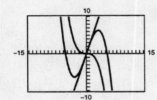

For $0 \le x \le 2$, $f(x)$ contributes most to the magnitude.
For $x > 6$, $g(x)$ contributes most to the magnitude.

31. (a) $T(x) = R(x) + B(x) = \frac{3}{4}x + \frac{1}{15}x^2$

(b)

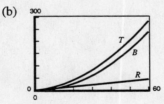

33.

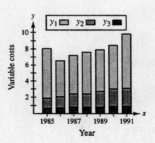

(c) $B(x)$ contributes more to $T(x)$ at higher speeds.

35. $f(x) = x^2,\ g(x) = x - 1$

(a) $(f \circ g)(x) = f(g(x)) = f(x - 1) = (x - 1)^2$

(b) $(g \circ f)(x) = g(f(x)) = g(x^2) = x^2 - 1$

(c) $(f \circ f)(x) = f(f(x)) = f(x^2) = (x^2)^2 = x^4$

37. $f(x) = 3x + 5,\ g(x) = 5 - x$

(a) $(f \circ g)(x) = f(g(x)) = f(5 - x) = 3(5 - x) + 5 = 20 - 3x$

(b) $(g \circ f)(x) = g(f(x)) = g(3x + 5) = 5 - (3x + 5) = -3x$

(c) $(f \circ f)(x) = f(f(x)) = f(3x + 5) = 3(3x + 5) + 5 = 9x + 20$

39. (a) $(f \circ g)(x) = f(g(x)) = f(x^2) = \sqrt{x^2 + 4}$

$(g \circ f)(x) = g(f(x)) = g(\sqrt{x + 4}) = (\sqrt{x + 4})^2$
$= x + 4, \ x \geq -4$

(b)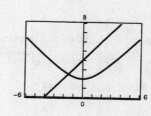

They are not equal.

41. (a) $(f \circ g)(x) = f(g(x)) = f(3x + 1)$
$= \frac{1}{3}(3x + 1) - 3 = x - \frac{8}{3}$

(b) $(g \circ f)(x) = g(f(x)) = g(\frac{1}{3}x - 3)$
$= 3(\frac{1}{3}x - 3) + 1 = x - 8$

(b)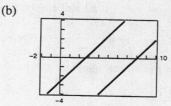

They are not equal.

43. (a) $(f \circ g)(x) = f(g(x)) = f(x^6) = (x^6)^{2/3} = x^4$

$(g \circ f)(x) = g(f(x)) = g(x^{2/3}) = (x^{2/3})^6 = x^4$

(b)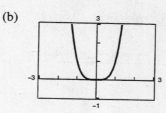

They are equal.

45. (a) $(f + g)(3) = f(3) + g(3) = 2 + 1 = 3$

(b) $\left(\dfrac{f}{g}\right)(2) = \dfrac{f(2)}{g(2)} = \dfrac{0}{2} = 0$

47. (a) $(f \circ g)(2) = f(g(2)) = f(2) = 0$

(b) $(g \circ f)(2) = g(f(2)) = g(0) = 4$

49. Let $f(x) = x^2$ and $g(x) = 2x + 1$, then $(f \circ g)(x) = h(x)$. This is not a unique solution. For example, if $f(x) = (x + 1)^2$ and $g(x) = 2x$, then $(f \circ g)(x) = h(x)$ as well.

51. Let $f(x) = \sqrt[3]{x}$ and $g(x) = x^2 - 4$, then $(f \circ g)(x) = h(x)$. This answer is not unique. Other possibilities may be:

$f(x) = \sqrt[3]{x - 4}$ and $g(x) = x^2$ or
$f(x) = \sqrt[3]{-x}$ and $g(x) = 4 - x^2$ or
$f(x) = \sqrt[9]{x}$ and $g(x) = (4 - x^2)^3$

53. Let $f(x) = 1/x$ and $g(x) = x + 2$, then $(f \circ g)(x) = h(x)$. Again, this is not a unique solution. Other possibilities may be:

$f(x) = \dfrac{1}{x + 2}$ and $g(x) = x$ or

$f(x) = \dfrac{1}{x + 1}$ and $g(x) = x + 1$ or

$f(x) = \dfrac{1}{x^2 + 2}$ and $g(x) = \sqrt{x}$

55. Let $f(x) = x^2 + 2x$ and $g(x) = x + 4$. Then $(f \circ g)(x) = h(x)$. (Answer is not unique.)

57. (a) The domain of $f(x) = \sqrt{x}$ is $x \geq 0$.

(b) The domain of $g(x) = x^2 + 1$ is all real numbers.

(c) $(f \circ g)(x) = f(g(x)) = f(x^2 + 1) = \sqrt{x^2 + 1}$

The domain of $f \circ g$ is all real numbers.

59. (a) The domain of $f(x) = 3/(x^2 - 1)$ is all real numbers except $x = \pm 1$.

(b) The domain of $g(x) = x + 1$ is all real numbers.

(c) $f \circ g = f(g(x)) = f(x + 1) = \dfrac{3}{(x + 1)^2 - 1} = \dfrac{3}{x^2 + 2x} = \dfrac{3}{x(x + 2)}$

This domain of $f \circ g$ is all real numbers except $x = 0$ and $x = -2$.

61. $f(x) = 3x - 4$

$$\frac{f(x + h) - f(x)}{h} = \frac{[3(x + h) - 4] - (3x - 4)}{h}$$

$$= \frac{3x + 3h - 4 - 3x + 4}{h}$$

$$= \frac{3h}{h}$$

$$= 3, \ h \neq 0$$

63. $f(x) = \dfrac{4}{x}$

$$\frac{f(x + h) - f(x)}{h} = \frac{\dfrac{4}{x + h} - \dfrac{4}{x}}{h} = \frac{\dfrac{4x - 4(x + h)}{x(x + h)}}{\dfrac{h}{1}}$$

$$= \frac{4x - 4x - 4h}{x(x + h)} \cdot \frac{1}{h}$$

$$= \frac{-4h}{x(x + h)} \cdot \frac{1}{h}$$

$$= \frac{-4}{x(x + h)}, \ h \neq 0$$

65. (a) $(A \circ r)(t)$ gives the area of the circle as a function of time.

$$(A \circ r)(t) = A(r(t))$$

$$= A(0.6t)$$

$$= \pi(0.6t)^2 = 0.36\pi t^2$$

67. $(C \circ x)(t) = C(x(t))$

$$= 60(50t) + 750$$

$$= 3000t + 750$$

$C \circ x$ represents the cost after t production hours.

69. $g(f(x)) = g(x - 500{,}000) = 0.03(x - 500{,}000)$

represents 3 percent of the amount over $500,000.

(b)

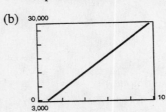

The cost increases to $15,000 when $t = 4.75$ hours.

71. Let $f(x)$ and $g(x)$ be odd functions, and define $h(x) = f(x)g(x)$. Then,

$$h(-x) = f(-x)g(-x)$$
$$= [-f(x)][-g(x)] \quad \text{since } f \text{ and } g \text{ are both odd}$$
$$= f(x)g(x) = h(x).$$

Thus, h is even.

Let $f(x)$ and $g(x)$ be even functions, and define $h(x) = f(x)g(x)$. Then,

$$h(-x) = f(-x)g(-x)$$
$$= f(x)g(x) \quad \text{since } f \text{ and } g \text{ are both even}$$
$$= h(x).$$

Thus, h is even.

73. $g(-x) = \frac{1}{2}[f(-x) + f(-(-x))] = \frac{1}{2}[f(-x) + f(x)] = g(x),$

which shows that g is even.

$$h(-x) = \frac{1}{2}[f(-x) - f(-(-x))] = \frac{1}{2}[f(-x) - f(x)]$$
$$= -\frac{1}{2}[f(x) - f(-x)] = -h(x),$$

which shows that h is odd.

75. (a) $f(x) = g(x) + h(x)$

$$= \frac{1}{2}[f(x) + f(-x)] + \frac{1}{2}[f(x) - f(-x)]$$

$$= \frac{1}{2}[(x^2 - 2x + 1) + (x^2 + 2x + 1)] + \frac{1}{2}[(x^2 - 2x + 1) - (x^2 + 2x + 1)]$$

$$= (x^2 + 1) + (-2x)$$

(b) $f(x) = g(x) + h(x)$

$$= \frac{1}{2}[f(x) + f(-x)] + \frac{1}{2}[f(x) - f(-x)]$$

$$= \frac{1}{2}\left[\left(\frac{1}{x+1} + \frac{1}{-x+1}\right)\right] + \frac{1}{2}\left[\left(\frac{1}{x+1} - \frac{1}{-x+1}\right)\right] = \frac{1}{1-x^2} + \frac{-x}{1-x^2}$$

$$= \frac{-1}{x^2-1} + \frac{x}{x^2-1}$$

77.

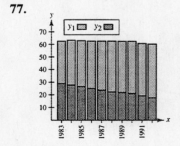

79.

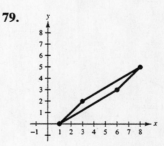

81. $xy > 0$ means $x > 0$ and $y > 0$, or $x < 0$ and $y < 0$. Thus, it lies in Quadrant I or III.

Section 1.5   Inverse Functions

- ■ Two functions f and g are inverses of each other if $f(g(x)) = x$ for every x in the domain of g and $g(f(x)) = x$ for every x in the domain of f.

- ■ Be able to find the inverse of a function, if it exists.

  1. Replace $f(x)$ with y.

  2. Interchange x and y.

  3. Solve for y. If this equation represents y as a function of x, then you have found $f^{-1}(x)$. If this equation does not represent y as a function of x, then f does not have an inverse function.

- ■ A function f has an inverse function if and only if no **horizontal** line crosses the graph of f at more than one point.

- ■ A function f has an inverse function if and only if f is one-to-one.

Solutions to Odd-Numbered Exercises

1. The inverse is a line through $(-1, 0)$.
Matches graph (c).

3. The inverse is half a parabola starting at $(1, 0)$.
Matches graph (a).

5. $f^{-1}(x) = \dfrac{x}{8} = \dfrac{1}{8}x$

$$f(f^{-1}(x)) = f\left(\dfrac{x}{8}\right) = 8\left(\dfrac{x}{8}\right) = x$$

$$f^{-1}(f(x)) = f^{-1}(8x) = \dfrac{8x}{8} = x$$

7. $f^{-1}(x) = x - 10$

$$f(f^{-1}(x)) = f(x - 10) = (x - 10) + 10 = x$$

$$f^{-1}(f(x)) = f^{-1}(x + 10) = (x + 10) - 10 = x$$

9. $f^{-1}(x) = x^3$

$$f(f^{-1}(x)) = f(x^3) = \sqrt[3]{x^3} = x$$

$$f^{-1}(f(x)) = f^{-1}(\sqrt[3]{x}) = \left(\sqrt[3]{x}\right)^3 = x$$

11. (a) $f(g(x)) = f\left(\dfrac{x}{2}\right) = 2\left(\dfrac{x}{2}\right) = x$

  $g(f(x)) = g(2x) = \dfrac{2x}{2} = x$

13. (a) $f(g(x)) = f\left(\dfrac{x - 1}{5}\right) = 5\left(\dfrac{x - 1}{5}\right) + 1 = x$

  $g(f(x)) = g(5x + 1) = \dfrac{(5x + 1) - 1}{5} = x$

(b)

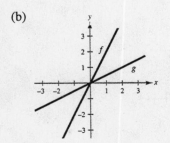

(b)

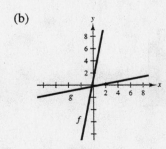

15. (a) $f(g(x)) = f(\sqrt[3]{x}) = (\sqrt[3]{x})^3 = x$

$g(f(x)) = g(x^3) = \sqrt[3]{x^3} = x$

(b)

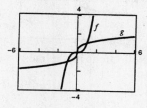

Reflections in the line $y = x$

17. (a) $f(g(x)) = f(x^2 + 4), \; x \geq 0$

$= \sqrt{(x^2 + 4) - 4} = x$

$g(f(x)) = g(\sqrt{x - 4})$

$= (\sqrt{x - 4})^2 + 4 = x$

(b)

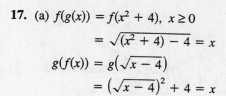

Reflections in the line $y = x$

19. (a) $f(g(x)) = f(\sqrt[3]{1 - x}) = 1 - (\sqrt[3]{1 - x})^3 = 1 - (1 - x) = x$

$g(f(x)) = g(1 - x^3) = \sqrt[3]{1 - (1 - x^3)} = \sqrt[3]{x^3} = x$

(b)

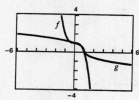

Reflections in the line $y = x$

21. Since no horizontal line crosses the graph of f at more than one point, f **has** an inverse.

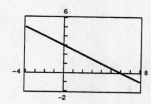

23. Since some horizontal lines cross the graph of f twice, f does **not** have an inverse.

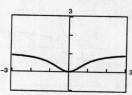

25. $g(x) = \dfrac{4 - x}{6}$

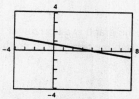

g passes the horizontal line test, so g **has** an inverse.

27. $h(x) = |x + 4| - |x - 4|$

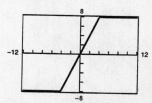

h does not pass the horizontal line test, so h does **not** have an inverse.

29. $f(x) = -2x\sqrt{16 - x^2}$

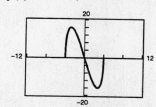

f does not pass the horizontal line test, so f does **not** have an inverse.

31. $f(x) = 2x - 3$

$y = 2x - 3$

$x = 2y - 3$

$y = \dfrac{x + 3}{2}$

$f^{-1}(x) = \dfrac{x + 3}{2}$

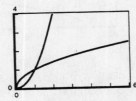

Reflections in the line $y = x$

33. $f(x) = x^5$

$y = x^5$

$x = y^5$

$y = \sqrt[5]{x}$

$f^{-1}(x) = \sqrt[5]{x}$

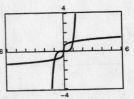

Reflections in the line $y = x$

35. $f(x) = \sqrt{x}$

$y = \sqrt{x}$

$x = \sqrt{y}$

$y = x^2$

$f^{-1}(x) = x^2, \ x \geq 0$

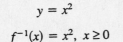

Reflections in the line $y = x$

37. $f(x) = \sqrt{4 - x^2}, \ 0 \leq x \leq 2$

$y = \sqrt{4 - x^2}$

$x = \sqrt{4 - y^2}$

$f^{-1}(x) = \sqrt{4 - x^2}, \ 0 \leq x \leq 2$

Reflections in the line $y = x$

39. $f(x) = \sqrt[3]{x - 1}$

$y = \sqrt[3]{x - 1}$

$x = \sqrt[3]{y - 1}$

$x^3 = y - 1$

$y = x^3 + 1$

$f^{-1}(x) = x^3 + 1$

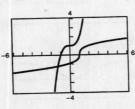

Reflections in the line $y = x$

41. $f(x) = x^4$

$y = x^4$

$x = y^4$

$y = \pm \sqrt[4]{x}$

This does not represent y as a function of x.
f does not have an inverse.

43. $g(x) = \dfrac{x}{8}$

$y = \dfrac{x}{8}$

$x = \dfrac{y}{8}$

$y = 8x$

This is a function of x, so g has an inverse.
$g^{-1}(x) = 8x$

45. $p(x) = -4$

$y = -4$

Since $y = -4$ for all x, the graph is a horizontal line and fails the horizontal line test. p does not have an inverse.

47. $f(x) = (x + 3)^2,\ x \geq -3 \implies y \geq 0$

$y = (x + 3)^2,\ x \geq -3,\ y \geq 0$

$x = (y + 3)^2,\ y \geq -3,\ x \geq 0$

$\sqrt{x} = y + 3\ \ ,\ y \geq -3,\ x \geq 0$

$y = \sqrt{x} - 3,\ x \geq 0,\ y \geq -3$

This is a function of x, so f has an inverse.

$f^{-1}(x) = \sqrt{x} - 3,\ x \geq 0$

49. $h(x) = \dfrac{4}{x^2}$ is not one-to-one, and does not have an inverse. For example, $h(1) = g(-1) = 4$.

51. $f(x) = \sqrt{2x + 3} \implies x \geq -\dfrac{3}{2},\ y \geq 0$

$y = \sqrt{2x + 3},\ x \geq -\dfrac{3}{2},\ y \geq 0$

$x = \sqrt{2y + 3},\ y \geq -\dfrac{3}{2},\ x \geq 0$

$x^2 = 2y + 3,\ x \geq 0,\ y \geq -\dfrac{3}{2}$

$y = \dfrac{x^2 - 3}{2},\ x \geq 0,\ y \geq -\dfrac{3}{2}$

This is a function of x, so f has an inverse.

$f^{-1}(x) = \dfrac{x^2 - 3}{2},\ x \geq 0$

53. $g(x) = x^2 - x^4$

The graph fails the horizontal line test, so g does not have an inverse.

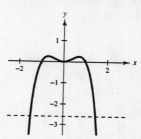

55. $f(x) = 25 - x^2,\ x \leq 0 \implies y \leq 25$

$y = 25 - x^2,\ x \leq 0,\ y \leq 25$

$x = 25 - y^2,\ y \leq 0,\ x \leq 25$

$y^2 = 25 - x,\ x \leq 25,\ y \leq 0$

$y = -\sqrt{25 - x},\ x \leq 25,\ y \leq 0$

This is a function of x, so f has an inverse.

$f^{-1}(x) = -\sqrt{25 - x},\ x \leq 25$

57. If we let $f(x) = (x - 2)^2,\ x \geq 2$, then f has an inverse. [Note: we could also let $x \leq 2$.]

$f(x) = (x - 2)^2,\ x \geq 2 \implies y \geq 0$

$y = (x - 2)^2,\ x \geq 2,\ y \geq 0$

$x = (y - 2)^2,\ x \geq 0,\ y \geq 2$

$\sqrt{x} = y - 2,\qquad x \geq 0,\ y \geq 2$

$\sqrt{x} + 2 = y,\qquad x \geq 0,\ y \geq 2$

Thus, $f^{-1}(x) = \sqrt{x} + 2,\ x \geq 0$.

59. If we let $f(x) = |x + 2|,\ x \geq -2$, then f has an inverse. [Note: we could also let $x \leq -2$.]

$f(x) = |x + 2|,\ x \geq -2$

$f(x) = x + 2$ when $x \geq -2$

$y = x + 2,\ x \geq -2,\ y \geq 0$

$x = y + 2,\ x \geq 0,\ y \geq -2$

$x - 2 = y,\qquad x \geq 0,\ y \geq -2$

Thus, $f^{-1}(x) = x - 2,\ x \geq 0$.

61.

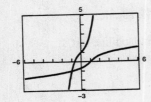

x	$f(x)$
-2	-4
-1	-2
1	2
3	3

x	$f^{-1}(x)$
-4	-2
-2	-1
2	1
3	3

63. $f(x) = x^3 + x + 1$

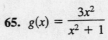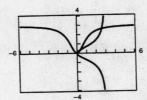

65. $g(x) = \dfrac{3x^2}{x^2 + 1}$

The graph of the inverse relation is an inverse function since it satisfies the vertical line test.

The graph of the inverse relation is not an inverse function since it does not satisfy the vertical line test.

67. False, $f(x) = x^2$ is even and does not have an inverse.

69. True

In Exercises 71, 73, and 75, $f(x) = \frac{1}{8}x - 3$, $f^{-1}(x) = 8(x + 3)$, $g(x) = x^3$, $g^{-1}(x) = \sqrt[3]{x}$.

71. $(f^{-1} \circ g^{-1})(1) = f^{-1}(g^{-1}(1)) = f^{-1}(\sqrt[3]{1}) = 8(\sqrt[3]{1} + 3) = 32$

73. $(f^{-1} \circ f^{-1})(6) = f^{-1}(f^{-1}(6)) = f^{-1}(8[6 + 3]) = f^{-1}(72) = 8(72 + 3) = 600$

75. $(f \circ g)(x) = f(g(x)) = f(x^3) = \frac{1}{8}x^3 - 3$

$$y = \frac{1}{8}x^3 - 3$$

$$x = \frac{1}{8}y^3 - 3$$

$$x + 3 = \frac{1}{8}y^3$$

$$8(x + 3) = y^3$$

$$\sqrt[3]{8(x + 3)} = y$$

$$(f \circ g)^{-1}(x) = 2\sqrt[3]{x + 3}$$

In Exercises 77 and 79, $f(x) = x + 4$, $f^{-1}(x) = x - 4$, $g(x) = 2x - 5$, $g^{-1}(x) = \dfrac{x + 5}{2}$.

77. $(g^{-1} \circ f^{-1})(x) = g^{-1}(f^{-1}(x)) = g^{-1}(x - 4) = \dfrac{(x - 4) + 5}{2} = \dfrac{x + 1}{2}$

79. $(f \circ g)(x) = f(g(x)) = f(2x - 5) = (2x - 5) + 4 = 2x - 1$

$$(f \circ g)^{-1}(x) = \dfrac{x + 1}{2}$$

Note that $(f \circ g)^{-1}(x) = (g^{-1} \circ f^{-1})(x)$.

81. Let $(f \circ g)^{-1}(x) = y$ which implies that $(f \circ g)(y) = x$. Hence, $f(g(y)) = x$ or $f^{-1}(x) = g(y)$ and
$g^{-1}(f^{-1}(x)) = (g^{-1} \circ f^{-1}) = y$. Thus, $(f \circ g)^{-1}(x) = (g^{-1} \circ f^{-1})(x)$.

83. (a)

$$y = 8 + 0.75x$$

$$x = 8 + 0.75y$$

$$x - 8 = 0.75y$$

$$\frac{x - 8}{0.75} = y$$

$$f^{-1}(x) = \frac{x - 8}{0.75}$$

$x =$ hourly wage
$y =$ number of units produced

(b)

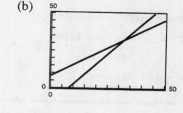

(c) If 10 units are produced, then $y = 8 + 0.75(10) = \$15.50$.

(d) If the hourly wage is \$22.25, then

$$y = \frac{22.25 - 8}{0.75} = 19 \text{ units.}$$

85. (a)

$$y = 0.03x^2 + 254.50, \ 0 < x < 100$$

$$x = 0.03y^2 + 254.50$$

$$x - 254.50 = 0.03y^2$$

$$\frac{x - 254.50}{0.03} = y^2$$

$$\sqrt{\frac{x - 254.50}{0.03}} = y, \ x > 254.50$$

$$f^{-1}(x) = \sqrt{\frac{x - 254.50}{0.03}}$$

$x =$ temperature in degrees Fahrenheit
$y =$ percent load for a diesel engine

(b)

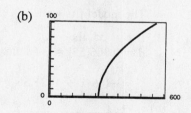

(c) $0.03x^2 + 254.50 < 500$

$$0.03x^2 < 245.5$$

$$x^2 < 8183\tfrac{1}{3}$$

$$x < 90.46$$

Thus, $0 < x < 90.46$.

Section 1.6   Exploring Data: Linear Models and Scatter Plots

---

- ■ You should know how to construct a scatter plot.
- ■ You should know how to fit a line to data by
  - (a) Visually determining the line.
  - (b) Finding the least squares regression line.

Solutions to Odd-Numbered Exercises

1. Negative correlation—y decreases as x increases.

3. No correlation.

5. (a)

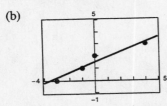

(b) Yes. The cancer mortality increases linearly with increased exposure to the carcinogenic substance.

7. (a) $y = 0.42x + 1.50$

(b)

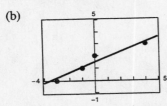

(c) Yes, the model appears valid.

9. (a) $y = 0.95x + 0.92$

(b)

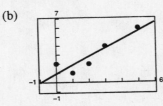

(c) Yes, the model appears valid.

11. (a) $d = 0.066F$ or $F = 15.15d$

(b)

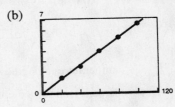

The model fits well.

(c) If $F = 55$, then $d \approx 0.066(55) = 3.63$ cm.

13. (a) $R = 1.21t + 4.73$

(b)

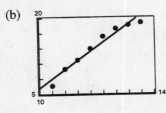

(c) The slope is the average increase in the monthly basic rate per year.

(d) For the year 2000, $t = 20$, and $R \approx \$28.84$.

15. (a) $y = 47.77x + 103.77$

(b)

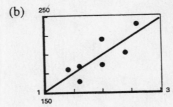

(c) The slope is the average increase in sales for increases in advertising expenditures.

(d) If $x = \$1500$, then $y \approx \$175,000$.

17. (a) $y = -0.64x + 1321.56$

(b)

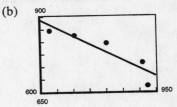

(c) The negative slope indicates that the amount of mortgage debt held by commercial banks is increasing as the amount held by savings institutions decreases.

❑ Review Exercises for Chapter 1

Solutions to Odd-Numbered Exercises

1. No, the table does not describe a function. The input value 10 has two distinct outputs, 5 and -5. (So does the input value 5.)

3. $16x - y^4 = 0$

$$y^4 = 16x$$

$$y = \pm 2\sqrt[4]{x}$$

y is **not** a function of x. Some x-values correspond to two y-values.

5. $y = \sqrt{1 - x}$

Each x-value, $x \leq 1$, corresponds to only one y-value so y **is** a function of x.

7. $f(x) = x^2 + 1$

(a) $f(2) = (2)^2 + 1 = 5$

(b) $f(t^2) = (t^2)^2 + 1 = t^4 + 1$

(c) $-f(x) = -(x^2 + 1) = -x^2 - 1$

9. $h(x) = 6 - 5x^2$

(a) $h(2) = 6 - 5(2)^2 = 6 - 20 = -14$

(b) $h(x + 3) = 6 - 5(x + 3)^2$

$$= 6 - 5(x^2 + 6x + 9)$$

$$= -5x^2 - 30x - 39$$

(c) $\dfrac{h(x + t) - h(x)}{t} = \dfrac{6 - 5(x + t)^2 - (6 - 5x^2)}{t}$

$$= \dfrac{-5(x^2 + 2tx + t^2) + 5x^2}{t}$$

$$= \dfrac{-10tx - 5t^2}{t}$$

$$= -10x - 5t, \ t \neq 0$$

11. $g(x) = \begin{cases} \frac{1}{2}x + 1, & x \leq 2 \\ x - 2, & x > 2 \end{cases}$

(a) $g(-2) = \frac{1}{2}(-2) + 1 = 0$

(b) $g(2) = \frac{1}{2}(2) + 1 = 2$

(c) $g(10) = 10 - 2 = 8$

13. $f(x) = \sqrt{25 - x^2}$

Domain: $25 - x^2 \geq 0$

$$(5 + x)(5 - x) \geq 0$$

Critical numbers: $x = \pm 5$

Test intervals: $(-\infty, -5), \ (-5, 5), \ (5, \infty)$

Solution set: $[-5, 5]$

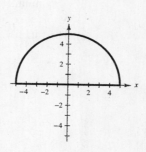

15. $g(s) = \dfrac{5}{3s - 9} = \dfrac{5}{3(s - 3)}$

Domain: All real numbers except $s = 3$

17. $h(x) = \dfrac{x}{x^2 - x - 6}$

$$= \dfrac{x}{(x - 3)(x + 2)}$$

Domain: $(-\infty, -2), (-2, 3), (3, \infty)$

19. $f(x) = \dfrac{3x}{2(3 - x)}$

The second setting shows the most complete graph.

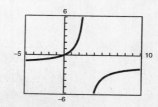

21.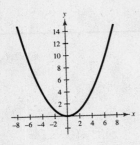

23.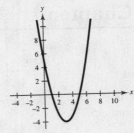

25.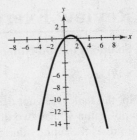

27.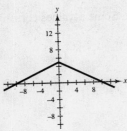

29.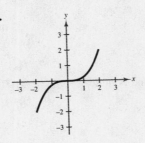

31.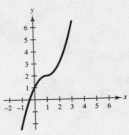

33. $h(x) = -x^4$ is a reflection in the x-axis.

35. $h(x) = (x - 2)^4$ is a horizontal shift to the right.

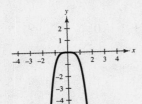

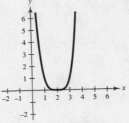

37. Vertical shift downward 2 units: $y = \sqrt{x} - 2$

39. Reflection in the x-axis: $y = -\sqrt{x}$

41.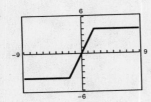

(a) Constant on: $(-\infty, -2), (2, \infty)$
    Increasing on: $(-2, 2)$
(b) Maximum: 4
    Minimum: -4
(c) The function is odd $g(x) = -g(-x)$.

43. (a) Increasing: $(-\infty, 3)$
        Decreasing: $(3, \infty)$
    (b) Relative maximum: $(3, 27)$
    (c) Neither even nor odd

45. (a) $C = 5.35x + 16,000$
    (b) $P = R - C = 8.20x - (5.35x + 16,000)$
        $= 2.85x - 16,000$

47. (a) $\quad f(x) = \frac{1}{2}x - 3$
    $y = \frac{1}{2}x - 3$
    $x = \frac{1}{2}y - 3$
    $x + 3 = \frac{1}{2}y$
    $2(x + 3) = y$
    $f^{-1}(x) = 2x + 6$

(b)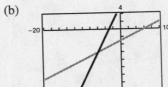

(c) $f^{-1}(f(x)) = f^{-1}\left(\frac{1}{2}x - 3\right)$
    $= 2\left(\frac{1}{2}x - 3\right) + 6$
    $= x - 6 + 6$
    $= x$
    $f(f^{-1}(x)) = f(2x + 6)$
    $= \frac{1}{2}(2x + 6) - 3$
    $= x + 3 - 3$
    $= x$

49. (a)
$$f(x) = \sqrt{x + 1}$$
$$y = \sqrt{x + 1}$$
$$x = \sqrt{y + 1}$$
$$x^2 = y + 1$$
$$x^2 - 1 = y$$
$$f^{-1}(x) = x^2 - 1, \; x \geq 0$$

(b)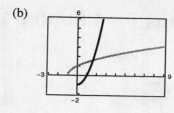

(c)
$$f^{-1}(f(x)) = f^{-1}\left(\sqrt{x + 1}\right)$$
$$= \left(\sqrt{x + 1}\right)^2 - 1$$
$$= x + 1 - 1$$
$$= x$$
$$f(f^{-1}(x)) = f(x^2 - 1)$$
$$= \sqrt{(x^2 - 1) + 1}$$
$$= \sqrt{x^2}$$
$$= x \text{ for } x \geq 0$$

Note: The inverse must have a restricted domain.

51. (a)
$$f(x) = x^2 - 5, \; x \geq 0$$
$$y = x^2 - 5$$
$$x = y^2 - 5$$
$$x + 5 = y^2$$
$$y = \sqrt{x + 5} \quad \text{(Choose positive square root)}$$
$$f^{-1}(x) = \sqrt{x + 5}$$

(b)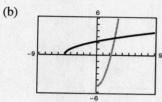

(c)
$$f^{-1}(f(x)) = f^{-1}(x^2 - 5)$$
$$= \sqrt{x^2 - 5 + 5}$$
$$= \sqrt{x^2} = x$$
$$f(f^{-1}(x)) = f\left(\sqrt{x + 5}\right)$$
$$= (x + 5) - 5 = x$$

53. $f(x) = 2(x - 4)^2$ is increasing on $[4, \infty)$.

Let $f(x) = 2(x - 4)^2, \; x \geq 4$ and $y \geq 0$.
$$y = 2(x - 4)^2$$
$$x = 2(y - 4)^2, \; x \geq 0, \; y \geq 4$$
$$\frac{x}{2} = (y - 4)^2$$
$$\sqrt{\frac{x}{2}} = y - 4$$
$$\sqrt{\frac{x}{2}} + 4 = y$$
$$f^{-1}(x) = \sqrt{\frac{x}{2}} + 4, \; x \geq 0$$

55.
$$f(x) = \sqrt{x^2 - 4}, \; x \geq 2$$
$$y = \sqrt{x^2 - 4}$$
$$x = \sqrt{y^2 - 4}$$
$$x^2 = y^2 - 4$$
$$x^2 + 4 = y^2$$
$$y = \sqrt{x^2 + 4}$$
$$f^{-1}(x) = \sqrt{x^2 + 4}, \; x \geq 0$$

57.
$$(f - g)(4) = f(4) - g(4)$$
$$= [3 - 2(4)] - \sqrt{4}$$
$$= -5 - 2$$
$$= -7$$

59. $(fh)(1) = f(1)h(1) = (3 - 2(1))(3(1)^2 + 2)$
$$= (1)(5) = 5$$

61.
$$(h \circ g)(7) = h(g(7))$$
$$= h\left(\sqrt{7}\right)$$
$$= 3\left(\sqrt{7}\right)^2 + 2$$
$$= 23$$

63. $g^{-1}(3) = 9$ because $g(9) = 3$.

65. (a)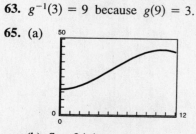

(b) $S = 34.4$

(c) The model will continue to decrease.

67. c, b, d, a

❑ Chapter Test Solutions for Chapter 1

1. $f(-6) = 10 - \sqrt{3 - (-6)} = 10 - 3 = 7$

2. $f(t - 3) = 10 - \sqrt{3 - (t - 3)} = 10 - \sqrt{6 - t}$

3. $\dfrac{f(x) - f(2)}{x - 2} = \dfrac{10 - \sqrt{3 - x} - 9}{x - 2} = \dfrac{\sqrt{3 - x} - 1}{2 - x}$

4. $3 - x \geq 0 \implies$ domain is all $x \leq 3$.

5. No, for some x there corresponds more than one value of y. For instance, if $x = 1$, $y = \pm 1/\sqrt{3}$.

6. (a)

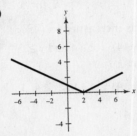

(b)

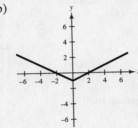

7. $C = 5.60x + 24,000$

$P = R - C = 9.20x - (5.60x + 24,000)$

$= 3.60x - 24,000$

8.

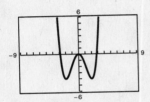

Even function
Increasing: $(-2, 0)$, $(2, \infty)$
Decreasing: $(-\infty, -2)$, $(0, 2)$

9.

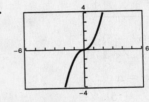

Odd function
Increasing: $(-\infty, \infty)$

10. (a) $(f - g)(x) = x^2 - \sqrt{2 - x}$, Domain: $x \leq 2$

(b) $\left(\dfrac{f}{g}\right)(x) = \dfrac{x^2}{\sqrt{2 - x}}$, Domain: $x < 2$

(c) $(f \circ g)(x) = f\left(\sqrt{2 - x}\right) = 2 - x$, Domain $x \leq 2$

(d) $y = \sqrt{2 - x}$

$x = \sqrt{2 - y}$ Interchange x and y

$x^2 = 2 - y$

$y = g^{-1}(x) = 2 - x^2$, $x \geq 0$

11. The graph of f must satisfy the horizontal line test.

12. f and f^{-1} are reflections in the line $y = x$.

13. Using your graphing utility, you obtain $y = 0.0568x - 1.5808$. For $x = 88$, $y \approx 3.4$.

❑ Practice Test for Chapter 1

1. Use a graphing utility to graph the equation $y = 4/x^2 - 5$. Approximate any x-intercepts of the graph.

2. Use a graphing utility to graph the equation $y = |x - 3| + 2$. Approximate any x-intercepts of the graph.

3. Graph $3x - 5y = 15$ by hand.

4. Graph $y = \sqrt{9 - x}$ by hand.

5. Solve $5x + 4 = 7x - 8$.

6. Solve $\dfrac{x}{3} - 5 = \dfrac{x}{5} + 1$.

7. Does the equation $x^4 + y^4 = 16$ represent y as a function of x?

8. Evaluate the function $f(x) = |x - 2|/(x - 2)$ at the points $x = 0$, $x = 2$, and $x = 4$.

9. Find the domain of the function $f(x) = 5/(x^2 - 16)$.

10. Find the domain of the function $g(t) = \sqrt{4 - t}$.

11. Use a graphing utility to sketch the graph of the function $f(x) = 3 - x^6$ and determine if the function is even, odd, or neither.

12. Use a graphing utility to approximate any relative minimum or maximum values of the function $y = 4 - x + x^3$.

13. Compare the graph of $f(x) = x^3 - 3$ with the graph of $y = x^3$.

14. Compare the graph of $f(x) = \sqrt{x - 6}$ with the graph of $y = \sqrt{x}$.

15. Find $g \circ f$ if $f(x) = \sqrt{x}$ and $g(x) = x^2 - 2$. What is the domain of $g \circ f$?

16. Find f/g if $f(x) = 3x^2$ and $g(x) = 16 - x^4$. What is the domain of f/g?

17. Show that $f(x) = 3x + 1$ and $g(x) = \dfrac{x - 1}{3}$ are inverse functions algebraically and graphically.

18. Find the inverse of $f(x) = \sqrt{9 - x^2}$, $0 \le x \le 3$. Graph f and f^{-1} in the same viewing rectangle.

19. Use a graphing utility to find the least squares regression line for the points $(-1, 0)$, $(0, 1)$, $(3, 3)$ and $(4, 5)$. Graph the points and the line.

20. The following ordered pairs give the scores of two consecutive 10-point quizzes for a class of 12 students. Create a scatter plot for the data. Does the relationship between consecutive quiz scores appear to be linear? Why or why not?

$(6, 4)$, $(7, 5)$, $(7, 6)$, $(10, 9)$, $(10, 10)$, $(9, 7)$, $(4, 2)$, $(8, 4)$, $(10, 8)$, $(6, 5)$, $(7. 8)$, $(9, 5)$

CHAPTER 2
Polynomial and Rational Functions

CHAPTER 2
Polynomial and Rational Functions

Section 2.1   Quadratic Functions

You should know the following facts about parabolas.

■ $f(x) = ax^2 + bx + c$, $a \neq 0$, is a quadratic function, and its graph is a parabola.

■ If $a > 0$, the parabola opens upward and the vertex is the minimum point. If $a < 0$, the parabola opens downward and the vertex is the maximum point.

■ The vertex is $(-b/2a, f(-b/2a))$.

■ To find the x-intercepts (if any), solve
   $$ax^2 + bx + c = 0.$$

■ The standard form of the equation of a parabola is
   $$f(x) = a(x - h)^2 + k$$
   where $a \neq 0$.

   (a) The vertex is (h, k).

   (b) The axis is the vertical line $x = h$.

Solutions to Odd-Numbered Exercises

1. $f(x) = (x - 2)^2$ opens upward and has vertex $(2, 0)$. Matches graph (g).

3. $f(x) = x^2 - 2$ opens upward and has vertex $(0, -2)$. Matches graph (b).

5. $f(x) = 4 - (x - 2)^2 = -(x - 2)^2 + 4$ opens downward and has vertex $(2, 4)$. Matches graph (f).

7. $f(x) = x^2 + 3$ opens upward and has vertex $(0, 3)$. Matches graph (e).

9. (a) $y = \frac{1}{2}x^2$

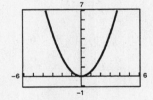

Vertical shrink

(b) $y = -\frac{1}{8}x^2$

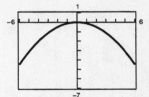

Vertical shrink and reflection in the x-axis

(c) $y = \frac{3}{2}x^2$

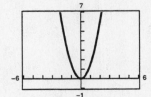

Vertical stretch

(d) $y = -3x^2$

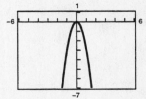

Vertical stretch and reflection in the x-axis

11. (a) $y = (x - 1)^2$

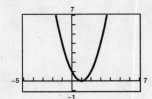

Horizontal translation one unit to the right

(b) $y = (x + 1)^2$

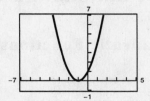

Horizontal translation one unit to the left.

(c) $y = (x - 3)^2$

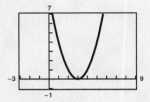

Horizontal translation three units to the right

(d) $y = (x + 3)^2$

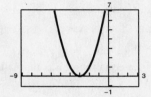

Horizontal translation three units to the left

13. $f(x) = 16 - x^2$
Vertex: $(0, 16)$
Intercepts: $(-4, 0)$, $(0, 16)$, $(4, 0)$

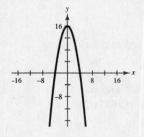

15. $h(x) = x^2 - 8x + 16 = (x - 4)^2$
Vertex: $(4, 0)$
Intercepts: $(0, 16)$, $(4, 0)$

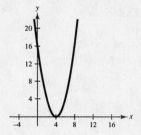

17. $f(x) = x^2 - x + \frac{5}{4} = \left(x - \frac{1}{2}\right)^2 + 1$
Vertex: $\left(\frac{1}{2}, 1\right)$
Intercepts: $\left(0, \frac{5}{4}\right)$

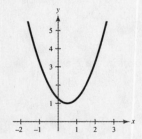

19. $f(x) = -x^2 + 2x + 5 = -(x - 1)^2 + 6$
Vertex: $(1, 6)$
Intercepts: $\left(1 - \sqrt{6}, 0\right)$, $(0, 5)$, $\left(1 + \sqrt{6}, 0\right)$

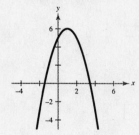

21. $h(x) = 4x^2 - 4x + 21 = 4\left(x - \frac{1}{2}\right)^2 + 20$

Vertex: $\left(\frac{1}{2}, 20\right)$

Intercept: $(0, 21)$

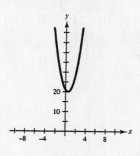

25. $f(x) = 2x^2 - 16x + 31$

$= 2(x - 4)^2 - 1$

Vertex: $(4, -1)$

Intercepts: $\left(4 \pm \frac{1}{2}\sqrt{2}, 0\right), \ (0, 31)$

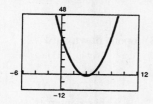

23. $f(x) = -(x^2 + 2x - 3) = -(x + 1)^2 + 4$

Vertex: $(-1, 4)$

Intercepts: $(-3, 0), \ (0, 3), \ (1, 0)$

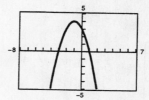

27. $(1, 0)$ is the vertex.

$f(x) = a(x - 1)^2 + 0 = a(x - 1)^2$

Since the graph passes through the point $(0, 1)$ we have:

$1 = a(0 - 1)^2$

$1 = a$

$f(x) = 1(x - 1)^2 = (x - 1)^2$

29. $(-2, 2)$ is the vertex.

$f(x) = a(x + 2)^2 + 2$

Since the graph passes through the point $(-1, 0)$, we have:

$0 = a(-1 + 2)^2 + 2$

$-2 = a$

$f(x) = -2(x + 2)^2 + 2$

33. $(3, 4)$ is the vertex.

$f(x) = a(x - 3)^2 + 4$

Since the graph passes through the point $(1, 2)$, we have:

$2 = a(1 - 3)^2 + 4$

$-2 = 4a$

$-\frac{1}{2} = a$

$f(x) = -\frac{1}{2}(x - 3)^2 + 4$

31. $(-2, 5)$ is the vertex.

$f(x) = a(x + 2)^2 + 5$

Since the graph passes through the point $(0, 9)$, we have:

$9 = a(0 + 2)^2 + 5$

$4 = 4a$

$1 = a$

$f(x) = 1(x + 2)^2 + 5 = (x + 2)^2 + 5$

35. $(5, 12)$ is the vertex.

$f(x) = a(x - 5)^2 + 12$

Since the graph passes through the point $(7, 15)$, we have:

$15 = a(7 - 5)^2 + 12$

$3 = 4a \ \Rightarrow \ a = \frac{3}{4}$

$f(x) = \frac{3}{4}(x - 5)^2 + 12$

37. $y = x^2 - 16$

$0 = x^2 - 16$

x-intercepts: $(\pm 4, 0)$

$x^2 = 16$

$x = \pm 4$

39. $y = x^2 - 4x - 5$

$0 = x^2 - 4x - 5$

x-intercepts: $(5, 0), (-1, 0)$

$0 = (x - 5)(x + 1)$

$x = 5$ or $x = -1$

41. $y = x^2 - 4x$

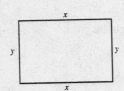

x-intercepts: $(0, 0), (4, 0)$

$0 = x^2 - 4x$

$0 = x(x - 4)$

$x = 0$ or $x = 4$

43. $y = 2x^2 - 7x - 30$

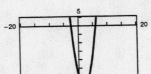

x-intercepts: $\left(-\frac{5}{2}, 0\right), (6, 0)$

$0 = 2x^2 - 7x - 30$

$0 = (2x + 5)(x - 6)$

$x = -\frac{5}{2}$ or $x = 6$

45. $f(x) = [x - (-1)](x - 3)$ opens upward

$= (x + 1)(x - 3)$

$= x^2 - 2x - 3$

$f(x) = -[x - (-1)](x - 3)$

$= -(x + 1)(x - 3)$ opens

$= -(x^2 - 2x - 3)$ downward

$= -x^2 + 2x + 3$

Note: $f(x) = a(x + 1)(x - 3)$ has x-intercepts $(-1, 0)$ and $(3, 0)$ for all real numbers $a \neq 0$.

47. $f(x) = [x - (-3)]\left[x - \left(-\frac{1}{2}\right)\right](2)$ opens upward

$= (x + 3)\left(x + \frac{1}{2}\right)(2)$

$= (x + 3)(2x + 1)$

$= 2x^2 + 7x + 3$

$f(x) = -(2x^2 + 7x + 3)$ opens downward

$= -2x^2 - 7x - 3$

Note: $f(x) = a(x + 3)(2x + 1)$ has x-intercepts $(-3, 0)$ and $\left(-\frac{1}{2}, 0\right)$ for all real numbers $a \neq 0$.

49. Let $x =$ the first number and $y =$ the second number. Then the sum is

$x + y = 110 \implies y = 110 - x$.

The product is $P(x) = xy = x(110 - x) = 110x - x^2$.

$P(x) = -x^2 + 110x$

$= -(x^2 - 110x + 3025 - 3025)$

$= -[(x - 55)^2 - 3025]$

$= -(x - 55)^2 + 3025$

The maximum value of the product occurs at the vertex of $P(x)$ and is 3025. This happens when $x = y = 55$.

51.

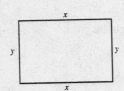

$2x + 2y = 100$

$y = 50 - x$

(a) $A(x) = xy = x(50 - x)$

Domain: $0 < x < 50$

—CONTINUED—

51. —CONTINUED—

(b)

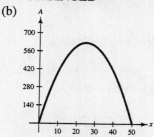

(c) The area is maximum (625 square feet) when $x = y = 25$. The rectangle has dimensions 25 ft × 25 ft. Algebraically, you have:

$$A(x) = -(x^2 - 50x)$$
$$= -(x^2 - 50x + 625) + 625$$
$$= -(x - 25)^2 + 625$$

53. $R = 900x - 0.1x^2 = -0.1x^2 + 900x$

The vertex occurs at

$$x = -\frac{b}{2a} = -\frac{900}{2(-0.1)} = 4500.$$

The revenue is maximum when $x = 4500$ units.

55. (a) $4x + 3y = 200 \implies y = \frac{1}{3}(200 - 4x)$

x	y	Area
2	$\frac{1}{3}[200 - 4(2)]$	$2xy = 256$
4	$\frac{1}{3}[200 - 4(4)]$	$2xy \approx 490$
6	$\frac{1}{3}[200 - 4(6)]$	$2xy = 704$
8	$\frac{1}{3}[200 - 4(8)]$	$2xy = 896$
10	$\frac{1}{3}[200 - 4(10)]$	$2xy \approx 1067$
12	$\frac{1}{3}[200 - 4(12)]$	$2xy = 1216$

(b)

x	y	Area
20	$\frac{1}{3}[200 - 4(20)]$	$2xy = 1600$
22	$\frac{1}{3}[200 - 4(22)]$	$2xy \approx 1643$
24	$\frac{1}{3}[200 - 4(24)]$	$2xy \approx 1664$
26	$\frac{1}{3}[200 - 4(26)]$	$2xy \approx 1664$
28	$\frac{1}{3}[200 - 4(28)]$	$2xy \approx 1643$
30	$\frac{1}{3}[200 - 4(30)]$	$2xy = 1600$

(c) $A = 2xy = 2x\left(\dfrac{200 - 4x}{3}\right) = \dfrac{2x(4)(50 - x)}{3}$

$$= \frac{8x(50 - x)}{3}$$

(d)

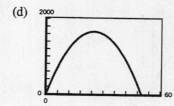

This area is maximum when $x = 25$ feet and $y = \frac{100}{3} = 33\frac{1}{3}$ feet.

(e) $A = \dfrac{8}{3}x(50 - x)$

$$= -\frac{8}{3}(x^2 - 50x)$$
$$= -\frac{8}{3}(x^2 - 50x + 625 - 625)$$
$$= -\frac{8}{3}[(x - 25)^2 - 625]$$
$$= -\frac{8}{3}(x - 25)^2 + \frac{5000}{3}$$

The maximum area occurs at the vertex and is 5000/3 square feet. This happens when $x = 25$ feet and $y = (200 - 4(25))/3 = 100/3$ feet. The dimensions are $2x = 50$ feet by $33\frac{1}{3}$ feet.

57. $y = -\dfrac{1}{12}x^2 + 2x + 4$

(a)
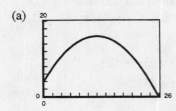

(b) When $x = 0$, $y = 4$ feet.

(c) The vertex occurs at
$$x = -\frac{b}{2a} = -\frac{2}{2(-1/12)} = 12.$$

The maximum height is

$$y = -\frac{1}{12}(12)^2 + 2(12) + 4$$

$$= 16 \text{ feet.}$$

(d) You can solve this part graphically by finding the x-intercept of the graph:
$$x \approx 25.856.$$
When the ball strikes the ground, $y = 0$.

$$0 = -\frac{1}{12}x^2 + 2x + 4$$

$$0 = x^2 - 24x - 48 \quad \text{Multiply both sides by } -12.$$

$$x = \frac{-(-24) \pm \sqrt{(-24)^2 - 4(1)(-48)}}{2(1)}$$

$$= \frac{24 \pm \sqrt{768}}{2} = \frac{24 \pm 16\sqrt{3}}{2} = 12 \pm 8\sqrt{3}$$

Using the positive value for x, we have
$$x = 12 + 8\sqrt{3} \approx 25.86 \text{ feet.}$$

59. $V = 0.77x^2 - 1.32x - 9.31, \ 5 \le x \le 40$

(a)
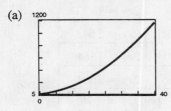

(b) $V(16) = 166.69$ board feet

(c) $500 = 0.77x^2 - 1.32x - 9.31$

$$0 = 0.77x^2 - 1.32x - 509.31$$

Using the Quadratic Formula and selecting the positive value for x, we have $x \approx 26.6$ inches in diameter. Or, use a graphing utility.

61. $C = 4024.5 + 51.4t - 3.1t^2, \ -10 \le t \le 30$

(a)
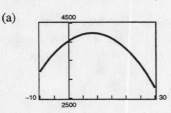

(b) $-\dfrac{b}{2a} = \dfrac{-51.4}{2(-3.1)} \approx 8.29$

The vertex occurs when $x \approx 8.29$ which corresponds to 1968. The warnings may not have had an immediate effect, but over time they and other findings about the health risks of cigarettes have had an effect.

(c) $C(0) = 4024.5$ annually; $\dfrac{4024.5}{366} \approx 11$ daily

63. (a) If $f(x) = ax^2 + bx + c$ has two real zeros, then by the Quadratic Formula they are

$$x = \frac{-b \pm \sqrt{b^2 - 4ac}}{2a}.$$

The average of the zeros of f is

$$\frac{\dfrac{-b - \sqrt{b^2 - 4ac}}{2a} + \dfrac{-b + \sqrt{b^2 - 4ac}}{2a}}{2} = \frac{\dfrac{-2b}{2a}}{2} = -\frac{b}{2a}.$$

This is the x-coordinate of the vertex of the graph.

(b) The zeros of $f(x) = \frac{1}{2}(x - 3)^2 - 2$

$$= \frac{1}{2}(x^2 - 6x + 5)$$

$$= \frac{1}{2}(x - 15)(x - 1) \text{ are } x = 1, 5.$$

The x-coordinate of the vertex is 3, the average of 1 and 5.

Section 2.2   Polynomial Functions of Higher Degree

- You should know the following basic principles about polynomials.
- $f(x) = a_n x^n + a_{n-1} x^{n-1} + \cdots + a_2 x^2 + a_1 x + a_0$ is a polynomial function of degree n.
- If f is of odd degree and
  (a) $a_n > 0$, then
     1. $f(x) \to \infty$ as $x \to \infty$.
     2. $f(x) \to -\infty$ as $x \to -\infty$.
  (b) $a_n < 0$, then
     1. $f(x) \to -\infty$ as $x \to \infty$.
     2. $f(x) \to \infty$ as $x \to -\infty$.
- If f is of even degree and
  (a) $a_n > 0$, then
     1. $f(x) \to \infty$ as $x \to \infty$.
     2. $f(x) \to \infty$ as $x \to -\infty$.
  (b) $a_n < 0$, then
     1. $f(x) \to -\infty$ as $x \to \infty$.
     2. $f(x) \to -\infty$ as $x \to -\infty$.
- The following are equivalent for a polynomial function.
  (a) $x = a$ is a zero of a function.
  (b) $x = a$ is a solution of the polynomial equation $f(x) = 0$.
  (c) $(x - a)$ is a factor of the polynomial.
  (d) $(a, 0)$ is an x-intercept of the graph of f.
- A polynomial of degree n has at most n distinct zeros.
- If f is a polynomial function such that $a < b$ and $f(a) \neq f(b)$, then f takes on every value between $f(a)$ and $f(b)$ in the interval $[a, b]$.
- If you can find a value where a polynomial is positive and another value where it is negative, then there is at least one real zero between the values.

Solutions to Odd-Numbered Exercises

1. $f(x) = -2x + 3$ is a line with y-intercept $(0, 3)$. Matches graph (f).

3. $f(x) = -2x^2 - 5x$ is a parabola with x-intercepts $(0, 0)$ and $\left(-\frac{5}{2}, 0\right)$ and opens downward. Matches graph (c).

5. $f(x) = -\frac{1}{4}x^4 + 3x^2$ has intercepts $(0, 0)$ and $\left(\pm 2\sqrt{3}, 0\right)$. Matches graph (e).

7. $f(x) = x^4 + 2x^3$ has intercepts $(0, 0)$ and $(-2, 0)$. Matches graph (g).

9. $y = x^3$

(a) $f(x) = (x - 2)^3$

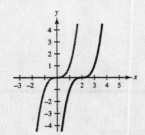

Horizontal shift two units to the right

(b) $f(x) = x^3 - 2$

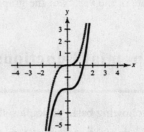

Vertical shift two units downward

(c) $f(x) = -\frac{1}{2}x^3$

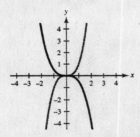

Reflection in the x-axis and a vertical shrink

(d) $f(x) = (x - 2)^3 - 2$

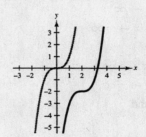

Horizontal shift two units to the right and a vertical shift two units downward

11. $y = x^4$

(a) $f(x) = (x + 3)^4$

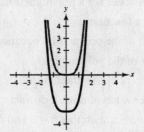

Horizontal shift three units to the left

(b) $f(x) = x^4 - 3$

Vertical shift three units downward

—**CONTINUED**—

11. —CONTINUED—

(c) $f(x) = 4 - x^4$

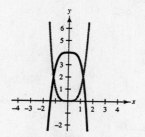

Reflection in the *x*-axis and then a vertical shift four units upward

(d) $f(x) = \frac{1}{2}(x - 1)^4$

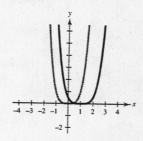

Horizontal shift one unit to the right and a vertical shrink

13. $f(x) = 3x^3 - 9x + 1;\ g(x) = 3x^3$

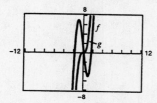

15. $f(x) = -(x^4 - 4x^3 + 16x);\ g(x) = -x^4$

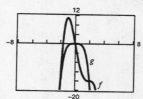

17. $f(x) = 2x^2 - 3x + 1$
Degree: 2
Leading coefficient: 2
The degree is even and the leading coefficient is positive. The graph rises to the left and right.

19. $g(x) = 5 - \frac{7}{2}x - 3x^2$
Degree: 2
Leading coefficient: -3
The degree is even and the leading coefficient is negative. The graph falls to the left and right.

21. $f(x) = 2x^5 - 5x^3 + 7.5$
Degree: 5
Leading coefficient: 2
The degree is odd and the leading coefficient is positive. The graph falls to the left and rises to the right.

23. $f(x) = 6 - 2x + 4x^2 - 5x^3$
Degree: 3
Leading coefficient: -5
The degree is odd and the leading coefficient is negative. The graph rises to the left and falls to the right.

25. $h(t) = -\frac{2}{3}(t^2 - 5t + 3)$
Degree: 2

Leading coefficient: $-\frac{2}{3}$

The degree is even and the leading coefficient is negative. The graph falls to the left and right.

27. $f(x) = x^2 - 25$
$\quad = (x + 5)(x - 5)$
$\quad x = \pm 5$

29. $h(t) = t^2 - 6t + 9$
$\quad = (t - 3)^2$
$\quad t = 3$

31. $f(x) = x^2 + x - 2$
$\quad = (x + 2)(x - 1)$
$\quad x = -2, 1$

33. $f(t) = t^3 - 4t^2 + 4t$
$\quad = t(t - 2)^2$
$\quad t = 0, 2$

35. $f(x) = 3x^2 - 12x + 3$

$= 3(x^2 - 4x + 1)$

$x = \dfrac{4 \pm \sqrt{16 - 4}}{2} = 2 \pm \sqrt{3}$

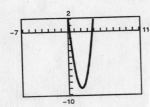

37. $g(t) = \dfrac{1}{2}t^4 - \dfrac{1}{2}$

$= \dfrac{1}{2}(t + 1)(t - 1)(t^2 + 1)$

$t = \pm 1$

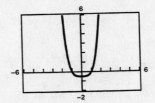

39. $f(x) = 2x^4 - 2x^2 - 40$

$= 2(x^2 + 4)(x + \sqrt{5})(x - \sqrt{5})$

$x = \pm\sqrt{5}$

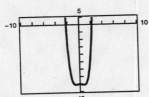

41. $f(x) = 5x^4 + 15x^2 + 10$

$= 5(x^4 + 3x^2 + 2)$

$= 5(x^2 + 2)(x^2 + 1)$

No real zeros

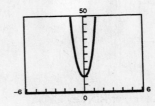

43. $y = 4x^3 - 20x^2 + 25x$

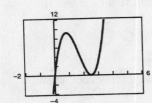

$0 = 4x^3 - 20x^2 + 25x$

$0 = x(2x - 5)^2$

$x = 0 \ \text{ or } \ x = \dfrac{5}{2}$

x-intercepts: $(0, 0), \left(\dfrac{5}{2}, 0\right)$

45. $y = x^5 - 5x^3 + 4x$

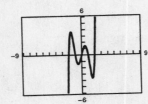

$0 = x^5 - 5x^3 + 4x$

$0 = x(x^2 - 1)(x^2 - 4)$

$0 = x(x + 1)(x - 1)(x + 2)(x - 2)$

$x = 0, \pm 1, \pm 2$

47. $f(x) = (x - 0)(x - 10)$

$f(x) = x^2 - 10x$

Note: $f(x) = a(x - 0)(x - 10) = ax(x - 10)$
has zeros 0 and 10 for all real numbers a.

49. $f(x) = (x - 2)(x - (-6))$

$= (x - 2)(x + 6)$

$= x^2 + 4x - 12$

Note: $f(x) = a(x - 2)(x + 6)$ has zeros 2 and -6
for all real numbers a.

51. $f(x) = (x - 0)(x - (-2))(x - (-3))$

$\quad = x(x + 2)(x + 3)$

$\quad = x^3 + 5x^2 + 6x$

Note: $f(x) = ax(x + 2)(x + 3)$ has zeros
$0, -2, -3$ for all real numbers a.

53. $f(x) = (x - 4)(x + 3)(x - 3)(x - 0)$

$\quad = (x - 4)(x^2 - 9)x$

$\quad = x^4 - 4x^3 - 9x^2 + 36x$

Note: $f(x) = a(x^4 - 4x^3 - 9x^2 + 36x)$ has these
zeros for all real numbers a.

55. $f(x) = \left[x - \left(1 + \sqrt{3}\right)\right]\left[x - \left(1 - \sqrt{3}\right)\right]$

$\quad = \left[(x - 1) - \sqrt{3}\right]\left[(x - 1) + \sqrt{3}\right]$

$\quad = (x - 1)^2 - \left(\sqrt{3}\right)^2$

$\quad = x^2 - 2x + 1 - 3$

$\quad = x^2 - 2x - 2$

Note: $f(x) = a(x^2 - 2x - 2)$ has these zeros for all real numbers a.

57. $f(x) = x^3 - 3x^2 + 3$

(a)

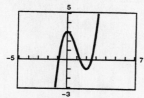

The functions has three zeros. They are
in the intervals $(-1, 0)$, $(1, 2)$ and $(2, 3)$.

(b) $-0.879, 1.347, 2.532$

59. $g(x) = 3x^4 + 4x^3 - 3$

(a)

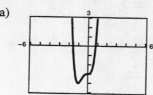

The function has two zeros. They are in the
intervals $(-2, -1)$ and $(0, 1)$.

(b) $-1.585, 0.779$

61. $f(x) = -\frac{3}{2}$

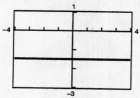

Xmin=-4
Xmax=4
Xscl=1
Ymin=-3
Ymax=1
Yscl=1

Horizontal line

63. $f(t) = \frac{1}{4}(t^2 - 2t + 15)$

$\quad = \frac{1}{4}(t - 1)^2 + \frac{7}{2}$

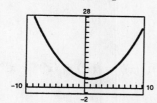

Xmin=-10
Xmax=10
Xscl=1
Ymin=-2
Ymax=28
Yscl=2

Parabola; opens upward

Vertex: $\left(1, \frac{7}{2}\right)$

65. $f(x) = x^2(x - 4)$

Two x-intercepts

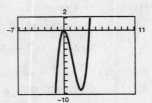

67. $g(t) = -\frac{1}{4}(t - 2)^2(t + 2)^2$

Two x-intercepts
y-axis symmetry

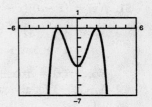

69. $f(x) = x^3 - 4x = x(x + 2)(x - 2)$

Symmetric to origin
Three x-intercepts

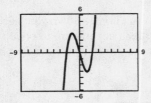

71. $g(x) = \frac{1}{5}(x + 1)^2(x - 3)(2x - 9)$

Three x-intercepts
No symmetry

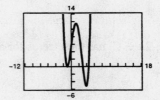

73. $f(x) = x^4$; $f(x)$ is even.

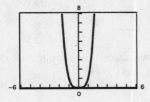

(a) $g(x) = f(x) + 2$

Vertical shift two units upward
$g(-x) = f(-x) + 2$

$\qquad = f(x) + 2$

$\qquad = g(x)$

Even

(b) $g(x) = f(x + 2)$

Horizontal shift two units to the left
Neither odd nor even

(c) $g(x) = f(-x) = (-x)^4 = x^4$

Reflection in the y-axis
The graph looks the same.
Even

(d) $g(x) = -f(x) = -x^4$

Reflection in the x-axis
Even

(e) $g(x) = f(\frac{1}{2}x) = \frac{1}{16}x^4$

Horizontal shrink
Even

(f) $g(x) = \frac{1}{2}f(x) = \frac{1}{2}x^4$

Vertical shrink
Even

(g) $g(x) = f(x^{3/4}) = (x^{3/4}) = x^3$

Odd

(h) $g(x) = (f \circ f)(x) = f(f(x)) = f(x^4) = (x^4)^4 = x^{16}$

Even

75. (a) $V(x) =$ length $\times$ width $\times$ height

$$= (24 - 2x)(24 - 4x)x$$

$$= 8x(6 - x)(12 - x)$$

(b) Domain $0 < x < 6$

(c)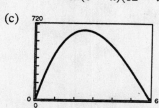

Maximum occurs at $x \approx 2.54$.

77. The point of diminishing returns (where the graph changes from curving upward to curving downward) occurs when $x = 200$. The point is $(200, 160)$ which corresponds to spending \$2,000,000 on advertising to obtain a revenue of \$160 million.

79. (a) $y_1 = -0.158t^3 + 2.850t^2 - 3.814t + 74.703$

(b) $y_2 = -0.007t^3 + 0.196t^2 + 2.533t + 60.844$

(c)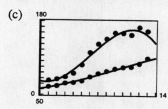

The median price of homes in the South is less than in the Northeast.

Section 2.3   Real Zeros of Polynomial Functions

You should know the following basic techniques and principles of polynomial division.

- The Division Algorithm (Long Division of Polynomials)
- Synthetic Division
- $f(k)$ is equal to the remainder of $f(x)$ divided by $(x - k)$.
- $f(k) = 0$ if and only if $(x - k)$ is a factor of $f(x)$.
- The Rational Zero Test
- The Upper and Lower Bound Rule

Solutions to Odd-Numbered Exercises

1. $y_2 = 4 + \dfrac{4}{x-1}$

$= \dfrac{4(x-1) + 4}{x-1}$

$= \dfrac{4x - 4 + 4}{x-1}$

$= \dfrac{4x}{x-1}$

$= y_1$

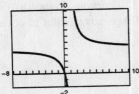

3. $y_2 = x - 2 + \dfrac{4}{x+2}$

$= \dfrac{(x-2)(x+2) + 4}{x+2}$

$= \dfrac{x^2 - 4 + 4}{x+2}$

$= \dfrac{x^2}{x+2}$

$= y_1$

5. $y_2 = x^3 - 4x + \dfrac{4x}{x^2 + 1}$

$= \dfrac{(x^3 - 4x)(x^2 + 1) + 4x}{x^2 + 1}$

$= \dfrac{x^5 + x^3 - 4x^3 - 4x + 4x}{x^2 + 1}$

$= \dfrac{x^5 - 3x^3}{x^2 + 1} = y_1$

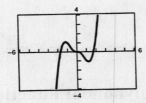

7.
$$
\begin{array}{r}
2x + 4 \\
x + 3 \overline{)\, 2x^2 + 10x + 12} \\
-(2x^2 + 6x) \\
\hline
4x + 12 \\
-(4x + 12) \\
\hline
0
\end{array}
$$

$$\frac{2x^2 + 10x + 12}{x + 3} = 2x + 4$$

9.
$$
\begin{array}{r}
x^2 - 3x + 1 \\
4x + 5 \overline{)\, 4x^3 - 7x^2 - 11x + 5} \\
-(4x^3 + 5x^2) \\
\hline
-12x^2 - 11x \\
-(-12x^2 - 15x) \\
\hline
4x + 5 \\
-(4x + 5) \\
\hline
0
\end{array}
$$

$$\frac{4x^3 - 7x^2 - 11x + 5}{4x + 5} = x^2 - 3x + 1$$

11.

$$
\begin{array}{r}
x^3 + 3x^2 \qquad\quad - 1 \\
x + 2\overline{)\ x^4 + 5x^3 + 6x^2\ -x - 2\ } \\
- (x^4 + 2x^3) \\
\hline
3x^3 + 6x^2 \\
- (3x^3 + 6x^2) \\
\hline
-x - 2 \\
-(-x - 2) \\
\hline
0
\end{array}
$$

$$\frac{x^4 + 5x^3 + 6x^2 - x - 2}{x + 2} = x^3 + 3x^2 - 1$$

13.

$$
\begin{array}{r}
7 \\
x + 2\overline{)\ 7x + 3\ } \\
- (7x + 14) \\
\hline
- 11
\end{array}
$$

$$\frac{7x + 3}{x + 2} = 7 - \frac{11}{x + 2}$$

15.

$$
\begin{array}{r}
3x + 5 \\
2x^2 + 0x + 1\overline{)\ 6x^3 + 10x^2 + x + 8\ } \\
- (6x^3 + 0x^2 + 3x) \\
\hline
10x^2 - 2x + 8 \\
- (10x^2 + 0x + 5) \\
\hline
- 2x + 3
\end{array}
$$

$$\frac{6x^3 + 10x^2 + x + 8}{2x^2 + 1} = 3x + 5 - \frac{2x - 3}{2x^2 + 1}$$

17.

$$
\begin{array}{r}
x^2 + 2x + 4 \\
x^2 - 2x + 3\overline{)\ x^4 + 0x^3 + 3x^2 + 0x + 1\ } \\
- (x^4 - 2x^3 + 3x^2) \\
\hline
2x^3 + 0x^2 + 0x \\
- (2x^3 - 4x^2 + 6x) \\
\hline
4x^2 - 6x + 1 \\
- (4x^2 - 8x + 12) \\
\hline
2x - 11
\end{array}
$$

$\Rightarrow$

$$\frac{x^4 + 3x^2 + 1}{x^2 - 2x + 3} = x^2 + 2x + 4 + \frac{2x - 11}{x^2 - 2x + 3}$$

19.

$$
\begin{array}{r}
2x \\
x^2 - 2x + 1\overline{)\ 2x^3 - 4x^2 - 15x + 5\ } \\
-(2x^3 - 4x^2 + 2x) \\
\hline
- 17x + 5
\end{array}
$$

$$\frac{2x^3 - 4x^2 - 15x + 5}{(x - 1)^2} = 2x - \frac{17x - 5}{(x - 1)^2}$$

21.

$$
\begin{array}{r}
x^{2n} + 6x^n + 9 \\
x^n + 3\overline{)\ x^{3n} + 9x^{2n} + 27x^n + 27\ } \\
- (x^{3n} + 3x^{2n}) \\
\hline
6x^{2n} + 27x^n \\
- (6x^{2n} + 18x^n) \\
\hline
9x^n + 27 \\
- (9x^n + 27) \\
\hline
0
\end{array}
$$

$$\frac{x^{3n} + 9x^{2n} + 27x^n + 27}{x^n + 3} = x^{2n} + 6x^n + 9$$

23.

$$
\begin{array}{r|rrrr}
-2 & 4 & 8 & -9 & -18 \\
   &   & -8 & 0 & 18 \\
\hline
   & 4 & 0 & -9 & 0
\end{array}
$$

$$\frac{4x^3 + 8x^2 - 9x - 18}{x + 2} = 4x^2 - 9$$

27.

$$
\begin{array}{r|rrrr}
-8 & 1 & 0 & 0 & 512 \\
   &   & -8 & 64 & -512 \\
\hline
   & 1 & -8 & 64 & 0
\end{array}
$$

$$\frac{x^3 + 512}{x + 8} = x^2 - 8x + 64$$

31.

$$
\begin{array}{r|rrrr}
-\frac{1}{2} & 4 & 16 & -23 & -15 \\
             &   & -2 & -7 & 15 \\
\hline
             & 4 & 14 & -30 & 0
\end{array}
$$

$$\frac{4x^3 + 16x^2 - 23x - 15}{x + \frac{1}{2}} = 4x^2 + 14x - 30$$

35. $f(x) = x^3 - x^2 - 14x + 11, \; k = 4$

$$
\begin{array}{r|rrrr}
4 & 1 & -1 & -14 & 11 \\
  &   & 4 & 12 & -8 \\
\hline
  & 1 & 3 & -2 & 3
\end{array}
$$

$16 + 12 - 2$

$28 - 2 = 26$

$f(x) = (x - 4)(x^2 + 3x - 2) + 3/(x - 4)$

$f(4) = (0)(26) + 3 = 3$

39. $f(x) = 4x^3 - 13x + 10$

(a)

$$
\begin{array}{r|rrrr}
1 & 4 & 0 & -13 & 10 \\
  &   & 4 & 4 & -9 \\
\hline
  & 4 & 4 & -9 & \underline{1} = f(1)
\end{array}
$$

(c)

$$
\begin{array}{r|rrrr}
\frac{1}{2} & 4 & 0 & -13 & 10 \\
            &   & 2 & 1 & -6 \\
\hline
            & 4 & 2 & -12 & \underline{4} = f\left(\frac{1}{2}\right)
\end{array}
$$

25.

$$
\begin{array}{r|rrrr}
4 & 5 & -6 & 0 & 8 \\
  &   & 20 & 56 & 224 \\
\hline
  & 5 & 14 & 56 & 232
\end{array}
$$

$$\frac{5x^3 - 6x^2 + 8}{x - 4} = 5x^2 + 14x + 56 + \frac{232}{x - 4}$$

29.

$$
\begin{array}{r|rrrrr}
2 & -3 & 0 & 0 & 0 & 0 \\
  &   & -6 & -12 & -24 & -48 \\
\hline
  & -3 & -6 & -12 & -24 & -48
\end{array}
$$

$$\frac{-3x^4}{x - 2} = -3x^3 - 6x^2 - 12x - 24 - \frac{48}{x - 2}$$

33. A divisor divides evenly into the dividend if the remainder is zero.

37. $f(x) = x^3 + 3x^2 - 2x - 14, \; k = \sqrt{2}$

$$
\begin{array}{r|rrrr}
\sqrt{2} & 1 & 3 & -2 & -14 \\
         &   & \sqrt{2} & 2 + 3\sqrt{2} & 6 \\
\hline
         & 1 & 3 + \sqrt{2} & 3\sqrt{2} & -8
\end{array}
$$

$f(x) = (x - \sqrt{2})\left[x^2 + (3 + \sqrt{2})x + 3\sqrt{2}\right] - 8/(\sqrt{2})$

$f(\sqrt{2}) = (0)(4 + 6\sqrt{2}) - 8 = -8$

(b)

$$
\begin{array}{r|rrrr}
-2 & 4 & 0 & -13 & 10 \\
   &   & -8 & 16 & -6 \\
\hline
   & 4 & -8 & 3 & \underline{4} = f(-2)
\end{array}
$$

(d)

$$
\begin{array}{r|rrrr}
8 & 4 & 0 & -13 & 10 \\
  &   & 32 & 256 & 1944 \\
\hline
  & 4 & 32 & 243 & \underline{1954} = f(8)
\end{array}
$$

41.

$$
\begin{array}{r|rrrr}
2 & 1 & 0 & -7 & 6 \\
  &   & 2 & 4 & -6 \\
\hline
  & 1 & 2 & -3 & 0
\end{array}
$$

$$
\begin{aligned}
x^3 - 7x + 6 &= (x-2)(x^2 + 2x - 3) \\
&= (x-2)(x+3)(x-1)
\end{aligned}
$$

Zeros: $2, -3, 1$

43.

$$
\begin{array}{r|rrrr}
\frac{1}{2} & 2 & -15 & 27 & -10 \\
  &   & 1 & -7 & 10 \\
\hline
  & 2 & -14 & 20 & 0
\end{array}
$$

$$
\begin{aligned}
2x^3 - 15x^2 + 27x - 10 &= \left(x - \tfrac{1}{2}\right)(2x^2 - 14x + 20) \\
&= (2x-1)(x-2)(x-5)
\end{aligned}
$$

Zeros: $\frac{1}{2}, 2, 5$

45.

$$
\begin{array}{r|rrrr}
\sqrt{3} & 1 & 2 & -3 & -6 \\
  &   & \sqrt{3} & 3 + 2\sqrt{3} & 6 \\
\hline
  & 1 & 2 + \sqrt{3} & 2\sqrt{3} & 0
\end{array}
$$

$$
\begin{array}{r|rrr}
-\sqrt{3} & 1 & 2 + \sqrt{3} & 2\sqrt{3} \\
  &   & -\sqrt{3} & -2\sqrt{3} \\
\hline
  & 1 & 2 & 0
\end{array}
$$

$$
x^3 + 2x^2 - 3x - 6 = \left(x - \sqrt{3}\right)\left(x + \sqrt{3}\right)(x + 2)
$$

Zeros: $\pm\sqrt{3}, -2$

47.

$$
\begin{array}{r|rrrr}
1 + \sqrt{3} & 1 & -3 & 0 & 2 \\
  &   & 1 + \sqrt{3} & 1 - \sqrt{3} & -2 \\
\hline
  & 1 & -2 + \sqrt{3} & 1 - \sqrt{3} & 0
\end{array}
$$

$$
\begin{array}{r|rrr}
1 - \sqrt{3} & 1 & -2 + \sqrt{3} & 1 - \sqrt{3} \\
  &   & 1 - \sqrt{3} & -1 + \sqrt{3} \\
\hline
  & 1 & -1 & 0
\end{array}
$$

$$
\begin{aligned}
x^3 - 3x^2 + 2 &= \left[x - \left(1 + \sqrt{3}\right)\right]\left[x - \left(1 - \sqrt{3}\right)\right](x - 1) \\
&= (x - 1)\left(x - 1 - \sqrt{3}\right)\left(x - 1 + \sqrt{3}\right)
\end{aligned}
$$

Zeros: $1, 1 \pm \sqrt{3}$

49. (a) $R = 16.823 + 1.415t - 0.115t^2 - 0.023t^3$

(b)

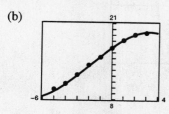

(c)

t	-5	-4	-3	-2	-1	0	1	2	3
R	9.75	10.80	12.16	13.72	15.32	16.82	18.10	19.01	19.41

—CONTINUED—

49. **—CONTINUED—**

(d) 6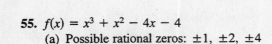

6	-0.023	-0.115	1.415	16.823
		-0.138	-1.518	-0.618
	-0.023	-0.253	-0.103	16.205

$R(6) \approx 16.205$. Since the cubic falls to the right, it is not accurate for values past 1996.

51. $f(x) = x^3 + 3x^2 - x - 3$
Possible rational zeros: $\pm 1, \pm 3$
Zeros shown on graph: $-3, -1, 1$

53. $f(x) = 2x^4 - 17x^3 + 35x^2 + 9x - 45$
Possible rational zeros: $\pm 1, \pm 3, \pm 5, \pm 9, \pm 15, \pm 45,$
$\pm \frac{1}{2}, \pm \frac{3}{2}, \pm \frac{5}{2}, \pm \frac{9}{2}, \pm \frac{15}{2}, \pm \frac{45}{2}$
Zeros shown of graph: $-1, \frac{3}{2}, 3, 5$

55. $f(x) = x^3 + x^2 - 4x - 4$
(a) Possible rational zeros: $\pm 1, \pm 2, \pm 4$
(b)
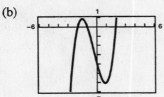
(c) $-2, -1, 2$ on graph

57. $f(x) = -4x^3 + 15x^2 - 8x - 3$
(a) Possible rational zeros: $\pm \frac{1}{4}, \pm \frac{1}{2}, \pm \frac{3}{4}, \pm 1, \pm \frac{3}{2}, \pm 3$
(b)
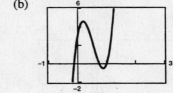
(c) $-\frac{1}{4}, 1, 3$ on graph

59. $f(x) = -2x^4 + 13x^3 - 21x^2 + 2x + 8$
(a) Possible rational zeros:
$\pm \frac{1}{2}, \pm 1, \pm 2, \pm 4, \pm 8$
(b)
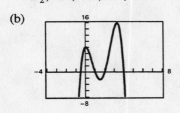
(c) $-\frac{1}{2}, 1, 2, 4$ on graph

61. $f(x) = 32x^3 - 52x^2 + 17x + 3$
(a) Possible rational zeros: $\pm 1, \pm 3, \pm \frac{1}{2}, \pm \frac{3}{2}, \pm \frac{1}{4}, \pm \frac{3}{4},$
$\pm \frac{1}{8}, \pm \frac{3}{8}, \pm \frac{1}{16}, \pm \frac{3}{16}, \pm \frac{1}{32}, \pm \frac{3}{32}$
(b)

(c) $-\frac{1}{8}, \frac{3}{4}, 1$ on graph

63. (a) $f(x) = 4x^3 + 7x^2 - 11x - 18$

Possible rational zeros: $\pm 1, \pm 2, \pm 3, \pm 6, \pm 9,$
$\pm 18, \pm \frac{1}{2}, \pm \frac{3}{2}, \pm \frac{9}{2}, \pm \frac{1}{4}, \pm \frac{3}{4}, \pm \frac{9}{4}$

(b)
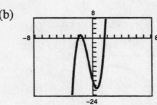

-2	4	7	-11	-18
		-8	2	18
	4	-1	-9	0

The zeros of $4x^2 - x - 9$ are $x = -2$ and
$$x = \frac{1 \pm \sqrt{1 + 4(4)(9)}}{8} = \frac{1}{8} \pm \frac{\sqrt{145}}{8}.$$

65. $z^4 - z^3 - 2z - 4 = 0$

Possible rational zeros: $\pm 1,\ \pm 2,\ \pm 4$

$$
\begin{array}{r|rrrrr}
-1 & 1 & -1 & 0 & -2 & -4 \\
   &   & -1 & 2 & -2 & 4 \\
\hline
   & 1 & -2 & 2 & -4 & 0 \\
\end{array}
$$

$$
\begin{array}{r|rrrr}
2 & 1 & -2 & 2 & -4 \\
  &   & 2 & 0 & 4 \\
\hline
  & 1 & 0 & 2 & 0 \\
\end{array}
$$

$z^4 - z^3 - 2z - 4 = (x + 1)(x - 2)(x^2 + 2) = 0$

The only real zeros are -1 and 2. You can verify this by graphing the function $f(z) = z^4 - z^3 - 2z - 4$.

67. $x^4 - 13x^2 - 12x = 0$

$x(x^3 - 13x - 12) = 0$

$\qquad\qquad x = 0$ is a solution.

Possible rational zeros of $x^3 - 13x - 12 = 0$ are $\pm 1, \pm 2, \pm 3, \pm 4, \pm 6$ and ± 12. Using a graphing utility or synthetic division, you find that the zeros are $0, -1, -3, 4$.

69. $2x^4 - 11x^3 - 6x^2 + 64x + 32 = 0$

Using a graphing utility, you can see that there are three zeros. Using synthetic division, you can verify that these zeros are $-2, -\frac{1}{2}, 4$.

71. $f(x) = x^4 - 3x^2 + 2$

(a) From the calculator we have
$x = \pm 1$ and $x \approx \pm 1.414$.

(b)
$$
\begin{array}{r|rrrrr}
1 & 1 & 0 & -3 & 0 & 2 \\
  &   & 1 & 1 & -2 & -2 \\
\hline
  & 1 & 1 & -2 & -2 & 0 \\
\end{array}
$$

$$
\begin{array}{r|rrrr}
-1 & 1 & 1 & -2 & -2 \\
   &   & -1 & 0 & 2 \\
\hline
   & 1 & 0 & -2 & 0 \\
\end{array}
$$

$f(x) = (x - 1)(x + 1)(x^2 - 2)$

$\qquad = (x - 1)(x + 1)(x - \sqrt{2})(x + \sqrt{2})$

The exact roots are $x = \pm 1,\ \pm\sqrt{2}$.

73. $h(x) = x^5 - 7x^4 + 10x^3 + 14x^2 - 24x$

(a) $h(x) = x(x^4 - 7x^3 + 10x^2 + 14x - 24)$
From the calculator we have $x = 0, 3, 4$
and $x \approx \pm 1.414$.

(b)
$$
\begin{array}{r|rrrrr}
3 & 1 & -7 & 10 & 14 & -24 \\
  &   & 3 & -12 & -6 & 24 \\
\hline
  & 1 & -4 & -2 & 8 & 0 \\
\end{array}
$$

$$
\begin{array}{r|rrrr}
4 & 1 & -4 & -2 & 8 \\
  &   & 4 & 0 & -8 \\
\hline
  & 1 & 0 & -2 & 0 \\
\end{array}
$$

$f(x) = x(x - 3)(x - 4)(x^2 - 2)$

$\qquad = x(x - 3)(x - 4)(x - \sqrt{2})(x + \sqrt{2})$

The exact roots are $x = 0,\ 3,\ 4,\ \pm\sqrt{2}$.

75. $f(x) = x^4 - 4x^3 + 15$

(a)

4	1	-4	0	0	15
		4	0	0	0
	1	0	0	0	15

4 is an upper bound.

(b)

-1	1	-4	0	0	15
		-1	5	-5	5
	1	-5	5	-5	20

-1 is a lower bound.

77. $f(x) = x^4 - 4x^3 + 16x - 16$

(a)

5	1	-4	0	16	-16
		5	5	25	205
	1	1	5	41	189

5 is an upper bound.

(b)

-3	1	-4	0	16	-16
		-3	21	-63	141
	1	-7	21	-47	125

-3 is a lower bound.

79. $P(x) = x^4 - \frac{25}{4}x^2 + 9$

$\quad = \frac{1}{4}(4x^4 - 25x^2 + 36)$

$\quad = \frac{1}{4}(4x^2 - 9)(x^2 - 4)$

$\quad = \frac{1}{4}(2x + 3)(2x - 3)(x + 2)(x - 2)$

The zeros are $\pm\frac{3}{2}$ and ± 2.

81. $f(x) = x^3 - \frac{1}{4}x^2 - x + \frac{1}{4}$

$\quad = \frac{1}{4}(4x^3 - x^2 - 4x + 1)$

$\quad = \frac{1}{4}[x^2(4x - 1) - 1(4x - 1)]$

$\quad = \frac{1}{4}(4x - 1)(x^2 - 1)$

$\quad = \frac{1}{4}(4x - 1)(x + 1)(x - 1)$

The zeros are $\frac{1}{4}$ and ± 1.

83. $f(x) = x^3 - 1 = (x - 1)(x^2 + x + 1)$
Rational zeros: 1 $(x = 1)$
Irrational zeros: 0
Matches (d).

85. $f(x) = x^3 - x = x(x + 1)(x - 1)$
Rational zeros: 3 $(x = 0, \pm 1)$
Irrational zeros: 0
Matches (b).

87. $g(x) = -f(x)$. This function would have the same zeros as $f(x)$ so r_1, r_2, and r_3 are also zeros of $g(x)$.

89. $g(x) = f(x - 5)$. The graph of $g(x)$ is a horizontal shift of the graph of $f(x)$ five units to the right so the zeros of $g(x)$ are $5 + r_1$, $5 + r_2$, and $5 + r_3$.

91. $g(x) = 3 + f(x)$. Since $g(x)$ is a vertical shift of the graph of $f(x)$, the zeros of $g(x)$ cannot be determined.

93. (a)

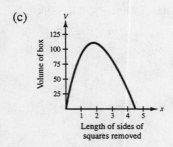

(b) $V = l \cdot w \cdot h = (15 - 2x)(9 - 2x)x$

$\quad = x(9 - 2x)(15 - 2x)$

Since length, width, and height cannot be negative, we have $0 < x < \frac{9}{2}$ for the domain.

(c)

The volume is maximum when $x \approx 1.82$.
The dimensions are: length $= 15 - 2(1.82) = 11.36$
width $= 9 - 2(1.82) = 5.36$
height $= x = 1.82$
$1.82 \text{ cm} \times 5.36 \text{ cm} \times 11.36 \text{ cm}$

(d) $56 = x(9 - 2x)(15 - 2x)$

$56 = 135x - 48x^2 + 4x^3$

$0 = 4x^3 - 48x^2 + 135x - 56$

The zeros of this polynomial are $\frac{1}{2}$, $\frac{7}{2}$, and 8.
x cannot equal 8 since it is not in the domain of V.
[The length cannot equal -1 and the width cannot equal -7. The product of $(8)(-1)(-7) = 56$ so it showed up as an extraneous solution.]

95. $y = -5.05x^3 + 3857x - 38,411.25,\ 13 \le x \le 18$

(a)

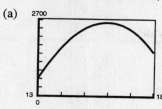

(b) The second air-fuel ration of 16.89 can be obtained by finding the second point where the curves y and $y_1 = 2400$ intersect.

(c) Solve $-5.05x^3 + 3857x - 38,411.25 = 2400$ or $-5.05x^3 + 3857x - 40,811.25 = 0$ by synthetic division.

$$
\begin{array}{r|rrrr}
15 & -5.05 & 0 & 3857 & -40811.25 \\
   &   & -75.75 & -1136.25 & 40811.25 \\
\hline
   & -5.05 & -75.75 & 2720.75 & 0
\end{array}
$$

(d) The positive zero of the quadratic $-5.05x^2 - 75.75x + 2720.75$ can be found by the Quadratic Formula.

$$
x \approx \frac{75.75 - \sqrt{(75.75)^2 - 4(-5.05)(2720.75)}}{2(-5.05)} \approx 16.89
$$

97. (a)

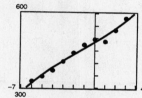

(b) By graphing I together with $y_2 = 525$, you see that the curves intersect at $x \approx 1.51$, or 1991.

(c) Yes, the curve is increasing to the right.

Section 2.4   Complex Numbers

- ■ You should know how to work with complex numbers.
- ■ Operations on complex numbers
  - (a) Addition: $(a + bi) + (c + di) = (a + c) + (b + d)i$
  - (b) Subtraction: $(a + bi) - (c + di) = (a - c) + (b - d)i$
  - (c) Multiplication: $(a + bi)(c + di) = (ac - bd) + (ad + bc)i$
  - (c) Division: $\dfrac{a + bi}{c + di} = \dfrac{a + bi}{c + di} \cdot \dfrac{c - di}{c - di} = \dfrac{ac + bd}{c^2 + d^2} + \dfrac{bc - ad}{c^2 + d^2}i$
- ■ The complex conjugate of $a + bi$ is $a - bi$:
  $$(a + bi)(a - bi) = a^2 + b^2$$
- ■ The additive inverse of $a + bi$ is $-a - bi$.
- ■ The multiplicative inverse of $a + bi$ is
  $$\frac{a - bi}{a^2 + b^2}.$$
- ■ $\sqrt{-a} = \sqrt{a}\,i$ for $a > 0$.

Solutions to Odd-Numbered Exercises

1. $a + bi = -10 + 6i$

    $a = -10$

    $b = 6$

3. $(a - 1) + (b + 3)i = 5 + 8i$

    $a - 1 = 5 \quad \Rightarrow \quad a = 6$

    $b + 3 = 8 \quad \Rightarrow \quad b = 5$

5. $4 + \sqrt{-9} = 4 + 3i$

7. $2 - \sqrt{-27} = 2 - \sqrt{27}\,i = 2 - 3\sqrt{3}\,i$

9. $\sqrt{-75} = \sqrt{75}\,i = 5\sqrt{3}\,i$

11. $-6i + i^2 = -6i - 1 = -1 - 6i$

13. $8 = 8 + 0i = 8$

15. $\sqrt{-0.09} = \sqrt{0.09}\,i = 0.3i$

17. $(5 + i) + (6 - 2i) = 11 - i$

19. $(8 - i) - (4 - i) = 8 - i - 4 + i = 4$

21. $\left(-2 + \sqrt{-8}\,\right) + \left(5 - \sqrt{-50}\right) = -2 + 2\sqrt{2}\,i + 5 - 5\sqrt{2}\,i$

                                $= 3 - 3\sqrt{2}\,i$

23. $13i - (14 - 7i) = 13i - 14 + 7i = -14 + 20i$

25. $-\left(\frac{3}{2} + \frac{5}{2}i\right) + \left(\frac{5}{3} + \frac{11}{3}i\right) = -\frac{3}{2} - \frac{5}{2}i + \frac{5}{3} + \frac{11}{3}i$

                                    $= -\frac{9}{6} - \frac{15}{6}i + \frac{10}{6} + \frac{22}{6}i$

                                    $= \frac{1}{6} + \frac{7}{6}i$

27. $\sqrt{-6} \cdot \sqrt{-2} = \left(\sqrt{6}\,i\right)\left(\sqrt{2}\,i\right) = \sqrt{12}\,i^2 = \left(2\sqrt{3}\right)(-1) = -2\sqrt{3}$

29. $\left(\sqrt{-10}\right)^2 = \left(\sqrt{10}\,i\right)^2 = 10i^2 = -10$

31. $(1 + i)(3 - 2i) = 3 - 2i + 3i - 2i^2$
$$= 3 + i + 2$$
$$= 5 + i$$

33. $6i(5 - 2i) = 30i - 12i^2 = 30i + 12 = 12 + 30i$

35. $\left(\sqrt{14} + \sqrt{10}\,i\right)\left(\sqrt{14} - \sqrt{10}\,i\right) = 14 - 10i^2 = 14 + 10 = 24$

37. $(4 + 5i)^2 = 16 + 40i + 25i^2 = 16 + 40i - 25$
$$= -9 + 40i$$

39. $(2 + 3i)^2 + (2 - 3i)^2 = 4 + 12i + 9i^2 + 4 - 12i + 9i^2$
$$= 4 + 12i - 9 + 4 - 12i - 9$$
$$= -10$$

41. $\sqrt{-6}\sqrt{-6} = \sqrt{6}\,i\,\sqrt{6}\,i = 6i^2 = -6 \quad \left(\sqrt{-6}\,\sqrt{-6} \neq \sqrt{(-6)(-6)}\right)$

43. The complex conjugate of $5 + 3i$ is $5 - 3i$.
$(5 + 3i)(5 - 3i) = 25 - 9i^2 = 25 + 9 = 34$

45. The complex conjugate of $-2 - \sqrt{5}\,i$ is $-2 + \sqrt{5}\,i$.
$\left(-2 - \sqrt{5}\,i\right)\left(-2 + \sqrt{5}\,i\right) = 4 - 5i^2 = 4 + 5 = 9$

47. The complex conjugate of $20i$ is $-20i$.
$(20i)(-20i) = -400i^2 = 400$

49. The complex conjugate of $\sqrt{8}$ is $\sqrt{8}$.
$\left(\sqrt{8}\right)\left(\sqrt{8}\right) = 8$

51. $\dfrac{6}{i} = \dfrac{6}{i} \cdot \dfrac{-i}{-i} = \dfrac{-6i}{-i^2} = \dfrac{-6i}{1} = -6i$

53. $\dfrac{4}{4 - 5i} = \dfrac{4}{4 - 5i} \cdot \dfrac{4 + 5i}{4 + 5i} = \dfrac{4(4 + 5i)}{16 + 25} = \dfrac{16 + 20i}{41} = \dfrac{16}{41} + \dfrac{20}{41}i$

55. $\dfrac{2 + i}{2 - i} = \dfrac{2 + i}{2 - i} \cdot \dfrac{2 + i}{2 + i} = \dfrac{4 + 4i + i^2}{4 + 1} = \dfrac{3 + 4i}{5} = \dfrac{3}{5} + \dfrac{4}{5}i$

57. $\dfrac{6 - 7i}{i} = \dfrac{6 - 7i}{i} \cdot \dfrac{-i}{-i} = \dfrac{-6i - 7}{1} = -7 - 6i$

59. $\dfrac{1}{(4 - 5i)^2} = \dfrac{1}{16 - 40i + 25i^2} = \dfrac{1}{-9 - 40i} \cdot \dfrac{-9 + 40i}{-9 + 40i}$
$$= \dfrac{-9 + 40i}{81 + 1600} = \dfrac{-9 + 40i}{1681} = -\dfrac{9}{1681} + \dfrac{40}{1681}i$$

61. $\dfrac{2}{1+i} - \dfrac{3}{1-i} = \dfrac{2(1-i) - 3(1+i)}{(1+i)(1-i)}$

$= \dfrac{2 - 2i - 3 - 3i}{1 + 1}$

$= \dfrac{-1 - 5i}{2}$

$= -\dfrac{1}{2} - \dfrac{5}{2}i$

63. $\dfrac{i}{3-2i} + \dfrac{2i}{3+8i} = \dfrac{i(3+8i) + 2i(3-2i)}{(3-2i)(3+8i)}$

$= \dfrac{3i + 8i^2 + 6i - 4i^2}{9 + 24i - 6i - 16i^2}$

$= \dfrac{4i^2 + 9i}{9 + 18i + 16}$

$= \dfrac{-4 + 9i}{25 + 18i} \cdot \dfrac{25 - 18i}{25 - 18i}$

$= \dfrac{-100 + 72i + 225i - 162i^2}{625 + 324}$

$= \dfrac{-100 + 297i + 162}{949}$

$= \dfrac{62 + 297i}{949}$

$= \dfrac{62}{949} + \dfrac{297}{949}i$

65. $\quad i = i \qquad i^5 = i \qquad i^9 = i \qquad i^{13} = i$

$\quad i^2 = -1 \qquad i^6 = -1 \qquad i^{10} = -1 \qquad i^{14} = -1$

$\quad i^3 = -i \qquad i^7 = -i \qquad i^{11} = -i \qquad i^{15} = -i$

$\quad i^4 = 1 \qquad i^8 = 1 \qquad i^{12} = 1 \qquad i^{16} = 1$

The numbers exhibit a periodic pattern: $i, -1, -i, 1$, etc.

67. $-6i^3 + i^2 = -6i^2i + i^2$

$= -6(-1)i + (-1)$

$= 6i - 1$

$= -1 + 6i$

69. $-5i^5 = -5i^2i^2i$

$= -5(-1)(-1)i$

$= -5i$

71. $\left(\sqrt{-75}\right)^3 = \left(5\sqrt{3}i\right)^3 = 5^3\left(\sqrt{3}\right)^3 i^3$

$= 125(3\sqrt{3})(-i)$

$= -375\sqrt{3}i$

73. $\dfrac{1}{i^3} = \dfrac{1}{-i} = \dfrac{1}{-i} \cdot \dfrac{i}{i} = \dfrac{1}{-i^2} = \dfrac{i}{1} = i$

75. $4 + 3i$

77. $0 + 6i = 6i$

79.

81.

83. The complex number 0 is in the Mandelbrot Set since for $c = 0$, the corresponding Mandelbrot sequence is 0, 0, 0, 0, 0, 0, . . . which is bounded.

85. The complex number $\frac{1}{2}i$, is in the Mandelbrot Set since for $c = \frac{1}{2}i$, the corresponding Mandelbrot sequence is

$\frac{1}{2}i, -\frac{1}{4} + \frac{1}{2}i, -\frac{3}{16} + \frac{1}{4}i, -\frac{7}{256} + \frac{13}{32}i, \frac{-10{,}767}{65{,}536} + \frac{1957}{4096}i, -\frac{864{,}513{,}055}{4{,}294{,}967{,}296} + \frac{46{,}037{,}845}{134{,}217{,}728}i$

which is bounded. Or in decimal form

$0.5i, -0.25 + 0.5i, -0.1875 + 0.25i, -0.02734 + 0.40625i,$

$-0.164291 + 0.477783i, -0.201285 + 0.343009i.$

87. The complex number 1 is not in the Mandelbrot Set since for $c = 1$, the corresponding Mandelbrot sequence is 1, 2, 5, 26, 677, 458,330 which is unbounded.

89. $(2)^3 = 8$

$$\left(-1 + \sqrt{3}i\right)^3 = (-1)^3 + 3(-1)^2\left(\sqrt{3}i\right) + 3(-1)\left(\sqrt{3}i\right)^2 + \left(\sqrt{3}i\right)$$
$$= -1 + 3\sqrt{3}i - 9i^2 + 3\sqrt{3}i^3$$
$$= -1 + 3\sqrt{3}i + 9 - 3\sqrt{3}i$$
$$= 8$$

$$\left(-1 - \sqrt{3}i\right)^3 = (-1)^3 + 3(-1)^2\left(-\sqrt{3}i\right) + 3(-1)\left(-\sqrt{3}i\right)^2\left(\sqrt{3}i\right)^3$$
$$= -1 - 3\sqrt{3}i - 9i^2 - 3\sqrt{3}i^3$$
$$= -1 - 3\sqrt{3}i + 9 + 3\sqrt{3}i$$
$$= 8$$

The three numbers are cube roots of 8.

91. $(a + bi) + (a - bi) = 2a$ which is a real number.
$(a + bi) - (a - bi) = 2bi$ is an imaginary number.

93. $\overline{(a_1 + b_1 i)(a_2 + b_2 i)} = \overline{(a_1 a_2 - b_1 b_2) + (a_1 b_2 + b_1 a_2)i}$
$$= (a_1 a_2 - b_1 b_2) - (a_1 b_2 + b_1 a_2)i$$

$\overline{(a_1 + b_1 i)}\ \overline{(a_2 + b_2 i)} = (a_1 - b_1 i)(a_2 - b_2 i)$
$$= (a_1 a_2 - b_1 b_2 i) - (a_1 b_2 + b_1 a_2)i,$$

which are equal.

95. $(x^3 - 3x^2) - (6 - 2x - 4x^2) = x^3 - 3x^2 - 6 + 2x + 4x^2 = x^3 + x^2 + 2x - 6$

97. $\left(3x - \frac{1}{2}\right)(x + 4) = 3x^2 - \frac{1}{2}x + 12x - \frac{1}{2}(4) = 3x^2 + \frac{23}{2}x - 2$

99. $[(x + y) + 3]^2 = (x + y)^2 + 6(x + y) + 9 = x^2 + 2xy + y^2 + 6x + 6y + 9$

101. $F = \alpha \dfrac{m_1 m_2}{r^2}$

$Fr^2 = \alpha\, m_1 m_2$

$r^2 = \dfrac{\alpha\, m_1 m_2}{F}$

$r = \sqrt{\dfrac{\alpha\, m_1 m_2}{F}}$ $(r > 0$ is distance$)$

103. $\text{Time} = \dfrac{\text{Distance}}{\text{Speed}} = \dfrac{200}{100} + \dfrac{200}{80} = \dfrac{1600 + 2000}{800} = \dfrac{9}{2}\ \text{hrs}$

$\text{Average speed} = \dfrac{\text{Distance}}{\text{Time}} = \dfrac{400}{9/2} = \dfrac{800}{9} \approx 88.9\ \text{km/hr}$

Section 2.5   The Fundamental Theorem of Algebra

- ■ You should know that if f is a polynomial of degree $n > 0$, then f has at least one zero in the complex number system. (Fundamental Theorem of Algebra)
- ■ You should know that if $a + bi$ is a complex zero of a polynomial f, with real coefficients, then $a - bi$ is also a complex zero of f.
- ■ You should know the difference between a factor that is irreducible over the rationals (such as $x^2 - 7$) and a factor that is irreducible over the reals (such as $x^2 + 9$).

Solutions to Odd-Numbered Exercises

1. $f(x) = x(x - 6)^2 = x(x - 6)(x - 6)$
The three zeros are $x = 0$, $x = 6$, and $x = 6$.

3. $h(t) = (t - 3)(t - 2)(t - 3i)(t + 3i)$
The four zeros are $t = 3$, $t = 2$, $t = 3i$, and $t = -3i$.

5. $f(x) = x^3 - 4x^2 + x - 4 = x^2(x - 4) + 1(x - 4)$
$= (x - 4)(x^2 + 1)$

The only real zero of $f(x)$ is $x = 4$. This corresponds to the x-intercept of $(4, 0)$ on the graph.

7. $f(x) = x^4 + 4x^2 + 4 = (x^2 + 2)^2$

$f(x)$ has no real zeros and the graph of $f(x)$ has no x-intercepts.

9. $f(x) = x^2 + 25 = (x + 5i)(x - 5i)$
The zeros of $f(x)$ are $x = \pm 5i$.

11. $h(x) = x^2 - 4x + 1$

h has no rational zeros. By the Quadratic Formula, the zeros are $x = \dfrac{4 \pm \sqrt{16 - 4}}{2} = 2 \pm \sqrt{3}$.

$$h(x) = \left[x - \left(2 + \sqrt{3}\right)\right]\left[x - \left(2 - \sqrt{3}\right)\right] = \left(x - 2 - \sqrt{3}\right)\left(x - 2 + \sqrt{3}\right)$$

13. $f(x) = x^4 - 81$

$$= (x^2 - 9)(x^2 + 9)$$

$$= (x + 3)(x - 3)(x + 3i)(x - 3i)$$

The zeros of $f(x)$ are $x = \pm 3$ and $x = \pm 3i$.

15. $f(z) = z^2 - 2z + 2$

f has no rational zeros. By the Quadratic Formula, the zeros are $z = \dfrac{2 \pm \sqrt{4 - 8}}{2} = 1 \pm i$.

$$f(z) = [z - (1 + i)][z - (1 - i)] = (z - 1 - i)(x - 1 + i)$$

17. $f(t) = t^3 - 3t^2 - 15t + 125$

Possible rational zeros: $\pm 1, \pm 5, \pm 25, \pm 125$

-5	1	-3	-15	125
		-5	40	-125
	1	-8	25	0

By the Quadratic Formula, the zeros of $t^2 - 8t + 25$ are $t = \dfrac{8 \pm \sqrt{64 - 100}}{2} = 4 \pm 3i$.

The zeros of $f(t)$ are $t = -5$ and $t = 4 \pm 3i$.

$$f(t) = [t - (-5)][t - (4 + 3i)][t - (4 - 3i)]$$

$$= (t + 5)(t - 4 - 3i)(t - 4 + 3i)$$

19. $f(x) = 16x^3 - 20x^2 - 4x + 15$

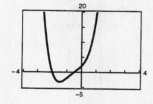

The graph reveals one zero at $x = -\dfrac{3}{4}$.

$-\dfrac{3}{4}$	16	-20	-4	15
		-12	24	-15
	16	-32	20	0

By the Quadratic Formula, the zeros of $16x^2 - 32x + 20 = 4(4x^2 - 8x + 5)$ are $x = \dfrac{8 \pm \sqrt{64 - 80}}{8} = 1 \pm \dfrac{1}{2}i$.

The zeros of $f(x)$ are $x = -\dfrac{3}{4}$ and $x = 1 \pm \dfrac{1}{2}i$.

$$(4x + 3)\left(x - 1 + \dfrac{1}{2}i\right)\left(x - 1 - \dfrac{1}{2}i\right)$$

21. $f(x) = 5x^3 - 9x^2 + 28x + 6$

Possible rational zeros: $\pm 1, \pm 2, \pm 3, \pm 6, \pm\frac{1}{5}, \pm\frac{2}{5}, \pm\frac{3}{5}, \pm\frac{6}{5}$

$$
\begin{array}{r|rrrr}
-\dfrac{1}{5} & 5 & -9 & 28 & 6 \\
 & & -1 & 2 & -6 \\
\hline
 & 5 & -10 & 30 & 0
\end{array}
$$

By the Quadratic Formula, the zeros of $5x^2 - 10x + 30 = 5(x^2 - 2x + 6)$ are $x = \dfrac{2 \pm \sqrt{4 - 24}}{2} = 1 \pm \sqrt{5}i$.

The zeros of $f(x)$ are $x = -\frac{1}{5}$ and $x = 1 \pm \sqrt{5}i$.

$$
\begin{aligned}
f(x) &= \left[x - \left(-\frac{1}{5}\right)\right](5)\left[x - \left(1 + \sqrt{5}i\right)\right]\left[x - \left(1 - \sqrt{5}i\right)\right] \\
&= (5x + 1)(x - 1 - \sqrt{5}i)(x - 1 + \sqrt{5}i)
\end{aligned}
$$

23. $f(x) = x^4 + 10x^2 + 9$

$\qquad = (x^2 + 1)(x^2 + 9)$

$\qquad = (x + i)(x - i)(x + 3i)(x - 3i)$

The zeros of $f(x)$ are $x = \pm i$ and $x = \pm 3i$.

25. $g(x) = x^4 - 4x^3 + 8x^2 - 16x + 16$

Possible rational zeros: $\pm 1, \pm 2, \pm 4, \pm 8, \pm 16$

$$
\begin{array}{r|rrrrr}
2 & 1 & -4 & 8 & -16 & 16 \\
 & & 2 & -4 & 8 & -16 \\
\hline
2 & 1 & -2 & 4 & -8 & 0 \\
 & & 2 & 0 & 8 & \\
\hline
 & 1 & 0 & 4 & 0 &
\end{array}
$$

$$
\begin{aligned}
g(x) &= (x - 2)(x - 2)(x^2 + 4) \\
&= (x - 2)^2(x + 2i)(x - 2i)
\end{aligned}
$$

The zeros of g are 2 and $\pm 2i$.

27. $f(x) = 2x^4 + 5x^3 + 4x^2 + 5x + 2$

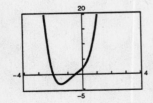

The graph reveals one zero at $x = -2$ and $x = -\frac{1}{2}$.

$$
\begin{array}{r|rrrrr}
-2 & 2 & 5 & 4 & 5 & 2 \\
 & & -4 & -2 & -4 & -2 \\
\hline
 & 2 & 1 & 2 & 1 & 0
\end{array}
$$

$$
\begin{array}{r|rrrr}
-\dfrac{1}{2} & 2 & 1 & 2 & 1 \\
 & & -1 & 0 & -1 \\
\hline
 & 2 & 0 & 2 & 0
\end{array}
$$

The zeros of $2x^2 + 2 = 2(x^2 + 1)$ are $x = \pm i$.

The zeros of $f(x)$ are $(x + 2)(2x + 1)(x - i)(x + i)$.

29. $f(x) = (x - 1)(x - 5i)(x + 5i)$

$\qquad = (x - 1)(x^2 + 25)$

$\qquad = x^3 - x^2 + 25x - 25$

Note: $f(x) = a(x^3 - x^2 + 25x - 25)$, where a is any real number, has the zero 1 and $\pm 5i$.

31. $f(x) = (x - 2)(x - 4 - i)(x - 4 + i)$

$= (x - 2)(x^2 - 8x + 17)$

$= x^3 - 10x^2 + 33x - 34$

33. $f(x) = (x - i)(x + i)(x - 6i)(x + 6i)$

$= (x^2 + 1)(x^2 + 36)$

$= x^4 + 37x^2 + 36$

Note: $f(x) = a(x^4 + 37x^2 + 36)$, where a is any real number, has the zeros $\pm i$ and $\pm 6i$.

35. If $1 + \sqrt{3}i$ is a zero, so is its conjugate $1 - \sqrt{3}i$.

$f(x) = (x + 5)^2(x - 1 + \sqrt{3}i)(x - 1 - \sqrt{3}i)$

$= (x^2 + 10x + 25)(x^2 - 2x + 4)$

$= x^4 + 8x^3 + 9x^2 - 10x + 100$

37. $f(x) = x^4 + 6x^2 - 27$

(a) $f(x) = (x^2 + 9)(x^2 - 3)$

(b) $f(x) = (x^2 + 9)(x + \sqrt{3})(x - \sqrt{3})$

(c) $f(x) = (x + 3i)(x - 3i)(x + \sqrt{3})(x - \sqrt{3})$

39.

$$
\begin{array}{r}
x^2 - 2x + 3 \\
x^2 - 2x - 2\overline{)x^4 - 4x^3 + 5x^2 - 2x - 6} \\
\underline{x^4 - 2x^3 - 2x^2} \\
-2x^3 + 7x^2 - 2x \\
\underline{-2x^3 + 4x^2 + 4x} \\
3x^2 - 6x - 6 \\
\underline{3x^2 - 6x - 6} \\
0
\end{array}
$$

$f(x) = (x^2 - 2x + 3)(x^2 - 2x - 2)$

(a) $f(x) = (x^2 - 2x - 2)(x^2 - 2x + 3)$

(b) $f(x) = (x - 1 + \sqrt{3})(x - 1 - \sqrt{3})(x^2 - 2x + 3)$

(c) $f(x) = (x - 1 + \sqrt{3})(x - 1 - \sqrt{3})(x - 1 + \sqrt{2}i)(x - 1 - \sqrt{2}i)$

Note: Use the Quadratic Formula for (b) and (c).

41. $f(x) = 2x^3 + 3x^2 + 50x + 75$

Since $5i$ is a zero, so is $-5i$.

$$
\begin{array}{r|rrrr}
5i & 2 & 3 & 50 & 75 \\
& & 10i & -50 + 15i & -75 \\
\hline
& 2 & 3 + 10i & 15i & 0
\end{array}
$$

$$
\begin{array}{r|rrr}
-5i & 2 & 3 + 10i & 15i \\
& & -10i & -15i \\
\hline
& 2 & 3 & 0
\end{array}
$$

The zero of $2x + 3$ is $x = -\frac{3}{2}$. The zeros of f are $x = -\frac{3}{2}$ and $x = \pm 5i$.

Alternate Solution

Since $x = \pm 5i$ are zeros of $f(x)$, $(x + 5i)(x - 5i) = x^2 + 25$ is a factor of $f(x)$. By long division we have:

$$
\begin{array}{r}
2x + 3 \\
x^2 + 0x + 25\overline{)2x^3 + 3x^2 + 50x + 75} \\
\underline{2x^3 + 0x^2 + 50x} \\
3x^2 + 0x + 75 \\
\underline{3x^2 + 0x + 75} \\
0
\end{array}
$$

Thus, $f(x) = (x^2 + 25)(2x + 3)$ and the zeros of f are $x = \pm 5i$ and $x = -\frac{3}{2}$.

43. $f(x) = 2x^4 - x^3 + 7x^2 - 4x - 4$
Since $2i$ is a zero, so is $-2i$.

$$
\begin{array}{r|rrrrr}
2i & 2 & -1 & 7 & -4 & -4 \\
   &   & 4i & -8-2i & 4-2i & 4 \\
\hline
   & 2 & -1+4i & -1-2i & -2 & 0
\end{array}
$$

$$
\begin{array}{r|rrrr}
-2i & 2 & -1+4i & -1-2i & -2i \\
    &   & -4i & 2i & 2i \\
\hline
    & 2 & -1 & -1 & 0
\end{array}
$$

The zeros of $2x^2 - x - 1 = (2x + 1)(x - 1)$ are $x = -\frac{1}{2}$ and $x = 1$. The zeros of f are $x = \pm 2i$, $x = -\frac{1}{2}$, and $x = 1$.

Alternate Solution

Since $x = \pm 2i$ are zeros of $f(x)$, $(x + 2i)(x - 2i) = x^2 + 4$ is a factor of $f(x)$. By long division we have:

$$
\require{enclose}
\begin{array}{r}
2x^2 - x - 1 \\
x^2 + 0x + 4 \enclose{longdiv}{2x^4 - x^3 + 7x^2 - 4x - 4} \\
\underline{2x^4 + 0x^3 + 8x^2 } \\
-x^3 - x^2 - 4x \\
\underline{-x^3 + 0x^2 - 4x } \\
-x^2 + 0x - 4 \\
\underline{-x^2 + 0x - 4} \\
0
\end{array}
$$

Thus, $f(x) = (x^2 + 4)(2x^2 - x - 1)$
$$= (x + 2i)(x - 2i)(2x + 1)(x - 1)$$

and the zeros of f are $x = \pm 2i$, $x = -\frac{1}{2}$, and $x = 1$.

45. $g(x) = 4x^3 + 23x^2 + 34x - 10$
Since $-3 + i$ is a zero, so is $-3 - i$.

$$
\begin{array}{r|rrrr}
-3+i & 4 & 23 & 34 & -10 \\
     &   & -12+4i & -37-i & 10 \\
\hline
     & 4 & 11+4i & -3-i & 0
\end{array}
$$

$$
\begin{array}{r|rrr}
-3-i & 4 & 11+4i & -3-i \\
     &   & -12-4i & 3+i \\
\hline
     & 4 & -1 & 0
\end{array}
$$

The zeros of $4x - 1$ is $x = \frac{1}{4}$. The zeros of $g(x)$ are $x = -3 \pm i$ and $x = \frac{1}{4}$.

—CONTINUED—

45. —CONTINUED—

Alternate Solution

Since $-3 \pm i$ are zeros of $g(x)$,

$$[x - (-3 + i)][x - (-3 - i)] = [(x - 3) - i][(x + 3) + i]$$
$$= (x + 3)^2 - i^2 = x^2 + 6x + 10$$

is a factor of $g(x)$. By long division we have:

$$
\begin{array}{r}
4x - 1 \\
x^2 + 6x + 10 \overline{\smash{\big)}\ 4x^3 + 23x^2 + 34x - 10} \\
\underline{4x^3 + 24x^2 + 40x} \\
-x^2 - 6x - 10 \\
\underline{-x^2 - 6x - 10} \\
0
\end{array}
$$

Thus, $g(x) = (x^2 + 6x + 10)(4x - 1)$ and the zeros of g are $x = -3 \pm i$ and $x = \frac{1}{4}$.

47. (a) The root feature yields the real roots 1 and 2, and the complex roots $-3 \pm 1.414i$.

(b) By synthetic division,

$$
\begin{array}{r|rrrrr}
1 & 1 & 3 & -5 & -21 & 22 \\
  &   & 1 & 4 & -1 & -22 \\
\hline
  & 1 & 4 & -1 & -22 & 0 \\
\end{array}
$$

$$
\begin{array}{r|rrrr}
2 & 1 & 4 & -1 & -22 \\
  &   & 2 & 12 & 22 \\
\hline
  & 1 & 6 & 11 & 0 \\
\end{array}
$$

The complex roots of $x^2 + 6x + 11$ are

$$x = \frac{-6 \pm \sqrt{6^2 - 4(11)}}{2} = -3 \pm \sqrt{2}i.$$

49. (a) The root feature yields the real root 0.75, and the complex roots $0.5 \pm 1.118i$.

(b) By synthetic division,

$$
\begin{array}{r|rrrr}
\frac{3}{4} & 8 & -14 & 18 & -9 \\
  &   & 6 & -6 & 9 \\
\hline
  & 8 & -8 & 12 & 0 \\
\end{array}
$$

The complex roots of $8x^2 - 8x + 12$ are

$$x = \frac{8 \pm \sqrt{64 - 4(8)(12)}}{2(8)} = \frac{1}{2} \pm \frac{\sqrt{5}}{2}i.$$

51. $f(x) = x^3 + ix^2 + ix - 1$

(a)
$$
\begin{array}{r|rrrr}
i & 1 & i & i & -1 \\
  &   & i & -2 & -1 - 2i \\
\hline
  & 1 & 2i & -2 + i & -2 - 2i \\
\end{array}
$$

Since the remainder is not zero, $x = i$ is not a zero of f.

(b) The theorem that states that complex zeros occur in conjugate pairs has the condition that the coefficients of f must be real numbers. This polynomial has complex coefficients for x^2 and x.

53. (a) No, the answers will not change if the graph is shifted to the right 2 units.

(b) No, the answer will not change.

55. No. Setting $P = R - C = xp - C = x(140 - 0.0001x) - (80x + 150,000) = 9,000,000$ yields a quadratic with no real roots:

$$-0.0001x^2 + 60x - 9,150,000 = 0$$

57. $f(x) = [x - (a + bi)][x - (a - bi)]$
$= [(x - a) - bi][(x - a) + bi]$
$= (x - a)^2 - (bi)^2$
$= x^2 - 2ax + a^2 + b^2$

59.

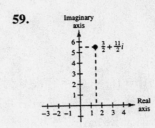

Section 2.6   Rational Functions and Asymptotes

■ You should know the following basic facts about rational functions.

(a) A function of the form $f(x) = P(x)/Q(x)$, $Q(x) \neq 0$, where $P(x)$ and $Q(x)$ are polynomials, is called a rational function.

(b) The domain of a rational function is the set of all real numbers except those which make the denominator zero.

(c) If $f(x) = P(x)/Q(x)$ is in reduced form, and a is a value such that $Q(a) = 0$, then the line $x = a$ is a vertical asymptote of the graph of f. $f(x) \to \infty$ or $f(x) \to -\infty$ as $x \to a$.

(d) The line $y = b$ is a horizontal asymptote of the graph of f if $f(x) \to b$ as $x \to \infty$ or $x \to -\infty$.

(e) Let $f(x) = \dfrac{P(x)}{Q(x)} = \dfrac{a_n x^n + a_{n-1} x^{n-1} + \cdots + a_1 x + a_0}{b_m x^m + b_{m-1} x^{m-1} + \cdots + b_1 x + b_0}$ where $P(x)$ and $Q(x)$ have no common factors.

1. If $n < m$, then the x-axis ($y = 0$) is a horizontal asymptote.

2. If $n = m$, then $y = \dfrac{a_n}{b_m}$ is a horizontal asymptote.

3. If $n > m$, then there are no horizontal asymptotes.

Solutions to Odd-Numbered Exercises

1. $f(x) = \dfrac{1}{x - 1}$

(a)

x	$f(x)$
0.5	-2
0.9	-10
0.99	-100
0.999	-1000

x	$f(x)$
1.5	2
1.1	10
1.01	100
1.001	1000

x	$f(x)$
5	0.25
10	$0.\overline{1}$
100	$0.\overline{01}$
1000	$0.\overline{001}$

(b) The zero of the denominator is $x = 1$, so $x = 1$ is a vertical asymptote. The degree of the numerator is less than the degree of the denominator so the x-axis, or $y = 0$ is a horizontal asymptote.

(c) The domain is all real numbers except $x = 1$.

3. $f(x) = \dfrac{3x}{|x-1|}$

(a)

x	$f(x)$
0.5	3
0.9	27
0.99	297
0.999	2997

x	$f(x)$
1.5	9
1.1	33
1.01	303
1.001	3003

x	$f(x)$
5	3.75
10	$3.\overline{33}$
100	$3.\overline{03}$
1000	$3.\overline{003}$

(b) The zero of the denominator is $x = 1$, so $x = 1$ is a vertical asymptote. Since $f(x) \to 3$ as $x \to \infty$ and $f(x) \to -3$ as $x \to -\infty$, both $y = 3$ and $y = -3$ are horizontal asymptotes.

(c) The domain is all real numbers except $x = 1$.

5. $f(x) = \dfrac{3x^2}{x^2 - 1}$

(a)

x	$f(x)$
0.5	-1
0.9	-12.79
0.99	-147.8
0.999	-1498

x	$f(x)$
1.5	5.4
1.1	17.29
1.01	152.3
1.001	1502.3

x	$f(x)$
5	3.125
10	$3.\overline{03}$
100	$3.\overline{0003}$
1000	3

(b) The zeros of the denominator are $x = \pm 1$ so both $x = 1$ and $x = -1$ are vertical asymptotes.

Since the degree of the numerator equals the degree of the denominator, $y = \frac{3}{1} = 3$ is a horizontal asymptote.

(c) The domain is all real numbers except $x = \pm 1$.

7. $f(x) = \dfrac{2}{x + 2}$

Vertical asymptote: $y = -2$
Horizontal asymptote: $y = 0$
Matches graph (a).

9. $f(x) = \dfrac{4x + 1}{x}$

Vertical asymptote: $x = 0$
Horizontal asymptote: $y = 4$
Matches graph (c).

11. $f(x) = \dfrac{x - 2}{x - 4}$

Vertical asymptote: $x = 4$
Horizontal asymptote: $y = 1$
Matches graph (b).

13. $f(x) = \dfrac{1}{x^2}$

Domain: all real numbers except $x = 0$
Vertical asymptote: $x = 0$
Horizontal asymptote: $y = 0$
[Degree of $p(x)$ < degree of $q(x)$]

15. $f(x) = \dfrac{2 + x}{2 - x} = \dfrac{x + 2}{-x + 2}$

Domain: all real numbers except $x = 2$
Vertical asymptote: $x = 2$
Horizontal asymptote: $y = -1$
[Degree of $p(x)$ = degree of $q(x)$]

17. $f(x) = \dfrac{x^3}{x^2 - 1}$

Domain: all real numbers except $x = \pm 1$
Vertical asymptote: $x = \pm 1$
Horizontal asymptotes: None
[Degree of $p(x)$ > degree of $q(x)$]

19. $f(x) = \dfrac{3x^2 + 1}{x^2 + x + 9}$

Domain: All real numbers. The denominator has no real zeros. [Try the Quadratic Formula on the denominator.]

Vertical asymptote: None

Horizontal asymptote: $y = 3$

[Degree of $p(x)$ = degree of $q(x)$]

21. $f(x) = \dfrac{x^2 - 4}{x + 2}$, $g(x) = x - 2$

(a) Domain of f: all real numbers except -2

Domain of g: all real numbers

(b) Since $x + 2$ is a common factor of both the numerator and the denominator of $f(x)$, $x = -2$ is not a vertical asymptote of f. f has no vertical asymptotes.

(c)

x	-4	-3	-2.5	-2	-1.5	-1	0
$f(x)$	-6	-5	-4.5	undef.	-3.5	-3	-2
$g(x)$	-6	-5	-4.5	-4	-3.5	-3	-2

(d) f and g differ only where f is undefined.

23. $f(x) = \dfrac{x - 3}{x^2 - 3x}$, $g(x) = \dfrac{1}{x}$

(a) Domain of f: all real number except 0 and 3

Domain of g: all real numbers except 0

(b) Since $x - 3$ is a common factor of both the numerator and the denominator of f, $x = 3$ is not a vertical asymptote of f. The only vertical asymptote is $x = 0$.

(c)

x	-1	-0.5	0	0.5	2	3	4
$f(x)$	-1	-2	undef.	2	$\frac{1}{2}$	undef.	$\frac{1}{4}$
$g(x)$	-1	-2	undef	2	$\frac{1}{2}$	$\frac{1}{3}$	$\frac{1}{4}$

(d) They differ only at $x = 3$, where f is undefined and g is defined.

25. $f(x) = \dfrac{1}{(x + 2)(x - 1)} = \dfrac{1}{x^2 + x - 2}$

27. $f(x) = \dfrac{2x^2}{x^2 + 1}$

29. $f(x) = 4 - \dfrac{1}{x}$

(a) As $x \to \pm\infty$, $f(x) \to 4$.

(b) As $x \to \infty$, $f(x) \to 4$ but is less than 4.

(c) As $x \to -\infty$, $f(x) \to 4$ but is greater than 4.

31. $f(x) = \dfrac{2x - 1}{x - 3}$

(a) As $x \to \pm\infty$, $f(x) \to 2$.

(b) As $x \to \infty$, $f(x) \to 2$ but is greater than 2.

(c) As $x \to -\infty$, $f(x) \to 2$ but is less than 2.

33. $f(x) = \dfrac{x^2 - 4}{x + 1} = \dfrac{(x + 2)(x - 2)}{x + 1}$

The zeros of f correspond to the zeros of the numerator and are $x = \pm 2$.

35. $f(x) = 1 - \dfrac{2}{x - 3} = \dfrac{x - 5}{x - 3}$

The zero of f corresponds to the zero of the numerator and is $x = 5$.

37. $C = \dfrac{255p}{100 - p}$, $0 \le p < 100$

(a) $C(10) = \dfrac{255(10)}{100 - 10} \approx 28.33$ million dollars

(b) $C(40) = \dfrac{255(40)}{100 - 40} = 170$ million dollars

(c) $C(75) = \dfrac{255(75)}{100 - 75} = 765$ million dollars

(d) $C \to \infty$ as $x \to 100$. No, it would not be possible to remove 100% of the pollutants.

39. (a)

M	200	400	600	800	1000	1200	1400	1600	1800	2000
t	0.472	0.596	0.710	0.817	0.916	1.009	1.096	1.178	1.255	1.328

The greater the mass, the more time required per oscillation. The model is a good fit to the actual data.

(b) You can find M corresponding to $t = 1.056$ by finding the point of intersection of

$$t = \dfrac{38M + 16{,}965}{10(M + 500)} \quad \text{and} \quad t = 1.056.$$

If you do this, you obtain $M \approx 1306$ grams.

41. $N = \dfrac{20(5 + 3t)}{1 + 0.04t}$, $0 \le t$

(a) $N(5) \approx 333$ deer

$N(10) = 500$ deer

$N(25) = 800$ deer

(b) The herd is limited by the horizontal asymptote: $N = \dfrac{60}{0.04} = 1500$ deer

43. $P = \dfrac{0.5 + 0.9(n - 1)}{1 + 0.9(n - 1)}$, $0 < n$

(a)

n	1	2	3	4	5	6	7	8	9	10
P	0.50	0.74	0.82	0.86	0.89	0.91	0.92	0.93	0.94	0.95

P approaches 1 as n increases.

(b) $P = \dfrac{0.9n - 0.4}{0.9n + 0.1}$

The percentage of correct responses is limited by a horizontal asymptote:

$$P = \dfrac{0.9}{0.9} = 1 = 100\%$$

45.
$$225 - 50x = 0$$
$$-50x = -225$$
$$x = \dfrac{-225}{(-50)} = \dfrac{9}{2}$$

47.
$$2z^2 - 3z - 35 = 0$$
$$(2z + 7)(z - 5) = 0$$
$$z = -\dfrac{7}{2}, \, 5$$

Section 2.7   Graphs of Rational Functions

■ You should be able to graph $f(x) = \dfrac{p(x)}{q(x)}$.

(a) Find the x- and y-intercepts.

(b) Find any vertical or horizontal asymptotes.

(c) Plot additional points.

(d) If the degree of the numerator is one more than the degree of the denominator, use long division to find the slant asymptote.

Solutions to Odd-Numbered Exercises

1. $g(x) = \dfrac{2}{x} + 1$

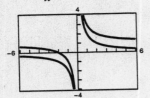

Vertical shift one unit upward

3. $g(x) = -\dfrac{2}{x}$

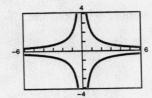

Reflection in the x-axis

5. $g(x) = \dfrac{2}{x^2} - 2$

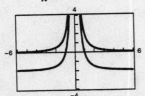

Vertical shift two units downward

7. $g(x) = \dfrac{2}{(x-2)^2}$

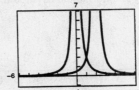

Horizontal shift two units to the right

9. $g(x) = \dfrac{4}{(x+2)^3}$

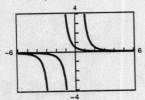

Horizontal shift two units to the left

11. $g(x) = -\dfrac{4}{x^3}$

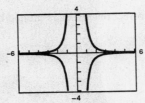

Reflection in the x-axis the left

13. $f(x) = \dfrac{1}{x + 2}$

y-intercept: $\left(0, \dfrac{1}{2}\right)$

Vertical asymptote: $x = -2$
Horizontal asymptote: $y = 0$

x	−4	−3	−1	0	1
y	$-\frac{1}{2}$	−1	1	$\frac{1}{2}$	$\frac{1}{3}$

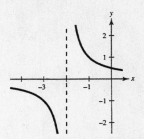

15. $h(x) = -\dfrac{1}{x + 2}$

y-intercept: $\left(0, -\dfrac{1}{2}\right)$

Vertical asymptote: $x = -2$
Horizontal asymptote: $y = 0$

x	−4	−3	−1	0
y	$\frac{1}{2}$	1	−1	$-\frac{1}{2}$

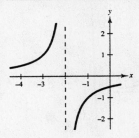

Note: This is the graph of $f(x) = \dfrac{1}{x + 2}$
(Exercise 13) reflected about the *x*-axis.

17. $C(x) = \dfrac{5 + 2x}{1 + x} = \dfrac{2x + 5}{x + 1}$

x-intercept: $\left(-\dfrac{5}{2}, 0\right)$

y-intercept: $(0, 5)$

Vertical asymptote: $x = -1$
Horizontal asymptote: $y = 2$

x	−4	−3	−2	0	1	2
C(x)	1	$\frac{1}{2}$	−1	5	$\frac{7}{2}$	3

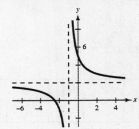

19. $g(x) = \dfrac{1}{x + 2} + 2 = \dfrac{2x + 5}{x + 2}$

Intercepts: $\left(-\dfrac{5}{2}, 0\right), \left(0, \dfrac{5}{2}\right)$

Vertical asymptote: $x = -2$
Horizontal asymptote: $y = 2$

x	−4	−3	−1	0	1
y	$\frac{3}{2}$	1	3	$\frac{5}{2}$	$\frac{7}{3}$

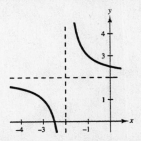

Note: This is the graph of $f(x) = \dfrac{1}{x + 2}$
(Exercise 13) shifted upward two units.

21. $f(x) = \dfrac{x^2}{x^2 + 9}$

Intercept: $(0, 0)$

Horizontal asymptote: $y = 1$

y-axis symmetry

x	±1	±2	±3
y	$\frac{1}{10}$	$\frac{4}{13}$	$\frac{1}{2}$

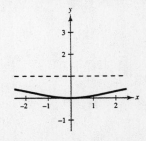

23. $h(x) = \dfrac{x^2}{x^2 - 9}$

Intercept: $(0, 0)$

Vertical asymptotes: $x = \pm 3$
Horizontal asymptote: $y = 1$
y-axis symmetry

x	± 5	± 4	± 2	± 1	0
y	$\frac{25}{16}$	$\frac{16}{7}$	$-\frac{4}{5}$	$-\frac{1}{8}$	0

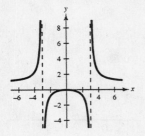

25. $g(s) = \dfrac{s}{s^2 + 1}$

Intercept: $(0, 0)$

Horizontal asymptote: $y = 0$
Origin symmetry

s	-2	-1	0	0	1
$g(s)$	$-\frac{2}{5}$	$-\frac{1}{2}$	0	$\frac{1}{2}$	$\frac{2}{5}$

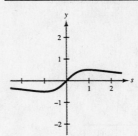

27. $g(x) = \dfrac{4(x + 1)}{x(x - 4)}$

Intercept: $(-1, 0)$

Vertical asymptotes: $x = 0$ and $x = 4$

Horizontal asymptote: $y = 0$

x	-2	-1	1	2	3	5	6
y	$-\frac{1}{3}$	0	$-\frac{8}{3}$	-3	$-\frac{16}{3}$	$\frac{24}{5}$	$\frac{7}{3}$

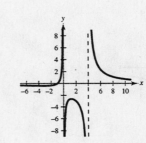

29. $f(x) = \dfrac{3x}{x^2 - x - 2} = \dfrac{3x}{(x + 1)(x - 2)}$

Intercept: $(0, 0)$

Vertical asymptotes: $x = -1, 2$

Horizontal asymptote: $y = 0$

x	-3	0	1	3	4
y	$-\frac{9}{10}$	0	$-\frac{3}{2}$	$\frac{9}{4}$	$\frac{6}{5}$

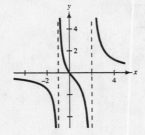

31. $f(x) = \dfrac{2 + x}{1 - x} = -\dfrac{x + 2}{x - 1}$

x-intercept: $(-2, 0)$

y-intercept: $(0, 2)$

Vertical asymptote: $x = 1$

Horizontal asymptote: $y = -1$

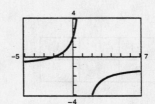

Domain: $x \neq 1$ or $(-\infty, 1) \cup (1, \infty)$

33. $f(t) = \dfrac{3t + 1}{t}$

t-intercept: $\left(-\dfrac{1}{3}, 0\right)$

Vertical asymptote: $t = 0$

Horizontal asymptote: $y = 3$

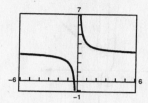

Domain: $t \neq 0$ or $(-\infty, 0) \cup (0, \infty)$

35. $h(t) = \dfrac{4}{t^2 + 1}$

Domain: all real numbers OR $(-\infty, \infty)$

Horizontal asymptote: $y = 0$

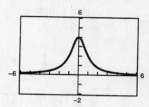

37. $f(t) = \dfrac{2t^2}{t^2 - 4}$

Domain: all real numbers except ± 2,
OR $(-\infty, -2) \cup (-2, 2) \cup (2, \infty)$

Vertical asymptote: $x = \pm 2$
Horizontal asymptote: $y = 2$

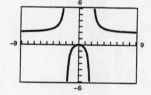

39. $f(x) = \dfrac{20x}{x^2 + 1} - \dfrac{1}{x} = \dfrac{19x^2 - 1}{x(x^2 + 1)}$

Domain: all real numbers except 0,
OR $(-\infty, 0) \cup (0, \infty)$

Vertical asymptote: $x = 0$
Horizontal asymptote: $y = 0$
Origin Symmetry

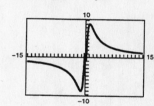

41. $h(x) = \dfrac{6x}{\sqrt{x^2 + 1}}$

There are two horizontal asymptotes: $y = \pm 6$

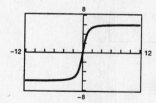

43. $f(x) = \dfrac{4(x - 1)^2}{x^2 - 4x + 5}$

The graph crosses its horizontal asymptote: $y = 4$

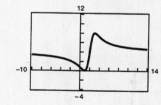

45. $h(x) = \dfrac{6 - 2x}{3 - x} = \dfrac{2(3 - x)}{3 - x}$

Since $h(x)$ is not reduced and $(3 - x)$ is a factor of both the numerator and the denominator, $x = 3$ is not a horizontal asymptote. There is a hole in the graph at $x = 3$.

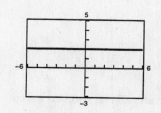

47. False. The graph would have two distinct branches that are separated by the vertical asymptote.

49. $f(x) = \dfrac{2x^2 + 1}{x} = 2x + \dfrac{1}{x}$

Vertical asymptote: $x = 0$
Slant asymptote: $y = 2x$
Origin symmetry

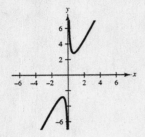

51. $g(x) = \dfrac{x^2 + 1}{x} = x + \dfrac{1}{x}$

Vertical asymptote: $x = 0$
Slant asymptote: $y = x$
Origin symmetry

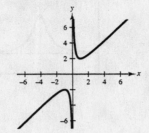

53. $f(x) = \dfrac{x^3}{x^2 - 1} = x + \dfrac{x}{x^2 - 1}$

Intercept: $(0, 0)$
Vertical asymptotes: $x = \pm 1$
Slant asymptote: $y = x$
Origin symmetry

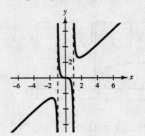

55. $f(x) = \dfrac{x^2 - x + 1}{x - 1} = x + \dfrac{1}{x - 1}$

y-intercept: $(0, -1)$
Vertical asymptote: $x = 1$
Slant asymptote: $y = x$

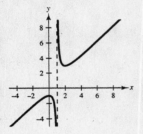

57. (a) x-intercept: $(-1, 0)$

(b) $0 = \dfrac{x + 1}{x - 3}$

$0 = x + 1$

$-1 = x$

59. (a) x-intercepts: $(\pm 1, 0)$

(b) $0 = \dfrac{1}{x} - x$

$x = \dfrac{1}{x}$

$x^2 = 1$

$x = \pm 1$

61. $y = \dfrac{2x^2 + x}{x + 1} = 2x - 1 + \dfrac{1}{x + 1}$

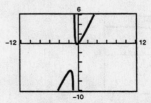

Domain: all real numbers except $x = -1$
Vertical asymptote: $x = -1$
Slant asymptote: $y = 2x - 1$

63. $g(x) = \dfrac{1 + 3x^2 - x^3}{x^2} = \dfrac{1}{x^2} + 3 - x = -x + 3 + \dfrac{1}{x^2}$

Domain: all real numbers except 0
OR $(-\infty, 0) \cup (0, \infty)$
Vertical asymptote: $x = 0$
Slant asymptote: $y = -x + 3$

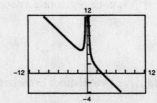

65. $y = \dfrac{1}{x + 5} + \dfrac{4}{x}$

(a)

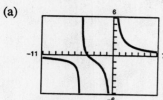

x-intercept: $(-4, 0)$

(b)
$$0 = \frac{1}{x + 5} + \frac{4}{x}$$
$$-\frac{4}{x} = \frac{1}{x + 5}$$
$$-4(x + 5) = x$$
$$-4x - 20 = x$$
$$-5x = 20$$
$$x = -4$$

67. $y = x - \dfrac{6}{x - 1}$

(a)

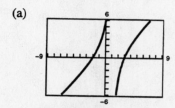

x-intercept: $(-2, 0), (3, 0)$

(b)
$$0 = x - \frac{6}{x - 1}$$
$$\frac{6}{x - 1} = x$$
$$6 = x(x - 1)$$
$$0 = x^2 - x - 6$$
$$0 = (x + 2)(x - 3)$$
$$x = -2, \quad x = 3$$

69. (a) $0.25(50) + 0.75(x) = C(50 + x)$

$$C = \frac{12.50 + 0.75x}{50 + x} \cdot \frac{4}{4}$$
$$C = \frac{50 + 3x}{4(50 + x)}$$
$$= \frac{3x + 50}{4(x + 50)}$$

(b) Domain: $x > 0$ and $x \le 1000 - 50$
Thus, $0 \le x \le 950$ OR $[0, 950]$.

(c)

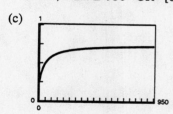

As the tank is filled, the rate that the concentration is increasing slows down. It approaches the horizontal asymptote of $C = \frac{3}{4} = 0.75$.

71. $f(x) = \dfrac{3(x + 1)}{x^2 + x + 1}$

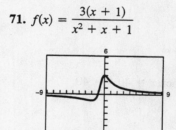

Minimum: $(-2, -1)$
Maximum: $(0, 3)$

73. (a) $A = xy$ and

$$(x - 2)(y - 4) = 30$$

$$y - 4 = \frac{30}{x - 2}$$

$$y = 4 + \frac{30}{x - 2} = \frac{4x + 22}{x - 2}$$

Thus, $A = xy = x\left(\frac{4x + 22}{x - 2}\right) = \frac{2x(x + 11)}{x - 2}$.

(b) Domain: Since the margins on the left and right are each 1 inch, $x > 2$, OR $(2, \infty)$.

(c)

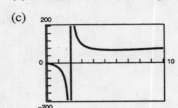

The area is minimum when $x \approx 5.87$ in. and $y \approx 11.75$ in.

75. $C = 100\left(\frac{200}{x^2} + \frac{x}{x + 30}\right)$, $1 \leq x$

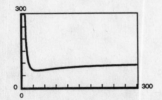

The minimum occurs when $x \approx 40.4 \approx 40$.

77. $C = \frac{3t^2 + t}{t^3 + 50}$, $0 \leq t$

(a) The horizontal asymptote is the t-axis, or $C = 0$. This indicates that the chemical eventually dissipates.

(b)

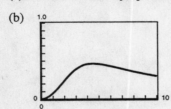

The maximum occurs when $t \approx 4.5$.

79. (a)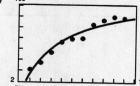

(b)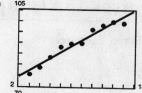

$$y = 2.81t + 68.77$$

(c)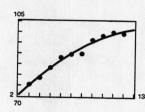

(d) The quadratic and rational models fit the data better than the line. The rational model may be a better predictor since the parabola is near its maximum.

$$y = -0.18t^2 + 5.52t + 60.12$$

81. $y = x + 1 + \dfrac{a}{x - 2}$

This has a slant asymptote of $x + 1$ and a vertical asymptote of $x = 2$.

$$0 = -2 + 1 + \frac{a}{-2 - 2}$$

Since $x = -2$ is a zero, $(-2, 0)$ is on the graph. Use this point to solve for a.

$$1 = \frac{a}{-4}$$

$$-4 = a$$

Hence, $y = x + 1 + \dfrac{-4}{x - 2} = \dfrac{x^2 - x - 6}{x - 2}$.

❏ Review Exercises for Chapter 2

Solutions to Odd-Numbered Exercises

1. $f(x) = \left(x + \frac{3}{2}\right)^2 + 1$

Vertex: $\left(-\frac{3}{2}, 1\right)$

y-intercept: $\left(0, \frac{13}{4}\right)$

No x-intercepts

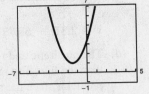

3. $f(x) = \frac{1}{3}(x^2 + 5x - 4)$

$\quad = \frac{1}{3}\left(x^2 + 5x + \frac{25}{4} - \frac{25}{4} - 4\right)$

$\quad = \frac{1}{3}\left[\left(x - \frac{5}{2}\right)^2 - \frac{41}{4}\right]$

$\quad = \frac{1}{3}\left(x - \frac{5}{2}\right)^2 - \frac{41}{12}$

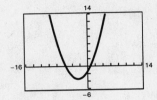

Vertex: $\left(\frac{5}{2}, -\frac{41}{12}\right)$

y-intercept: $\left(0, -\frac{4}{3}\right)$

x-intercepts: $0 = \frac{1}{3}(x^2 + 5x - 4)$

$\qquad\qquad 0 = x^2 + 5x - 4$

$\qquad\qquad x = \dfrac{-5 \pm \sqrt{41}}{2}$ Use the Quadratic Formula.

$\qquad\qquad \left(\dfrac{-5 \pm \sqrt{41}}{2}, 0\right)$

5. Vertex: $(1, -4) \implies f(x) = a(x - 1)^2 - 4$

Point: $(2, -3) \implies -3 = a(2 - 1)^2 - 4$

$\qquad\qquad\qquad\qquad 1 = a$

Thus, $f(x) = (x - 1)^2 - 4$.

7. (a) $y = 2x^2$
Vertical stretch

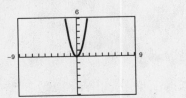

(b) $y = -2x^2$
Vertical stretch and a reflection in the x-axis

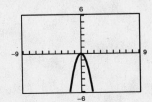

–CONTINUED–

7. –CONTINUED–

(c) $y = x^2 + 2$
Vertical shift two units upward

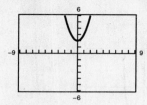

(d) $y = (x + 2)^2$
Horizontal shift two units to the left

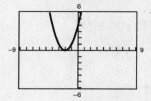

9. $g(x) = x^2 - 2x$
$= x^2 - 2x + 1 - 1$
$= (x - 1)^2 - 1$

The minimum occurs at the vertex $(1, -1)$.

11. $f(x) = 6x - x^2$
$= -(x^2 - 6x + 9 - 9)$
$= -(x - 3)^2 + 9$

The maximum occurs at the vertex $(3, 9)$.

13. $f(t) = -2t^2 + 4t + 1$
$= -2(t^2 - 2t + 1 - 1) + 1$
$= -2[(t - 1)^2 - 1] + 1$
$= -2(t - 1)^2 + 3$

The maximum occurs at the vertex $(1, 3)$.

15. $h(x) = x^2 + 5x - 4$
$= x^2 + 5x + \frac{25}{4} - \frac{25}{4} - 4$
$= \left(x + \frac{5}{2}\right)^2 - \frac{25}{4} - \frac{16}{4}$
$= \left(x + \frac{5}{2}\right)^2 - \frac{41}{4}$

The minimum occurs at the vertex $\left(-\frac{5}{2}, -\frac{41}{4}\right)$.

17. (a)

x	y	Area
1	$4 - \frac{1}{2}(1)$	$(1)[4 - \frac{1}{2}(1)] = \frac{7}{2}$
2	$4 - \frac{1}{2}(2)$	$(2)[4 - \frac{1}{2}(2)] = 6$
3	$4 - \frac{1}{2}(3)$	$(3)[4 - \frac{1}{2}(3)] = \frac{15}{2}$
4	$4 - \frac{1}{2}(4)$	$(4)[4 - \frac{1}{2}(4)] = 8$
5	$4 - \frac{1}{2}(5)$	$(5)[4 - \frac{1}{2}(5)] = \frac{15}{2}$
6	$4 - \frac{1}{2}(6)$	$(6)[4 - \frac{1}{2}(6)] = 6$

(b) The dimensions that will produce a maximum area are $x = 4$ and $y = 2$.

(c) $A = xy = x\left(\dfrac{8 - x}{2}\right)$ since

$x + 2y - 8 = 0 \implies y = \dfrac{8 - x}{2}$.

Since the figure is in the first quadrant and x and y must be positive, the domain of

$A = x\left(\dfrac{8 - x}{2}\right)$ is $0 < x < 8$.

(d)

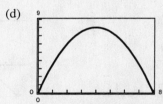

The maximum area of 8 occurs at the vertex when

$x = 4$ and $y = \dfrac{8 - 4}{2} = 2$.

(e) $A = x\left(\dfrac{8 - x}{2}\right)$

$= \dfrac{1}{2}(8x - x^2)$

$= -\dfrac{1}{2}(x^2 - 8x)$ or $-\dfrac{1}{2}x^2 + 4x$

$= -\dfrac{1}{2}(x^2 - 8x + 16 - 16)$

$= -\dfrac{1}{2}[(x - 4)^2 - 16]$

$= -\dfrac{1}{2}(x - 4)^2 + 8$

The maximum area of 8 occurs when $x = 4$ and

$y = \dfrac{8 - 4}{2} = 2$.

19. $f(x) = -x^2 + 6x + 9$

The degree is even and the leading coefficient is negative. The graph falls to the left and right.

21. $f(x) = \frac{3}{4}(x^4 + 3x^2 + 2)$

The degree is even and the leading coefficient is positive. The graph rises to the left and right.

23. $f(x) = \frac{1}{2}x^3 - 2x + 1$; $g(x) = \frac{1}{2}x^3$

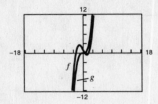

25. $g(x) = x^4 - x^3 - 2x^2$

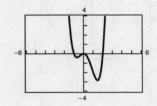

27. $f(t) = t^3 - 3t = t(t^2 - 3)$

Intercepts: $(0, 0), \left(\pm\sqrt{3}, 0\right)$

The graph rises to the right and falls to the left.

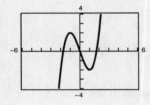

29. $f(x) = x(x + 3)^2$

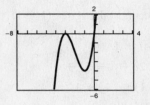

31. $y_1 = \dfrac{x^2}{x - 2}$

$$y_2 = x + 2 + \frac{4}{x - 2}$$

$$= \frac{(x + 2)(x - 2)}{x - 2} + \frac{4}{x - 2}$$

$$= \frac{x^2 - 4}{x - 2} + \frac{4}{x - 2}$$

$$= \frac{x^2}{x - 2}$$

$$= y_1$$

33.

$$
\begin{array}{r}
8x + 5 \\
3x - 2\overline{)24x^2 - x - 8} \\
\underline{24x^2 - 16x} \\
15x - 8 \\
\underline{15x - 10} \\
2
\end{array}
$$

Thus, $\dfrac{24x^2 - x - 8}{3x - 2} = 8x + 5 + \dfrac{2}{3x - 2}$.

35.

$$
\begin{array}{r}
x^2 - 2 \\
x^2 - 1\overline{)x^4 - 3x^2 + 2} \\
\underline{x^4 - x^2} \\
-2x^2 + 2 \\
\underline{-2x^2 + 2} \\
0
\end{array}
$$

Thus, $\dfrac{x^4 - 3x^2 + 2}{x^2 - 1} = x^2 - 2,\qquad (x \neq \pm 1)$.

37. 2 | 0.25   −4   0   0   0

$$
\begin{array}{r|rrrrr}
2 & 0.25 & -4 & 0 & 0 & 0 \\
  &   & \frac{1}{2} & -7 & -14 & -28 \\
\hline
  & \frac{1}{4} & -\frac{7}{2} & -7 & -14 & -28
\end{array}
$$

Hence, $\dfrac{0.25x^4 - 4x^3}{x - 2} = 0.25x^4 - 3.5x^2 - 7x - 14 - \dfrac{28}{x - 2}$.

39. $\frac{2}{3}$

$$
\begin{array}{r|rrrrr}
\frac{2}{3} & 6 & -4 & -27 & 18 & 0 \\
  &   & 4 & 0 & -18 & 0 \\
\hline
  & 6 & 0 & -27 & 0 & 0
\end{array}
$$

Thus, $\dfrac{6x^4 - 4x^3 - 27x^2 + 18x}{x - (2/3)} = 6x^3 - 27x$.

41. $(7 + 5i) + (-4 + 2i) = (7 - 4) + (5i + 2i)$
$$= 3 + 7i$$

43. $5i(13 - 8i) = 65i - 40i^2 = 40 + 65i$

45. $\dfrac{6 + i}{i} = \dfrac{6 + i}{i} \cdot \dfrac{-i}{-i} = \dfrac{-6i - i^2}{-i^2}$

$$= \dfrac{-6i + 1}{1} = 1 - 6i$$

47. $f(x) = 6(x + 1)^2 \left(x - \dfrac{1}{3}\right)\left(x + \dfrac{1}{2}\right)$ Multiply by 6 to clear the fractions.

$$= (x + 1)^2 \, 3\left(x - \dfrac{1}{3}\right) 2\left(x + \dfrac{1}{2}\right)$$
$$= (x^2 + 2x + 1)(3x - 1)(2x + 1)$$
$$= (x^2 + 2x + 1)(6x^2 + x - 1)$$
$$= 6x^4 + 13x^3 + 7x^2 - x - 1$$

Note: $f(x) = a(6x^4 + 13x^3 + 7x^2 - x - 1)$, where a is any real number, has zeros $-1, -1, \frac{1}{3}$, and $-\frac{1}{2}$.

49. $f(x) = 4x^3 - 11x^2 + 10x - 3$

Possible rational zeros: $\pm 1, \pm 3, \pm \frac{1}{2}, \pm \frac{3}{2}, \pm \frac{1}{4}, \pm \frac{3}{4}$. Use a graphing utility to see that $x = 1$ is probably a zero.

$$
\begin{array}{r|rrrr}
1 & 4 & -11 & 10 & -3 \\
  &   & 4 & -7 & 3 \\
\hline
  & 4 & -7 & 3 & 0
\end{array}
$$

$4x^3 - 11x^2 + 10x - 3 = (x - 1)(4x^2 - 7x + 3) = (x - 1)^2(4x - 3)$

Thus, the zeros of f are $x = 1$ and $x = \frac{3}{4}$.

51. $f(x) = 6x^3 - 5x^2 + 24x - 20$

Graphing $f(x)$ with a graphing utility suggests that $x = \frac{5}{6}$ is a zero.

$$
\begin{array}{r|rrrr}
\frac{5}{6} & 6 & -5 & 24 & -20 \\
  &   & 5 & 0 & 20 \\
\hline
  & 6 & 0 & 24 & 0
\end{array}
$$

The quadratic $6x^2 + 24 = 0$ has complex zeros $x = \pm 2i$. Thus, the zeros are $\frac{5}{6}, 2i, -2i$.

53. $f(x) = 6x^4 - 25x^3 + 14x^2 + 27x - 18$

Possible rational zeros: $\pm 1, \pm 2, \pm 3, \pm 6, \pm 9, \pm 18, \pm\frac{1}{2}, \pm\frac{3}{2}, \pm\frac{9}{2}, \pm\frac{1}{3}, \pm\frac{2}{3}, \pm\frac{1}{6}$. Use a graphing utility to see that $x = -1$ and $x = 3$ are probably zeros.

$$
\begin{array}{r|rrrrr}
-1 & 6 & -25 & 14 & 27 & -18 \\
   &   & -6 & 31 & -45 & 18 \\
\hline
   & 6 & -31 & 45 & -18 & 0
\end{array}
$$

$$
\begin{array}{r|rrrr}
3 & 6 & -31 & 45 & -18 \\
  &   & 18 & -39 & 18 \\
\hline
  & 6 & -13 & 6 & 0
\end{array}
$$

$$6x^4 - 25x^3 + 14x^2 + 27x - 18 = (x + 1)(x - 3)(6x^2 - 13x + 6)$$
$$= (x + 1)(x - 3)(3x - 2)(2x - 3)$$

Thus, the zeros of f are $x = -1$, $x = 3$, $x = \frac{2}{3}$, and $x = \frac{3}{2}$.

55. $f(x) = x^4 + 2x + 1$

(a)

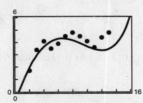

(b) The graph has two x-intercepts, so there are two real zeros.

(c) The zeros are $x = -1$ and $x \approx -0.54$.

57. $h(x) = x^3 - 6x^2 + 12x - 10$

(a)

(b) The graph has one x-intercept, so there is one real zero.

(c) $x \approx 3.26$

59. (a) $S = -0.8195 + 1.6762t - 0.1347t^2 - 0.0028t^3 + 0.0004t^4$

The model is a fairly "good fit."

(b) One explanation may be a recession. The model also shows a downturn in sales.

(c) $S(8) - S(11) \approx 0.72$
The actual decrease of 0.48 was more than this.

(d) $S(15) \approx 4.82$

61. $f(x) = \dfrac{-5}{x^2}$

y-axis symmetry
Vertical asymptote: $x = 0$
Horizontal asymptote: $y = 0$

x	± 3	± 2	± 1
y	$-\frac{5}{9}$	$-\frac{5}{4}$	-5

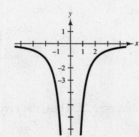

63. $p(x) = \dfrac{x^2}{x^2 + 1}$

Intercept: $(0, 0)$
y-axis symmetry
Horizontal asymptote: $y = 1$

x	± 3	± 2	± 1	0
y	$\frac{9}{10}$	$\frac{4}{5}$	$\frac{1}{2}$	0

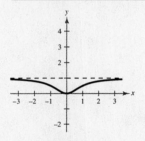

65. $f(x) = \dfrac{x}{x^2 + 1}$

Intercept: $(0, 0)$
Origin symmetry
Horizontal asymptote: $y = 0$

x	-2	-1	0	1	2
y	$-\frac{2}{5}$	$-\frac{1}{2}$	0	$\frac{1}{2}$	$\frac{2}{5}$

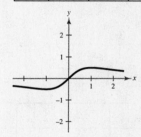

67. $f(x) = \dfrac{2x^3}{x^2 + 1} = 2x - \dfrac{2x}{x^2 + 1}$

Intercept: $(0, 0)$
Origin symmetry
Slant asymptote: $y = 2x$

x	-2	-1	0	1	2
y	$-\frac{16}{5}$	-1	0	1	$\frac{16}{5}$

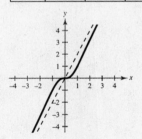

69. $s(x) = \dfrac{8x^2}{x^2 + 4}$

Intercept: $(0, 0)$
Horizontal asymptote: $y = 8$

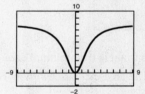

71. $g(x) = \dfrac{x^2 + 1}{x + 1} = x - 1 + \dfrac{2}{x + 1}$

Intercept: $(0, 1)$
Slant asymptote: $y = x - 1$

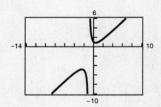

73. $f(x) = \dfrac{2x^2}{(x + 3)(x - 4)} = \dfrac{2x^2}{x^2 - x - 12}$

This answer is not unique.

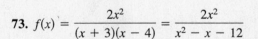

75. (a)

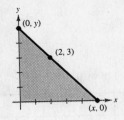

(b) $A = \frac{1}{2}bh = \frac{1}{2}xy.$

From the slope, $\dfrac{y - 3}{-2} = \dfrac{0 - 3}{x - 2} \Longrightarrow y = \dfrac{6}{x - 2} + 3 = \dfrac{3x}{x - 2}.$

Hence, $A = \dfrac{1}{2}x\left(\dfrac{3x}{x - 2}\right) = \dfrac{3x^2}{2(x - 2)}.$

(c)

x	2.5	3	3.5	4	4.5
A	18.75	13.50	12.25	12	12.15

(d)

$x = 4$ yields the triangle of minimum area.

(e) $y = \frac{3}{2}(x + 2)$. The area increases without bound as x increases.

❑ Chapter Test Solutions for Chapter 2

1. (a) $g(x) = 2 - x^2$ is a reflection in the x-axis followed by a vertical translation 2 units upward.

(b) $g(x) = \left(x - \frac{3}{2}\right)^2$ is a horizontal translation $\frac{3}{2}$ units to the right.

2. $y = x^2 + 4x + 3 = x^2 + 4x + 4 - 1 = (x + 2)^2 - 1$

Vertex: $(-2, -1)$

$y = 0 \implies y = 3$

$y = 0 \implies x^2 + 4x + 3 = 0 \implies (x + 3)(x + 1) = 0 \implies x = -1, -3$

Intercepts: $(0, 3),\ (-1, 0),\ (-3, 0)$

3. Let $y = a(x - h)^2 + k$. The vertex $(3, -6)$ implies that $y = a(x - 3)^2 - 6$. For $(0, 3)$ you obtain

$3 = a(0 - 3)^2 - 6 = 9a - 6 \implies a = 1.$

Thus, $y = (x - 3)^2 - 6 = x^2 - 6x + 3.$

4. (a) $y = -\frac{1}{20}x^2 + 3x + 5 = -\frac{1}{20}(x^2 - 60x + 900) + 5 + 45$

$= -\frac{1}{20}(x - 30)^2 + 50$

Maximum height: $y = 50$ feet

(b) The term 5 determines the height at which the ball was thrown. Changing the constant term results in a vertical translation of the graph and therefore changes the maximum height.

5.
$$
\begin{array}{r}
3x \\
x^2 + 1\overline{)\ 3x^3 + 0x^2 + 4x - 1} \\
\underline{3x^3 + 3x} \\
x - 1
\end{array}
$$

$3x + \dfrac{x - 1}{x^2 + 1}$

6. $(10 - 8i)$

$(2 - 3i) = 20 - 16i - 30i - 24$

$= -4 - 46i$

7. Possible rational zeros:

$\pm 24, \pm 12, \pm 8, \pm 6, \pm 4, \pm 3, \pm 2, \pm 1, \pm\frac{3}{2}, \pm\frac{1}{2}$

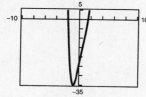

Rational zeros: $-2, \frac{3}{2}$

8. Possible rational zeros:

$\pm 2, \pm 1, \pm\frac{2}{3}, \pm\frac{1}{3}$

Rational zeros: $\pm 1, -\frac{2}{3}$

9. Real zeros: $1.380, -0.819$

10. Real zeros: $-1.414, -0.667, 1.414$

11. $(x - 0)(x - 3)(x - (3 + i))(x - (3 - i))$

$x(x - 3)(x^2 - 6x + 10)$

$x^4 - 9x^3 + 28x^2 - 30x$

12. $\left(x - (1 + \sqrt{3}i)\right)\left(x - (1 - \sqrt{3}i)\right)(x - 2)(x - 2)$

$(x^2 - 2x + 4)(x^2 - 4x + 4)$

$x^4 - 6x^3 + 16x^2 - 24x + 16$

13.

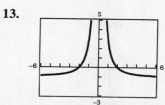

14.

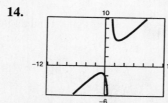

15. $f(x) = \dfrac{4x^2}{(x - 3)(x + 3)}$

$= \dfrac{4x^2}{x^2 - 9}$

❑ Practice Test for Chapter 2

1. Sketch the graph of $f(x) = x^2 - 6x + 5$ by hand and identify the vertex and the intercepts.

2. Find the number of units x that produce a minimum cost C if $C = 0.01x^2 - 90x + 15,000$.

3. Find the quadratic function that has a maximum at $(1, 7)$ and passes through the point $(2, 5)$.

4. Find two quadratic functions that have x-intercepts $(2, 0)$ and $\left(\frac{4}{3}, 0\right)$.

5. Use the leading Coefficient Test to determine the right-hand and left-hand behavior of the graph of the polynomial function $f(x) = -3x^5 + 2x^3 - 17$.

6. Find all the real zeros of $f(x) = x^5 - 5x^3 + 4x$. Verify your answer with a graphing utility.

7. Use a graphing utility to approximate any points of intersection of $y = 3x^2 - 4$ and $y = 2 - x$.

8. Use a graphing utility to approximate any points of intersection of $y = 2x^2 + 3$ and $y = 5 + \sqrt{x}$.

9. Write $\dfrac{2}{1 + i}$ in standard form.

10. Write $\dfrac{3 + i}{2} - \dfrac{i + 1}{4}$ in standard form.

11. Solve $28 + 5x - 3x^2 = 0$ by factoring.

12. Solve $(x - 2)^2 = 24$ by taking the square root of both sides.

13. Solve $x^2 - 4x - 9 = 0$ by completing the square.

14. Solve $x^2 + 5x - 1 = 0$ by the Quadratic Formula.

15. Solve $3x^2 - 2x + 4 = 0$ by the Quadratic Formula.

16. The perimeter of a rectangle is 1100 feet. Find the dimensions so that the enclosed area will be 60,000 square feet.

17. Find two consecutive even positive integers whose product is 624.

18. Solve $x^3 - 10x^2 + 24x = 0$ by factoring.

19. Solve $\sqrt[3]{6 - x} = 4$.

20. Solve $(x^2 - 8)^{2/5} = 4$.

21. Solve $x^4 - x^2 - 12 = 0$.

22. Find a polynomial function with 0, 3, and -2 as zeros.

23. Divide $3x^4 - 7x^2 + 2x - 10$ by $x - 3$ using long division.

24. Use synthetic division to divide $3x^5 + 13x^4 + 12x - 1$ by $x + 5$.

25. List all possible rational zeros of the function $f(x) = 6x^3 - 5x^2 + 4x - 15$.

26. Find a polynomial with real coefficients that has 2, $3 + i$, and $3 - i$ as zeros.

27. Sketch the graph of $f(x) = \dfrac{x - 1}{2x}$ and label all intercepts and asymptotes.

28. Sketch the graph of $f(x) = \dfrac{3x^2 - 4}{x}$ and label all intercepts and asymptotes.

29. Find all the asymptotes of $f(x) = \dfrac{8x^2 - 9}{x^2 + 1}$.

30. Find all the asymptotes of $f(x) = \dfrac{4x^2 - 2x + 7}{x - 1}$.

CHAPTER 3
Exponential and Logarithmic Functions

C H A P T E R 3
Exponential and Logarithmic Functions

Section 3.1   Exponential Functions and Their Graphs

- You should know that a function of the form $y = a^x$, where $a > 0$, $a \neq 1$, is called an exponential function with base a.
- You should be able to graph exponential functions.
- You should be familiar with the number e and the natural exponential function $f(x) = e^x$.
- You should know formulas for compound interest.

    (a) For n compoundings per year: $A = P\left(1 + \dfrac{r}{n}\right)^{nt}$.

    (b) For continuous compoundings: $A = Pe^{rt}$.

Solutions to Odd-Numbered Exercises

1. $(3.4)^{5.6} \approx 946.852$

3. $(1.005)^{400} \approx 7.352$

5. $5^{-\pi} \approx 0.006$

7. $100^{\sqrt{2}} \approx 673.639$

9. $e^{-3/4} \approx 0.472$

11. $\begin{aligned} f(x) &= 3^{x-2} \\ &= 3^x 3^{-2} \\ &= 3^x\left(\frac{1}{3^2}\right) \\ &= \frac{1}{9}(3^x) \\ &= h(x) \end{aligned}$

Thus, $f(x) \neq g(x)$, but $f(x) = h(x)$. You can confirm your answer graphically by graphing f, g, and h in the same viewing rectangle.

13. $\begin{aligned} f(x) &= 16(4^{-x}) \\ &= 4^2(4^{-x}) \\ &= 4^{2-x} \\ &= \left(\frac{1}{4}\right)^{-(2-x)} \\ &= \left(\frac{1}{4}\right)^{x-2} \\ &= g(x) \end{aligned}$

and $\begin{aligned} f(x) &= 16(4^{-x}) \\ &= 16(2^2)^{-x} \\ &= 16(2^{-2x}) \\ &= h(x) \end{aligned}$

Thus, $f(x) = g(x) = h(x)$. You can confirm your answer graphically by graphing f, g, and h in the same viewing rectangle.

15. $g(x) = 5^x$

x	-2	-1	0	1	2
$g(x)$	$\frac{1}{25}$	$\frac{1}{5}$	1	5	25

Asymptote: $y = 0$
Intercept: $(0, 1)$
Increasing

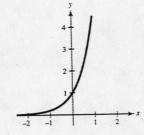

17. $f(x) = \left(\frac{1}{5}\right)^x = 5^{-x}$

x	-2	-1	0	1	2
y	25	5	1	$\frac{1}{5}$	$\frac{1}{25}$

Asymptote: $y = 0$
Intercepts: $(0, 1)$
Decreasing

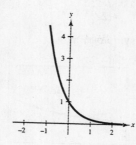

19. $h(x) = 5^{x-2}$

x	-1	0	1	2	3
y	$\frac{1}{125}$	$\frac{1}{25}$	$\frac{1}{5}$	1	5

Asymptote: $y = 0$
Intercepts: $\left(0, \frac{1}{25}\right)$
Increasing

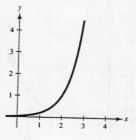

21. $g(x) = 5^{-x} - 3$

x	-1	0	1	2
y	2	-2	$-2\frac{4}{5}$	$-2\frac{24}{25}$

Asymptote: $y = -3$
Intercepts: $(1, -2)$, $(-0.683, 0)$
Decreasing

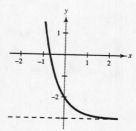

23. $f(x) = 2^x$ rises to the right.

Asymptote: $y = 0$
Intercept: $(0, 1)$
Matches graph (c).

25. $f(x) = 2^{-x}$ falls to the right.

Asymptote: $y = 0$
Intercept: $(0, 1)$
Matches graph (e).

27. $f(x) = 2^x - 4$ rises to the right.

Asymptote: $y = -4$
Intercept: $(0, -3)$
Matches graph (g).

29. $f(x) = -2^{x-2} = -(2^{x-2})$ falls to the right.

Asymptote: $y = 0$
Intercept: $(0, -2^{-2}) = \left(0, -\frac{1}{4}\right)$
Matches graph (a).

31. $y = 2^{-x^2}$
Asymptote: $y = 0$

33. $f(x) = 3^{x-2} + 1$
Asymptote: $y = 1$

35. $y = 1.08^{-5x}$

Asymptote: $y = 0$

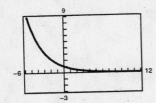

37. $s(t) = 2e^{0.12t}$

Asymptote: $y = 0$

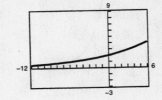

39. $g(x) = 1 + e^{-x}$

Asymptote: $y = 1$

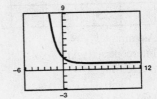

41. $y = 3^x$ and $y = 4^x$

x	-2	-1	0	1	2
3^x	$\frac{1}{9}$	$\frac{1}{3}$	1	3	9
4^x	$\frac{1}{16}$	$\frac{1}{4}$	1	4	16

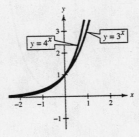

(a) $4^x < 3^x$ when $x < 0$.

(b) $4^x > 3^x$ when $x > 0$.

43. $f(x) = 3^x$

(a) $g(x) = f(x - 2) = 3^{x-2}$

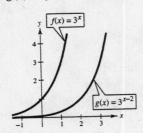

Horizontal shift two units to the right

(b) $h(x) = -\frac{1}{2}f(x) = -\frac{1}{2}(3^x)$

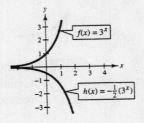

Vertical shrink and a reflection about the x-axis

(c) $q(x) = f(-x) + 3 = 3^{-x} + 3$

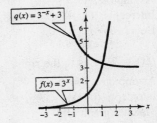

Reflection about the y-axis and a vertical translation three units upward

45. (a) $f(x) = x^2 e^{-x}$

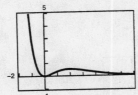

Decreasing: $(-\infty, 0)$, $(2, \infty)$

Increasing: $(0, 2)$

Relative maximum: $(2, 4e^{-2})$

Relative minimum: $(0, 0)$

(b) $g(x) = x2^{3-x}$

Decreasing: $(1.44, \infty)$

Increasing: $(-\infty, 1.44)$

Relative maximum: $(1.44, 4.25)$

47. The exponential function, $y = e^x$, increases at a faster rate than the polynomial function, $y = x^n$.

49. $f(x) = \left(1 + \dfrac{0.5}{x}\right)^x$ and $g(x) = e^{0.5}$

(Horizontal line)

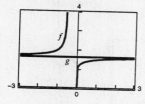

As $x \rightarrow \infty, f(x) \rightarrow g(x)$.

51. $A = 5000e^{(0.075)(50)} \approx \$212,605.51$

53. $P = \$2500$, $r = 12\%$, $t = 10$ years

Compounded n times per year: $A = 2500\left(1 + \dfrac{0.12}{n}\right)^{10n}$

Compounded continuously: $A = 2500e^{0.12(10)}$

n	1	2	4	12	365	Continuous compounding
A	\$7,764.62	\$8,017.84	\$8,155.09	\$8250.97	\$8298.66	\$8,300.29

55. $P = \$2500$, $r = 12\%$, $t = 20$ years

Compounded n times per year: $A = 2500\left(1 + \dfrac{0.12}{n}\right)^{20n}$

Compounded continuously: $A = 2500e^{0.12(20)}$

n	1	2	4	12	365	Continuous compounding
A	\$24,115.73	\$25,714.29	\$26,602.23	\$27,231.38	\$27,547.07	\$27,557.94

57.

$A = Pe^{rt}$

$100,000 = Pe^{0.09t}$

$\dfrac{100,000}{e^{0.09t}} = P$

$P = 100,000e^{-0.09t}$

t	1	10	20	30	40	50
P	\$91,393.12	\$40,656.97	\$16,529.89	\$6,720.55	\$2,732.37	\$1,110.90

59. $P = 100,000\left(1 + \dfrac{0.10}{12}\right)^{-12t}$

t	1	10	20	30	40	50
P	\$90,521.24	\$36,940.70	\$13,646.15	\$5,040.98	\$1,862.17	\$687.90

61. $P = 5000\left(1 - \dfrac{4}{4 + e^{-0.002x}}\right)$

(a)

(b) If $x = 500, p \approx \$421.12$

(c) For $x = 600, p \approx \$350.13$

63. $P(t) = 100e^{0.2197t}$

  (a) $P(0) \approx 100$

  (b) $P(5) \approx 300$

  (c) $P(10) \approx 900$

67. $P = 10{,}958e^{-0.15h}$

  (a), (b)

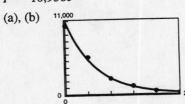

69. (a) $T = -1.239t + 73.021$

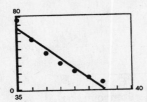

The temperature decreases at a slower rate as it approaches the room temperature.

  (b) $T = 0.034t^2 - 2.264t + 77.295$
The parabola is increasing when $t = 60$.

71.

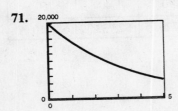

When $t = 2$, $v(2) \approx \$11{,}250$.

75.

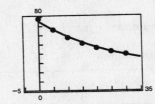

65. $Q = 25\left(\frac{1}{2}\right)^{t/1620}$

  (a) When $t = 0$, $Q = 25\left(\frac{1}{2}\right)^{0/1620} = 25(1) = 25$ units.

  (b) When $t = 1000$, $Q = 25\left(\frac{1}{2}\right)^{1000/1620} \approx 16.30$ units.

  (c)

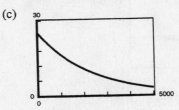

  (c)

h	0	5	10	15	20
P	10,958	5176	2445	1155	546

The model is a "good fit".

  (d) $P(8) \approx 3330 \text{ kg/m}^2$

  (e) $2000 = 10{,}958e^{-0.15h}$ when $x \approx 11.3$.

  (c) $T = 54.438(0.964)^t$

  (d) The horizontal asymptote of the exponential is $T = 0$.

73. Since $\sqrt{2} \approx 1.414$ we know that $1 < \sqrt{2} < 2$.

Thus:

$2^1 < 2^{\sqrt{2}} < 2^2$

$2 \ \ < 2^{\sqrt{2}} < 4$

77. (a) $f(u + v) = a^{u+v} = a^u \cdot a^v = f(u)f(v)$

  (b) $f(2x) = a^{2x} = (a^x)^2 = [f(x)]^2$

Section 3.2   Logarithmic Functions and Their Graphs

- ■ You should know that a function of the form $y = \log_a x$, where $a > 0$, $a \neq 1$, and $x > 0$, is called a logarithm of x to base a.

- ■ You should be able to convert from logarithmic form to exponential form and vice versa.

  $$y = \log_a x \iff a^y = x$$

- ■ You should know the following properties of logarithms.

  (a) $\log_a 1 = 0$ since $a^0 = 1$.

  (b) $\log_a a = 1$ since $a^1 = a$.

  (c) $\log_a a^x = x$ since $a^x = a^x$.

  (d) If $\log_a x = \log_a y$, then $x = y$.

- ■ You should know the definition of the natural logarithmic function.

  $$\log_e x = \ln x, x > 0$$

- ■ You should know the properties of the natural logarithmic function.

  (a) $\ln 1 = 0$ since $e^0 = 1$.

  (b) $\ln e = 1$ since $e^1 = e$.

  (c) $\ln e^x = x$ since $e^x = e^x$.

  (d) If $\ln x = \ln y$, then $x = y$.

- ■ You should be able to graph logarithmic functions.

Solutions to Odd-Numbered Exercises

1. $\log_4 64 = 3 \implies 4^3 = 64$

3. $\log_7 \frac{1}{49} = -2 \implies 7^{-2} = \frac{1}{49}$

5. $\log_{32} 4 = \frac{2}{5} \implies 32^{2/5} = 4$

7. $\ln 1 = 0 \implies e^0 = 1$

9. $5^3 = 125 \implies \log_5 125 = 3$

11. $81^{1/4} = 3 \implies \log_{81} 3 = \frac{1}{4}$

13. $6^{-2} = \frac{1}{36} \implies \log_6 \frac{1}{36} = -2$

15. $e^3 = 20.0855\ldots \implies \ln 20.0855\ldots = 3$

17. $e^x = 4 \implies \ln 4 = x$

19. $\log_2 16 = \log_2 2^4 = 4$

21. $\log_{16} 4 = \log_{16} 16^{1/2} = \frac{1}{2}$

23. $\log_7 1 = \log_7 7^0 = 0$

25. $\log_{10} 0.01 = \log_{10} 10^{-2} = -2$

27. $\ln e^3 = 3$

29. $\log_a a^2 = 2$

31. $\log_{10} 345 \approx 2.538$

33. $\log_{10} 145 \approx 2.161$

35. $\ln 18.42 \approx 2.913$

37. $\ln(1 + \sqrt{3}) \approx 1.005$

39. $\ln 0.32 \approx -1.139$

41. $f(x) = 3^x$, $g(x) = \log_3 x$

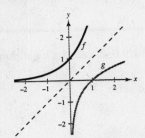

f and *g* are inverses. Their graphs are reflected about the line $y = x$.

43. $f(x) = e^x$, $g(x) = \ln x$

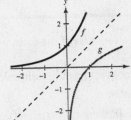

f and *g* are inverses. Their graphs are reflected about the line $y = x$.

45. $f(x) = \log_3 x + 2$
Asymptote: $x = 0$
Point on graph: $(1, 2)$
Matches graph (c).

47. $f(x) = -\log_3(x + 2)$
Asymptote: $x = -2$
Point on graph: $(-1, 0)$
Matches graph (d).

49. $f(x) = \log_3(1 - x)$
Asymptote: $x = 1$
Point on graph: $(0, 0)$
Matches graph (b).

51. $f(x) = \log_4 x$
Domain: $x > 0 \implies$ The domain is $(0, \infty)$.
Vertical asymptote: $x = 0$
x-intercept: $(1, 0)$
$y = \log_4 x \implies 4^y = x$

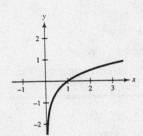

x	$\frac{1}{4}$	1	4	2
y	-1	0	1	$\frac{1}{2}$

53. $h(x) = \log_4(x - 3)$
Domain: $x - 3 > 0$ or $(3, \infty)$
Vertical asymptote: $x = 3$
Intercept: $(4, 0)$

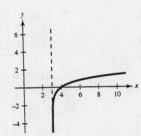

55. $y = -\log_3 x + 2$
Domain: $(0, \infty)$
Vertical asymptote: $x = 0$
x-intercept: $-\log_3 x + 2 = 0$
$$2 = \log_3 x$$
$$3^2 = x$$
$$9 = x$$

The *x*-intercept is $(9, 0)$.
$y = -\log_2 x + 2$
$\log_3 x = 2 - y \implies 3^{2-y} = x$

x	27	9	3	1	$\frac{1}{3}$
y	-1	0	1	2	3

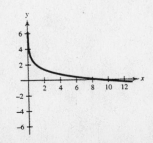

57. $y = \log_{10}\left(\dfrac{x}{5}\right)$

Domain: $\dfrac{x}{5} > 0 \implies x > 0$

The domain is $(0, \infty)$.

Vertical asymptote: $\dfrac{x}{5} = 0 \implies x = 0$

The vertical asymptote is the y-axis.

x-intercept: $\log_{10}\left(\dfrac{x}{5}\right) = 0$

$$\dfrac{x}{5} = 10^0$$

$$\dfrac{x}{5} = 1 \implies x = 5$$

The x-intercept is $(5, 0)$.

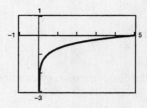

59. $f(x) = \ln(x - 2)$

Domain: $x - 2 > 0 \implies x > 2$

The domain is $(2, \infty)$.

Vertical asymptote: $x - 2 = 0 \implies x = 2$

x-intercept: $0 = \ln(x - 2)$

$$e^0 = x - 2$$

$$3 = x$$

The x-intercept is $(3, 0)$.

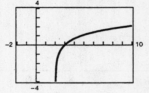

61. $g(x) = \ln(-x)$

Domain: $-x > 0 \implies x < 0$

The domain is $(-\infty, 0)$.

Vertical asymptote: $-x = 0 \implies x = 0$

x-intercept: $0 = \ln(-x)$

$$e^0 = -x$$

$$-1 = x$$

The x-intercept is $(-1, 0)$.

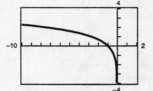

63. $f(x) = \dfrac{x}{2} - \ln\dfrac{x}{4}$

Domain: $(0, \infty)$

Increasing on $(2, \infty)$

Decreasing on $(0, 2)$

Relative minimum: $(2, 1.693)$

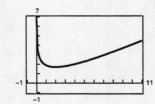

65. $h(x) = 4x \ln x$

Domain: $(0, \infty)$

Increasing on $(0.368, \infty)$

Decreasing on $(0, 0.368)$

Relative minimum: $(0.368, -1.472)$

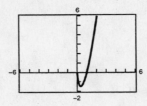

67. $t = \dfrac{10 \ln 2}{\ln 67 - \ln 50} \approx 23.68$ years

69. (a) False, y is not an exponential function of x. (y can never be 0.)

(b) True, y could be $\log_2 x$.

(c) True, x could be 2^y.

(d) False, y is not linear. (The points are not collinear.)

71. $y = (x - 1) - \frac{1}{2}(x - 1)^2 + \frac{1}{3}(x - 1)^3 - \frac{1}{4}(x - 1)^4$

The pattern implies that as we take more terms, the graph of y will more closely resemble that of $\ln x$ on the interval $(0, 2)$.

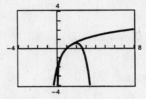

73. $t = \dfrac{\ln K}{0.095}$

(a)

K	1	2	4	6	8	10	12
t	0	7.3	14.6	18.9	21.9	24.2	26.2

(b)

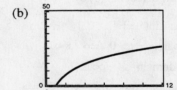

75. (a)

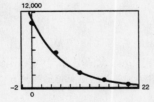

(b)

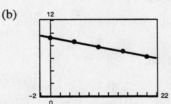

(c) $P = e^{-0.1499h + 9.3018} \approx 10{,}957.7 e^{-0.1499h}$

$\ln P = -0.1499h + 9.3018$

77. $y = 10 \ln\left(\dfrac{10 + \sqrt{100 - x^2}}{x}\right) - \sqrt{100 - x^2}$

(a)

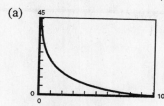

Domain: $0 < x \leq 10$

(b) Asymptote: $x = 0$

(c) When $x = 2$, $y \approx 13.126$. Since the rope is 10 feet long, the third side of the shaded right triangle is $\sqrt{100 - 2^2} = \sqrt{96}$. Thus, the person is at $13.126 + \sqrt{96} \approx 22.924$.

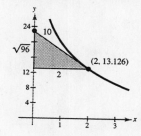

(d) $p = y + \sqrt{100 - x^2} = 10 \ln\left(\dfrac{10 + \sqrt{100 - x^2}}{x}\right)$

(e)

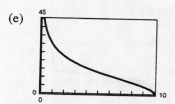

The position of the person changes most at the beginning.

79. $y = 80.4 - 11 \ln x$

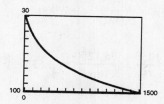

$y(300) = 80.4 - 11 \ln 300 \approx 17.66 \text{ ft}^3/\text{min}$

81. $w = 19{,}440(\ln 9 - \ln 3) \approx 21{,}357 \text{ ft-lb}$

83. $t = 10.042 \ln\left(\dfrac{1316.35}{1316.35 - 1250}\right) \approx 30 \text{ years}$

85. Total amount $= (1316.35)(30)(12) = \$473{,}886$
Interest $= 473{,}886 - 150{,}000 = \$323{,}886$

87. (a) $(0, \infty)$

(b) $y = \log_{10} x$
$x = \log_{10} y$
$10^x = y$
$f^{-1}(x) = 10^x$

(c) Since $\log_{10} 1000 = 3$ and $\log_{10} 10{,}000 = 4$, the interval in which $f(x)$ will be found is $(3, 4)$.

(d) When $f(x)$ is negative, x is in the interval $(0, 1)$.

(e) $0 = \log_{10} 1$
$1 = \log_{10} 10$
$2 = \log_{10} 100$
$3 = \log_{10} 1000$

When $f(x)$ is increased by one unit, x is increased by a factor of 10.

(f) $f(x_1) = 3n \qquad\qquad f(x_2) = n$
$\log_{10} x_1 = 3n \qquad \log_{10} x_2 = n$
$x_1 = 10^{3n} \qquad\qquad x_2 = 10^n$
$x_1 : x_2 = 10^{3n} : 10^n = 10^{2n} : 1$

Section 3.3 Properties of Logarithms

■ You should know the following properties of logarithms.

(a) $\log_a x = \dfrac{\log_b x}{\log_b a}$

(b) $\log_a (uv) = \log_a u + \log_a v$ $\qquad$ $\ln (uv) = \ln u + \ln v$

(c) $\log_a (u/v) = \log_a u - \log_a v$ $\qquad$ $\ln (u/v) = \ln u - \ln v$

(d) $\log_a u^n = n \log_a u$ $\qquad\qquad$ $\ln u^n = n \ln u$

■ You should be able to rewrite logarithmic expressions using these properties.

Solutions to Odd-Numbered Exercises

1. $f(x) = \log_{10} x$

$g(x) = \dfrac{\ln x}{\ln 10}$

$f(x) = g(x)$

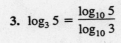

3. $\log_3 5 = \dfrac{\log_{10} 5}{\log_{10} 3}$

5. $\log_2 x = \dfrac{\log_{10} x}{\log_{10} 2}$

7. $\log_3 5 = \dfrac{\ln 5}{\ln 3}$

9. $\log_2 x = \dfrac{\ln x}{\ln 2}$

11. $\log_3 7 = \dfrac{\log_{10} 7}{\log_{10} 3} = \dfrac{\ln 7}{\ln 3} \approx 1.771$

13. $\log_{1/2} 4 = \dfrac{\log_{10} 4}{\log_{10}(1/2)} = \dfrac{\ln 4}{\ln(1/2)} = -2.000$

15. $\log_9(0.4) = \dfrac{\log_{10} 0.4}{\log_{10} 9} = \dfrac{\ln 0.4}{\ln 9} \approx -0.417$

17. $\log_{15} 1250 = \dfrac{\log_{10} 1250}{\log_{10} 15} = \dfrac{\ln 1250}{\ln 15} \approx 2.633$

19. $\log_{10} 5x = \log_{10} 5 + \log_{10} x$

21. $\log_{10} \dfrac{5}{x} = \log_{10} 5 - \log_{10} x$

23. $\log_8 x^4 = 4 \log_8 x$

25. $\ln \sqrt{z} = \ln z^{1/2} = \frac{1}{2} \ln z$

27. $\ln xyz = \ln x + \ln y + \ln z$

29. $\ln \sqrt{a - 1} = \frac{1}{2} \ln(a - 1)$

31. $\ln z(z - 1)^2 = \ln z + \ln(z - 1)^2$
$\qquad\qquad\qquad = \ln z + 2\ln(z - 1)$

33. $\ln \sqrt[3]{\dfrac{x}{y}} = \dfrac{1}{3} \ln \dfrac{x}{y}$

$\qquad\quad = \dfrac{1}{3}[\ln x - \ln y]$

$\qquad\quad = \dfrac{1}{3} \ln x - \dfrac{1}{3} \ln y$

35. $\ln \left(\dfrac{x^4 \sqrt{y}}{z^5} \right) = \ln x^4 \sqrt{y} - \ln z^5$

$\qquad\qquad\quad = \ln x^4 + \ln \sqrt{y} - \ln z^5$

$\qquad\qquad\quad = 4 \ln x + \dfrac{1}{2} \ln y - 5 \ln z$

37. $\log_b\left(\dfrac{x^2}{y^2z^3}\right) = \log_b x^2 - \log_b y^2z^3$

$= \log_b x^2 - [\log_b y^2 + \log_b z^3]$

$= 2\log_b x - 2\log_b y - 3\log_b z$

39. $y_1 = \ln[x^3(x+4)]$

$y_2 = 3\ln x + \ln(x+4)$

$y_1 = y_2$

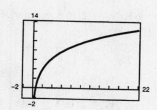

41. $\ln x + \ln 2 = \ln 2x$

43. $\log_4 z - \log_4 y = \log_4 \dfrac{z}{y}$

45. $2\log_2(x+4) = \log_2(x+4)^2$

47. $\frac{1}{3}\ln 5x = \ln(5x)^{1/3} = \ln\sqrt[3]{5x}$

49. $\ln x - 3\ln(x+1) = \ln x - \ln(x+1)^3$

$= \ln\dfrac{x}{(x+1)^3}$

51. $\ln(x-2) - \ln(x+2) = \ln\left(\dfrac{x-2}{x+2}\right)$

53. $\ln x - 2[\ln(x+2) + \ln(x-2)] = \ln x - 2\ln[(x+2)(x-2)]$

$= \ln x - 2\ln(x^2 - 4)$

$= \ln x - \ln(x^2 - 4)^2$

$= \ln\dfrac{x}{(x^2 - 4)^2}$

55. $\frac{1}{3}[2\ln(x+3) + \ln x - \ln(x^2 - 1)] = \frac{1}{3}[\ln(x+3)^2 + \ln x - \ln(x^2 - 1)]$

$= \frac{1}{3}[\ln[x(x+3)^2] - \ln(x^2 - 1)]$

$= \frac{1}{3}\ln\dfrac{x(x+3)^2}{x^2 - 1}$

$= \ln\sqrt[3]{\dfrac{x(x+3)^2}{x^2 - 1}}$

57. $\frac{1}{3}[\ln y + 2\ln(y+4)] - \ln(y-1) = \frac{1}{3}[\ln y + \ln(y+4)^2] - \ln(y-1)$

$= \frac{1}{3}\ln[y(y+4)^2] - \ln(y-1)$

$= \ln\sqrt[3]{y(y+4)^2} - \ln(y-1)$

$= \ln\dfrac{\sqrt[3]{y(y+4)^2}}{y-1}$

59. $2\ln 3 - \dfrac{1}{2}\ln(x^2 + 1) = \ln 3^2 - \ln\sqrt{x^2 + 1}$

$= \ln\dfrac{9}{\sqrt{x^2 + 1}}$

61. $y_1 = 2[\ln 8 - \ln(x^2 + 1)]$

$y_2 = \ln\left[\dfrac{64}{(x^2 + 1)^2}\right]$

$y_1 = y_2$

$y_1 = 2[\ln 8 - \ln(x^2 + 1)]$

$\quad = 2 \ln\left(\dfrac{8}{x^2 + 1}\right)$

$\quad = \ln\left[\dfrac{64}{(x^2 + 1)^2}\right] = y_2$

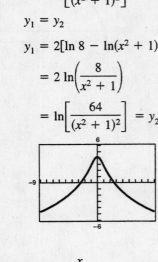

63. $y_1 = \ln x^2$

$y_2 = 2 \ln x$

$y_1 = y_2$ for $x > 0$.

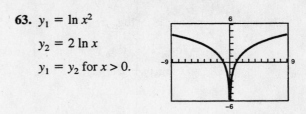

They are not equivalent. The domain of $f(x)$ is all real numbers except 0. The domain of $g(x)$ is $x > 0$.

65. $f(x) = \ln \dfrac{x}{2}$

$g(x) = \dfrac{\ln x}{\ln 2}$

$h(x) = \ln x - \ln 2$

$f(x) = h(x)$ by Property 2.

67. $\log_3 9 = 2 \log_3 3 = 2$

69. $\log_4 16^{1.2} = 1.2(\log_4 16) = 1.2(2) = 2.4$

71. $\log_3 (-9)$ is undefined. -9 is not in the domain of $\log_3 x$.

73. $\log_5 75 - \log_5 3 = \log_5 \frac{75}{3} = \log_5 25$

75. $\ln e^2 - \ln e^5 = 2 - 5 = -3$

77. $\log_{10} 0$ is undefined. 0 is not in the domain of $\log_{10} x$.

79. $\ln e^{4.5} = 4.5$

81. $\log_4 8 = \log_4 2^3 = 3 \log_4 2$

$\qquad = 3 \log_4 \sqrt{4} = 3\log_4 4^{1/2}$

$\qquad = 3\left(\tfrac{1}{2}\right) \log_4 4 = \tfrac{3}{2}$

83. $\log_7 \sqrt{70} = \frac{1}{2} \log_7 70 = \frac{1}{2} \log_7(10 \cdot 7)$

$\qquad = \frac{1}{2} \log_7 10 + \frac{1}{2} \log_7 7 = \frac{1}{2} \log_7 10 + \frac{1}{2}$

85. $\log_5 \frac{1}{250} = \log_5 1 - \log_5 250 = 0 - \log_5 (125 \cdot 2)$

$\qquad = -\log_5(5^3 \cdot 2) = -[\log_5 5^3 + \log_5 2]$

$\qquad = -[3 \log_5 5 + \log_5 2] = -3 - \log_5 2$

87. $\ln(5e^6) = \ln 5 + \ln e^6 = \ln 5 + 6 = 6 + \ln 5$

89. $\beta = 10 \log_{10}\left(\dfrac{I}{10^{-16}}\right) = 10[\log_{10} I - \log_{10} 10^{-16}]$

$\qquad = 10[\log_{10} I - (-16) \log_{10} 10]$

$\qquad = 10[\log_{10} I + 16]$

If $I = 10^{-10}$, then $\log_{10} I = \log_{10} 10^{-10} = -10$
and $\beta = 10[-10 + 16] = 60$ db.

91. (a)

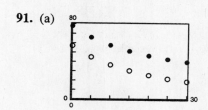

(b) The data $(t, T - 21)$ fits the exponential model
$T - 21 = 54.4380(0.9635)^t$. For the original data the
model is $T = 54.4380(0.9635)^t$.

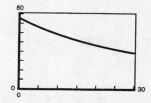

(c) $\ln(T - 21) = -0.03721t + 3.9971$

$\qquad T - 21 = e^{-0.0377t + 3.9971}$

$\qquad\quad T = 21 + 54.44(e^{-0.0377t})$

$\qquad\qquad = 21 + 54.44(0.9635)^t$

(d) $T = \dfrac{4960}{6t + 80} + 21$

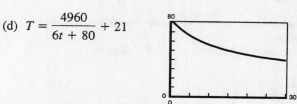

93. $f(x) = \log_2 x = \dfrac{\log_{10} x}{\log_{10} 2}$

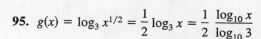

95. $g(x) = \log_3 x^{1/2} = \dfrac{1}{2} \log_3 x = \dfrac{1}{2} \dfrac{\log_{10} x}{\log_{10} 3}$

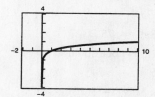

97. $f(x) = \ln x$

False, $f(0) \neq 0$ since 0 is not in the domain of $f(x)$. $f(1) = \ln 1 = 0$

99. False, $f(x) - f(2) = \ln x - \ln 2 = \ln \dfrac{x}{2} \neq \ln(x - 2)$.

101. False, $f(u) = 2f(v) \implies \ln u = 2 \ln v \implies \ln u = \ln v^2 \implies u = v^2$.

103. Let $x = \log_b u$ and $y = \log_b v$, then $b^x = u$ and $b^y = v$.

$\dfrac{u}{v} = \dfrac{b^x}{b^y} = b^{x-y}$

$\log_b\left(\dfrac{u}{v}\right) = \log_b(b^{x-y}) = x - y = \log_b u - \log_b v$

105. $\dfrac{24xy^{-2}}{16x^{-3}y} = \dfrac{24xx^3}{16yy^2} = \dfrac{3x^4}{2y^3}$

107. $(18x^3y^4)^{-3}(18x^3y^4)^3 = \dfrac{(18x^3y^4)^3}{(18x^3y^4)^3} = 1$ if $x \neq 0, y \neq 0$.

Section 3.4   Solving Exponential and Logarithmic Equations

> ■ To solve an exponential equation, isolate the exponential expression, then take the logarithm of both sides. Then solve for the variable.
>
>   1. $\log_a a^x = x$
>
>   2. $\ln e^x = x$
>
> ■ To solve a logarithmic equation, rewrite it in exponential form. Then solve for the variable.
>
>   1. $a^{\log_a x} = x$
>
>   2. $e^{\ln x} = x$
>
> ■ If $a > 0$ and $a \neq 1$ we have the following:
>
>   1. $\log_a x = \log_a y \implies x = y$
>
>   2. $a^x = a^y \implies x = y$
>
> ■ Use your graphing utility to approximate solutions.

Solutions to Odd-Numbered Exercises

1. $4^{2x-7} = 64$

    (a) $x = 5$

        $4^{2(5)-7} = 4^3 = 64$

        Yes, $x = 5$ is a solution.

    (b) $x = 2$

        $4^{2(2)-7} = 4^{-3} = \frac{1}{64} \neq 64$

        No, $x = 2$ is not a solution.

3. $3e^{x+2} = 75$

    (a) $x = -2 + e^{25}$

        $3e^{(-2+e^{25})+2} = 3e^{e^{25}} \neq 75$

        No, $x = -2 + e^{25}$ is not a solution.

    (b) $x = -2 + \ln 25$

        $3e^{(-2+\ln 25)+2} = 3e^{\ln 25} = 3(25) = 75$

        Yes, $x = -2 + \ln 25$ is a solution.

    (c) $x \approx 1.2189$

        $3e^{1.2189+2} = 3e^{3.2189} \approx 75$

        Yes, $x \approx 1.2189$ is a solution.

5. $\log_4(3x) = 3 \implies 3x = 4^3 \implies 3x = 64$

    (a) $x \approx 20.3560$

        $3(20.3560) = 61.0680 \neq 64$

        No, $x \approx 20.3560$ is not a solution.

    (b) $x = -4$

        $3(-4) = -12 \neq 64$

        No, $x = -4$ is not a solution.

    (c) $x = \frac{64}{3}$

        $3\left(\frac{64}{3}\right) = 64$

        Yes, $x = \frac{64}{3}$ is a solution.

7. $f(x) = g(x)$

    $2^x = 8$

    $2^x = 2^3$

    $x = 3$

    Point of intersection: $(3, 8)$

9. $f(x) = g(x)$

    $\log_3 x = 2$

    $x = 3^2$

    $x = 9$

    Point of intersection: $(9, 2)$

11. $4^x = 16$

    $4^x = 4^2$

    $x = 2$

13. $7^x = \frac{1}{49}$

    $7^x = 7^{-2}$

    $x = -2$

15. $\left(\frac{3}{4}\right)^x = \frac{27}{64}$

    $\left(\frac{3}{4}\right)^x = \left(\frac{3}{4}\right)^3$

    $x = 3$

17. $\log_4 x = 3$

    $x = 4^3$

    $x = 64$

19. $\log_{10} x = -1$

    $x = 10^{-1}$

    $x = \frac{1}{10}$

21. $\log_{10} 10^{x^2} = x^2$

23. $e^{\ln(5x+2)} = 5x + 2$

25. $e^{\ln x^2} = x^2$

27. $e^x = 10$

    $x = \ln 10 \approx 2.303$

29. $7 - 2e^x = 5$

    $-2e^x = -2$

    $e^x = 1$

    $x = \ln 1 = 0$

31. $500e^{-x} = 300$

    $e^{-x} = \frac{3}{5}$

    $-x = \ln \frac{3}{5}$

    $x = -\ln \frac{3}{5} = \ln \frac{5}{3} \approx 0.511$

33. $10^x = 42$

    $x = \log_{10} 42 \approx 1.623$

35.

x	0.6	0.7	0.8	0.9	1.0
$f(x)$	6.05	8.17	11.02	14.88	20.09

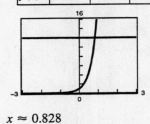

$x \approx 0.828$

37.

x	5	6	7	8	9
$f(x)$	1756	1598	1338	908	200

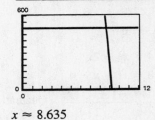

$x \approx 8.635$

39. $e^{2x} - 4e^x - 5 = 0$

  $(e^x - 5)(e^x + 1) = 0$

    $e^x = 5$ or $e^x = -1$ (No solution)

        $x = \ln 5 \approx 1.609$

41. $3^{2x} = 80$

  $\ln 3^{2x} = \ln 80$

  $2x \ln 3 = \ln 80$

     $x = \dfrac{\ln 80}{2 \ln 3} \approx 1.994$

43. $5^{-t/2} = 0.20$

  $5^{-t/2} = \frac{1}{5}$

  $5^{-t/2} = 5^{-1}$

    $-\dfrac{t}{2} = -1$

    $t = 2$

45. $2^{3-x} = 565$

    $\ln 2^{3-x} = \ln 565$

  $(3 - x) \ln 2 = \ln 565$

  $3 \ln 2 - x \ln 2 = \ln 565$

      $-x \ln 2 = \ln 565 - \ln 2^3$

      $-x \ln 2 = \ln 8 - \ln 5656$

        $x = \dfrac{\ln 8 - \ln 565}{\ln 2} \approx -6.142$

47. $8(10^{3x}) = 12$

   $10^{3x} = \frac{12}{8}$

    $3x = \log_{10}\left(\frac{3}{2}\right)$

     $x = \frac{1}{3} \log_{10}\left(\frac{3}{2}\right) \approx 0.059$

49. Using the root feature for

$$y = \left(1 + \frac{0.10}{12}\right)^{12t} - 2 = 0$$

you obtain $t \approx 6.960$.

51.

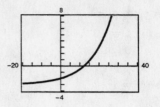

zero at $x = -0.427$

53.

zero at $t = 12.207$

55. $\ln x = -3$

$\qquad x = e^{-3} \approx 0.050$

57. $\ln\sqrt{x + 2} = 1$

$\qquad \sqrt{x + 2} = e^1$

$\qquad x + 2 = e^2$

$\qquad x = e^2 - 2 \approx 5.389$

59. $\log_{10}(x + 4) - \log_{10} x = \log_{10}(x + 2)$

$$\log_{10}\left(\frac{x + 4}{x}\right) = \log_{10}(x + 2)$$

$$\frac{x + 4}{x} = x + 2$$

$$x + 4 = x^2 + 2x$$

$$0 = x^2 + x - 4$$

$$x = \frac{-1 \pm \sqrt{17}}{2} \quad \text{Quadratic Formula}$$

Choosing the positive value of x (the negative value is extraneous), we have $\dfrac{-1 + \sqrt{17}}{2}$.

61.

x	2	3	4	5	6
$f(x)$	1.39	1.79	2.08	12.30	2.49

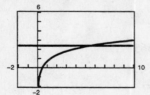

$x \approx 5.512$

63.

x	12	13	14	15	16
$f(x)$	9.79	10.22	10.63	11.00	11.36

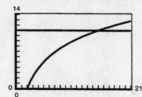

$x \approx 14.988$

65. $\log_{10}(z - 3) = 2$

$\qquad z - 3 = 10^2$

$\qquad z = 10^2 + 3 = 103$

67. $2 \ln x = 7$

$\qquad \ln x = \frac{7}{2}$

$\qquad x = e^{7/2} \approx 33.115$

69. $\ln x + \ln(x - 2) = 1$

$$\ln[x(x - 2)] = 1$$

$$x(x - 2) = e^1$$

$$x^2 - 2x - e = 0$$

$$x = \frac{2 \pm \sqrt{4 + 4e}}{2}$$

$$= \frac{2 \pm 2\sqrt{1 + e}}{3}$$

Using the positive value for x, we have
$x = 1 + \sqrt{1 + e} \approx 2.928$.

73. $\ln(x + 5) = \ln(x - 1) - \ln(x + 1).$

$$\ln(x + 5) = \ln\left(\frac{x - 1}{x + 1}\right)$$

$$x + 5 = \frac{x - 1}{x + 1}$$

$$(x + 5)(x + 1) = x - 1$$

$$x^2 + 6x + 5 = x - 1$$

$$x^2 + 5x + 6 = 0$$

$$(x + 2)(x + 3) = 0$$

$$x = -2 \text{ or } x = -3$$

Both of these solutions are extraneous, so the equation has no solution.

77. $y_1 = 7$

$y_2 = 2^x$

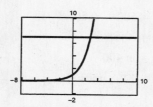

From the graph we have $x \approx 2.807$.

81.
$$A = Pe^{rt}$$

$$2000 = 1000e^{0.085t}$$

$$2 = e^{0.085t}$$

$$\ln 2 = 0.085t$$

$$\frac{\ln 2}{0.085} = t$$

$$t \approx 8.2 \text{ years}$$

71. $\log_3 x + \log_3(x^2 - 8) = \log_3 8x$

$$\log_3 x(x^2 - 8) = \log_3 8x$$

$$x(x^2 - 8) = 8x$$

$$x^3 - 8x = 8x$$

$$x^3 - 16x = 0$$

$$x(x + 4)(x - 4) = 0$$

$$x = 0, \ x = -4 \text{ or } x = 4$$

The only solution that is in the domain is $x = 4$.
Both $x = 0$ and $x = -4$ are extraneous.

75. $\ln x + \ln(x^2 + 1) = 8$

$$\ln x(x^2 + 1) = 8$$

$$x(x^2 + 1) = e^8$$

$$x^3 + x - e^8 = 0$$

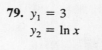

From the graph we have $x \approx 14.369$.

79. $y_1 = 3$

$y_2 = \ln x$

From the graph
we have
$x \approx 20.806$.

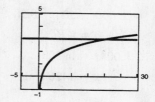

83. *Doubling Time*

$$2P = Pe^{rt}$$

$$2 = e^{rt}$$

$$\ln 2 = rt$$

$$\frac{\ln 2}{r} = t$$

Quadrupling Time

$$4P = Pe^{rt}$$

$$4 = e^{rt}$$

$$\ln 4 = rt$$

$$\frac{\ln 4}{r} = t$$

$$\frac{\ln 2^2}{r} = t$$

$$\frac{2\ln 2}{r} = t$$

$$2\left(\frac{\ln 2}{r}\right) = t$$

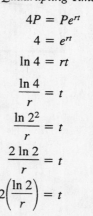

Yes, it takes twice as long to quadruple.

85. $A = Pe^{rt}$

$3000 = 1000e^{0.085t}$

$3 = e^{0.085t}$

$\ln 3 = 0.085t$

$\dfrac{\ln 3}{0.085} = t$

$t \approx 12.9$ years

87. (a)

(b) From the graph we see horizontal asymptotes at $y = 0$ and $y = 100$. These represent the lower and upper percent bounds.

(c) Males:

$$50 = \frac{100}{1 + e^{-0.6114(x-69.71)}}$$

$1 + e^{-0.6114(x-69.71)} = 2$

$e^{-0.6114(x-69.71)} = 1$

$-0.6114(x - 69.71) = \ln 1$

$-0.6114(x - 69.71) = 0$

$x = 69.71$ inches

Females:

$$50 = \frac{100}{1 + e^{-0.66607(x-64.51)}}$$

$1 + e^{-0.66607(x-64.51)} = 2$

$e^{-0.66607(x-64.51)} = 1$

$-0.66607(x - 64.51) = \ln 1$

$-0.66607(x - 64.51) = 0$

$x = 64.51$ inches

89. $p = 500 - 0.5(e^{0.004x})$

(a) $p = 350$

$350 = 500 - 0.5(e^{0.004x})$

$300 = e^{0.004x}$

$0.004x = \ln 300$

$x \approx 1426$ units

(b) $p = 300$

$300 = 500 - 0.5(e^{0.004x})$

$400 = e^{0.004x}$

$0.004x = \ln 400$

$x \approx 1498$ units

91. $V = 6.7e^{-48.1/t}, \ t \ge 0$

(a)

(b) As $x \to \infty$, $V \to 6.7$.

Horizontal asymptote: $y = 6.7$
The yield will approach
6.7 million cubic feet per acre.

(c) $1.3 = 6.7e^{-48.1/t}$

$\dfrac{1.3}{6.7} = e^{-48.1/t}$

$\ln\left(\dfrac{13}{67}\right) = \dfrac{-48.1}{t}$

$t = \dfrac{-48.1}{\ln(13/67)} \approx 29.3$ years

93. $T = 20[1 + 7(2^{-h})]$

(a)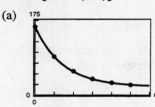

(b) We see a horizontal asymptote at $y = 20$. This represents the room temperature.

(c) $100 = 20[1 + 7(2^{-h})]$

$5 = 1 + 7(2^{-h})$

$4 = 7(2^{-h})$

$\dfrac{4}{7} = 2^{-h}$

$\ln\left(\dfrac{4}{7}\right) = \ln 2^{-h}$

$\ln\left(\dfrac{4}{7}\right) = -h \ln 2$

$\dfrac{\ln(4/7)}{-\ln 2} = h$

$h \approx 0.81$ hour

95. $4x - 3y - 9 = 0$

$-3y = -4x + 9$

$y = \frac{4}{3}x - 3$

Line with slope $\frac{4}{3}$

y-intercept -3

Matches (b).

97. $y = 25 - 2.25x$

Slope: -2.25

y-intercept: 25

Matches (f).

99. $y - 3 = 0$

$y = 3$

Horizontal line

Matches (d).

Section 3.5   **Exponential and Logarithmic Models**

■ You should be able to solve compound interest problems.

1. $A = P\left(1 + \dfrac{r}{n}\right)^{nt}$

2. $A = Pe^{rt}$

■ You should be able to solve growth and decay problems.

(a) Exponential growth if $b > 0$ and $y = ae^{bx}$.

(b) Exponential decay if $b > 0$ and $y = ae^{-bx}$.

■ You should be able to use the Gaussian model

$y = ae^{-(x-b)^2/c}$.

■ You should be able to use the logistics growth model

$y = \dfrac{a}{1 + be^{-(x-c)/d}}$.

■ You should be able to use the logarithmic models

$y = \ln(ax + b)$ and $y = \log_{10}(ax + b)$.

Solutions to Odd-Numbered Exercises

1. $y = 2e^{x/4}$

This is an exponential growth model. Matches graph (b).

3. $y = \frac{1}{16}(x^2 + 8x + 32)$

This is a quadratic function. It's graph is a parabola. Matches graph (e).

5. $y = \ln(x + 1)$

This is a logarithmic model. Matches graph (f).

7. Since $A = 1000e^{0.12t}$, the time to double is given by $2000 = 1000e^{0.12t}$ and we have

$$t = \frac{\ln 2}{0.12} \approx 5.78 \text{ years.}$$

Amount after 10 years: $A = 1000e^{1.2} \approx \$3320.12$

9. Since $A = 750e^{rt}$ and $A = 1500$ when $t = 7.75$, we have the following.

$$15000 = 750e^{7.75r}$$

$$r = \frac{\ln 2}{7.75} \approx 0.0894 = 8.94\%$$

Amount after 10 years: $A = 750e^{0.0894(10)} \approx \$1833.67$

11. Since $A = 500e^{rt}$ and $A = 1292.85$ when $t = 10$, we have the following.

$$1292.85 = 500e^{10r}$$

$$r = \frac{\ln(1292.85/500)}{10} \approx 0.9095 = 9.5\%$$

The time to double is given by

$$1000 = 500e^{0.095t}$$

$$t = \frac{\ln 2}{0.095} \approx 7.30 \text{ years.}$$

13. Since $A = Pe^{0.045t}$ and $A = 10,000.00$ when $t = 10$, we have the following.

$$10,000.00 = Pe^{0.045(10)}$$

$$\frac{10,000.00}{e^{0.045(10)}} = P \approx 6376.28$$

The time to double is given by

$$t = \frac{\ln 2}{0.045} \approx 15.40 \text{ years.}$$

15.

$$500,000 = P\left(1 + \frac{0.075}{12}\right)^{12(20)}$$

$$\frac{500,000}{\left(1 + \frac{0.075}{12}\right)^{12(20)}} = P = \$112,087.09$$

17. $P = 1000, r = 11\%$

(a) $n = 1$

$$t = \frac{\ln 2}{\ln(1 + 0.11)} \approx 6.642 \text{ years}$$

(b) $n = 12$

$$t = \frac{\ln 2}{12 \ln\left(1 + \frac{0.11}{12}\right)} \approx 6.330 \text{ years}$$

(c) $n = 365$

$$t = \frac{\ln 2}{365 \ln\left(1 + \frac{0.11}{365}\right)} \approx 6.302 \text{ years}$$

(d) Continuously

$$t = \frac{\ln 2}{0.11} \approx 6.301 \text{ years}$$

19.
$$3P = Pe^{rt}$$
$$3 = e^{rt}$$
$$\ln 3 = rt$$
$$\frac{\ln 3}{r} = t$$

r	2%	4%	6%	8%	10%	12%
$t = \dfrac{\ln 3}{r}$	54.93	27.47	18.31	13.73	10.99	9.16

21.
$$3P = P(1 + r)^t$$
$$3 = (1 + r)^t$$
$$\ln 3 = \ln(1 + r)^t$$
$$\ln 3 = t\ln(1 + r)$$
$$\frac{\ln 3}{\ln(1 + r)} = t$$

r	2%	4%	6%	8%	10%	12%
$t = \dfrac{\ln 3}{\ln(1 + r)}$	55.47	28.01	18.85	14.27	11.53	9.69

23. Continuous compounding results in faster growth.
$$A = 1 + 0.075[\![t]\!]$$
and $A = e^{0.07t}$

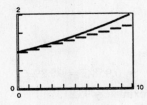

25. $\dfrac{1}{2}C = Ce^{k(1620)}$

$$k = \frac{\ln 0.5}{1620}$$

Given $C = 10$ grams, after 1000 years, we have

$$y = Ce^{[(\ln 0.5)/1620](1000)}$$
$$C \approx 6.52 \text{ grams.}$$

27. $\dfrac{1}{2}C = Ce^{k(5730)}$

$$k = \frac{\ln 0.5}{5730}$$

Given $y = 3$ grams, after 1000 years, we have

$$3 = Ce^{[(\ln 0.5)/5730](1000)}$$
$$C \approx 2.66 \text{ grams.}$$

29.
$$y = ae^{bx}$$
$$1 = ae^{b(0)} \implies 1 = a$$
$$10 = e^{b(3)}$$
$$\ln 10 = 3b$$
$$\frac{\ln 10}{3} = b \qquad \implies b \approx 0.7675$$

Thus, $y = e^{0.7675x}$.

31.
$$y = ae^{bx}$$
$$1 = ae^{b(0)} \implies 1 = a$$
$$\frac{1}{4} = e^{b(3)}$$
$$\ln\left(\frac{1}{4}\right) = 3b$$
$$\frac{\ln(1/4)}{3} = b \qquad \implies b \approx -0.4621$$

Thus, $y = e^{-0.4621x}$.

33.
$$P = 105,300e^{0.015t}$$
$$150,000 = 105,300e^{0.015t}$$
$$\ln\frac{1500}{1053} = 0.015t$$
$$t \approx 23.59$$

The population will reach 150,000 during 2013. [Note: 1990 + 23.59.]

35. For 1945, use $t = -45$.

$$1350 = 2500d^{k(-45)}$$
$$\ln\left(\frac{1350}{2500}\right) = -45k \approx 0.0137$$

For 2010, use $t = 20$.

$$P = 2500e^{0.0137(20)} \approx 3288 \text{ people}$$

37.
$$y = ae^{bt}$$
$$4.22 = ae^{b(0)} \implies a = 4.22$$
$$6.49 = 4.22e^{b(10)}$$
$$\frac{6.49}{4.22} = e^{10b}$$
$$\ln\left(\frac{6.49}{4.22}\right) = 10b \implies b \approx 0.0430$$
$$y = 4.22e^{0.0430t}$$

When $t = 20$,
$y = 4.22e^{0.0430(20)} \approx 9.97$ million.

39.
$$y = ae^{bt}$$
$$3.00 = ae^{b(0)} \implies a = 3$$
$$2.74 = 3e^{b(10)}$$
$$\frac{2.74}{3} = e^{10b}$$
$$\ln\left(\frac{2.74}{3}\right) = 10b \implies b \approx 0.0091$$
$$y = 3e^{-0.0091t}$$

When $t = 20$,
$y = 3e^{-0.0091(20)} \approx 2.50$ million.

41. b is determined by the growth rate. The greater the rate of growth, the greater the value of b.

43.
$$N = 100e^{kt}$$
$$300 = 100e^{5k}$$
$$k = \frac{\ln 3}{5} \approx 0.2197$$
$$N = 100e^{0.2197t}$$
$$200 = 100e^{0.2197t}$$
$$t = \frac{\ln 2}{0.2197} \approx 3.15 \text{ hours}$$

45.
$$y = Ce^{kt}$$
$$\frac{1}{2}C = Ce^{(1620)k}$$
$$\ln\frac{1}{2} = 1620k$$
$$k = \frac{\ln(1/2)}{1620}$$

When $t = 100$, we have
$$y = Ce^{[\ln(1/2)/1620](100)} \approx 0.958C = 95.8\%C.$$

After 100 years, approximately 95.8% of the radioactive radium will remain.

47. $y = ae^{bt}$.
At time $t = 0$,
$$y = 4600 = ae^{b(0)} \implies a = 4600.$$
At time $t = 2$,
$$3000 = 4600e^{b(2)}$$
$$\frac{30}{46} = e^{2b}$$
$$b = \frac{1}{2}\ln\left(\frac{30}{46}\right) \approx -0.2137.$$
After 3 years,
$$y = 4600e^{b(3)}$$
$$= 4600e^{3/2\ln(30/46)}$$
$$= 4600 \cdot \left(\frac{30}{46}\right)^{3/2} \approx \$2423.$$

49. $S(t) = 100(1 - e^{kt})$

(a)
$$15 = 100(1 - e^{k(1)})$$
$$-85 = -100e^{k}$$
$$k = \ln 0.85$$
$$k \approx -0.1625$$
$$S(t) = 100(1 - e^{-0.1625t})$$

(b)

(c) $S(5) = 100(1 - e^{-0.1625(5)})$
$$\approx 55.625 = 55,625 \text{ units}$$

51. $S = 10(1 - e^{kx})$

$x = 5$ (in hundreds), $S = 2.5$ (in thousands)

(a) $2.5 = 10(1 - e^{k(5)})$

$\qquad 0.25 = 1 - e^{5k}$

$\qquad e^{5k} = 0.75$

$\qquad 5k = \ln 0.75$

$\qquad k \approx -0.0575$

$\qquad S = 10(1 - e^{-0.0575x})$

(b) When $x = 7$,

$\qquad S = 10(1 - e^{-0.0575(7)}) \approx 3.314$
which corresponds to 3314 units.

53. $N = 30(1 - e^{kt})$

(a) $N = 19$, $t = 20$

$\qquad 19 = 30(1 - e^{20k})$

$\qquad 20k = \ln\dfrac{11}{30}$

$\qquad k \approx -0.050$

$\qquad N = 30(1 - e^{-0.050t})$

(b) $N = 25$

$\qquad 25 = 30(1 - e^{-0.05t})$

$\qquad \dfrac{5}{30} = e^{-0.05t}$

$\qquad t = -\dfrac{1}{0.05}\ln\dfrac{5}{30} \approx 36$ days

(c) No, this is not a linear function.

55. $R = \log_{10}\dfrac{I}{I_0} = \log_{10} I$ since $I_0 = 1$.

(a) $R = \log_{10} 80{,}500{,}000 \approx 7.91$

(b) $R = \log_{10} 48{,}275{,}000 \approx 7.68$

57. $\beta(I) = \log_{10}\dfrac{I}{I_0}$ where $I_0 = 10^{-16}$ watt/cm^2.

(a) $\beta(10^{-14}) = 10\log_{10}\dfrac{10^{-14}}{10^{-16}} = 10\log_{10}10^2 = 20$ decibels

(b) $\beta(10^{-9}) = 10\log_{10}\dfrac{10^{-9}}{10^{-16}} = 10\log_{10}10^7 = 70$ decibels

(c) $\beta(10^{-6.5}) = 10\log_{10}\dfrac{10^{-6.5}}{10^{-16}} = 10\log_{10}10^{9.5} = 95$ decibels

(d) $\beta(10^{-4}) = 10\log_{10}\dfrac{10^{-4}}{10^{-16}} = 10\log_{10}10^{12} = 120$ decibels

59. $\beta = 10\log_{10}\dfrac{I}{I_0}$

$10^{\beta/10} = \dfrac{I}{I_0}$

$I = I_0 10^{\beta/10}$

% decrease $= \dfrac{I_0 10^{9.3} - I_0 10^{8.0}}{I_0 10^{9.3}} \times 100 \approx 95\%$

61. pH $= -\log_{10}[H^+] = -\log_{10}[2.3 \times 10^{-5}] \approx 4.64$

63. $5.8 = -\log_{10}[H^+]$

$10^{-5.8} = H^+$

$H^+ \approx 1.58 \times 10^{-6}$ moles per liter

65. pH $= -\log_{10}[H^+]$

$-$pH $= \log_{10}[H^+]$

$10^{-\text{pH}} = [H^+]$

$\dfrac{\text{Hydrogen ion concentration of fruit}}{\text{Hydrogen ion concentration of tablet}} = \dfrac{10^{-2.5}}{10^{-9.5}} = 10^7$

67. Interest: $u = M - \left(M - \dfrac{Pr}{12}\right)\left(1 + \dfrac{r}{12}\right)^{12t}$

Principle: $v = \left(M - \dfrac{Pr}{12}\right)\left(1 + \dfrac{r}{12}\right)^{12t}$

(a) $P = 120{,}000$, $t = 35$, $r = 0/095$, $M = 985.93$

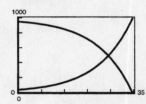

(b) In the early years of the mortgage, the majority of the monthly payment goes toward interest. The principle and interest are nearly equal when $t \approx 27.676 \approx 28$ years.

(c) $P = 120{,}000$, $t = 20$, $r = 0.095$, $M = 1118.56$

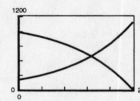

The interest is still the majority of the monthly payment in the early years. Now the principle and interest are nearly equal when $t \approx 12.675 \approx 12.7$ years.

69. $t_1 = 40.757 + 0.556s - 15.817 \ln s$

$t_2 = 1.2259 + 0.0023s^2$

(a) Linear Model: $t_3 \approx 0.2729s - 6.0143$

Exponential Model: $t_4 \approx 1.5385e^{1.0291s}$

(b)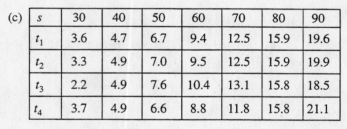

(c)

s	30	40	50	60	70	80	90
t_1	3.6	4.7	6.7	9.4	12.5	15.9	19.6
t_2	3.3	4.9	7.0	9.5	12.5	15.9	19.9
t_3	2.2	4.9	7.6	10.4	13.1	15.8	18.5
t_4	3.7	4.9	6.6	8.8	11.8	15.8	21.1

(d) Model t_1: $S_1 = |3.4 - 3.6| + |5 - 4.7| + |7 - 6.7| + |9.3 - 9.4| + |12 - 12.5| +$
$\qquad |15.8 - 15.9| + |20 - 19.6| = 1.9$

Model t_2: $S_2 = |3.4 - 3.3| + |5 - 4.9| + |7 - 7| + |9.3 - 9.5| + |12 - 12.5| +$
$\qquad |15.8 - 15.9| + |20 - 19.9| = 1.2$

Model t_3: $S_3 = |3.4 - 2.2| + |5 - 4.9| + |7 - 7.6| + |9.3 - 10.4| + |12 - 13.1| +$
$\qquad |15.8 - 15.8| + |20 - 18.5| = 5.6$

Model t_4: $S_4 = |3.4 - 3.7| + |5 - 4.9| + |7 - 6.6| + |9.3 - 8.8| + |12 - 11.8| +$
$\qquad |15.8 - 15.8| + |20 - 21.1| = 2.6$

t_2, the Quadratic model, is the best fit with the data.

71. Each essay will vary.

73. $t = -2.5 \ln\left(\dfrac{T - 70}{98.6 - 70}\right)$

At 9:00 A.M. we have: $t = -2.5 \ln\left(\dfrac{85.7 - 70}{98.6 - 70}\right) \approx 1.5$ hours.

From this we can conclude that the person died at 7:30 A.M.

75. $\dfrac{3}{2}$ | 8   −36   54   −27

                 12   −36   27

        8   −24   18   0

$\dfrac{8x^3 - 36x^2 + 54x - 27}{x - 3/2} = 8x^2 - 24x + 18$

$\left(x \neq \dfrac{3}{2}\right)$

77. −5 | 1   0   0   −3   1

                   −5   25   −125   640

        1   −5   25   −128   641

$\dfrac{x^4 - 3x + 1}{x + 5} = x^3 - 5x^2 + 25x - 128 + \dfrac{641}{x + 5}$

Section 3.6   Exploring Data: Nonlinear Models

- ■ You should know how to classify scatter plots.
- ■ You should know how to use your calculator or computer to fit a nonlinear model to data.

Solutions to Odd-Numbered Exercises

1. A logarithmic model seems best.

3. A Gaussian model seems best.

5. An exponential model seems best.

7. A Gaussian model seems best.

9. Logarithmic model

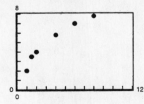

11. Exponential model

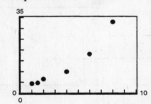

13. Linear model

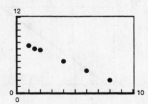

15. $y = 3.807(13,057)^x$

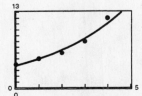

17. $y = 8.463(0.7775)^x$

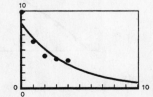

19. $y = 2.083 + 1.257 \ln x$

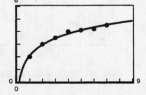

21. $y = 9.826 - 4.097 \ln x$

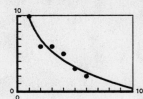

23. $y = 1985x^{0.760}$

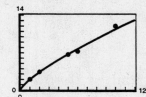

25. $y = 16.103x^{-3.174}$

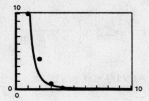

27. (a) $y = 38.233d^{1.955}$

(b)

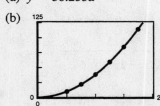

(c) $y = 38.233(2)^{1.955} \approx 148.2$ tons

29. (a) $y = 0.088x + 4.413$

(b) $y = 4.456(1.017)^x$

(c) The linear model is better.

(d) Linear: $y = 0.088(21) + 4.413$

$= 6.261$ billion

Exponential: $y = 4.456(1.017)^{21}$

≈ 6.349 billion

31. (a) $y = 5.088x^{0.645}$

(b)

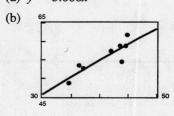

(c) $y = 5.088(50)^{0.645}$

≈ 63.44 million board feet

33. (a) $f = 396.48T^{0.05055}$

(b) 487

(c) $x^{1/2}$

35. (a) $y_1 = 7.01x + 41.14$

$y_2 = 5.31 + 45.83 \ln x$

$y_3 = 51.26(1.08)^x$

$y_4 = 34.00x^{0.513}$

(b)

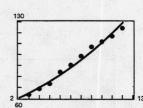

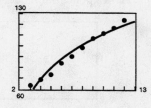

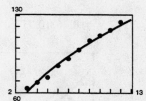

The linear model seems to best fit the data.

—CONTINUED—

35. —CONTINUED—

(c)

x	y	$y - y_1$	$(y - y_1)^2$	$y - y_2$	$(y - y_2)^2$	$y - y_3$	$(y - y_3)^2$	$y - y_4$	$(y - y_4)^2$
3	63.2	1.03	1.06	7.54	56.86	-1.37	1.88	3.46	11.99
4	68.6	-0.58	0.34	-0.24	0.06	-1.14	1.30	-0.64	0.41
5	73.2	-2.99	8.94	-5.87	34.46	-2.12	4.48	-4.43	19.66
6	83.9	0.70	0.49	-3.53	12.44	2.56	6.54	-1.35	1.81
7	90.3	0.09	0.01	-4.19	17.57	2.45	6.00	-1.96	3.84
8	98.4	1.18	1.39	-2.21	4.89	3.52	12.40	-0.40	0.16
9	107.0	2.77	7.67	0.99	0.98	4.53	20.53	2.04	4.18
10	111.7	0.46	0.21	0.86	0.74	1.03	1.07	0.92	0.84
11	116.8	-1.45	2.10	1.59	2.54	-2.72	7.40	0.46	0.22
12	124.3	-0.96	0.92	5.11	26.08	-4.78	22.86	2.65	7.04

(d) Linear Model

y_1: 23.14, y_2: 156.62, y_3: 84.46, y_4: 50.15

(e) Sum of the squares of the errors

❏ Review Exercises for Chapter 3

Solutions to Odd-Numbered Exercises

1. $f(x) = 4^x$
Intercept: $(0, 1)$
Horizontal asymptote: x-axis
Increasing on: $(-\infty, \infty)$
Matches graph (e).

3. $f(x) = -4^x$
Intercept: $(0, -1)$
Horizontal asymptote: x-axis
Decreasing on: $(-\infty, \infty)$
Matches graph (b).

5. $f(x) = \log_4 x$
Intercept: $(1, 0)$
Vertical asymptote: y-axis
Increasing on: $(0, \infty)$
Matches graph (a).

7. $f(x) = 6^x$
Intercept: $(0, 1)$
Increasing horizontal asymptote: x-axis

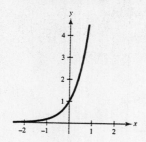

9. $g(x) = 6^{-x} = \left(\frac{1}{6}\right)^x$
Intercept: $(0, 1)$
Decreasing horizontal asymptote: x-axis

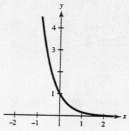

11. $h(x) = e^{-x/2}$

x	-2	-1	0	1	2
y	2.72	1.65	1	0.61	0.37

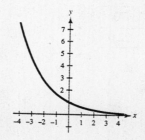

13. $f(x) = e^{x+2}$

x	-3	-2	-1	0	1
y	0.37	1	2.72	7.39	20.09

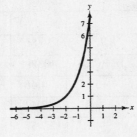

15.

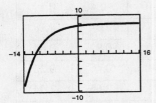

Horizontal asymptote: $y = 8$

17.

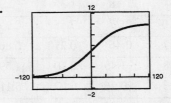

Horizontal asymptotes: $y = 0$, $y = 10$

19. $A = 3500\left(1 + \dfrac{0.105}{n}\right)^{10n}$ or $A = 3500e^{(0.105)(10)}$

n	1	2	4	12	365	Continuous Compounding
A	\$9,499.28	\$9,738.91	\$9,867.22	\$9,956.20	\$10,000.27	\$10,001.78

21. $200{,}000 = Pe^{0.08t}$

$$P = \dfrac{200{,}000}{e^{0.08t}}$$

t	1	10	20	30	40	50
P	\$184,623.27	\$89,865.79	\$40,379.30	\$18,143.59	\$8,152.44	\$3,663.13

23. (a)

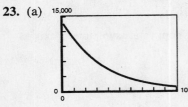

(b) $v(2) = 14{,}000\left(\frac{3}{4}\right)^2 = \$7875$

(c) The car depreciates most rapidly at the beginning, which is realistic.

25. $A = 500e^{-0.013t}$

$A = 500e^{-0.013(60)}$

≈ 229.2 units per milliliter

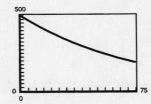

27. $y = 28e^{0.6 - 0.012s}$, $s \geq 50$

s	50	55	60	65	70
y	28	26.4	24.8	23.4	22.0

29. $g(x) = \log_2 x \implies 2^y = x$

Domain: $(0, \infty)$

Vertical asymptote: $x = 0$

x	$\frac{1}{4}$	$\frac{1}{2}$	1	2	4
y	-2	-1	0	1	2

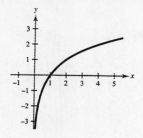

31. $f(x) = \ln x + 3$

Domain: $(0, \infty)$

Vertical asymptote: $x = 0$

x	1	2	3	$\frac{1}{2}$	$\frac{1}{4}$
$f(x)$	3	3.69	4.10	2.31	1.61

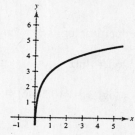

33. $h(x) = \ln(e^{x-1})$

$\quad\quad = (x - 1) \ln e$

$\quad\quad = x - 1$

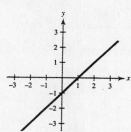

35. $y = \log_{10}(x^2 + 1)$

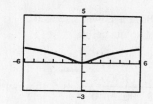

37. $\quad 4^3 = 64$

$\log_4 64 = 3$

39. $\log_{10} 1000 = \log_{10} 10^3 = 3$

41. $\log_3 \frac{1}{9} = \log_3 1 - \log_3 9 = -\log_3 3^2 = -2 \log_3 3 = -2$

43. $\ln e^7 = 7$

45. $\ln 1 = 0$ since $e^0 = 1$.

47. $\log_4 9 = \dfrac{\log_{10} 9}{\log_{10} 4} \approx 1.585$

$\log_4 9 = \dfrac{\ln 9}{\ln 4} \approx 1.585$

49. $\log_{12} 200 = \dfrac{\log_{10} 200}{\log_{10} 12} \approx 2.132$

$\log_{12} 200 = \dfrac{\ln 200}{\ln 12} \approx 2.132$

51. $\log_5 5x^2 = \log_5 5 + \log_5 x^2$

$\quad\quad\quad\quad = 1 + 2 \log_5 x$

53. $\log_{10} \dfrac{5\sqrt{y}}{x^2} = \log_{10} 5\sqrt{y} - \log_{10} x^2$

$\quad\quad\quad\quad\quad = \log_{10} 5 + \log_{10} \sqrt{y} - \log_{10} x^2$

$\quad\quad\quad\quad\quad = \log_{10} 5 + \dfrac{1}{2} \log_{10} y - 2 \log_{10} x$

55. $\ln[(x^2 + 1)(x - 1)] = \ln(x^2 + 1) + \ln(x - 1)$

57. $\log_2 5 + \log_2 x = \log_2 5x$

59. $\dfrac{1}{2}\ln|2x - 1| - 2\ln|x + 1| = \ln\sqrt{|2x - 1|} - \ln|x + 1|^2$

$$= \ln \dfrac{\sqrt{|2x - 1|}}{(x + 1)^2}$$

61. $\ln 3 + \dfrac{1}{3}\ln(4 - x^2) - \ln x = \ln\left[\dfrac{3(4 - x^2)^{1/3}}{x}\right] = \ln\left[\dfrac{3\sqrt[3]{4 - x^2}}{x}\right]$

63. True; by the inverse properties, $\log_b b^{2x} = 2x$.

65. False, the domain of $f(x) = \ln x$ is $(0, \infty)$.

67. False; $\ln x + \ln y = \ln(xy) \neq \ln(x + y)$

69. $\log_b 25 = \log_b 5^2 = 2\log_b 5 \approx 2(0.8271) = 1.6542$

71. $\log_b \sqrt{3} = \log_b 3^{1/2} = \frac{1}{2}\log_b 3 \approx \frac{1}{2}(0.5646) = 0.2823$

73. $t = 50\log_{10}\dfrac{18{,}000}{18{,}000 - h}$

(a) $0 \leq h < 18{,}000$

(b)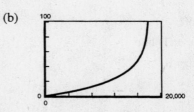

(c) The plane climbs at a slower rate as it approaches its absolute ceiling.

(d) If $h = 4000$,

$$t = 50\log_{10}\dfrac{18{,}000}{18{,}000 - 4000} \approx 5.46 \text{ minutes.}$$

Vertical asymptote: $h = 18{,}000$

75. $e^x = 12$

$x = \ln 12 \approx 2.485$

77. $3e^{-5x} = 132$

$e^{-5x} = 44$

$-5x = \ln 44$

$x = \dfrac{\ln 44}{5} \approx -0.757$

81. $\ln 3x = 8.2$

$3x = e^{8.2}$

$x = \dfrac{e^{8.2}}{3} \approx 1213.650$

79.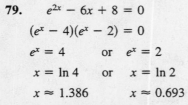

$e^{2x} - 6x + 8 = 0$

$(e^x - 4)(e^x - 2) = 0$

$e^x = 4$ or $e^x = 2$

$x = \ln 4$ or $x = \ln 2$

$x \approx 1.386$ $\qquad x \approx 0.693$

Extraneous solutions, no solution

83. $\ln x - \ln 3 = 2$

$\ln\dfrac{x}{3} = 2$

$\dfrac{x}{3} = e^2$

$x = 3e^2 \approx 22.167$

85.
$$\log(x - 1) = \log(x - 2) - \log(x + 2)$$

$$\log(x - 1) = \log\left(\frac{x - 2}{x + 2}\right)$$

$$x - 1 = \frac{x - 2}{x + 2}$$

$$(x - 1)(x + 2) = x - 2$$

$$x^2 + x - 2 = 2 - 2$$

$$x^2 = 0$$

$$x = 0$$

Since $x = 0$ is not in the domain of $\ln(x - 1)$ or of $\ln(x - 2)$, it is an extraneous solution. The equation has no solution. You can verify this by graphing each side of the equation and observing that the two curves do not intersect.

87. $2^{0.6x} - 3x = 0$

Graph $y_1 = 2^{0.6x} - 3x$.

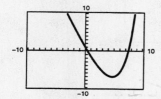

The x-intercepts are at $x \approx 0.39$ and at $x \approx 7.48$.

89. $2 \ln(x + 3) + 3x = 8$

Graph $y_1 = 2 \ln(x + 3) + 3x - 8$.

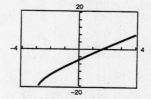

The x-intercept is at $x \approx 1.64$.

91.
$$y = ae^{bx}$$

$$2 = ae^{b(0)} \implies a = 2$$

$$3 = 2e^{b(4)}$$

$$1.5 = e^{4b}$$

$$\ln 1.5 = 4b \implies b \approx 0.1014$$

Thus, $y \approx 2e^{0.1014x}$.

93.
$$y = ae^{bx}$$

$$4 = ae^{b(0)} = a \implies a = 4$$

$$\frac{1}{2} = 4e^{b(5)}$$

$$\frac{1}{8} = e^{5b}$$

$$\ln \frac{1}{8} = 5b \implies b = -\frac{\ln 8}{5} \approx -0.4159$$

95. $p = 500 - 0.5e^{0.004x}$

(a)
$$p = 450$$

$$450 = 500 - 0.5e^{0.004x}$$

$$0.5e^{0.004x} = 50$$

$$e^{0.004x} = 100$$

$$0.004x = \ln 100$$

$$x \approx 1151 \text{ units}$$

(b)
$$p = 400$$

$$400 = 500 - 0.5e^{0.004x}$$

$$0.5e^{0.004x} = 100$$

$$e^{0.004x} = 200$$

$$0.004x = \ln 200$$

$$x \approx 1325 \text{ units}$$

97. (a)
$$t = \frac{\ln 2}{r}$$

$$7.75 = \frac{\ln 2}{r}$$

$$r = \frac{\ln 2}{7.75}$$

$$\approx 0.0894$$

$$\text{or } 8.94\%$$

(b) $A = Pe^{rt}$

$$A = 750e^{[(\ln 2)/7.75](10)}$$

$$\approx \$1834.37$$

(c) $A = Pe^{[(\ln 2)/7.75](1)}$

$$\approx P(1.0936)$$

$$= P(1 + 0.0936)$$

$$\text{Effective yield} \approx 0.0936$$

$$= 9.36\%$$

99.
$$\beta = 10 \log_{10}\left(\frac{I}{10^{-16}}\right)$$

$$125 = 10 \log_{10}\left(\frac{I}{10^{-16}}\right)$$

$$\frac{125}{10} = \log_{10}\left(\frac{I}{10^{-16}}\right)$$

$$10^{12.5} = \frac{I}{10^{-16}}$$

$$10^{12.5} = 10^{16}I$$

$$10^{12.5}10^{-16} = I$$

$$10^{-3.5} = I$$

101. $y = 234.6839(0.8746)^x$
$= 234.684e^{-0.134x}$

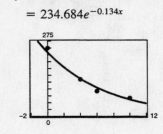

103. (a) $y = 12.907x^{0.5102}$

   (b) For 2001, $y = 12.907(21)^{0.5102} \approx 61.0$ billion.

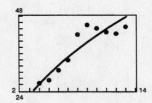

❑ Cumulative Test for Chapters P–3

1.
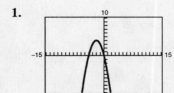

2.
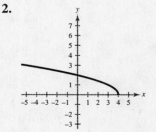

3.
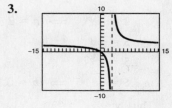

4.
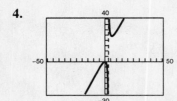

5.
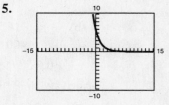

6.
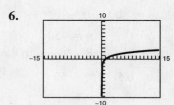

7. Slope: $\dfrac{8-1}{3+(1/2)} = \dfrac{7}{7/2} = 2$

   Line: $y - 8 = 2(x - 3)$
   $$y = 2x + 2$$
   $$y - 2x - 2 = 0$$

8. It does not pass the vertical line test.

9. (a) $r(x) = \frac{1}{2}\sqrt[3]{x}$ is a vertical shrink of $y = \sqrt[3]{x}$.

(b) $h(x) = \sqrt[3]{x} + 2$ is a vertical shift 2 units upward.

(c) $g(x) = \sqrt[3]{x + 2}$ is a horizontal shift 2 units to the left.

11.
$$2x - 3(x - 4) = 5$$
$$2x - 3x + 12 = 5$$
$$-x = -7$$
$$x = 7$$

13. $3y^2 + 6y + 2 = 0$
$$y = \frac{-6 \pm \sqrt{6^2 - 4(3)(2)}}{2(3)}$$
$$= \frac{-6 \pm \sqrt{12}}{6}$$
$$= -1 \pm \frac{1}{3}\sqrt{3}$$

15. $6e^{2x} = 72$
$$e^{2x} = 12$$
$$2x = \ln 12$$
$$x = \frac{1}{2}\ln 12 \approx 1.242$$

17. $P = 230 + 20x - \frac{1}{2}x^2$
$$= -\frac{1}{2}(x^2 - 40x + 400) + 230 + 200$$
$$-\frac{1}{2}(x - 20)^2 + 430$$

$x = 20$ yields a maximum profit of 430. Thus $2000 is the amount.

19. $x \approx 1.20$

21. $y = 5.280\,(1.4455)^x$

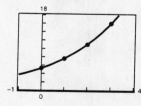

10. $h(x) = 5x - 2$ is one-to-one, since it is a nonvertical line.
$$y = 5x - 2$$
$$x = 5y - 2 \qquad \text{Interchange } x \text{ and } y.$$
$$h^{-1}(x) = y = \frac{1}{5}(x + 2) \qquad \text{Solve for } y.$$

12. $\dfrac{2}{t - 3} + \dfrac{2}{t - 2} = \dfrac{10}{t^2 + 5t + 6} = \dfrac{10}{(t - 3)(t - 2)}$
$$2(t - 2) + 2(t - 3) = 10$$
$$4t - 10 = 10$$
$$t = 5$$

14.
$$\sqrt{x + 10} = x - 2$$
$$x + 10 = (x - 2)^2 = x^2 - 4x + 4$$
$$x^2 - 5x - 6 = 0$$
$$(x - 6)(x + 1) = 0$$
$$x = 6 \qquad (x = -1 \text{ is extraneous.})$$

16. $\log_2 x + \log_2 5 = 6$
$$\log_2(5x) = 6$$
$$5x = 2^6 = 64$$
$$x = \frac{64}{5}$$

18. $f(x) = x^3 + 2x^2 + 4x + 8 \qquad x = -2$ is a zero.
$$= (x + 2)(x^2 + 4)$$
$$= (x + 2)(x + 2i)(x - 2i)$$
$$-2, \pm 2i$$

20. $2 \ln x - \dfrac{1}{2}\ln(x + 5) = \ln x^2 - \ln(x + 5)^{1/2}$
$$= \ln \frac{x^2}{\sqrt{x + 5}}$$

❑ Practice Test for Chapter 3

1. Solve for x: $x^{3/5} = 8$

2. Solve for x: $3^{x-1} = \frac{1}{81}$

3. Graph $f(x) = 2^{-x}$ by hand.

4. Graph $g(x) = e^x + 1$ by hand.

5. If $5000 is invested at 9% interest, find the amount after three years if the interest is compounded

   (a) monthly   (b) quarterly   (c) continuously.

6. Write the equation in logarithmic form: $7^{-2} = \frac{1}{49}$

7. Solve for x: $x - 4 = \log_2 \frac{1}{64}$

8. Given $\log_b 2 = 0.3562$ and $\log_b 5 = 0.8271$, evaluate $\log_b \sqrt[4]{8/25}$.

9. Write $5 \ln x - \frac{1}{2} \ln y + 6 \ln z$ as a single logarithm.

10. Using your calculator and the change of base formula, evaluate $\log_9 28$.

11. Use your calculator to solve for N: $\log_{10} N = 0.6646$

12. Graph $y = \log_4 x$ by hand.

13. Determine the domain of $f(x) = \log_3(x^2 - 9)$.

14. Graph $y = \ln(x - 2)$ by hand.

15. True or false: $\dfrac{\ln x}{\ln y} = \ln(x - y)$

16. Solve for x: $5^x = 41$

17. Solve for x: $x - x^2 = \log_5 \frac{1}{25}$

18. Solve for x: $\log_2 x + \log_2(x - 3) = 2$

19. Solve for x: $\dfrac{e^x + e^{-x}}{3} = 4$

20. Six thousand dollars is deposited into a fund at an annual percentage rate of 13%. Find the time required for the investment to double if the interest is compounded continuously.

21. Use a graphing utility to find the points of intersection of the graphs of $y = \ln(3x)$ and $y = e^x - 4$.

22. Use a graphing utility to find the power model $y = ax^b$ for the data $(1, 1)$, $(2, 5)$, $(3, 8)$, and $(4, 17)$.

CHAPTER 4
Trigonometric Functions

CHAPTER 4
Trigonometric Functions

Section 4.1 Radian and Degree Measure

You should know the following basic facts about angles, their measurement, and their applications.

- ■ Types of Angles:
  - (a) Acute: Measure between $0°$ and $90°$.
  - (b) Right: Measure $90°$.
  - (c) Obtuse: Measure between $90°$ and $180°$.
  - (d) Straight: Measure $180°$.
- ■ α and β are complementary if $\alpha + \beta = 90°$. They are supplementary if $\alpha + \beta = 180°$.
- ■ Two angles in standard position that have the same terminal side are called coterminal angles.
- ■ To convert degrees to radians, use $1° = \pi/180$ radians.
- ■ To convert radians to degrees, use 1 radian $= (180/\pi)°$.
- ■ $1' =$ one minute $= 1/60$ of $1°$.
- ■ $1'' =$ one second $= 1/60$ of $1' = 1/3600$ of $1°$.
- ■ The length of a circular arc is $s = r\theta$ where θ is measured in radians.
- ■ Speed $=$ distance/time
- ■ Angular speed $= \theta/t = s/rt$

Solutions to Odd-Numbered Exercises

1.

The angle shown is approximately 2 radians.

3.

The angle shown is approximately -3 radians.

5. (a) Since $0 < \dfrac{\pi}{5} < \dfrac{\pi}{2}$; $\dfrac{\pi}{5}$ lies in Quadrant I.

   (b) Since $\pi < \dfrac{7\pi}{5} < \dfrac{3\pi}{2}$; $\dfrac{7\pi}{5}$ lies in Quadrant III.

7. (a) Since $-\dfrac{\pi}{2} < -\dfrac{\pi}{12} < 0$; $-\dfrac{\pi}{12}$ lies in Quadrant IV.

   (b) Since $-\dfrac{3\pi}{2} < -\dfrac{11\pi}{9} < -\pi$; $-\dfrac{11\pi}{9}$ lies in Quadrant II.

9. (a) Since $\pi < 3.5 < \dfrac{3\pi}{2}$; 3.5 lies in Quadrant III.

   (b) Since $\dfrac{\pi}{2} < 2.25 < \pi$; 2.25 lies in Quadrant II.

11. (a)

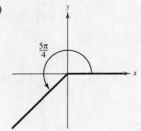

(b)

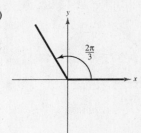

13. (a)

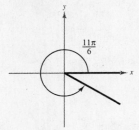

(b)

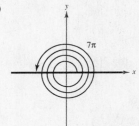

15. (a) Coterminal angles for $\dfrac{\pi}{12}$

$$\frac{\pi}{12} + 2\pi = \frac{25\pi}{12}$$

$$\frac{\pi}{12} - 2\pi = -\frac{23\pi}{12}$$

(b) Coterminal angles for $\dfrac{2\pi}{3}$

$$\frac{2\pi}{3} + 2\pi = \frac{8\pi}{3}$$

$$\frac{2\pi}{3} - 2\pi = -\frac{4\pi}{3}$$

17. (a) Coterminal angles for $-\dfrac{9\pi}{4}$

$$-\frac{9\pi}{4} + 4\pi = \frac{7\pi}{4}$$

$$\frac{7\pi}{4} - 2\pi = -\frac{\pi}{4}$$

(b) Coterminal angles for $-\dfrac{2\pi}{15}$

$$-\frac{2\pi}{15} + 2\pi = \frac{28\pi}{15}$$

$$-\frac{2\pi}{15} - 2\pi = -\frac{32\pi}{15}$$

19.

The angle shown is approximately 210°.

21.

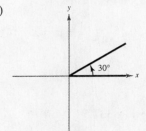

The angle shown is approximately −45°.

23. (a) Since $90° < 130° < 180°$; $130°$ lies in Quadrant II.

   (b) Since $270° < 285° < 360°$; $285°$ lies in Quadrant IV.

25. (a) Since $-180° < -132°50' < -90°$; $-132°\ 50'$ lies in Quadrant III.

   (b) Since $-360° < -336° < -270°$; $-336°$ lies in Quadrant I.

27. (a)

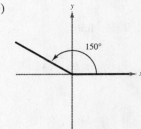

(b)

29. (a)

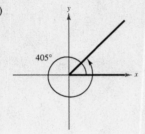

(b)

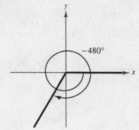

31. (a) Coterminal angles for 45°

$45° + 360° = 405°$

$45° - 360° = -315°$

(b) Coterminal angles for $-36°$

$-36° + 360° = 324°$

$-36° - 360° = -396°$

33. (a) Coterminal angles for 300°

$300° + 360° = 660°$

$300° - 360° = -60°$

(b) Coterminal angles for 740°

$740° - 2(360°) = 20°$

$20° - 360° = -340°$

35. (a) $54° \, 45' = 54° + \left(\frac{45}{60}\right)° = 54.75°$

(b) $-128° \, 30' = -128° - \left(\frac{30}{60}\right)° = -128.5°$

37. (a) $85° \, 18' \, 30'' = \left(85 + \frac{18}{60} + \frac{30}{3600}\right)° \approx 85.308°$

(b) $330° \, 25'' = \left(330 + \frac{25}{3600}\right)° \approx 330.007°$

39. (a) $240.6° = 240° + 0.6(60)' = 240° \, 36'$

(b) $-145.8° = -[145° + 0.8(60')] = -145° \, 48'$

41. (a) $2.5 = 2.5\left(\frac{180}{\pi}\right)° \approx 143.23945° \approx 143° \, 14' \, 22''$

(b) $-3.58 = -3.58\left(\frac{180}{\pi}\right)° \approx -205.11889° \approx -205° \, 7' \, 8''$

43. (a) Complement: $90° - 18° = 72°$

Supplement: $180° - 18° = 162°$

(b) Complement: Not possible; 115° is greater than 90°.

Supplement: $180° - 115° = 65°$

45. (a) Complement: $\frac{\pi}{2} - \frac{\pi}{3} = \frac{\pi}{6}$

Supplement: $\pi - \frac{\pi}{3} = \frac{2\pi}{3}$

(b) Complement: Not possible; $\frac{3\pi}{4}$ is greater than $\frac{\pi}{2}$.

Supplement: $\pi - \frac{3\pi}{4} = \frac{\pi}{4}$

47. (a) $30° = 30\left(\frac{\pi}{180}\right) = \frac{\pi}{6}$

(b) $150° = 150\left(\frac{\pi}{180}\right) = \frac{5\pi}{6}$

49. (a) $-20° = -20\left(\frac{\pi}{180}\right) = -\frac{\pi}{9}$

(b) $-240° = -240\left(\frac{\pi}{180}\right) = -\frac{4\pi}{3}$

51. (a) $\frac{3\pi}{2} = \frac{3\pi}{2}\left(\frac{180}{\pi}\right)° = 270°$

(b) $\frac{7\pi}{6} = \frac{7\pi}{6}\left(\frac{180}{\pi}\right)° = 210°$

53. (a) $\frac{7\pi}{3} = \frac{7\pi}{3}\left(\frac{180}{\pi}\right)° = 420°$

(b) $-\frac{11\pi}{30} = -\frac{11\pi}{30}\left(\frac{180}{\pi}\right)° = -66°$

55. $115° = 115\left(\frac{\pi}{180}\right) \approx 2.007$ radians

57. $-216.35° = -216.35\left(\frac{\pi}{180}\right) \approx -3.776$ radians

59. $532° = 532\left(\frac{\pi}{180}\right) \approx 9.285$ radians

61. $-0.83° = -0.83\left(\frac{\pi}{180}\right) \approx -0.014$ radian

63. $\frac{\pi}{7} = \frac{\pi}{7}\left(\frac{180}{\pi}\right) \approx 25.714°$

65. $\frac{15\pi}{8} = \frac{15\pi}{8}\left(\frac{180}{\pi}\right) = 337.5°$

67. $-4.2\pi = -4.2\pi\left(\frac{180}{\pi}\right) = -756°$

69. $-2 = -2\left(\frac{180}{\pi}\right) \approx -114.592°$

71. $s = r\theta$

$6 = 5\theta$

$\theta = \frac{6}{5}$ radians

73. $s = r\theta$

$32 = 7\theta$

$\theta = \frac{32}{7} = 4\frac{4}{7}$ radians

75. $s = r\theta$

$4 = 15\theta$

$\theta = \frac{4}{15}$ radian

77. $s = r\theta$

$25 = 14.5\theta$

$\theta = \frac{25}{14.5} \approx 1.724$ radians

79. $s = r\theta$, θ in radians

$s = 15(180)\left(\dfrac{\pi}{180}\right) = 15\pi$ inches

≈ 47.12 inches

81. $s = r\theta$, θ in radians

$s = 6(2) = 12$ meters

83. $\theta = 41°\ 15'\ 42'' - 32°\ 47'\ 9'' = 8°\ 28'\ 33'' \approx 8.47583° \approx 0.14793$ radian

$s = r\theta = 4000(0.14793) \approx 591.72$ miles

85. $\theta = 42°\ 7'\ 15'' - 25°\ 46'\ 37'' = 16°\ 20'\ 38'' \approx 0.285255$ radian

$s = r\theta = 4000(0.285255) \approx 1141.02$ miles

87. $\theta = \dfrac{s}{r} = \dfrac{600}{6378} \approx 0.094$ radian $\approx 5.39°$

89. $\theta = \dfrac{s}{r} = \dfrac{2.5}{6} = \dfrac{25}{60} = \dfrac{5}{12}$ radian

91. (a) 50 miles per hour $= 50(5280)/60 = 4400$ feet per minute

The circumference of the tire is $C = 2.5\pi$ feet.

The number of revolutions per minute is $r = 4400/2.5\pi \approx 560.2$ rev/min.

(b) The angular speed is θ/t.

$\theta = \dfrac{4400}{2.5\pi}(2\pi) = 3520$ radians

Angular speed $= \dfrac{3520 \text{ radians}}{1 \text{ minute}} = 3520$ rad/min

93. 1 Radian $= \left(\dfrac{180}{\pi}\right)° \approx 57.3°$, so one radian is much larger than one degree.

95. (a) $\dfrac{\text{Revolutions}}{\text{Second}} = \dfrac{2400}{60} = 40$ rev/sec

Angular speed $= (2\pi)(40) = 80\pi$ rad/sec

(b) Radius of saw blade $= \dfrac{7.5}{2} = 3.75$ in.

Radius in feet $= \dfrac{3.75 \text{ in.}}{12 \text{ in.}} = 0.3125$ ft

Speed $= \dfrac{s}{t} = \dfrac{r\theta}{t} = r\dfrac{\theta}{t}$

$= r(\text{angular speed})$

$= 0.3125(80\pi) = 78.54$ ft/sec

97. (a) Arc length of larger sprocket in feet: $s = r\theta = \left(\dfrac{4}{12}\right)(2\pi) = \dfrac{2\pi}{3}$ feet

The angle θ of the smaller sprocket is $\theta = \dfrac{s}{r} = \dfrac{(2\pi/3) \text{ ft}}{(2/12) \text{ ft}} = 4\pi$.

The arc length of the tire is $s = r\theta = \left(\dfrac{14}{12}\right)(4\pi) = \dfrac{14\pi}{3}$ feet.

Speed $= \dfrac{s}{t} = \dfrac{(14\pi/3) \text{ ft}}{1 \text{ sec}} = \dfrac{14\pi}{3}$ ft/sec

(b) $\left(\dfrac{14\pi}{3} \text{ ft/sec}\right)(3600 \text{ sec/hr})\left(\dfrac{1 \text{ mile}}{5280 \text{ ft}}\right) \approx 10$ mi/hr

99. $A = \frac{1}{2}r^2\theta = \frac{1}{2}(10)^2 \cdot \frac{\pi}{3} = \frac{50}{3}\pi \ m^2$

101. Two angles in standard position are coterminal angles if they have the same initial and terminal sides.

103. $s = r\theta \implies \theta = \frac{s}{r} = \frac{1.2}{4000} = 0.0003$ radians or $0.0172°$

Section 4.2   Trigonometric Functions: The Unit Circle

- ■ You should know the definition of the trigonometric functions in terms of the unit circle. Let t be a real number and (x, y) the point on the unit circle corresponding to t.

$$\sin t = y \qquad\qquad \csc t = \frac{1}{y}, \quad y \neq 0$$

$$\cos t = x \qquad\qquad \sec t = \frac{1}{x}, \quad x \neq 0$$

$$\tan t = \frac{y}{x}, \quad x \neq 0 \qquad\qquad \cot t = \frac{x}{y}, \quad y \neq 0$$

- ■ The cosine and secant functions are even.

$$\cos(-t) = \cos t \qquad\qquad \sec(-t) = \sec t$$

- ■ The other four trigonometric functions are odd.

$$\sin(-t) = -\sin t \qquad\qquad \csc(-t) = -\csc t$$

$$\tan(-t) = -\tan t \qquad\qquad \cot(-t) = -\cot t$$

- ■ Be able to evaluate the trigonometric functions with a calculator.

Solutions to Odd-Numbered Exercises

1. $x = -\frac{3}{5}, \quad y = \frac{4}{5}$

$\sin t = y = \frac{4}{5}$ $\qquad\qquad$ $\csc t = \frac{1}{y} = \frac{5}{4}$

$\cos t = x = -\frac{3}{5}$ $\qquad\qquad$ $\sec t = \frac{1}{x} = -\frac{5}{3}$

$\tan t = \frac{y}{x} = -\frac{4}{3}$ $\qquad\qquad$ $\cot t = \frac{x}{y} = -\frac{3}{4}$

3. $x = \frac{8}{17}, \quad y = -\frac{15}{17}$

$\sin t = y = -\frac{15}{17}$ $\qquad\qquad$ $\csc t = \frac{1}{y} = -\frac{17}{15}$

$\cos t = x = \frac{8}{17}$ $\qquad\qquad$ $\sec t = \frac{1}{x} = \frac{17}{8}$

$\tan t = \frac{y}{x} = -\frac{15}{8}$ $\qquad\qquad$ $\cot t = \frac{x}{y} = -\frac{8}{15}$

5. $x = -\frac{\sqrt{3}}{2}, \quad y = -\frac{1}{2}$

$\sin t = y = -\frac{1}{2}$ $\qquad\qquad$ $\csc t = \frac{1}{y} = -2$

$\cos t = x = -\frac{\sqrt{3}}{2}$ $\qquad\qquad$ $\sec t = \frac{1}{x} = -\frac{2}{\sqrt{3}} = -\frac{2\sqrt{3}}{3}$

$\tan t = \frac{y}{x} = \frac{1}{\sqrt{3}} = \frac{\sqrt{3}}{3}$ $\qquad\qquad$ $\cot t = \frac{x}{y} = \sqrt{3}$

7. $t = \dfrac{\pi}{4}$ corresponds to $\left(\dfrac{\sqrt{2}}{2}, \dfrac{\sqrt{2}}{2}\right)$.

9. $t = \dfrac{5\pi}{6}$ corresponds to $\left(-\dfrac{\sqrt{3}}{2}, \dfrac{1}{2}\right)$.

11. $t = \dfrac{4\pi}{3}$ corresponds to $\left(-\dfrac{1}{2}, -\dfrac{\sqrt{3}}{2}\right)$.

13. $t = \dfrac{3\pi}{2}$ corresponds to $(0, -1)$.

15. $t = \dfrac{\pi}{4}$ corresponds to $\left(\dfrac{\sqrt{2}}{2}, \dfrac{\sqrt{2}}{2}\right)$.

$$\sin t = y = \dfrac{\sqrt{2}}{2}$$

$$\cos t = x = \dfrac{\sqrt{2}}{2}$$

$$\tan t = \dfrac{y}{x} = 1$$

17. $t = -\dfrac{\pi}{6}$ corresponds to $\left(\dfrac{\sqrt{3}}{2}, -\dfrac{1}{2}\right)$.

$$\sin t = y = -\dfrac{1}{2}$$

$$\cos t = x = \dfrac{\sqrt{3}}{2}$$

$$\tan t = \dfrac{y}{x} = -\dfrac{\sqrt{3}}{3}$$

19. $t = -\dfrac{5\pi}{4}$ corresponds to $\left(-\dfrac{\sqrt{2}}{2}, \dfrac{\sqrt{2}}{2}\right)$.

$$\sin t = y = \dfrac{\sqrt{2}}{2}$$

$$\cos t = x = -\dfrac{\sqrt{2}}{2}$$

$$\tan t = \dfrac{y}{x} = -1$$

21. $t = \dfrac{11\pi}{6}$ corresponds to $\left(\dfrac{\sqrt{3}}{2}, -\dfrac{1}{2}\right)$.

$$\sin t = y = -\dfrac{1}{2}$$

$$\cos t = x = \dfrac{\sqrt{3}}{2}$$

$$\tan t = \dfrac{y}{x} = -\dfrac{\sqrt{3}}{3}$$

23. $t = \dfrac{4\pi}{3}$ corresponds to $\left(-\dfrac{1}{2}, -\dfrac{\sqrt{3}}{2}\right)$.

$$\sin t = y = -\dfrac{\sqrt{3}}{2}$$

$$\cos t = x = -\dfrac{1}{2}$$

$$\tan t = \dfrac{y}{x} = \sqrt{3}$$

25. $t = -\dfrac{3\pi}{2}$ corresponds to $(0, 1)$.

$$\sin t = y = 1$$
$$\cos t = x = 0$$
$$\tan t = \dfrac{y}{x} \text{ is undefined.}$$

27. $t = \dfrac{3\pi}{4}$ corresponds to $\left(-\dfrac{\sqrt{2}}{2}, \dfrac{\sqrt{2}}{2}\right)$.

$$\sin t = y = \dfrac{\sqrt{2}}{2} \qquad \csc t = \dfrac{1}{y} = \sqrt{2}$$

$$\cos t = x = -\dfrac{\sqrt{2}}{2} \qquad \sec t = \dfrac{1}{x} = -\sqrt{2}$$

$$\tan t = \dfrac{y}{x} = -1 \qquad \cot t = \dfrac{x}{y} = -1$$

29. $t = \dfrac{\pi}{2}$ corresponds to $(0, 1)$.

$$\sin t = y = 1 \qquad \csc t = \dfrac{1}{y} = 1$$

$$\cos t = x = 0 \qquad \sec t = \dfrac{1}{x} \text{ is undefined.}$$

$$\tan t = \dfrac{y}{x} \text{ is undefined.} \qquad \cot t = \dfrac{x}{y} = 0$$

31. $t = -\dfrac{4\pi}{3}$ corresponds to $\left(-\dfrac{1}{2}, \dfrac{\sqrt{3}}{2}\right)$.

$$\sin t = y = \dfrac{\sqrt{3}}{2} \qquad \csc t = \dfrac{1}{y} = \dfrac{2\sqrt{3}}{3}$$

$$\cos t = x = -\dfrac{1}{2} \qquad \sec t = \dfrac{1}{x} = -2$$

$$\tan t = \dfrac{y}{x} = -\sqrt{3} \qquad \cot t = \dfrac{x}{y} = -\dfrac{\sqrt{3}}{3}$$

33. $\sin 3\pi = \sin \pi = 0$

35. $\cos\dfrac{8\pi}{3} = \cos\dfrac{2\pi}{3} = -\dfrac{1}{2}$

37. $\cos\dfrac{19\pi}{6} = \cos\dfrac{7\pi}{6} = -\dfrac{\sqrt{3}}{2}$

39. $\sin\left(-\dfrac{9\pi}{4}\right) = \sin\left(-\dfrac{\pi}{4}\right) = -\dfrac{\sqrt{2}}{2}$

41. $\sin t = \dfrac{1}{3}$

    (a) $\sin(-t) = -\sin t = -\dfrac{1}{3}$

    (b) $\csc(-t) = -\csc t = -3$

43. $\cos(-t) = -\dfrac{7}{8}$

    (a) $\cos t = \cos(-t) = -\dfrac{7}{8}$

    (b) $\sec(-t) = \dfrac{1}{\cos(-t)} = -\dfrac{8}{7}$

45. $\sin t = \dfrac{4}{5}$

    (a) $\sin(\pi - t) = \sin t = \dfrac{4}{5}$

    (b) $\sin(t + \pi) = -\sin t = -\dfrac{4}{5}$

47. $\sin\dfrac{\pi}{4} \approx 0.7071$

49. $\cos(-3) \approx -0.9900$

51. $\cos(-1.7) \approx -0.1288$

53. $\csc 0.8 = \dfrac{1}{\sin 0.8} \approx 1.3940$

55. $\sec 22.8 = \dfrac{1}{\cos 22.8} \approx -1.4486$

57. (a) $\sin 5 \approx -1$

    (b) $\cos 2 \approx -0.4$

59. (a) $\sin t = 0.25$

       $t \approx 0.25$ or 2.89

    (b) $\cos t = -0.25$

       $t \approx 1.82$ or 4.46

61. Let $t = \pi/4$

    $\cos(2\pi/4) = 0$

    $2\cos(\pi/4) = \sqrt{2}$

    $\cos 2t \neq 2\cos t$

63. (a) The points have y-axis symmetry.

    (b) $\sin t_1 = \sin(\pi - t_1)$ since they have the same y-value.

    (c) $-\cos t_1 = \cos(\pi - t_1)$ since the x-values have the opposite signs.

65. $y(t) = \dfrac{1}{4}\cos 6t$

    (a) $y(0) = \dfrac{1}{4}\cos 0 = 0.2500$ feet

    (b) $y\left(\dfrac{1}{4}\right) = \dfrac{1}{4}\cos\dfrac{3}{2} \approx 0.0177$ feet

    (c) $y\left(\dfrac{1}{2}\right) = \dfrac{1}{4}\cos 3 \approx -0.2475$ feet

67. $I = 5e^{-2(0.7)}\sin(0.7) \approx 0.794$

69. Let $h(t) = f(t)g(t)$

          $= \sin t \cos t.$

    Then, $h(-t) = \sin(-t)\cos(-t)$

                $= -\sin t \cos t$

                $= -h(t).$

    Thus, $h(t)$ is odd.

71. $f(x) = \dfrac{1}{2}(3x - 2)$

         $y = \dfrac{1}{2}(3x - 2)$

         $x = \dfrac{1}{2}(3y - 2)$

       $2x = 3y - 2$

    $2x + 2 = 3y$

    $\dfrac{2}{3}(x + 1) = y$

    $f^{-1}(x) = \dfrac{2}{3}(x + 1)$

73. $f(x) = \sqrt{x^2 - 4}, \quad x \geq 2, \quad y \geq 0$

$$y = \sqrt{x^2 - 4}$$

$$x = \sqrt{y^2 - 4}$$

$$x^2 = y^2 - 4$$

$$x^2 + 4 = y^2$$

$$\sqrt{x^2 + 4} = y, \quad x \geq 0$$

$$f^{-1}(x) = \sqrt{x^2 + 4}, \quad x \geq 0$$

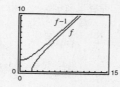

Note: The domain of $f^{-1}(x)$ equals the range of $f(x)$.

Section 4.3   Right Triangle Trigonometry

■ You should know the right triangle definition of trigonometric functions.

(a) $\sin \theta = \dfrac{\text{opp}}{\text{hyp}}$ 　　 (b) $\cos \theta = \dfrac{\text{adj}}{\text{hyp}}$ 　　 (c) $\tan \theta = \dfrac{\text{opp}}{\text{adj}}$

(d) $\csc \theta = \dfrac{\text{hyp}}{\text{opp}}$ 　　 (e) $\sec \theta = \dfrac{\text{hyp}}{\text{adj}}$ 　　 (f) $\cot \theta = \dfrac{\text{adj}}{\text{opp}}$

■ You should know the following identities.

(a) $\sin \theta = \dfrac{1}{\csc \theta}$ 　　 (b) $\csc \theta = \dfrac{1}{\sin \theta}$ 　　 (c) $\cos \theta = \dfrac{1}{\sec \theta}$

(d) $\sec \theta = \dfrac{1}{\cos \theta}$ 　　 (e) $\tan \theta = \dfrac{1}{\cot \theta}$ 　　 (f) $\cot \theta = \dfrac{1}{\tan \theta}$

(g) $\tan \theta = \dfrac{\sin \theta}{\cos \theta}$ 　　 (h) $\cot \theta = \dfrac{\cos \theta}{\sin \theta}$ 　　 (i) $\sin^2 \theta + \cos^2 \theta = 1$

(j) $1 + \tan^2 \theta = \sec^2 \theta$ 　　 (k) $1 + \cot^2 \theta = \csc^2 \theta$

■ You should know that two acute angles α and β are complementary if $\alpha + \beta = 90°$, and cofunctions of complementary angles are equal.

■ You should know the trigonometric function values of $30°$, $45°$, and $60°$, or be able to construct triangles from which you can determine them.

Solutions to Odd-Numbered Exercises

1.

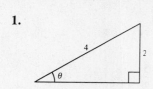

$\text{adj} = \sqrt{4^2 - 2^2} = \sqrt{12} = 2\sqrt{3}$

$\sin \theta = \dfrac{\text{opp}}{\text{hyp}} = \dfrac{2}{4} = \dfrac{1}{2}$ 　　　　 $\csc \theta = \dfrac{\text{hyp}}{\text{opp}} = \dfrac{4}{2} = 2$

$\cos \theta = \dfrac{\text{adj}}{\text{hyp}} = \dfrac{2\sqrt{3}}{4} = \dfrac{\sqrt{3}}{2}$ 　　　　 $\sec \theta = \dfrac{\text{hyp}}{\text{adj}} = \dfrac{4}{2\sqrt{3}} = \dfrac{2\sqrt{3}}{3}$

$\tan \theta = \dfrac{\text{opp}}{\text{adj}} = \dfrac{2}{2\sqrt{3}} = \dfrac{\sqrt{3}}{3}$ 　　　　 $\cot \theta = \dfrac{\text{adj}}{\text{opp}} = \dfrac{2\sqrt{3}}{2} = \sqrt{3}$

3.

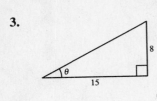

$$\text{hyp} = \sqrt{8^2 + 15^2} = 17$$

$$\sin \theta = \frac{\text{opp}}{\text{hyp}} = \frac{8}{17} \qquad\qquad \csc \theta = \frac{\text{hyp}}{\text{opp}} = \frac{17}{8}$$

$$\cos \theta = \frac{\text{adj}}{\text{hyp}} = \frac{15}{17} \qquad\qquad \sec \theta = \frac{\text{hyp}}{\text{adj}} = \frac{17}{15}$$

$$\tan \theta = \frac{\text{opp}}{\text{adj}} = \frac{8}{15} \qquad\qquad \cot \theta = \frac{\text{adj}}{\text{opp}} = \frac{15}{8}$$

5.

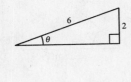

$$\text{adj} = \sqrt{3^2 - 1^2} = \sqrt{8} = 2\sqrt{2}$$

$$\sin \theta = \frac{\text{opp}}{\text{hyp}} = \frac{1}{3} \qquad\qquad \csc \theta = \frac{\text{hyp}}{\text{opp}} = 3$$

$$\cos \theta = \frac{\text{adj}}{\text{hyp}} = \frac{2\sqrt{2}}{3} \qquad\qquad \sec \theta = \frac{\text{hyp}}{\text{adj}} = \frac{3}{2\sqrt{2}} = \frac{3\sqrt{2}}{4}$$

$$\tan \theta = \frac{\text{opp}}{\text{adj}} = \frac{1}{2\sqrt{2}} = \frac{\sqrt{2}}{4} \qquad\qquad \cot \theta = \frac{\text{adj}}{\text{opp}} = 2\sqrt{2}$$

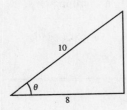

$$\text{adj} = \sqrt{6^2 - 2^2} = \sqrt{32} = 4\sqrt{2}$$

$$\sin \theta = \frac{\text{opp}}{\text{hyp}} = \frac{2}{6} = \frac{1}{3} \qquad\qquad \csc \theta = \frac{\text{hyp}}{\text{opp}} = \frac{6}{2} = 3$$

$$\cos \theta = \frac{\text{adj}}{\text{hyp}} = \frac{4\sqrt{2}}{6} = \frac{2\sqrt{2}}{3} \qquad\qquad \sec \theta = \frac{\text{hyp}}{\text{adj}} = \frac{6}{4\sqrt{2}} = \frac{3}{2\sqrt{2}} = \frac{3\sqrt{2}}{4}$$

$$\tan \theta = \frac{\text{opp}}{\text{adj}} = \frac{2}{4\sqrt{2}} = \frac{1}{2\sqrt{2}} = \frac{\sqrt{2}}{4} \qquad\qquad \cot \theta = \frac{\text{adj}}{\text{opp}} = \frac{4\sqrt{2}}{2} = 2\sqrt{2}$$

The function values are the same since the triangles are similar and the corresponding sides are proportional.

7.

$$\text{opp} = \sqrt{10^2 - 8^2} = 6$$

$$\sin \theta = \frac{\text{opp}}{\text{hyp}} = \frac{6}{10} = \frac{3}{5} \qquad\qquad \csc \theta = \frac{\text{hyp}}{\text{opp}} = \frac{10}{6} = \frac{5}{3}$$

$$\cos \theta = \frac{\text{adj}}{\text{hyp}} = \frac{8}{10} = \frac{4}{5} \qquad\qquad \sec \theta = \frac{\text{hyp}}{\text{adj}} = \frac{10}{8} = \frac{5}{4}$$

$$\tan \theta = \frac{\text{opp}}{\text{adj}} = \frac{6}{8} = \frac{3}{4} \qquad\qquad \cot \theta = \frac{\text{adj}}{\text{opp}} = \frac{8}{6} = \frac{4}{3}$$

$$\text{opp} = \sqrt{2.5^2 - 2^2} = 1.5$$

$$\sin \theta = \frac{\text{opp}}{\text{hyp}} = \frac{1.5}{2.5} = \frac{3}{5} \qquad\qquad \csc \theta = \frac{\text{hyp}}{\text{opp}} = \frac{2.5}{1.5} = \frac{5}{3}$$

$$\cos \theta = \frac{\text{adj}}{\text{hyp}} = \frac{2}{2.5} = \frac{4}{5} \qquad\qquad \sec \theta = \frac{\text{hyp}}{\text{adj}} = \frac{2.5}{2} = \frac{5}{4}$$

$$\tan \theta = \frac{\text{opp}}{\text{adj}} = \frac{1.5}{2} = \frac{3}{4} \qquad\qquad \cot \theta = \frac{\text{adj}}{\text{opp}} = \frac{2}{1.5} = \frac{4}{3}$$

The function values are the same since the triangles are similar and the corresponding sides are proportional.

9. Given: $\sin \theta = \dfrac{2}{3} = \dfrac{\text{opp}}{\text{hyp}}$

$$2^2 + (\text{adj})^2 = 3^2$$

$$\text{adj} = \sqrt{5}$$

$$\cos \theta = \frac{\sqrt{5}}{3}$$

$$\tan \theta = \frac{2\sqrt{5}}{5}$$

$$\cot \theta = \frac{\sqrt{5}}{2}$$

$$\sec \theta = \frac{3\sqrt{5}}{5}$$

$$\csc \theta = \frac{3}{2}$$

11. Given: $\sec \theta = 2 = \dfrac{2}{1} = \dfrac{\text{hyp}}{\text{adj}}$

$$(\text{opp})^2 + 1^2 = 2^2$$

$$\text{opp} = \sqrt{3}$$

$$\sin \theta = \frac{\sqrt{3}}{2}$$

$$\cos \theta = \frac{1}{2}$$

$$\tan \theta = \sqrt{3}$$

$$\cot \theta = \frac{\sqrt{3}}{3}$$

$$\csc \theta = \frac{2\sqrt{3}}{3}$$

13. Given: $\tan \theta = 3 = \dfrac{3}{1} = \dfrac{\text{opp}}{\text{adj}}$

$$3^2 + 1^2 = (\text{hyp})^2$$

$$\text{hyp} = \sqrt{10}$$

$$\sin \theta = \frac{3\sqrt{10}}{10}$$

$$\cos \theta = \frac{\sqrt{10}}{10}$$

$$\cot \theta = \frac{1}{3}$$

$$\sec \theta = \sqrt{10}$$

$$\csc \theta = \frac{\sqrt{10}}{3}$$

15. Given: $\cot \theta = \dfrac{3}{2} = \dfrac{\text{adj}}{\text{opp}}$

$$2^2 + 3^2 = (\text{hyp})^2$$

$$\text{hyp} = \sqrt{13}$$

$$\sin \theta = \frac{2}{\sqrt{13}} = \frac{2\sqrt{13}}{13}$$

$$\cos \theta = \frac{3}{\sqrt{13}} = \frac{3\sqrt{13}}{13}$$

$$\tan \theta = \frac{2}{3}$$

$$\csc \theta = \frac{\sqrt{13}}{2}$$

$$\sec \theta = \frac{\sqrt{13}}{3}$$

17. $\sin 60° = \dfrac{\sqrt{3}}{2}$, $\cos 60° = \dfrac{1}{2}$

(a) $\tan 60° = \dfrac{\sin 60°}{\cos 60°} = \sqrt{3}$

(b) $\sin 30° = \cos 60° = \dfrac{1}{2}$

(c) $\cos 30° = \sin 60° = \dfrac{\sqrt{3}}{2}$

(d) $\cot 60° = \dfrac{\cos 60°}{\sin 60°} = \dfrac{1}{\sqrt{3}} = \dfrac{\sqrt{3}}{3}$

19. $\csc \theta = 3$, $\sec \theta = \dfrac{3\sqrt{2}}{4}$

(a) $\sin \theta = \dfrac{1}{\csc \theta} = \dfrac{1}{3}$

(b) $\cos \theta = \dfrac{1}{\sec \theta} = \dfrac{2\sqrt{2}}{3}$

(c) $\tan \theta = \dfrac{\sin \theta}{\cos \theta} = \dfrac{1/3}{(2\sqrt{2})/3} = \dfrac{\sqrt{2}}{4}$

(d) $\sec(90° - \theta) = \csc \theta = 3$

21. $\cos \alpha = \dfrac{1}{4}$

   (a) $\sec \alpha = \dfrac{1}{\cos \alpha} = 4$

   (b) $\sin^2 \alpha + \cos^2 \alpha = 1$

$$\sin^2 \alpha + \left(\dfrac{1}{4}\right)^2 = 1$$

$$\sin^2 \alpha = \dfrac{15}{16}$$

$$\sin \alpha = \pm \dfrac{\sqrt{15}}{4}$$

   (c) $\cot \alpha = \dfrac{\cos \alpha}{\sin \alpha} = \pm \dfrac{1/4}{\sqrt{15}/4} = \pm \dfrac{1}{\sqrt{15}} = \pm \dfrac{\sqrt{15}}{15}$

   (d) $\sin(90° - \alpha) = \cos \alpha = \dfrac{1}{4}$

23. $\tan \theta \cot \theta = \tan \theta \left(\dfrac{1}{\tan \theta}\right) = 1$

25. $\tan \alpha \cos \alpha = \left(\dfrac{\sin \alpha}{\cos \alpha}\right) \cos \alpha = \sin \alpha$

27. $(1 + \cos \theta)(1 - \cos \theta) = 1 - \cos^2 \theta$
$$= (\sin^2 \theta + \cos^2 \theta) - \cos^2 \theta$$
$$= \sin^2 \theta$$

29. $(\sec \theta + \tan \theta)(\sec \theta - \tan \theta) = \sec^2 \theta - \tan^2 \theta$
$$= (1 + \tan^2 \theta) - \tan^2 \theta$$
$$= 1$$

31. $\dfrac{\sin \theta}{\cos \theta} + \dfrac{\cos \theta}{\sin \theta} = \dfrac{\sin^2 \theta + \cos^2 \theta}{\sin \theta \cos \theta}$

$$= \dfrac{1}{\sin \theta \cos \theta}$$

$$= \dfrac{1}{\sin \theta} \cdot \dfrac{1}{\cos \theta}$$

$$= \csc \theta \sec \theta$$

33. (a) $\cos 60° = \dfrac{1}{2}$

   (b) $\tan \dfrac{\pi}{6} = \dfrac{1}{\sqrt{3}} = \dfrac{\sqrt{3}}{3}$

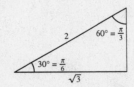

35. (a) $\cot 45° = 1$

   (b) $\cos 45° = \dfrac{1}{\sqrt{2}} = \dfrac{\sqrt{2}}{2}$

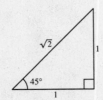

37. (a) $\sin 10° \approx 0.1736$

   (b) $\cos 80° \approx 0.1736$

   Note: $\cos 80° = \sin(90° - 80°) = \sin 10°$

39. (a) $\sin 16.35° \approx 0.2815$

(b) $\csc 16.35° = \dfrac{1}{\sin 16.35°} \approx 3.5523$

41. (a) $\sec 42° \, 12' = \sec 42.2° = \dfrac{1}{\cos 42.2°} \approx 1.3499$

(b) $\csc 48° \, 7' = \dfrac{1}{\sin\left(48 + \frac{7}{60}\right)°} \approx 1.3432$

43. Make sure that your calculator is in radian mode.

(a) $\cot \dfrac{\pi}{16} = \dfrac{1}{\tan(\pi/16)} \approx 5.0273$

(b) $\tan \dfrac{\pi}{16} \approx 0.1989$

45. Make sure that your calculator is in radian mode.

(a) $\csc 1 = \dfrac{1}{\sin 1} \approx 1.1884$

(b) $\tan \dfrac{1}{2} \approx 0.5463$

47. (a) $\sin \theta = \dfrac{1}{2} \implies \theta = 30° = \dfrac{\pi}{6}$

(b) $\csc \theta = 2 \implies \theta = 30° = \dfrac{\pi}{6}$

49. (a) $\sec \theta = 2 \implies \theta = 60° = \dfrac{\pi}{3}$

(b) $\cot \theta = 1 \implies \theta = 45° = \dfrac{\pi}{4}$

51. (a) $\csc \theta = \dfrac{2\sqrt{3}}{3} \implies \theta = 60° = \dfrac{\pi}{3}$

(b) $\sin \theta = \dfrac{\sqrt{2}}{2} \implies \theta = 45° = \dfrac{\pi}{4}$

53. (a) $\sin \theta = 0.8191 \implies \theta \approx 55° \approx 0.960$ radian

(b) $\cos \theta = 0.0175 \implies \theta \approx 89° \approx 1.553$ radians

55. (a) $\tan \theta = 1.1920 \implies \theta \approx 50° \approx 0.873$ radian

(b) $\tan \theta = 0.4663 \implies \theta \approx 25° \approx 0.436$ radian

57. $\tan 30° = \dfrac{y}{75}$

$\dfrac{\sqrt{3}}{3} = \dfrac{y}{75}$

$75\left(\dfrac{\sqrt{3}}{3}\right) = y$

$25\sqrt{3} = y$

59. $\cot 60° = \dfrac{x}{32}$

$\dfrac{\sqrt{3}}{3} = \dfrac{x}{32}$

$\dfrac{32\sqrt{3}}{3} = x$

61. $\sin 40° = \dfrac{15}{r}$

$r = \dfrac{15}{\sin 40°} \approx 23.3$

63. $\sin 50° = \dfrac{y}{8}$

$y = 8 \sin 50° \approx 6.1$

65. $\dfrac{h}{23} = \dfrac{6}{8}$

$h = \dfrac{138}{8} = \dfrac{69}{4} = 17\frac{1}{4}$ ft

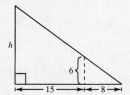

67. (a)

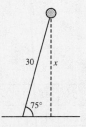

(b) $\sin 75° = \dfrac{x}{30}$

(c) $x = 30 \sin 75° \approx 28.98$ meters

69. Let $x =$ distance from the boat to the shoreline

$$\tan 3° = \frac{60}{x}$$

$$x = \frac{60}{\tan 3°} \approx 1144.87 \text{ feet}$$

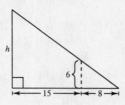

71.

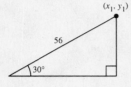

$$\sin 30° = \frac{y_1}{56}$$

$$y_1 = (\sin 30°)(56) = \left(\frac{1}{2}\right)(56) = 28$$

$$\cos 30° = \frac{x_1}{56}$$

$$x_1 = \cos 30°(56) = \frac{\sqrt{3}}{2}(56) = 28\sqrt{3}$$

$$(x_1, y_1) = \left(28\sqrt{3}, 28\right)$$

$$\sin 60° = \frac{y_2}{56}$$

$$y_2 = \sin 60°(56) = \left(\frac{\sqrt{3}}{2}\right)(56) = 28\sqrt{3}$$

$$\cos 60° = \frac{x_2}{56}$$

$$x_2 = (\cos 60°)(56) = \left(\frac{1}{2}\right)(56) = 28$$

$$(x_2, y_2) = \left(28, 28\sqrt{3}\right)$$

73. $x \approx 9.397,\ y \approx 3.420$

$$\sin \theta = \frac{y}{10} \approx 0.34$$

$$\cos \theta = \frac{x}{10} \approx 0.94$$

$$\tan \theta = \frac{y}{x} \approx 0.36$$

$$\cot \theta = \frac{x}{y} \approx 2.75$$

$$\sec \theta = \frac{10}{x} \approx 1.06$$

$$\csc \theta = \frac{10}{y} \approx 2.92$$

75. (a)

θ	0	0.1	0.2	0.3	0.4	0.5
$\sin \theta$	0	0.0998	0.1987	0.2955	0.3894	0.4794

(b) In the interval $(0, 0.5]$, $\theta > \sin \theta$.

(c) As $\theta \to 0$, $\sin \theta \to 0$.

77. (a)

θ	0°	20°	40°	60°	80°
$\sin \theta$	0	0.3420	0.6428	0.8660	0.9848
$\cos \theta$	1	0.9397	0.7660	0.5000	0.1736
$\tan \theta$	0	0.3640	0.8391	1.7321	5.6713

(b) Sine and tangent are increasing; cosine is decreasing.

(c) In each case, $\tan \theta = \dfrac{\sin \theta}{\cos \theta}$.

79. (a)

θ	20°	30°	40°	50°	60°	70°
R	170.3	229.5	261.0	261.0	229.5	170.3

(b) The range increases as θ increases from 20° through 40°, and decreases as θ increases from 50° through 70°.

(c) The maximum range should be halfway between 40° and 50°: 45°. You can verify this by analyzing tabular values near 45°.

(d)

θ	43°	44°	45°	46°	47°
R	264.4	264.8	265.0	264.8	264.4

81. True, $\csc x = \dfrac{1}{\sin x} \implies \sin 60° \csc 60° = \sin 60°\left(\dfrac{1}{\sin 60°}\right) = 1$

83. False, $\sin 45° + \cos 45° = \dfrac{\sqrt{2}}{2} + \dfrac{\sqrt{2}}{2} = \sqrt{2} \neq 1$

85. False, $\dfrac{\sin 60°}{\sin 30°} = \dfrac{\cos 30°}{\sin 30°} = \cot 30° \approx 1.7321 \neq \sin 2° = 0.0349$

Section 4.4   Trigonometric Functions of Any Angle

- Know the Definitions of Trigonometric Functions of Any Angle.

  If θ is in standard position, (x, y) a point on the terminal side and $r = \sqrt{x^2 + y^2} \neq 0$, then

  $\sin \theta = \dfrac{y}{r}$ $\csc \theta = \dfrac{r}{y}, \ y \neq 0$

  $\cos \theta = \dfrac{x}{r}$ $\sec \theta = \dfrac{r}{x}, \ x \neq 0$

  $\tan \theta = \dfrac{y}{x}, \ x \neq 0$ $\cot \theta = \dfrac{x}{y}, \ y \neq 0.$

- You should know the signs of the trigonometric functions in each quadrant.
- You should know the trigonometric function values of the quadrant angles 0, $\dfrac{\pi}{2}$, π, and $\dfrac{3\pi}{2}$.
- You should be able to find reference angles.
- You should be able to evaluate trigonometric functions of any angle. (Use reference angles.)
- You should know that the period of sine and cosine is 2π.

Solutions to Odd-Numbered Exercises

1. (a) $(x, y) = (4, 3)$

$r = \sqrt{16 + 9} = 5$

$\sin \theta = \dfrac{y}{r} = \dfrac{3}{5}$ $\csc \theta = \dfrac{r}{y} = \dfrac{5}{3}$

$\cos \theta = \dfrac{x}{r} = \dfrac{4}{5}$ $\sec \theta = \dfrac{r}{x} = \dfrac{5}{4}$

$\tan \theta = \dfrac{y}{x} = \dfrac{3}{4}$ $\cot \theta = \dfrac{x}{y} = \dfrac{4}{3}$

(b) $(x, y) = (-8, -15)$

$r = \sqrt{64 + 225} = 17$

$\sin \theta = \dfrac{y}{r} = -\dfrac{15}{17}$ $\csc \theta = \dfrac{r}{y} = -\dfrac{17}{15}$

$\cos \theta = \dfrac{x}{r} = -\dfrac{8}{17}$ $\sec \theta = \dfrac{r}{x} = -\dfrac{17}{8}$

$\tan \theta = \dfrac{y}{x} = \dfrac{15}{8}$ $\cot \theta = \dfrac{x}{y} = \dfrac{8}{15}$

3. (a) $(x, y) = \left(-\sqrt{3}, -1\right)$

$r = \sqrt{3 + 1} = 2$

$\sin \theta = \dfrac{y}{r} = -\dfrac{1}{2}$ $\qquad \csc \theta = \dfrac{r}{y} = -2$

$\cos \theta = \dfrac{x}{r} = -\dfrac{\sqrt{3}}{2}$ $\qquad \sec \theta = \dfrac{r}{x} = -\dfrac{2\sqrt{3}}{3}$

$\tan \theta = \dfrac{y}{x} = \dfrac{\sqrt{3}}{3}$ $\qquad \cot \theta = \dfrac{x}{y} = \sqrt{3}$

(b) $(x, y) = (-2, 2)$

$r = \sqrt{4 + 4} = 2\sqrt{2}$

$\sin \theta = \dfrac{y}{r} = \dfrac{\sqrt{2}}{2}$ $\qquad \csc \theta = \dfrac{r}{y} = \sqrt{2}$

$\cos \theta = \dfrac{x}{r} = -\dfrac{\sqrt{2}}{2}$ $\qquad \sec \theta = \dfrac{r}{x} = -\sqrt{2}$

$\tan \theta = \dfrac{y}{x} = -1$ $\qquad \cot \theta = \dfrac{x}{y} = -1$

5. (a) $(x, y) = (7, 24)$

$r = \sqrt{49 + 576} = 25$

$\sin \theta = \dfrac{y}{r} = \dfrac{24}{25}$ $\qquad \csc \theta = \dfrac{r}{y} = \dfrac{25}{24}$

$\cos \theta = \dfrac{x}{r} = \dfrac{7}{25}$ $\qquad \sec \theta = \dfrac{r}{x} = \dfrac{25}{7}$

$\tan \theta = \dfrac{y}{x} = \dfrac{24}{7}$ $\qquad \cot \theta = \dfrac{x}{y} = \dfrac{7}{24}$

(b) $(x, y) = (7, -24)$

$r = \sqrt{49 + 576} = 25$

$\sin \theta = \dfrac{y}{r} = -\dfrac{24}{25}$ $\qquad \csc \theta = \dfrac{r}{y} = -\dfrac{25}{24}$

$\cos \theta = \dfrac{x}{r} = \dfrac{7}{25}$ $\qquad \sec \theta = \dfrac{r}{x} = \dfrac{25}{7}$

$\tan \theta = \dfrac{y}{x} = -\dfrac{24}{7}$ $\qquad \cot \theta = \dfrac{x}{y} = -\dfrac{7}{24}$

7. (a) $(x, y) = (-4, 10)$

$r = \sqrt{16 + 100} = 2\sqrt{29}$

$\sin \theta = \dfrac{y}{r} = \dfrac{5\sqrt{29}}{29}$ $\qquad \csc \theta = \dfrac{r}{y} = \dfrac{\sqrt{29}}{5}$

$\cos \theta = \dfrac{x}{r} = -\dfrac{2\sqrt{29}}{29}$ $\qquad \sec \theta = \dfrac{r}{x} = -\dfrac{\sqrt{29}}{2}$

$\tan \theta = \dfrac{y}{x} = -\dfrac{5}{2}$ $\qquad \cot \theta = \dfrac{x}{y} = -\dfrac{2}{5}$

(b) $(x, y) = (3, -5)$

$r = \sqrt{9 + 25} = \sqrt{34}$

$\sin \theta = \dfrac{y}{r} = -\dfrac{5\sqrt{34}}{34}$ $\qquad \csc \theta = \dfrac{r}{y} = -\dfrac{\sqrt{34}}{5}$

$\cos \theta = \dfrac{x}{r} = \dfrac{3\sqrt{34}}{34}$ $\qquad \sec \theta = \dfrac{r}{x} = \dfrac{\sqrt{34}}{3}$

$\tan \theta = \dfrac{y}{x} = -\dfrac{5}{3}$ $\qquad \cot \theta = \dfrac{x}{y} = -\dfrac{3}{5}$

9. (a) $\sin \theta < 0 \implies \theta$ lies in Quadrant III or in Quadrant IV.

$\cos \theta < 0 \implies \theta$ lies in Quadrant II or in Quadrant III.

$\sin \theta < 0$ *and* $\cos \theta < 0 \implies \theta$ lies in Quadrant III.

(b) $\sin \theta > 0 \implies \theta$ lies in Quadrant I or in Quadrant II.

$\cos \theta < \theta \implies \theta$ lies in Quadrant II or in Quadrant III.

$\sin \theta > 0$ *and* $\cos \theta < 0 \implies \theta$ lies in Quadrant II.

11. (a) $\sin \theta > 0 \implies \theta$ lies in Quadrant I or in Quadrant II.

$\tan \theta < 0 \implies \theta$ lies in Quadrant II or in Quadrant IV.

$\sin \theta > 0$ *and* $\tan \theta < 0 \implies \theta$ lies in Quadrant II.

(b) $\cos \theta > 0 \implies \theta$ lies in Quadrant I or in Quadrant IV.

$\tan \theta < 0 \implies \theta$ lies in Quadrant II or in Quadrant IV.

$\cos \theta > 0$ *and* $\tan \theta < 0 \implies \theta$ lies in Quadrant IV.

13. $\sin \theta = \dfrac{y}{r} = \dfrac{3}{5} \implies x^2 = 25 - 9 = 16$

θ in Quadrant II $\implies x = -4$

$\sin \theta = \dfrac{y}{r} = \dfrac{3}{5}$ $\qquad$ $\csc \theta = \dfrac{r}{y} = \dfrac{5}{3}$

$\cos \theta = \dfrac{x}{r} = -\dfrac{4}{5}$ $\qquad$ $\sec \theta = \dfrac{r}{x} = -\dfrac{5}{4}$

$\tan \theta = \dfrac{y}{x} = -\dfrac{3}{4}$ $\qquad$ $\cot \theta = \dfrac{x}{y} = -\dfrac{4}{3}$

15. $\sin \theta < 0 \implies y < 0$

$\tan \theta = \dfrac{y}{x} = \dfrac{-15}{8} \implies r = 17$

$\sin \theta = \dfrac{y}{r} = -\dfrac{15}{17}$ $\qquad$ $\csc \theta = \dfrac{r}{y} = -\dfrac{17}{15}$

$\cos \theta = \dfrac{x}{r} = \dfrac{8}{17}$ $\qquad$ $\sec \theta = \dfrac{r}{x} = \dfrac{17}{8}$

$\tan \theta = \dfrac{y}{x} = -\dfrac{15}{8}$ $\qquad$ $\cot \theta = \dfrac{x}{y} = -\dfrac{8}{15}$

17. $\cot \theta = \dfrac{x}{y} = -\dfrac{3}{1} = \dfrac{3}{-1}$

$\cos \theta > 0 \implies x$ is positive; $x = 3, y = -1, r = \sqrt{10}$

$\sin \theta = \dfrac{y}{r} = -\dfrac{\sqrt{10}}{10}$ $\qquad$ $\csc \theta = \dfrac{r}{y} = -\sqrt{10}$

$\cos \theta = \dfrac{x}{r} = \dfrac{3\sqrt{10}}{10}$ $\qquad$ $\sec \theta = \dfrac{r}{x} = \dfrac{\sqrt{10}}{3}$

$\tan \theta = \dfrac{y}{x} = -\dfrac{1}{3}$ $\qquad$ $\cot \theta = \dfrac{x}{y} = -3$

19. $\sec \theta = \dfrac{r}{x} = \dfrac{2}{-1} \implies y^2 = 4 - 1 = 3$

$\sin \theta > 0 \implies y = \sqrt{3}$

$\sin \theta = \dfrac{y}{r} = \dfrac{\sqrt{3}}{2}$ $\qquad$ $\csc \theta = \dfrac{r}{y} = \dfrac{2\sqrt{3}}{3}$

$\cos \theta = \dfrac{x}{r} = -\dfrac{1}{2}$ $\qquad$ $\sec \theta = \dfrac{r}{x} = -2$

$\tan \theta = \dfrac{y}{x} = -\sqrt{3}$ $\qquad$ $\cot \theta = \dfrac{x}{y} = -\dfrac{\sqrt{3}}{3}$

21. $\sin \theta = 0 \implies \theta = n\pi$

$\sec \theta = -1 \implies \theta = \pi$

$\sin \theta = \dfrac{y}{r} = \dfrac{0}{r} = 0$ $\qquad$ $\csc \theta = \dfrac{r}{y}$ is undefined.

$\cos \theta = \dfrac{x}{r} = \dfrac{-r}{r} = -1$ $\qquad$ $\sec \theta = \dfrac{r}{x} = -1$

$\tan \theta = \dfrac{y}{x} = \dfrac{0}{x} = 0$ $\qquad$ $\cot \theta = \dfrac{x}{y}$ is undefined.

23. To find a point on the terminal side of θ, use any point on the line $y = -x$ that lies in Quadrant II. $(-1, 1)$ is one such point.

$x = -1, y = 1, r = \sqrt{2}$

$\sin \theta = \dfrac{1}{\sqrt{2}} = \dfrac{\sqrt{2}}{2}$ $\qquad$ $\csc \theta = \sqrt{2}$

$\cos \theta = -\dfrac{1}{\sqrt{2}} = -\dfrac{\sqrt{2}}{2}$ $\qquad$ $\sec \theta = -\sqrt{2}$

$\tan \theta = -1$ $\qquad$ $\cot \theta = -1$

25. To find a point on the terminal side of θ, use any point on the line $y = 2x$ that lies in Quadrant III. $(-1, -2)$ is one such point.

$x = -1, y = -2, r = \sqrt{5}$

$\sin \theta = -\dfrac{2}{\sqrt{5}} = -\dfrac{2\sqrt{5}}{5}$ $\qquad$ $\csc \theta = \dfrac{\sqrt{5}}{-2} = -\dfrac{\sqrt{5}}{2}$

$\cos \theta = -\dfrac{1}{\sqrt{5}} = -\dfrac{\sqrt{5}}{5}$ $\qquad$ $\sec \theta = \dfrac{\sqrt{5}}{-1} = -\sqrt{5}$

$\tan \theta = \dfrac{-2}{-1} = 2$ $\qquad$ $\cot \theta = \dfrac{-1}{-2} = \dfrac{1}{2}$

27. $(x, y) = (-1, 0)$

$\cos \pi = \dfrac{x}{r} = \dfrac{-1}{1} = -1$

29. $(x, y) = (-1, 0)$

$\sec \pi = \dfrac{r}{x} = \dfrac{1}{-1} = -1$

31. $(x, y) = (0, 1)$

$\tan \dfrac{\pi}{2} = \dfrac{y}{x} = \dfrac{1}{0}$ undefined

33. $(x, y) = (0, 1)$

$\cot \dfrac{\pi}{2} = \dfrac{x}{y} = \dfrac{0}{1} = 0$

35. (a) $\theta = 203°$

$\theta' = 203° - 180° = 23°$

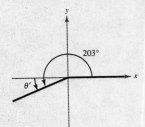

(b) $\theta = 127°$

$\theta' = 180° - 127° = 53°$

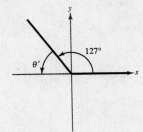

37. (a) $\theta = -245°$

$360° - 245° = 115°$ (coterminal angle)

$\theta' = 180° - 115° = 65°$

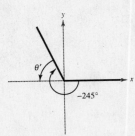

(b) $\theta = -72°$

$\theta' = 72°$

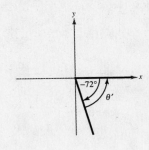

39. (a) $\theta = \dfrac{2\pi}{3}$

$\theta' = \pi - \dfrac{2\pi}{3} = \dfrac{\pi}{3}$

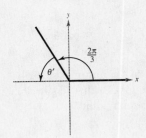

(b) $\theta = \dfrac{7\pi}{6}$

$\theta' = \dfrac{7\pi}{6} - \pi = \dfrac{\pi}{6}$

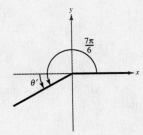

41. (a) $\theta = 3.5$

$\theta' = 3.5 - \pi$

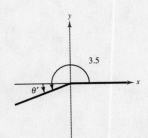

(b) $\theta = 5.8$

$\theta' = 2\pi - 5.8$

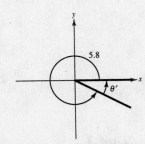

43. (a) $\theta' = 45°$, Quadrant III

$$\sin 225° = -\sin 45° = -\frac{\sqrt{2}}{2}$$

$$\cos 225° = -\cos 45° = -\frac{\sqrt{2}}{2}$$

$$\tan 225° = \tan 45° = 1$$

(b) $\theta' = 45°$, Quadrant II

$$\sin(-225°) = \sin 45° = \frac{\sqrt{2}}{2}$$

$$\cos(-225°) = -\cos 45° = -\frac{\sqrt{2}}{2}$$

$$\tan(-225°) = -\tan 45° = -1$$

45. (a) $\theta' = 30°$, Quadrant I

$$\sin 750° = \sin 30° = \frac{1}{2}$$

$$\cos 750° = \cos 30° = \frac{\sqrt{3}}{2}$$

$$\tan 750° = \tan 30° = \frac{\sqrt{3}}{3}$$

(b) $\theta' = 30°$, Quadrant II

$$\sin 510° = \sin 30° = \frac{1}{2}$$

$$\cos 510° = -\cos 30° = -\frac{\sqrt{3}}{2}$$

$$\tan 510° = -\tan 30° = -\frac{\sqrt{3}}{3}$$

47. (a) $\theta' = \frac{\pi}{3}$, Quadrant III

$$\sin \frac{4\pi}{3} = -\sin \frac{\pi}{3} = -\frac{\sqrt{3}}{2}$$

$$\cos \frac{4\pi}{3} = -\cos \frac{\pi}{3} = -\frac{1}{2}$$

$$\tan \frac{4\pi}{3} = \tan \frac{\pi}{3} = \sqrt{3}$$

(b) $\theta' = \frac{\pi}{3}$, Quadrant II

$$\sin \frac{2\pi}{3} = \sin \frac{\pi}{3} = \frac{\sqrt{3}}{2}$$

$$\cos \frac{2\pi}{3} = -\cos \frac{\pi}{3} = -\frac{1}{2}$$

$$\tan \frac{2\pi}{3} = -\tan \frac{\pi}{3} = -\sqrt{3}$$

49. (a) $\theta' = \frac{\pi}{6}$, Quadrant IV

$$\sin\left(-\frac{\pi}{6}\right) = -\sin \frac{\pi}{6} = -\frac{1}{2}$$

$$\cos\left(-\frac{\pi}{6}\right) = \cos \frac{\pi}{6} = \frac{\sqrt{3}}{2}$$

$$\tan\left(-\frac{\pi}{6}\right) = -\tan \frac{\pi}{6} = -\frac{\sqrt{3}}{3}$$

(b) $\theta' = \frac{\pi}{6}$, Quadrant II

$$\sin \frac{5\pi}{6} = \sin \frac{\pi}{6} = \frac{1}{2}$$

$$\cos \frac{5\pi}{6} = -\cos \frac{\pi}{6} = -\frac{\sqrt{3}}{2}$$

$$\tan \frac{5\pi}{6} = -\tan \frac{\pi}{6} = -\frac{\sqrt{3}}{3}$$

51. (a) $\theta' = \frac{\pi}{4}$, Quadrant II

$$\sin \frac{11\pi}{4} = \sin \frac{\pi}{4} = \frac{\sqrt{2}}{2}$$

$$\cos \frac{11\pi}{4} = -\cos \frac{\pi}{4} = -\frac{\sqrt{2}}{2}$$

$$\tan \frac{11\pi}{4} = -\tan \frac{\pi}{4} = -1$$

(b) $\theta' = \frac{\pi}{6}$, Quadrant IV

$$\sin\left(-\frac{13\pi}{6}\right) = -\sin \frac{\pi}{6} = -\frac{1}{2}$$

$$\cos\left(-\frac{13\pi}{6}\right) = \cos \frac{\pi}{6} = \frac{\sqrt{3}}{2}$$

$$\tan\left(-\frac{13\pi}{6}\right) = -\tan \frac{\pi}{6} = -\frac{\sqrt{3}}{3}$$

53. (a) $\sin 10° \approx 0.1736$

(b) $\csc 10° = \dfrac{1}{\sin 10°} \approx 5.7588$

55. (a) $\cos(-110°) \approx -0.3420$

(b) $\cos 250° \approx -0.3420$

57. (a) $\tan 240° \approx 1.7321$ (b) $\cot 210° = \dfrac{1}{\tan 210°} \approx 1.7321$

59. (a) $\tan \dfrac{\pi}{9} \approx 0.3640$ (b) $\tan \dfrac{10\pi}{9} \approx 0.3640$

61. (a) $\sin 0.65 \approx 0.6052$ (b) $\sin(-5.63) \approx 0.6077$

63. (a) $\sin \theta = \dfrac{1}{2} \implies$ reference angle is $30°$ or $\dfrac{\pi}{6}$ and θ is in Quadrant I or Quadrant II.

Values in degrees: $30°, 150°$

Values in radian: $\dfrac{\pi}{6}, \dfrac{5\pi}{6}$

(b) $\sin \theta = -\dfrac{1}{2} \implies$ reference angle is $30°$ or $\dfrac{\pi}{6}$ and θ is in Quadrant III or Quadrant IV.

Values in degrees: $210°, 330°$

Values in radians: $\dfrac{7\pi}{6}, \dfrac{11\pi}{6}$

65. (a) $\csc \theta = \dfrac{2\sqrt{3}}{3} \implies$ reference angle is $60°$ or $\dfrac{\pi}{3}$ and θ is in Quadrant I or Quadrant II.

Values in degrees: $60°, 120°$

Values in radians: $\dfrac{\pi}{3}, \dfrac{2\pi}{3}$

(b) $\cot \theta = -1 \implies$ reference angle is $45°$ or $\dfrac{\pi}{4}$ and θ is in Quadrant II or Quadrant IV.

Values in degrees: $135°, 315°$

Values in radians: $\dfrac{3\pi}{4}, \dfrac{7\pi}{4}$

67. (a) $\tan \theta = 1 \implies$ reference angle is $45°$ or $\dfrac{\pi}{4}$ and θ is in Quadrant I or Quadrant III.

Values in degrees: $45°, 225°$

Values in radians: $\dfrac{\pi}{4}, \dfrac{5\pi}{4}$

(b) $\cot \theta = -\sqrt{3} \implies$ reference angle is $30°$ or $\dfrac{\pi}{6}$ and θ is in Quadrant II or Quadrant IV.

Values in degrees: $150°, 330°$

Values in radians: $\dfrac{5\pi}{6}, \dfrac{11\pi}{6}$

69. (a) $\sin \theta = 0.8191 \implies \theta' \approx 54.99°$

Quadrant I: $\theta = \sin^{-1} 0.8191 \approx 54.99°$

Quadrant II: $\theta = 180° - \sin^{-1} 0.8191 \approx 125.01°$

(b) $\theta' = \sin^{-1} 0.2589 \approx 15.00°$

Quadrant III: $\theta = 180° + 15° = 195°$

Quadrant IV: $\theta = 360° - 15° = 345°$

71. (a) $\cos \theta = 0.9848 \implies \theta' \approx 0.175$

Quadrant I: $\theta = \cos^{-1}(0.9848) \approx 0.175$

Quadrant IV: $\theta = 2\pi - \theta' \approx 6.109$

(b) $\theta' = \cos^{-1} 0.5890 \approx 0.941$

Quadrant II: $\theta = \pi - 0.941 \approx 2.201$

Quadrant III: $\theta = \pi + 0.941 \approx 4.083$

73. (a) $\tan \theta = 1.192 \implies \theta' \approx 0.873$

Quadrant I: $\theta = \tan^{-1} 1.192 \approx 0.873$

Quadrant III: $\theta = \pi + \theta' \approx 4.014$

(b) $\theta' = \tan^{-1} 8.144 \approx 1.4486$

Quadrant II: $\theta = \pi - 1.4486 \approx 1.693$

Quadrant IV: $\theta = 2\pi - 1.4486 \approx 4.835$

75.
$$\sin \theta = -\frac{3}{5}$$

$$\sin^2 \theta + \cos^2 \theta = 1$$

$$\cos^2 \theta = 1 - \sin^2 \theta$$

$$\cos^2 \theta = 1 - \left(-\frac{3}{5}\right)^2$$

$$\cos^2 \theta = 1 - \frac{9}{25}$$

$$\cos^2 \theta = \frac{16}{25}$$

$\cos \theta > 0$ in Quadrant IV.

$$\cos \theta = \frac{4}{5}$$

77. $\tan \theta = \frac{3}{2}$

$$\sec^2 \theta = 1 + \tan^2 \theta$$

$$\sec^2 \theta = 1 + \left(\frac{3}{2}\right)^2$$

$$\sec^2 \theta = 1 + \frac{9}{4}$$

$$\sec^2 \theta = \frac{13}{4}$$

$\sec \theta < 0$ in Quadrant III.

$$\sec \theta = -\frac{\sqrt{13}}{2}$$

79. $\cos \theta = \frac{5}{8}$

$$\cos \theta = \frac{1}{\sec \theta} \implies \sec \theta = \frac{1}{\cos \theta}$$

$$\sec \theta = \frac{1}{5/8} = \frac{8}{5}$$

81. (a) $t = 1$

$$T = 45 - 23 \cos \left[\frac{2\pi}{365}(1 - 32)\right] \approx 25.2° \text{ F}$$

(b) $t = 185$

$$T = 45 - 23 \cos \left[\frac{2\pi}{365}(185 - 32)\right] \approx 65.1° \text{ F}$$

(c) $t = 291$

$$T = 45 - 23 \cos \left[\frac{2\pi}{365}(291 - 32)\right] \approx 50.8° \text{ F}$$

83. $\sin \theta = \frac{6}{d} \implies d = \frac{6}{\sin \theta}$

(a) $d = \dfrac{6}{\sin 30°} = 12$ miles

(b) $d = \dfrac{6}{\sin 90°} = 6$ miles

(c) $d = \dfrac{6}{\sin 120°} \approx 6.9$ miles

85. Selecting the point $(-0.8, 0.6)$ on the terminal side, you obtain $\cos \theta \approx -0.8$.

87. $y = 2^{x-1}$

x	-1	0	1	2	3
y	$\frac{1}{4}$	$\frac{1}{2}$	1	2	4

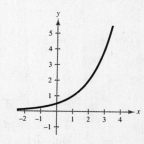

89. $y = \ln(x - 1)$

Domain: $x - 1 > 0 \Rightarrow x > 1$

x	1.1	1.5	2	3	4
y	-2.30	-0.69	0	0.69	1.10

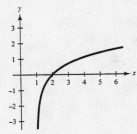

Section 4.5   Graphs of Sine and Cosine Functions

- You should be able to graph $y = a \sin(bx - c)$ and $y = a \cos(bx - c)$.

- Amplitude: $|a|$

- Period: $\dfrac{2\pi}{|b|}$

- Shift: Solve $bx - c = 0$ and $bx - c = 2\pi$.

- Key increments: $\dfrac{1}{4}$ (period)

Solutions to Odd-Numbered Exercises

1. $y = 3 \sin 2x$

Period: $\dfrac{2\pi}{2} = \pi$

Amplitude: $|3| = 3$

```
Xmin = -2π
Xmax = 2π
Xscl = π/2
Ymin = -4
Ymax = 4
Yscl = 1
```

3. $y = \dfrac{5}{2} \cos \dfrac{x}{2}$

Period: $\dfrac{2\pi}{1/2} = 4\pi$

Amplitude $\left|\dfrac{5}{2}\right| = \dfrac{5}{2}$

```
Xmin = -4π
Xmax = 4π
Xscl = π
Ymin = -3
Ymax = 3
Yscl = 1
```

5. $y = \dfrac{2}{3} \sin \pi x$

Period: $\dfrac{2\pi}{\pi} = 2$

Amplitude: $\left|\dfrac{2}{3}\right| = \dfrac{2}{3}$

```
Xmin = -π
Xmax = π
Xscl = π/2
Ymin = -1
Ymax = 1
Yscl = .5
```

7. $y = -2 \sin x$

Period: $\dfrac{2\pi}{1} = 2\pi$

Amplitude: $|-2| = 2$

9. $y = 3 \sin 10x$

Period: $\dfrac{2\pi}{10} = \dfrac{\pi}{5}$

Amplitude: $|3| = 3$

11. $y = \dfrac{1}{2} \cos \dfrac{2\pi}{3}$

Period: $\dfrac{2\pi}{2/3} = 3\pi$

Amplitude: $\left|\dfrac{1}{2}\right| = \dfrac{1}{2}$

13. $y = 3 \sin 4\pi x$

Period: $\dfrac{2\pi}{4\pi} = \dfrac{1}{2}$

Amplitude: $|3| = 3$

15. $f(x) = \sin x$

$g(x) = \sin(x - \pi)$

The graph of g is a horizontal shift to the right π units of the graph of f (a phase shift).

17. $f(x) = \cos 2x$

$g(x) = -\cos 2x$

The graph of g is a reflection in the x-axis of the graph of f.

19. $f(x) = \cos x$

$g(x) = -3 \cos x$

The graph of g is three times the amplitude of f, and reflected in the x-axis.

23. The graph of g has twice the amplitude as the graph of f. The period is the same.

27. $y_1 = \frac{1}{2} \sin x$; $y_2 = \frac{3}{2} \sin x$; $y_3 = -3 \sin x$

Changing the value of a changes the amplitude.

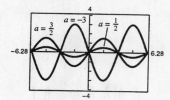

31. $f(x) = -2 \sin x$
Period: 2π
Amplitude: 2
$g(x) = 4 \sin x$
Period: 2π
Amplitude: 4 upward.

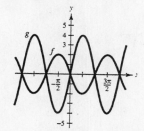

35. $f(x) = -\dfrac{1}{2} \sin \dfrac{x}{2}$

Period: 4π

Amplitude: $\dfrac{1}{2}$

$g(x) = 3 - \dfrac{1}{2} \sin \dfrac{x}{2}$ is the graph of $f(x)$ shifted vertically three units upward.

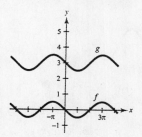

21. $f(x) = \sin x$

$f(x) = 2 + \sin x$

The graph of g is a vertical shift upward of 2 units of the graph of f.

25. The graph of g is a horizontal shift π units to the right of the graph of f.

29. $y_1 = \sin\left(\frac{1}{2}x\right)$; $y_2 = \sin\left(\frac{3}{2}x\right)$; $y_3 = \sin(4x)$

Changing the value of b changes the period.

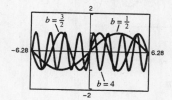

33. $f(x) = \cos x$
Period: 2π
Amplitude: 1
$g(x) = 1 + \cos x$ is a vertical shift of the graph of $f(x)$ one unit upward.

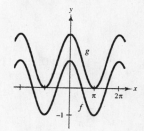

37. $f(x) = 2 \cos x$
Period: 2π
Amplitude: 2
$g(x) = 2 \cos(x + \pi)$ is the graph of $f(x)$ shifted π units to the left.

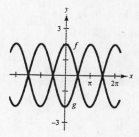

39. Since sine and cosine are cofunctions and x and $x - (\pi/2)$ are complementary, we have

$$\sin x = \cos\left(x - \frac{\pi}{2}\right).$$

Period: 2π

Amplitude: 1

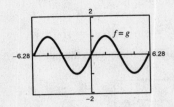

43. $y = -2\sin 6x$; $a = -2$, $b = 6$, $c = 0$

Period: $\dfrac{2\pi}{6} = \dfrac{\pi}{3}$

Amplitude: $|-2| = 2$

Key points: $(0, 0)$, $\left(\dfrac{\pi}{12}, -2\right)$, $\left(\dfrac{\pi}{6}, 0\right)$, $\left(\dfrac{\pi}{4}, 2\right)$, $\left(\dfrac{\pi}{3}, 0\right)$

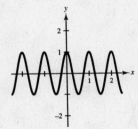

45. $y = \cos 2\pi x$

Period: $\dfrac{2\pi}{2\pi} = 1$

Amplitude: 1

Key points: $(1, 0)$, $\left(0, \dfrac{1}{4}\right)$, $\left(-1, \dfrac{1}{2}\right)$, $\left(0, \dfrac{3}{4}\right)$, $(1, 1)$

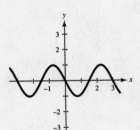

47. $y = -\sin \dfrac{2\pi x}{3}$; $a = -1$, $b = \dfrac{2\pi}{3}$, $c = 0$

Period: $\dfrac{2\pi}{2\pi/3} = 3$

Amplitude: 1

Key points: $(0, 0)$, $\left(\dfrac{3}{4}, -1\right)$, $\left(\dfrac{3}{2}, 0\right)$, $\left(\dfrac{9}{4}, 1\right)$, $(3, 0)$

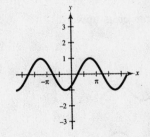

41. $f(x) = \cos x$

$g(x) = -\sin\left(x - \dfrac{\pi}{2}\right) = \sin\left(\dfrac{\pi}{2} - x\right) = \cos x$

Thus, $f(x) = g(x)$.

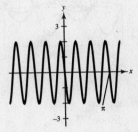

49. $y = \sin\left(x - \dfrac{\pi}{4}\right)$; $a = 1$, $b = 1$, $c = \dfrac{\pi}{4}$

Period: 2π

Amplitude: 1

Shift: Set $x - \dfrac{\pi}{4} = 0$ and $x - \dfrac{\pi}{4} = 2\pi$

$$x = \frac{\pi}{4} \qquad\qquad x = \frac{9\pi}{4}$$

Key points: $\left(\dfrac{\pi}{4}, 0\right)$, $\left(\dfrac{3\pi}{4}, 1\right)$, $\left(\dfrac{5\pi}{4}, 0\right)$, $\left(\dfrac{7\pi}{4}, -1\right)$, $\left(\dfrac{9\pi}{4}, 0\right)$

51. $y = 3\cos(x + \pi)$

Period: 2π

Amplitude: 3

Shift: Set $x + \pi = 0$ and $x + \pi = 2\pi$

$\qquad\qquad x = -\pi \qquad\qquad x = \pi$

Key points: $(3, -\pi)$, $\left(0, -\dfrac{\pi}{2}\right)$, $(-3, 0)$, $\left(0, \dfrac{\pi}{2}\right)$, $(3, \pi)$

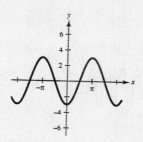

53. $y = \dfrac{1}{10}\cos(60\pi x)$; $a = \dfrac{1}{10}$, $b = 60\pi$, $c = 0$

Period: $\dfrac{2\pi}{60\pi} = \dfrac{1}{30}$

Amplitude: $\dfrac{1}{10}$

Key points: $\left(0, \dfrac{1}{10}\right)$, $\left(\dfrac{1}{120}, 0\right)$, $\left(\dfrac{1}{60}, -\dfrac{1}{10}\right)$, $\left(\dfrac{1}{40}, 0\right)$, $\left(\dfrac{1}{30}, \dfrac{1}{10}\right)$

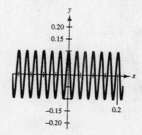

55. $y = 2 - \sin\dfrac{2\pi x}{3}$

Vertical shift 2 units upward of the graph in Exercise #47.

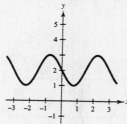

57. $y = 3\cos(x + \pi) - 3$

Vertical shift 2 units downward of the graph in Exercise #51.

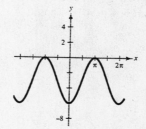

59. $y = \dfrac{2}{3}\cos\left(\dfrac{x}{2} - \dfrac{\pi}{4}\right)$; $a = \dfrac{2}{3}$, $b = \dfrac{1}{2}$, $c = \dfrac{\pi}{4}$

Period: 4π

Amplitude: $\dfrac{2}{3}$

Shift: Set $\dfrac{x}{2} - \dfrac{\pi}{4} = 0$ and $\dfrac{x}{2} - \dfrac{\pi}{4} = 2\pi$

$\qquad\qquad x = \dfrac{\pi}{2} \qquad\qquad x = \dfrac{9\pi}{2}$

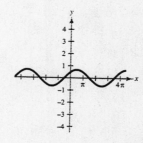

Key points: $\left(\dfrac{\pi}{2}, \dfrac{2}{3}\right)$, $\left(\dfrac{3\pi}{2}, 0\right)$, $\left(\dfrac{5\pi}{2}, \dfrac{-2}{3}\right)$, $\left(\dfrac{7\pi}{2}, 0\right)$, $\left(\dfrac{9\pi}{2}, \dfrac{2}{3}\right)$

61. $y = -2 \sin(4x + \pi)$

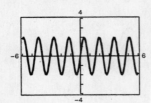

63. $y = \cos\left(2\pi x - \dfrac{\pi}{2}\right) + 1$

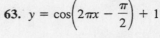
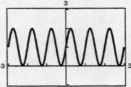

65. $y = -0.1 \sin\left(\dfrac{\pi x}{10} + \pi\right)$

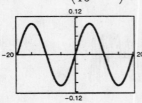

67. $y = 5 \cos(\pi - 2x) + 2$

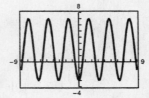

69. $f(x) = a \cos x + d$

Amplitude: $\frac{1}{2}[8 - 0] = 4$

Since $f(x)$ is the graph of $g(x) = 4 \cos x$ reflected about the x-axis and shifted vertically 4 units upward, we have $a = -4$ and $d = 4$. Thus, $f(x) = -4 \cos x + 4 = 4 - 4 \cos x$.

71. $y = a \sin(bx - c)$

Amplitude: $|a| = |3|$.

Since the graph is reflected about the x-axis, we have $a = -3$.

Period: $\dfrac{2\pi}{b} = \pi \implies b = 2$

Phase shift: $c = 0$

Thus, $y = -3 \sin 2x$.

73. $y = a \sin(bx + c)$

Amplitude: $a = 1$

Period: $2\pi \implies b = 1$

Phase shift: $bx + c = 0$ when $x = \dfrac{\pi}{4}$

$$(1) \; \dfrac{\pi}{4} + c = 0 \implies c = -\dfrac{\pi}{4}$$

Thus, $y = \sin\left(x - \dfrac{\pi}{4}\right)$.

75. $y_1 = \sin x$

$y_2 = -\dfrac{1}{2}$

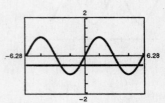

In the interval $[-2\pi, 2\pi]$, $\sin x = -\dfrac{1}{2}$ when

$$x = -\dfrac{5\pi}{6}, \; -\dfrac{\pi}{6}, \; \dfrac{7\pi}{6}, \; \dfrac{11\pi}{6}.$$

77. $y_1 = \cos x$

$y_2 = \dfrac{\sqrt{2}}{2}$

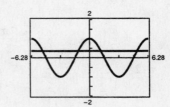

In the interval $[-2\pi, 2\pi]$, $\cos x = \dfrac{\sqrt{2}}{2}$ when $x = \pm\dfrac{\pi}{4}, \; \pm\dfrac{7\pi}{4}$.

79. (a) $h(x) = \cos^2 x$ is even.   (b) $g(x) = \sin^2 x$ is even.   (c) $h(x) = \sin x \cos x$ is odd.

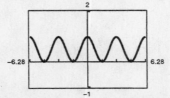
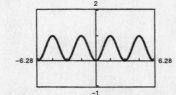
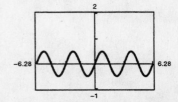

81. $y = 0.85 \sin \dfrac{\pi t}{3}$

(a)

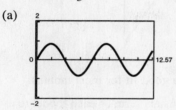

(b) Time for one cycle = one period = $\dfrac{2\pi}{\pi/3} = 6$ sec

(c) Cycles per min = $\dfrac{60}{6} = 10$ cycles per min

83. (a) $\sin x \approx x - \dfrac{x^3}{3!} + \dfrac{x^5}{5!}$

Near $x = 0$ the graphs are approximately the same.

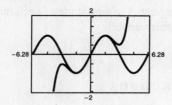

(b) $\cos x \approx 1 - \dfrac{x^2}{2!} + \dfrac{x^4}{4!}$

Near $x = 0$ the graphs are approximately the same.

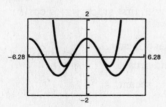

(c) $\sin x \approx x - \dfrac{x^3}{3!} + \dfrac{x^5}{5!} - \dfrac{x^7}{7!}$

$\cos x \approx 1 - \dfrac{x^2}{2!} + \dfrac{x^4}{4!} - \dfrac{x^6}{6!}$

The accuracy is increased.

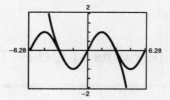

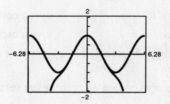

85. $S = 22.3 - 3.4 \cos \dfrac{\pi t}{6}, \ 1 \le t \le 12$

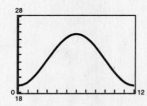

Maximum sales: June

Minimum sales: December

87. $C = 30.3 + 21.6 \sin \left(\dfrac{2\pi t}{365} + 10.9 \right)$

(a) period = $\dfrac{2\pi}{b} = \dfrac{2\pi}{(2\pi/365)} = 365$ days

This is to be expected: 365 days = 1 year

(b) The constant 30.3 gallons is the average daily fuel consumption.

(c)

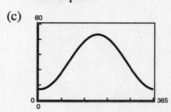

Consumption exceeds 40 gallons/day when $124 \le x \le 252$. (Graph C together with $y = 40$).

89. (a) A model for Chicago is

$$C(t) = 56.35 + 27.35 \sin\left(\frac{\pi t}{6} + 4.19\right).$$

(c)

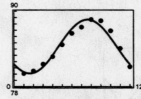

The model is a good fit.

(e) Each model has a period of 12. This corresponds to the 12 months in a year.

(b)

The model is a good fit for most months.

(d) Use the constant term of each model to estimate the average annual temperature.

Honolulu: 84.40°

Chicago: 56.35°

(f) Chicago has a greater variability in temperatures during the year. The amplitude of each model indicates this variability.

Section 4.6   Graphs of Other Trigonometric Functions

■ You should be able to graph:

$$y = a \tan(bx - c) \qquad\qquad y = a \cot(bx - c)$$
$$y = a \sec(bx - c) \qquad\qquad y = a \csc(bx - c)$$

■ When graphing $y = a \sec(bx - c)$ or $y = a \csc(bx - c)$ you should know to first graph $y = a \cos(bx - c)$ or $y = a \sin(bx - c)$ since

   (a) The intercepts of sine and cosine are vertical asymptotes of cosecant and secant.

   (b) The maximums of sine and cosine are local minimums of cosecant and secant.

   (c) The minimums of sine and cosine are local maximums of cosecant and secant.

■ You should be able to graph using a damping factor.

Solutions to Odd-Numbered Exercises

1. $y = \sec \dfrac{x}{2}$

Period: $\dfrac{2\pi}{1/2} = 4\pi$

Matches graph (g).

5. $y = \cot \dfrac{\pi x}{2}$

Period: $\dfrac{\pi}{\pi/2} = 2$

Matches graph (b).

3. $y = \tan 2x$

Period: $\dfrac{\pi}{2}$

Matches graph (f).

7. $y = -\csc x$

Period: 2π

Matches graph (e).

9. $y = \frac{1}{3} \tan x$.

Period: π

Two consecutive asymptotes:

$x = -\frac{\pi}{x}$ and $x = \frac{\pi}{2}$

x	$-\frac{\pi}{4}$	0	$\frac{\pi}{4}$
y	$-\frac{1}{3}$	0	$\frac{1}{3}$

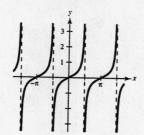

13. $y = -\frac{1}{2} \sec x$

Graph $y = -\frac{1}{2} \cos x$ first.

Period: 2π

One cycle: 0 to 27

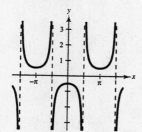

17. $y = \sec \pi x - 1$

Reflect the graph in Exercise #15 about the
x-axis and then shift it vertically down one unit.

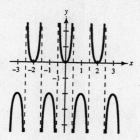

11. $y = \tan 2x$

Period: $\frac{\pi}{2}$

Two consecutive asymptotes:

$2x = -\frac{\pi}{2} \Rightarrow x = -\frac{\pi}{4}$

$2x = \frac{\pi}{2} \Rightarrow x = \frac{\pi}{4}$

x	$-\frac{\pi}{8}$	0	$\frac{\pi}{8}$
y	-1	0	1

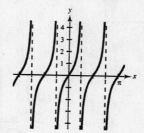

15. $y = -\sec \pi x$

Graph $y = -\cos \pi x$ first.

Period: $\frac{2\pi}{\pi} = 2$

One cycle: 0 to 2

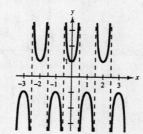

19. $y = \csc \frac{x}{2}$

Graph $y = \sin \frac{x}{2}$ first.

Period: $\frac{2\pi}{1/2} = 4\pi$

One cycle: 0 to 4π

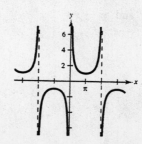

21. $y = \cot \dfrac{x}{2}$

Period: $\dfrac{\pi}{1/2} = 2\pi$

Two consecutive asymptotes:

$\dfrac{x}{2} = 0 \implies x = 0$

$\dfrac{x}{2} = \pi \implies x = 2\pi$

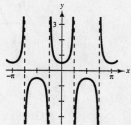

x	$\dfrac{\pi}{2}$	π	$\dfrac{3\pi}{2}$
y	1	0	-1

23. $y = \dfrac{1}{2} \sec 2x$

Graph $y = \dfrac{1}{2} \cos 2x$ first.

Period: $\dfrac{2\pi}{2} = \pi$

One cycle: 0 to π

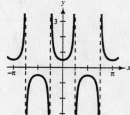

25. $y = \tan \dfrac{\pi x}{4}$

Period: $\dfrac{\pi}{\pi/4} = 4$

Two consecutive asymptotes:

$\dfrac{\pi x}{4} = -\dfrac{\pi}{2} \implies x = -2$

$\dfrac{\pi x}{4} = \dfrac{\pi}{2} \implies x = 2$

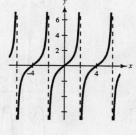

x	-1	0	1
y	-1	0	1

27. $y = \csc(\pi - x)$

Graph $y = \sin(\pi - x)$ first.

Period: 2π

Shift: Set $\pi - x = 0$ and $\pi - x = 2\pi$

$\qquad\qquad x = \pi \qquad\qquad x = -\pi$

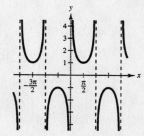

29. $y = 2 \cot\left(x + \dfrac{\pi}{2}\right)$

Graph $y = \dfrac{2}{\tan\left(x + \dfrac{\pi}{2}\right)}$

Period: π

Shift: Set $x + \dfrac{\pi}{2} = 0$ and $x + \dfrac{\pi}{2} = \pi$

$$x = -\dfrac{\pi}{2} \qquad x = \dfrac{\pi}{2}$$

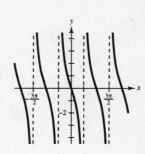

31. $y = \tan\dfrac{x}{3}$

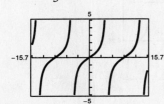

33. $y = -2 \sec 4x$

$$= \dfrac{-2}{\cos 4x}$$

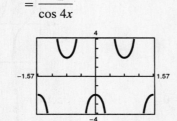

35. $y = \tan\left(x - \dfrac{\pi}{4}\right)$

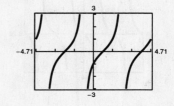

37. $y = \dfrac{1}{4} \cot\left(x - \dfrac{\pi}{2}\right)$

$$= \dfrac{1}{4 \tan(x - \pi/2)}$$

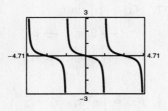

39. $y = 2 \sec(2x - \pi)$

$$y = \dfrac{2}{\cos(2x - \pi)}$$

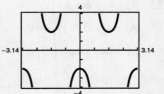

41. $\tan x = 1$

$$x = -\dfrac{7\pi}{4}, \ -\dfrac{3\pi}{4}, \ \dfrac{\pi}{4}, \ \dfrac{5\pi}{4}$$

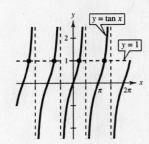

43. $\sec x = -2$

$$x = \pm\dfrac{2\pi}{3}, \ \pm\dfrac{4\pi}{3}$$

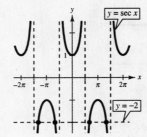

45. The graph of $f(x) = \sec x$ has y-axis symmetry. Thus, the function is even.

47. As $x \to \dfrac{\pi}{2}$ from the left, $f(x) = \tan x \to \infty$.

As $x \to \dfrac{\pi}{2}$ from the right, $f(x) = \tan x \to -\infty$.

49. $f(x) = 2 \sin x$

$g(x) = \dfrac{1}{2} \csc x$

(a)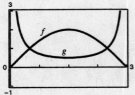

(b) $f > g$ on the interval, $\dfrac{\pi}{6} < x < \dfrac{5\pi}{6}$

(c) As $x \to \pi, f(x) = 2 \sin x \to 0$ and $g(x) = \dfrac{1}{2} \csc x \to \infty$ since $g(x)$ is the reciprocal of $f(x)$.

51. $y_1 = \sin x \csc x$ and $y_2 = 1$

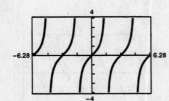

$\sin x \csc x = \sin x \left(\dfrac{1}{\sin x} \right) = 1, \sin x \neq 0$

53. $y_1 = \dfrac{\cos x}{\sin x}$ and $y_2 = \cot x = \dfrac{1}{\tan x}$

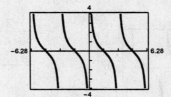

$\cot x = \dfrac{\cos x}{\sin x}$

55. $f(x) = x \cos x$

As $x \to 0, f(x) \to 0$.

Matches graph (d).

57. $g(x) = |x| \sin x$

As $x \to 0, g(x) \to 0$.

Matches graph (b).

59. $f(x) = \sin x + \cos\left(x + \dfrac{\pi}{2}\right), g(x) = 0$

$f(x) = g(x)$

The graph is the line $y = 0$.

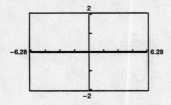

61. $f(x) = \sin^2 x, g(x) = \dfrac{1}{2}(1 - \cos 2x)$

$f(x) = g(x)$

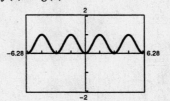

63. $f(x) = 2^{-x/4} \cos \pi x$

$-2^{-x/4} \leq f(x) \leq 2^{-2x/4}$

The damping factor is $y = 2^{-x/4}$.

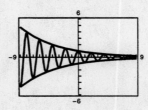

As $x \to \infty, f \to 0$

65. $g(x) = e^{-x^2/2} \sin x$

$-e^{-x^2/2} \leq g(x) \leq e^{-x^2/2}$

The damping factor is $y = e^{-x^2/2}$.

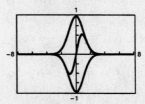

As $x \to \infty, g \to 0$

67. $\tan x = \dfrac{5}{d}$

$d = \dfrac{5}{\tan x} = 5 \cot x$

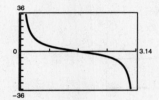

69. As the predator population increases, the number of prey decrease. When the number of prey is small, the number of predators decreases.

71. (a)

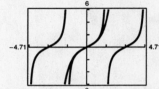

(b) The displacement function is periodic, but damped. It approaches 0 as t increases.

73. $\tan x \approx x + \dfrac{2x^3}{3!} + \dfrac{16x^5}{5!}$

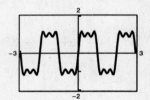

The graphs are approximately the same when x is near zero. As x gets larger, the graphs are further apart.

75. (a) $y_1 = \dfrac{4}{\pi}\left(\sin(\pi x) + \dfrac{1}{3} \sin 3(\pi x)\right)$

$y_2 = \dfrac{4}{\pi}\left(\sin(\pi x) + \dfrac{1}{3} \sin 3(\pi x) + \dfrac{1}{5} \sin 5(\pi x)\right)$

(b) $y_3 = \dfrac{4}{\pi}\left(\sin(\pi x) + \dfrac{1}{3} \sin 3(\pi x) + \dfrac{1}{5} \sin 5(\pi x) + \dfrac{1}{7} \sin 7(\pi x)\right)$

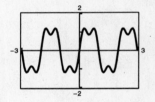

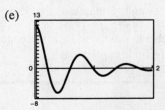

(c) $y_4 = \dfrac{4}{\pi}\left(\sin(\pi x) + \dfrac{1}{3} \sin 3(\pi x) + \dfrac{1}{5} \sin 5(\pi x) + \dfrac{1}{7} \sin 7(\pi x) + \dfrac{1}{9} \sin 9(\pi x)\right)$

77. $y = \dfrac{6}{x} + \cos x, \quad x > 0$

As $x \to 0, \ y \to \infty$.

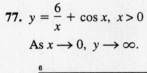

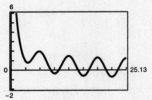

79. (a) Yes. For each t there corresponds one and only one value of y.

(b) 1.27 oscillations per sec

(c) $y = 12(0.2231)^t \cos 8t$

(d) $y = 12e^{-1.5t} \cos 8t$

(e)

81. (a) 850 rev/min

(b) The direction of the saw is reversed.

(c) $L = 60\left[\left(\dfrac{\pi}{2} + \phi\right) + \cot \phi\right], \ 0 < f < \dfrac{\pi}{2}$

(e) Straight line lengths change faster.

(d)

ϕ	0.3	0.6	0.9	1.2	1.5
L	306.2	217.9	195.9	189.6	188.5

(f)

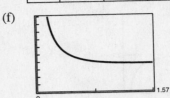

83. Graphing $y = \dfrac{300}{1 + e^{-x}} - 100$, you see that $x \approx -0.693$ is the only root. Analytically,

$$\dfrac{300}{1 + e^{-x}} = 100$$

$$3 = 1 + e^{-x}$$

$$2 = e^{-x}$$

$$\ln 2 = -x \implies x = -\ln 2 \approx -0.693.$$

85. $\log_8 x + \log_8(x - 1) = \frac{1}{3}$

$$\log_8 x(x - 1) = \tfrac{1}{3}$$

$$8^{1/3} = x(x - 1)$$

$$x^2 - x - 2 = 0$$

$$(x - 2)(x + 1) = 0 \implies x = 2 \ (x = -1 \text{ is not in the domain.})$$

Section 4.7 Inverse Trigonometric Functions

- You should know the definitions, domains, and ranges of $y = \arcsin x$, $y = \arccos x$, and $y = \arctan x$.

Function	Domain	Range
$y = \arcsin x \implies x = \sin y$	$-1 \le x \le 1$	$-\dfrac{\pi}{2} \le y \le \dfrac{\pi}{2}$
$y = \arccos x \implies x = \cos y$	$-1 \le x \le 1$	$0 \le y \le \pi$
$y = \arctan x \implies x = \tan y$	$-\infty < x < \infty$	$-\dfrac{\pi}{2} < x < \dfrac{\pi}{2}$

- You should know the inverse properties of the inverse trigonometric functions.

  $\sin(\arcsin x) = x, \ -1 \le x \le 1$ and $\arcsin(\sin y) = y, \ -\dfrac{\pi}{2} \le y \le \dfrac{\pi}{2}$

  $\cos(\arccos x) = x, \ -1 \le x \le 1$ and $\arccos(\cos y) = y, \ 0 \le y \le \pi$

  $\tan(\arctan x) = x$ and $\arctan(\tan y) = y, \ -\dfrac{\pi}{2} < y < \dfrac{\pi}{2}$

- You should be able to use the triangle technique to convert trigonometric functions of inverse trigonometric functions into algebraic expressions.

Solutions to Odd-Numbered Exercises

1. (a)

x	-1.0	-0.8	-0.6	-0.4	-0.2
y	-1.5708	-0.9273	-0.6435	-0.4115	-0.2014

x	0	0.2	0.4	0.6	0.8	1
y	0	0.2014	0.4115	0.6435	0.9273	1.5708

(b)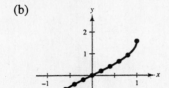

(c)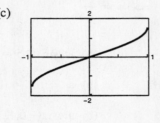

(d) $(0, 0)$, Symmetric to the origin

3. (a)

x	-10	-8	-6	-4	-2
y	-1.4711	-1.4464	-1.4056	-1.3258	-1.1071

x	0	2	4	6	8	10
y	0	1.1071	1.3258	1.4056	1.4464	1.4711

(b)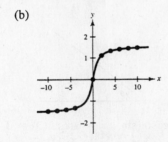

(c)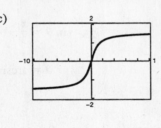

(d) Horizontal asymptotes: $y = \pm\dfrac{\pi}{2}$

5. $y = \arctan x$

$$\tan\left(-\frac{\pi}{6}\right) = -\frac{\sqrt{3}}{3} \implies \left(-\frac{\sqrt{3}}{3}, -\frac{\pi}{6}\right)$$

$$\arctan\left(-\sqrt{3}\right) = -\frac{\pi}{3} \implies \left(-\sqrt{3}, -\frac{\pi}{3}\right)$$

$$\tan\left(\frac{\pi}{4}\right) = 1 \implies \left(1, \frac{\pi}{4}\right)$$

7. (a) $y = \arcsin\dfrac{1}{2} \implies \sin y = \dfrac{1}{2}$ for $-\dfrac{\pi}{2} \le y \le \dfrac{\pi}{2} \implies y = \dfrac{\pi}{6}$

(b) $y = \arcsin 0 \implies \sin y = 0 \implies y = 0$

9. (a) $y = \arctan\dfrac{\sqrt{3}}{3} \implies \tan y = \dfrac{\sqrt{3}}{3}$ for $-\dfrac{\pi}{2} < y < \dfrac{\pi}{2} \implies y = \dfrac{\pi}{6}$

(b) $y = \arctan(-1) \implies \tan y = -1 \implies y = -\dfrac{\pi}{4}$

11. (a) $y = \arctan(-\sqrt{3}) \implies \tan y = \sqrt{3}$ for $-\frac{\pi}{2} < y < \frac{\pi}{2} \implies y = -\frac{\pi}{3}$

   (b) $y = \arctan\sqrt{3} \implies \tan y = \sqrt{3} \implies y = \frac{\pi}{3}$

13. (a) $y = \arcsin\frac{\sqrt{3}}{2} \implies \sin y = \frac{\sqrt{3}}{2}$ for $-\frac{\pi}{2} \le y \le \frac{\pi}{2} \implies y = \frac{\pi}{3}$

   (b) $y = \arctan\left(\frac{-\sqrt{3}}{3}\right) \implies \tan y = \frac{-\sqrt{3}}{3} \implies y = -\frac{\pi}{6}$

15. (a) $\arccos 0.28 = \cos^{-1} 0.28 \approx 1.29$ (b) $\arcsin 0.45 \approx 0.47$

17. (a) $\arctan(-3) = \tan^{-1}(-3) \approx -1.25$ (b) $\arctan 15 \approx 1.50$

19. (a) $\arccos(-0.41) = \cos^{-1}(-0.41) \approx 1.99$ (b) $\arcsin(-0.125) \approx -0.13$

21. $f(x) = \tan x$ and $g(x) = \arctan x$

   Graph: $y_1 = \tan x$

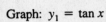

   $\qquad\quad y_2 = \tan^{-1} x$

   $\qquad\quad y_3 = x$

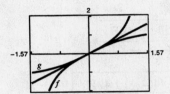

23. $\tan\theta = \frac{x}{4}$

   $\theta = \arctan\frac{x}{4}$

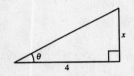

25. $\sin\theta = \frac{x+2}{5}$

   $\theta = \arcsin\left(\frac{x+2}{5}\right)$

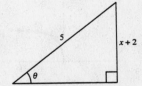

27. $\sin(\arcsin 0.3) = 0.3$

29. $\cos[\arccos(-0.1)] = -0.1$

31. $\arcsin(\sin 3\pi) = \arcsin(0) = 0$

   Note: 3π is not in the range of the arcsine function.

33. Let $y = \arctan\frac{3}{4}$. Then,

   $\tan y = \frac{3}{4}$, $0 < y < \frac{\pi}{2}$

   and $\sin y = \frac{3}{5}$.

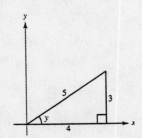

35. Let $y = \arctan 2$. Then,

   $\tan y = 2 = \frac{2}{1}$, $0 < y < \frac{\pi}{2}$

   and $\cos y = \frac{1}{\sqrt{5}} = \frac{\sqrt{5}}{5}$.

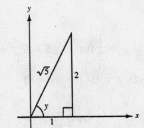

37. Let $y = \arcsin \dfrac{5}{13}$. Then,

$\sin y = \dfrac{5}{13}, \; 0 < y < \dfrac{\pi}{2}$

and $\cos y = \dfrac{12}{13}$.

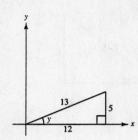

39. Let $y = \arctan\left(-\dfrac{3}{5}\right)$. Then,

$\tan y = -\dfrac{3}{5}, \; -\dfrac{\pi}{2} < y < 0$

and $\sec y = \dfrac{\sqrt{34}}{5}$.

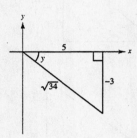

41. Let $y = \arccos\left(-\dfrac{2}{3}\right)$. Then,

$\cos y = -\dfrac{2}{3}, \; -\dfrac{\pi}{2} < y < \pi$

and $\sin y = \dfrac{\sqrt{5}}{3}$.

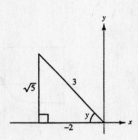

43. Let $y = \arctan x$. Then,

$\tan y = x$

and $\cot y = \dfrac{1}{x}$.

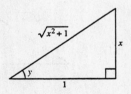

45. Let $y = \arcsin(2x)$. Then,

$\sin y = 2x = \dfrac{2x}{1}$

and $\cos y = \sqrt{1 - 4x^2}$.

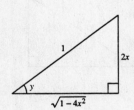

47. Let $y = \arccos x$. Then,

$$\cos y = x = \frac{x}{1}$$

and $\sin y = \sqrt{1 - x^2}$.

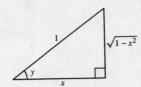

49. Let $y = \arccos\left(\dfrac{x}{3}\right)$. Then,

$$\cos y = \frac{x}{3}$$

and $\tan y = \dfrac{\sqrt{9 - x^2}}{x}$.

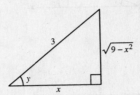

51. Let $y = \arctan \dfrac{x}{\sqrt{2}}$. Then,

$$\tan y = \frac{x}{\sqrt{2}}$$

and $\csc y = \dfrac{\sqrt{x^2 + 2}}{x}$.

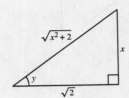

53. $f(x) = \sin(\arctan 2x)$, $\quad g(x) = \dfrac{2x}{\sqrt{1 + 4x^2}}$

Let $y = \arctan 2x$. Then,

$$\tan y = 2x = \frac{2x}{1}$$

and $\sin y = \dfrac{2x}{\sqrt{1 + 4x^2}}$.

$$g(x) = \frac{2x}{\sqrt{1 + 4x^2}} = f(x)$$

The graph has horizontal asymptotes at $y = \pm 1$.

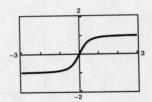

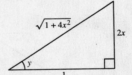

55. Let $y = \arctan \dfrac{9}{x}$. Then,

$$\tan y = \frac{9}{x} \text{ and } \sin y = \frac{9}{\sqrt{x^2 + 81}}.$$

Thus, $y = \arcsin \dfrac{9}{\sqrt{x^2 + 81}}$.

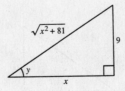

57. Let $y = \arccos \dfrac{3}{\sqrt{x^2 - 2x + 10}}$. Then,

$$\cos y = \frac{3}{\sqrt{x^2 - 2x + 10}} = \frac{3}{\sqrt{(x - 1)^2 + 9}}$$

and $\sin y = \dfrac{|x - 1|}{\sqrt{(x - 1)^2 + 9}}$.

Thus, $y = \arcsin \dfrac{|x - 1|}{\sqrt{(x - 1)^2 + 9}} = \arcsin \dfrac{|x - 1|}{\sqrt{x^2 - 2x + 10}}$.

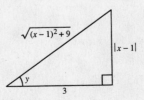

59. $y = 2 \arccos x$

Domain: $-1 \leq x \leq 1$

Range: $0 \leq y \leq 2\pi$

Vertical stretch of $f(x) = \arccos x$

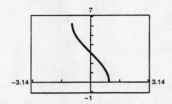

61. The graph of $f(x) = \arcsin(x - 1)$ is a horizontal translation of the graph of $y = \arcsin x$ by one unit.

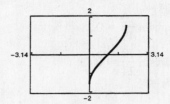

63. $f(x) = \arctan 2x$

Domain: all real numbers

Range: $-\dfrac{\pi}{2} < y < \dfrac{\pi}{2}$

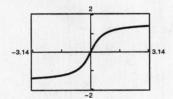

65. $h(v) = \tan(\arccos v) = \dfrac{\sqrt{1 - v^2}}{v}$

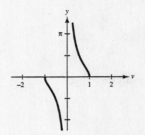 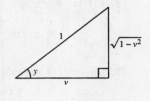

Domain: $-1 \leq v \leq 1,\ v \neq 0$

Range: all real numbers

67. $f(t) = 3 \cos 2t + 3 \sin 2t = \sqrt{3^2 + 3^2} \sin\left(2t + \arctan \dfrac{3}{3}\right)$

$$= 3\sqrt{2} \sin(2t + \arctan 1)$$

$$= 3\sqrt{2} \sin\left(2t + \dfrac{\pi}{4}\right)$$

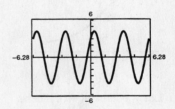

The graphs are the same.

69. $f(x) = \sin x,\ f^{-1}(x) = \arcsin x$

(a) $f \circ f^{-1} = f(f^{-1}(x)) = f(\arcsin x) = \sin(\arcsin x)$

$f^{-1} \circ f = f^{-1}(f(x)) = f^{-1}(\sin x) = \arcsin(\sin x)$

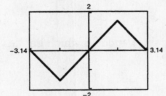 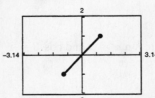

(b) Both the domain and range of $f \circ f^{-1} = \sin(\arcsin x)$ are the intervals of $[-1, 1]$.

The domain of $f^{-1} \circ f$ is all real numbers. The range is still the interval $[-1, 1]$.

Neither graph is the line $y = x$ because of these domain/range restrictions.

71. (a) $\sin \theta = \dfrac{10}{s}$

$\theta = \arcsin \dfrac{10}{s}$

(b) $s = 48$: $\theta = \arcsin \dfrac{10}{48} \approx 0.21$

$s = 24$: $\theta = \arcsin \dfrac{10}{24} \approx 0.43$

73. $\beta = \arctan \dfrac{3x}{x^2 + 4}$

(a)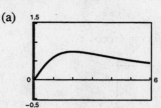

(b) β is maximum when $x = 2$.

(c) The graph has a horizontal asymptote at $\beta = 0$.
As x increases, β decreases.

75. (a) $\tan \theta = \dfrac{5}{x}$

$\theta = \arctan \dfrac{5}{x}$

(b) $x = 10$: $\theta = \arctan \dfrac{5}{10} \approx 26.6° = 0.46$ rad

$x = 3$: $\theta = \arctan \dfrac{5}{3} \approx 59.0° = 1.03$ rad

77. $y = \text{arccot } x$ if and only if $\cot y = x$.

Domain: $-\infty < x < \infty$

Range: $0 < x < \pi$

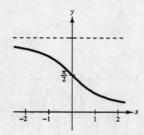

79. $y = \text{arccsc } x$ if and only if $\csc y = x$.

Domain: $(-\infty, -1] \cup [1, \infty)$

Range: $\left[-\dfrac{\pi}{2}, 0 \right) \cup \left(0, \dfrac{\pi}{2} \right]$

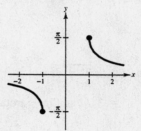

81. Let $y = \arcsin(-x)$. Then,

$\sin y = -x$

$-\sin y = x$

$\sin(-y) = x$

$-y = \arcsin x$

$y = -\arcsin x.$

Therefore, $\arcsin(-x) = -\arcsin x.$

83. $y = \pi - \arccos x$

$\cos y = \cos(\pi - \arccos x)$

$\cos y = \cos \pi \cos(\arccos x) + \sin \pi \sin(\arccos x)$

$\cos y = -x$

$y = \arccos(-x)$

85. Let $\alpha = \arcsin x$ and $\beta = \arccos x$. Then, $\sin \alpha = x$ and $\cos \beta = x$. Thus, $\sin \alpha = \cos \beta$ which implies that α and β are complementary angles and we have

$$\alpha + \beta = \frac{\pi}{2}$$

$$\arcsin x + \arccos x = \frac{\pi}{2}.$$

87. Now: Cost = 23,500 + 725 = $24,225

Wait a month: Cost = 23,500 (1.04) = $24,440

The customer should buy now and save $215.

89. Let x = the number of people presently in the group. Each person's share is now $\dfrac{250,000}{x}$. If two more join the group, each person's share would then be $\dfrac{250,000}{x + 2}$.

$$\boxed{\begin{array}{c}\text{Share per person with}\\\text{two more people}\end{array}} = \boxed{\begin{array}{c}\text{Original share}\\\text{per person}\end{array}} - 6250$$

$$\frac{250,000}{x + 2} = \frac{250,000}{x} - 6250$$

$$250,000x = 250,000(x + 2) - 6250x(x + 2)$$

$$250,000x = 250,000x + 500,000 - 6250x^2 - 12500x$$

$$6250x^2 + 12500x - 500,000 = 0$$

$$6250(x^2 + 2x - 80) = 0$$

$$6250(x + 10)(x - 8) = 0$$

$$x = -10 \quad \text{or} \quad x = 8$$

Not possible
There were 8 people in the original group.

Section 4.8   Applications and Models

- ■ You should be able to solve right triangles.
- ■ You should be able to solve right triangle applications.
- ■ You should be able to solve applications of simple harmonic motion: $d = a \sin wt$ or $d = a \cos wt$

Solutions to Odd-Numbered Exercises

1. Given: $A = 20°$, $b = 10$

$$\tan A = \frac{a}{b} \implies a = b \tan A = 10 \tan 20° \approx 3.64$$

$$\cos A = \frac{a}{c} \implies c = \frac{a}{\cos A} = \frac{10}{\cos 20°} \approx 10.64$$

$$B = 90° - 20° = 70°$$

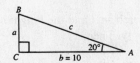

3. Given: $B = 71°$, $b = 24$

$$\tan B = \frac{b}{a} \implies a = \frac{b}{\tan B} = \frac{24}{\tan 71°} \approx 8.26$$

$$\sin B = \frac{b}{c} \implies c = \frac{b}{\sin B} = \frac{24}{\sin 71°} \approx 25.38$$

$$A = 90° - 71° = 19°$$

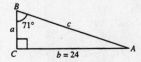

5. Given: $a = 6$, $b = 10$

$c^2 = a^2 + b^2 \implies c = \sqrt{36 + 100}$
$$= 2\sqrt{34} \approx 11.66$$

$\tan A = \dfrac{a}{b} = \dfrac{6}{10} \implies A = \arctan \dfrac{3}{5} \approx 30.96°$

$B = 90° - 30.964° = 59.04°$

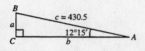

7. $b = 16$, $c = 52$

$a = \sqrt{52^2 - 16^2}$
$$= \sqrt{2448} = 12\sqrt{17} \approx 49.48$$

$\cos A = \dfrac{16}{52}$

$A = \arccos \dfrac{16}{52} \approx 72.08°$

$B = 90 - 72.08 \approx 17.92$

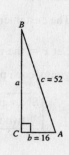

9. $A = 12° \ 15'$, $c = 430.5$

$B = 90° - 12° \ 15' = 77° \ 45'$

$\sin 12° \ 15' = \dfrac{a}{430.5}$
$$a = 430.5 \sin 12° \ 15' \approx 91.34$$

$\cos 12° \ 15' = \dfrac{b}{430.5}$
$$b = 430.5 \cos 12° \ 15' \approx 420.70$$

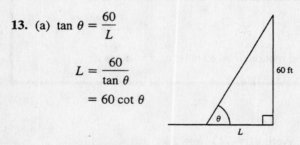

11. $\tan \theta = \dfrac{h}{\frac{1}{2}b} \implies h = \dfrac{1}{2}b \tan\theta$

$h = \dfrac{1}{2}b \tan \theta$

$h = \dfrac{1}{2}(4) \tan 52° \approx 2.56$ in.

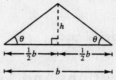

13. (a) $\tan \theta = \dfrac{60}{L}$

$L = \dfrac{60}{\tan \theta}$
$$= 60 \cot \theta$$

60 ft

(b)

θ	10°	20°	30°	40°	50°
L	340	165	104	72	50

(c) No, the shadow lengths do not increase in equal increments. The cotangent function is not linear.

15. (a) $\sin \theta = \dfrac{h}{20}$

$h = 20 \sin \theta$

20 ft

(b)

θ	60°	65°	70°	75°	80°
h	17.3	18.1	18.8	19.3	19.7

17. (a)

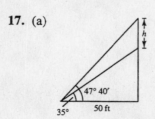

(b) Let the height of the church $= x$ and the height of the church and steeple $= y$. Then:

$\tan 35° = \dfrac{x}{50}$ and $\tan 47° \ 40' = \dfrac{y}{50}$

$x = 50 \tan 35°$ and $y = 50 \tan 47° \ 40'$

$h = y - x = 50(\tan 47° \ 40' - \tan 35°)$

(c) $h \approx 19.9$ feet

19. $\sin 34° = \dfrac{x}{4000}$

$x = 4000 \sin 34°$

≈ 2236.8 feet

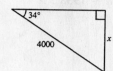

23. $\sin \theta = \dfrac{4000}{4150}$

$\theta = \arcsin\left(\dfrac{4000}{4150}\right)$

$\theta \approx 74.5°$

$\alpha = 90° - 74.5° = 15.5°$

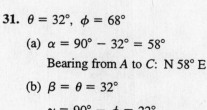

27. $\sin 10.5° = \dfrac{x}{4}$

$x = 4 \sin 10.5°$

≈ 0.73 mile

21. $\tan \theta = \dfrac{75}{50}$

$\theta = \arctan \dfrac{3}{2} \approx 56.3°$

25. Since the airplane speed is

$$\left(275 \,\dfrac{\text{ft}}{\text{sec}}\right)\left(60 \,\dfrac{\text{sec}}{\text{min}}\right) = 16{,}500 \,\dfrac{\text{ft}}{\text{min}},$$

after one minute its distance travelled is 16,500 feet.

$\sin 18° = \dfrac{a}{16{,}500}$

$a = 16{,}500 \sin 18°$

≈ 5099 ft

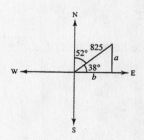

29. The plane has traveled $1.5(550) = 825$ miles.

$\sin 38° = \dfrac{a}{825} \implies a \approx 508$ miles north

$\cos 38° = \dfrac{b}{825} \implies b \approx 650$ miles east

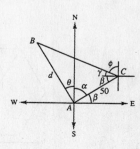

31. $\theta = 32°, \;\; \phi = 68°$

(a) $\alpha = 90° - 32° = 58°$

Bearing from A to C: N 58° E

(b) $\beta = \theta = 32°$

$\gamma = 90° - \phi = 22°$

$C = \beta + \gamma = 54°$

$\tan C = \dfrac{d}{50} \implies \tan 54° = \dfrac{d}{50} \implies d \approx 68.82$ yd

33. $\tan \theta = \dfrac{45}{30} \implies \theta \approx 56.3°$

Bearing: N 56.3° W

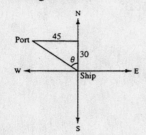

35. $\tan 6.5° = \dfrac{350}{d} \implies d \approx 3071.91$ ft

$\tan 4° = \dfrac{350}{D} \implies D \approx 5005.23$ ft

Distance between ships: $D - d \approx 1933.3$ ft

37. $\tan 57° = \dfrac{a}{x} \implies x = a \cot 57°$

$\tan 16° = \dfrac{a}{x + (55/6)}$

$\tan 16° = \dfrac{a}{a \cot 57° + (55/6)}$

$\cot 16° = \dfrac{a \cot 57° + (55/6)}{a}$

$a \cot 16° - a \cot 57° = \dfrac{55}{6} \implies a \approx 3.23$ miles

$\approx 17{,}054$ ft

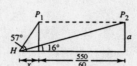

39. $L_1: 3x - 2y = 5 \implies y = \dfrac{3}{2}x - \dfrac{5}{2} \implies m_1 = \dfrac{3}{2}$

$L_2: \ x - \ y = 1 \implies y = -x + 1 \implies m_2 = -1$

$\tan \alpha = \left| \dfrac{-1 - (3/2)}{1 + (-1)(3/2)} \right| = \left| \dfrac{-5/2}{-1/2} \right| = 5$

$\alpha = \arctan 5 \approx 78.7°$

41. The diagonal of the base has a length of $\sqrt{a^2 + a^2} = \sqrt{2}a$.

Now, we have:

$\tan \theta = \dfrac{a}{\sqrt{2}a} = \dfrac{1}{\sqrt{2}}$

$\theta = \arctan \dfrac{1}{\sqrt{2}}$

$\theta \approx 35.3°$

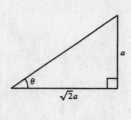

43. $\cos 30° = \dfrac{b}{r}$

$b = \cos 30° r$

$b = \dfrac{\sqrt{3}r}{2}$

$y = 2b = 2\left(\dfrac{\sqrt{3}r}{2} \right) = \sqrt{3}r$

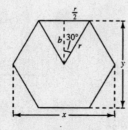

45. $\sin 36° = \dfrac{d}{25} \implies d \approx 14.6946$

Length of side: $2d \approx 29.4$ inches

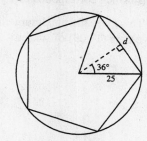

47. $\tan 35° = \dfrac{a}{10}$

$a = 10 \tan 35° \approx 7$

$\cos 33° = \dfrac{10}{c}$

$c = \dfrac{10}{\cos 35°} \approx 12.2$

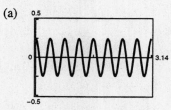

49. $d = 4 \cos 8\pi t$

(a) Maximum displacement = amplitude = 4

(b) Frequency $= \dfrac{\omega}{2\pi} = \dfrac{8\pi}{2\pi}$

$= 4$ cycles per unit of time

(c) $8\pi t = \dfrac{\pi}{2} \implies t = \dfrac{1}{16}$

51. $d = \dfrac{1}{16} \sin 120 \pi t$

(a) Maximum displacement = amplitude $= \dfrac{1}{16}$

(b) Frequency $= \dfrac{\omega}{2\pi} = \dfrac{120\pi}{2\pi}$

$= 60$ cycles per unit of time

(c) $120\pi t = \pi \implies t = \dfrac{1}{120}$

53. $d = 0$ when $t = 0$, $a = 4$, Period = 2

Use $d = a \sin \omega t$ since $d = 0$ when $t = 0$

$\dfrac{2\pi}{\omega} = 2 \implies \omega = \pi$

Thus, $d = 4 \sin \pi t$.

55. $d = 3$ when $t = 0$, $a = 3$, Period = 1.5

Use $d = a \cos \omega t$ since $d = 3$ when $t = 0$

$\dfrac{2\pi}{\omega} = 1.5 \implies \omega = \dfrac{4\pi}{3}$

Thus, $d = 3 \cos\left(\dfrac{4\pi}{3} t\right) = 3 \cos\left(\dfrac{4\pi t}{3}\right)$.

57. $d = a \sin \omega t$

Period $= \dfrac{2\pi}{\omega} = \dfrac{1}{\text{frequency}}$

$\dfrac{2\pi}{\omega} = \dfrac{1}{264}$

$\omega = 2\pi(264) = 528\pi$

59. $y = \dfrac{1}{4} \cos 16t, \ t > 0$

(a)

0.5

0

3.14

−0.5

(b) Period: $\dfrac{2\pi}{16} = \dfrac{\pi}{8}$ seconds

(c) $\dfrac{1}{4} \cos 16t = 0$ when $16t = \dfrac{\pi}{2} \implies t = \dfrac{\pi}{32}$ seconds.

61. $S(t) = 18.09 + 1.41 \sin\left(\dfrac{\pi t}{6} + 4.60\right)$

(a)

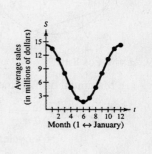

(b) The period is 12 months, which corresponds to 1 year.

(c) The amplitude is 1.41. This gives the maximum change in time from the average time (18.09) of sunset.

63. (a)

(b) $a = \dfrac{1}{2}(14.30 - 1.70) = 6.3$

$\dfrac{2\pi}{b} = 12 \implies b = \dfrac{\pi}{6}$

Shift: $d = 14.3 - 6.3 = 8$

$S = d + a \cos bt$

$S = 8 + 6.3 \cos\left(\dfrac{\pi t}{6}\right)$

The model is a good fit.

(c) Period: $\dfrac{2\pi}{\pi/6} = 12$

This corresponds to the 12 months in a year. Since the sales of outerwear is seasonal, this is reasonable.

(d) The amplitude represents the maximum displacement from the average sale of 8 million dollars. Sales are greatest in December (cold weather + Christmas) and least in June.

❑ **Review Exercises for Chapter 4**

Solutions to Odd-Numbered Exercises

1. $\theta = \dfrac{11\pi}{4}$

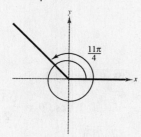

Coterminal angles: $\dfrac{11\pi}{4} - 2\pi = \dfrac{3\pi}{4}$

$\dfrac{3\pi}{4} - 2\pi = -\dfrac{5\pi}{4}$

3.

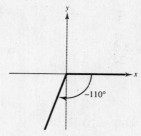

Coterminal angles: $-110° + 360° = 250°$

$-110° - 360° = -470°$

5. $135° \, 16' \, 45'' = \left(135 + \frac{16}{60} + \frac{45}{3600}\right)° \approx 135.28°$

7. $5° \, 22' \, 53'' = \left(5 + \frac{22}{60} + \frac{53}{3600}\right)° \approx 5.38°$

9. $135.27° = 135° + (0.27)(60)'$

$\qquad = 135° + 16' + 0.2(60)''$

$\qquad = 135° \, 16' \, 12''$

11. $-85.15° = -[85° + (0.15)(60)'] = -85° \, 9'$

13. $\dfrac{5\pi \text{ rad}}{7} = \dfrac{5\pi \text{ rad}}{7} \cdot \dfrac{180°}{\pi \text{ rad}} \approx 128.57°$

15. $-3.5 \text{ rad} = -3.5 \text{ rad} \cdot \dfrac{180°}{\pi \text{ rad}} \approx -200.54°$

17. $480° = 480° \cdot \dfrac{\pi \text{ rad}}{180°} = \dfrac{8\pi}{3} \text{ rad} \approx 8.3776 \text{ rad}$

19. $-33° \, 45' = -33.75° = -33.75° \cdot \dfrac{\pi \text{ rad}}{180°} = -\dfrac{3\pi}{16} \text{ rad} \approx -0.5890 \text{ rad}$

21. $252°$ is in Quadrant III.

Reference angle $= 252° - 180° = 72°$

23. $-\dfrac{6\pi}{5}$ is in Quadrant II and is coterminal to $\dfrac{4\pi}{5}$.

Reference angle $= \pi - \dfrac{4\pi}{5} = \dfrac{\pi}{5}$

25. $s = r\theta \implies \theta = \dfrac{s}{r} = \dfrac{25}{12}$

27. $s = r\theta = (20)(2.41) = 48.20$

29. The hypotenuse is $\sqrt{12^2 + 10^2} = \sqrt{244} = 2\sqrt{61}$.

$\sin \theta = \dfrac{\text{opp}}{\text{hyp}} = \dfrac{10}{2\sqrt{61}} = \dfrac{5}{\sqrt{61}} = \dfrac{5\sqrt{61}}{61}$

$\cos \theta = \dfrac{\text{adj}}{\text{hyp}} = \dfrac{12}{2\sqrt{61}} = \dfrac{6}{\sqrt{61}} = \dfrac{6\sqrt{61}}{61}$

$\csc \theta = \dfrac{1}{\sin \theta} = \dfrac{\sqrt{61}}{5}$

$\sec \theta = \dfrac{1}{\cos \theta} = \dfrac{\sqrt{61}}{6}$

$\tan \theta = \dfrac{\text{opp}}{\text{adj}} = \dfrac{10}{12} = \dfrac{5}{6}$

$\cot \theta = \dfrac{1}{\tan \theta} = \dfrac{6}{5}$

31. The opposite side is $\sqrt{6^2 - 4^2} = \sqrt{36 - 16} = \sqrt{20} = 2\sqrt{5}$.

$$\sin \theta = \frac{\text{opp}}{\text{hyp}} = \frac{2\sqrt{5}}{6} = \frac{\sqrt{5}}{3} \qquad\qquad \cos \theta = \frac{\text{adj}}{\text{hyp}} = \frac{4}{6} = \frac{2}{3}$$

$$\csc \theta = \frac{1}{\sin \theta} = \frac{3}{\sqrt{5}} = \frac{3\sqrt{5}}{5} \qquad\qquad \sec \theta = \frac{1}{\cos \theta} = \frac{3}{2}$$

$$\tan \theta = \frac{\text{opp}}{\text{adj}} = \frac{2\sqrt{5}}{4} = \frac{\sqrt{5}}{2} \qquad\qquad \cot \theta = \frac{1}{\tan \theta} = \frac{2}{\sqrt{5}} = \frac{2\sqrt{5}}{5}$$

33. $x = 12, y = 16$

$$r = \sqrt{144 + 256} = \sqrt{400} = 20$$

$$\sin \theta = \frac{y}{r} = \frac{4}{5} \qquad\qquad \csc \theta = \frac{r}{y} = \frac{5}{4}$$

$$\cos \theta = \frac{x}{r} = \frac{3}{5} \qquad\qquad \sec \theta = \frac{r}{x} = \frac{5}{3}$$

$$\tan \theta = \frac{y}{x} = \frac{4}{3} \qquad\qquad \cot \theta = \frac{x}{y} = \frac{3}{4}$$

35. $x = -7, y = 2, r = \sqrt{49 + 4} = \sqrt{53}$

$$\sin \theta = \frac{y}{r} = \frac{2}{\sqrt{53}} = \frac{2\sqrt{53}}{53} \qquad\qquad \csc \theta = \frac{\sqrt{53}}{2}$$

$$\cos \theta = \frac{x}{r} = -\frac{7}{\sqrt{53}} = -\frac{7\sqrt{53}}{53} \qquad\qquad \sec \theta = -\frac{\sqrt{53}}{7}$$

$$\tan \theta = \frac{y}{x} = -\frac{2}{7} \qquad\qquad \cot \theta = -\frac{7}{2}$$

37. $x = -4, y = -6, r = \sqrt{16 + 36} = 2\sqrt{13}$

$$\sin \theta = \frac{y}{r} = \frac{-6}{2\sqrt{13}} = -\frac{3\sqrt{13}}{13} \qquad\qquad \csc \theta = -\frac{\sqrt{13}}{3}$$

$$\cos \theta = \frac{x}{r} = \frac{-4}{2\sqrt{13}} = -\frac{2\sqrt{13}}{13} \qquad\qquad \sec \theta = -\frac{\sqrt{13}}{2}$$

$$\tan \theta = \frac{y}{x} = \frac{-6}{-4} = \frac{3}{2} \qquad\qquad \cot \theta = \frac{2}{3}$$

39. $\sec \theta = \frac{6}{5}$, $\tan \theta < 0 \implies \theta$ is in Quadrant IV.

$$r = 6, x = 5, y = -\sqrt{36 - 25} = -\sqrt{11}$$

$$\sin \theta = \frac{y}{r} = -\frac{\sqrt{11}}{6} \qquad\qquad \csc \theta = -\frac{6\sqrt{11}}{11}$$

$$\cos \theta = \frac{x}{r} = \frac{5}{6} \qquad\qquad \sec \theta = \frac{6}{5}$$

$$\tan \theta = \frac{y}{x} = -\frac{\sqrt{11}}{5} \qquad\qquad \cot \theta = -\frac{5\sqrt{11}}{11}$$

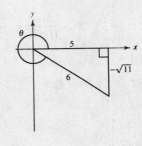

41. $\sin \theta = \frac{3}{8}$, $\cos \theta < 0 \implies \theta$ is in Quadrant II.

$$y = 3, r = 8, x = -\sqrt{55}$$

$$\sin \theta = \frac{y}{r} = \frac{3}{8} \qquad\qquad \csc \theta = \frac{8}{3}$$

$$\cos \theta = \frac{x}{r} = -\frac{\sqrt{55}}{8} \qquad\qquad \sec \theta = -\frac{8}{\sqrt{55}} = -\frac{8\sqrt{55}}{55}$$

$$\tan \theta = \frac{y}{x} = -\frac{3}{\sqrt{55}} = -\frac{3\sqrt{55}}{55} \qquad\qquad \cot \theta = -\frac{\sqrt{55}}{3}$$

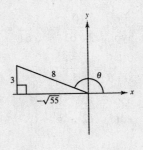

43. $\tan \dfrac{\pi}{3} = \sqrt{3}$

45. $\sin \dfrac{5\pi}{3} = -\sin \dfrac{\pi}{3} = -\dfrac{\sqrt{3}}{2}$

47. $\cos 495° = -\cos 45° = -\dfrac{\sqrt{2}}{2}$

49. $\tan 33° \approx 0.65$

51. $\sec \dfrac{12\pi}{5} = \dfrac{1}{\cos(12\pi/5)} \approx 3.24$

53. $\cos \theta = -\dfrac{\sqrt{2}}{2} \implies \theta$ is in Quadrant II or III.

Reference angle: $\dfrac{\pi}{4}$

$\theta = \dfrac{3\pi}{4}, \dfrac{5\pi}{4}$ or $\theta = 135°, 225°$

55. $\csc \theta = -2 \implies \theta$ is in Quadrant III or IV.

Reference angle: $\dfrac{\pi}{6}$

$\theta = \dfrac{7\pi}{6}, \dfrac{11\pi}{6}$ or $\theta = 210°, 330°$

57. $\sin \theta = 0.8387 \implies \theta$ is in Quadrant I or II.

$= \arcsin 0.8387$

$\theta \approx 0.9949$ rad or $0.9949 \cdot \dfrac{180}{\pi} = 57.0°$

$\theta \approx \pi - 0.9949 \approx 2.1467$ rad or

$2.1467 \cdot \dfrac{180}{\pi} = 123.0°$

59. $\sec \theta = -1.0353$, θ is in Quadrant II or III.

Reference angle: $15.0°$

$\theta = 165.0°$ or $165° \cdot \dfrac{\pi}{180°} \approx 2.8798$ rad

$\theta = 195.0°$ or $195° \cdot \dfrac{\pi}{180°} \approx 3.4034$ rad

61. Period: 2,
Amplitude: 5

63. Period: $3.14 \approx \pi$,
Amplitude: 3.4

65. $y = 3 \cos 2\pi x$

Amplitude: 3

Period: $\dfrac{2\pi}{2\pi} = 1$

67. $f(x) = 5 \sin \dfrac{2x}{5}$

Amplitude: 5

Period: $\dfrac{2\pi}{2/5} = 5\pi$

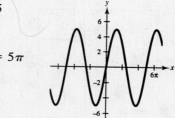

69. $f(x) = -\dfrac{1}{4} \cos \dfrac{\pi x}{4}$

Amplitude: $\left| -\dfrac{1}{4} \right| = \dfrac{1}{4}$

Period: $\dfrac{2\pi}{\pi/4} = 8$

71. $g(t) = \dfrac{5}{2} \sin(t - \pi)$

Amplitude: $\dfrac{5}{2}$

Period: 2π

Shift:
$t - \pi = 0$ and $t - \pi = 2\pi$
$\quad\quad t = \pi \quad\quad\quad\quad t = 3\pi$

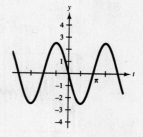

73. $h(t) = \tan\left(t - \dfrac{\pi}{4} \right)$

Period: π

Two consecutive asymptotes: $t - \dfrac{\pi}{4} = -\dfrac{\pi}{2}$ and $t - \dfrac{\pi}{4} = \dfrac{\pi}{2}$

$\quad\quad\quad\quad\quad\quad\quad t = -\dfrac{\pi}{4} \quad\quad\quad\quad t = \dfrac{3\pi}{4}$

t	0	$\pi/4$	$\pi/2$
$h(t)$	-1	0	1

75. $f(t) = \csc\left(3t - \dfrac{\pi}{2}\right)$

Graph: $y = \sin\left(3t - \dfrac{\pi}{2}\right)$ first.

Period: $\dfrac{2\pi}{3}$

Shift: $3t - \dfrac{\pi}{2} = 0$ and $3t - \dfrac{\pi}{2} = 2\pi$

$t = \dfrac{\pi}{6}$ $\qquad$ $t = \dfrac{5\pi}{6}$

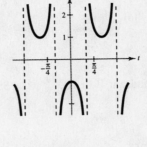

77. $y = \arcsin\dfrac{x}{2}$

Domain: $-2 \le x \le 2$

Range: $-\dfrac{\pi}{2} \le y \le \dfrac{\pi}{2}$

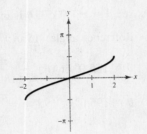

79. $f(x) = \dfrac{x}{4} - \sin x$

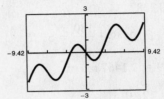

Not periodic

81. $f(x) = \dfrac{\pi}{2} + \arctan x$

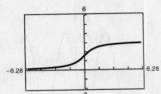

Not periodic

83. $h(\theta) = \theta \sin \pi\theta$

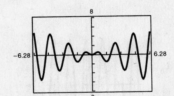

Not periodic

85. $f(t) = 2.5e^{-t/4} \sin 2\pi t$

$-2.5e^{-t/4} \le f(t) \le 2.5e^{-t/4}$

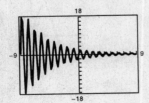

Not periodic

87. $g(x) = 3\left(\sin\dfrac{\pi x}{3} + 1\right)$

Period: $\dfrac{2\pi}{\pi/3} = 6$

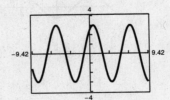

89. $f(x) = e^{\sin x}$

The graph is periodic.

Maximum: $\left(\dfrac{\pi}{2}, e\right)$

Minimum: $\left(\dfrac{3\pi}{2}, e^{-1}\right)$

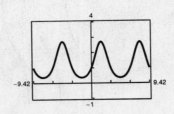

91. $g(x) = 2 \sin x \cos^2 x$

The graph is periodic.

Relative minimum: $\left(\dfrac{\pi}{2}, 0\right)$, $(3.76, -0.77)$, $(5.67, -0.77)$

Relative maximum: $(0.61, 0.77)$, $(2.53, 0.77)$, $\left(\dfrac{3\pi}{2}, 0\right)$

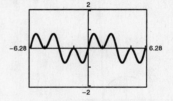

93. $f(x) = -2 \cos\left(x - \dfrac{\pi}{4}\right)$ **95.** $f(x) = -4 \cos\left(2x - \dfrac{\pi}{2}\right)$ **97.** $f(x) = \dfrac{1}{2} \tan \dfrac{x}{2}$

99. Let $y = \arcsin(x - 1)$. Then,

$$\sin y = (x - 1) = \frac{x - 1}{1} \text{ and}$$

$$\sec y = \frac{1}{\sqrt{-x^2 + 2x}} = \frac{\sqrt{-x^2 + 2x}}{-x^2 + 2x}.$$

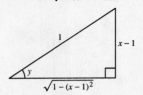

101. Let $y = \arccos \dfrac{x^2}{4 - x^2}$. Then,

$$\cos y = \frac{x}{4 - x^2} \text{ and}$$

$$\sin y = \frac{\sqrt{(4 - x^2)^2 - (x^2)^2}}{4 - x^2}$$

$$= \frac{\sqrt{16 - 8x^2}}{4 - x^2}$$

$$= \frac{2\sqrt{4 - 2x^2}}{4 - x^2}.$$

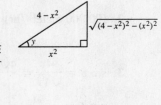

103. $\sin 50° = \dfrac{h}{12}$

$h = 12 \sin 50°$

≈ 9.2 m

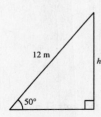

105. $\tan 25° = \dfrac{h}{2.5}$

$h = 2.5 \tan 25°$

≈ 1.2 miles

107. $\tan 1° \, 10' = \dfrac{a}{3.5}$

$a = 3.5 \tan 1° \, 10'$

≈ 0.071 km

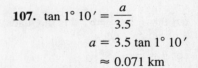

109. (a) $\tan \theta = \dfrac{x}{12}$

$x = 12 \tan \theta$

Area = Area of triangle − Area of sector

$$= \left(\frac{1}{2} bh\right) - \left(\frac{1}{2} r^2 \theta\right)$$

$$= \frac{1}{2} (12)(12 \tan \theta) - \frac{1}{2} (12^2)(\theta)$$

$$= 72 \tan \theta - 72\theta$$

$$= 72(\tan \theta - \theta)$$

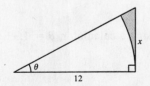

(b)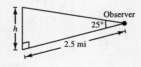

As $\theta \to \dfrac{\pi}{2}$, $A \to \infty$. The area increases without bound as θ approaches $\dfrac{\pi}{2}$.

❑ Chapter Test for Chapter 4

1. (a)

(b) $\dfrac{5\pi}{4} + 2\pi = \dfrac{13\pi}{4}$; $\dfrac{5\pi}{4} - 2\pi = -\dfrac{3\pi}{4}$

(c) $\dfrac{5\pi}{4} \cdot \dfrac{180}{\pi} = 225°$

2. $(90{,}000 \text{ meters/hr})\left(\dfrac{1}{60} \text{ hr/min}\right)\left(\dfrac{2\pi \text{ rad}}{2\pi\left(\frac{1}{2}\right) \text{ meters}}\right) = 3000 \text{ rad/min}$

3. $\sin \theta = \dfrac{4}{\sqrt{17}} = \dfrac{4\sqrt{17}}{17}$ $\csc \theta = \dfrac{\sqrt{17}}{4}$

$\cos \theta = -\dfrac{1}{\sqrt{17}} = -\dfrac{\sqrt{17}}{17}$ $\sec \theta = -\sqrt{17}$

$\tan \theta = -4$ $\cot \theta = -\dfrac{1}{4}$

4. $\tan \theta = \dfrac{3}{2} > 0 \implies \theta$ is in Quadrant I or III.

$\cot \theta = \dfrac{2}{3}$

$\sin \theta = \pm\dfrac{3}{\sqrt{13}} = \pm\dfrac{3\sqrt{13}}{13}$ $\cos \theta = \pm\dfrac{2}{\sqrt{13}} = \pm\dfrac{2\sqrt{13}}{13}$

$\csc \theta = \pm\dfrac{\sqrt{13}}{3}$ $\sec \theta = \pm\dfrac{\sqrt{13}}{2}$

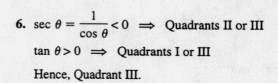

5. $\theta = 290° \implies \theta' = 70°$

6. $\sec \theta = \dfrac{1}{\cos \theta} < 0 \implies$ Quadrants II or III

$\tan \theta > 0 \implies$ Quadrants I or III

Hence, Quadrant III.

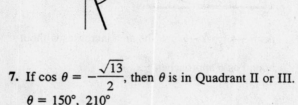

7. If $\cos \theta = -\dfrac{\sqrt{13}}{2}$, then θ is in Quadrant II or III.

$\theta = 150°,\ 210°$

8. $\csc \theta = \dfrac{1}{\sin \theta} = 1.030 \implies \sin \theta = \dfrac{1}{1.030}$ and θ in Quadrant I or II. Using a calculator, $\theta = 1.33$, 1.81 radians.

9. Amplitude: 2, shifted $\dfrac{\pi}{4}$ to the right.

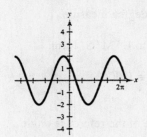

10.

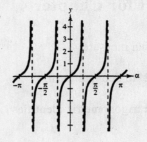

Period: $\dfrac{\pi}{2}$

11.

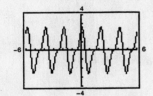

Period is 2

12.

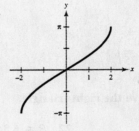

Not periodic

13. Amplitude $= 2$, reflected in x-axis $\implies a = -2$.

Period 4π and shifted to the right: $y = -2\sin\!\left(\dfrac{x}{2} - \dfrac{\pi}{4}\right)$

14. Let $u = \arccos\dfrac{2}{3} \implies \cos u = \dfrac{2}{3}$.

Then $\tan\!\left(\arccos\dfrac{2}{3}\right) = \tan u = \dfrac{\sqrt{5}}{2}$.

15.

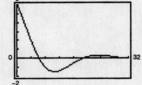

16. $(18)(3) = 54$ miles from port

$\sin 16° = -\dfrac{x}{54} \implies x = -14.88$

$\cos 16° = \dfrac{y}{54} \implies y = 51.91$

Position: $(-14.88, 51.91)$

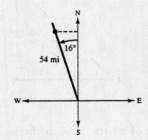

❑ Practice Test for Chapter 4

1 Express 350° in radian measure.

2. Express $(5\pi)/9$ in degree measure.

3. Convert 135° 14′ 12″ to decimal form.

4. Convert −22.569° to D° M′ S″ form.

5. If $\cos\theta = \frac{2}{3}$, use the trigonometric identities to find $\tan\theta$.

6. Find θ given $\sin\theta = 0.9063$.

7. Solve for x in the figure below.

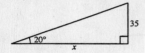

8. Find the magnitude of the reference angle for $\theta = (6\pi)/5$.

9. Evaluate csc 3.92.

10. Find $\sec\theta$ given that θ lies in Quadrant III and $\tan\theta = 6$.

11. Graph $y = 3\sin\dfrac{\pi}{2}$.

12. Graph $y = -2\cos(x - \pi)$.

13. Graph $y = \tan 2x$.

14. Graph $y = -\csc\left(x + \dfrac{\pi}{4}\right)$.

15. Graph $y = 2x + \sin x$, using a graphing calculator.

16. Graph $y = 3x\cos x$, using a graphing calculator.

17. Evaluate arcsin 1.

18. Evaluate arctan(−3).

19. Evaluate $\sin\left(\arccos\dfrac{4}{\sqrt{35}}\right)$.

20. Write an algebraic expression for $\cos\left(\arcsin\dfrac{x}{4}\right)$.

For Exercises 21–23, solve the right triangle.

21. $A = 40°$, $c = 12$

22. $B = 6.84°$, $a = 21.3$

23. $a = 5$, $b = 9$

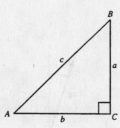

24. A 20-foot ladder leans against the side of a barn. Find the height of the top of the ladder if the angle of elevation of the ladder is 67°.

25. An observer in a lighthouse 250 feet above sea level spots a ship off the shore. If the angle of depression to the ship is 5°, how far out is the ship?

CHAPTER 5
Analytic Trigonometry

CHAPTER 5
Analytic Trigonometry

Section 5.1 Using Fundamental Identities

■ You should know the fundamental trigonometric identities.

(a) Reciprocal Identities

$$\sin u = \frac{1}{\csc u} \qquad\qquad \csc u = \frac{1}{\sin u}$$

$$\cos u = \frac{1}{\sec u} \qquad\qquad \sec u = \frac{1}{\cos u}$$

$$\tan u = \frac{1}{\cot u} = \frac{\sin u}{\cos u} \qquad\qquad \cot u = \frac{1}{\tan u} = \frac{\cos u}{\sin u}$$

(b) Pythagorean Identities

$$\sin^2 u + \cos^2 u = 1$$
$$1 + \tan^2 u = \sec^2 u$$
$$1 + \cot^2 u = \csc^2 u$$

(c) Cofunction Identities

$$\sin\left(\frac{\pi}{2} - u\right) = \cos u \qquad\qquad \cos\left(\frac{\pi}{2} - u\right) = \sin u$$

$$\tan\left(\frac{\pi}{2} - u\right) = \cot u \qquad\qquad \cot\left(\frac{\pi}{2} - u\right) = \tan u$$

$$\sec\left(\frac{\pi}{2} - u\right) = \csc u \qquad\qquad \csc\left(\frac{\pi}{2} - u\right) = \sec u$$

(d) Negative Angle Identities

$$\sin(-x) = -\sin x \qquad \csc(-x) = -\csc x$$
$$\cos(-x) = \cos x \qquad \sec(-x) = \sec x$$
$$\tan(-x) = -\tan x \qquad \cot(-x) = -\cot x$$

■ You should be able to use these fundamental identities to find function values.

■ You should be able to convert trigonometric expressions to equivalent forms by using the fundamental identities.

■ You should be able to check your answers with a graphing utility.

Solutions to Odd-Numbered Exercises

1. $\sin x = \dfrac{1}{2}, \ \cos x = \dfrac{\sqrt{3}}{2} \implies x$ is in

Quadrant I.

$\tan x = \dfrac{\sin x}{\cos x} = \dfrac{1/2}{\sqrt{3}/2} = \dfrac{1}{\sqrt{3}} = \dfrac{\sqrt{3}}{3}$

$\cot x = \dfrac{1}{\tan x} = \sqrt{3}$

$\sec x = \dfrac{1}{\cos x} = \dfrac{2}{\sqrt{3}} = \dfrac{2\sqrt{3}}{3}$

$\csc x = \dfrac{1}{\sin x} = 2$

3. $\sec\theta = \sqrt{2}, \ \sin\theta = -\dfrac{\sqrt{2}}{2} \implies \theta$ is in

Quadrant IV.

$\cos\theta = \dfrac{1}{\sec\theta} = \dfrac{1}{\sqrt{2}} = \dfrac{\sqrt{2}}{2}$

$\tan\theta = \dfrac{\sin\theta}{\cos\theta} = \dfrac{-\sqrt{2}/2}{\sqrt{2}/2} = -1$

$\cot\theta = \dfrac{1}{\tan\theta} = -1$

$\csc\theta = -\sqrt{2}$

5. $\tan x = \dfrac{5}{12}, \ \sec x = -\dfrac{13}{12} \implies x$ is in

Quadrant III.

$\cos x = \dfrac{1}{\sec x} = -\dfrac{12}{13}$

$\sin x = -\sqrt{1 - \cos^2 x} = -\sqrt{1 - \dfrac{144}{169}} = -\dfrac{5}{13}$

$\cot x = \dfrac{1}{\tan x} = \dfrac{12}{5}$

$\csc x = \dfrac{1}{\sin x} = -\dfrac{13}{5}$

7. $\sec\phi = -1, \ \sin\phi = 0 \implies \phi = \pi$

$\cos\phi = -1$

$\tan\phi = 0$

$\cot\phi$ is undefined.

$\csc\phi$ is undefined.

9. $\sin(-x) = -\sin x = -\dfrac{2}{3} \implies \sin x = \dfrac{2}{3}$

$\sin x = \dfrac{2}{3}, \ \tan x = -\dfrac{2\sqrt{5}}{5} \implies x$ is in

Quadrant II.

$\cos x = -\sqrt{1 - \sin^2 x} = -\sqrt{1 - \dfrac{4}{9}} = -\dfrac{\sqrt{5}}{3}$

$\cot x = \dfrac{1}{\tan x} = -\dfrac{\sqrt{5}}{2}$

$\sec x = \dfrac{1}{\cos x} = -\dfrac{3\sqrt{5}}{5}$

$\csc x = \dfrac{1}{\sin x} = \dfrac{3}{2}$

11. $\tan\theta = 2, \ \sin\theta < 0 \implies \theta$ is in Quadrant III.

$\sec\theta = -\sqrt{\tan^2\theta + 1} = -\sqrt{4 + 1} = -\sqrt{5}$

$\cos\theta = \dfrac{1}{\sec\theta} = -\dfrac{1}{\sqrt{5}} = -\dfrac{\sqrt{5}}{5}$

$\sin\theta = -\sqrt{1 - \cos^2\theta}$

$\quad = -\sqrt{1 - \dfrac{1}{5}} = -\dfrac{2}{\sqrt{5}} = -\dfrac{2\sqrt{5}}{5}$

$\csc\theta = \dfrac{1}{\sin\theta} = -\dfrac{\sqrt{5}}{2}$

$\cot\theta = \dfrac{1}{\tan\theta} = \dfrac{1}{2}$

13. $\sin\theta = -1, \ \cot\theta = 0 \implies \theta = \dfrac{3\pi}{2}$

$\cos\theta = \sqrt{1 - \sin^2\theta} = 0$

$\sec\theta$ is undefined.

$\tan\theta$ is undefined.

$\csc\theta = -1$

15. By looking at the basic graphs of $\sin x$ and $\csc x$, we

see that as $x \to \dfrac{\pi^-}{2}$, $\sin x \to 1$ and $\csc x \to 1$.

17. By looking at the basic graphs of $\tan x$ and $\cot x$, we see that as $x \to \dfrac{\pi^-}{2}$, $\tan x \to \infty$ and $\cot x \to 0$.

19. $\sec x \cos x = \sec x \cdot \dfrac{1}{\sec x} = 1$

The expression is matched with (d).

21. $\tan^2 x - \sec^2 x = \tan^2 x - (\tan^2 x + 1) = -1$

The expression is matched with (a).

23. $\dfrac{\sin(-x)}{\cos(-x)} = \dfrac{-\sin x}{\cos x} = -\tan x$

The expression is matched with (e).

25. $\sin x \sec x = \sin x \cdot \dfrac{1}{\cos x} = \tan x$

The expression is matched with (b).

27. $\sec^4 x - \tan^4 x = (\sec^2 x + \tan^2 x)(\sec^2 x - \tan^2 x)$

$\qquad\qquad = (\sec^2 x + \tan^2 x)(1) = \sec^2 x + \tan^2 x$

The expression is matched with (f).

29. $\dfrac{\sec^2 x - 1}{\sin^2 x} = \dfrac{\tan^2 x}{\sin^2 x} = \dfrac{\sin^2 x}{\cos^2 x} \cdot \dfrac{1}{\sin^2 x} = \sec^2 x$

The expression is matched with (e).

31. $\tan \phi \csc \phi = \dfrac{\sin \phi}{\cos \phi} \cdot \dfrac{1}{\sin \phi} = \dfrac{1}{\cos \phi} = \sec \phi$

33. $\cos \beta \tan \beta = \cos \beta \left(\dfrac{\sin \beta}{\cos \beta} \right)$

$\qquad\qquad = \sin \beta$

35. $\dfrac{\cot x}{\csc x} = \dfrac{\cos x / \sin x}{1/\sin x}$

$\qquad = \dfrac{\cos x}{\sin x} \cdot \dfrac{\sin x}{1} = \cos x$

37. $\sec \alpha \, \dfrac{\sin \alpha}{\tan \alpha} = \dfrac{1}{\cos \alpha}(\sin \alpha) \cot \alpha$

$\qquad = \dfrac{1}{\cos \alpha}(\sin \alpha) \left(\dfrac{\cos \alpha}{\sin \alpha} \right) = 1$

39. $\dfrac{\sin(-x)}{\cos x} = -\dfrac{\sin x}{\cos x} = -\tan x$

41. $\cos\left(\dfrac{\pi}{2} - x \right) \sec x = (\sin x)(\sec x)$

$\qquad\qquad = (\sin x) \left(\dfrac{1}{\cos x} \right)$

$\qquad\qquad = \dfrac{\sin x}{\cos x}$

$\qquad\qquad = \tan x$

43. $\dfrac{\cos^2 y}{1 - \sin y} = \dfrac{1 - \sin^2 y}{1 - \sin y}$

$\qquad = \dfrac{(1 + \sin y)(1 - \sin y)}{1 - \sin y}$

$\qquad = 1 + \sin y$

45. $\tan^2 x - \tan^2 x \sin^2 x = \tan^2 x(1 - \sin^2 x)$

$\qquad\qquad = \tan^2 x \cos^2 x$

$\qquad\qquad = \dfrac{\sin^2 x}{\cos^2 x} \cdot \cos^2 x$

$\qquad\qquad = \sin^2 x$

47. $\sin^2 x \sec^2 x - \sin^2 x = \sin^2 x(\sec^2 x - 1)$
$$= \sin^2 x \tan^2 x$$

49. $\tan^4 x + 2 \tan^2 x + 1 = (\tan^2 x + 1)^2$
$$= (\sec^2 x)^2$$
$$= \sec^4 x$$

51. $\sin^4 x - \cos^4 x = (\sin^2 x + \cos^2 x)(\sin^2 x - \cos^2 x)$
$$= (1)(\sin^2 x - \cos^2 x)$$
$$= \sin^2 x - \cos^2 x$$

53. $(\sin x + \cos x)^2 = \sin^2 x + 2 \sin x \cos x + \cos^2 x$
$$= (\sin^2 x + \cos^2 x) + 2 \sin x \cos x$$
$$= 1 + 2 \sin x \cos x$$

55. $(\sec x + 1)(\sec x - 1) = \sec^2 x - 1 = \tan^2 x$

57. $\dfrac{1}{1 + \cos x} + \dfrac{1}{1 - \cos x} = \dfrac{1 - \cos x + 1 + \cos x}{(1 + \cos x)(1 - \cos x)}$
$$= \dfrac{2}{1 - \cos^2 x}$$
$$= \dfrac{2}{\sin^2 x}$$
$$= 2 \csc^2 x$$

59. $\dfrac{\cos x}{1 + \sin x} + \dfrac{1 + \sin x}{\cos x} = \dfrac{\cos^2 x + (1 + \sin x)^2}{\cos x(1 + \sin x)}$
$$= \dfrac{2 + 2 \sin x}{\cos x(1 + \sin x)}$$
$$= \dfrac{2(1 + \sin x)}{\cos x(1 + \sin x)}$$
$$= \dfrac{2}{\cos x}$$
$$= 2 \sec x$$

61. $\dfrac{\sin^2 y}{1 - \cos y} = \dfrac{1 - \cos^2 y}{1 - \cos y}$
$$= \dfrac{(1 + \cos y)(1 - \cos y)}{1 - \cos y}$$
$$= 1 + \cos y$$

63. $\dfrac{3}{\sec x - \tan x} \cdot \dfrac{\sec x + \tan x}{\sec x + \tan x} = \dfrac{3(\sec x + \tan x)}{\sec^2 x - \tan^2 x}$
$$= \dfrac{3(\sec x + \tan x)}{1}$$
$$= 3(\sec x + \tan x)$$

65. $y_1 = \cos\left(\dfrac{\pi}{2} - x\right)$, $y_2 = \sin x$

x	0.2	0.4	0.6	0.8	1.0	1.2	1.4
y_1	0.1987	0.3894	0.5646	0.7174	0.8415	0.9320	0.9854
y_2	0.1987	0.3894	0.5646	0.7174	0.8415	0.9320	0.9854

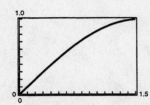

Conclusion: $y_1 = y_2$

67. $y_1 = \dfrac{\cos x}{1 - \sin x}$, $y_2 = \dfrac{1 + \sin x}{\cos x}$

x	0.2	0.4	0.6	0.8	1.0	1.2	1.4
y_1	1.2230	1.5085	1.8958	2.4650	3.4082	5.3319	11.6814
y_2	1.2230	1.5085	1.8958	2.4650	3.4082	5.3319	11.6814

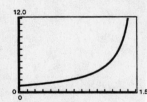

Conclusion: $y_1 = y_2$

69. $y_1 = \cos x \cot x + \sin x = \csc x$

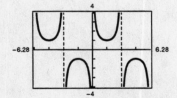

71. $\sqrt{25 - x^2} = \sqrt{25 - (5 \sin \theta)^2}$, $x - 5 \sin \theta$
$$= \sqrt{25 - 25 \sin^2 \theta}$$
$$= \sqrt{25(1 - \sin^2 \theta)}$$
$$= \sqrt{25 \cos^2 \theta}$$
$$= 5 \cos \theta$$

73. $\sqrt{x^2 - 9} = \sqrt{(3 \sec \theta)^2 - 9}$, $x = 3 \sec \theta$
$$= \sqrt{9 \sec^2 \theta - 9}$$
$$= \sqrt{9 (\sec^2 \theta - 1)}$$
$$= \sqrt{9 \tan^2 \theta}$$
$$= 3 \tan \theta$$

75. $\sqrt{x^2 + 25} = \sqrt{(5 \tan \theta)^2 + 25}$, $x = 5 \tan \theta$
$$= \sqrt{25 \tan^2 \theta + 25}$$
$$= \sqrt{25(\tan^2 \theta + 1)}$$
$$= \sqrt{25 \sec^2 \theta}$$
$$= 5 \sec \theta$$

77. $\sin \theta = \sqrt{1 - \cos^2 \theta}$

Let $y_1 = \sin x$ and $y_2 = \sqrt{1 - \cos^2 x}$, $0 \le x \le 2\pi$.

$y_1 = y_2$ for $0 \le x \le \pi$, so we have

$\sin \theta = \sqrt{1 - \cos^2 \theta}$ for $0 \le \theta \le \pi$.

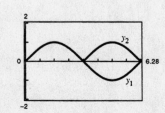

79. $\sec \theta = \sqrt{1 + \tan^2 \theta}$

Let $y_1 = \dfrac{1}{\cos x}$ and $y_2 = \sqrt{1 + \tan^2 x}$, $0 \leq x \leq 2\pi$.

$y_1 = y_2$ for $0 \leq x < \dfrac{\pi}{2}$ and $\dfrac{3\pi}{2} < x < 2\pi$, so we have

$\sin \theta = \sqrt{1 + \tan^2 \theta}$ for $0 \leq \theta < \dfrac{\pi}{2}$ and $\dfrac{3\pi}{2} < \theta < 2\pi$.

81. $\ln|\cos \theta| - \ln|\sin \theta| = \ln \dfrac{|\cos \theta|}{|\sin \theta|} = \ln|\cot \theta|$

83. False; $\dfrac{\sin k\theta}{\cos k\theta} = \tan k\theta$

85. True; $\sin \theta \csc \theta = \sin \theta \left(\dfrac{1}{\sin \theta} \right) = 1$,

provided $\sin \theta \neq 0$.

87. (a) $\csc^2 132° - \cot^2 132° \approx 1.8107 - 0.8107 = 1$

(b) $\csc^2 \dfrac{2\pi}{7} - \cot^2 \dfrac{2\pi}{7} \approx 1.6360 - 0.6360 = 1$

89. $\cos\left(\dfrac{\pi}{2} - \theta \right) = \sin \theta$

(a) $\theta = 80°$

$\cos(90° - 80°) = \sin 80°$

$0.9848 = 0.9848$

(b) $\theta = 0.8$

$\cos\left(\dfrac{\pi}{2} - 0.8 \right) = \sin 0.8$

$0.7174 = 0.7174$

91. Since $\sin^2 \theta + \cos^2 \theta = 1$ and $\cos^2 \theta = 1 - \sin^2 \theta$:

$\cos \theta = \pm\sqrt{1 - \sin^2 \theta}$

$\tan \theta = \dfrac{\sin \theta}{\cos \theta} = \pm\dfrac{\sin \theta}{\sqrt{1 - \sin^2 \theta}}$

$\cot \theta = \dfrac{1}{\tan \theta} = \pm\dfrac{\sqrt{1 - \sin^2 \theta}}{\sin \theta}$

$\sec \theta = \dfrac{1}{\cos \theta} = \pm\dfrac{1}{\sqrt{1 - \sin^2 \theta}}$

$\csc \theta = \dfrac{1}{\sin \theta}$

The sign depends on the choice of θ.

93. Seward has the greater variation in daylight hours because the coefficient of cosine is larger. The coefficients 6.4 and 1.9 determine the difference between the greatest and least number of daylight hours.

Section 5.2   Verifying Trigonometric Identities

- ■ You should know the difference between an expression, a conditional equation, and an identity.
- ■ You should be able to solve trigonometric identities, using the following techniques.
  - (a) Work with *one* side at a time. Do not "cross" the equal sign.
  - (b) Use algebraic techniques such as combining fractions, factoring expressions, rationalizing denominators, and squaring binomials.
  - (c) Use the fundamental identities.
  - (d) Convert all the terms into sines and cosines.

Solutions to Odd-Numbered Exercises

1. $\sin t \csc t = \sin t \left(\dfrac{1}{\sin t} \right) = 1$

3. $\dfrac{\sec^2 x}{\tan x} = \dfrac{1}{\cos^2 x} \cdot \dfrac{\cos x}{\sin x}$

$\qquad = \dfrac{1}{\cos x} \cdot \dfrac{1}{\sin x}$

$\qquad = \sec x \csc x$

5. $\cos^2 \beta - \sin^2 \beta = (1 - \sin^2 \beta) - \sin^2 \beta$

$\qquad\qquad = 1 - 2 \sin^2 \beta$

7. $\tan^2 \theta + 4 = (\sec^2 \theta - 1) + 4$

$\qquad\qquad = \sec^2 \theta + 3$

9. $\cos x + \sin x \tan x = \cos x + \sin x \cdot \dfrac{\sin x}{\cos x}$

$\qquad = \dfrac{\cos^2 x + \sin^2 x}{\cos x}$

$\qquad = \dfrac{1}{\cos x}$

$\qquad = \sec x$

11.

x	0.2	0.4	0.6	0.8	1.0	1.2	1.4
y_1	4.835	2.1785	1.2064	0.6767	0.3469	0.1409	0.0293
y_2	4.835	2.1785	1.2064	0.6767	0.3469	0.1409	0.0293

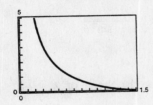

$\dfrac{1}{\sec x \tan x} = \cos x \cdot \dfrac{\cos x}{\sin x}$

$\qquad = \dfrac{\cos^2 x}{\sin x}$

$\qquad = \dfrac{1 - \sin^2 x}{\sin x}$

$\qquad = \dfrac{1}{\sin x} - \sin x$

$\qquad = \csc x - \sin x$

13.

x	0.2	0.4	0.6	0.8	1.0	1.2	1.4
y_1	4.835	2.1785	1.2064	0.6767	0.3469	0.1409	0.0293
y_2	4.835	2.1785	1.2064	0.6767	0.3469	0.1409	0.0293

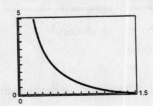

$\csc x - \sin x = \dfrac{1}{\sin x} - \sin x$

$\qquad = \dfrac{1 - \sin^2 x}{\sin x}$

$\qquad = \dfrac{\cos^2 x}{\sin x}$

$\qquad = \cos x \cdot \dfrac{\cos x}{\sin x}$

$\qquad = \cos x \cdot \cot x$

15.

x	0.2	0.4	0.6	0.8	1.0	1.2	1.4
y_1	1.0203	1.0857	1.2116	1.4353	1.8508	2.7597	5.8835
y_2	1.0203	1.0857	1.2116	1.4353	1.8508	2.7597	5.8835

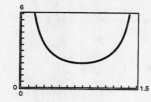

$\cos x + \sin x \tan x$

$$= \cos x + \sin x \cdot \frac{\sin x}{\cos x}$$

$$= \frac{\cos^2 x + \sin^2 x}{\cos x}$$

$$= \frac{1}{\cos x}$$

$$= \sec x$$

17.

x	0.2	0.4	0.6	0.8	1.0	1.2	1.4
y_1	5.1359	2.7880	2.1458	2.0009	2.1995	2.9609	5.9704
y_2	5.1359	2.7880	2.1458	2.0009	2.1995	2.9609	5.9704

$$\frac{1}{\tan x} + \frac{1}{\cot x} = \frac{\cot x + \tan x}{\tan x \cdot \cot x}$$

$$= \cot x + \tan x$$

19. $\sin^{1/2} x \cos x - \sin^{5/2} x \cos x = \sin^{1/2} x \cos x (1 - \sin^2 x) = \sin^{1/2} x \cos x \cdot \cos^2 x = \cos^3 x \sqrt{\sin x}$

21. $\cos\left(\dfrac{\pi}{2} - x\right)\csc x = \sin x\left(\dfrac{1}{\sin x}\right) = 1$

23. $\dfrac{\csc(-x)}{\sec(-x)} = \dfrac{1/\sin(-x)}{1/\cos(-x)}$

$$= \frac{\cos(-x)}{\sin(-x)}$$

$$= \frac{\cos x}{-\sin x}$$

$$= -\cot x$$

25. $\dfrac{\cos(-\theta)}{1 + \sin(-\theta)} = \dfrac{\cos \theta}{1 - \sin \theta} \cdot \dfrac{1 + \sin \theta}{1 + \sin \theta}$

$$= \frac{\cos \theta(1 + \sin \theta)}{1 - \sin^2 \theta}$$

$$= \frac{\cos \theta(1 + \sin \theta)}{\cos^2 \theta}$$

$$= \frac{1 + \sin \theta}{\cos \theta}$$

$$= \frac{1}{\cos \theta} + \frac{\sin \theta}{\cos \theta}$$

$$= \sec \theta + \tan \theta$$

27. $\dfrac{\sin x \cos y + \cos x \sin y}{\cos x \cos y - \sin x \sin y} = \dfrac{\dfrac{\sin x \cos y}{\cos x \cos y} + \dfrac{\cos x \sin y}{\cos x \cos y}}{\dfrac{\cos x \cos y}{\cos x \cos y} - \dfrac{\sin x \sin y}{\cos x \cos y}} = \dfrac{\tan x + \tan y}{1 - \tan x \tan y}$

29. $\dfrac{\tan x + \cot y}{\tan x \cot y} = \dfrac{\dfrac{1}{\cot x} + \dfrac{1}{\tan y}}{\dfrac{1}{\cot x} \cdot \dfrac{1}{\tan y}} = \dfrac{\dfrac{\tan y + \cot x}{\cot x \cdot \tan y}}{\dfrac{1}{\cot x \cdot \tan y}} = \tan y + \cot x$

31. $\sqrt{\dfrac{1 + \sin \theta}{1 - \sin \theta}} = \sqrt{\dfrac{1 + \sin \theta}{1 - \sin \theta} \cdot \dfrac{1 + \sin \theta}{1 + \sin \theta}}$ Note: Check your answer with a graphing utility. What happens if you leave off the absolute value?

$\qquad\qquad = \sqrt{\dfrac{(1 + \sin \theta)^2}{1 - \sin^2 \theta}}$

$\qquad\qquad = \sqrt{\dfrac{(1 + \sin \theta)^2}{\cos^2 \theta}}$

$\qquad\qquad = \dfrac{1 + \sin \theta}{|\cos \theta|}$

33. $\sin^2 x + \sin^2\left(\dfrac{\pi}{2} - x\right) = \sin^2 x + \cos^2 x = 1$ **35.** $\csc x \cos\left(\dfrac{\pi}{2} - x\right) = \dfrac{1}{\sin x} \cdot \sin x = 1$

37. $2\sec^2 x - 2\sec^2 x \sin^2 x - \sin^2 x - \cos^2 x = 2\sec^2 x(1 - \sin^2 x) - (\sin^2 x + \cos^2 x)$

$\qquad\qquad = 2\sec^2 x(\cos^2 x) - 1$

$\qquad\qquad = 2 \cdot \dfrac{1}{\cos^2 x} \cdot \cos^2 x - 1$

$\qquad\qquad = 2 - 1$

$\qquad\qquad = 1$

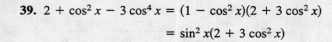
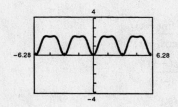

39. $2 + \cos^2 x - 3\cos^4 x = (1 - \cos^2 x)(2 + 3\cos^2 x)$

$\qquad\qquad = \sin^2 x(2 + 3\cos^2 x)$

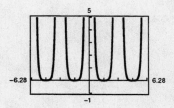

41. $\csc^4 x - 2\csc^2 x + 1 = (\csc^2 x - 1)^2$

$\qquad\qquad = (\cot^2 x)^2 = \cot^4 x$

43. $\sec^4 \theta - \tan^4 \theta = (\sec^2 \theta + \tan^2 \theta)(\sec^2 \theta - \tan^2 \theta)$

$\qquad\qquad = (1 + \tan^2 \theta + \tan^2 \theta)(1)$

$\qquad\qquad = 1 + 2\tan^2 \theta$

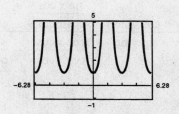

45. $\dfrac{\sin \beta}{1 - \cos \beta} \cdot \dfrac{1 + \cos \beta}{1 + \cos \beta} = \dfrac{\sin \beta(1 + \cos \beta)}{1 - \cos^2 \beta}$

$$= \dfrac{\sin \beta(1 + \cos \beta)}{\sin^2 \beta} = \dfrac{1 + \cos \beta}{\sin \beta}$$

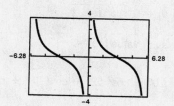

47. $\dfrac{\tan^3 \alpha - 1}{\tan \alpha - 1} = \dfrac{(\tan \alpha - 1)(\tan^2 \alpha + \tan \alpha + 1)}{\tan \alpha - 1} = \tan^2 \alpha + \tan \alpha + 1$

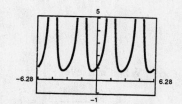

49. It appears that $y_1 = 1$. Analytically,

$$\dfrac{1}{\cot x + 1} + \dfrac{1}{\tan x + 1} = \dfrac{\tan x + 1 + \cot x + 1}{(\cot x + 1)(\tan x + 1)}$$

$$= \dfrac{\tan x + \cot x + 2}{\cot x \tan x + \cot x + \tan x + 1}$$

$$= \dfrac{\tan x + \cot x + 2}{\tan x + \cot x + 2}$$

$$= 1.$$

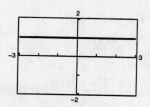

51. It appears that $y_1 = \sin x$. Analytically,

$$\dfrac{1}{\sin x} - \dfrac{\cos^2 x}{\sin x} = \dfrac{1 - \cos^2 x}{\sin x} = \dfrac{\sin^2 x}{\sin x} = \sin x.$$

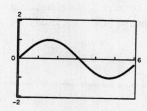

53. $\ln|\tan \theta| = \ln\left|\dfrac{\sin \theta}{\cos \theta}\right|$

$$= \ln \dfrac{|\sin \theta|}{|\cos \theta|}$$

$$= \ln|\sin \theta| - \ln|\cos \theta|$$

55. $-\ln(1 + \cos \theta) = \ln(1 + \cos \theta)^{-1}$

$$= \ln \dfrac{1}{1 + \cos \theta} \cdot \dfrac{1 - \cos \theta}{1 - \cos \theta}$$

$$= \ln \dfrac{1 - \cos \theta}{1 - \cos^2 \theta}$$

$$= \ln \dfrac{1 - \cos \theta}{\sin^2 \theta}$$

$$= \ln(1 - \cos \theta) - \ln \sin^2 \theta$$

$$= \ln(1 - \cos \theta) - 2 \ln|\sin \theta|$$

57. $\sqrt{\tan^2 \theta} = |\tan \theta|$

$\sqrt{\tan^2 \theta} \neq \tan \theta$ if θ lies in Quadrant II or IV.

One such angle is $\theta = \dfrac{3\pi}{4}$.

59. $|\tan \theta| = \sqrt{\sec^2 \theta - 1}$.

One such value is $x = \dfrac{3\pi}{4}$.

61. $\sin^2 25 + \sin^2 65 = \sin^2 25 + \cos^2 25° = 1$

63. $\cos^2 20° + \cos^2 52° + \cos^2 38° + \cos^2 70° = \cos^2 20° + \cos^2 52^2 + \sin^2(90° - 38°) + \sin^2(90° - 70°)$

$$= \cos^2 20° + \cos^2 52^2 + \sin^2 52° + \sin^2 20°$$

$$= (\cos^2 20° + \sin^2 20°) + (\cos^2 52° + \sin^2 52°)$$

$$= 1 + 1$$

$$= 2$$

65. When n is even,

$$\cos\left[\frac{(2n + 1)\pi}{2}\right] = \cos\frac{\pi}{2} = 0.$$

When n is odd,

$$\cos\left[\frac{(2n + 1)\pi}{2}\right] = \cos\frac{3\pi}{2} = 0.$$

Thus, $\cos\left[\dfrac{(2n + 1)\pi}{2}\right] = 0$ for all integers n.

67. $\mu W \cos \theta = W \sin \theta$

$$\mu = \frac{W \sin \theta}{W \cos \theta} = \frac{\sin \theta}{\cos \theta} = \tan \theta$$

Section 5.3   Solving Trigonometric Equations

- You should be able to identify and solve trigonometric equations.
- A trigonometric equation is a conditional equation. It is true for a specific set of values.
- To solve trigonometric equations, use algebraic techniques such as collecting like terms, taking square roots, factoring, squaring, converting to quadratic form, using formulas, and using inverse functions. Study the examples in this section.
- Use your graphing utility to calculate solutions and verify results.

Solutions to Odd-Numbered Exercises

1. $y = \sin\dfrac{\pi x}{2} + 1$

From the graph in the textbook we see that the curve has x-intercepts at $x = -1$ and at $x = 3$.

3. $y = \tan^2\left(\dfrac{\pi x}{6}\right) - 3$

From the graph in the textbook we see that the curve has x-intercepts at $x = \pm 2$.

5. $2 \cos x - 1 = 0$

(a) $2 \cos \dfrac{\pi}{3} - 1 = 2\left(\dfrac{1}{2}\right) - 1 = 0$

(b) $2 \cos \dfrac{5\pi}{3} - 1 = 2\left(\dfrac{1}{2}\right) - 1 = 0$

7. $3 \tan^2 2x - 1 = 0$

(a) $3\left[\tan 2\left(\dfrac{\pi}{12}\right)\right]^2 - 1 = 3 \tan^2 \dfrac{\pi}{6} - 1$

$\qquad = 3\left(\dfrac{1}{\sqrt{3}}\right)^2 - 1$

$\qquad = 0$

(b) $3\left[\tan 2\left(\dfrac{5\pi}{12}\right)\right]^2 - 1 = 3 \tan^2 \dfrac{5\pi}{6} - 1$

$\qquad = 3\left(-\dfrac{1}{\sqrt{3}}\right)^2 - 1$

$\qquad = 0$

9. $2 \sin^2 x - \sin x - 1 = 0$

(a) $2 \sin^2 \dfrac{\pi}{2} - \sin \dfrac{\pi}{2} - 1 = 2(1)^2 - 1 - 1$

$\qquad = 0$

(b) $2 \sin^2 \dfrac{7\pi}{6} - \sin \dfrac{7\pi}{6} - 1 = 2\left(-\dfrac{1}{2}\right)^2 - \left(-\dfrac{1}{2}\right) - 1$

$\qquad = \dfrac{1}{2} + \dfrac{1}{2} - 1$

$\qquad = 0$

11. $2 \cos x + 1 = 0$

$\qquad 2 \cos x = -1$

$\qquad \cos x = -\dfrac{1}{2}$

$\qquad x = \dfrac{2\pi}{3} + 2n\pi$

$\qquad \text{or } x = \dfrac{4\pi}{3} + 2n\pi$

13. $\sqrt{3} \csc x - 2 = 0$

$\qquad \sqrt{3} \csc x = 2$

$\qquad \csc x = \dfrac{2}{\sqrt{3}}$

$\qquad x = \dfrac{\pi}{3} + 2n\pi$

$\qquad \text{or } x = \dfrac{2\pi}{3} + 2n\pi$

15. $3 \sec^2 x - 4 = 0$

$\qquad \sec x = \pm \dfrac{2}{\sqrt{3}}$

$\qquad x = \dfrac{\pi}{6} + n\pi$

$\qquad \text{or } x = \dfrac{5\pi}{6} + n\pi$

17. $2 \sin^2 2x = 1$

$\sin 2x = \pm\dfrac{1}{\sqrt{2}} = \pm\dfrac{\sqrt{2}}{2}$

$2x = \dfrac{\pi}{4} + 2n\pi, \ 2x = \dfrac{3\pi}{4} + 2n\pi, \ 2x = \dfrac{5\pi}{4} + 2n\pi, \ 2x = \dfrac{7\pi}{4} + 2n\pi,$

$2x = \dfrac{9\pi}{4} + 2n\pi, \ 2x = \dfrac{11\pi}{4} + 2n\pi, \ 2x = \dfrac{13\pi}{4} + 2n\pi, \ 2x = \dfrac{15\pi}{4} + 2n\pi$

Thus, $x = \dfrac{\pi}{8} + n\pi, \ \dfrac{3\pi}{8} + n\pi, \ \dfrac{5\pi}{8} + n\pi, \ \dfrac{7\pi}{8} + n\pi.$

19. $4 \sin^2 - 3 = 0$

$$\sin x = \pm \frac{\sqrt{3}}{2}$$

$$x = \frac{\pi}{3} + n\pi$$

$$\text{or } x = \frac{2\pi}{3} + n\pi$$

21.
$$\sin^2 x = 3 \cos^2 x$$

$$\sin^2 x - 3(1 - \sin^2 x) = 0$$

$$4 \sin^2 x = 3$$

$$\sin x = \pm \frac{\sqrt{3}}{2}$$

$$x = \frac{\pi}{3} + n\pi$$

$$\text{or } x = \frac{2\pi}{3} + n\pi$$

23. $(3 \tan^2 x - 1)(\tan^2 x - 3) = 0$

$3 \tan^2 x - 1 = 0 \qquad \text{or} \quad \tan^2 x - 3 = 0$

$\tan x = \pm \frac{1}{\sqrt{3}} \qquad\qquad \tan x = \pm \sqrt{3}$

$x = \frac{\pi}{6} + n\pi \qquad\qquad x = \frac{\pi}{3} + n\pi$

$\text{or } x = \frac{5\pi}{6} + n\pi \qquad \text{or } x = \frac{2\pi}{3} + n\pi$

25.
$$\cos^3 x = \cos x$$

$$\cos^3 x - \cos x = 0$$

$$\cos x(\cos^2 x - 1) = 0$$

$\cos x = 0 \qquad \text{or} \quad \cos^2 x - 1 = 0$

$x = \frac{\pi}{2}, \frac{3\pi}{2} \qquad\qquad \cos x = \pm 1$

$$x = 0, \ \pi$$

27. $3 \tan^3 x - \tan x = 0$

$\tan x(3 \tan^2 x - 1) = 0$

$\tan x = 0 \qquad \text{or} \quad 3 \tan^2 x - 1 = 0$

$x = 0, \ \pi \qquad\qquad \tan x = \pm \frac{\sqrt{3}}{3}$

$$x = \frac{\pi}{6}, \frac{5\pi}{6}, \frac{7\pi}{6}, \frac{11\pi}{6}$$

29.
$$\sec^2 x - \sec x - 2 = 0$$

$(\sec x - 2)(\sec x + 1) = 0$

$\sec x - 2 = 0 \qquad \text{or} \quad \sec x + 1 = 0$

$\sec x = 2 \qquad\qquad \sec x = -1$

$x = \frac{\pi}{3}, \frac{5\pi}{3} \qquad\qquad x = \pi$

31. $2 \sin x + \csc x = 0$

$$2 \sin x + \frac{1}{\sin x} = 0$$

$$2 \sin^2 x + 1 = 0$$

Since $2 \sin^2 x + 1 > 0$, there are no solutions.

33.
$$\csc x + \cot x = 1$$

$$\frac{1}{\sin x} + \frac{\cos x}{\sin x} = 1$$

$$1 + \cos x = \sin x$$

$$(1 + \cos x)^2 = \sin^2 x$$

$$1 + 2 \cos x + \cos^2 x = 1 - \cos^2 x$$

$$2 \cos^2 x + 2 \cos x = 0$$

$$2 \cos x(\cos x + 1) = 0$$

$\cos x = 0 \qquad \text{or} \qquad \cos x = -1$

$x = \frac{\pi}{2}, \frac{3\pi}{2} \qquad\qquad x = \pi$

($3\pi/2$ is extraneous.)   (π is extraneous.)

$x = \pi/2$ is the only solution.

35. $\cos\left(\dfrac{x}{2}\right) = \dfrac{\sqrt{2}}{2}$

$\dfrac{x}{2} = \dfrac{\pi}{4} + 2n\pi$

$x = \dfrac{\pi}{2} + 4n\pi$

$x = \dfrac{\pi}{2}$

37. $\dfrac{1 + \cos x}{1 - \cos x} = 0$

$1 + \cos x = 0$

$\cos x = -1$

$x = \pi$

39. $2\sec^2 x + \tan^2 x - 3 = 0$

$2(\tan^2 x + 1) + \tan^2 x - 3 = 0$

$3\tan^2 x - 1 = 0$

$\tan x = \pm\dfrac{\sqrt{3}}{3}$

$x = \dfrac{\pi}{6},\ \dfrac{5\pi}{6},\ \dfrac{7\pi}{6},\ \dfrac{11\pi}{6}$

41. $\dfrac{\cos x \cot x}{1 - \sin x} = 3$

Graph $y = \dfrac{\cos x}{(1 - \sin x)\tan x} - 3$.

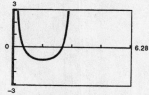

The solutions are approximately $x \approx 0.5236$, $x \approx 2.6180$

43. $4\sin^3 x + 2\sin^2 x - 2\sin x - 1 = 0$

Graph $y = 4\sin^3 x + 2\sin^2 x - 2\sin x - 1$.

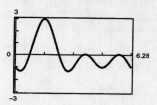

By altering the y-range to $\text{Ymin} = -.5$ and $\text{Ymax} = .5$, you see that there are 6 solutions: 0.7854, 2.3562, 3.6652, 3.9270, 5.4978, 5.7596.

45. $2\sin x - x = 0$

Graph $y_1 = 2\sin x - x$.

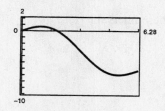

The x-intercepts occur at $x = 0$ and $x \approx 1.8955$.

47. $\sec^2 x + 0.5\tan x - 1 = 0$

Graph $y_1 = \dfrac{1}{(\cos x)^2} + 0.5\tan x - 1$.

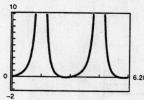

The x-intercepts occur at $x = 0$, $x \approx 2.6779$, $x = 3.1416$ and $x \approx 5.8195$.

49. $12 \sin^2 x - 13 \sin x + 3 = 0$

$(3 \sin x - 1)(4 \sin x - 3) = 0$

$3 \sin x - 1 = 0$ or $4 \sin x - 3 = 0$

$\sin x = \frac{1}{3}$ $\sin x = \frac{3}{4}$

$x = 0.3398,\ 2.8018$ $x = 0.8481,\ 2.2935$

Graph $y_1 = 12 \sin^2 x - 13 \sin x + 3$.

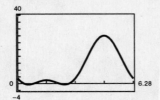

The x-intercepts occur at $x \approx 0.3398$, $x \approx 0.8481$, $x \approx 2.2935$, and $x \approx 2.8018$.

51. (a)

x	0	1	2	3	4	5	6
$f(x)$	Undef.	0.83	-1.36	-2.93	-4.46	-6.34	-13.02

The zero is in the interval $(1, 2)$ since f changes signs in the interval.

(b)

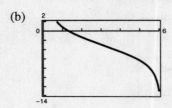

The interval is the same as part (a).

(c) 1.3065

53. (a)

x	0	1	2	3	4	5	6
$f(x)$	-1	1.39	1.65	-0.70	-1.94	-2.00	-1.48

The zeros are in the intervals $(0, 1)$ and $(2, 3)$ since f changes signs in these intervals.

(b)

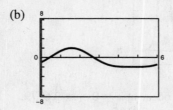

The interval is the same as part (a).

(c) 0.4271, 2.7145

55. (a) $f(x) = \sin x + \cos x$

Maximum: $\left(\dfrac{\pi}{4}, \sqrt{2}\right)$

Minimum: $\left(\dfrac{5\pi}{4}, -\sqrt{2}\right)$

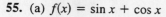

(b) $\cos x - \sin x = 0$

$\cos x = \sin x$

$1 = \dfrac{\sin x}{\cos x}$

$\tan x = 1$

$x = \dfrac{\pi}{4}, \dfrac{5\pi}{4}$

$f\left(\dfrac{\pi}{4}\right) = \sin\dfrac{\pi}{4} + \cos\dfrac{\pi}{4} = \dfrac{\sqrt{2}}{2} + \dfrac{\sqrt{2}}{2} = \sqrt{2}$

$f\left(\dfrac{5\pi}{4}\right) = \sin\dfrac{5\pi}{4} + \cos\dfrac{5\pi}{4} = -\sin\dfrac{\pi}{4} + \left(-\cos\dfrac{\pi}{4}\right) = -\dfrac{\sqrt{2}}{2} - \dfrac{\sqrt{2}}{2} = -\sqrt{2}$

Therefore, the maximum point in the interval $[0, 2\pi)$ is $\left(\pi/4, \sqrt{2}\right)$ and the minimum point is $\left(5\pi/4, -\sqrt{2}\right)$.

57. $f(x) = \tan\dfrac{\pi x}{4}$. $\tan 0 = 0$, but 0 is not positive. By graphing $y = \tan\dfrac{\pi x}{4} - x$, you see that the smallest positive fixed point is $x = 1$.

59. $f(x) = \cos\dfrac{1}{x}$

(a) The domain of $f(x)$ is all real numbers except 0.

(b) The graph has y-axis symmetry and a horizontal asymptote at $y = 1$.

(c) As $x \rightarrow 0$, $f(x)$ oscillates between -1 and 1.

(d) There are an infinite number of solutions in the interval $[-1, 1]$.

(e) The greatest solution appears to occur at $x \approx 0.6366$.

61.
$$y = \dfrac{1}{12}(\cos 8t - 3\sin 8t)$$

$\dfrac{1}{12}(\cos 8t - 3\sin 8t) = 0$

$\cos 8t = 3\sin 8t$

$\dfrac{1}{3} = \tan 8t$

$8t = 0.32175 + n\pi$

$t = 0.04 + \dfrac{n\pi}{8}$

In the interval $0 \le t \le 1$, $t = 0.04, 0.43$, and 0.83.

63. $r = \dfrac{1}{32}v_0{}^2 \sin 2\theta$, $r = 300$, $v_0 = 100$

$300 = \dfrac{1}{32}(100)^2 \sin 2\theta$

$\sin 2\theta = 0.96$

$2\theta \approx 1.287$ or $2\theta = \pi - 1.287 \approx 1.855$

$\theta \approx 0.6435 \approx 37°$ $\theta \approx 0.928 \approx 53°$

65. $A = 2x \cos x$, $-\dfrac{\pi}{2} < x < \dfrac{\pi}{2}$

(a)

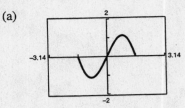

(b) $A \geq 1$ for $0.6 < x < 1.1$

The maximum area of $A \approx 1.12$ occurs when $x \approx 0.86$.

67. (a)

(b) By checking the graphs we see that

   (*iii*) $r = 1.05 \sin[0.95(t + 6.32)] + 6.20$ best fits the data.

(c) The constant term gives the rate of 6.20%.

(d) Period: $\dfrac{2\pi}{0.95} \approx 7$ years

(e) $r \approx 6.00$ when $t \approx 10.4$ which corresponds to late 2000.

Section 5.4   Sum and Difference Formulas

■ You should memorize the sum and difference formulas.

$\sin(u \pm v) = \sin u \cos v \pm \cos u \sin v$

$\cos(u \pm v) = \cos u \cos v \mp \sin u \sin v$

$\tan(u \pm v) = \dfrac{\tan u \pm \tan v}{1 \mp \tan u \tan v}$

■ You should be able to use these formulas to find the values of the trigonometric functions of angles whose sums or differences are special angles.

■ You should be able to use these formulas to solve trigonometric equations.

Solutions to Odd-Numbered Exercises

1. (a) $\cos\left(\dfrac{\pi}{4} + \dfrac{\pi}{3}\right) = \cos\dfrac{\pi}{4}\cos\dfrac{\pi}{3} - \sin\dfrac{\pi}{4}\sin\dfrac{\pi}{3}$

$= \dfrac{\sqrt{2}}{2} \cdot \dfrac{1}{2} - \dfrac{\sqrt{2}}{2} \cdot \dfrac{\sqrt{3}}{2}$

$= \dfrac{\sqrt{2} - \sqrt{6}}{4}$

(b) $\cos\dfrac{\pi}{4} + \cos\dfrac{\pi}{3} = \dfrac{\sqrt{2}}{2} + \dfrac{1}{2} = \dfrac{\sqrt{2} + 1}{2}$

3. (a) $\sin\left(\dfrac{7\pi}{6} - \dfrac{\pi}{3}\right) = \sin\dfrac{5\pi}{6} = \sin\dfrac{\pi}{6} = \dfrac{1}{2}$

(b) $\sin\dfrac{7\pi}{6} - \sin\dfrac{\pi}{3} = -\dfrac{1}{2} - \dfrac{\sqrt{3}}{2} = \dfrac{-1 - \sqrt{3}}{2}$

5. Both statements are false. Parts (a) and (b) are unequal in Exercises 1–4.

7. $\sin 75° = \sin(30° + 45°)$

$= \sin 30° \cos 45° + \sin 45° \cos 30°$

$= \dfrac{1}{2} \cdot \dfrac{\sqrt{2}}{2} + \dfrac{\sqrt{2}}{2} \cdot \dfrac{\sqrt{3}}{2}$

$= \dfrac{\sqrt{2}}{4}(1 + \sqrt{3})$

$\cos 75° = \cos(30° + 45°)$

$= \cos 30° \cos 45° - \sin 30° \sin 45°$

$= \dfrac{\sqrt{3}}{2} \cdot \dfrac{\sqrt{2}}{2} - \dfrac{1}{2} \cdot \dfrac{\sqrt{2}}{2}$

$= \dfrac{\sqrt{2}}{4}(\sqrt{3} - 1)$

$\tan 75° = \tan(30° + 45°)$

$= \dfrac{\tan 30° + \tan 45°}{1 - \tan 30° \tan 45°}$

$= \dfrac{(\sqrt{3}/3) + 1}{1 - (\sqrt{3}/3)} = \dfrac{\sqrt{3} + 3}{3 - \sqrt{3}} \cdot \dfrac{3 + \sqrt{3}}{3 + \sqrt{3}}$

$= \dfrac{6\sqrt{3} + 12}{6} = \sqrt{3} + 2$

11. $\sin 195° = \sin(225° - 30°)$

$= \sin 225° \cos 30° - \sin 30° \cos 225°$

$= -\sin 45° \cos 30° + \sin 30° \cos 45°$

$= -\dfrac{\sqrt{2}}{2} \cdot \dfrac{\sqrt{3}}{2} + \dfrac{\sqrt{2}}{2} \cdot \dfrac{1}{2}$

$= \dfrac{\sqrt{2}}{4}(1 - \sqrt{3})$

$\cos 195° = \cos(225° - 30°)$

$= \cos 225° \cos 30° + \sin 225° \sin 30°$

$= -\cos 45° \cos 30° - \sin 45° \sin 30°$

$= \dfrac{\sqrt{2}}{2} \cdot \dfrac{\sqrt{2}}{2} - \dfrac{\sqrt{2}}{2} \cdot \dfrac{1}{2}$

$= -\dfrac{\sqrt{2}}{4}(\sqrt{3} + 1)$

$\tan 195° = \tan(225° - 30°)$

$= \dfrac{\tan 225° - \tan 30°}{1 + \tan 225° \tan 30°}$

$= \dfrac{\tan 45° - \tan 30°}{1 + \tan 45° \tan 30°}$

$= \dfrac{1 - (\sqrt{3}/3)}{1 + (\sqrt{3}/3)} = \dfrac{\sqrt{3}}{3 + \sqrt{3}} \cdot \dfrac{3 - \sqrt{3}}{3 - \sqrt{3}}$

$= \dfrac{12 - 6\sqrt{3}}{6} = 2 - \sqrt{3}$

9. $\sin 105° = \sin(60° + 45°)$

$= \sin 60° \cos 45° + \sin 45° \cos 60°$

$= \dfrac{\sqrt{3}}{2} \cdot \dfrac{\sqrt{2}}{2} + \dfrac{\sqrt{2}}{2} \cdot \dfrac{1}{2}$

$= \dfrac{\sqrt{2}}{4}(\sqrt{3} + 1)$

$\cos 105° = \cos(60° + 45°)$

$= \cos 60° \cos 45° - \sin 60° \sin 45°$

$= \dfrac{1}{2} \cdot \dfrac{\sqrt{2}}{2} - \dfrac{\sqrt{3}}{2} \cdot \dfrac{\sqrt{2}}{2}$

$= \dfrac{\sqrt{2}}{4}(1 - \sqrt{3})$

$\tan 105° = \tan(60° + 45°)$

$= \dfrac{\tan 60° + \tan 45°}{1 - \tan 60° \tan 45°}$

$= \dfrac{\sqrt{3} + 1}{1 - \sqrt{3}} = \dfrac{\sqrt{3} + 1}{1 - \sqrt{3}} \cdot \dfrac{1 + \sqrt{3}}{1 + \sqrt{3}}$

$= \dfrac{4 + 2\sqrt{3}}{-2} = -2 - \sqrt{3}$

13. $\sin \dfrac{11\pi}{12} = \sin\left(\dfrac{3\pi}{4} + \dfrac{\pi}{6}\right)$

$\qquad = \sin \dfrac{3\pi}{4} \cos \dfrac{\pi}{6} + \sin \dfrac{\pi}{6} \cos \dfrac{3\pi}{4}$

$\qquad = -\dfrac{\sqrt{2}}{2} \cdot \dfrac{\sqrt{3}}{2} + \dfrac{1}{2}\left(-\dfrac{\sqrt{2}}{2}\right)$

$\qquad = \dfrac{\sqrt{2}}{4}(\sqrt{3} - 1)$

$\cos \dfrac{11\pi}{12} = \cos\left(\dfrac{3\pi}{4} + \dfrac{\pi}{6}\right)$

$\qquad = \cos \dfrac{3\pi}{4} \cos \dfrac{\pi}{6} - \sin \dfrac{3\pi}{4} \sin \dfrac{\pi}{6}$

$\qquad = \dfrac{\sqrt{2}}{2} \cdot \dfrac{\sqrt{3}}{2} - \dfrac{\sqrt{2}}{2} \cdot \dfrac{1}{2}$

$\qquad = -\dfrac{\sqrt{2}}{4}(\sqrt{3} + 1)$

$\tan \dfrac{11\pi}{4} = \tan\left(\dfrac{3\pi}{4} + \dfrac{\pi}{6}\right)$

$\qquad = \dfrac{\tan(3\pi/4) - \tan(\pi/6)}{1 - \tan(3\pi/4)\tan(\pi/6)}$

$\qquad = \dfrac{-1 + (\sqrt{3}/3)}{1 - (-1)(\sqrt{3}/3)}$

$\qquad = \dfrac{-3 + \sqrt{3}}{3 + \sqrt{3}} \cdot \dfrac{3 - \sqrt{3}}{3 - \sqrt{3}}$

$\qquad = \dfrac{-12 + 6\sqrt{3}}{6} = -2 + \sqrt{3}$

15. $\sin \dfrac{17\pi}{12} = \sin\left(\dfrac{9\pi}{4} - \dfrac{5\pi}{6}\right)$

$\qquad = \sin \dfrac{9\pi}{4} \cos \dfrac{5\pi}{6} - \sin \dfrac{5\pi}{6} \cos \dfrac{9\pi}{4}$

$\qquad = \dfrac{\sqrt{2}}{2}\left(-\dfrac{\sqrt{3}}{2}\right) - \left(\dfrac{1}{2}\right)\left(\dfrac{\sqrt{2}}{2}\right)$

$\qquad = -\dfrac{\sqrt{2}}{4}(\sqrt{3} + 1)$

$\cos \dfrac{17\pi}{12} = \cos\left(\dfrac{9\pi}{4} - \dfrac{5\pi}{6}\right)$

$\qquad = \cos \dfrac{9\pi}{4} \cos \dfrac{5\pi}{6} + \sin \dfrac{9\pi}{4} \sin \dfrac{5\pi}{6}$

$\qquad = \dfrac{\sqrt{2}}{2}\left(-\dfrac{\sqrt{3}}{2}\right) + \dfrac{\sqrt{2}}{2}\left(\dfrac{1}{2}\right)$

$\qquad = \dfrac{\sqrt{2}}{4}\left(1 - \sqrt{3}\right)$

$\tan \dfrac{17\pi}{4} = \tan\left(\dfrac{9\pi}{4} - \dfrac{5\pi}{6}\right)$

$\qquad = \dfrac{\tan(9\pi/4) - \tan(5\pi/6)}{1 + \tan(9\pi/4)\tan(5\pi/6)}$

$\qquad = \dfrac{1 - (-\sqrt{3}/3)}{1 + (-\sqrt{3}/3)}$

$\qquad = \dfrac{3 + \sqrt{3}}{3 - \sqrt{3}} \cdot \dfrac{3 + \sqrt{3}}{3 + \sqrt{3}}$

$\qquad = \dfrac{12 + 6\sqrt{3}}{6} = 2 + \sqrt{3}$

17. $\cos 25° \cos 15° - \sin 25° \sin 15° = \cos(25° + 15°) = \cos 40°$

19. $\sin 230° \cos 30° - \cos 230° \sin 30° = \sin(230° - 30°) = \sin 200°$

21. $\dfrac{\tan 325° - \tan 86°}{1 + \tan 325° \tan 86°} = \tan(325° - 86°) = \tan 239°$

23. $\sin 3 \cos 1.2 - \cos 3 \sin 1.2 = \sin(3 - 1.2) = \sin 1.8$

25. $\dfrac{\tan 2x + \tan x}{1 - \tan 2x \tan x} = \tan(2x + x) = \tan 3x$

27.

x	0.2	0.4	0.6	0.8	1.0	1.2	1.4
y_1	0.9801	0.9211	0.8253	0.6967	0.5403	0.3624	0.1700
y_2	0.9801	0.9211	0.8253	0.6967	0.5403	0.3624	0.1700

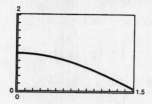

$$y_1 = \sin\left(\frac{\pi}{2} + x\right)$$
$$= \sin\frac{\pi}{2}\cos x + \sin x \cdot \cos\frac{\pi}{2}$$
$$= \cos x$$
$$= y_2$$

29.

x	0.2	0.4	0.6	0.8	1.0	1.2	1.4
y_1	0.6621	0.7978	0.9017	0.9696	0.9989	0.9883	0.9384
y_2	0.6621	0.7978	0.9017	0.9696	0.9989	0.9883	0.9384

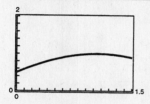

$$y_1 = \sin\left(\frac{\pi}{6} + x\right)$$
$$= \sin\frac{\pi}{6}\cos x + \sin x \cdot \cos\frac{\pi}{6}$$
$$= \frac{1}{2}\cos x + \frac{\sqrt{3}}{2}\sin x$$
$$= \frac{1}{2}(\cos x + \sqrt{3}\sin x)$$
$$= y_2$$

31.

x	0.2	0.4	0.6	0.8	1.0	1.2	1.4
y_1	0.9605	0.8484	0.6812	0.4854	0.2919	0.1313	0.0289
y_2	0.9605	0.8484	0.6812	0.4854	0.2919	0.1313	0.0289

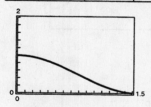

$$y_1 = \cos(x + \pi)\cos(x - \pi)$$
$$= (\cos x \cdot \cos\pi - \sin x \cdot \sin\pi)$$
$$\quad [\cos x \cos\pi + \sin x \sin\pi]$$
$$= [-\cos x][-\cos x]$$
$$= \cos^2 x$$
$$= y_2$$

For Exercises 33–35, we have:

$\sin u = \frac{5}{13}$, u **is in Quadrant I** $\implies \cos u = \frac{12}{13}$

$\cos v = -\frac{3}{5}$, v **is in Quadrant II** $\implies \sin v = \frac{4}{5}$

33. $\sin(u + v) = \sin u \cos v + \cos u \sin v$
$$= \left(\tfrac{5}{13}\right)\left(-\tfrac{3}{5}\right) + \left(\tfrac{12}{13}\right)\left(\tfrac{4}{5}\right)$$
$$= \frac{33}{65}$$

35. $\cos(u + v) = \cos u \cos v - \sin u \sin v$
$$= \left(\tfrac{12}{13}\right)\left(-\tfrac{3}{5}\right) - \left(\tfrac{5}{13}\right)\left(\tfrac{4}{5}\right)$$
$$= \frac{-56}{65}$$

For Exercises 37–39, we have:

$\sin u = \frac{7}{25}$, u is in Quadrant II $\implies$ $\cos u = \frac{-24}{25}$

$\cos v = \frac{4}{5}$, v is in Quadrant IV $\implies$ $\sin v = -\frac{3}{5}$

37. $\cos(u + v) = \cos u \cdot \cos v - \sin u \cdot \sin v$

$$= \left(-\frac{24}{25}\right)\left(\frac{4}{5}\right) - \left(\frac{7}{25}\right)\left(-\frac{3}{5}\right)$$

$$= \frac{-96 + 21}{125} = \frac{-75}{125} = \frac{-3}{5}$$

39. $\sin(v - u) = \sin u \cdot \cos v - \sin u \cdot \cos v$

$$= \left(-\frac{3}{5}\right)\left(-\frac{24}{25}\right) - \left(\frac{7}{25}\right)\left(\frac{4}{5}\right)$$

$$= \frac{72 - 28}{125} = \frac{44}{125}$$

41. $\cos(\pi - \theta) + \sin\left(\frac{\pi}{2} + \theta\right) = \cos \pi \cos \theta + \sin \pi \sin \theta + \sin \frac{\pi}{2} \cos \theta + \sin \theta \cos \frac{\pi}{2}$

$$= (-1)(\cos \theta) + (0)(\sin \theta) + (1)(\cos \theta) + (\sin \theta)(0) = -\cos \theta + \cos \theta = 0$$

43. $\sin(x + y) + \sin(x - y) = \sin x \cos y + \sin y \cos x + \sin x \cos y - \sin y \cos x = 2 \sin x \cos y$

45. $\cos(n\pi + \theta) = \cos n\pi \cos \theta - \sin n\pi \sin \theta$

$$= (-1)^n (\cos \theta) - (0)(\sin \theta)$$

$$= (-1)^n (\cos \theta), \text{ where } n \text{ is an integer.}$$

47. $C = \arctan \frac{b}{a} \implies \tan C = \frac{b}{a} \implies \sin C = \frac{b}{\sqrt{a^2 + b^2}}, \cos C = \frac{a}{\sqrt{a^2 + b^2}}$

$$\sqrt{a^2 + b^2} \sin(B\theta + C) = \sqrt{a^2 + b^2}\left(\sin B\theta \cdot \frac{a}{\sqrt{a^2 + b^2}} + \frac{b}{\sqrt{a^2 + b^2}} \cdot \cos B\theta\right) = a \sin B\theta + b \cos B\theta$$

49. The graph of $g(x) = \cos(\pi + x)$ looks like that of $f(x) = -\cos x$. Analytic

$$g(x) = \cos(\pi + x) = \cos \pi \cdot \cos x - \sin \pi \cdot \sin x = -\cos x.$$

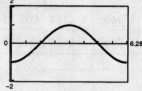

51. $\sin \theta + \cos \theta$

$a = 1$, $b = 1$, $B = 1$

(a) $C = \arctan \frac{b}{a} = \arctan 1 = \frac{\pi}{4}$

$\sin \theta + \cos \theta = \sqrt{a^2 + b^2} \sin(B\theta + C)$

$$= \sqrt{2} \sin\left(\theta + \frac{\pi}{4}\right)$$

(b) $C = \arctan \frac{a}{b} = \arctan 1 = \frac{\pi}{4}$

$\sin \theta + \cos \theta = \sqrt{a^2 + b^2} \cos(B\theta - C)$

$$= \sqrt{2} \cos\left(\theta - \frac{\pi}{4}\right)$$

53. $12 \sin 3\theta + 5 \cos 3\theta$

$a = 12$, $b = 5$, $B = 3$

(a) $C = \arctan \frac{b}{a} = \arctan \frac{5}{12} \approx 0.3948$

$12 \sin 3\theta + 5 \cos 3\theta = \sqrt{a^2 + b^2} \sin(B\theta + C)$

$$\approx 13 \sin(3\theta + 0.3948)$$

(b) $C = \arctan \frac{a}{b} = \arctan \frac{12}{5} \approx 1.1760$

$12 \sin 3\theta + 5 \cos 3\theta = \sqrt{a^2 + b^2} \cos(B\theta - C)$

$$\approx 13 \cos(3\theta - 1.1760)$$

55. $C = \arctan \dfrac{b}{a} = \dfrac{\pi}{2} \;\Longrightarrow\; a = 0$

$\sqrt{a^2 + b^2} = 2 \;\Longrightarrow\; b = 2$

$B = 1$

$2 \sin\!\left(\theta + \dfrac{\pi}{2}\right) = (0)(\sin\theta) + (2)(\cos\theta) = 2\cos\theta$

57. $\sin(\arcsin x + \arccos x) = \sin(\arcsin x)\cos(\arccos x) + \sin(\arccos x)\cos(\arcsin x)$

$$= x \cdot x + \sqrt{1 - x^2} \cdot \sqrt{1 - x^2}$$

$$= x^2 + 1 - x^2$$

$$= 1$$

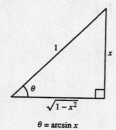

$\theta = \arcsin x$ $\qquad\qquad\qquad$ $\theta = \arccos x$

59.
$$\sin\!\left(x + \dfrac{\pi}{3}\right) + \sin\!\left(x - \dfrac{\pi}{3}\right) = 1$$

$$\sin x \cos\dfrac{\pi}{3} + \cos x \sin\dfrac{\pi}{3} + \sin x \cos\dfrac{\pi}{3} - \cos x \sin\dfrac{\pi}{3} = 1$$

$$2\sin x(0.5) = 1$$

$$\sin x = 1$$

$$x = \dfrac{\pi}{2}$$

61.
$$\cos\!\left(x + \dfrac{\pi}{4}\right) - \cos\!\left(x - \dfrac{\pi}{4}\right) = 1$$

$$\cos x \cos\dfrac{\pi}{4} - \sin x \sin\dfrac{\pi}{4} - \left(\cos x \cos\dfrac{\pi}{4} + \sin x \sin\dfrac{\pi}{4}\right) = 1$$

$$-2\sin x\!\left(\dfrac{\sqrt{2}}{2}\right) = 1$$

$$-\sqrt{2}\,\sin x = 1$$

$$\sin x = -\dfrac{1}{\sqrt{2}}$$

$$\sin x = -\dfrac{\sqrt{2}}{2}$$

$$x = \dfrac{5\pi}{4},\ \dfrac{7\pi}{4}$$

63. Graph $y_1 = \cos\left(x + \dfrac{\pi}{4}\right) + \cos\left(x - \dfrac{\pi}{4}\right)$ and $y_2 = 1$.

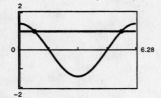

The points of intersection occur at $x = \dfrac{\pi}{4}$ and $x = \dfrac{7\pi}{4}$.

65. $y_1 + y_2 = A \cos 2\pi\left(\dfrac{t}{T} - \dfrac{x}{\lambda}\right) + A \cos 2\pi\left(\dfrac{t}{T} + \dfrac{x}{\lambda}\right)$

$\qquad = A\left[\cos\left(\dfrac{2\pi t}{T}\right)\cos\left(\dfrac{2\pi x}{\lambda}\right) + \sin\left(\dfrac{2\pi t}{T}\right)\sin\left(\dfrac{2\pi x}{\lambda}\right)\right] + A\left[\cos\left(\dfrac{2\pi t}{T}\right)\cos\left(\dfrac{2\pi x}{\lambda}\right) - \sin\left(\dfrac{2\pi t}{T}\right)\sin\left(\dfrac{2\pi x}{\lambda}\right)\right]$

$\qquad = 2A \cos\left(\dfrac{2\pi t}{T}\right)\cos\left(\dfrac{2\pi x}{\lambda}\right)$

67. $y = \dfrac{1}{3}\sin 2t + \dfrac{1}{4}\cos 2t$

(a)

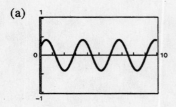

(b) $a = \dfrac{1}{3}$, $b = \dfrac{1}{4}$, $B = 2$

$\qquad C = \arctan\dfrac{b}{a} = \arctan\dfrac{3}{4} \approx 0.6435$

$\qquad y \approx \sqrt{\left(\tfrac{1}{3}\right)^2 + \left(\tfrac{1}{4}\right)^2}\,\sin(2t + 0.6435)$

$\qquad = \dfrac{5}{12}\sin(2t + 0.6435)$

(c) Amplitude: $\dfrac{5}{12}$

(d) Frequency: $\dfrac{1}{\text{period}} = \dfrac{b}{2\pi} = \dfrac{2}{2\pi} = \dfrac{1}{\pi}$

69. $f(h) = \dfrac{\cos\left(\dfrac{\pi}{6} + h\right) - \cos\dfrac{\pi}{6}}{h}$

$g(h) = \cos\dfrac{\pi}{6}\left(\dfrac{\cos h - 1}{h}\right) - \sin\dfrac{\pi}{6}\left(\dfrac{\sin h}{h}\right)$

(a) The domains are both $(-\infty, 0)$, $(0, \infty)$.

(b)

h	0.01	0.02	0.05	0.1	0.2	0.5
$f(h)$	-0.5043	-0.5086	-0.5214	-0.5424	-0.5830	-0.6915
$g(h)$	-0.5043	-0.5086	-0.5214	-0.5424	-0.5830	-0.6915

—CONTINUED—

69. —CONTINUED—

(c)
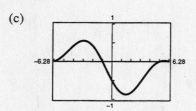

(d) The values tend to $y = -\dfrac{1}{2}$.

Section 5.5   Multiple-Angle and Product-to-Sum Formulas

■ You should know the following double-angle formulas.

(a) $\sin 2u = 2 \sin u \cos u$

(b) $\cos 2u = \cos^2 u - \sin^2 u$

$\qquad = 2\cos^2 u - 1$

$\qquad = 1 - 2\sin^2 u$

(c) $\tan 2u = \dfrac{2 \tan u}{1 - \tan^2 u}$

■ You should be able to reduce the power of a trigonometric function.

(a) $\sin^2 u = \dfrac{1 - \cos 2u}{2}$ (b) $\cos^2 u = \dfrac{1 + \cos 2u}{2}$ (c) $\tan^2 u = \dfrac{1 - \cos 2u}{1 + \cos 2u}$

■ You should be able to use the half-angle formulas.

(a) $\sin \dfrac{u}{2} = \pm \sqrt{\dfrac{1 - \cos u}{2}}$ (b) $\cos \dfrac{u}{2} = \pm \sqrt{\dfrac{1 + \cos u}{2}}$ (c) $\tan \dfrac{u}{2} = \dfrac{1 - \cos u}{\sin u} = \dfrac{\sin u}{1 + \cos u}$

■ You should be able to use the product-sum formulas.

(a) $\sin u \sin v = \dfrac{1}{2}[\cos(u - v) - \cos(u + v)]$ (b) $\cos u \cos v = \dfrac{1}{2}[\cos(u - v) + \cos(u + v)]$

(c) $\sin u \cos v = \dfrac{1}{2}[\sin(u + v) + \sin(u - v)]$ (d) $\cos u \sin v = \dfrac{1}{2}[\sin(u + v) - \sin(u - v)]$

■ You should be able to use the sum-product formulas.

(a) $\sin x + \sin y = 2 \sin\left(\dfrac{x + y}{2}\right)\cos\left(\dfrac{x - y}{2}\right)$ (b) $\sin x - \sin y = 2 \cos\left(\dfrac{x + y}{2}\right)\sin\left(\dfrac{x - y}{2}\right)$

(c) $\cos x + \cos y = 2 \cos\left(\dfrac{x + y}{2}\right)\cos\left(\dfrac{x - y}{2}\right)$ (d) $\cos x - \cos y = -2 \sin\left(\dfrac{x + y}{2}\right)\sin\left(\dfrac{x - y}{2}\right)$

Solutions to Odd-Numbered Exercises

Figure for Exercises 1–7

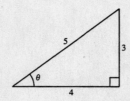

$$\sin \theta = \tfrac{3}{5}$$
$$\cos \theta = \tfrac{4}{5}$$
$$\tan \theta = \tfrac{3}{4}$$

1. $\sin \theta = \tfrac{3}{5}$

3. $\cos 2\theta = 2 \cos^2 \theta - 1$

$$= 2\left(\tfrac{4}{5}\right)^2 - 1$$
$$= \tfrac{32}{25} - \tfrac{25}{25}$$
$$= \tfrac{7}{25}$$

5. $\tan 2\theta = \dfrac{2 \tan \theta}{1 - \tan^2 \theta}$

$$= \dfrac{2(3/4)}{1 - (3/4)^2}$$
$$= \dfrac{3/2}{1 - (9/16)}$$
$$= \dfrac{3}{2} \cdot \dfrac{16}{7}$$
$$= \dfrac{24}{7}$$

7. $\csc 2\theta = \dfrac{1}{\sin 2\theta}$

$$= \dfrac{1}{2 \sin \theta \cos \theta}$$
$$= \dfrac{1}{2(3/5)(4/5)}$$
$$= \dfrac{25}{24}$$

9.
$$\sin 2x - \sin x = 0$$
$$2 \sin x \cos x - \sin x = 0$$
$$\sin x(2 \cos x - 1) = 0$$

$\sin x = 0 \quad$ or $\quad 2 \cos x - 1 = 0$

$x = 0, \pi \qquad\qquad \cos x = \dfrac{1}{2}$

$$x = \dfrac{\pi}{3}, \dfrac{5\pi}{3}$$

$$x = 0, \dfrac{\pi}{3}, \pi, \dfrac{5\pi}{3}$$

11. $4 \sin x \cos x = 1$

$$2 \sin 2x = 1$$
$$\sin 2x = \dfrac{1}{2}$$

$2x = \dfrac{\pi}{6} + 2n\pi \quad$ or $\quad 2x = \dfrac{5\pi}{6} + 2n\pi$

$x = \dfrac{\pi}{12} + n\pi \qquad\qquad x = \dfrac{5\pi}{12} + n\pi$

$x = \dfrac{\pi}{12}, \dfrac{13\pi}{12} \qquad\qquad x = \dfrac{5\pi}{12}, \dfrac{17\pi}{12}$

13.
$$\cos 2x = \cos x$$
$$\cos^2 x - \sin^2 x = \cos x$$
$$\cos^2 x(1 - \cos^2 x) - \cos x = 0$$
$$2\cos^2 x - \cos x - 1 = 0$$
$$(2\cos x + 1)(\cos x - 1) = 0$$
$$2\cos x + 1 = 0 \qquad \text{or} \quad \cos x - 1 = 0$$
$$\cos x = -\frac{1}{2} \qquad\qquad \cos x = 1$$
$$x = \frac{2\pi}{3}, \frac{4\pi}{3} \qquad\qquad x = 0$$

15.
$$\tan 2x - \cot x = 0$$
$$\frac{2\tan x}{1 - \tan^2 x} = \cot x$$
$$2\tan x = \cot x(1 - \tan^2 x)$$
$$2\tan x = \cot x - \cot x \tan^2 x$$
$$2\tan x = \cot x - \tan x$$
$$3\tan x = \cot x$$
$$3\tan x - \cot x = 0$$
$$3\tan x - \frac{1}{\tan x} = 0$$
$$\frac{3\tan^2 x - 1}{\tan x} = 0$$
$$\frac{1}{\tan x}(3\tan^2 x - 1) = 0$$
$$\cot x(3\tan^2 x - 1) = 0$$
$$\cot x = 0 \qquad \text{or} \quad 3\tan^2 x - 1 = 0$$
$$x = \frac{\pi}{2}, \frac{3\pi}{2} \qquad\qquad \tan^2 x = \frac{1}{3}$$
$$\tan x = \pm\frac{\sqrt{3}}{3}$$
$$x = \frac{\pi}{6}, \frac{5\pi}{6}, \frac{7\pi}{6}, \frac{11\pi}{6}$$
$$x = \frac{\pi}{6}, \frac{\pi}{2}, \frac{5\pi}{6}, \frac{7\pi}{6}, \frac{3\pi}{2}, \frac{11\pi}{6}$$

17.
$$\sin 4x = -2\sin 2x$$
$$\sin 4x + 2\sin 2x = 0$$
$$2\sin 2x \cos 2x + 2\sin 2x = 0$$
$$2\sin 2x(\cos 2x + 1) = 0$$
$$2\sin 2x = 0 \qquad \text{or} \quad \cos 2x + 1 = 0$$
$$\sin 2x = 0 \qquad\qquad \cos 2x = -1$$
$$2x = n\pi \qquad\qquad 2x = \pi + 2n\pi$$
$$x = \frac{n}{2}\pi \qquad\qquad x = \frac{\pi}{2} + n\pi$$
$$x = 0, \frac{\pi}{2}, \pi, \frac{3\pi}{2} \qquad\qquad x = \frac{\pi}{2}, \frac{3\pi}{2}$$

19. $f(x) = 6\sin x \cos x$
$$= 3(2\sin x \cos x)$$
$$= 3\sin 2x$$

21. $g(x) = 4 - 8\sin^2 x$
$$= 4(1 - 2\sin^2 x)$$
$$= 4\cos 2x$$

23. $\sin u = \dfrac{3}{5}, \ 0 < u < \dfrac{\pi}{2} \ \Rightarrow \ \cos v = \dfrac{4}{5}$

$\sin 2u = 2 \sin u \cos u = 2 \cdot \dfrac{3}{5} \cdot \dfrac{4}{5} = \dfrac{24}{25}$

$\cos 2u = \cos^2 u - \sin^2 u = \dfrac{16}{25} - \dfrac{9}{25} = \dfrac{7}{25}$

$\tan 2u = \dfrac{2 \tan u}{1 - \tan^2 u} = \dfrac{2(3/4)}{1 - (9/16)} = \dfrac{24}{7}$

25. $\tan u = \dfrac{1}{2}, \ \pi < u < \dfrac{3\pi}{2} \ \Rightarrow \ \sin u = -\dfrac{1}{\sqrt{5}}$ and

$$\cos u = -\dfrac{2}{\sqrt{5}}$$

$\sin 2u = 2 \sin u \cos u = 2\left(-\dfrac{1}{\sqrt{5}}\right)\left(-\dfrac{2}{\sqrt{5}}\right) = \dfrac{4}{5}$

$\cos 2u = \cos^2 u - \sin^2 u = \left(-\dfrac{2}{\sqrt{5}}\right)^2 - \left(-\dfrac{1}{\sqrt{5}}\right)^2 = \dfrac{3}{5}$

$\tan 2u = \dfrac{2 \tan u}{1 - \tan^2 u} = \dfrac{2(1/2)}{1 - (1/4)} = \dfrac{4}{3}$

27. $\cos^4 x = (\cos^2 x)(\cos^2 x) = \left(\dfrac{1 + \cos 2x}{2}\right)\left(\dfrac{1 + \cos 2x}{2}\right) = \dfrac{1 + 2\cos 2x + \cos^2 2x}{4}$

$$= \dfrac{1 + 2\cos 2x + (1 + \cos 4x)/2}{4}$$

$$= \dfrac{2 + 4\cos 2x + 1 + \cos 4x}{8}$$

$$= \dfrac{3 + 4\cos 2x + \cos 4x}{8}$$

$$= \dfrac{1}{8}(3 + 4\cos 2x + \cos 4x)$$

29. $(\sin^2 x)(\cos^2 x) = \left(\dfrac{1 - \cos 2x}{2}\right)\left(\dfrac{1 + \cos 2x}{2}\right)$

$$= \dfrac{1 - \cos^2 2x}{4}$$

$$= \dfrac{1}{4}\left(1 - \dfrac{1 + \cos 4x}{2}\right)$$

$$= \dfrac{1}{8}(2 - 1 - \cos 4x)$$

$$= \dfrac{1}{8}(1 - \cos 4x)$$

31. $\sin^2 x \cos^4 x = \sin^2 x \cos^2 x \cos^2 x = \left(\dfrac{1 - \cos 2x}{2}\right)\left(\dfrac{1 + \cos 2x}{2}\right)\left(\dfrac{1 + \cos 2x}{2}\right)$

$$= \frac{1}{8}(1 - \cos 2x)(1 + \cos 2x)(1 + \cos 2x)$$

$$= \frac{1}{8}(1 - \cos^2 2x)(1 + \cos 2x)$$

$$= \frac{1}{8}(1 + \cos 2x - \cos^2 2x - \cos^3 2x)$$

$$= \frac{1}{8}\left[1 + \cos 2x - \left(\frac{1 + \cos 4x}{2}\right) - \cos 2x\left(\frac{1 + \cos 4x}{2}\right)\right]$$

$$= \frac{1}{16}[2 + 2\cos 2x - 1 - \cos 4x - \cos 2x - \cos 2x \cos 4x]$$

$$= \frac{1}{16}\left[1 + \cos 2x - \cos 4x - \left(\frac{1}{2}\cos 2x + \frac{1}{2}\cos 6x\right)\right]$$

$$= \frac{1}{32}(2 + 2\cos 2x - 2\cos 4x - \cos 2x - \cos 6x)$$

$$= \frac{1}{32}(2 + \cos 2x - 2\cos 4x - \cos 6x)$$

Figure for Exercises 33 – 37

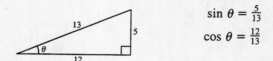

$\sin \theta = \frac{5}{13}$

$\cos \theta = \frac{12}{13}$

33. $\cos \dfrac{\theta}{2} = \sqrt{\dfrac{1 + \cos \theta}{2}} = \sqrt{\dfrac{1 + (12/13)}{2}} = \sqrt{\dfrac{25}{26}} = \dfrac{5}{\sqrt{26}}$

35. $\tan \dfrac{\theta}{2} = \dfrac{\sin \theta}{1 + \cos \theta} = \dfrac{5/13}{1 + (12/13)} = \dfrac{5}{25} = \dfrac{1}{5}$

37. $\csc \dfrac{\theta}{2} = \dfrac{1}{\sin(\theta/2)} = \dfrac{1}{\sqrt{(1 - \cos \theta/2)}} = \dfrac{1}{\sqrt{\left(1 - \frac{12}{13}\right)/2}} = \dfrac{1}{\sqrt{1/26}} = \sqrt{26}$

39. $\sin 105° = \sin\left(\dfrac{1}{2} \cdot 210°\right) = \sqrt{\dfrac{1 - \cos 210°}{2}} = \sqrt{\dfrac{1 + (\sqrt{3}/2)}{2}} = \dfrac{1}{2}\sqrt{2 + \sqrt{3}}$

$\cos 105° = \cos\left(\dfrac{1}{2} \cdot 210°\right) = \sqrt{\dfrac{1 + \cos 210°}{2}} = \sqrt{\dfrac{1 - (\sqrt{3}/2)}{2}} = -\dfrac{1}{2}\sqrt{2 - \sqrt{3}}$

$\tan 105° = \tan\left(\dfrac{1}{2} \cdot 210°\right) = \dfrac{\sin 210°}{1 + \cos 210°} = \dfrac{-1/2}{1 - (\sqrt{3}/2)} = -2 - \sqrt{3}$

41. $\sin 112° 30' = \sin\left(\dfrac{1}{2} \cdot 225°\right) = \sqrt{\dfrac{1 - \cos 225°}{2}} = \sqrt{\dfrac{1 + (\sqrt{2}/2)}{2}} = \dfrac{1}{2}\sqrt{2 + \sqrt{2}}$

$\cos 112° 30' = \cos\left(\dfrac{1}{2} \cdot 225°\right) = -\sqrt{\dfrac{1 + \cos 225°}{2}} = -\sqrt{\dfrac{1 - (\sqrt{2}/2)}{2}} = -\dfrac{1}{2}\sqrt{2 - \sqrt{2}}$

$\tan 112° 30' = \tan\left(\dfrac{1}{2} \cdot 225°\right) = \dfrac{\sin 225°}{1 + \cos 225°} = \dfrac{-\sqrt{2}/2}{1 - (\sqrt{2}/2)} = -1 - \sqrt{2}$

43. $\sin \dfrac{\pi}{8} = \sin\left[\dfrac{1}{2}\left(\dfrac{\pi}{4}\right)\right] = \sqrt{\dfrac{1 - \cos(\pi/4)}{2}} = \dfrac{1}{2}\sqrt{2 - \sqrt{2}}$

$\cos \dfrac{\pi}{8} = \cos\left[\dfrac{1}{2}\left(\dfrac{\pi}{4}\right)\right] = \sqrt{\dfrac{1 + \cos(\pi/4)}{2}} = \dfrac{1}{2}\sqrt{2 + \sqrt{2}}$

$\tan \dfrac{\pi}{8} = \tan\left[\dfrac{1}{2}\left(\dfrac{\pi}{4}\right)\right] = \dfrac{\sin(\pi/4)}{1 + \cos(\pi/4)} = \dfrac{\sqrt{2}/2}{1 + (\sqrt{2}/2)} = \sqrt{2} - 1$

45. $\sin u = \dfrac{5}{13}, \dfrac{\pi}{2} < u < \pi \implies \cos u = -\dfrac{12}{13}$

$\sin\left(\dfrac{u}{2}\right) = \sqrt{\dfrac{1 - \cos u}{2}} = \sqrt{\dfrac{1 + (12/13)}{2}} = \dfrac{5\sqrt{26}}{26}$

$\cos\left(\dfrac{u}{2}\right) = \sqrt{\dfrac{1 + \cos u}{2}} = \sqrt{\dfrac{1 - (12/13)}{2}} = \dfrac{\sqrt{26}}{26}$

$\tan\left(\dfrac{u}{2}\right) = \dfrac{\sin u}{1 + \cos u} = \dfrac{5/13}{1 - (12/13)} = \dfrac{5}{1} = 5$

47. $\tan u = -\dfrac{5}{8}, \dfrac{3\pi}{2} < u < 2\pi \implies \sin u = -\dfrac{8}{\sqrt{89}}$ and $\cos u = \dfrac{5}{\sqrt{89}}$

$\sin\left(\dfrac{u}{2}\right) = \sqrt{\dfrac{1 - \cos u}{2}} = \sqrt{\dfrac{1 - (8/\sqrt{89})}{2}} \sqrt{\dfrac{\sqrt{89} - 8}{2\sqrt{89}}} = \sqrt{\dfrac{89 - 8\sqrt{89}}{178}}$

$\cos\left(\dfrac{u}{2}\right) = -\sqrt{\dfrac{1 + \cos u}{2}} = -\sqrt{\dfrac{1 + (8/\sqrt{89})}{2}} = -\sqrt{\dfrac{\sqrt{89} + 8}{2\sqrt{89}}} = -\sqrt{\dfrac{89 + 8\sqrt{89}}{178}}$

$\tan\left(\dfrac{u}{2}\right) = \dfrac{1 - \cos u}{\sin u} = \dfrac{1 - (8/\sqrt{89})}{-5/\sqrt{89}} = \dfrac{8 - \sqrt{89}}{5}$

49. $\sqrt{\dfrac{1 - \cos 6x}{2}} = |\sin 3x|$

51. $-\sqrt{\dfrac{1 - \cos 8x}{1 + \cos 8x}} = -\dfrac{\sqrt{(1 - \cos 8x)/2}}{\sqrt{(1 + \cos 8x)/2}}$

$= -\left|\dfrac{\sin 4x}{\cos 4x}\right|$

$= -|\tan 4x|$

53. $\sin\dfrac{x}{2} + \cos x = 0$

$$\pm\sqrt{\dfrac{1-\cos x}{2}} = -\cos x$$

$$\dfrac{1-\cos x}{2} = \cos^2 x$$

$$0 = 2\cos^2 x + \cos x - 1$$

$$= (2\cos x - 1)(\cos x + 1)$$

$$\cos x = \dfrac{1}{2} \quad \text{or} \quad \cos x = -1$$

$$x = \dfrac{\pi}{3}, \dfrac{5\pi}{3} \qquad x = \pi$$

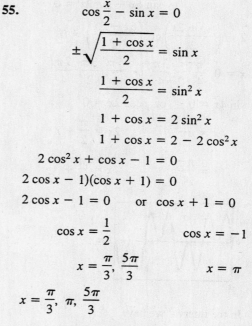

By checking these values in the original equations, we see that $x = \pi/3$ and $x = 5\pi/3$ are extraneous, and $x = \pi$ is the only solution.

55. $\cos\dfrac{x}{2} - \sin x = 0$

$$\pm\sqrt{\dfrac{1+\cos x}{2}} = \sin x$$

$$\dfrac{1+\cos x}{2} = \sin^2 x$$

$$1 + \cos x = 2\sin^2 x$$

$$1 + \cos x = 2 - 2\cos^2 x$$

$$2\cos^2 x + \cos x - 1 = 0$$

$$2\cos x - 1)(\cos x + 1) = 0$$

$$2\cos x - 1 = 0 \quad \text{or} \quad \cos x + 1 = 0$$

$$\cos x = \dfrac{1}{2} \qquad\qquad \cos x = -1$$

$$x = \dfrac{\pi}{3}, \dfrac{5\pi}{3} \qquad\qquad x = \pi$$

$$x = \dfrac{\pi}{3}, \ \pi, \ \dfrac{5\pi}{3}$$

$\pi/3$, π, and $5\pi/3$ are all solutions to the equation.

57. $6\sin\dfrac{\pi}{4}\cos\dfrac{\pi}{4} = 6\cdot\dfrac{1}{2}\left[\sin\left(\dfrac{\pi}{4}+\dfrac{\pi}{4}\right) + \sin\left(\dfrac{\pi}{4}-\dfrac{\pi}{4}\right)\right] = 3\left(\sin\dfrac{\pi}{2} + \sin 0\right)$

59. $\sin 5\theta \cos 3\theta = \dfrac{1}{2}[\sin(5\theta + 3\theta) + \sin(5\theta - 3\theta) = \dfrac{1}{2}(\sin 8\theta + \sin 2\theta)$

61. $5\cos(-5\beta)\cos 3\beta = 5\cdot\dfrac{1}{2}[\cos(-5\beta - 3\beta) + \cos(-5\beta + 3\beta)] = \dfrac{5}{2}[\cos(-8\beta) + \cos(-2\beta)]$

$$= \dfrac{5}{2}(\cos 8\beta + \cos 2\beta)$$

63. $\sin 60° + \sin 30° = 2\sin\left(\dfrac{60° + 30°}{2}\right)\cos\left(\dfrac{60° - 30°}{2}\right) = 2\sin 45° \cos 15°$

65. $\cos\dfrac{3\pi}{4} - \cos\dfrac{\pi}{4} = -2\sin\left(\dfrac{(3\pi/4) + (\pi/4)}{2}\right) + \left(\dfrac{(3\pi/4) - (\pi/4)}{2}\right) = -2\sin\dfrac{\pi}{2}\sin\dfrac{\pi}{4}$

67. $\cos 6x + \cos 2x = 2\cos\left(\dfrac{6x + 2x}{2}\right)\cos\left(\dfrac{6x - 2x}{2}\right) = 2\cos 4x \cos 2x$

69. $\sin(\alpha + \beta) - \sin(\alpha - \beta) = 2\cos\left(\dfrac{\alpha + \beta + \alpha - \beta}{2}\right)\sin\left(\dfrac{\alpha + \beta - \alpha + \beta}{2}\right) = 2\cos\alpha\sin\beta$

71. $\cos(\phi + 2\pi) + \cos\phi = 2\cos\left(\dfrac{\phi + 2\pi + \phi}{2}\right)\cos\left(\dfrac{\phi + 2\pi - \phi}{2}\right) = 2\cos(\phi + \pi)\cos\pi$

73.
$$\sin 6x + \sin 2x = 0$$
$$2\sin\left(\frac{6x+2x}{2}\right)\cos\left(\frac{6x-2x}{2}\right) = 0$$
$$x = 0, \frac{\pi}{4}, \frac{\pi}{2}, \frac{3\pi}{4}, \pi, \frac{5\pi}{4}, \frac{3\pi}{2}, \frac{7\pi}{4}$$
$$\sin 4x = 0 \quad \text{or} \quad \cos 2x = 0$$
$$4x = n\pi \qquad\qquad 2x = \frac{\pi}{2} + n\pi$$
$$x = \frac{n\pi}{4} \qquad\qquad x = \frac{\pi}{4} + \frac{n\pi}{2}$$

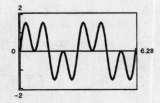

In the interval we have
$$x = 0, \frac{\pi}{4}, \frac{\pi}{2}, \frac{3\pi}{4}, \pi, \frac{5\pi}{4}, \frac{3\pi}{2}, \frac{7\pi}{4}.$$

75. $\dfrac{\cos 2x}{\sin 3x - \sin x} - 1 = 0$
$$\frac{\cos 2x}{\sin 3x - \sin x} = 1$$
$$\frac{\cos 2x}{2\cos 2x \sin x} = 1$$
$$2\sin x = 1$$
$$\sin x = \frac{1}{2}$$
$$x = \frac{\pi}{6}, \frac{5\pi}{6}$$

77. $\sin^2 \alpha = \left(\frac{5}{13}\right)^2 = \frac{25}{169}$
$$\sin^2 \alpha = 1 - \cos^2 \alpha = 1 - \left(\frac{12}{13}\right)^2 = 1 - \frac{144}{169} = \frac{25}{169}$$

79. $\sin \alpha \cos \beta = \left(\frac{5}{13}\right)\left(\frac{4}{5}\right) = \frac{4}{13}$
$$\sin \alpha \cos \beta = \cos\left(\frac{\pi}{2} - \alpha\right) \sin\left(\frac{\pi}{2} - \beta\right) = \left(\frac{5}{13}\right)\left(\frac{4}{5}\right) = \frac{4}{13}$$

81. $\csc 2\theta = \dfrac{1}{\sin 2\theta}$
$$= \frac{1}{2\sin\theta\cos\theta}$$
$$= \frac{1}{\sin 2\theta} \cdot \frac{1}{2\cos\theta}$$
$$= \frac{\csc\theta}{2\cos\theta}$$

83. $\cos^2 2\alpha - \sin^2 2\alpha = \cos[2(2\alpha)]$
$$= \cos 4\alpha$$

85. $(\sin x + \cos x)^2 = \sin^2 x + 2\sin x \cos x + \cos^2 x$
$$= (\sin^2 x + \cos^2 x) + 2\sin x \cos x$$
$$= 1 + \sin 2x$$

87. $1 + \cos 10y = 1 + \cos^2 5y - \sin^2 5y$
$$= 1 + \cos^2 5y - (1 - \cos^2 5y)$$
$$= 2\cos^2 5y$$

89. $\sec \dfrac{u}{2} = \pm \sqrt{\dfrac{1}{\cos(u/2)}}$

$\quad = \pm \sqrt{\dfrac{2}{1 + \cos u}}$

$\quad = \pm \sqrt{\dfrac{2 \sin u}{\sin u(1 + \cos u)}}$

$\quad = \pm \sqrt{\dfrac{2 \sin u}{\sin u + \sin u \cos u}}$

$\quad = \pm \sqrt{\dfrac{(2 \sin u)/(\cos u)}{(\sin u)/(\cos u) + (\sin u \cos u)/(\cos u)}}$

$\quad = \pm \sqrt{\dfrac{2 \tan u}{\tan u + \sin u}}$

91. $\cos 3\beta = \cos(2\beta + \beta)$

$\quad = \cos 2\beta \cos \beta - \sin 2\beta \sin \beta$

$\quad = (\cos^2 \beta - \sin^2 \beta) \cos \beta - 2 \sin \beta \cos \beta \sin \beta$

$\quad = \cos^3 \beta - \sin^2 \beta \cos \beta - 2 \sin^2 \beta \cos \beta$

$\quad = \cos^3 \beta - 3 \sin^2 \beta \cos \beta$

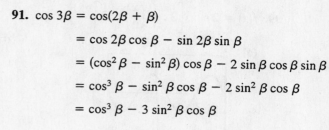

93 $\dfrac{\cos 4x - \cos 2x}{2 \sin 3x} = \dfrac{-2 \sin\left(\dfrac{4x + 2x}{2}\right) \sin\left(\dfrac{4x - 2x}{2}\right)}{2 \sin 3x}$

$\quad = \dfrac{-2 \sin 3x \sin x}{2 \sin 3x}$

$\quad = -\sin x$

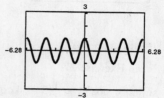

95. $\sin^2 x = \dfrac{1 - \cos 2x}{2} = \dfrac{1}{2} - \dfrac{\cos 2x}{2}$

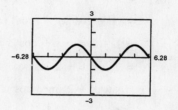

97. (a) $y = 4 \sin \dfrac{x}{2} + \cos x$

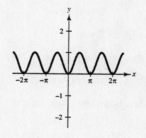

Maximum: $(\pi, 3)$

(b) $2 \cos \dfrac{x}{2} - \sin x = 0$

$2\left(\pm \sqrt{\dfrac{1 + \cos x}{2}} \right) = \sin x$

$4\left(\dfrac{1 + \cos x}{2} \right) = \sin^2 x$

$2(1 + \cos x) = 1 - \cos^2 x$

$\cos^2 x + 2 \cos x + 1 = 0$

$(\cos x + 1)^2 = 0$

$\cos x = -1$

$x = \pi$

99. $f(x) = 2 \sin x \left[2 \cos^2\left(\dfrac{x}{2}\right) - 1 \right]$

(a)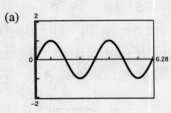

(b) The graph appears to be that of $y = \sin 2x$.

(c) $2 \sin x \left[2 \cos^2\left(\dfrac{x}{2}\right) - 1 \right] = 2 \sin x \left[2\, \dfrac{1 + \cos x}{2} - 1 \right]$

$= 2 \sin x \,[\cos x]$

$= \sin 2x$

101. $f(x) = \sin^4 x + \cos^4 x$

(a) From Exercise 27:

$\sin^4 x + \cos^4 x = \tfrac{1}{8}(3 - 4 \cos 2x + \cos 4x) + \tfrac{1}{8}(3 + 4 \cos 2x + \cos 4x)$

$= \tfrac{1}{8}(6 + 2 \cos 4x)$

$= \tfrac{1}{4}(3 + \cos 4x)$

(b) $\sin^4 x + \cos^4 x = (\sin^2 x)^2 + \cos^4 x$

$= (1 - \cos^2 x)^2 + \cos^4 x$

$= 1 - 2 \cos^2 x + \cos^4 x + \cos^4 x$

$= 2 \cos^4 x - 2 \cos^2 x + 1$

(c) $\sin^4 x + \cos^4 x = \sin^4 x + 2 \sin^2 x \cos^2 x + \cos^4 x - 2 \sin^2 x \cos^2 x$

$= (\sin^2 x + \cos^2 x)^2 - 2 \sin^2 x \cos^2 x$

$= 1 - 2 \sin^2 x \cos^2 x$

(d) $\sin^4 x + \cos^4 x = 1 - 2 \sin^2 x \cos^2 x$

$= 1 - (2 \sin x \cos x)^2 \left(\tfrac{1}{2}\right)$

$= 1 - \tfrac{1}{2} \sin^2(2x)$

(e) The expression has been rewritten four different ways. Hence, there is often more than one way to write a correct answer.

103. $\sin(2 \arcsin x) = 2 \sin(\arcsin x) \cos(\arcsin x) = 2x\sqrt{1 - x^2}$

❑ Review Exercises for Chapter 5

Solutions to Odd-Numbered Exercises

1. $\dfrac{1}{\cot^2 x + 1} = \dfrac{1}{\csc^2 x} = \sin^2 x$

3. $\dfrac{\sin^2 \alpha - \cos^2 \alpha}{\sin^2 \alpha - \sin \alpha \cos \alpha} = \dfrac{(\sin \alpha + \cos \alpha)(\sin \alpha - \cos \alpha)}{\sin \alpha(\sin \alpha - \cos \alpha)}$

$= \dfrac{\sin \alpha + \cos \alpha}{\sin \alpha}$

$= 1 + \cot \alpha$

5. $\tan^2 \theta(\csc^2 \theta - 1) = \tan^2 \theta(\cot^2 \theta)$

$$= \tan^2 \theta \left(\frac{1}{\tan^2 \theta} \right)$$

$$= 1$$

7. $\dfrac{2 \tan(x + 1)}{1 - \tan^2(x + 1)} = \tan[2(x + 1)]$

$$= \tan(2x + 2)$$

9. $\tan x(1 - \sin^2 x) = \tan x \cos^2 x$

$$= \frac{\sin x}{\cos x} \cdot \cos^2 x$$

$$= \sin x \cos x$$

$$= \frac{1}{2}(2 \sin x \cos x)$$

$$= \frac{1}{2} \sin 2x$$

11. $\sec^2 x \cot x - \cot x = \cot x(\sec^2 x - 1)$

$$= \cot x \tan^2 x$$

$$= \frac{1}{\tan x} \tan^2 x$$

$$= \tan x$$

13. $\sin^5 x \cos^2 x = \sin^4 x \cos^2 x \sin x$

$$= (1 - \cos^2 x)^2 \cos^2 x \sin x$$

$$= (1 - 2 \cos^2 x + \cos^4 x) \cos^2 x \sin x$$

$$= (\cos^2 x - 2 \cos^4 x + \cos^6 x) \sin x$$

15. $\sin 3\theta \sin \theta = \frac{1}{2}[\cos(3\theta - \theta) - \cos(3\theta + \theta)]$

$$= \frac{1}{2}(\cos 2\theta - \cos 4\theta)$$

17. $\sqrt{\dfrac{1 - \sin \theta}{1 + \sin \theta}} = \sqrt{\dfrac{1 - \sin \theta}{1 + \sin \theta} \cdot \dfrac{1 - \sin \theta}{1 - \sin \theta}}$

$$= \sqrt{\frac{(1 - \sin \theta)^2}{1 - \sin^2 \theta}} = \sqrt{\frac{(1 - \sin \theta)^2}{\cos^2 \theta}} = \frac{|1 - \sin \theta|}{|\cos \theta|} = \frac{1 - \sin \theta}{|\cos \theta|}$$

Note: We can drop the absolute value on $1 - \sin \theta$ since it is always nonnegative.

19. $\cos 3x = \cos(2x + x)$

$$= \cos 2x \cos x - \sin 2x \sin x$$

$$= (\cos^2 x - \sin^2 x) \cos x - 2 \sin x \cos x \sin x$$

$$= \cos^3 x - 3 \sin^2 x \cos x$$

$$= \cos^3 x - 3 \cos x(1 - \cos^2 x)$$

$$= \cos^3 x - 3 \cos x + 3 \cos^3 x$$

$$= 4 \cos^3 x - 3 \cos x$$

21. $\cot\left(\dfrac{\pi}{2} - x \right) = \dfrac{\cos[(\pi/2) - x]}{\sin[(\pi/2) - x]}$

$$= \frac{\cos(\pi/2) \cos x + \sin(\pi/2) \sin x}{\sin(\pi/2) \cos x - \sin x \cos(\pi/2)}$$

$$= \frac{\sin x}{\cos x}$$

$$= \tan x$$

23. $\dfrac{\sec x - 1}{\tan x} = \dfrac{(1/\cos x) - 1}{\sin x/\cos x} = \dfrac{1 - \cos x}{\sin x} = \tan \dfrac{x}{2}$

25. $2 \sin y \cos y \sec 2y = (\sin 2y)(\sec 2y)$

$\qquad\qquad\qquad\quad = \dfrac{\sin 2y}{\cos 2y}$

$\qquad\qquad\qquad\quad = \tan 2y$

27. $\sin\left(x - \dfrac{3\pi}{2}\right) = \sin x \cos \dfrac{3\pi}{2} - \sin \dfrac{3\pi}{2} \cos x$

$\qquad\qquad\qquad\; = (\sin x)(0) - (-1)(\cos x)$

$\qquad\qquad\qquad\; = \cos x$

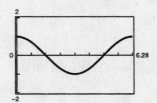

29. $\dfrac{1 - \cos 2x}{1 + \cos 2x} = \dfrac{1 - (1 - 2 \sin^2 x)}{1 + (2 \cos x^2 - 1)}$

$\qquad\qquad\quad = \dfrac{2 \sin^2 x}{2 \cos^2 x}$

$\qquad\qquad\quad = \tan^2 x$

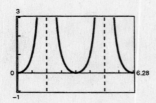

31. $\sin \dfrac{5\pi}{12} = \sin\left(\dfrac{2\pi}{3} - \dfrac{\pi}{4}\right)$

$\qquad\quad = \sin\left(\dfrac{2\pi}{3}\right)\cos\left(\dfrac{\pi}{4}\right) - \cos \dfrac{2\pi}{3} \sin\left(\dfrac{\pi}{4}\right)$

$\qquad\quad = \left(\dfrac{\sqrt{3}}{2}\right)\left(\dfrac{\sqrt{2}}{2}\right) - \left(-\dfrac{1}{2}\right)\left(\dfrac{\sqrt{2}}{2}\right)$

$\qquad\quad = \dfrac{\sqrt{2}}{4}\left(\sqrt{3} + 1\right)$

33. $\cos(157° 30') = \cos \dfrac{315°}{2} = -\sqrt{\dfrac{1 + \cos 315°}{2}} = -\sqrt{\dfrac{1 + \cos 45°}{2}}$

$\qquad\qquad\qquad = -\sqrt{\dfrac{1 + \sqrt{2}/2}{2}} = -\sqrt{\dfrac{2 + \sqrt{2}}{4}} = -\dfrac{1}{2}\sqrt{2 + \sqrt{2}}$

For Exercises 35–39:

$\sin u = \dfrac{3}{4},\; u \text{ in Quadrant II} \;\Rightarrow\; \cos u = -\dfrac{\sqrt{7}}{4}$

$\cos v = -\dfrac{5}{13},\; v \text{ in Quadrant II} \;\Rightarrow\; \sin v = \dfrac{12}{13}$

35. $\sin(u + v) = \sin u \cos v + \cos u \sin v$

$\qquad\qquad = \left(\dfrac{3}{4}\right)\left(-\dfrac{5}{13}\right) + \left(-\dfrac{\sqrt{7}}{4}\right)\left(\dfrac{12}{13}\right)$

$\qquad\qquad = -\dfrac{15}{52} - \dfrac{12\sqrt{7}}{52}$

$\qquad\qquad = \dfrac{-3(5 + 4\sqrt{7})}{52}$

37. $\cos(u - v) = \cos u \cos v + \sin u \sin v$

$\qquad\qquad = \left(-\dfrac{\sqrt{7}}{4}\right)\left(-\dfrac{5}{13}\right) + \left(\dfrac{3}{4}\right)\left(\dfrac{12}{13}\right)$

$\qquad\qquad = \dfrac{5\sqrt{7} + 36}{52}$

39. $\cos\dfrac{u}{2} = \sqrt{\dfrac{1+\cos u}{2}}$

$\qquad = \sqrt{\dfrac{1+(-\sqrt{7}/4)}{2}}$

$\qquad = \sqrt{\dfrac{4-\sqrt{7}}{8}}$

$\qquad = \dfrac{1}{4}\sqrt{2(4-\sqrt{7})}$

41. If $\dfrac{\pi}{2} < \theta < \pi$, then $\cos\dfrac{\theta}{2} < 0$. False, if

$\dfrac{\pi}{2} < \theta < \pi \;\Rightarrow\; \dfrac{\pi}{4} < \dfrac{\theta}{2} < \dfrac{\pi}{2}$,

which is in Quadrant I $\;\Rightarrow\; \cos(\theta/2) > 0$.

43. $4\sin(-x)\cos(-x) = -2\sin 2x$. True.

$4\sin(-x)\cos(-x) = 4(-\sin x)(\cos x) = -4\sin x\cos x = -2(2\sin x\cos x) = -2\sin 2x$

45. $\qquad \sin x - \tan x = 0$

$\qquad \sin x - \dfrac{\sin x}{\cos x} = 0$

$\qquad \sin x\cos x - \sin x = 0$

$\qquad \sin x(\cos x - 1) = 0$

$\sin x = 0 \quad$ or $\quad \cos x - 1 = 0$

$x = 0, \pi \qquad\qquad \cos x = 1$

$\qquad\qquad\qquad\qquad x = 0$

47. $\qquad \sin 2x + \sqrt{2}\sin x = 0$

$\qquad 2\sin x\cos x + \sqrt{2}\sin x = 0$

$\qquad \sin x(2\cos x + \sqrt{2}) = 0$

$\sin x = 0 \quad$ or $\quad 2\cos x + \sqrt{2} = 0$

$x = 0, \pi \qquad\qquad \cos x = -\dfrac{\sqrt{2}}{2}$

$\qquad\qquad\qquad\qquad x = \dfrac{3\pi}{4}, \dfrac{5\pi}{4}$

49. $\qquad \cos^2 x + \sin x = 1$

$\qquad 1 - \sin^2 x + \sin x = 1$

$\qquad \sin x(\sin x - 1) = 0$

$\sin x = 0 \qquad$ or $\quad \sin x = 1$

$x = 0, \pi \qquad\qquad x = \dfrac{\pi}{2}$

51. $\qquad \dfrac{1+\sin x}{\cos x} + \dfrac{\cos x}{1+\sin x} = 4$

$\qquad (1+\sin x)^2 + \cos^2 x = 4\cos x(1+\sin x)$

$\qquad 1 + 2\sin x + \sin^2 x + \cos^2 x = 4\cos x(1+\sin x)$

$\qquad 2 + 2\sin x = 4\cos x(1+\sin x) = 0$

$\qquad 2(1+\sin x)(1-2\cos x) = 0$

$1 + \sin x = 0 \quad$ or $\quad 1 - 2\cos x = 0$

$\sin x = -1 \qquad\qquad \cos x = \dfrac{1}{2}$

$x = \dfrac{3\pi}{2} \qquad\qquad x = \dfrac{\pi}{3}, \dfrac{5\pi}{3}$

(extraneous solution)

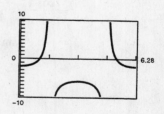

53. $\tan^3 x - \tan^2 x + 3 \tan x - 3 = 0$

$\tan^2 x(\tan x - 1) + 3(\tan x - 1) = 0$

$(\tan^2 x + 3)(\tan x - 1) = 0$

$\tan^2 x + 3 = 0$ or $\tan x - 1 = 0$

(No solution)   or $\tan x = 1$

$$x = \frac{\pi}{4}, \frac{5\pi}{4}$$

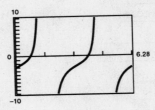

55. False, $\sin \theta = \frac{1}{2}$ has an infinite number of solutions but is not an identity.

57. $\cos 3\theta + \cos 2\theta = 2 \cos\left(\dfrac{3\theta + 2\theta}{2}\right) \cos\left(\dfrac{3\theta - 2\theta}{2}\right)$

$$= 2 \cos \frac{5\theta}{2} \cos \frac{\theta}{2}$$

59. $\sin 3\alpha \sin 2\alpha = \frac{1}{2}[\cos(3\alpha - 2\alpha) - \cos(3\alpha + 2\alpha)$

$$= \frac{1}{2}(\cos \alpha - \cos 5\alpha)$$

61. $\cos(2 \arccos 2x) = \cos 2\theta$

$$= \cos^2 \theta - \sin^2 \theta$$

$$= (2x)^2 - \left(\sqrt{1 - 4x^2}\right)^2$$

$$= 4x^2 - (1 - 4x^2)$$

$$= 8x^2 - 1$$

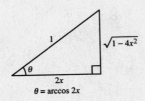

$\theta = \arccos 2x$

63. $\sin^{-1/2} \cos x = \dfrac{\cos x}{\sin^{1/2} x}$

$$= \frac{\cos x}{\sqrt{\sin x}} \cdot \frac{\sqrt{\sin x}}{\sqrt{\sin x}}$$

$$= \frac{\cos x}{\sin x} \sqrt{\sin x}$$

$$= \cot x \sqrt{\sin x}$$

65. $y = 1.5 \sin 8t - 0.5 \cos 8t$

(a) $a = \dfrac{3}{2}$, $b = -\dfrac{1}{2}$, $B = 8$, $C = \arctan\left(-\dfrac{1/2}{3/2}\right)$

$$y = \sqrt{(3/2)^2 + (1/2)^2} \sin\left(8t + \arctan -\frac{1}{3}\right)$$

$$y = \frac{1}{2} \sqrt{10} \sin\left(8t - \arctan \frac{1}{3}\right)$$

(b)

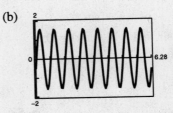

(c) The amplitude is $\dfrac{\sqrt{10}}{2}$.

(d) Frequency $= \dfrac{1}{\text{period}} = \dfrac{4}{\pi}$

❑ Chapter Test for Chapter 5

1. θ is in Quadrant III.

$$\tan^2 \theta + 1 = \sec^2 \theta \implies \frac{9}{4} + 1 = \sec^2 \theta \implies \sec \theta = -\frac{\sqrt{13}}{2}$$

$$\cos \theta = -\frac{2}{\sqrt{13}}$$

$$\sin \theta = (\tan \theta)(\cos \theta) = \frac{3}{2}\left(-\frac{2}{\sqrt{13}}\right) = \frac{-3}{\sqrt{13}}$$

$$\csc \theta = \frac{-\sqrt{13}}{3}$$

$$\cot \theta = \frac{1}{\tan \theta} = \frac{2}{3}$$

2. $\csc^2 \beta (1 - \cos^2 \beta) = \dfrac{1}{\sin^2 \beta} \cdot \sin^2 \beta = 1$

3. $\dfrac{\sec^4 x - \tan^4 x}{\sec^2 x + \tan^2 x} = \dfrac{[(\sec^2 x) + (\tan^2 x)][\sec^2 x - \tan^2 x]}{\sec^2 x + \tan^2 x} = \sec^2 x - \tan^2 x = 1$

4. $\dfrac{\cos \theta}{\sin \theta} + \dfrac{\sin \theta}{\cos \theta} = \dfrac{\cos^2 \theta + \sin^2 \theta}{\sin \theta \cos \theta} = \dfrac{1}{\sin \theta \cos \theta}$

5. Since $\tan^2 \theta = \sec^2 \theta - 1$ for all θ, then $\tan \theta = -\sqrt{\sec^2 \theta - 1}$ in Quadrants II and IV. Thus, $\pi/2 < \theta \le \pi$ and $3\pi/2 < \theta < 2\pi$.

6.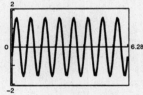

The graph appears equal.

Analytically,
$$\begin{aligned} y_1 &= \cos x + \sin x \tan x \\ &= \cos x + \sin x (\sin x / \cos x) \\ &= \frac{\cos^2 x + \sin^2 x}{\cos x} \\ &= \frac{1}{\cos x} \\ &= \sec x = y_2 \end{aligned}$$

7. $\sin \theta \cdot \sec \theta = \sin \theta \dfrac{1}{\cos \theta} = \tan \theta$

8. $\sec^2 \theta \tan^2 \theta + \sec^2 \theta = \sec^2 \theta (\tan^2 \theta + 1) = \sec^4 \theta$

9. $\dfrac{\csc \alpha + \sec \alpha}{\sin \alpha + \cos \alpha} = \dfrac{\dfrac{1}{\sin \alpha} + \dfrac{1}{\cos \alpha}}{\sin \alpha + \cos \alpha} = \dfrac{\dfrac{\cos \alpha + \sin \alpha}{\sin \alpha \cdot \cos \alpha}}{(\sin \alpha + \cos \alpha)}$

$$= \dfrac{1}{\sin \alpha \cos \alpha} = \dfrac{\cos^2 \alpha + \sin^2 \alpha}{\sin \alpha \cos \alpha}$$

$$= \dfrac{\cos \alpha}{\sin \alpha} + \dfrac{\sin \alpha}{\cos \alpha} = \cot \alpha + \tan \alpha$$

10. $\cos\left(x + \dfrac{\pi}{2}\right) = \cos x \cos \dfrac{\pi}{2} - \sin x \sin \dfrac{\pi}{2}$

$$= 0 - \sin x = -\sin x$$

11. $\sin(n\pi + \theta) = \sin n\pi \cdot \cos \theta + \sin \theta \cdot \cos n\pi$

$$= \cos n\pi \cdot \sin \theta$$

$$= (-1)^n \sin \theta$$

12. $(\sin x + \cos x)^2 = \sin^2 x + \cos^2 x + 2 \sin x \cos x$

$$= 1 + \sin 2x$$

13. $\tan^2 x + \tan x = 0$

   $\tan x(\tan x + 1) = 0$

   $\tan x = 0 \implies x = 0, \pi$

   $\tan x + 1 = 0 \implies \tan x = -1 \implies x = \dfrac{3\pi}{4}, \dfrac{7\pi}{4}$

14. $\sin 2\alpha - \cos \alpha = 0$

   $2 \sin \alpha \cos \alpha - \cos \alpha = 0$

   $\cos \alpha(2 \sin \alpha - 1) = 0$

   $\cos \alpha = 0 \implies \alpha = \dfrac{\pi}{2}, \dfrac{3\pi}{2}$

   $2 \sin \alpha - 1 = 0 \implies \sin \alpha = \dfrac{1}{2} \implies \alpha = \dfrac{\pi}{6}, \dfrac{5\pi}{6}$

15. $4 \cos^2 x - 3 = 0$

$$\cos^2 x = \dfrac{3}{4}$$

$$\cos x = \pm\dfrac{\sqrt{3}}{2}$$

$$x = \dfrac{\pi}{6}, \dfrac{5\pi}{6}, \dfrac{7\pi}{6}, \dfrac{11\pi}{6}$$

16. $\csc^2 x - \csc x - 2 = 0$

   $(\csc x - 2)(\csc x + 1) = 0$

   $\csc x - 2 = 0 \implies \csc x = 2 \implies \sin x = \dfrac{1}{2} \implies x = \dfrac{\pi}{6}, \dfrac{5\pi}{6}$

   $\csc x + 1 = 0 \implies \csc x = -1 \implies \sin x = -1 \implies x = \dfrac{3\pi}{2}$

17. Let $y = 3 \cos x - x$. The zeros are $-2.938, -2.663, 1.170$.

18. Since $|\cos x| \leq 1$, $|\cos^2 x + \cos x| \leq 2$ for all x.

19. $\cos 105° = \cos(135° - 30°) = \cos 135° \cos 30° + \sin 135° \sin 30°$

$$= \left(-\dfrac{\sqrt{2}}{2}\right)\left(\dfrac{\sqrt{3}}{2}\right) + \left(\dfrac{\sqrt{2}}{2}\right)\dfrac{1}{2}$$

$$= -\dfrac{\sqrt{6}}{4} + \dfrac{\sqrt{2}}{4} = \dfrac{\sqrt{2} - \sqrt{6}}{4}$$

20. $\sin 2u = 2 \sin u \cos u = 2\dfrac{2}{\sqrt{5}} \cdot \dfrac{1}{\sqrt{5}} = \dfrac{4}{5}$

$$\tan 2u = \dfrac{2 \tan u}{1 - \tan^2 u} = \dfrac{2(2)}{1 - 2^2} = -\dfrac{4}{3}$$

❏ Practice Test for Chapter 5

1 Find the value of the other five trigonometric functions, given $\tan x = \frac{4}{11}$, $\sec x < 0$.

2. Simplify $\dfrac{\sec^2 x + \csc^2 x}{\csc^2 x(1 + \tan^2 x)}$.

3. Rewrite as a single logarithm and simplify $\ln|\tan \theta| - \ln|\cot \theta|$.

4. True or false:
$$\cos\left(\frac{\pi}{2} - x\right) = \frac{1}{\csc x}$$

5. Factor and simplify: $\sin^4 x + (\sin^2 x)\cos^2 x$

6. Multiply and simplify: $(\csc x + 1)(\csc x - 1)$

7. Rationalize the denominator and simplify:
$$\frac{\cos^2 x}{1 - \sin x}$$

8. Verify:
$$\frac{1 + \cos \theta}{\sin \theta} + \frac{\sin \theta}{1 + \cos \theta} = 2\csc \theta$$

9. Verify:
$$\tan^4 x + 2\tan^2 x + 1 = \sec^4 x$$

10. Use the sum or difference formulas to determine:

(a) $\sin 105°$ (b) $\tan 15°$

11. Simplify: $(\sin 42°)\cos 38° - (\cos 42°)\sin 38°$

12. Verify $\tan\left(\theta + \dfrac{\pi}{4}\right) = \dfrac{1 + \tan \theta}{1 - \tan \theta}$.

13. Write $\sin(\arcsin x - \arccos x)$ as an algebraic expression in x.

14. Use the double-angle formulas to determine:

(a) $\cos 120°$ (b) $\tan 300°$

15. Use the half-angle formulas to determine:

(a) $\sin 22.5°$ (b) $\tan \dfrac{\pi}{12}$

16. Given $\sin = 4/5$, θ lies in Quadrant II, find $\cos \theta/2$.

17. Use the power-reducing identities to write $(\sin^2 x)\cos^2 x$ in terms of the first power of cosine.

18. Rewrite as a sum: $6(\sin 5\theta)\cos 2\theta$.

19. Rewrite as a product:
$$\sin(x + \pi) + \sin(x - \pi).$$

20. Verify $\dfrac{\sin 9x + \sin 5x}{\cos 9x - \cos 5x} = -\cot 2x$.

21. Verify:
$$(\cos u)\sin v = \tfrac{1}{2}[\sin(u + v) - \sin(u - v)].$$

22. Find all solutions in the interval $[0, 2\pi)$:
$$4\sin^2 x = 1$$

23. Find all solutions in the interval $[0, 2\pi)$:
$$\tan^2 \theta + \left(\sqrt{3} - 1\right)\tan \theta - \sqrt{3} = 0$$

24. Find all solutions in the interval $[0, 2\pi)$:
$$\sin 2x = \cos x$$

25. Use the Quadratic Formula to find all solutions in the interval $[0, 2\pi)$:
$$\tan^2 x - 6\tan x + 4 = 0$$

CHAPTER 6
Additional Topics in Trigonometry

CHAPTER 6
Additional Topics in Trigonometry

Section 6.1   Law of Sines

- If ABC is any oblique triangle with sides a, b, and c, then the Law of Sines says
$$\frac{a}{\sin A} = \frac{b}{\sin B} = \frac{c}{\sin C}.$$

- You should be able to use the Law of Sines to solve an oblique triangle for the remaining three parts, given:
  (a) Two angles and any side (AAS or ASA)
  (b) Two sides and an angle opposite one of them (SSA)
   1. If A is acute and $h = b \sin A$:
      (a) $a < h$, no triangle is possible.
      (b) $a = h$ or $a > b$, one triangle is possible.
      (c) $h < a < b$, two triangles are possible.
   2. If A is obtuse and $h = b \sin A$:
      (a) $a \le b$, no triangle is possible.
      (b) $a > b$, one triangle is possible.

- The area of any triangle equals one-half the product of the lengths of two sides times the sine of their included angle.
$$A = \tfrac{1}{2}ab \sin C = \tfrac{1}{2}ac \sin B = \tfrac{1}{2}bc \sin A$$

Solutions to Odd-Numbered Exercises

1. Given: $A = 30°$, $B = 45°$, $a = 20$

$C = 180° - A - B = 105°$

$b = \dfrac{a}{\sin A}(\sin B) = \dfrac{20 \sin 45°}{\sin 30°} = 20\sqrt{2} \approx 28.28$

$c = \dfrac{a}{\sin A}(\sin C) = \dfrac{20 \sin 105°}{\sin 30°} \approx 38.64$

3. Given: $A = 10°$, $B = 60°$, $a = 7.5$

$C = 180° - A - B = 110°$

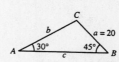

$b = \dfrac{a}{\sin A}(\sin B) = \dfrac{7.5}{\sin 10°} = (\sin 60°) \approx 37.40$

$c = \dfrac{a}{\sin A}(\sin C) = \dfrac{7.5}{\sin 10°}(\sin 110°) \approx 40.59$

5. Given: $A = 36°$, $a = 8$, $b = 5$

$\sin B = \dfrac{b \sin A}{a} = \dfrac{5 \sin 36°}{8} \approx 0.36737 \implies B \approx 21.55°$

$C = 180 - A - B \approx 180° - 36° - 21.55 = 122.45°$

$c = \dfrac{a}{\sin A}(\sin C) = \dfrac{8}{\sin 36°}(\sin 122.45°) \approx 11.49$

7. Given: $A = 150°$, $C = 20°$, $a = 200$

   $B = 180° - A - C = 180° - 150° - 20° = 10°$

   $b = \dfrac{a}{\sin A}(\sin B) = \dfrac{200}{\sin 150°}(\sin 10°) \approx 69.46$

   $c = \dfrac{a}{\sin A}(\sin C) = \dfrac{200}{\sin 150°}(\sin 20°) \approx 136.81$

9. Given: $A = 83° \, 20'$, $C = 54.6°$, $c = 18.1$

   $B = 180° - A - C = 180° - 80° \, 20' - 54° \, 36' = 42° \, 4'$

   $a = \dfrac{c}{\sin C}(\sin A) = \dfrac{18.1}{\sin 54.6°}(\sin 83° \, 20') \approx 22.05$

   $b = \dfrac{c}{\sin C}(\sin B) = \dfrac{18.1}{\sin 54.6°}(\sin 42° \, 4') \approx 14.88$

11. Given: $B = 15° \, 30'$, $a = 4.5$, $b = 6.8$

    $\sin A = \dfrac{a \sin B}{b} = \dfrac{4.5 \sin 15° \, 30'}{6.8} \approx 0.17685 \implies A \approx 10° \, 11'$

    $C = 180° - A - B \approx 180° - 10° \, 11' - 15.5° = 154° \, 19'$

    $c = \dfrac{b}{\sin B}(\sin C) = \dfrac{6.8}{\sin 15° \, 30'}(\sin 154° \, 19') \approx 11.03$

13. Given: $A = 110° \, 15'$, $a = 48$, $b = 16$

    $\sin B = \dfrac{b \sin A}{a} = \dfrac{16 \sin 110° \, 15'}{48} \approx 0.31273 \implies B \approx 18° \, 13'$

    $C = 180° - A - B \approx 180° - 110° \, 15' - 18° \, 13' = 51° \, 32'$

    $c = \dfrac{a}{\sin A}(\sin C) = \dfrac{48}{\sin 110° \, 15'}(\sin 51° \, 32') \approx 40.05$

15. Given: $a = 4.5$, $b = 12.8$, $A = 58°$

    $h = 12.8 \sin 58° \approx 10.86$

    Since $a < h$, no triangle is formed.

17. Given: $a = 4.5$, $b = 5$, $A = 58°$

    $\sin B = \dfrac{b \sin A}{a} = \dfrac{5 \sin 58°}{4.5} \approx 0.9423 \implies B = 70.4° \text{ or } B = 109.6°$

Case 1	Case 2
$B \approx 70.4°$	$B \approx 109.6°$
$C \approx 180° - 70.4° - 58° = 51.6°$	$C \approx 180° - 109.6° - 58° = 12.4°$
$c \approx \dfrac{4.5}{\sin 58°}(\sin 51.6°) \approx 4.16$	$c \approx \dfrac{4.5}{\sin 58°}(\sin 12.4°) \approx 1.14$

19. Given: $A = 36°$, $a = 5$

    (a) One solution if $b \le 5$ or $b = \dfrac{5}{\sin 36°}$.

    (b) Two solutions if $5 < b < \dfrac{5}{\sin 36°}$.

    (c) No solution if $b > \dfrac{5}{\sin 36°}$.

21. (a)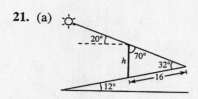

(b) $\dfrac{16}{\sin 70°} = \dfrac{h}{\sin 32°}$

(c) $h = \dfrac{16 \sin 32°}{\sin 70°} \approx 9$ meters

25. Given: $A = 74° - 28° = 46°$,

$B = 180° - 41° - 74° = 65°$, $c = 100$

$C = 180° - 46° - 65° = 69°$

$a = \dfrac{c}{\sin C}(\sin A) = \dfrac{100}{\sin 69°}(\sin 46°) \approx 77$ meters

27. (a)

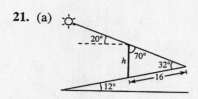

(c) $\dfrac{y}{\sin 71.2°} = \dfrac{x}{\sin 90°}$

$y = x \sin 71.2° \approx 119289.1261 \sin 71.2°$

≈ 112924.963 feet ≈ 21.4 miles

29. $A = 65° - 28° = 37°$

$c = 30$

$B = 180° - 16.5° - 65° = 98.5°$

$C = 180° - 37° - 98.5° = 44.5°$

$a = \dfrac{c}{\sin C}(\sin A) = \dfrac{30}{\sin 44.5°}(\sin 37°) \approx 25.8$ km to B

$b = \dfrac{c}{\sin C}(\sin B) = \dfrac{30}{\sin 44.5°}(\sin 98.5°) \approx 42.3$ km to A

31. $A = 90° - 62° = 28°$,

$B = 90° + 38° = 128°$, $c = 5$

$C = 180° - 128° - 28° = 24°$

$a = \dfrac{c}{\sin C}(\sin A) = \dfrac{5}{\sin 24°}(\sin 28°) \approx 5.77$

$d = a \sin(90° - 38°) \approx 5.77 \sin 52° \approx 4.55$ miles

23. $\dfrac{\sin(42° - \theta)}{10} = \dfrac{\sin 48°}{17}$

$\sin(42° - \theta) \approx 0.43714$

$\theta \approx 16.1°$

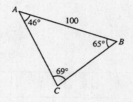

(b) $\dfrac{x}{\sin 17.5°} = \dfrac{9000}{\sin 1.3°}$

$x \approx 1198289.1261$ feet ≈ 22.6 miles

(d) $z = x \sin 18.8° \approx 119289.1261 \sin 18.8°$

$\approx 38,442.8$ feet

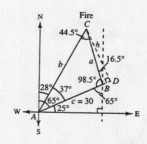

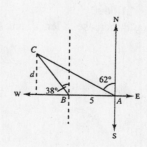

33. (a) $\dfrac{\sin \alpha}{9} = \dfrac{\sin \beta}{18}$

$\sin \alpha = 0.5 \sin \beta$

$\alpha = \arcsin(0.5 \sin \beta)$

(b)

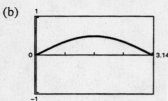

Domain: $0 < \beta < \pi$

Range: $0 < c \le \pi/6$

(c) $\gamma = \pi - \alpha - \beta - \pi - \beta - \arcsin(0.5 \sin \beta)$

$\dfrac{c}{\sin \gamma} = \dfrac{18}{\sin \beta}$

$c = \dfrac{18 \sin \gamma}{\sin \beta} = \dfrac{18 \sin[\pi - \beta - \arcsin(0.5 \sin \beta)]}{\sin \beta}$

(d)

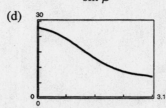

Domain: $0 < \beta < \pi$

Range: $9 < c < 27$

(e)

β	0	0.4	0.8	1.2	1.6	2.0	2.4	2.8
α	0	0.1960	0.3669	0.4848	0.5234	0.4720	0.3445	0.1683
c	Undef.	25.95	23.07	19.19	15.33	12.29	10.31	9.27

As $\beta \to 0$, $c \to 27$.

As $\beta \to \pi$, $c \to 9$.

35. Area $= \frac{1}{2}ab \sin C = \frac{1}{2}(4)(6) \sin 120° \approx 10.4$

37. Area $= \frac{1}{2}bc \sin A = \frac{1}{2}(57)(85) \sin 43° \, 45' \approx 1675.2$

39. Area $= \frac{1}{2}ac \sin B = \frac{1}{2}(62)(20) \sin 130° \approx 474.9$

41. (a) $A = \dfrac{1}{2}(30)(20) \sin\left(\theta + \dfrac{\theta}{2}\right) - \dfrac{1}{2}(8)(20) \sin \dfrac{\theta}{2} - \dfrac{1}{2}(8)(30) \sin \theta$

$= 300 \sin \dfrac{3\theta}{2} - 80 \sin \dfrac{\theta}{2} - 120 \sin \theta$

$= 20\left[15 \sin \dfrac{3\theta}{2} - 4 \sin \dfrac{\theta}{2} - 6 \sin \theta \right]$

(b)

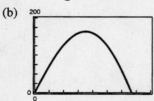

(c) Domain: $0 \le \theta \le 1.6690$

The domain would increase in length and the area would increase if the 8 centimeter line segment were decreased.

Section 6.2   Law of Cosines

> ■ If *ABC* is any oblique triangle with sides *a*, *b*, and *c*, then the Law of Cosines says:
>
> (a) $a^2 = b^2 + c^2 - 2bc \cos A$ or $\cos A = \dfrac{b^2 + c^2 - a^2}{2bc}$
>
> (b) $b^2 = a^2 + c^2 - 2ac \cos B$ or $\cos B = \dfrac{a^2 + c^2 - b^2}{2ac}$
>
> (c) $c^2 = a^2 + b^2 - 2ab \cos C$ or $\cos C = \dfrac{a^2 + b^2 - c^2}{2ab}$
>
> ■ You should be able to use the Law of Cosines to solve an oblique triangle for the remaining three parts, given:
>
> (a) Three sides (SSS)
>
> (b) Two sides and their included angle (SAS)
>
> ■ Given any triangle with sides of length *a*, *b*, and *c*, then the area of the triangle is
>
> $$\text{Area} = \sqrt{s(s - a)(s - b)(s - c)}, \quad \text{where } s = \frac{a + b + c}{2}. \quad \text{(Heron's Formula)}$$

Solutions to Odd-Numbered Exercises

1. Given: $a = 6$, $b = 8$, $c = 12$

$\cos A = \dfrac{b^2 + c^2 - a^2}{2bc} = \dfrac{64 + 144 - 36}{2(8)(12)} \approx 0.8958 \implies A \approx 26.4°$

$\sin B = \dfrac{b \sin A}{a} \approx \dfrac{8 \sin 26.4°}{6} \approx 0.5928 \implies B \approx 36.3°$

$C \approx 180° - 26.4° - 36.3° = 117.3°$

3. Given: $A = 30°$, $b = 15$, $c = 30$

$a^2 = b^2 + c^2 - 2bc \cos A$

$\quad = 225 + 900 - 2(15)(30) \cos 30° \approx 18.5897$

$a \approx 18.6$

$\cos B = \dfrac{a^2 + c^2 - b^2}{2ac} \approx \dfrac{(18.6)^2 + 900 - 225}{2(18.6)(30)} \approx 0.9148$

$B \approx 23.8°$

$C \approx 180° - 30° - 23.8° = 126.2°$

5. Given: $a = 9$, $b = 12$, $c = 15$

$\cos C = \dfrac{a^2 + b^2 - c^2}{2ab} = \dfrac{81 + 144 - 225}{2(9)(12)} = 0 \implies C = 90°$

$\sin A = \dfrac{9}{15} = \dfrac{3}{5} \implies A \approx 36.9°$

$B \approx 180° - 90° - 36.9° = 53.1°$

7. Given: $a = 75.4$, $b = 52$, $c = 52$

$$\cos A = \frac{b^2 + c^2 - a^2}{2bc} = \frac{52^2 + 52^2 - 75.4^2}{2(52)(52)} = -0.05125 \implies A = 92.9°$$

$$\sin B = \frac{b \sin A}{a} = \frac{52(0.9987)}{75.4} \approx 0.68875 \implies B \approx 43.53°$$

$$C = B \approx 43.53°$$

9. Given: $B = 8° \, 45'$, $a = 25$, $c = 15$

$$b^2 = a^2 + c^2 - 2ac \cos B \approx 625 + 225 - 2(25)(15)(0.9884) \approx 108.7 \implies b \approx 10.4$$

$$\sin C = \frac{c \sin B}{b} = \frac{15(0.1521)}{10.43} \approx 0.2188 \implies C \approx 12.64° \approx 12° \, 38'$$

$$A \approx 180° - 8° \, 45' - 12° \, 38' = 158° \, 37'$$

11.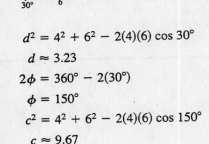

$$d^2 = 4^2 + 6^2 - 2(4)(6) \cos 30°$$

$$d \approx 3.23$$

$$2\phi = 360° - 2(30°)$$

$$\phi = 150°$$

$$c^2 = 4^2 + 6^2 - 2(4)(6) \cos 150°$$

$$c \approx 9.67$$

13.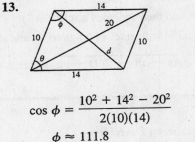

$$\cos \phi = \frac{10^2 + 14^2 - 20^2}{2(10)(14)}$$

$$\phi \approx 111.8$$

$$2\theta \approx 360° - 2(111.80°)$$

$$\theta = 68.2°$$

$$d^2 = 10^2 + 14^2 - 2(10)(14) \cos 68.2°$$

$$d \approx 13.86$$

15. $\cos \alpha = \dfrac{(9)^2 + (10)^2 - (6)^2}{2(9)(10)}$

$$\alpha = 36.3°$$

$$\cos \beta = \frac{6^2 + 10^2 - 9^2}{2(6)(10)}$$

$$\beta \approx 62.7°$$

$$z = 180° - \alpha - \beta \approx 80.9$$

$$\mu = 180° - z \approx 99.1$$

$$b^2 = 9^2 + 6^2 - 2(9)(6)(\cos 99.0°)$$

$$b \approx 11.58$$

$$\cos \omega = \frac{9^2 + 11.58^2 - 6^2}{2(9)(11.58)}$$

$$\omega \approx 30.8°$$

$$\theta = \alpha + \omega \approx 67.1°$$

$$\cos x = \frac{6^2 + 11.58^2 - 9^2}{2(6)(11.58)}$$

$$x \approx 50.1$$

$$\phi = \beta + x \approx 112.8°$$

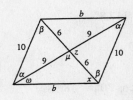

17.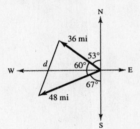

$B = 105° + 32° = 137°$

$b^2 = a^2 + c^2 - 2ac \cdot \cos B$

$\qquad = 648^2 + 810^2 - 2(648)(810)\cos(137°)$

$\qquad = 1{,}843{,}749.862$

$b = 1357.8$ miles

From the Law of Sines, $\dfrac{a}{\sin A} = \dfrac{b}{\sin B} \implies \sin A = \dfrac{a}{b}\sin B = \dfrac{648}{1357.8}\sin(137°) \approx 0.32548$

$\implies A \approx 19° \implies$ Bearing N 56° E

19.

$b^2 = 250^2 + 300^2 - 2(250)(300)\cos 100°$

$b \approx 422.5$ meters

21. $C = 180° - 53° - 67° = 60°$

$c^2 = a^2 + b^2 - 2ab\cos C = 36^2 + 48^2 - 2(36)(48)(0.5) = 1872$

$c \approx 43.3$ mi

23. (a) $\cos \theta = \dfrac{273^2 + 178^2 - 235^2}{2(273)(178)}$

$\qquad \theta \approx 58.4°$

$\qquad$ Bearing: N 58.4° W

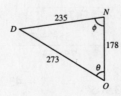

(b) $\cos \phi = \dfrac{235^2 + 178^2 - 273^2}{2(235)(178)}$

$\qquad \phi \approx 81.5°$

$\qquad$ Bearing: S 81.5° W

25. $d^2 = 60.5^2 + 90^2 - 2(60.5)(90)\cos 45° \approx 4059.9 \implies d \approx 63.7$ ft

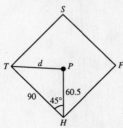

27. $\overline{RS} = \sqrt{8^2 + 10^2} = \sqrt{164} = 2\sqrt{41} \approx 12.8$ ft

$\overline{PQ} = \frac{1}{2}\sqrt{16^2 + 10^2} = \frac{1}{2}\sqrt{356} = \sqrt{89} \approx 9.4$ ft

$\tan P = \frac{10}{16}$

$\quad P = \arctan \frac{5}{8} \approx 32.0°$

$\quad \overline{QS} = \sqrt{8^2 + 9.4^2 - 2(8)(9.4)\cos 32°} \approx \sqrt{24.81} \approx 5.0$ ft

29. (a) $7^2 = 1.5^2 + x^2 - 2(1.5)(x)\cos\theta$

$\qquad 49 = 2.25 + x^2 - 3x\cos\theta$

(b) $\qquad\qquad x^2 - 3x\cos\theta = 46.75$

$x^2 - 3x\cos\theta + \left(\dfrac{3\cos\theta}{2}\right)^2 = 46.75 + \left(\dfrac{3\cos\theta}{2}\right)^2$

$\left[x - \dfrac{3\cos\theta}{2}\right]^2 = \dfrac{187}{4} - \dfrac{9\cos^2\theta}{4}$

$x - \dfrac{3\cos\theta}{2} = \pm\sqrt{\dfrac{187 + 9\cos^2\theta}{4}}$

Choosing the positive values of x, we have
$x = \frac{1}{2}\left(3\cos\theta + \sqrt{9\cos^2\theta + 187}\right)$.

(c)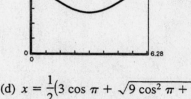

(d) $x = \dfrac{1}{2}\left(3\cos\pi + \sqrt{9\cos^2\pi + 187}\right)$

$\quad = 5.5$

$\quad \approx 6$ inches

31. $A = 180° - 40° - 20° = 120°$

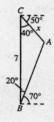

$\dfrac{x}{\sin 20°} = \dfrac{7}{\sin 120°}$

$\quad x = \dfrac{7\sin 20°}{\sin 120°}$

$\quad x = 2.76$ feet

33. $a = 25$, $b = 55$, $c = 72$, $s = \dfrac{25 + 55 + 72}{2} = 76$

(a) $A = \sqrt{(76)(76 - 25)(76 - 55)(76 - 72)} \approx 570.60$

(b) $\cos C = \dfrac{a^2 + b^2 - c^2}{2ab} = \dfrac{25^2 + 55^2 - 72^2}{2(25)(55)} \implies C \approx 123.905°$

$\quad 2R = \dfrac{c}{\sin C} \approx \dfrac{72}{\sin 123.905°} \implies R \approx 43.3754$

$\quad A = \pi R^2 \approx 5910.67$

(c) $r = \sqrt{\dfrac{(s - a)(s - b)(s - c)}{s}} = \sqrt{\dfrac{(51)(21)(4)}{76}} \approx 7.5079$

$\quad A = \pi r^2 \approx 177.09$

35. $a = 5$, $b = 7$, $c = 10 \implies s = \dfrac{a + b + c}{2} = 11$

Area $= \sqrt{s(s - a)(s - b)(s - c)} = \sqrt{11(6)(4)(1)} \approx 16.25$

37. $a = 12$, $b = 15$, $c = 9 \implies s = \dfrac{12 + 15 + 9}{2} = 18$

Area $= \sqrt{18(6)(3)(9)} = 54$

39. $a = 20,\ b = 20,\ c = 10 \implies s = \dfrac{20 + 20 + 10}{2} = 25$

Area $= \sqrt{25(5)(5)(15)} \approx 96.82$

41.
$$\frac{1}{2}bc(1 + \cos A) = \frac{1}{2}bc\left[1 + \frac{b^2 + c^2 - a^2}{2bc}\right]$$

$$= \frac{1}{2}bc\left[\frac{2bc + b^2 + c^2 - a^2}{2bc}\right]$$

$$= \frac{1}{4}[(b + c)^2 - a^2]$$

$$= \frac{1}{4}[(b + c) + a][(b + c) - a]$$

$$= \frac{b + c + a}{2} \cdot \frac{b + c - a}{2}$$

$$= \frac{a + b + c}{2} \cdot \frac{-a + b + c}{2}$$

Section 6.3   Vectors in the Plane

- A vector **v** is the collection of all directed line segments that are equivalent to a given directed line segment $\overrightarrow{PQ}$.

- You should be able to *geometrically* perform the operations of vector addition and scalar multiplication.

- The component form of the vector with initial point $P = (p_1, p_2)$ and terminal point $Q = (q_1, q_2)$ is

  $\overrightarrow{PQ} = \langle q_1 - p_1, q_2 - p_2 \rangle = \langle v_1, v_2 \rangle = \mathbf{v}.$

- The magnitude of $\mathbf{v} = \langle v_1, v_2 \rangle$ is given by $\|\mathbf{v}\| = \sqrt{v_1{}^2 + v_2{}^2}$.

- You should be able to perform the operations of scalar multiplication and vector addition in component form.

- You should know the following properties of vector addition and scalar multiplication.

  (a) $\mathbf{u} + \mathbf{v} = \mathbf{v} + \mathbf{u}$

  (b) $(\mathbf{u} + \mathbf{v}) + \mathbf{w} = \mathbf{u} + (\mathbf{v} + \mathbf{w})$

  (c) $\mathbf{u} + \mathbf{0} = \mathbf{u}$

  (d) $\mathbf{u} + (-\mathbf{u}) = \mathbf{0}$

  (e) $c(d\mathbf{u}) = (cd)\mathbf{u}$

  (f) $(c + d)\mathbf{u} = c\mathbf{u} + d\mathbf{u}$

  (g) $c(\mathbf{u} + \mathbf{v}) = c\mathbf{u} + c\mathbf{v}$

  (h) $1(\mathbf{u}) = \mathbf{u},\ 0\mathbf{u} = \mathbf{0}$

  (i) $\|c\mathbf{v}\| = |c|\,\|\mathbf{v}\|$

- A unit vector in the direction of **v** is given $\mathbf{u} = \dfrac{\mathbf{v}}{\|\mathbf{v}\|}$.

- The standard unit vectors are $\mathbf{i} = \langle 1, 0 \rangle$ and $\mathbf{j} = \langle 0, 1 \rangle$. $\mathbf{v} = \langle v_1, v_2 \rangle$ can be written as $\mathbf{v} = v_1\mathbf{i} + v_2\mathbf{j}$.

- A vector **v** with magnitude $\|\mathbf{v}\|$ and direction θ can be written as $\mathbf{v} = a\mathbf{i} + b\mathbf{j} = \mathbf{v}(\cos \theta)\mathbf{i} + \mathbf{v}(\sin \theta)\mathbf{j}$ where $\tan \theta = b/a$.

Solutions to Odd-Numbered Exercises

1. Initial point:$(0, 0)$
Terminal point: $(4, 3)$
$\mathbf{v} = \langle 4 - 0, 3 - 0 \rangle = \langle 4, 3 \rangle$
$\|\mathbf{v}\| = \sqrt{4^2 + 3^2} = 5$

3. Initial point: $(2, 2)$
Terminal point: $(-1, 4)$
$\mathbf{v} = \langle -1 - 2, 4 - 2 \rangle = \langle -3, 2 \rangle$
$\|\mathbf{v}\| = \sqrt{(-3)^2 + 2^2} = \sqrt{13}$

5. Initial point: $(3, -2)$
Terminal point: $(3, 3)$
$\mathbf{v} = \langle 3 - 3, 3 - (-2) \rangle = \langle 0, 5 \rangle$
$\|\mathbf{v}\| = 5$

7. Initial point: $\left(\frac{5}{2}, 1\right)$

Terminal point: $\left(-2, -\frac{3}{2}\right)$

$\mathbf{v} = \left\langle -2 - \frac{5}{2}, -\frac{3}{2} - 1 \right\rangle = \left\langle -\frac{9}{2}, -\frac{5}{2} \right\rangle$

$\|\mathbf{v}\| = \sqrt{\left(-\frac{9}{2}\right)^2 + \left(-\frac{5}{2}\right)^2} = \sqrt{\frac{81 + 25}{4}} = \frac{1}{2}\sqrt{106}$

9. Initial point: $(-3, -5)$
Terminal point: $(5, 1)$
$\mathbf{v} = \langle 5 - (-3), 1 - (-5) \rangle = \langle 8, 6 \rangle$
$\|\mathbf{v}\| = \sqrt{8^2 + 6^2} = \sqrt{100} = 10$

11.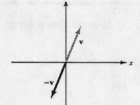

13.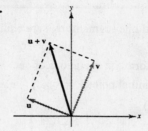

15.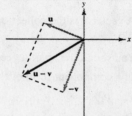

17. $\mathbf{u} = \langle 1, 2 \rangle, \mathbf{v} = \langle 3, 1 \rangle$
(a) $\mathbf{u} + \mathbf{v} = \langle 4, 3 \rangle$
(b) $\mathbf{u} - \mathbf{v} = \langle -2, 1 \rangle$
(c) $2\mathbf{u} - 3\mathbf{v} = \langle 2, 4 \rangle - \langle 9, 3 \rangle = \langle -7, 1 \rangle$

19. $\mathbf{u} = \mathbf{i} + \mathbf{j}, \mathbf{v} = 2\mathbf{i} - 3\mathbf{j}$
(a) $\mathbf{u} + \mathbf{v} = 3\mathbf{i} - 2\mathbf{j}$
(b) $\mathbf{u} - \mathbf{v} = -\mathbf{i} + 4\mathbf{j}$
(c) $2\mathbf{u} - 3\mathbf{v} = (2\mathbf{i} + 2\mathbf{j}) - (6\mathbf{i} - 9\mathbf{j})$
$= -4\mathbf{i} + 11\mathbf{j}$

21. $\mathbf{u} = \frac{1}{\|\mathbf{v}\|}\mathbf{v}$
$= \frac{1}{\sqrt{5^2 + 0^2}}\langle 5, 0 \rangle$
$= \frac{1}{5}\langle 5, 0 \rangle$
$= \langle 1, 0 \rangle$

23. $\mathbf{u} = \frac{1}{\|\mathbf{v}\|}\mathbf{v}$
$= \frac{1}{\sqrt{(-2)^2 + 2^2}}\langle -2, 2 \rangle$
$= \frac{1}{2\sqrt{2}}\langle -2, 2 \rangle$
$= \left\langle -\frac{1}{\sqrt{2}}, \frac{1}{\sqrt{2}} \right\rangle$

25. $\mathbf{u} = \frac{1}{\|\mathbf{v}\|}\mathbf{v}$
$= \frac{1}{\sqrt{16 + 9}}(4\mathbf{i} - 3\mathbf{j}) = \frac{1}{5}(4\mathbf{i} - 3\mathbf{j})$
$= \frac{4}{5}\mathbf{i} - \frac{3}{5}\mathbf{j}$

27. $\mathbf{u} = \frac{1}{\|\mathbf{v}\|}\mathbf{v} = \frac{1}{2}(2\mathbf{j}) = \mathbf{j}$

29. $5\left(\dfrac{1}{\|\mathbf{v}\|}\mathbf{v}\right) = 5\left(\dfrac{1}{\sqrt{3^2+3^2}}\langle 3, 3\rangle\right)$

$\qquad = 5\left(\dfrac{1}{3\sqrt{2}}\langle 3, 3\rangle\right)$

$\qquad = \left\langle \dfrac{5}{\sqrt{2}}, \dfrac{5}{\sqrt{2}}\right\rangle$

31. $7\left(\dfrac{1}{\|\mathbf{v}\|}\mathbf{v}\right) = 7\left(\dfrac{1}{\sqrt{(-3)^2+4^2}}\langle -3, 4\rangle\right)$

$\qquad = \dfrac{7}{5}\langle -3, 4\rangle$

$\qquad = \left\langle -\dfrac{21}{5}, \dfrac{28}{5}\right\rangle$

33. $\mathbf{v} = \dfrac{3}{2}\mathbf{u}$

$\qquad = \dfrac{3}{2}(2\mathbf{i} - \mathbf{j})$

$\qquad = 3\mathbf{i} - \dfrac{3}{2}\mathbf{j} = \langle 3, -\dfrac{3}{2}\rangle$

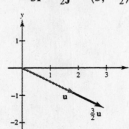

35. $\mathbf{v} = \mathbf{u} + 2\mathbf{w}$

$\qquad = (2\mathbf{i} - \mathbf{j}) + 2(\mathbf{i} + 2\mathbf{j})$

$\qquad = 4\mathbf{i} + 3\mathbf{j} = \langle 4, 3\rangle$

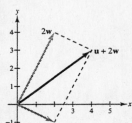

37. $\mathbf{v} = \dfrac{1}{2}(3\mathbf{u} + \mathbf{w})$

$\qquad = \dfrac{1}{2}(6\mathbf{i} - 3\mathbf{j} + \mathbf{i} + 2\mathbf{j})$

$\qquad = \dfrac{7}{2}\mathbf{i} - \dfrac{1}{2}\mathbf{j} = \langle \dfrac{7}{2}, -\dfrac{1}{2}\rangle$

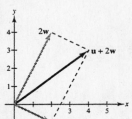

39. $\mathbf{v} = 5(\cos 30°\mathbf{i} + \sin 30°\mathbf{j})$

$\qquad \|\mathbf{v}\| = 5, \ \theta = 30°$

41. $\mathbf{v} = 6\mathbf{i} - 6\mathbf{j}$

$\qquad \|\mathbf{v}\| = \sqrt{6^2 + (-6)^2} = \sqrt{72} = 6\sqrt{2}$

$\qquad \tan \theta = \dfrac{-6}{6} = -1$

Since $\mathbf{v}$ lies in Quadrant IV, $\theta = 315°$.

43. $\mathbf{v} = \langle 3\cos 0°, \ 3\sin 0°\rangle$

$\qquad = \langle 3, 0\rangle$

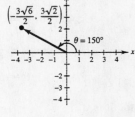

45. $\mathbf{v} = \langle 3\sqrt{2}\cos 150°, \ 3\sqrt{2}\sin 150°\rangle$

$\qquad = \left\langle -\dfrac{3\sqrt{6}}{2}, \dfrac{3\sqrt{2}}{2}\right\rangle$

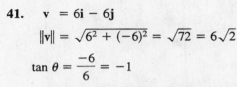

47. $\mathbf{v} = 2\left(\dfrac{1}{\sqrt{3^2+1^2}}\right)(\mathbf{i} + 3\mathbf{j})$

$\qquad = \dfrac{2}{\sqrt{10}}(\mathbf{i} + 3\mathbf{j})$

$\qquad = \dfrac{\sqrt{10}}{5}\mathbf{i} + \dfrac{3\sqrt{10}}{5}\mathbf{j} = \left\langle \dfrac{\sqrt{10}}{5}, \dfrac{3\sqrt{10}}{5}\right\rangle$

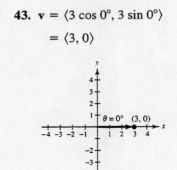

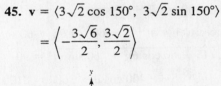

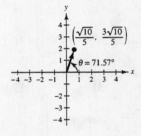

49. $\mathbf{u} = \langle 5\cos 0°, 5\sin 0° \rangle = \langle 5, 0 \rangle$

$\mathbf{v} = \langle 5\cos 90°, 5\sin 90° \rangle = \langle 0, 5 \rangle$

$\mathbf{u} + \mathbf{v} = \langle 5, 5 \rangle$

51. $\mathbf{u} = \langle 20\cos 45°, 20\sin 45° \rangle = \langle 10\sqrt{2}, 10\sqrt{2} \rangle$

$\mathbf{v} = \langle 50\cos 180°, 50\sin 180° \rangle = \langle -50, 0 \rangle$

$\mathbf{u} + \mathbf{v} = \langle 10\sqrt{2} - 50, 10\sqrt{2} \rangle$

53. $\mathbf{v} = \mathbf{i} + \mathbf{j}$

$\mathbf{w} = 2(\mathbf{i} - \mathbf{j})$

$\mathbf{u} = \mathbf{v} - \mathbf{w} = -\mathbf{i} + 3\mathbf{j}$

$\|\mathbf{v}\| = \sqrt{2}$

$\|\mathbf{w}\| = 2\sqrt{2}$

$\|\mathbf{v} - \mathbf{w}\| = \sqrt{10}$

$\cos \alpha = \dfrac{\|\mathbf{v}\|^2 + \|\mathbf{w}\|^2 - \|\mathbf{v} - \mathbf{w}\|^2}{2\|\mathbf{v}\|\,\|\mathbf{w}\|} = \dfrac{2 + 8 - 10}{2\sqrt{2}\cdot 2\sqrt{2}} = 0$

$\alpha = 90°$

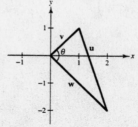

55. $\mathbf{v} = \mathbf{i} + \mathbf{j}$

$\mathbf{w} = 3\mathbf{i} - \mathbf{j}$

$\mathbf{u} = \mathbf{v} - \mathbf{w} = -2\mathbf{i} + 2\mathbf{j}$

$\cos \alpha = \dfrac{\|\mathbf{v}\|^2 + \|\mathbf{w}\|^2 - \|\mathbf{v} - \mathbf{w}\|^2}{2\|\mathbf{v}\|\,\|\mathbf{w}\|} = \dfrac{2 + 10 - 8}{2\sqrt{2}\,\sqrt{10}} \approx 0.4472$

$\alpha = 63.4°$

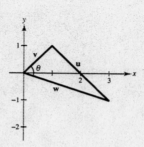

57. Force One: $\mathbf{u} = 45\mathbf{i}$

Force Two: $\mathbf{v} = 60\cos\theta\,\mathbf{i} + 60\sin\theta\,\mathbf{j}$

Resultant Force: $\mathbf{u} + \mathbf{v} = (45 + 60\cos\theta)\mathbf{i} + 60\sin\theta\,\mathbf{j}$

$\|\mathbf{u} + \mathbf{v}\| = \sqrt{(45 + 60\cos\theta)^2 + (60\sin\theta)^2} = 90$

$2025 + 5400\cos\theta + 3600 = 8100$

$5400\cos\theta = 2475$

$\cos\theta = \dfrac{2475}{5400} \approx 0.4583$

$\theta \approx 62.7°$

59. (a) The angle between them is $0°$.

(b) The angle between them is $180°$.

(c) No. At most it can be equal to the sum when the angle between them is $0°$.

61. (a) $\mathbf{u} = 220\mathbf{i}, \quad \mathbf{v} = 150\cos 30°\mathbf{i} + 150\sin 30°\mathbf{j}$

$\mathbf{u} + \mathbf{v} = (220 + 75\sqrt{3})\mathbf{i} + 75\mathbf{j}$

$\|\mathbf{u} + \mathbf{v}\| = \sqrt{(220 + 75\sqrt{3})^2 + 75^2} \approx 357.85$ newtons

$\tan\theta = \dfrac{75}{220 + 75\sqrt{3}} \implies \theta \approx 12.1°$

—CONTINUED—

61. **—CONTINUED—**

(b) $\mathbf{u} + \mathbf{v} = 220\mathbf{i} + (150 \cos \theta\mathbf{i} + 150 \sin \theta\mathbf{j})$

$$M = \|\mathbf{u} + \mathbf{v}\| = \sqrt{(220^2 + 150^2(\cos^2 \theta + \sin^2 \theta) + 2(220)(150) \cos \theta}$$

$$= \sqrt{70{,}900 + 66{,}000 \cos \theta}$$

$$= 10\sqrt{709 + 660 \cos \theta}$$

$$\alpha = \arctan\left(\frac{15 \sin \theta}{22 + 15 \cos \theta}\right)$$

(c)

θ	0°	30°	60°	90°	120°	150°	180°
M	370.0	357.9	322.3	266.3	194.7	117.2	70.0
α	0°	12.1°	23.8°	34.3°	41.9°	39.8°	0°

(d)

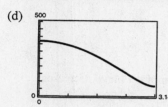 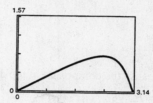

(e) For increasing θ the two vectors tend to work against each other resulting in a decrease in the magnitude of the resultant.

63.

$$\mathbf{u} = (75 \cos 30°)\mathbf{i} + (75 \sin 30)\mathbf{j} \approx 64.95\mathbf{i} + 37.5\mathbf{j}$$

$$\mathbf{v} = (100 \cos 45°)\mathbf{i} + (100 \sin 45°)\mathbf{j} \approx 70.71\mathbf{i} + 70.71\mathbf{j}$$

$$\mathbf{w} = (125 \cos 120°)\mathbf{i} + (125 \sin 120°)\mathbf{j} \approx -62.5\mathbf{i} + 108.3\mathbf{j}$$

$$\mathbf{u} + \mathbf{v} + \mathbf{w} = 73.16\mathbf{i} + 216.5\mathbf{j}$$

$$\|\mathbf{u} + \mathbf{v} + \mathbf{w}\| = 228.5 \text{ pounds}$$

$$\tan \theta = \frac{216.5}{73.16} \approx 2.9592 \approx 71.3°$$

65.

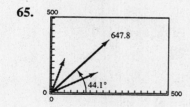

$$\mathbf{u} = 400 \cos 25°\mathbf{i} + 400 \sin 25°\mathbf{j}$$

$$\mathbf{v} = 300 \cos 70°\mathbf{i} + 300 \sin 70°\mathbf{j}$$

$$\mathbf{u} + \mathbf{v} \approx 465.13\mathbf{i} + 450.96\mathbf{j}$$

$$\|\mathbf{u} + \mathbf{v}\| \approx \sqrt{(465.13)^2 + (450.96)^2} \approx 647.8$$

$$\alpha = \arctan\left(\frac{450.96}{465.13}\right) \approx 44.1°$$

67. Horizontal component of velocity:

$80 \cos 40° \approx 61.28$ ft/sec

Vertical component of velocity:

$80 \sin 40° \approx 51.42$ ft/sec

69. Cable $\overrightarrow{AC}$: $\mathbf{u} = \|\mathbf{u}\|(\cos 50°\mathbf{i} - \sin 50°\mathbf{j})$

Cable $\overrightarrow{BC}$: $\mathbf{u} = \|\mathbf{u}\|(\cos 30°\mathbf{i} - \sin 30°\mathbf{j})$

Resultant: $\mathbf{u} + \mathbf{v} = -1000\mathbf{j}$

$\|\mathbf{u}\| \cos 50° - \|\mathbf{v}\| \cos 30° = 0$

$-\|\mathbf{u}\| \sin 50° - \|\mathbf{v}\| \sin 30° = -2000$

Solving this system of equations yields:

$T_{AC} = \|\mathbf{u}\| \approx 1758.8$ pounds

$T_{BC} = \|\mathbf{v}\| \approx 1305.4$ pounds

71. (a) Tow line 1: $\mathbf{u} = \|\mathbf{u}\| (\cos 20°\mathbf{i} + \sin 20°\mathbf{j})$

Tow line 2: $\mathbf{v} = \|\mathbf{u}\| (\cos(-20°)\mathbf{i} + \sin(-20°)\mathbf{j})$

Resultant: $\mathbf{u} + \mathbf{v} = 6000\mathbf{i} = [\|\mathbf{u}\| \cos 20° + \|\mathbf{u}\| \cos(-20°)]\mathbf{i}$

$$\implies 6000 = 2\|\mathbf{u}\| \cos 20°$$

$$\implies \|\mathbf{u}\| \approx 3192.5 \text{ lb}$$

(b) $\mathbf{u} + \mathbf{v} = 6000\mathbf{i} = 2\|\mathbf{u}\| \cos \theta \implies T = \|\mathbf{u}\| = 3000 \sec \theta$

(c)

θ	10°	20°	30°	40°	50°	60°
T	3046.3	3192.5	3464.1	2916.2	4667.2	6000.0

(d)

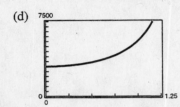

(e) The tension increases because the component in the direction of the motion of the barge decreases.

73. Airspeed: $\mathbf{u} = (875 \cos 32°)\mathbf{i} - (875 \sin 32°)\mathbf{j}$

Groundspeed: $\mathbf{v} = (800 \cos 40°)\mathbf{i} - (800 \sin 40°)\mathbf{j}$

Wind: $\mathbf{w} = \mathbf{v} - \mathbf{u} = (800 \cos 40° - 875 \cos 32°)\mathbf{j} + (-800 \sin 40° + 875 \sin 32°)\mathbf{j}$

$$\approx -129.2065\mathbf{i} - 50.5507\mathbf{j}$$

Wind speed: $\|\mathbf{w}\| \approx \sqrt{(-129.2065)^2 + (-50.5507)^2}$

$$\approx 138.7 \text{ kilometers per hour}$$

Wind direction: $\tan \theta = \dfrac{-50.5507}{-129.2065}$

$$\theta \approx 21.4°$$

$$\text{N } 21.4° \text{ E}$$

75. $W = FD = (85 \cos 60°)(20)$

$= 850 \text{ ft/lb}$

77. True

79. False, $a = b = 0$.

81. Let $\mathbf{v} = (\cos \theta)\mathbf{i} + (\sin \theta)\mathbf{j}$

$\|\mathbf{v}\| = \sqrt{\cos^2 \theta + \sin^2 \theta} = \sqrt{1} = 1$

Therefore, $\mathbf{v}$ is a unit vector for any value of θ.

83. $\mathbf{u} = \langle 5 - 1, 2 - 6 \rangle = \langle 4, -4 \rangle$

$\mathbf{v} = \langle 9 - 4, 4 - 5 \rangle = \langle 5, -1 \rangle$

$\mathbf{u} - \mathbf{v} = \langle -1, -3 \rangle$

$\mathbf{v} - \mathbf{u} = \langle 1, 3 \rangle$

Section 6.4   Vectors and Dot Products

■ Know the definition of the dot product of $\mathbf{u} = \langle u_1, u_2 \rangle$ and $\mathbf{v} = \langle v_1, v_2 \rangle$.

$$\mathbf{u} \cdot \mathbf{v} = u_1 v_1 + u_2 v_2$$

■ Know the following properties of the dot product:

1. $\mathbf{u} \cdot \mathbf{v} = \mathbf{v} \cdot \mathbf{u}$
2. $\mathbf{0} \cdot \mathbf{v} = 0$
3. $\mathbf{u} \cdot (\mathbf{v} + \mathbf{w}) = \mathbf{u} \cdot \mathbf{v} + \mathbf{u} \cdot \mathbf{w}$
4. $\mathbf{v} \cdot \mathbf{v} = \|\mathbf{v}\|^2$
5. $c(\mathbf{u} \cdot \mathbf{v}) = c\mathbf{u} \cdot \mathbf{v} = \mathbf{u} \cdot c\mathbf{v}$

■ If θ is the angle between two nonzero vectors $\mathbf{u}$ and $\mathbf{v}$, then

$$\cos \theta = \frac{\mathbf{u} \cdot \mathbf{v}}{\|\mathbf{u}\| \, \|\mathbf{v}\|}.$$

■ The vectors $\mathbf{u}$ and $\mathbf{v}$ are orthogonal if $\mathbf{u} \cdot \mathbf{v} = 0$.

■ Know the definition of vector components. $\mathbf{u} = \mathbf{w}_1 + \mathbf{w}_2$ where $\mathbf{w}_1$ and $\mathbf{w}_2$ are orthogonal, and $\mathbf{w}_1$ is parallel to $\mathbf{v}$. $\mathbf{w}_1$ is called the projection of $\mathbf{u}$ onto $\mathbf{v}$ and is denoted by

$$\mathbf{w}_1 = \text{proj}_{\mathbf{v}}\mathbf{u} = \left(\frac{\mathbf{u} \cdot \mathbf{v}}{\|\mathbf{v}\|^2} \right) \mathbf{v}.$$

Then we have $\mathbf{w}_2 = \mathbf{u} - \mathbf{w}_1$.

■ Know the definition of work.

1. Projection form: $W = \|\text{proj}_{\overrightarrow{PQ}} \mathbf{F}\| \, \|\overrightarrow{PQ}\|$
2. Dot product form: $W = \mathbf{F} \cdot \overrightarrow{PQ}$

Solutions to Odd-Numbered Exercises

1. $\mathbf{u} = \langle 3, 4 \rangle$, $\mathbf{v} = \langle 2, -3 \rangle$

$\mathbf{u} \cdot \mathbf{v} = 3(2) + 4(-3) = -6$

3. $\mathbf{u} = 4\mathbf{i} - 2\mathbf{j}$, $\mathbf{v} = \mathbf{i} - \mathbf{j}$

$\mathbf{u} \cdot \mathbf{v} = 4(1) + (-2)(-1) = 6$

5. $\mathbf{u} = \langle 2, 2 \rangle$

$\mathbf{u} \cdot \mathbf{u} = 2(2) + 2(2) = 8$

The result is a scalar.

7. $\mathbf{u} = \langle 2, 2 \rangle$, $\mathbf{v} = \langle -3, 4 \rangle$

$(\mathbf{u} \cdot \mathbf{v})\mathbf{v} = [(2)(-3) + 2(4)] \langle -3, 4 \rangle = 2\langle -3, 4 \rangle = \langle -6, 8 \rangle$

The result is a vector.

9. $\mathbf{u} = \langle -5, 12 \rangle$

$\|\mathbf{u}\| = \sqrt{\mathbf{u} \cdot \mathbf{u}} = \sqrt{(-5)^2 + 12^2} = 13$

11. $\mathbf{u} = 20\mathbf{i} + 25\mathbf{j}$

$\|\mathbf{u}\| = \sqrt{(20)^2 + (25)^2} = \sqrt{1025} = 5\sqrt{41}$

13. $\mathbf{u} = \langle 1245, 2600 \rangle$, $\mathbf{v} = \langle 12.20, 8.50 \rangle$

$\mathbf{u} \cdot \mathbf{v} = 1245(12.20) + 2600(8.50) = \$37,289$

This gives the total revenue that can be earned by selling all of the units.

15. $\mathbf{u} = \langle 1, 0 \rangle$, $\mathbf{v} = \langle 0, -2 \rangle$

$$\cos \theta = \frac{\mathbf{u} \cdot \mathbf{v}}{\|\mathbf{u}\| \|\mathbf{v}\|} = \frac{0}{(1)(2)} = 0$$

$$\theta = 90°$$

17. $\mathbf{u} = 3\mathbf{i} + 4\mathbf{j}$, $\mathbf{v} = -2\mathbf{j}$

$$\cos \theta = \frac{\mathbf{u} \cdot \mathbf{v}}{\|\mathbf{u}\| \|\mathbf{v}\|} = -\frac{8}{(5)(2)}$$

$$\theta = \arccos\left(-\frac{4}{5}\right)$$

$$\theta \approx 143.13°$$

19. $\mathbf{u} = \left(\cos \frac{\pi}{3}\right)\mathbf{i} + \left(\sin \frac{\pi}{3}\right)\mathbf{j} = \frac{1}{2}\mathbf{i} + \frac{\sqrt{3}}{2}\mathbf{j}$

$\mathbf{v} = \left(\cos \frac{3\pi}{4}\right)\mathbf{i} + \left(\sin \frac{3\pi}{4}\right)\mathbf{j} = -\frac{\sqrt{2}}{2}\mathbf{i} + \frac{\sqrt{2}}{2}\mathbf{j}$

$\|\mathbf{u}\| = \|\mathbf{v}\| = 1$

$$\cos \theta = \frac{\mathbf{u} \cdot \mathbf{v}}{\|\mathbf{u}\| \|\mathbf{v}\|} = \mathbf{u} \cdot \mathbf{v} = \left(\frac{1}{2}\right)\left(-\frac{\sqrt{2}}{2}\right) + \left(\frac{\sqrt{3}}{2}\right)\left(\frac{\sqrt{2}}{2}\right) = \frac{-\sqrt{2} + \sqrt{6}}{4}$$

$$\theta = \arccos\left(\frac{-\sqrt{2} + \sqrt{6}}{4}\right) = 75° = \frac{5\pi}{12}$$

21. $\mathbf{u} = 3\mathbf{i} + 4\mathbf{j}$, $\mathbf{v} = -7\mathbf{i} + 5\mathbf{j}$

$$\cos \theta = \frac{\mathbf{u} \cdot \mathbf{v}}{\|\mathbf{u}\| \|\mathbf{v}\|} = -\frac{1}{(5)(\sqrt{74})} \implies \theta \approx 91.33°$$

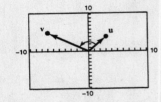

23. $\mathbf{u} = 5\mathbf{i} + 5\mathbf{j}$, $\mathbf{v} = -6\mathbf{i} + 6\mathbf{j}$

$$\cos \theta = \frac{\mathbf{u} \cdot \mathbf{v}}{\|\mathbf{u}\| \|\mathbf{v}\|} = 0 \implies \theta = 90°$$

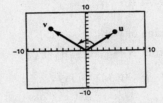

25. $P = (1, 2)$, $Q = (3, 4)$, $R = (2, 5)$

$\overrightarrow{PQ} = \langle 2, 2 \rangle$, $\overrightarrow{PR} = \langle 1, 3 \rangle$, $\overrightarrow{QR} = \langle -1, -1 \rangle$

$$\cos \alpha = \frac{\overrightarrow{PQ} \cdot \overrightarrow{PR}}{\|\overrightarrow{PQ}\| \|\overrightarrow{PR}\|} = \frac{8}{(2\sqrt{2})(\sqrt{10})} \implies \alpha = \arccos \frac{2}{\sqrt{5}} \approx 26.6°$$

$$\cos \beta = \frac{\overrightarrow{PQ} \cdot \overrightarrow{QR}}{\|\overrightarrow{PQ}\| \|\overrightarrow{QR}\|} = 0 \implies \beta = 90°. \text{ Thus, } \gamma = 180° - 26.6° - 90° = 63.4°.$$

27. $\mathbf{u} \cdot \mathbf{v} = \|\mathbf{u}\| \|\mathbf{v}\| \cos \theta$

$$= (4)(10) \cos \frac{2\pi}{3}$$

$$= 40\left(-\frac{1}{2}\right)$$

$$= -20$$

29. $\mathbf{u} = \langle -12, 30 \rangle$, $\mathbf{v} = \left\langle \frac{1}{2}, -\frac{5}{4} \right\rangle$

$\mathbf{u} = -24\mathbf{v} \implies \mathbf{u}$ and $\mathbf{v}$ are parallel.

31. $\mathbf{u} = \frac{1}{4}(3\mathbf{i} - \mathbf{j})$, $\mathbf{v} = 5\mathbf{i} + 6\mathbf{j}$

$\mathbf{u} \neq k\mathbf{v} \implies$ Not parallel

$\mathbf{u} \cdot \mathbf{v} \neq 0 \implies$ Not orthogonal

Neither

33. $\mathbf{u} = 2\mathbf{i} - 2\mathbf{j}$, $\mathbf{v} = -\mathbf{i} - \mathbf{j}$

$\mathbf{u} \cdot \mathbf{v} = 0 \implies \mathbf{u}$ and $\mathbf{v}$ are orthogonal.

35. $\mathbf{u} = \langle 3, 4 \rangle$, $\mathbf{v} = \langle 8, 2 \rangle$

$$\mathbf{w}_1 = \text{proj}_\mathbf{v}\mathbf{u} = \left(\frac{\mathbf{u} \cdot \mathbf{v}}{\|\mathbf{v}\|^2}\right)\mathbf{v} = \left(\frac{32}{68}\right)\mathbf{v} = \frac{8}{17}\langle 8, 2 \rangle = \frac{16}{17}\langle 4, 1 \rangle$$

$$\mathbf{w}_2 = \mathbf{u} - \mathbf{w}_1 = \langle 3, 4 \rangle - \frac{16}{17}\langle 4, 1 \rangle = \frac{13}{17}\langle -1, 4 \rangle$$

37. $\mathbf{u} = \langle 0, 3 \rangle$, $\mathbf{v} = \langle 2, 15 \rangle$

$$\mathbf{w}_1 = \text{proj}_\mathbf{v}\mathbf{u} = \left(\frac{\mathbf{u} \cdot \mathbf{v}}{\|\mathbf{v}\|^2}\right)\mathbf{v} = \frac{45}{229}\langle 2, 15 \rangle$$

$$\mathbf{w}_2 = \mathbf{u} - \mathbf{w}_1 = \langle 0, 3 \rangle - \frac{45}{229}\langle 2, 15 \rangle = \left\langle -\frac{90}{229}, \frac{12}{229} \right\rangle = \frac{6}{229}\langle -15, 2 \rangle$$

39. $\text{proj}_\mathbf{v}\mathbf{u} = \mathbf{u}$ since they are parallel.

$$\text{proj}_\mathbf{v}\mathbf{u} = \frac{\mathbf{u} \cdot \mathbf{v}}{\|\mathbf{v}\|^2}\mathbf{v} = \frac{18 + 18}{36 + 16}\mathbf{v} = \frac{26}{52}\langle 6, 4 \rangle = \langle 3, 2 \rangle = \mathbf{u}$$

41. $\text{proj}_\mathbf{v}\mathbf{u} = \mathbf{0}$ since they are perpendicular.

$$\text{proj}_\mathbf{v}\mathbf{u} = \frac{\mathbf{u} \cdot \mathbf{v}}{\|\mathbf{v}\|^2}\mathbf{v} = \mathbf{0}, \text{ since } \mathbf{u} \cdot \mathbf{v} = 0.$$

43. $\mathbf{u} = \langle 3, 5 \rangle$

For $\mathbf{v}$ to be orthogonal to $\mathbf{u}$, $\mathbf{u} \cdot \mathbf{v}$ must equal 0.

Two possibilities: $\langle -5, 3 \rangle$ and $\langle 5, -3 \rangle$.

45. $\mathbf{u} = \frac{1}{2}\mathbf{i} - \frac{2}{3}\mathbf{j}$

For $\mathbf{u}$ and $\mathbf{v}$ to be orthogonal, $\mathbf{u} \cdot \mathbf{v}$ must equal 0.

Two possibilities: $\frac{2}{3}\mathbf{i} + \frac{1}{2}\mathbf{j}$ and $-\frac{2}{3}\mathbf{i} - \frac{1}{2}\mathbf{j}$.

47. (a) $\mathbf{F} = -36{,}000\mathbf{j}$ Gravitational force

$$\mathbf{v} = (\cos 10°)\mathbf{i} + (\sin 10°)\mathbf{j}$$

$$\mathbf{w}_1 = \text{proj}_\mathbf{v}\mathbf{F} = \left(\frac{\mathbf{F} \cdot \mathbf{v}}{\|\mathbf{v}\|^2}\right)\mathbf{v} = (\mathbf{F} \cdot \mathbf{v})\mathbf{v} \approx -6251.3\mathbf{v}$$

The magnitude of this force is 6251.3, therefore a force of 6251.3 pounds is needed to keep the truck from rolling down the hill.

(b) $\mathbf{w}_2 = \mathbf{F} - \mathbf{w}_1 = -36{,}000\mathbf{j} + 6251.3(\cos 10°\mathbf{i} + \sin 10°\mathbf{j})$

$\quad = [(6251.3 \cos 10°)\mathbf{i} + (6251.3 \sin 10° - 36{,}000)\mathbf{j}]$

$\|\mathbf{w}_2\| \approx 35{,}453.1$ pounds

49. (a) $\mathbf{u} \cdot \mathbf{v} = 0 \implies \mathbf{u}$ and $\mathbf{v}$ are orthogonal and $\theta = \frac{\pi}{2}$.

(b) $\mathbf{u} \cdot \mathbf{v} > 0 \implies \cos \theta > 0 \implies 0 \le \theta < \frac{\pi}{2}$

(c) $\mathbf{u} \cdot \mathbf{v} < 0 \implies \cos \theta < 0 \implies \frac{\pi}{2} < \theta \le \pi$

51. $w = (245)(3) = 735$ Newton-meters

53. $w = (\cos 30°)(45)(20) \approx 779.4$ foot-pounds

55. $w = \|\text{proj}_{\overrightarrow{PQ}}\mathbf{v}\| \, \|\overrightarrow{PQ}\|$ where $\overrightarrow{PQ} = \langle 4, 7 \rangle$ and $\mathbf{v} = \langle 1, 4 \rangle$.

$$\text{proj}_{\overrightarrow{PQ}}\mathbf{v} = \left(\frac{\mathbf{v} \cdot \overrightarrow{PQ}}{\|\overrightarrow{PQ}\|^2}\right)\overrightarrow{PQ} = \left(\frac{32}{65}\right)\langle 4, 7 \rangle$$

$$w = \|\text{proj}_{\overrightarrow{PQ}}\mathbf{v}\| \, \|\overrightarrow{PQ}\| = \left(\frac{32\sqrt{65}}{65}\right)(\sqrt{65}) = 32$$

57. Since $\text{proj}_v\mathbf{u}$ is a scalar multiple of $\mathbf{v}$, you can write $\text{proj}_v\mathbf{u} = \alpha\mathbf{v}$. If $\alpha > 0$, then $\cos\theta > 0$ and

$$\|\text{proj}_v\mathbf{u}\| = \|\alpha\mathbf{v}\| = \|\mathbf{u}\|\cos\theta$$

$$= \frac{\|\mathbf{u}\|\,\|\mathbf{v}\|\cos\theta}{\|\mathbf{v}\|}$$

$$= \frac{\mathbf{u}\cdot\mathbf{v}}{\|\mathbf{v}\|}$$

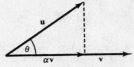

Thus, $\alpha = \dfrac{\mathbf{u}\cdot\mathbf{v}}{\|\mathbf{v}\|^2}$ and $\text{proj}_v\mathbf{u} = \dfrac{\mathbf{u}\cdot\mathbf{v}}{\|\mathbf{v}\|^2}\mathbf{v}$.

59. Use the Law of Cosines on the triangle.

$$\|\mathbf{u}-\mathbf{v}\|^2 = \|\mathbf{u}\|^2 + \|\mathbf{v}\|^2 - 2\|\mathbf{u}\|\,\|\mathbf{v}\|\cos\theta$$

$$= \|\mathbf{u}\|^2 + \|\mathbf{v}\|^2 - 2\,\mathbf{u}\cdot\mathbf{v}$$

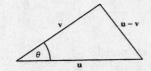

61. $\mathbf{u}\cdot(c\mathbf{v}+d\mathbf{w}) = \mathbf{u}\cdot(c\mathbf{v}) + \mathbf{u}\cdot(d\mathbf{w})$

$$= c(\mathbf{u}\cdot\mathbf{v}) + d(\mathbf{u}\cdot\mathbf{w})$$

$$= c0 + d0 = 0$$

63. (a) $1 - \dfrac{1}{x^2} = \dfrac{x^2-1}{x^2}$

(b) $1 - \dfrac{1}{\sec^2 x} = \dfrac{\sec^2 x - 1}{\sec^2 x} = \dfrac{\tan^2 x}{\sec^2 x} = \sin^2 x$

65. (a) $\dfrac{y}{1+z} + \dfrac{1+z}{y} = \dfrac{y^2 + (1+z)^2}{(1+z)y}$

(b) $\dfrac{\sin x}{1+\cos x} + \dfrac{1+\cos x}{\sin x} = \dfrac{\sin^2 x + (1+\cos x)^2}{\sin x(1+\cos x)}$

$$= \dfrac{\sin^2 x + 1 + 2\cos x + \cos^2 x}{\sin x(1+\cos x)} = \dfrac{2+2\cos x}{\sin x(1+\cos x)} = \dfrac{2}{\sin x} = 2\csc x$$

Section 6.5   DeMoivre's Theorem

- You should be able to graphically represent complex numbers.
- The absolute value of the complex numbers $z = a + bi$ is $|z| = \sqrt{a^2 + b^2}$.
- The trigonometric form of the complex number $z = a + bi$ is $z = r(\cos\theta + i\sin\theta)$ where

  (a) $a = r\cos\theta$ (b) $b = r\sin\theta$

  (c) $r = \sqrt{a^2 + b^2}$; r is called the modulus of z.   (d) $\tan\theta = b/a$; θ is called the argument of z.
- Given $z_1 = r_1(\cos\theta_1 + i\sin\theta_1)$ and $z_2 = r_2(\cos\theta_2 + i\sin\theta_2)$:

  (a) $z_1 z_2 = r_1 r_2[\cos(\theta_1 + \theta_2) + i\sin(\theta_1 + \theta_2)]$

  (b) $\dfrac{z_1}{z_2} = \dfrac{r_1}{r_2}[\cos(\theta_1 - \theta_2) + i\sin(\theta_1 - \theta_2)]$, $z_2 \neq 0$
- You should know DeMoivre's Theorem: If $z = r(\cos\theta + i\sin\theta)$, then for any positive integer n,

  $$z^n = r^n(\cos n\theta + i\sin n\theta).$$
- You should know that for any positive integer n, $z = r(\cos\theta + i\sin\theta)$ has n distinct nth roots given by

  $$\sqrt[n]{r}\left[\cos\left(\frac{\theta + 2\pi k}{n}\right) + i\sin\left(\frac{\theta + 2\pi k}{n}\right)\right]$$

  where $k = 0, 1, 2, \ldots, n-1$.

Solutions to Odd-Numbered Exercises

1. $|-5i| = \sqrt{0^2 + (-5)^2}$
$\qquad\quad = \sqrt{25} = 5$

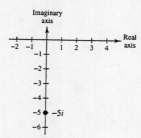

3. $|-4 + 4i| = \sqrt{(-4)^2 + (4)^2}$
$\qquad\qquad = \sqrt{32} = 4\sqrt{2}$

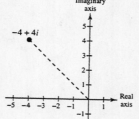

5. $|6 - 7i| = \sqrt{6^2 + (-7)^2}$
$\qquad\qquad = \sqrt{85}$

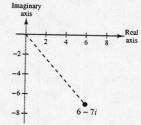

7. $z = 3i$
$\quad r = \sqrt{0^2 + 3^2} = \sqrt{9} = 3$
$\quad \tan \theta = \dfrac{3}{0},\text{ undefined} \implies \theta = \dfrac{\pi}{2}$
$\quad z = 3\left(\cos \dfrac{\pi}{2} + i \sin \dfrac{\pi}{2}\right)$

9. $z = -2 - 2i$
$\quad r = \sqrt{(-2)^2 + (-2)^2} = \sqrt{8} = 2\sqrt{2}$
$\quad \tan \theta = \dfrac{-2}{-2} = 1,\ \theta \text{ is in Quadrant III.}$
$\quad \theta = \dfrac{5\pi}{4}$
$\quad z = 2\sqrt{2}\left(\cos \dfrac{5\pi}{4} + i \sin \dfrac{5\pi}{4}\right)$

11. $z = 3 - 3i$
$\quad r = \sqrt{3^2 + (-3)^2} = \sqrt{18} = 3\sqrt{2}$
$\quad \tan \theta = \dfrac{-3}{3} = -1 \implies \theta = \dfrac{7\pi}{4}$
$\quad z = 3\sqrt{2}\left(\cos \dfrac{7\pi}{4} + i \sin \dfrac{7\pi}{4}\right)$

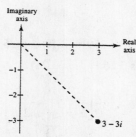

13. $z = \sqrt{3} + i$
$\quad r = \sqrt{(\sqrt{3})^2 + 1^2} = \sqrt{4} = 2$
$\quad \tan \theta = \dfrac{1}{\sqrt{3}} = \dfrac{\sqrt{3}}{3} \implies \theta = \dfrac{\pi}{6}$
$\quad z = 2\left(\cos \dfrac{\pi}{6} + i \sin \dfrac{\pi}{6}\right)$

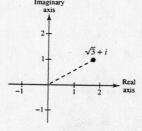

15. $z = -2(1 + \sqrt{3}i)$
$\quad r = \sqrt{(-2)^2 + (-2\sqrt{3})^2} = \sqrt{16} = 4$
$\quad \tan \theta = \dfrac{\sqrt{3}}{1} = \sqrt{3} \implies \theta = \dfrac{4\pi}{3}$
$\quad z = 4\left(\cos \dfrac{4\pi}{3} + i \sin \dfrac{4\pi}{3}\right)$

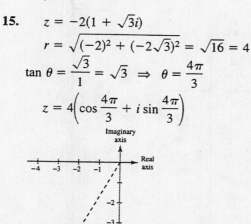

17. $z = 0 + 6i$
$\quad r = \sqrt{0^2 + (6)^2} = \sqrt{36} = 6$
$\quad \tan \theta = \dfrac{6}{0},\text{ undefined} \implies \theta = \dfrac{\pi}{2}$
$\quad z = 6\left(\cos \dfrac{\pi}{2} + i \sin \dfrac{\pi}{2}\right)$

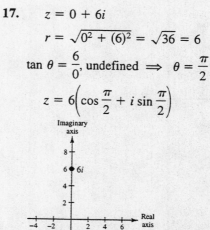

19. $z = -7 + 4i$

$r = \sqrt{(-7)^2 + (4)^2} = \sqrt{65}$

$\tan \theta = \dfrac{4}{-7} \;\Rightarrow\; \theta = 2.62$

$z \approx \sqrt{65}(\cos 2.62 + i \sin 2.62)$

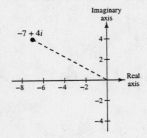

21. $z = 7 + 0i$

$r = \sqrt{(7)^2 + (0)^2} = \sqrt{49} = 7$

$\tan \theta = \dfrac{0}{7} = 0 \;\Rightarrow\; \theta = 0$

$z = 7(\cos 0 + i \sin 0)$

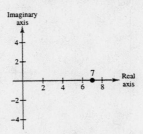

23. $z = 1 + 6i$

$r = \sqrt{1^2 + (6)^2} = \sqrt{37}$

$\tan \theta = \dfrac{6}{1} = 6 \;\Rightarrow\; \theta = 1.41$

$z \approx \sqrt{37}(\cos 1.41 + i \sin 1.41)$

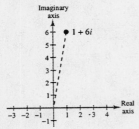

25. $z = -3 - i$

$r = \sqrt{(-3)^2 + (-1)^2} = \sqrt{10}$

$\tan \theta = \dfrac{-1}{-3} = \dfrac{1}{3} \;\Rightarrow\; \theta \approx 3.46$

$z \approx \sqrt{10}(\cos 3.46 + i \sin 3.46)$

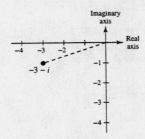

27. $5 + 2i \approx 5.39 \angle 0.38$

$= 5.39(\cos 0.38 + i \sin 0.38)$

29. $3\sqrt{2} - 7i \approx 8.19 \angle -1.0259 \approx 8.19 \angle 5.26$

$= 8.19(\cos 5.26 + i \sin 5.26)$

31. $2(\cos 150° + i \sin 150°) = \left[-\dfrac{\sqrt{3}}{2} + i\left(\dfrac{1}{2}\right) \right]$

$= -\sqrt{3} + i$

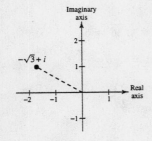

33. $\dfrac{3}{2}(\cos 300° + i \sin 300°) = \dfrac{3}{2}\left[\dfrac{1}{2} + i\left(-\dfrac{\sqrt{3}}{2} \right) \right]$

$= \dfrac{3}{4} - \dfrac{3\sqrt{3}}{4}i$

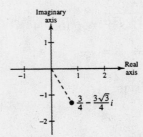

35. $3.75\left(\cos\dfrac{3\pi}{4} + i\sin\dfrac{3\pi}{4}\right) = -\dfrac{15\sqrt{2}}{8} + \dfrac{15\sqrt{2}}{8}i$

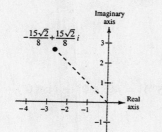

37. $4\left(\cos\dfrac{3\pi}{2} + i\sin\dfrac{3\pi}{2}\right) = 4(0 - i) = -4i$

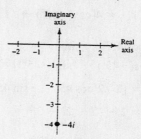

39. $3[\cos(18°\ 45') + i\sin(18°\ 45')] \approx 2.8408 + 0.9643i$

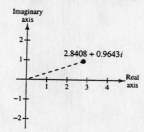

41. $5\left(\cos\dfrac{\pi}{9} + i\sin\dfrac{\pi}{9}\right) \approx 4.70 + 1.71i$

43. $12\left(\cos\dfrac{3\pi}{5} + i\sin\dfrac{3\pi}{5}\right) \approx -3.71 + 11.41i$

45.

The absolute value of each power is 1.

47. $\left[3\left(\cos\dfrac{\pi}{3} + i\sin\dfrac{\pi}{3}\right)\right]\left[4\left(\cos\dfrac{\pi}{6} + i\sin\dfrac{\pi}{6}\right)\right] = (3)(4)\left[\cos\left(\dfrac{\pi}{3} + \dfrac{\pi}{6}\right) + i\sin\left(\dfrac{\pi}{6} + \dfrac{\pi}{3}\right)\right]$

$$= 12\left(\cos\dfrac{\pi}{2} + i\sin\dfrac{\pi}{2}\right)$$

49. $\left[\tfrac{5}{3}(\cos 140° + i\sin 140°)\right]\left[\tfrac{2}{3}(\cos 60° + i\sin 60°)\right] = \left(\tfrac{5}{3}\right)\left(\tfrac{2}{3}\right)[\cos(140° + 60°) + i\sin(140° + 60°)]$

$$= \tfrac{10}{9}(\cos 200° + i\sin 200°)$$

51. $[0.45(\cos 310° + i\sin 310°)][0.60(\cos 200° + i\sin 200°)] = (0.45)(0.60)[\cos(310° + 200°) + i\sin(310° + 200°)]$

$$= 0.27(\cos 510° + i\sin 510°)$$

$$= 0.27(\cos 150° + i\sin 150°)$$

53. $\dfrac{\cos 40° + i\sin 40°}{\cos 10° + i\sin 10°} = \cos(40° - 10°) + i\sin(40° - 10°) = \cos 30° + i\sin 30°$

55. $\dfrac{2(\cos 120° + i\sin 120°)}{4(\cos 40° + i\sin 40°)} = \dfrac{1}{2}[\cos(120° - 40°) + i\sin(120° - 40°)]$

$$= \dfrac{1}{2}(\cos 80° + i\sin 80°)$$

57. $\dfrac{12(\cos 52° + i \sin 52°)}{3(\cos 110° + i \sin 110°)} = 4[\cos(52° - 110°) + i \sin(52° - 110°)]$

$$= 4[\cos(-58°) + i \sin(-58°)]$$

59. (a) $2 + 2i = 2\sqrt{2}(\cos 45° + i \sin 45°)$

$1 - i = \sqrt{2}[\cos(-45°) + i \sin(-45°)]$

(b) $(2 + 2i)(1 - i) = [2\sqrt{2}(\cos 45° + i \sin 45°)][\sqrt{2}(\cos(-45°) + i \sin(-45°))] = 4(\cos 0° + i \sin 0°) = 4.$

(c) $(2 + 2i)(1 - i) = 2 - 2i + 2i - 2i^2 = 2 + 2 = 4$

61. (a) $-2i = 2[\cos(-90°) + i \sin(-90°)]$

$1 + i = \sqrt{2}(\cos 45° + i \sin 45°)$

(b) $-2i(1 + i) = 2[\cos(-90°) + i \sin(-90°)][\sqrt{2}(\cos 45° + i \sin 45°)]$

$$= 2\sqrt{2}[\cos(-45°) + i \sin(-45°)]$$

$$= 2\sqrt{2}\left[\frac{1}{\sqrt{2}} - \frac{1}{\sqrt{2}}i\right] = 2 - 2i$$

(c) $-2i(1 + i) = -2i - 2i^2 = -2i + 2 = 2 - 2i$

63. (a) $5 = 5(\cos 0° + i \sin 0°)$

$2 + 3i \approx \sqrt{13}(\cos 56.31° + i \sin 56.31°)$

(b) $\dfrac{5}{2 + 3i} \approx \dfrac{5(\cos 0° + i \sin 0°)}{\sqrt{13}(\cos 56.31° + i \sin 56.31°)} = \dfrac{5\sqrt{13}}{13}(\cos -56.31° + i \sin -56.31) \approx 0.7692 - 1.154i$

(c) $\dfrac{5}{2 + 3i} = \dfrac{5}{2 + 3i} \cdot \dfrac{2 - 3i}{2 - 3i} = \dfrac{10 - 15i}{13} = \dfrac{10}{13} - \dfrac{15}{13}i \approx 0.7692 - 1.154i$

65. $\dfrac{z_1}{z_2} = \dfrac{r_1(\cos \theta_1 + i \sin \theta_1)}{r_2(\cos \theta_2 + i \sin \theta_2)} \cdot \dfrac{\cos \theta_2 - i \sin \theta_2}{\cos \theta_2 - i \sin \theta_2}$

$$= \dfrac{r_1}{r_2(\cos^2 \theta_2 + \sin^2 \theta_2)}[\cos \theta_1 \cos \theta_2 + \sin \theta_1 \sin \theta_2 + i(\sin \theta_1 \cos \theta_2 - \sin \theta_2 \cos \theta_1)]$$

$$= \dfrac{r_1}{r_2}[\cos(\theta_1 - \theta_2) + i \sin(\theta_1 - \theta_2)]$$

67. (a) $z\bar{z} = [r(\cos \theta + i \sin \theta)][r(\cos(-\theta) + i \sin(-\theta)]$

$$= r^2[\cos(\theta - \theta) + i \sin(\theta - \theta)]$$

$$= r^2[\cos 0 + i \sin 0]$$

$$= r^2$$

(b) $\dfrac{z}{\bar{z}} = \dfrac{r(\cos \theta + i \sin \theta)}{r[\cos(-\theta) + i \sin(-\theta)]}$

$$= \dfrac{r}{r}[\cos(\theta - (-\theta)) + i \sin(\theta - (-\theta))]$$

$$= \cos 2\theta + i \sin 2\theta$$

69. Let $z = x + iy$ such that:

$$|z| = 2 \implies 2 = \sqrt{x^2 + y^2}$$
$$\implies 4 = x^2 + y^2: \text{circle with radius of 2}$$

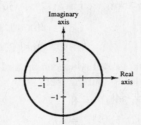

71. $(1 + i)^5 = \left[\sqrt{2}\left(\cos\dfrac{\pi}{4} + i\sin\dfrac{\pi}{4}\right)\right]^5$

$$= \left(\sqrt{2}\right)^5\left(\cos\dfrac{5\pi}{4} + i\sin\dfrac{5\pi}{4}\right)$$

$$= 4\sqrt{2}\left(-\dfrac{\sqrt{2}}{2} - \dfrac{\sqrt{2}}{2}i\right)$$

$$= -4 - 4i$$

73. $(-1 + i)^{10} = \left[\sqrt{2}\left(\cos\dfrac{3\pi}{4} + i\sin\dfrac{3\pi}{4}\right)\right]^{10}$

$$= \left(\sqrt{2}\right)^{10}\left(\cos\dfrac{30\pi}{4} + i\sin\dfrac{30\pi}{4}\right)$$

$$= 32\left[\cos\left(\dfrac{3\pi}{2} + 6\pi\right) + i\sin\left(\dfrac{3\pi}{2} + 6\pi\right)\right]$$

$$= 32\left(\cos\dfrac{3\pi}{2} + i\sin\dfrac{3\pi}{2}\right)$$

$$= 32[0 + i(-1)]$$

$$= -32i$$

75. $2\left(\sqrt{3} + i\right)^7 = 2\left[2\left(\cos\dfrac{\pi}{6} + i\sin\dfrac{\pi}{6}\right)\right]^7$

$$= 2\left[2^7\left(\cos\dfrac{7\pi}{6} + i\sin\dfrac{7\pi}{6}\right)\right]$$

$$= 256\left(-\dfrac{\sqrt{3}}{2} - \dfrac{1}{2}i\right)$$

$$= -128\sqrt{3} - 128i$$

77. $[5(\cos 20° + i\sin 20°)]^3 = 5^3(\cos 60° + i\sin 60°) = \dfrac{125}{2} + \dfrac{125\sqrt{3}}{2}i$

79. $\left(\cos\dfrac{5\pi}{4} + i\sin\dfrac{5\pi}{4}\right)^{10} = \cos\dfrac{25\pi}{2} + i\sin\dfrac{25\pi}{2}$

$$= \cos\left(12\pi + \dfrac{\pi}{2}\right) + i\sin\left(12\pi + \dfrac{\pi}{2}\right) = \cos\dfrac{\pi}{2} + i\sin\dfrac{\pi}{2} = i$$

81. $[5(\cos 3.2 + i\sin 3.2)]^4 = 5^4(\cos 12.8 + i\sin 12.8)$

$$\approx 608.02 + 144.69i$$

83. $(3 - 2i)^5 = -597 - 122i$

85. $[3(\cos 15° + i\sin 15°)]^4 = 81(\cos 60° + i\sin 60°)$

$$= \dfrac{81}{2} + \dfrac{81\sqrt{3}}{2}i$$

87. $\left[-\dfrac{1}{2}(1 + \sqrt{3}i)\right]^6 = \left[\cos\dfrac{4\pi}{3} + i\sin\dfrac{4\pi}{3}\right]^6$

$$= \cos 8\pi + i\sin 8\pi$$

$$= 1$$

89. (a) In trigonometric form we have: $2(\cos 30° + i \sin 30°)$

$$2(\cos 150° + i \sin 150°)$$
$$2(\cos 270° + i \sin 270°)$$

(b) There are three roots evenly spaced around a circle of radius 2. Therefore, they represent the cube roots of some number of modulus 8. Cubing them shows that they are all cube roots of $8i$.

(c) $[2(\cos 30° + i \sin 30°)]^3 = 8i$

$[2(\cos 150° + i \sin 150°)]^3 = 8i$

$[2(\cos 270° + i \sin 270°)]^3 = 8i$

91. (a) Square roots of $5(\cos 120° + i \sin 120°)$:

$$\sqrt{5}\left[\cos\left(\frac{120° + 360°k}{2}\right) + i \sin\left(\frac{120° + 360°k}{2}\right)\right], \ k = 0, \ 1$$

$$\sqrt{5}(\cos 60° + i \sin 60°)$$
$$\sqrt{5}(\cos 240° + i \sin 240°)$$

(c) $\dfrac{\sqrt{5}}{2} + \dfrac{\sqrt{15}}{2}i, \ -\dfrac{\sqrt{5}}{2} - \dfrac{\sqrt{15}}{2}i$

(b)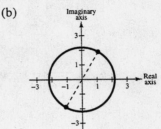

93. (a) Fourth roots of $16\left(\cos\dfrac{4\pi}{3} + i \sin\dfrac{4\pi}{3}\right)$:

$$\sqrt[4]{16}\left[\cos\left(\frac{(4\pi/3) + 2k\pi}{4}\right) + i \sin\left(\frac{(4\pi/3) + 2k\pi}{4}\right)\right], \ k = 0, 1, 2, 3$$

$$2\left(\cos\frac{\pi}{3} + i \sin\frac{\pi}{3}\right)$$

$$2\left(\cos\frac{5\pi}{6} + i \sin\frac{5\pi}{6}\right)$$

$$2\left(\cos\frac{4\pi}{3} + i \sin\frac{4\pi}{3}\right)$$

$$2\left(\cos\frac{11\pi}{6} + i \sin\frac{11\pi}{6}\right)$$

(c) $1 + \sqrt{3}i, \ -\sqrt{3} + i, \ -1 - \sqrt{3}i, \ \sqrt{3} - i$

(b)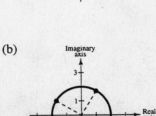

95. (a) Square roots of $-25i = 25\left(\cos\dfrac{3\pi}{2} + i \sin\dfrac{3\pi}{2}\right)$:

$$\sqrt{25}\left[\cos\left(\frac{(3\pi/2) + 2k\pi}{2}\right) + i \sin\left(\frac{(3\pi/2) + 2k\pi}{2}\right)\right], \ k = 0, 1$$

$$5\left(\cos\frac{3\pi}{4} + i \sin\frac{3\pi}{4}\right)$$

$$5\left(\cos\frac{7\pi}{4} + i \sin\frac{7\pi}{4}\right)$$

(c) $-\dfrac{5\sqrt{2}}{2} + \dfrac{5\sqrt{2}}{2}i, \ \dfrac{5\sqrt{2}}{2} - \dfrac{5\sqrt{2}}{2}i$

(b)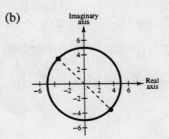

97. (a) Cube roots of $-\dfrac{125}{3}(1 + \sqrt{3}i) = 125\left(\cos \dfrac{4\pi}{3} + i \sin \dfrac{4\pi}{3}\right)$:

$$\sqrt[3]{125}\left[\cos\left(\dfrac{(4\pi/3) + 2k\pi}{3}\right) + i\sin\left(\dfrac{(4\pi/3) + 2k\pi}{3}\right)\right], \ k = 0, 1, 2$$

$$5\left(\cos \dfrac{4\pi}{9} + i \sin \dfrac{4\pi}{9}\right)$$

$$5\left(\cos \dfrac{10\pi}{9} + i \sin \dfrac{10\pi}{9}\right)$$

$$2\left(\cos \dfrac{16\pi}{9} + i \sin \dfrac{16\pi}{9}\right)$$

(b)

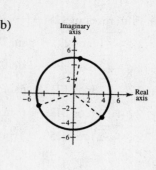

(c) $0.8682 + 4.924i, \ -4.698 - 1.710i, \ 3.830 - 3.214i$

99. (a) Cube roots of $8 = 8(\cos 0 + i \sin 0)$:

$$\sqrt[3]{8}\left[\cos\left(\dfrac{2k\pi}{3}\right) + i\sin\left(\dfrac{2k\pi}{3}\right)\right], \ k = 0, 1, 2$$

$$2(\cos 0 + i \sin 0)$$

$$2\left(\cos \dfrac{2\pi}{3} + i \sin \dfrac{2\pi}{3}\right)$$

$$2\left(\cos \dfrac{4\pi}{3} + i \sin \dfrac{4\pi}{3}\right)$$

(b)

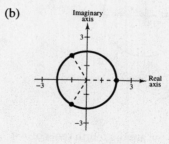

(c) $2, \ -1 + \sqrt{3}i, \ -1 - \sqrt{3}i$

101. (a) Fifth roots of $1 = \cos 0 + i \sin 0$:

$$\cos \dfrac{2k\pi}{5} + i \sin \dfrac{2k\pi}{5}, k = 0, 1, 2, 3, 4$$

$$\cos 0 + i \sin 0$$

$$\cos \dfrac{2\pi}{5} + i \sin \dfrac{2\pi}{5}$$

$$\cos \dfrac{4\pi}{5} + i \sin \dfrac{4\pi}{5}$$

$$\cos \dfrac{6\pi}{5} + i \sin \dfrac{6\pi}{5}$$

$$\cos \dfrac{8\pi}{5} + i \sin \dfrac{8\pi}{5}$$

(b)

(c) $1, 0.3090 + 0.9511i, -0.8090 + 0.5878i, -0.8090 - 0.5878i, 0.3090 - 0.9511i$

103. (a) Cube roots of $-125 = 125(\cos 180° + i \sin 180°)$ are:

$$5(\cos 60° + i \sin 60°)$$

$$5(\cos 180° + i \sin 180°)$$

$$5(\cos 300° + i \sin 300°)$$

(b)

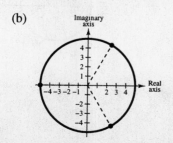

(c) $\dfrac{5}{2} + \dfrac{5\sqrt{3}}{2}i, \ -5, \ \dfrac{5}{2} - \dfrac{5\sqrt{3}}{2}i$

105. (a) Fifth roots of $128(-1 + i) = 128\sqrt{2}(\cos 135° + i \sin 135°)$ are:

$$2\sqrt[5]{4\sqrt{2}}\,(\cos 27° + i \sin 27°)$$

$$2\sqrt[5]{4\sqrt{2}}\,(\cos 99° + i \sin 99°)$$

$$2\sqrt[5]{4\sqrt{2}}\,(\cos 171° + i \sin 171°)$$

$$2\sqrt[5]{4\sqrt{2}}\,(\cos 243° + i \sin 243°)$$

$$2\sqrt[5]{4\sqrt{2}}\,(\cos 315° + i \sin 315°)$$

(b)

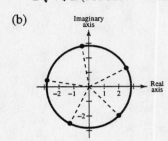

(c) $2.52 + 1.28i, -0.44 + 2.79i, -2.79 + 0.44i,$
$-1.28 - 252i, 2 - 2i$

107. $x^4 - i = 0$

$$x^4 = i$$

The solutions are the fourth roots of $i = \cos\dfrac{\pi}{2} + i \sin\dfrac{\pi}{2}$:

$$\sqrt[4]{1}\left[\cos\left(\frac{(\pi/2) + 2k\pi}{4}\right) + i \sin\left(\frac{(\pi/2) + 2k\pi}{4}\right)\right],\ k = 0, 1, 2, 3$$

$$\cos\frac{\pi}{8} + i \sin\frac{\pi}{8}$$

$$\cos\frac{5\pi}{8} + i \sin\frac{5\pi}{8}$$

$$\cos\frac{9\pi}{8} + i \sin\frac{9\pi}{8}$$

$$\cos\frac{13\pi}{8} + i \sin\frac{13\pi}{8}$$

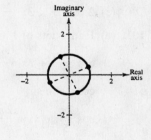

109. $x^5 + 243 = 0$

$$x^5 = -243$$

The solutions are the fifth roots of $-243 = 243(\cos \pi + i \sin \pi)$:

$$\sqrt[5]{243}\left[\cos\left(\frac{\pi + 2k\pi}{5}\right) + i \sin\left(\frac{\pi + 2k\pi}{5}\right)\right],\ k = 0, 1, 2, 3, 4$$

$$3\left(\cos\frac{\pi}{5} + i \sin\frac{\pi}{5}\right)$$

$$3\left(\cos\frac{3\pi}{5} + i \sin\frac{3\pi}{5}\right)$$

$$3(\cos \pi + i \sin \pi) = -3$$

$$3\left(\cos\frac{7\pi}{5} + i \sin\frac{7\pi}{5}\right)$$

$$3\left(\cos\frac{9\pi}{5} + i \sin\frac{9\pi}{5}\right)$$

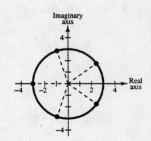

111. $x^3 + 64i = 0$

$$x^3 = -64i$$

The solutions are the fifth roots of $-64i$:

$$\sqrt[3]{64}\left[\cos\left(\frac{(3\pi/2) + 2k\pi}{3}\right) + i\sin\left(\frac{(3\pi/2) + 2k\pi}{3}\right)\right], \quad k = 0, 1, 2$$

$$4\left(\cos\frac{\pi}{2} + i\sin\frac{\pi}{2}\right) = 4i$$

$$4\left(\cos\frac{7\pi}{6} + i\sin\frac{7\pi}{6}\right) = -2\sqrt{3} - 2i$$

$$4\left(\cos\frac{11\pi}{6} + i\sin\frac{11\pi}{6}\right) = 2\sqrt{3} - 2i$$

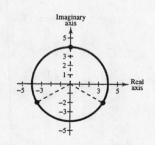

113. $x^3 - (1 - i) = 0$

$$x^3 = 1 - i = \sqrt{2}(\cos 315° + i\sin 315°)$$

The solutions are the cube roots of $1 - i$:

$$\sqrt[3]{\sqrt{2}}\left[\cos\left(\frac{315° + 360°k}{3}\right) + i\sin\left(\frac{315° + 360°k}{3}\right)\right], \quad k = 0, 1, 2$$

$$\sqrt[6]{2}(\cos 105° + i\sin 105°)$$

$$\sqrt[6]{2}(\cos 225° + i\sin 225°)$$

$$\sqrt[6]{2}(\cos 345° + i\sin 345°)$$

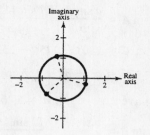

115. $\sin 28.1° = \dfrac{h}{18} \implies h = 18\sin 28.1° \approx 8.48$

❑ Review Exercises for Chapter 6

Solutions to Odd-Numbered Exercises

1. Given: $a = 5$, $b = 8$, $c = 10$

$$\cos C = \frac{a^2 + b^2 - c^2}{2ab} = \frac{25 + 64 - 100}{80} \approx -0.1375 \implies C \approx 97.9°$$

$$\sin A = \frac{a\sin C}{c} \approx \frac{5(0.9905)}{10} \approx 0.4953 \implies A \approx 29.7°$$

$$\sin B = \frac{b\sin C}{c} \approx \frac{8(0.9905)}{10} \approx 0.7924 \implies B \approx 52.4°$$

3. Given: $A = 12°$, $B = 58°$, $a = 5$

$$C = 180° - A - B = 180° - 12° - 58° = 110°$$

$$b = \frac{a\sin B}{\sin A} = \frac{5\sin 58°}{\sin 12°} \approx \frac{5(0.8480)}{0.2079} \approx 20.4$$

$$c = \frac{a\sin C}{\sin A} = \frac{5\sin 110°}{\sin 12°} \approx \frac{5(0.9397)}{0.2079} \approx 22.6$$

5. Given: $B = 110°$, $a = 4$, $c = 4$

$$b^2 = a^2 + c^2 - 2ac \cos B \approx 16 + 16 - 2(4)(4)(-0.3420) \approx 42.94 \implies b \approx 6.6$$

$$\sin A = \frac{a \sin B}{b} = \frac{4 \sin 110°}{6.6} \approx \frac{4(0.9397)}{6.6} \approx 0.5736 \implies A \approx 35°$$

$$c = a \implies C = A \approx 35°$$

7. Given: $A = 75°$, $a = 2.5$, $b = 16.5$

$$\sin B = \frac{b \sin A}{a} = \frac{16.5 \sin 75°}{2.5} \approx \frac{16.5(0.9659)}{2.5} \approx 5.375 \implies \text{no triangle formed}$$

No solution

9. Given: $B = 115°$, $a = 7$, $b = 14.5$

$$\sin A = \frac{a \sin B}{b} = \frac{7 \sin 115°}{14.5} \approx \frac{7(0.9063)}{14.5} \approx 0.4375 \implies A \approx 25.9°$$

$$C \approx 180° - 115° - 25.9° = 39.1°$$

$$c^2 = a^2 + b^2 - 2ab \cos C \approx 7^2 + 14.5^2 - 2(7)(14.5)(0.7760) \approx 101.7 \implies c \approx 10.1$$

11. Given: $A = 15°$, $a = 5$, $b = 10$

$$\sin B = \frac{b \sin A}{a} = \frac{10 \sin 15°}{5} \approx \frac{10(0.2588)}{5} \approx 0.5176 \implies B \approx 31.2° \text{ or } 148.8°$$

Case 1: $B \approx 31.2°$ | Case 2: $B \approx 148.8°$

$C \approx 180° - 15° - 31.2° = 133.8°$ | $C \approx 180° - 15° - 148.8° = 16.2°$

$c = \dfrac{a \sin C}{\sin A} \approx 13.9$ | $c = \dfrac{a \sin C}{\sin A} \approx 5.39$

13. Given: $B = 150°$, $a = 10$, $c = 20$

$$b^2 = a^2 + c^2 - 2ac \cos B \approx 100 + 400 - 400(-0.8660) \approx 846.4 \implies b \approx 29.1$$

$$\sin C = \frac{c \sin B}{b} \approx \frac{20(0.5)}{29.09} \approx 0.3437 \implies C \approx 20.1°$$

$$\sin A = \frac{a \sin B}{b} \approx \frac{10(0.5)}{29.09} \approx 0.1719 \implies A \approx 9.9°$$

15. Given: $B = 25°$, $a = 6.2$, $b = 4$

$$\sin A = \frac{a \sin B}{b} \approx 0.6551 \implies A \approx 40.9° \text{ or } 139.1°$$

Case 1: $A \approx 40.9°$ | Case 2: $A \approx 139.1°$

$C \approx 180° - 25° - 40.9° = 114.1°$ | $C \approx 180° - 25° - 139.1° = 15.9°$

$c \approx 8.6$ | $c \approx 2.6$

17. $a = 4$, $b = 5$, $c = 7$

$$s = \frac{a + b + c}{2} = \frac{4 + 5 + 7}{2} = 8$$

$$\text{Area} = \sqrt{s(s - a)(s - b)(s - c)}$$

$$= \sqrt{8(4)(3)(1)} \approx 9.798$$

19. $A = 27°$, $b = 5$, $c = 8$

$$\text{Area} = \frac{1}{2}bc \sin A = \frac{1}{2}(5)(8)(0.4540) = 9.08$$

21. $\alpha = 180° - 31° = 149°$

$\phi = 180° - 149° - 17° = 14°$

$x = \dfrac{50 \sin 17°}{\sin \phi} = \dfrac{50 \sin 17°}{\sin 14°} \approx 60.43$

$h = x \sin 31°$

$\approx 50.43(0.5150) \approx 31.1$ meters

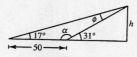

23. $\sin 28° = \dfrac{h}{75}$

$h = 75 \sin 28° \approx 35.21$ feet

$\cos 28° = \dfrac{x}{75}$

$x = 75 \cos 28° \approx 66.2211$ feet

$\tan 45° = \dfrac{H}{x}$

$H = x \tan 45° \approx 66.22$ feet

Height of tree: $H - h \approx 31$ feet

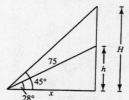

25. $d^2 = 850^2 + 1060^2 - 2(850)(1060) \cos 72°$

$\approx 1{,}289{,}251$

$d \approx 1135$ miles

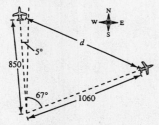

27. The height is given by

$h = a \cdot \sin 55° \approx 14.6$ m.

29. Initial point: $(-5, 4)$

Terminal point: $(2, -1)$

$\mathbf{v} = \langle 2 - (-5), -1, - 4 \rangle = \langle 7, -5 \rangle$

31. Initial point: $(0, 10)$

Terminal point: $(7, 3)$

$\mathbf{v} = \langle 7 - 0, 3 - 10 \rangle = \langle 7, -7 \rangle$

33. $\langle 8 \cos 120°, 8 \sin 120° \rangle = (-4, 4\sqrt{3})$

35. $\mathbf{v} = -10\mathbf{i} + 10\mathbf{j}$

$\|\mathbf{v}\| = \sqrt{(-10)^2 + (10)^2} = \sqrt{200} = 10\sqrt{2}$

$\tan \theta = \dfrac{10}{-10} = -1 \implies \theta = 135°$ since

$\mathbf{v}$ is in Quadrant II.

$\mathbf{v} = 10\sqrt{2}(\mathbf{i} \sin 135° + \mathbf{j} \cos 135°)$

37. $\mathbf{u} = 6\mathbf{i} - 5\mathbf{j}$

$\dfrac{1}{\|\mathbf{u}\|}\mathbf{u} = \dfrac{1}{\sqrt{6^2 + 5^2}}(6\mathbf{i} - 5\mathbf{j}) = \dfrac{6}{\sqrt{61}}\mathbf{i} - \dfrac{5}{\sqrt{61}}\mathbf{j}$

$= \left\langle \dfrac{6}{\sqrt{61}}, -\dfrac{5}{\sqrt{61}} \right\rangle$

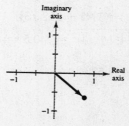

39. $\mathbf{u} = 6\mathbf{i} - 5\mathbf{j}$, $\mathbf{v} = 10\mathbf{i} + 3\mathbf{j}$

$4\mathbf{u} - 5\mathbf{v} = (24\mathbf{i} - 20\mathbf{j}) - (50\mathbf{i} + 15\mathbf{j}) = -26\mathbf{i} - 35\mathbf{j}$

$$= \langle -26, -35 \rangle$$

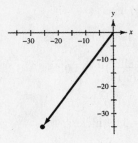

41. $\mathbf{u} = 15[(\cos 20°)\mathbf{i} + (\sin 20°)\mathbf{j}]$

$\mathbf{v} = 20[(\cos 63°)\mathbf{i} + (\sin 63°)\mathbf{j}]$

$\mathbf{u} + \mathbf{v} \approx 23.1752\mathbf{i} + 22.9504\mathbf{j}$

$\|\mathbf{u} + \mathbf{v}\| \approx 32.62$

$\tan \theta = \dfrac{22.9504}{23.1752} \implies \theta \approx 44.72°$

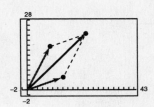

43. $\tan \alpha = \frac{12}{5} \implies \sin \alpha = \frac{12}{13}$ and $\cos \alpha = \frac{5}{13}$

$\tan \beta = \frac{3}{4} \implies \sin(180 - \beta) = \frac{3}{5}$ and $\cos(180° - \beta) = -\frac{4}{5}$

$\mathbf{u} = 250\left(\frac{5}{13}\mathbf{i} + \frac{12}{13}\mathbf{j}\right)$

$\mathbf{v} = 100\left(-\frac{4}{5}\mathbf{i} + \frac{3}{5}\mathbf{j}\right)$

$\mathbf{w} = 200(0\mathbf{i} - \mathbf{j})$

$\mathbf{r} = \mathbf{u} + \mathbf{v} + \mathbf{w} = \left(\frac{1250}{13} - 80 + 0\right)\mathbf{i} + \left(\frac{3000}{13} + 60 - 200\right)\mathbf{j} = \frac{210}{13}\mathbf{i} + \frac{1180}{13}\mathbf{j}$

$\|\mathbf{r}\| = \sqrt{\left(\frac{210}{13}\right)^2 + \left(\frac{1180}{13}\right)^2} \approx 92.2$

$\tan \theta = \frac{1180}{210} \implies \theta \approx 79.9°$

45. Rope One: $\mathbf{u} = \|\mathbf{u}\|(\cos 30°\mathbf{i} - \sin 30°\mathbf{j}) = \|\mathbf{u}\|\left(\frac{\sqrt{3}}{2}\mathbf{i} - \frac{1}{2}\mathbf{j}\right)$

Rope Two: $\mathbf{v} = \|\mathbf{u}\|(-\cos 30°\mathbf{i} - \sin 30°\mathbf{j}) - \|\mathbf{u}\|\left(-\frac{\sqrt{3}}{2}\mathbf{i} - \frac{1}{2}\mathbf{j}\right)$

Resultant: $\mathbf{u} + \mathbf{v} = -\|\mathbf{u}\|\mathbf{j} = -180\mathbf{j}$

$$\|\mathbf{u}\| = 180$$

Therefore, the tension on each rope is $\|\mathbf{u}\| = 180$ lb.

47. Force $= 500 \sin 12° \approx 104$ lbs

49. Airspeed: $\mathbf{u} = 724(\cos 60°\mathbf{i} + \sin 60°\mathbf{j})$

$$= 362(\mathbf{i} + \sqrt{3}\mathbf{j})$$

Wind: $\mathbf{w} = 32\mathbf{i}$

Groundspeed $= \mathbf{u} + \mathbf{w} = (394\mathbf{i} + 362\sqrt{3}\mathbf{j})$

$\|\mathbf{u} + \mathbf{w}\| = \sqrt{(394)^2 + \left(362\sqrt{3}\right)^2} \approx 740.5$ km/hr

$\tan \theta = \dfrac{362\sqrt{3}}{394} \implies \theta \approx 57.9°$

Bearing: N 32.1° E

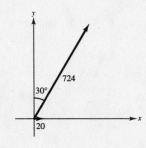

51. $P(7, -4), Q(-3, 2)$

$\overrightarrow{PQ} = \langle -3 - 7, 2 - (-4) \rangle = \langle -10, 6 \rangle$

$\|\overrightarrow{PQ}\| = \sqrt{(-10)^2 + (6)^2} = \sqrt{136} = 2\sqrt{34}$

$\dfrac{\overrightarrow{PQ}}{\|\overrightarrow{PQ}\|} = \dfrac{1}{2\sqrt{34}}\langle -10, 6 \rangle = \dfrac{1}{\sqrt{34}}\langle -5, 3 \rangle$

53. $\mathbf{u} = \langle 39, -12 \rangle, \mathbf{v} = \langle -26, 8 \rangle$

$\mathbf{u} \cdot \mathbf{v} = 39(-26) + (-12)(8) = -1110 \neq 0 \implies \mathbf{u}$ and $\mathbf{v}$ are not orthogonal.

$\mathbf{v} = -\tfrac{2}{3}\mathbf{u} \implies \mathbf{u}$ and $\mathbf{v}$ are parallel.

55. $\mathbf{u} = \cos\dfrac{7\pi}{4}\mathbf{i} + \sin\dfrac{7\pi}{4}\mathbf{j} = \left\langle \dfrac{1}{\sqrt{2}}, -\dfrac{1}{\sqrt{2}} \right\rangle$

$\mathbf{v} = \cos\dfrac{5\pi}{6}\mathbf{i} + \sin\dfrac{5\pi}{6}\mathbf{j} = \left\langle -\dfrac{\sqrt{3}}{2}, \dfrac{1}{2} \right\rangle$

$\cos\theta = \dfrac{\mathbf{u} \cdot \mathbf{v}}{\|\mathbf{u}\|\,\|\mathbf{v}\|} = \dfrac{\dfrac{-\sqrt{3}}{2\sqrt{2}} - \dfrac{1}{2\sqrt{2}}}{(1)(1)} \approx -0.966 \implies \theta \approx 165° \text{ or } \dfrac{11\pi}{12}$

57. $\mathbf{u} = \langle 2\sqrt{2}, -4 \rangle, \mathbf{v} = \langle -\sqrt{2}, 1 \rangle$

$\cos\theta = \dfrac{\mathbf{u} \cdot \mathbf{v}}{\|\mathbf{u}\|\,\|\mathbf{v}\|} = \dfrac{-8}{(\sqrt{24})(\sqrt{3})} \implies \theta \approx 160.5°$

59.

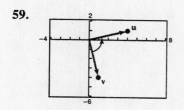

$\cos\theta = \dfrac{\mathbf{u} \cdot \mathbf{v}}{\|\mathbf{u}\|\,\|\mathbf{v}\|} = 0 \implies \theta = 90°$

61.

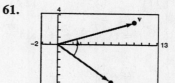

$\cos\theta = \dfrac{\mathbf{u} \cdot \mathbf{v}}{\|\mathbf{u}\|\,\|\mathbf{v}\|} = \dfrac{70 - 15}{\sqrt{74}\sqrt{109}} \approx 0.612 \implies \theta \approx 52.2°$

63. $\mathbf{u} = \langle -4, 3 \rangle, \mathbf{v} = \langle -8, -2 \rangle$

$\text{proj}_{\mathbf{v}}\mathbf{u} = \left(\dfrac{\mathbf{u} \cdot \mathbf{v}}{\|\mathbf{v}\|^2} \right)\mathbf{v} = \left(\dfrac{26}{68} \right)\langle -8, -2 \rangle$

$= -\dfrac{13}{17}\langle 4, 1 \rangle$

65. $\text{proj}_{\mathbf{v}}\mathbf{u} = \left(\dfrac{\mathbf{u} \cdot \mathbf{v}}{\|\mathbf{v}\|^2} \right)\mathbf{v} = \dfrac{-5}{2}\langle 1, -1 \rangle$

67. $z = -3 = 3(\cos 180° + i\sin 180°)$

69. $z = 5 - 2i, r = \sqrt{25 + 4} = \sqrt{29}, \theta = \arctan\left(-\tfrac{2}{5}\right) \approx 338.2°$

$z = \sqrt{29}(\cos 338.2° + i\sin 338.2°)$

71. $5 - 5i$

$$r = \sqrt{5^2 + (-5)^2} = \sqrt{50} = 5\sqrt{2}$$

$\tan \theta = \dfrac{-5}{5} = -1 \implies \theta \approx 315°$ since the complex number is in Quadrant IV.

$5 - 5i = 5\sqrt{2}(\cos 315° + i \sin 315°)$

73. $5 + 12i$

$$r = \sqrt{5^2 + 12^2} = \sqrt{169} = 13$$

$\tan \theta = \dfrac{12}{5} \implies \theta \approx 67.8°$ since the number is in Quadrant I.

$5 + 12i \approx 13(\cos 67.38° + i \sin 67.38°)$

75. $100(\cos 240° + i \sin 240°) = 100\left(-\dfrac{1}{2} - \dfrac{\sqrt{3}}{2}i\right)$

$$= -50 - 50\sqrt{3}i$$

77. $13(\cos 0 + i \sin 0) = 13(1 + 0i) = 13$

79. (a) $z_1 = 2\sqrt{3} - 2i = 4(\cos 330° + i \sin 330°)$

$z_2 = -10i = 10(\cos 270° + i \sin 270°)$

(b) $z_1 z_2 = [4(\cos 330° + i \sin 330°)][10(\cos 270° + i \sin 270°)]$

$= 40(\cos 600° + i \sin 600°)$

$= 40(\cos 240° + i \sin 240°)$

$\approx -20.00 - 34.64i$

$\dfrac{z_1}{z_2} = \dfrac{4(\cos 330° + i \sin 330°)}{10(\cos 270° + i \sin 270°)}$

$= \dfrac{2}{5}(\cos 60° + i \sin 60°)$

81. $\left[5\left(\cos \dfrac{\pi}{12} + i \sin \dfrac{\pi}{12}\right)\right]^4 = 5^4\left(\dfrac{4\pi}{12} + i \sin \dfrac{4\pi}{12}\right)$

$= 625\left(\cos \dfrac{\pi}{3} + i \sin \dfrac{\pi}{3}\right)$

$= 625\left(\dfrac{1}{2} + \dfrac{\sqrt{3}}{2}i\right)$

$= \dfrac{625}{2} + \dfrac{625\sqrt{3}}{2}i$

83. $(2 + 3i)^6 = [\sqrt{13}(\cos 56.3° + i \sin 56.3°)]^6$

$= 13^3(\cos 337.9° + i \sin 337.9°)$

$\approx 13^3(0.9263 - 0.3769i)$

$\approx 2035 - 828i$

85. (a) The trigonometric form of the three roots shown is:

$4(\cos 60° + i \sin 60°)$

$4(\cos 180° + i \sin 180°)$

$4(\cos 300° + i \sin 300°)$

(b) Since there are three evenly spaced roots on the circle of radius 4, they are cube roots of a complex number of modulus $4^3 = 64$. Cubing them yields -64.

(c) $[4(\cos 60° + i \sin 60°)]^3 = -64$

$[4(\cos 180° + i \sin 180°)]^3 = -64$

$[4(\cos 300° + i \sin 300°)]^3 = -64$

87. (a) 2

$2(\cos 72° + i \sin 72°)$

$2(\cos 144° + i \sin 144°)$

$2(\cos 216° + i \sin 216°)$

$2(\cos 288° + i \sin 288°)$

(b) Since there are five equally spaced roots on the circle of radius 2, they are fifth roots of a complex number of modulus $32 = 2^5$.

89. Sixth roots of $-729i = 729\left(\cos \dfrac{3\pi}{2} + i \sin \dfrac{3\pi}{2}\right)$:

$$\sqrt[6]{729}\left(\cos \frac{(3\pi/2) + 2k\pi}{6} + i \sin \frac{(3\pi/2) + 2k\pi}{6}\right), k = 1, 2, 3, 4, 5$$

$3\left(\cos \dfrac{\pi}{4} + i \sin \dfrac{\pi}{4}\right)$

$3\left(\cos \dfrac{7\pi}{12} + i \sin \dfrac{7\pi}{12}\right)$

$3\left(\cos \dfrac{11\pi}{12} + i \sin \dfrac{11\pi}{12}\right)$

$3\left(\cos \dfrac{5\pi}{4} + i \sin \dfrac{5\pi}{4}\right)$

$3\left(\cos \dfrac{19\pi}{12} + i \sin \dfrac{19\pi}{12}\right)$

$3\left(\cos \dfrac{23\pi}{12} + i \sin \dfrac{23\pi}{12}\right)$

91. $x^4 + 81 = 0$

$x^4 = -81$

$-81 = 81(\cos \pi + i \sin \pi)$

$$\sqrt[4]{-81} = \sqrt[4]{81}\left[\cos\left(\frac{\pi + 2\pi k}{4}\right) + i \sin\left(\frac{\pi + 2\pi k}{4}\right)\right], \ k = 0, 1, 2, 3$$

$3\left(\cos \dfrac{\pi}{4} + i \sin \dfrac{\pi}{4}\right) = \dfrac{3\sqrt{2}}{2} + \dfrac{3\sqrt{2}}{2}i$

$3\left(\cos \dfrac{3\pi}{4} + i \sin \dfrac{3\pi}{4}\right) = -\dfrac{3\sqrt{2}}{2} + \dfrac{3\sqrt{2}}{2}i$

$3\left(\cos \dfrac{5\pi}{4} + i \sin \dfrac{5\pi}{4}\right) = -\dfrac{3\sqrt{2}}{2} - \dfrac{3\sqrt{2}}{2}i$

$3\left(\cos \dfrac{7\pi}{4} + i \sin \dfrac{7\pi}{4}\right) = \dfrac{3\sqrt{2}}{2} - \dfrac{3\sqrt{2}}{2}i$

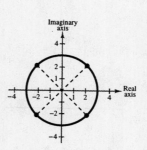

93. $x^3 + 8i = 0$

$$x^3 = -8i$$

$$-8i = 8\left(\cos\frac{3\pi}{2} + i\sin\frac{3\pi}{2}\right)$$

$$\sqrt[3]{-8i} = \sqrt[3]{8}\left[\cos\frac{(3\pi/2) + 2\pi k}{3} + i\sin\frac{(3\pi/2) + 2\pi k}{3}\right], \; k = 0, 1, 2$$

$$2\left(\cos\frac{\pi}{2} + i\sin\frac{\pi}{2}\right) = 2i$$

$$2\left(\cos\frac{7\pi}{6} + i\sin\frac{7\pi}{6}\right) = -\sqrt{3} - i$$

$$2\left(\cos\frac{11\pi}{6} + i\sin\frac{11\pi}{6}\right) = \sqrt{3} - i$$

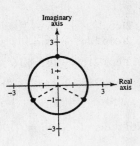

❏ Cumulative Test for Chapters 4-6

1. (a)

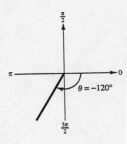

(b) $-120° + 360° = 240°$

(c) $-120° \cdot \dfrac{\pi}{180}\dfrac{\text{rad}}{\text{deg}} = -\dfrac{2\pi}{3}$

(d) $\theta' = 60°$

(e) $\sin\theta = -\dfrac{\sqrt{3}}{2}, \quad \cos\theta = -\dfrac{1}{2}, \quad \tan\theta = \sqrt{3}$

$\csc\theta = -\dfrac{2}{\sqrt{3}} = -\dfrac{2\sqrt{3}}{3}, \quad \sec\theta = -2,$

$\cot\theta = \dfrac{1}{\sqrt{3}} = \dfrac{\sqrt{3}}{3}$

2. 2.35 radians $\left(\dfrac{180}{\pi}\right) \approx 134.6°$

3. $\tan\theta = -\dfrac{4}{3} \implies \sec^2\theta = \tan^2\theta + 1 = \dfrac{16}{9} + 1 = \dfrac{25}{9}$

$\implies \sec\theta = \dfrac{5}{3}$ (Quadrant IV) $\implies \cos\theta = \dfrac{3}{5}$

4 (a)

(b)

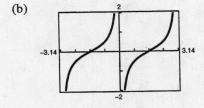

5. Amplitude: -3

Cosine curve reflected about the x-axis.

Period: $2 \implies h(x) = -3\cos(\pi x)$

Answer: $a = -3, b = \pi, c = 0$

6. Let $u = \arccos 2x \implies \cos u = 2x$. Then:

$$\sin(\arccos 2x) = \sin u$$
$$= \sqrt{1 - 4x^2}$$

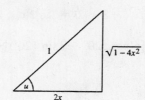

7. $\dfrac{\sin \theta - 1}{\cos \theta} - \dfrac{\cos \theta}{\sin \theta - 1} = \dfrac{\sin^2 \theta - 2 \sin \theta + 1 - \cos^2 \theta}{\cos \theta (\sin \theta - 1)}$

$$= \dfrac{\sin^2 \theta - 2 \sin \theta + \sin^2 \theta}{\cos \theta (\sin \theta - 1)}$$

$$= \dfrac{2 \sin \theta (\sin \theta - 1)}{\cos \theta (\sin \theta - 1)} = 2 \tan \theta$$

8. (a) $\cot^2 \alpha (\sec^2 \alpha - 1) = \cot^2 \alpha (\tan^2 \alpha) = 1$

(b) $\sin(x + y) \sin(x - y) = [\sin x \cos y + \cos x \sin y][\sin x \cos y - \sin y \cos x]$

$$= \sin^2 x \cos^2 y - \sin^2 y \cos^2 x$$
$$= \sin^2 x (1 - \sin^2 y) - \sin^2 y (1 - \sin^2 x)$$
$$= \sin^2 x - \sin^2 x \sin^2 y - \sin^2 y + \sin^2 y \sin^2 x$$
$$= \sin^2 x - \sin^2 y$$

9. (a) $\quad 2 \cos^2 \beta - \cos \beta = 0$

$$\cos \beta (2 \cos \beta - 1) = 0$$

$$\cos \beta = 0 \implies \beta = \frac{\pi}{2}, \frac{3\pi}{2}$$

$$2 \cos \beta - 1 = 0 \implies \cos \beta = \frac{1}{2} \implies \beta = \frac{\pi}{3}, \frac{5\pi}{3}$$

(b) $3 \tan^2 \theta - \cot \theta = 0$

$$3 \tan \theta - \frac{1}{\tan \theta} = 0$$

$$3 \tan^2 \theta - 1 = 0$$

$$\tan \theta = \pm \frac{1}{\sqrt{3}} \implies \theta = \frac{\pi}{6}, \frac{5\pi}{6}, \frac{7\pi}{6}, \frac{11\pi}{6}$$

10. (a) By the Law of Sines,

$$\frac{a}{\sin A} = \frac{b}{\sin B}$$

$$\frac{9}{\sin 30°} = \frac{8}{\sin B}$$

$$\sin B = \frac{8}{9} \sin 30° = \frac{8}{9}\left(\frac{1}{2}\right) = \frac{4}{9}.$$

Hence, $B \approx 26.4°$, and $C = 180° - 30° - 26.4° = 123.6°$. Finally,

$$\frac{c}{\sin C} = \frac{a}{\sin A} \implies c = 9 \frac{\sin(123.6°)}{\sin 30°} \approx 15.$$

(b) $a^2 = b^2 + c^2 - 2bc \cdot \cos A = 8^2 + 10^2 - 2(8)(10) \cos 30° = 25.44 \implies a \approx 5.0$. Then,

$$\frac{b}{\sin B} = \frac{a}{\sin A} \implies \sin B = \frac{b \sin A}{a} \approx \frac{8 \cdot \sin 30°}{5.0} = 0.793 \implies B \approx 52.5°.$$

(c) Finally, $C = 180° - A - B \approx 97.5°$.

11. $r = \sqrt{(-2)^2 + 2^2} = 2\sqrt{2} : -2 + 2i = 2\sqrt{2}(\cos 135° + i \sin 135°)$

12. $[4(\cos 30° + i \sin 30°)][6(\cos 120° + i \sin 120°)] = 24(\cos(30 + 120°) + i \sin(30° + 120°))$

$$= 24(\cos 150° + i \sin 150°)$$

$$= 24\left(-\frac{\sqrt{3}}{2} + i\frac{1}{2}\right) = -12\sqrt{3} + 12i$$

13. $1 = 1(\cos 0 + i \sin 0)$

$$\cos\left(\frac{0 + 2\pi k}{3}\right) + i \sin\left(\frac{0 + 2\pi k}{3}\right)$$

$k = 0: \cos 0 + i \sin 0 = 1$

$k = 1: \cos \dfrac{2\pi}{3} + i \sin \dfrac{2\pi}{3} = -\dfrac{1}{2} + \dfrac{\sqrt{3}}{2}i$

$k = 2: \cos \dfrac{4\pi}{3} + i \sin \dfrac{4\pi}{3} = -\dfrac{1}{2} - \dfrac{\sqrt{3}}{2}i$

14. $\tan 18° = \dfrac{h}{200}$

$\tan 16° 45' = \dfrac{k}{200}$

Hence, $f = h - k = 200 \tan 18° - 200 \tan 16° 45'$

$\approx 4.8 \approx 5$ feet.

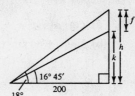

15. Add the two vectors:

$500(\cos 60\mathbf{i} + \sin 60\mathbf{j}) + 50(\cos 30\mathbf{i} + \sin 30\mathbf{j}) = \left(250 + 25\sqrt{3}\right)\mathbf{i} + \left(250\sqrt{3} + 25\right)\mathbf{j}$

$\tan \theta = \dfrac{250\sqrt{3} + 25}{250 + 25\sqrt{3}} \approx 1.56 \Longrightarrow \theta \approx 57.4$

Direction: N 32.6° E

Speed $= \sqrt{\left(250 + 25\sqrt{3}\right)^2 + \left(250\sqrt{3} + 25\right)^2} \approx 543.9$ km/hr

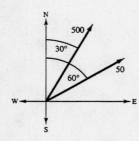

❑ Practice Test for Chapter 6

For Exercises 1 and 2, use the Law of Sines to find the remaining sides and angles of the triangle.

1. $A = 40°$, $B = 12°$, $b = 100$

2. $C = 150°$, $a = 5$, $c = 20$

3. Find the area of the triangle: $a = 3$, $b = 6$, $C = 130°$.

4. Determine the number of solutions to the triangle: $1 = 10$, $b = 35$, $A = 22.5°$.

For Exercises 5 and 6, use the Law of Cosines to find the remaining sides and angles off the triangle.

5. $a = 49$, $b = 53$, $c = 38$

6. $C = 29°$, $a = 100$, $b = 300$

7. Use Heron's Formula to find the area of the triangle: $a = 4.1$, $b = 6.8$, $c = 5.5$.

8. A ship travels 40 miles due east, then adjusts its course 12° southward. After traveling 70 miles in that direction, how far is the ship from its point of departure?

9. **w** is $4\mathbf{u} - 7\mathbf{v}$ where $\mathbf{u} = 3\mathbf{i} + \mathbf{j}$ and $\mathbf{v} = -\mathbf{i} + 2\mathbf{j}$. Find **w**.

10. Find a unit vector in the direction of $\mathbf{v} = 5\mathbf{i} - 3\mathbf{j}$.

11. Find the dot product and the angle between $\mathbf{u} = 6\mathbf{i} + 5\mathbf{j}$ and $\mathbf{v} = 2\mathbf{i} - 3\mathbf{j}$.

12. **v** is a vector of magnitude 4 making an angle of 30° with the positive *x*-axis. Find **v** in component form.

13. Find the projection of **u** onto **v** given $\mathbf{u} = \langle 3, -1 \rangle$ and $\mathbf{v} = \langle -2, 4 \rangle$.

14. Give the trigonometric form of $z = 5 - 5i$.

15. Give the standard form of $z = 6(\cos 225° + i \sin 225°)$.

16. Multiply $[7 \cos 23° + i \sin 23°)][4(\cos 7° + i \sin 7°)]$.

17. Divide $\dfrac{9\left(\cos \dfrac{5\pi}{4} + i \sin \dfrac{5\pi}{4}\right)}{3(\cos \pi + i \sin \pi)}$.

18. Find $(2 + 2i)^8$.

19. Find the cube roots of $8\left(\cos \dfrac{\pi}{3} + i \sin \dfrac{\pi}{3}\right)$.

20. Find all the solutions to $x^4 + i = 0$.

CHAPTER 7
Systems of Equations and Inequalities

C H A P T E R   7
Systems of Equations and Inequalities

Section 7.1   Solving Systems of Equations

■ You should be able to solve systems of equations by the method of substitution.

    1.   Solve one of the equations for one of the variables.

    2.   Substitute this expression into the other equation and solve.

    3.   Back-substitute into the first equation to find the value of the other variable.

    4.   Check your answer in each of the original equations.

■ You should be able to find solutions graphically. (See Example 5 in textbook.)

Solutions to Odd-Numbered Exercises

1. $2x + y = 6$ Equation 1

   $-x + y = 0$ Equation 2

Solve for y in Equation 1: $y = 6 - 2x$

Substitute for y in Equation 2: $-x + (6 - 2x) = 0$

Solve for x: $-3x + 6 = 0 \implies x = 2$

Back-substitute $x = 2$: $y = 6 - 2(2) = 2$

Answer: $(2, 2)$

3. $x - y = -4$ Equation 1

   $x^2 - y = -2$ Equation 2

Solve for y in Equation 1: $y = x + 4$

Substitute for y in Equation 2: $x^2 - (x + 4) = -2$

Solve for x: $x^2 - x - 2 = 0 \implies (x + 1)(x - 2) = 0 \implies x = -1, 2$

Back-substitute $x = -1$: $y = -1 + 4 = 3$

Back-substitute $x = 2$: $y = 2 + 4 = 6$

Answers: $(-1, 3), (2, 6)$

5. $x - 3y = 15$ Equation 1

   $x^2 + y^2 = 25$ Equation 2

Solve for x in Equation 1: $x = 3y + 15$

Substitute for x in Equation 2: $(3y + 15)^2 + y^2 = 25$

Solve for y: $10y^2 + 90y + 200 = 0 \implies y^2 + 9y + 20 = 0 \implies (y + 5)(y + 4) = 0 \implies y = -5, -4$

Back-substitute $y = -5$: $x = 3(-5) + 15 = 0$

Back-substitute $y = -4$: $x = 3(-4) + 15 = 3$

Answers: $(0, -5), (3, -4)$

7. $x^2 + y = 0$ Equation 1

$x^2 - 4x - y = 0$ Equation 2

Solve for y in Equation 1: $y = -x^2$

Substitute for y in Equation 2: $x^2 - 4x - (-x^2) = 0$

Solve for x: $2x^2 - 4x = 0 \implies 2x(x - 2) = 0 \implies x = 0, 2$

Back-substitute $x = 0$: $y = -0^2 = 0$

Back-substitute $x = 2$: $y = -2^2 = -4$

Answers: $(0, 0)$, $(2, -4)$

9. $x - 6y = -8$ Equation 1

$x^2 - 4y^3 = 0$ Equation 2

Solve for x in Equation 1: $x = 6y - 8$

Substitute for x in Equation 2: $(6y - 8)^2 - 4y^3 = 0$

Solve for y: $-4y^3 + 36y^2 - 96y + 64 = 0$

$$y^3 - 9y^2 + 24y - 16 = 0$$

$$(y - 1)(y - 4)^2 = 0 \implies y = 1, 4$$

Back-substitute $y = 1$: $x = 6(1) - 8 = -2$

Back-substitute $y = 4$: $x = 6(4) - 8 = 16$

Answers: $(-2, 1)$, $(16, 4)$

11. $x - y = 0$ Equation 1

$5x - 3y = 10$ Equation 2

Substitute for y in Equation 2: $5x - 3x = 10$

Solve for x: $2x = 10 \implies x = 5$

Back-substitute in Equation 1: $y = x = 5$

Answer: $(5, 5)$

13. $2x - y + 2 = 0$ Equation 1

$4x + y - 5 = 0$ Equation 2

Solve for y in Equation 1: $y = 2x + 2$

Substitute for y in Equation 2: $4x + (2x + 2) - 5 = 0$

Solve for x: $4x + (2x + 2) - 5 = 0 \implies 6x - 3 = 0 \implies x = \frac{1}{2}$

Back-substitute $x = \frac{1}{2}$: $y = 2x + 2 = 2\left(\frac{1}{2}\right) + 2 = 3$

Answer: $\left(\frac{1}{2}, 3\right)$

15. $1.5x + 0.8y = 2.3 \implies 15x + 8y = 23$

$\quad\; 0.3x - 0.2y = 0.1 \implies 3x - 2y = 1$

Solve for y in Equation 2: $-2y = 1 - 3x$

$$y = \frac{3x - 1}{2}$$

Substitute for y in Equation 1: $15x + 8\left(\frac{3x - 1}{2}\right) = 23$

$$15x + 12x - 4 = 23$$
$$27x = 27$$
$$x = 1$$

Then, $y = \dfrac{3x - 1}{2} = \dfrac{3(1) - 1}{2} = 1.$

17. $\frac{1}{5}x + \frac{1}{2}y = 8$ Equation 1

$\quad x + y = 20$ Equation 2

Solve for x in Equation 2: $x = 20 - y$

Substitute for x in Equation 1: $\frac{1}{5}(20 - y) + \frac{1}{2}y = 8$

Solve for y: $4 + \frac{3}{10}y = 8 \implies y = \frac{40}{3}$

Back-substitute $y = \frac{40}{3}$: $x = 20 - y = 20 - \frac{40}{3} = \frac{20}{3}$

Answer: $\left(\frac{20}{3}, \frac{40}{3}\right)$

19. $\quad 2x - y = 4$ Equation 1

$\quad -4x + 2y = -12$ Equation 2

Solve for y in Equation 1: $y = 2x - 4$

Substitute for y in Equation 2: $-4x + 2(2x - 4) = -12$

Solve for x: $-8 \neq -12$ Inconsistent

No Solution

21. $x - y = 0$ Equation 1

$\quad 2x + y = 0$ Equation 2

Solve for y in Equation 1: $y = x$

Substitute for y in Equation 2: $2x + x = 0$

Solve for x: $3x = 0 \implies x = 0$

Back-substitute $x = 0$: $y = x = 0$

Answer: $(0, 0)$

23. $-x + 2y = 2$

$\quad 3x + y = 15$

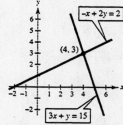

Point of intersection: $(4, 3)$

25. $x - 3y = -2$

 $5x + 3y = 17$

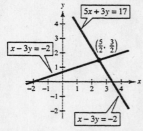

Point of intersection: $\left(\frac{5}{2}, \frac{3}{2}\right)$

27. $x + y = 4$

 $x^2 + y^2 - 4x = 0$

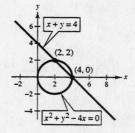

Points of intersection: $(2, 2)$, $(4, 0)$

29. $7x + 8y = 24 \implies y_1 = -\frac{7}{8}x + 3$

 $x - 8y = 8 \implies y_2 = \frac{1}{8}x - 1$

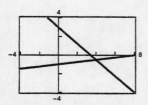

Point of intersection: $\left(4, -\frac{1}{2}\right)$

31. $2x - y + 3 = 0 \implies y_1 = 2x + 3$

 $x^2 + y^2 - 4x = 0 \implies y_2 = \sqrt{4x - x^2},\ y_3 = -\sqrt{4x - x^2}$

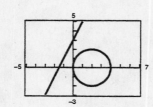   No points of intersection

33. $x^2 + y^2 = 8 \implies y_1 = \sqrt{8 - x^2}$ and $y_2 = -\sqrt{8 - x^2}$

 $y = x^2 \implies y_3 = x^2$

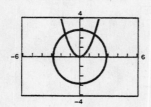

Points of intersection: $\left(\pm\sqrt{\dfrac{-1 + \sqrt{33}}{2}},\ \dfrac{-1 + \sqrt{33}}{2}\right) \approx (\pm 1.54,\ 2.37)$

35. $y = e^x$

$x - y + 1 = 0 \implies y = x + 1$

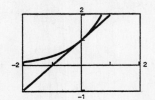

Point of intersection: $(0, 1)$

37. $y = \sqrt{x}$

$y = x$

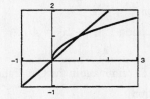

Points of intersection: $(0, 0)$, $(1, 1)$

39. $x^2 + y^2 = 169 \implies y_1 = \sqrt{169 - x^2}$ and $y_2 = -\sqrt{169 - x^2}$

$x^2 - 8y = 104 \implies y_3 = \frac{1}{8}x^2 - 13$

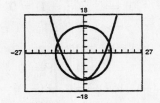

Points of intersection: $(0, -13)$, $(\pm 12, 5)$

41. $y = 2x$ Equation 1

$y = x^2 + 1$ Equation 2

Substitute for y in Equation 2: $2x = x^2 + 1$

Solve for x: $x^2 - 2x + 1 = (x - 1)^2 = 0 \implies x = 1$

Back-substitute $x = 1$ in Equation 1: $y = 2x = 2$

Answer: $(1, 2)$

43. $3x - 7y + 6 = 0$ Equation 1

     $x^2 - y^2 = 4$ Equation 2

Solve for y in Equation 1: $y = \dfrac{3x + 6}{7}$

Solve for y in Equation 2: $x^2 - \left(\dfrac{3x + 6}{7}\right)^2 = 4$

Solve for x: $x^2 - \left(\dfrac{9x^2 + 36x + 36}{49}\right) = 4$

$$49x^2 - (9x^2 + 36x + 36) = 196$$

$$40x^2 - 36x - 232 = 0$$

$10x^2 - 9x - 58 = 0 \implies x = \dfrac{9 \pm \sqrt{81 + 40(58)}}{20} \implies x = \dfrac{29}{10}, -2$

Back-substitute $x = \dfrac{29}{10}$: $y = \dfrac{3x + 6}{7} = \dfrac{3(29/10) + 6}{7} = \dfrac{21}{10}$

Back-substitute $x = -2$: $y = \dfrac{3x + 6}{7} = 0$

Answers: $\left(\dfrac{29}{10}, \dfrac{21}{10}\right)$, $(-2, 0)$

45. $x - 2y = 4$ Equation 1

$x^2 - y = 0$ Equation 2

Solve for y in Equation 2: $y = x^2$

Substitute for y in Equation 1: $x - 2x^2 = 4$

Solve for x: $0 = 2x^2 - x + 4$

No real solutions, the discriminant in the Quadratic Formula is negative.

Inconsistent, No solution

47. $y - e^{-x} = 1 \implies y = e^{-x} + 1$

$y - \ln x = 3 \implies y = \ln x + 3$

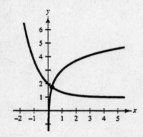

Point of intersection: Approximately $(0.287, 1.75)$

49. $y = x^4 - 2x^2 + 1$ Equation 1

$y = 1 - x^2$ Equation 2

Substitute for y in Equation 1: $1 - x^2 = x^4 - 2x^2 + 1$

Solve for x: $x^4 - x^2 = 0 \implies x^2(x^2 - 1) = 0$

$\implies x = 0, \pm 1$

Back-substitute $x = 0$: $1 - x^2 = 1$

Back-substitute $x = 1$: $1 - x^2 = 1 - 1^2 = 0$

Back-substitute $x = -1$: $1 - x^2 = 1 - (-1)^2 = 0$

Answers: $(0, 1), (\pm 1, 0)$

51. $xy - 1 = 0$ Equation 1

$2x - 4y + 7 = 0$ Equation 2

Solve for y in Equation 1: $y = \dfrac{1}{x}$

Substitute for y in Equation 2: $2x - 4\left(\dfrac{1}{x}\right) + 7 = 0$

Solve for x: $2x^2 - 4 + 7x = 0 \implies (2x - 1)(x + 4) = 0 \implies x = \dfrac{1}{2}, -4$

Back-substitute $x = \dfrac{1}{2}$: $y = \dfrac{1}{1/2} = 2$

Back-substitute $x = -4$: $y = \dfrac{1}{-4} = -\dfrac{1}{4}$

Answers: $\left(\dfrac{1}{2}, 2\right), \left(-4, -\dfrac{1}{4}\right)$

53. The system has no solution if you arrive at a false statement, ie. $4 = 8$, or you have a quadratic equation with a negative discriminant, which would yield imaginary roots.

55. (a) $C = 8650x + 250{,}000, \quad R = 9950x$

$R = C$

$9950x = 8650x + 250{,}000$

$1300x = 250{,}000$

$x \approx 192 \text{ units}$

(b)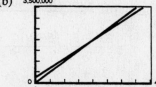

57. (a) $C = 2.65x + 350,000, \quad R = 4.15x$

$R = C$

$4.15x = 2.65x + 350,000$

$1.50x = 350,000$

$x \approx 233,333$ units

(b)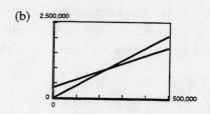

59. (a) $C = 3.45x + 16,000, \quad R = 5.95x$

(c) $R = C$

$5.95x = 3.45x + 16,000$

$2.50x = 16,000$

$x \approx 6400$ units

(b)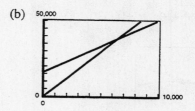

61. $x + \quad\; y = 20,000$

$0.065x + 0.085y = 1600$

Graphing $y_1 = 20,000 - x$ and

$y_2 = (1600 - 0.065x)\dfrac{1}{0.085}$

you will see that the point of intersection

occurs when $x = \$5000$.

63. $0.06x = 0.03x + 250$

$0.03x = 250$

     $x \approx \$8333.33$

To make the straight commission offer better, you
would have to sell more than $\$8333.33$ per week.

65. Supply: $p = 1.45 + 0.00014x^2$

Demand: $p = (2.388 - 0.007x)^2$

The two graphs intersect at $x = 100$ bushels.

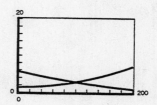

67. $2l + 2w = 30 \implies l + w = 15$

$\quad\quad l = w + 3 \implies (w + 3) + 2 = 15$

$\quad\quad\quad\quad\quad\quad\quad\quad\quad\quad 2w = 12$

$\quad\quad\quad\quad\quad\quad\quad\quad\quad\quad\; w = 6$

$l = w + 3 = 9$

Dimensions: 6 meters × 9 meters

69. $2l + 2w = 42 \implies l + w = 21$

$\quad\quad\quad w = \tfrac{3}{4}l \implies l + \tfrac{3}{4}l = 21$

$\quad\quad\quad\quad\quad\quad\quad\quad\quad \tfrac{7}{4}l = 21$

$\quad\quad\quad\quad\quad\quad\quad\quad\quad\quad l = 12$

$w = \tfrac{3}{4}, l = 9$

Dimensions: 9 inches × 12 inches

71. $2l + 2w = 40 \implies l + w = 20 \implies w = 20 - l$

$\quad\; lw = 96 \implies l(20 - l) = 96$

$\quad\quad\quad\quad\quad\quad\quad\; 20l - l^2 = 96$

$\quad\quad\quad\quad\quad\quad\quad 0 = l^2 - 20l + 96$

$\quad\quad\quad\quad\quad\quad\quad 0 = (l - 8)(l - 12)$

$\quad\quad\quad\quad\quad\quad\quad l = 8 \text{ or } l = 12$

$w = 12, w = 8$

Since the length is supposed to be greater than the width, we have $l = 12$ miles and $w = 8$ miles.

73. (a) The line $y = 2x$ intersects the parabola $y = x^2$ at two points, $(0, 0)$ and $(2, 4)$.

    (b) The line $y = 0$ intersects $y = x^2$ at $(0, 0)$ only.

    (c) The line $y = x - 2$ does not intersect $y = x^2$.

75. $(0, 6.6)$, $(1, 6.8)$, $(2, 7.1)$, $(3, 7.1)$

    (a) Linear model: $f(t) = 0.18t + 6.63$

        Quadratic model:
        $g(t) = -0.05t^2 + 0.33t + 6.58$

    (b)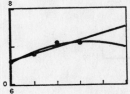

    (c) Points of intersection: $(0.382, 6.70)$, $(2.618, 7.10)$

    (d) Linear model when $t = 4$:
        $f(4) = 7.35$ million short tons

        Quadratic model when $t = 4$:
        $g(4) = 7.1$ million short tons

        Since the sale of newsprint has been slowly decreasing (more papers are on-line now and many people would rather listen to the news than read it) the Quadratic model is probably more accurate.

77. $(-2, 7)$, $(5, 5)$

$$m = \frac{5 - 7}{5 - (-2)} = -\frac{2}{7}$$

$$y - 7 = -\frac{2}{7}(x - (-2))$$

$$7y - 49 = -2x - 4$$

$$2x + 7y - 45 = 0$$

79. $(6, 3)$, $(10, 3)$

$$m = \frac{3 - 3}{10 - 6} = 0 \implies \text{The line is horizontal.}$$

$$y = 3 \implies y - 3 = 0$$

81. $\left(\frac{3}{5}, 0\right)$, $(4, 6)$

$$m = \frac{6 - 0}{4 - \frac{3}{5}} = \frac{6}{\frac{17}{5}} = \frac{30}{17}$$

$$y - 6 = \frac{30}{17}(x - 4)$$

$$17y - 102 = 30x - 120$$

$$0 = 30x - 17y - 18$$

Section 7.2   Systems of Linear Equations in Two Variables

■ You should be able to solve a linear system by the method of elimination.

1. Obtain coefficients for either x or y that differ only in sign. This is done by multiplying all the terms of one or both equations by appropriate constants.

2. Add the equations to eliminate one of the variables and then solve for the remaining variable.

3. Use back-substitution into either original equation and solve for the other variable.

4. Check your answer.

■ You should know that for a system of two linear equations, one of the following is true.

(a) There are infinitely many solutions; the lines are identical. The system is consistent.

(b) There is no solution; the lines are parallel. The system is inconsistent.

(c) There is one solution; the lines intersect at one point. The system is consistent.

Solutions to Odd-Numbered Exercises

1. $2x + y = 5$ Equation 1

    $x - y = 1$ Equation 2

    Add to eliminate y: $3x = 6 \implies x = 2$

    Substitute $x = 2$ in Equation 2: $2 - y = 1 \implies y = 1$

    Answer: $(2, 1)$

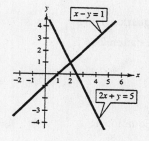

3. $x + y = 0$ Equation 1

    $3x + 2y = 1$ Equation 2

    Multiply Equation 1 by -2: $-2x - 2y = 0$

    Add this to Equation 2 to eliminate y: $x = 1$

    Substitute $x = 1$ in Equation 1: $1 + y = 0 \implies y = -1$

    Answer: $(1, -1)$

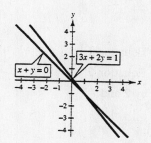

5. $x - y = 2$ Equation 1

    $-2x + 2y = 5$ Equation 2

    Multiply Equation 1 by 2: $2x - 2y = 4$

    Add this to Equation 2: $0 = 9$

    There are no solutions.

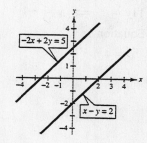

7. $3x - 2y = 5$ Equation 1

$-6x + 4y = -10$ Equation 2

Multiply Equation 1 by 2 and add to Equation 2: $0 = 0$

The equations are dependent. There are infinitely many solutions.

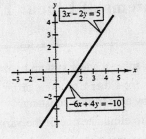

9. $9x + 3y = 1$ Equation 1

$3x - 6y = 5$ Equation 2

Multiply Equation 2 by (-3): $9x + 3y = 1$

$-9x + 18y = -15$

Add to eliminate x: $21y = -14 \implies y = -\frac{2}{3}$

Substitute $y = -\frac{2}{3}$ in Equation 1: $9x + 3\left(-\frac{2}{3}\right) = 1$

$x = \frac{1}{3}$

Answer: $\left(\frac{1}{3}, -\frac{2}{3}\right)$

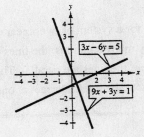

11. $x + 2y = 4$ Equation 1

$x - 2y = 1$ Equation 2

Add to eliminate y:

$2x = 5$

$x = \frac{5}{2}$

Substitute $x = \frac{5}{2}$ in Equation 1:

$\frac{5}{2} + 2y = 4 \implies y = \frac{3}{4}$

Answer: $\left(\frac{5}{2}, \frac{3}{4}\right)$

13. $2x + 3y = 18$ Equation 1

$5x - y = 11$ Equation 2

Multiply Equation 2 by 3: $15x - 3y = 33$

Add this to Equation 1 to eliminate y:

$17x = 51 \implies x = 3$

Substitute $x = 3$ in Equation 1:

$6 + 3y = 18 \implies y = 4$

Answer: $(3, 4)$

15. $3x + 2y = 10$ Equation 1

$2x + 5y = 3$ Equation 2

Multiply Equation 1 by 2 and Equation 2 by (-3):

$6x + 4y = 20$

$-6x - 15y = -9$

Add to eliminate x: $-11y = 1 \implies y = -1$

Substitute $y = -1$ in Equation 1:

$3x - 2 = 10 \implies x = 4$

Answer: $(4, -1)$

17. $2u + v = 120$ Equation 1

$u + 2v = 120$ Equation 2

Multiply Equation 2 by (-2):

$-2u - 4v = -240$

Add this to Equation 1 to eliminate u:

$-3v = -120$

$v = 40$

Substitute $v = 40$ in Equation 2:

$u + 80 = 120 \implies u = 40$

Answer: $(40, 40)$

19. $6r - 5s = 3$

$\quad 10r - 12s = 5$

Multiply Equation 1 by 5 and Equation 2 by (-3):

$$30r - 25s = \quad 15$$

$$-30r + 36s = -15$$

Add to eliminate r: $11s = 0$

$$s = 0$$

Substitute $s = 0$ in Equation 1: $6r - 5(0) = 3$

$$r = \tfrac{1}{2}$$

Answer: $\left(\tfrac{1}{2},\ 0\right)$

21. $\dfrac{x}{4} + \dfrac{y}{6} = 1$ Equation 1

$\quad x - y = 3$ Equation 2

Multiply Equation 1 by 6: $\dfrac{3}{2}x + y = 6$

Add this to Equation 2 to eliminate y:

$$\frac{5}{2}x = 9 \implies x = \frac{18}{5}$$

Substitute $x = \dfrac{18}{5}$ in Equation 2:

$$\frac{18}{5} - y = 3$$

$$y = \frac{3}{5}$$

Answer: $\left(\dfrac{18}{5}, \dfrac{3}{5}\right)$

23. $\dfrac{x+3}{4} + \dfrac{y-1}{3} = \quad 1$ Equation 1

$\quad\quad 2x - y = 12$ Equation 2

Multiply Equation 1 by 12 and Equation 2 by 4

$$3x + 4y = \quad 7$$

$$8x - 4y = 48$$

Add to eliminate y: $11x = 55 \implies x = 5$

Substitute $x = 5$ into Equation 2:

$$2(5) - y = 12 \implies y = -2$$

Answer: $(5, -2)$

25. $2.5x - 3y = 1.5$ Equation 1

$\quad 10x - 12y = 6$ Equation 2

Multiply Equation 1 by (-4):

$$-10x + 12y = -6$$

Add this to Equation 2 to eliminate x:

$$0 = 0 \text{ (Dependents)}$$

The solution set consists of all points lying
on the line

$$10x - 12y = 6.$$

Let $x = a$, then $y = \dfrac{5}{6}a - \dfrac{1}{2}$.

Answer: $\left(a, \dfrac{5}{6}a - \dfrac{1}{2}\right)$, where a is any
real number. Infinitely many solutions.

27. $0.05x - 0.03y = 0.21$ Equation 1

$\quad 0.07x + 0.02y = 0.16$ Equation 2

Multiply Equation 1 by 200 and
Equation 2 by 300:

$$10x - 6y = 42$$

$$21x + 6y = 48$$

Add to eliminate y: $31x = 90$

$$x = \tfrac{90}{31}$$

Substitute $x = \tfrac{90}{31}$ in Equation 2:

$$0.07\left(\tfrac{90}{31}\right) + 0.02y = 0.16$$

$$y = -\tfrac{67}{31}$$

Answer: $\left(\tfrac{90}{31}, -\tfrac{67}{31}\right)$

29. $4b + 3m = 3$ Equation 1

$3b + 11m = 13$ Equation 2

Multiply Equation 1 by 3 and Equation 2 by (-4):

$12b + 9m = 9$

$-12b - 44m = -52$

Add to eliminate b: $-35m = -43$

$m = \frac{43}{35}$

Substitute $m = \frac{43}{35}$ in Equation 1: $5b + 3\left(\frac{43}{35}\right) = 3 \implies b = -\frac{6}{35}$

Answer: $\left(-\frac{6}{35}, \frac{43}{35}\right)$

31. $\frac{1}{5}x - \frac{1}{3}y = 1$

$-3x + 5y = 9$

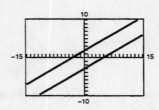

The lines are parallel.
The system is inconsistent.

33. $2x - 5y = 0$

$x - y = 3$

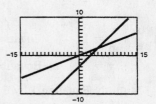

The system is consistent. There is one solution.

35. $8x + 9y = 42$

$6x - y = 16$

Solution: $(3, 2)$

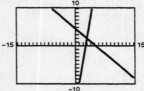

37. $4y = -8$

$7x - 2y = 25$

Solution: $(3, -2)$

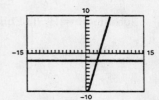

39. $3x - 5y = 7$ Equation 1

$2x + y = 9$ Equation 2

Multiply Equation 2 by 5:

$10x + 5y = 45$

Add this to Equation 1:

$13x = 52 \implies x = 4$

Back-substitute $x = 4$ into Equation 2:

$2(4) + y = 9 \implies y = 1$

Solution: $(4, 1)$

41. $y = 2x - 5$ Equation 1

$y = 5x - 11$ Equation 2

Since both equations are solved for y, set them equal to one another and solve for x.

$2x - 5 = 5x - 11$

$6 = 3x$

$2 = x$

Back-substitute $x = 2$ into Equation 1:

$y = 2(2) - 5 = -1$

Solution: $(2, -1)$

43. There are infinitely many systems that have the solution $\left(3, \frac{5}{2}\right)$. One possible system is:

$2(3) + 2\left(\frac{5}{2}\right) = 11 \implies 2x + 2y = 11$

$3 - 4\left(\frac{5}{2}\right) = -7 \implies x - 4y = -7$

45. $100y - x = \quad 200$ Equation 1

$\quad 99y - x = -198$ Equation 2

Subtract Equation 2 from Equation 1 to eliminate x: $y = 398$

Substitute $y = 398$ into Equation 1:

$\quad 100(398) - x = 200 \implies x = 39,600$

Solution: $(39,600, \ 398)$

The lines are not parallel. The scale on the axes must be changed to see the point of intersection.

47. No, it is not possible for a consistent system of linear equations to have exactly two solutions. Either the lines will intersect once or they will coincide and then the system would have infinite solutions.

49. $4x - 8y = -3$ Equation 1

$2x + ky = 16$ Equation 2

Multiply Equation 2 by -2: $-4x - 2ky = -32$

Add this to Equation 1: $-8y - 2ky = -35$

The system in inconsistent if $-8y - 2ky = 0$. This occurs when $k = -4$.
Note that for $k = -4$, the two original equations represent parallel lines.

51. Let $x =$ the ground speed and $y =$ the wind speed.

$3.6(x - y) = 1800$ Equation 1 $\qquad x - y = \quad 500$

$\quad 6(x + y) = 1800$ Equation 2 $\qquad x + y = \quad 600$

$\qquad\qquad\qquad\qquad\qquad\qquad\qquad \overline{\quad 2x \qquad\ = 1100}$

$\qquad\qquad\qquad\qquad\qquad\qquad\qquad\qquad x \quad\ \ = \quad 550$

$\qquad\qquad\qquad\qquad\qquad\qquad\quad 550 + y = \quad 600$

$\qquad\qquad\qquad\qquad\qquad\qquad\qquad\quad\ y = \quad\ \ 50$

Answer: $x = 550$ mph, $y = 50$ mph

53. Let $x =$ the number of liters at 20%, $y =$ the number of liters at 50%.

(a) $x + \quad y = \qquad 10$

$\quad 0.2x + 0.5y = 0.3(10)$

(c) -2 Equation 1 $\quad -2x - 2y = -20$

$\quad\ 10$ Equation 2 $\qquad 2x + 5y = \quad 30$

$\qquad\qquad\qquad\qquad\qquad \overline{\qquad\quad 3y = \quad 10}$

$\qquad\qquad\qquad\qquad\qquad\qquad y = \frac{10}{3}$

$\qquad\qquad\qquad\qquad\ x + \frac{10}{3} = 10$

$\qquad\qquad\qquad\qquad\qquad x = \frac{20}{3}$

(b)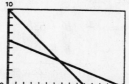

As x increases, y decreases.

Answer: $x = 6\frac{2}{3}$ liters at 20%, $y = 3\frac{1}{3}$ liters at 50%

55. Let x = amount invested at 10.5%, y = amount invested at 12%.

$$
\begin{aligned}
x + \quad y &= 12{,}000 \quad \text{Equation 1} \\
0.105x + 0.12y &= \quad 1350 \quad \text{Equation 2}
\end{aligned}
$$

$$
\begin{aligned}
-12x - 12y &= -144{,}000 \\
\underline{10.5x + 12y} &= \underline{\ 135{,}000} \\
-1.5x \quad\quad &= \quad -9000 \\
\\
x \quad &= \quad 6000 \\
6000 + \quad y &= 12{,}000 \\
y &= \quad 6000
\end{aligned}
$$

Answer: $y = \$6000$ at 12%, $x = \$6000$ at 10.5%.

57. Let x = number of adult tickets sold, y = number of child tickets sold.

$$
\begin{aligned}
x + \ y &= \quad 500 \quad \text{Equation 1} \\
7.5x + 4y &= \$3312.50 \quad \text{Equation 2}
\end{aligned}
$$

$$
\begin{aligned}
-4x - 4y &= -2000.00 \\
\underline{7.5x + 4y} &= \underline{\ 3312.50} \\
3.5x \quad &= \quad 1312.50 \\
\\
x \quad &= 375 \\
375 + \ y &= 500 \\
y &= 125
\end{aligned}
$$

Answer: $x = 375$ adult tickets, $y = 125$ child tickets

59. Let x = distance one person drives,
y = distance other person drives.

$$
\begin{aligned}
x + y &= 300 \quad \text{Equation 1} \\
y &= 3x \quad \text{Equation 2} \\
x + 3x &= 300 \quad \text{Use substitution.} \\
4x &= 300 \\
x &= 75 \\
y &= 3x = 225
\end{aligned}
$$

Answer: 75 km and 225 km

61.
$$
\begin{aligned}
5b + 10a &= 20.2 \implies -10b - 20a = -40.4 \\
10b + 30a &= 50.1 \implies \underline{\ 10b + 30a = \ \ 50.1} \\
&\qquad\qquad\qquad\qquad 10a = \quad 9.7 \\
&\qquad\qquad\qquad\qquad\ a = 0.97 \\
&\qquad\qquad\qquad\qquad\ b = 2.10
\end{aligned}
$$

Least squares regression line:
$y = 0.97x + 2.10$

63.
$$
\begin{aligned}
7b + 21a &= \ 35.1 \implies -21b - 63a = -105.3 \\
21b + 91a &= 114.2 \implies \underline{\ 21b + 91a = \ \ 114.2} \\
&\qquad\qquad\qquad\qquad 28a = \quad 8.9 \\
&\qquad\qquad\qquad\qquad\ a = \tfrac{89}{280} \\
&\qquad\qquad\qquad\qquad\ b = \tfrac{1137}{280}
\end{aligned}
$$

Least squares regression line: $y = \tfrac{1}{280}(89x + 1137) \approx 0.318x + 4.061$

65. $(-2, 0), (0, 1), (2, 3)$

$3b = 4 \implies b = \frac{4}{3}$

$8a = 6 \implies a = \frac{3}{4}$

Least squares regression line:

$y = \frac{3}{4}x + \frac{4}{3}$

67. $(0, 4), (1, 3), (1, 1), (2, 0)$

$$
\begin{aligned}
4b + 4a = 8 &\implies \quad 4b + 4a = 8 \\
4b + 6a = 4 &\implies \underline{-4b - 6a = -4} \\
& \qquad\qquad -2a = 4
\end{aligned}
$$

$a = -2$

$b = 4$

Least squares regression line: $y = -2x + 4$

69. $(1.0, \ 32), (1.5, \ 41), (2.0, \ 48), (2.5, \ 53)$

$n = 4, \quad \sum_{i=1}^{4} x_i = 7, \quad \sum_{i=1}^{4} y_i = 174, \quad \sum_{i=1}^{4} x_i^2 = 13.5, \quad \sum_{i=1}^{4} x_i y_i = 322$

$$
\begin{aligned}
4b + 7a = 174 &\implies 28b + 49a = 1218 \\
7b + 13.5a = 322 &\implies \underline{-28b - 54a = -1288} \\
& \qquad\qquad\quad -5a = -70
\end{aligned}
$$

$a = 14$

$b = 19$

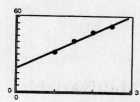

$y = ax + b$

$y = 14x + 19$

When $x = 1.6$, $y = 14(1.6) + 19 = 41.4$ bushels per acre.

71. Demand = Supply

$50 - 0.5 = 0.125x$

$50 = 0.625x$

$x = 80$ units

$p = \$10$

Answer: $(80, 10)$

73.   Demand = Supply

$140 - 0.00002x = 80 + 0.00001x$

$60 = 0.00003x$

$x = 2{,}000{,}000$ units

$p = \$100.00$

Answer: $(2{,}000{,}000, 100)$

75. Multiply the first equation by $(\cos x)$ and the second by $(-\sin x)$.

$u \sin x(\cos x) + v(\cos x)(\cos x) = 0(\cos x)$

$u \cos x(-\sin x) - v \sin x(-\sin x) = \sec x(-\sin x)$

Adding,

$v \cos^2 x + v \sin^2 x = -\sec x \cdot \sin x$

$v(\cos^2 x + \sin^2 x) = -\tan x$

$v = -\tan x.$

Finally,

$u \sin x + v \cos x = 0$

$\implies u \sin x - \tan x \cos x = 0$

$\implies u \sin x = \sin x \implies u = 1.$

77. Subtracting the two equations:

$$vxe^x - v(x + 1)e^x = -e^x \ln x$$
$$vxe^x - vxe^x - ve^x = -e^x \ln x$$
$$ve^x = e^x \ln x$$
$$v = \ln x$$

Finally, $ue^x + vxe^x = ue^x + \ln x \cdot x \cdot e^x = 0$

$$\Longrightarrow \quad ue^x = -x \ln x \cdot e^x$$
$$\Longrightarrow \quad u = -x \ln x$$

79. The domain of $f(x) = x^2 - 2x$ is all real numbers.

81. The domain of $h(x) = \sqrt{25 - x^2}$ is:

$$25 - x^2 \geq 0$$
$$x^2 \leq 25$$
$$-5 \leq x \leq 5$$

Section 7.3   Multivariable Linear Systems

■ You should know the operations that lead to equivalent systems of linear equations:

   (a) Interchange any two equations.

   (b) Multiply all terms of an equation by a nonzero constant.

   (c) Replace an equation by the sum of itself and a constant multiple of any other equation in the system.

■ You should be able to use the method of elimination.

Solutions to Odd-Numbered Exercises

1.
$$2x - y + 5z = 24 \quad \text{Equation 1}$$
$$y + 2z = 6 \quad \text{Equation 2}$$
$$z = 4 \quad \text{Equation 3}$$

Back-substitute $z = 4$ into Equation 2.

$$y + 2(4) = 6$$
$$z = -2$$

Back-substitute $y = -2$ and $z = 4$ into Equation 1.

$$2x - (-2) + 5(4) = 24$$
$$2x + 22 = 24$$
$$x = 1$$

Answer: $(1, -2, 4)$

3.
$$2x + y - 3z = 10 \quad \text{Equation 1}$$
$$y = 2 \quad \text{Equation 2}$$
$$y - z = 4 \quad \text{Equation 3}$$

Back-substitute $y = 2$ into Equation 3.

$$2 - z = 4$$
$$z = -2$$

Back-substitute $y = 2$ and $z = -2$ into Equation 1.

$$2x + 2 - 3(-2) = 10$$
$$2x + 8 = 10$$
$$x = 1$$

Answer: $(1, 2, -2)$

5. $4x - 2y + z = 8$ Equation 1

$\qquad\quad 2z = 4$ Equation 2

$\qquad -y + z = 4$ Equation 3

From Equation 2 we have $z = 2$. Back-substitute $z = 2$ into Equation 3.

$\qquad -y + 2 = 4$

$\qquad\qquad y = -2$

Back-substitute $y = -2$ and $z = 2$ into Equation 1.

$\qquad 4x - 2(-2) + 2 = 8$

$\qquad\quad 4x + 6 = 8$

$\qquad\qquad\quad x = \frac{1}{2}$

Answer: $\left(\frac{1}{2}, -2, 2\right)$

7. $\quad x - 2y + 3z = 5$ Equation 1

$\quad -x + 3y - 5z = 4$ Equation 2

$\quad 2x \qquad - 3z = 0$ Equation 3

Add Equation 1 to Equation 2.

$\qquad y - 2z = 9$

This is the first step in putting the system in row-echelon form.

9. $\quad x + y + z = \quad 6$ Equation 1

$\quad 2x - y + z = \quad 3$ Equation 2

$\quad 3x \quad - z = \quad 0$ Equation 3

$\quad x + y + z = \quad 6$

$\qquad -3y - z = -9$ -2Eq.1 $+$ Eq.2

$\quad x - 3y - 4z = -18$ -3Eq.1 $+$ Eq.3

$\quad x + y + z = \quad 6$

$\qquad -3y - z = -9$

$\qquad\qquad -3z = -9$ $-$Eq.2 $+$ Eq.3

$\qquad -3z = -9 \implies z = 3$

$\qquad -3y - 3 = -9 \implies y = 2$

$\quad x + 2 + 3 = 6 \implies x = 1$

Answer: $(1, 2, 3)$

11. $2x \qquad + 2z = 2$ Equation 1

$5x + 3y \qquad = 4$ Equation 2

$\qquad 3y - 4z = 4$ Equation 3

$x + 3y - 4z = 0$ -2Eq.1 $+$ Eq.2

$2x \qquad + 2z = 2$ Interchange

$\qquad 3y - 4z = 4$ the equations.

$x + 3y - 4z = 0$

$\qquad -6y + 10z = 2$ -2Eq.1 $+$ Eq.2

$\qquad 3y - 4z = 4$

$x + 3y - 4z = 0$

$\qquad -6y + 10z = 2$

$\qquad\qquad z = 5$ $\frac{1}{2}$Eq.2 $+$ Eq.3

$\qquad\qquad\qquad z = 5$

$-6y + 10(5) = 2 \implies y = 8$

$x + 3(8) - 4(5) = 0 \implies x = -4$

Answer: $(-4, 8, 5)$

13.
$$\begin{aligned} 3x + 3y \phantom{{}+ 3z} &= 9 \\ 2x \phantom{{}+ 3y} - 3z &= 10 \\ 6y + 4z &= -12 \end{aligned}$$ Interchange the equations.

$$\begin{aligned} x + y \phantom{{}+ 3z} &= 3 \\ 2x \phantom{{}+ 3y} - 3z &= 10 \\ 6y + 4z &= -12 \end{aligned}$$ $\frac{1}{3}$Eq.1

$$\begin{aligned} x + y \phantom{{}+ 3z} &= 3 \\ -2y - 3z &= 4 \\ 6y + 4z &= -12 \end{aligned}$$ -2Eq.1 + Eq.2

$$\begin{aligned} x + y \phantom{{}+ 3z} &= 3 \\ -2y - 3z &= 4 \\ -5z &= 0 \end{aligned}$$ 3Eq.2 + Eq.3

$$-5x = 0 \implies z = 0$$
$$-2y - 3(0) = 4 \implies y = -2$$
$$x - 2 = 3 \implies x = 5$$

Answer: $(5, -2, 0)$

17.
$$\begin{aligned} 3x + 3y + 5z &= 1 \\ 3x + 5y + 9z &= 0 \\ 5x + 9y + 17z &= 0 \end{aligned}$$

$$\begin{aligned} 6x + 6y + 10z &= 2 \\ 3x + 5y + 9z &= 0 \\ 5x + 9y + 17z &= 0 \end{aligned}$$ 2 Eq.1

$$\begin{aligned} x - 3y - 7z &= 2 \\ 3x + 5y + 9z &= 0 \\ 5x + 9y + 17z &= 0 \end{aligned}$$ $-$Eq.3 + Eq.1

$$\begin{aligned} x - 3y - 7z &= 2 \\ 14y + 30z &= -6 \\ 24y + 52z &= -10 \end{aligned}$$ -3Eq.1 + Eq.2
-5Eq.1 + Eq.3

15.
$$\begin{aligned} x + y - 2z &= 3 \\ 3x - 2y + 4z &= 1 \\ 2x - 3y + 6z &= 8 \end{aligned}$$ Interchange the equations.

$$\begin{aligned} x + y - 2z &= 3 \\ -5y + 10z &= -8 \\ -5y + 10z &= 2 \end{aligned}$$ -3Eq.1 + Eq.2
-2Eq.1 + Eq.3

$$\begin{aligned} x + y - 2z &= 3 \\ -5y + 10z &= -8 \\ 0 &= 10 \end{aligned}$$ $\to \leftarrow$ $-$Eq.2 + Eq.3

No solution, inconsistent

$$\begin{aligned} x - 3y - 7z &= 2 \\ 84y + 180z &= -36 \\ 84y + 182z &= -35 \end{aligned}$$ 6Eq.2
3.5Eq.3

$$\begin{aligned} x - 3y - 7z &= 2 \\ 84y + 180z &= -36 \\ 2z &= 1 \end{aligned}$$ $-$Eq.2 + Eq.3

$$2z = 1 \implies x = \tfrac{1}{2}$$
$$84y + 180\left(\tfrac{1}{2}\right) = -36 \implies y = -\tfrac{3}{2}$$
$$x - 3\left(-\tfrac{3}{2}\right) - 7\left(\tfrac{1}{2}\right) = 2 \implies x = 1$$

Answer: $\left(1, -\tfrac{3}{2}, \tfrac{1}{2}\right)$

19. $x + 2y - 7z = -4$
$2x + y + z = 13$
$3x + 9y - 36z = -33$

$x + 2y - 7z = -4$
$-3y + 15z = 21$ -2Eq.1 + Eq.2
$3y - 15z = -21$ -3Eq.1 + Eq.3

$x + 2y - 1z = -4$
$-3y + 15z = 21$
$0 = 0$ Eq.2 + Eq.3

$x + 2y - 7z = -4$
$y - 5z = -7$ $\frac{1}{3}$Eq.2

$x \quad + 3z = 10$ -2Eq.2 + Eq.1
$y - 5z = -7$

Let $z = a$, then:
$y = 5a - 7$
$x = -3a + 10$
Answer: $(-3a + 10, 5a - 7, a)$

23. $x - 2y + 5z = 2$
$4x \quad - z = 0$
Let $z = a$, then $x = \frac{1}{4}a$.
$\frac{1}{4}a - 2y + 5a = 2$
$a - 8y + 20a = 8$
$-8y = -21a + 8$
$y = \frac{21}{8}a - 1$
Answer: $\left(\frac{1}{4}a, \frac{21}{8}a - 1, a\right)$
To avoid fractions, we could go back and let
$z = 8a$, then $4x - 8a = 0 \implies x = 2a$.
$2a - 2y + 5(8a) = 2$
$-2y + 42a = 2$
$y = 21a - 1$
Answer: $(2a, 21a - 1, 8a)$

21. $3x - 3y + 6z = 6$
$x + 2y - z = 5$
$5x - 8y + 13z = 7$

$x - y + 2z = 2$ $\frac{1}{3}$Eq.1
$3y - 3z = 3$ $-$Eq.1 + Eq.2
$-3y + 3z = -3$ -5Eq.1 + Eq.3

$x - y + 2z = 2$
$y - z = 1$ $\frac{1}{3}$Eq.2
$0 = 0$ Eq.2 + Eq.3

$x \quad + z = 3$ Eq.2 + Eq.1
$y - z = 1$

Let $z = a$, then:
$y = a + 1$
$x = -a + 3$
Answer: $(-a + 3, a + 1, a)$

25. $2x - 3y + z = -2$
$-4x + 9y \quad = 7$

$2x - 3y + z = -2$
$3y + 2z = 3$ 2Eq.1 + Eq.2

$2x \quad + 3z = 1$ Eq.2 + Eq.1
$3y + 2z = 3$

Let $x = a$, then:
$y = -\frac{2}{3}a + 1$
$x = -\frac{3}{2}a + \frac{1}{2}$
Answer: $\left(-\frac{3}{2}a + \frac{1}{2}, -\frac{2}{3}a + 1, a\right)$

27.
$$\begin{aligned}
x \qquad\quad + 3w &= 4 \\
2y - z - w &= 0 \\
3y \qquad - 2w &= 1 \\
2x - y + 4z \qquad &= 5
\end{aligned}$$

$$\begin{aligned}
x \qquad\quad + 3w &= 4 \\
2y - z - w &= 0 \\
3y \qquad - 2w &= 1 \\
- y + 4z - 6w &= -3 \qquad -2\text{Eq.1} + \text{Eq.4}
\end{aligned}$$

$$\begin{aligned}
x \qquad\quad + 3w &= 4 \\
y - 4z + 6w &= 3 \qquad -\text{Eq.4 and} \\
2y - z - w &= 0 \qquad \text{interchange} \\
3y \qquad - 2w &= 1 \qquad \text{the equations.}
\end{aligned}$$

$$\begin{aligned}
x \qquad\quad + 3w &= 4 \\
y - 4z + 6w &= 3 \\
7z - 13w &= -6 \qquad -\text{Eq.2} + \text{Eq.3} \\
12z - 20w &= -8 \qquad -3\text{Eq.2} + \text{Eq.4}
\end{aligned}$$

$$\begin{aligned}
x \qquad\quad + 3w &= 4 \\
y - 4z + 6w &= 3 \\
z + 3w &= -2 \qquad -\tfrac{1}{2}\text{Eq.4} + \text{Eq.3} \\
12z - 20w &= -8
\end{aligned}$$

$$\begin{aligned}
x \qquad\quad + 3w &= 4 \\
y - 4z + 6w &= 3 \\
z - 3w &= -2 \\
16w &= 16 \qquad -12\text{Eq.3} + \text{Eq.4}
\end{aligned}$$

$$\begin{aligned}
16w &= 16 \implies w = 1 \\
z - 3(1) &= -2 \implies z = 1 \\
y - 4(1) + 6(1) &= 3 \implies y = 1 \\
x + 3(1) &= 4 \implies x = 1
\end{aligned}$$

Answer: $(1, 1, 1, 1)$

29.
$$\begin{aligned}
x \quad + 4z &= 1 \\
x + y + 10z &= 10 \\
2x - y + 2z &= -5
\end{aligned}$$

$$\begin{aligned}
x \quad + 4z &= 1 \\
y + 6z &= 9 \qquad -\text{Eq.1} + \text{Eq.2} \\
-y - 6z &= -7 \qquad -2\text{Eq.1} + \text{Eq.3}
\end{aligned}$$

$$\begin{aligned}
x \quad + 4z &= 1 \\
y + 6z &= 9 \\
0 &= 2 \qquad \to \leftarrow \text{Eq.2} + \text{Eq.3}
\end{aligned}$$

No solution, inconsistent.

31.
$$\begin{aligned}
2x + 3y \qquad &= 0 \\
4x + 3y - z &= 0 \\
8x + 3y + 3z &= 0
\end{aligned}$$

$$\begin{aligned}
2x + 3y \qquad &= 0 \\
-3y - z &= 0 \qquad -2\text{Eq.1} + \text{Eq.2} \\
-9y + 3z &= 0 \qquad -4\text{Eq.1} + \text{Eq.3}
\end{aligned}$$

$$\begin{aligned}
2x + 3y \qquad &= 0 \\
-3y - z &= 0 \\
6z &= 0 \qquad -3\text{Eq.2} + \text{Eq.3}
\end{aligned}$$

$$\begin{aligned}
6z &= 0 \implies z = 0 \\
-3y - 0 &= 0 \implies y = 0 \\
2x + 3(0) &= 0 \implies x = 0
\end{aligned}$$

Answer: $(0, 0, 0)$

33.
$$\begin{aligned}
23x + 4y - z &= 0 \qquad \text{Interchange} \\
12x + 5y + z &= 0 \qquad \text{the equations.}
\end{aligned}$$

$$\begin{aligned}
x + 6y + 3z &= 0 \qquad 2\text{Eq.2} - \text{Eq.1} \\
-67y - 35z &= 0 \qquad -12\text{Eq.1} + \text{Eq.2}
\end{aligned}$$

To avoid fractions, let $z = 67a$, then:
$$\begin{aligned}
-67y - 35(67a) &= 0 \\
y &= -35a \\
x + 6(-35a) + 3(67a) &= 0 \\
x &= 9a
\end{aligned}$$

Answer: $(9a, -35a, 67a)$

35. No, they are not equivalent. The constant in the second equation should be -11 and the coefficient of z in the third equation should be 2.

37. There are an infinite number of linear systems that has $(4, -1, 2)$ as their solution. One such system is as follows:

$$3(4) + (-1) - (2) = 9 \Longrightarrow 3x + y - z = 9$$
$$(4) + 2(-1) - (2) = 0 \Longrightarrow x + 2y - z = 0$$
$$-(4) + (-1) + 3(2) = 1 \Longrightarrow -x + y + 3z = 1$$

39. $y = ax^2 + bx + c$ passing through $(0, 0), (2, -2), (4, 0)$

$(0, \quad 0): \quad 0 = \qquad\qquad c$

$(2, -2): -2 = 4a + 2b + c \Longrightarrow -1 = 2a + b$

$(4, \quad 0): \quad 0 = 16a + 4b + c \Longrightarrow \quad 0 = 4a + b$

Answer: $a = \frac{1}{2}, b = -2, c = 0$

The equation of the parabola is $y = \frac{1}{2}x^2 - 2x$.

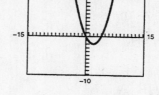

41. $y = ax^2 + bx + c$ passing through $(2, 0), (3, -1), (4, 0)$

$(2, \quad 0): \quad 0 = 4a + 2b + c$

$(3, -1): -1 = 9a + 3b + c \Longrightarrow -1 = 5a + b$

$(4, \quad 0): \quad 0 = 16a + 4b + c \Longrightarrow \quad 0 = 12a + 2b$

Answer: $a = 1, b = -6, c = 8$

The equation of the parabola is $y = x^2 - 6x + 8$.

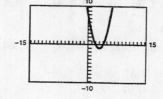

43. $x^2 + y^2 + Dx + Ey + F = 0$ passing through $(0, 0), (2, 2), (4, 0)$

$(0, 0): \qquad\qquad\qquad F = 0$

$(2, 2): \quad 8 + 2D + 2E + F = 0 \Longrightarrow D + E = -4$

$(4, 0): 16 + 4D \qquad + F = 0 \Longrightarrow D = -4 \text{ and } E = 0$

The equation of the circle is $x^2 + y^2 - 4x = 0$.

To graph, let $y_1 = \sqrt{4x - x^2}$ and $y_2 = -\sqrt{4x - x^2}$.

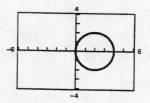

45. $x^2 + y^2 + Dx + Ey + F = 0$ passing through $(-3, -1), (2, 4), (-6, 8)$

$(-3, -1): \quad 10 - 3D - E + F = 0 \Longrightarrow \quad 10 = 3D + E - F$

$(\quad 2, \quad 4): \quad 20 + 2D + 4E + F = 0 \Longrightarrow \quad 20 = -2D - 4E - F$

$(-6, \quad 8): 100 - 6D + 8E + F = 0 \Longrightarrow 100 = 6D - 8E - F$

Answer: $D = 6, E = -8, F = 0$

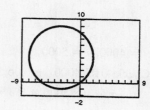

The equation of the circle is $x^2 + y^2 + 6x - 8y = 0$. To graph, complete the squares first, then solve for y.

$$(x^2 + 6x + 9) + (y^2 - 8y + 16) = 0 + 9 + 16$$
$$(x + 3)^2 + (y - 4)^2 = 25$$
$$(y - 4)^2 = 25 - (x + 3)^2$$
$$y - 4 = \pm\sqrt{25 - (x + 3)^2}$$
$$y = 4 \pm \sqrt{25 - (x + 3)^2}$$

Let $y_1 = 4 + \sqrt{25 - (x + 3)^2}$ and $y_2 = 4 - \sqrt{25 - (x + 3)^2}$.

47. $s = \frac{1}{2}at^2 + v_0t + s_0$

$(1, 128), (2, 80), (3, 0)$

$128 = \frac{1}{2}a + v_0 + s_0 \implies a + 2v_0 + 2s_0 = 256$

$80 = 2a + 2v_0 + s_0 \implies 2a + 2v_0 + s_0 = 80$

$0 = \frac{9}{2}a + 3v_0 + s_0 \implies 9a + 6v_0 + 2s_0 = 0$

Solving this system yields $a = -32, v_0 = 0, s_0 = 144$.

Thus, $s = \frac{1}{2}(-32)t^2 + (0)t + 144$

$= -16t^2 + 144.$

49. $s = \frac{1}{2}at^2 + v_0t + s_0$

$(1, 452), (2, 260), (3, 116)$

$452 = \frac{1}{2}a + v_0 + s_0 \implies a + 2v_0 + 2s_0 = 904$

$260 = 2a + 2v_0 + s_0 \implies 2a + 2v_0 + s_0 = 260$

$116 = \frac{9}{2}a + 3v_0 + s_0 \implies 9a + 6v_0 + 2s_0 = 232$

Solving this system yields $a = 48, v_0 = -264, s_0 = 692$.

Thus, $s = \frac{1}{2}(48)t^2 + (-264)t + 692$

$= 24t^2 - 264t + 692.$

51. Let x = amount at 5%.

Let y = amount at 6%.

Let z = amount at 7%.

$x + y + z = 16{,}000$

$0.05x + 0.06y + 0.07z = 990$

$x + 3000 = z$

$y + 2000 = z$

$(z - 3000) + (z - 2000) + z = 16{,}000$

$3z = 21{,}000$

$z = 7000$

$x = 4000, y = 5000$

Check: $0.05(4000) + 0.06(5000) + 0.07(7000) = 990$

Answer: $x = \$4000$ at 5%, $y = \$5000$ at 6%, $z = \$7000$ at 7%

53. Let x = amount at 8%.

Let y = amount at 9%.

Let z = amount at 10%.

$$x + \quad y + \quad z = 775{,}000$$
$$0.08x + 0.09y + 0.10z = \quad 67{,}000$$
$$x \qquad\qquad = \quad 4z$$

$$y + \quad 5z = 775{,}000$$
$$0.09y + 0.42z = \quad 67{,}000$$

$$z \approx 91{,}666.67$$
$$y = 775{,}000 - 5z = 316{,}666.67$$
$$x = 4z = 366{,}666.67$$

Answer: x = \$366,666.67 at 8%

y = \$316,666.67 at 9%

z = \$91,666.67 at 10%

55. Let C = amount in certificates of deposit.

Let M = amount in municipal bonds.

Let B = amount in blue-chip stocks.

Let G = amount in growth or speculative stocks.

$$C + M + B + G = 500{,}000$$
$$0.10C + 0.08M + 0.12B + 0.13G = 0.10(500{,}000)$$
$$B + G = \tfrac{1}{4}(500{,}000)$$

This system has infinitely many solutions.

Let $G = s$, then $B = 125{,}000 - s$

$$M = 125{,}000 + \tfrac{1}{2}s$$
$$C = 250{,}000 - \tfrac{1}{2}s.$$

Answer:

$\left(250{,}000 - \tfrac{1}{2}s,\ 125{,}000 + \tfrac{1}{2}s,\ 125{,}000 - s,\ s\right)$,

where $0 \le s \le 125{,}000$.

One possible solution is to let $s = 50{,}000$.

Certificates of deposit: \$225,000

Municipal bonds: \$150,000

Blue-chip stocks: \$75,000

Growth or speculative stocks: \$50,000

57. Let x = gallons of spray X.

Let y = gallons of spray Y.

Let z = gallons of spray Z.

Chemical A: $\tfrac{1}{5}x + \tfrac{1}{2}z = 12$ ⎫
Chemical B: $\tfrac{2}{5}x + \tfrac{1}{2}z = 16$ ⎬ $\Rightarrow x = 20,\ z = 16$

Chemical C: $\tfrac{2}{5}x + y = 26$ $\Rightarrow y = 18$

Answer: 20 liters of spray X

18 liters of spray Y

16 liters of spray Z

59.

	Product	
Truck	A	B
Large	6	3
Medium	4	4
Small	0	3

Possible solutions:

(1) 4 medium trucks

(2) 2 large trucks, 1 medium truck, 2 small trucks

(3) 3 large trucks, 1 medium truck, 1 small truck

(4) 3 large trucks, 3 small trucks

61. $t_1 - 2t_2 \qquad = \quad 0$

$t_1 \qquad\qquad 2a = 128 \Rightarrow \quad 2t_2 - 2a = \quad 128$

$\qquad t_2 + \quad a = \quad 32 \Rightarrow \quad -2t_2 - 2a = -64$

$$\overline{\qquad\qquad -4a = \quad 64}$$

$$a = -16$$
$$t_2 = \quad 48$$
$$t_1 = \quad 96$$

Answer: t_1 = 96 lb, t_2 = 48 lb, $a = -16$ ft/sec^2

63.

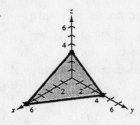

Four points are:
(6, 0, 0), (0, 4, 0), (0, 0, 3), (4, 0, 1)

65.

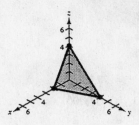

Four points are:
(2, 0, 0), (0, 4, 0), (0, 0, 4), (0, 2, 2)

67. Least squares regression parabola through $(-4, 5)$, $(-2, 6)$, $(2, 6)$, $(4, 2)$

$$n = 4$$

$\sum x_i = 0$ $\sum y_i = 19$

$\sum x_i^2 = 40$ $\sum x_i^3 = 0$

$\sum x_i^4 = 544$ $\sum x_i y_i = -12$

$\sum x_i^2 y_i = 160$

$$4c \quad + 40a = \quad 19$$
$$40b \qquad\quad = -12$$
$$40c \quad + 544a = 160$$

Solving this system yields $a = -\frac{5}{24}$, $b = -\frac{3}{10}$, and $c = \frac{41}{6}$. Thus, $y = -\frac{5}{24}x^2 - \frac{3}{10}x + \frac{41}{6}$.

69. Least squares regression parabola through $(0, 0)$, $(2, 2)$, $(3, 6)$, $(4, 12)$

$$n = 4$$

$\sum x_i = 9$ $\sum y_i = 20$

$\sum x_i^2 = 29$ $\sum x_i^3 = 99$

$\sum x_i^4 = 353$ $\sum x_i y_i = 70$

$\sum x_i^2 y_i = 254$

$$4c + \quad 9b + \quad 29a = \quad 20$$
$$9c + \quad 29b + \quad 99a = \quad 70$$
$$29c + \quad 99b + 353a = 254$$

Solving this system yields $a = 1$, $b = -1$, and $c = 0$. Thus, $y = x^2 - x$.

71. (a) Least squares regression parabola through $(20, 25)$, $(30, 55)$, $(40, 105)$, $(50, 188)$, $(60, 300)$

$$n = 5$$

$\sum x_i = 200$ $\sum y_i = 673$

$\sum x_i^2 = 9000$ $\sum x_i^3 = 440{,}000$

$\sum x_i^4 = 22{,}740{,}000$ $\sum x_i y_i = 33{,}750$

$\sum x_i^2 y_i = 1{,}777{,}500$

$$5c + \qquad 200b + \qquad 9000a = \qquad 673$$
$$200c + \quad 9000b + \quad 440{,}000a = \quad 33{,}750$$
$$9000c + 440{,}000b + 22{,}740{,}000a = 1{,}777{,}500$$

Solving this system yields $a \approx 0.14$, $b \approx -4.43$, and $c \approx 58.40$. Thus, $y = 0.14x^2 - 4.43x + 58.40$.

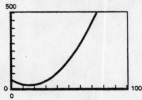

(b)   When $x = 70$, $y \approx 434.3$ feet.

73.
$$\left.\begin{array}{r} y + \quad \lambda = 0 \\ x + \lambda = 0 \end{array}\right\} \implies x = y = -\lambda$$

$x + y - 10 = 0 \implies 2x - 10 = 0$

$$x = 5$$
$$y = 5$$
$$\lambda = -5$$

75. $2x - 2x\lambda = 0 \implies x = x\lambda$

$-2y + \lambda = 0 \implies 2y = \lambda$

$y - x^2 = 0 \implies y = x^2$

From the first equation, $x = 0$ or $\lambda = 1$. If $x = 0$, then $y = 0^2 = 0$ and $\lambda = 0$. If $x \neq 0$, then $\lambda = 1 \implies y = \frac{1}{2}$ and $x = \pm\sqrt{\frac{1}{2}}$.

Thus, the solutions are:

(1) $x = y = \lambda = 0$

(2) $x = \dfrac{\sqrt{2}}{2}, \ y = \dfrac{1}{2}, \ \lambda = 1$

(3) $x = -\dfrac{\sqrt{2}}{2}, \ y = \dfrac{1}{2}, \ \lambda = 1$

77. The slope represents the average increase in sales per year.

79. $(0.075)(85) = 6.375$

81. $(0.005)(n) = 400$
$$n = 80,000$$

Section 7.4 Partial Fractions

■ You should know how to decompose a rational function $\dfrac{N(x)}{D(x)}$ into partial fractions.

(a) If the fraction is improper, divide to obtain

$$\frac{N(x)}{D(x)} = p(x) + \frac{N_1(x)}{D(x)}$$

where $p(x)$ is a polynomial.

(b) Factor the denominator completely into linear and irreducible (over the reals) quadratic factors.

(c) For each factor of the form $(px + q)^m$, the partial fraction decomposition includes the terms

$$\frac{A_1}{(px + q)} + \frac{A_2}{(px + q)^2} + \cdots + \frac{A_m}{(px + q)^m}.$$

(d) For each factor of the form $(ax^2 + bx + c)^n$, the partial fraction decomposition includes the terms

$$\frac{B_1 x + C_1}{ax^2 + bx + c} + \frac{B_2 x + C_2}{(ax^2 + bx + c)^2} + \cdots + \frac{B_n x + C_n}{(ax^2 + bx + c)^n}.$$

■ You should know how to determine the values of the constants in the numerators.

(a) Set $\dfrac{N_1(x)}{D(x)}$ = partial fraction decomposition.

(b) Multiply both sides by $D(x)$. This is called the basic equation.

(c) For distinct linear factors, substitute the roots of the distinct linear factors into the basic equation.

(d) For repeated linear factors, use the coefficients found in part (c) to rewrite the basic equation. Then use other values of x to solve for the remaining coefficients.

(e) For quadratic factors, expand the basic equation, collect like terms, and then equate the coefficients of like terms.

Solutions to Odd-Numbered Exercises

1. $\dfrac{7}{x^2 - 14x} = \dfrac{7}{x(x - 14)} = \dfrac{A}{x} + \dfrac{B}{x - 14}$

3. $\dfrac{12}{x^3 - 10x^2} = \dfrac{12}{x^2(x - 10)} = \dfrac{A}{x} + \dfrac{B}{x^2} + \dfrac{C}{x - 10}$

5. $\dfrac{2x - 3}{x^3 + 10x} = \dfrac{2x - 3}{x(x^2 + 10)} = \dfrac{A}{x} + \dfrac{Bx + C}{x^2 + 10}$

7 $\dfrac{1}{x^2 - 1} = \dfrac{A}{x + 1} + \dfrac{B}{x - 1}$

$$1 = A(x - 1) + B(x + 1)$$

Let $x = -1$: $1 = -2A \implies A = -\dfrac{1}{2}$

Let $x = 1$: $1 = 2B \implies B = \dfrac{1}{2}$

$$\frac{1}{x^2 - 1} = \frac{1/2}{x - 1} - \frac{1/2}{x + 1} = \frac{1}{2}\left[\frac{1}{x - 1} - \frac{1}{x + 1}\right]$$

9. $\dfrac{1}{x^2 + x} = \dfrac{A}{x} + \dfrac{B}{x + 1}$

$$1 = A(x + 1) + Bx$$

Let $x = 0$: $1 = A$

Let $x = -1$: $1 = -B \implies B = -1$

$$\frac{1}{x^2 + x} = \frac{1}{x} - \frac{1}{x + 1}$$

11. $\dfrac{1}{2x^2 + x} = \dfrac{A}{2x + 1} + \dfrac{B}{x}$

$$1 = Ax + B(2x + 1)$$

Let $x = -\dfrac{1}{2}$: $1 = -\dfrac{1}{2}A \implies A = -2$

Let $x = 0$: $1 = B$

$$\dfrac{1}{2x^2 + x} = \dfrac{1}{x} - \dfrac{2}{2x + 1}$$

13. $\dfrac{3}{x^2 + x - 2} = \dfrac{A}{x - 1} + \dfrac{B}{x + 2}$

$$3 = A(x + 2) + B(x - 1)$$

Let $x = 1$: $3 = 3A \implies A = 1$

Let $x = -2$: $3 = -3B \implies B = -1$

$$\dfrac{3}{x^2 + x - 2} = \dfrac{1}{x - 1} - \dfrac{1}{x + 2}$$

15. $\dfrac{x^2 + 12x + 12}{x^3 - 4x} = \dfrac{A}{x} + \dfrac{B}{x + 2} + \dfrac{C}{x - 2}$

$$x^2 + 12x + 12 = A(x + 2)(x - 2) + Bx(x - 2) + Cx(x + 2)$$

Let $x = 0$: $12 = -4A \implies A = -3$

Let $x = -2$: $-8 = 8B \implies B = -1$

Let $x = 2$: $40 = 8C \implies C = 5$

$$\dfrac{x^2 + 12x + 12}{x^3 - 4x} = -\dfrac{3}{x} - \dfrac{1}{x + 2} + \dfrac{5}{x - 2}$$

17. $\dfrac{4x^2 + 2x - 1}{x^2(x + 1)} = \dfrac{A}{x} + \dfrac{B}{x^2} + \dfrac{C}{x + 1}$

$$4x^2 + 2x - 1 = Ax(x + 1) + B(x + 1) + Cx^2$$

Let $x = 0$: $-1 = B$

Let $x = -1$: $1 = C$

Let $x = 1$: $5 = 2A + 2B + C$

$$5 = 2A - 2 + 1$$
$$6 = 2A$$
$$3 = A$$

$$\dfrac{4x^2 + 2x - 1}{x^2(x + 1)} = \dfrac{3}{x} - \dfrac{1}{x^2} + \dfrac{1}{x + 1}$$

19. $\dfrac{3x}{(x - 3)^2} = \dfrac{A}{x - 3} + \dfrac{B}{(x - 3)^2}$

$$3x = A(x - 3) + B$$

Let $x = 3$: $9 = B$

Let $x = 0$: $0 = -3A + B$

$$0 = -3A + 9$$
$$3 = A$$

$$\dfrac{3x}{(x - 3)^2} = \dfrac{3}{x - 3} + \dfrac{9}{(x - 3)^2}$$

21. $\dfrac{x^2 - 1}{x(x^2 + 1)} = \dfrac{A}{x} + \dfrac{Bx + C}{x^2 + 1}$

$$x^2 - 1 = A(x^2 + 1) + (Bx + C)x$$

Let $x = 0$: $-1 = A$

$$x^2 - 1 = Ax^2 + A + Bx^2 + Cx$$
$$= -x^2 - 1 + Bx^2 + Cx$$
$$= x^2(B - 1) + Cx - 1$$

Equating coefficients of like powers:

$$1 = B - 1, 2 = B, \text{ and } 0 = C$$

$$\dfrac{x^2 - 1}{x(x^2 + 1)} = -\dfrac{1}{x} + \dfrac{2x}{x^2 + 1}$$

23. $\dfrac{x^2}{x^4 - 2x^2 - 8} = \dfrac{x^2}{(x^2 - 4)(x^2 + 2)} = \dfrac{A}{x + 2} + \dfrac{B}{x - 2} + \dfrac{Cx + D}{x^2 + 2}$

$x^2 = A(x - 2)(x^2 + 2) + B(x + 2)(x^2 + 2) + (Cx + D)(x^2 - 4)$

Let $x = -2$: $4 = -24A \implies A = -\dfrac{1}{6}$

Let $x = 2$: $4 = 24B \implies B = \dfrac{1}{6}$

$x^2 = -\dfrac{1}{6}(x - 2)(x^2 + 2) + \dfrac{1}{6}(x + 2)(x^2 + 2) + (Cx + D)(x^2 - 4)$

$x^2 = -\dfrac{1}{6}x^3 + \dfrac{1}{3}x^2 - \dfrac{1}{3}x + \dfrac{2}{3} + \dfrac{1}{6}x^3 + \dfrac{1}{3}x^2 + \dfrac{1}{3}x + \dfrac{2}{3} + Cx^3 + Dx^2 - 4Cx - 4D$

$x^2 = Cx^3 + \left(\dfrac{2}{3} + D\right)x^2 - 4Cx + \left(\dfrac{4}{3} - 4D\right)$

Equating coefficients of like powers:

$C \implies 0$

$1 = \dfrac{2}{3} + D \implies D = \dfrac{1}{3}$

$\dfrac{x^2}{x^4 - 2x^2 - 8} = -\dfrac{1}{6(x - 2)} + \dfrac{1}{6(x + 2)} + \dfrac{1}{3(x^2 + 2)}$

25. $\dfrac{x}{16x^4 - 1} = \dfrac{A}{2x + 1} + \dfrac{B}{2x - 1} + \dfrac{Cx + D}{4x^2 + 1}$

$x = A(2x - 1)(4x^2 + 1) + B(2x + 1)(4x^2 + 1) + (Cx + D)(2x + 1)(2x - 1)$

Let $x = -\dfrac{1}{2}$: $-\dfrac{1}{2} = -4A \implies A = \dfrac{1}{8}$

Let $x = \dfrac{1}{2}$: $\dfrac{1}{2} - 4B \implies B = \dfrac{1}{8}$

Let $x = 0$: $0 = -A + B - D$

$0 = -\dfrac{1}{8} + \dfrac{1}{8} - D$

$0 = D$

Let $x = 1$: $1 = 5A + 15B + 3C + 3D$

$1 = \dfrac{5}{8} + \dfrac{15}{8} + 3C + 0$

$-\dfrac{1}{2} = C$

$\dfrac{x}{16x^4 - 1} = \dfrac{1/8}{2x + 1} + \dfrac{1/8}{2x - 1} - \dfrac{x/2}{4x^2 + 1} = \dfrac{1}{8(2x + 1)} + \dfrac{1}{8(2x - 1)} - \dfrac{x}{2(4x^2 + 1)}$

27. $\dfrac{x^2 + 5}{(x + 1)(x^2 - 2x + 3)} = \dfrac{A}{x + 1} + \dfrac{Bx + C}{x^2 - 2x + 3}$

$$x^2 + 5 = A(x^2 - 2x + 3) + (Bx + C)(x + 1)$$

Let $x = -1$: $6 = 6A \implies A = 1$

$x^2 + 5 = x^2 - 2x + 3 + Bx^2 + Bx + Cx + C$

$\qquad\quad = x^2(1 + B) + x(-2 + B + C) + (3 + C)$

Equating coefficients of like powers:

$1 = 1 + B, \quad 0 = -2 + B + C, \quad$ and $5 = 3 + C$

$0 = B \qquad\quad 0 = -2 + 0 + C \qquad 2 = C$

$\qquad\qquad\qquad 2 = C$

$\dfrac{x^2 + 5}{(x + 1)(x^2 - 2x + 3)} = \dfrac{1}{x + 1} + \dfrac{2}{x^2 - 2x + 3}$

29. $\dfrac{x^4}{(x - 1)^3} = \dfrac{x^4}{x^3 - 3x^2 + 3x - 1} = x + 3 + \dfrac{6x^2 - 8x + 3}{(x - 1)^3}$

$\dfrac{6x^2 - 8x + 3}{(x - 1)^3} = \dfrac{A}{x - 1} + \dfrac{B}{(x - 1)^2} + \dfrac{C}{(x - 1)^3}$

$6x^2 - 8x + 3 = A(x - 1)^2 + B(x - 1) + C$

Let $x = 1$: $1 = C$

$6x^2 - 8x + 3 = Ax^2 - 2Ax + A + Bx - B + 1$

$6x^2 - 8x + 3 = Ax^2 + (-2A + B)x + (A - B + 1)$

Equating coefficients of like powers:

$6 = A, \quad -8 = -2A + B$ and $3 = A - B + 1$

$\qquad\qquad -8 = -12 + B \qquad\quad 3 = 6 - B + 1$

$\qquad\qquad\quad 4 = B \qquad\qquad\quad 4 = B$

$\dfrac{x^4}{(x - 1)^3} = x + 3 + \dfrac{6}{x - 1} + \dfrac{4}{(x - 1)^2} + \dfrac{1}{(x - 1)^3}$

31. $\dfrac{5 - x}{2x^2 + x - 1} = \dfrac{A}{2x - 1} + \dfrac{B}{x + 1}$

$\qquad -x + 5 = A(x + 1) + B(2x - 1)$

Let $x = \dfrac{1}{2}$: $\dfrac{9}{2} = \dfrac{3}{2}A \implies A = 3$

Let $x = -1$: $6 = -3B \implies B = -2$

$\dfrac{5 - x}{2x^2 + x - 1} = \dfrac{3}{2x - 1} - \dfrac{2}{x + 1}$

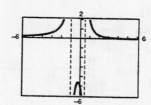

33. $\dfrac{x-1}{x^3+x^2} = \dfrac{A}{x} + \dfrac{B}{x^2} + \dfrac{C}{x+1}$

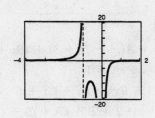

$x - 1 = Ax(x+1) + B(x+1) + Cx^2$

Let $x = -1$: $-2 = C$

Let $x = 0$: $-1 = B$

Let $x = 1$: $0 = 2A + 2B + C$

$\qquad\qquad 0 = 2A - 2 - 2$

$\qquad\qquad 2 = A$

$\dfrac{x-1}{x^3+x^2} = \dfrac{2}{x} - \dfrac{1}{x^2} - \dfrac{2}{x+1}$

35. $\dfrac{x^2+x+2}{(x^2+2)^2} = \dfrac{Ax+B}{x^2+2} + \dfrac{Cx+D}{(x^2+2)^2}$

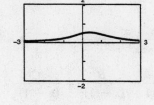

$x^2 + x + 2 = (Ax+B)(x^2+2) + Cx + D$

$x^2 + x + 2 = Ax^3 + Bx^2 + (2A+C)x + (2B+D)$

Equating coefficient of like powers:

$0 = A$

$1 = B$

$1 = 2A + C \implies C = 1$

$2 = 2B + D \implies D = 0$

$\dfrac{x^2+x+2}{(x^2+2)^2} = \dfrac{1}{x^2+2} + \dfrac{x}{(x^2+2)^2}$

37. $\dfrac{2x^3 - 4x^2 - 15x + 5}{x^2 - 2x - 8} = 2x + \dfrac{x+5}{(x+2)(x-4)}$

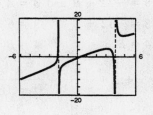

$\dfrac{x+5}{(x+2)(x-4)} = \dfrac{A}{x+2} + \dfrac{B}{x-4}$

$x + 5 = A(x-4) + B(x+2)$

Let $x = -2$: $3 = -6A \implies A = -\dfrac{1}{2}$

Let $x = 4$: $9 = 6B \implies B = \dfrac{3}{2}$

$\dfrac{2x^3 - 4x^2 - 15x + 5}{x^2 - 2x - 8} = 2x + \dfrac{1}{2}\left[\dfrac{3}{x-4} - \dfrac{1}{x+2}\right]$

39. $\dfrac{1}{a^2 - x^2} = \dfrac{A}{a+x} + \dfrac{B}{a-x}$, a is a constant

$\qquad\quad 1 = A(a-x) + B(a+x)$

Let $x = -a$: $1 = 2aA \implies A = \dfrac{1}{2a}$

Let $x = a$: $1 = 2aB \implies B = \dfrac{1}{2a}$

$\dfrac{1}{a^2 - x^2} = \dfrac{1}{2a}\left[\dfrac{1}{a+x} + \dfrac{1}{a-x}\right]$

41. $\dfrac{1}{y(a-y)} = \dfrac{A}{y} + \dfrac{B}{a-y}$

$\qquad\quad 1 = A(a-y) + By$

Let $y = 0$: $1 = aA \implies A = \dfrac{1}{a}$

Let $y = a$: $1 = aB \implies B = \dfrac{1}{a}$

$\dfrac{1}{y(a-y)} = \dfrac{1}{a}\left(\dfrac{1}{y} + \dfrac{1}{a-y}\right)$

43. $\dfrac{x-12}{x(x-4)} = \dfrac{A}{x} + \dfrac{B}{x-4}$

$x - 12 = A(x-4) + Bx$

Let $x = 0$: $-12 = -4A \implies A = 3$

Let $x = 4$: $-8 = 4B \implies B = -2$

$\dfrac{x-12}{x(x-4)} = \dfrac{3}{x} - \dfrac{2}{x-4}$

$y = \dfrac{x-12}{x(x-4)}$

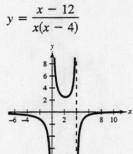

$y = \dfrac{3}{x}$

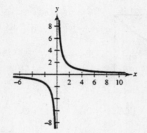

$y = -\dfrac{2}{x-4}$

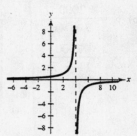

Vertical asymptotes: $x = 0$
and $x = 4$

Vertical asymptote: $x = 0$

Vertical asymptote: $x = 4$

The combination of the vertical asymptotes of the terms of the decompositions are the same as the vertical asymptotes of the rational function.

45. $\dfrac{2(4x-3)}{x^2-9} = \dfrac{A}{x-3} + \dfrac{B}{x+3}$

$2(4x-3) = A(x+3) + B(x-3)$

Let $x = 3$: $18 = 6A \implies A = 3$

Let $x = -3$: $-30 = -6B \implies B = 5$

$\dfrac{2(4x-3)}{x^2-9} = \dfrac{3}{x-3} + \dfrac{5}{x+3}$

$y = \dfrac{2(4x-3)}{x^2-9}$

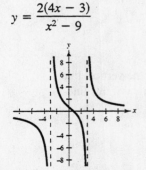

$y = \dfrac{3}{x-3}$

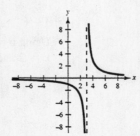

$y = \dfrac{5}{x+3}$

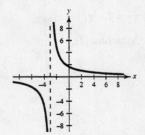

Vertical asymptotes: $x = \pm 3$

Vertical asymptote: $x = 3$

Vertical asymptote: $y = -3$

The combination of the vertical asymptotes of the terms of the decompositions are the same as the vertical asymptotes of the rational function.

47. (a) $\dfrac{2000(4 - 3x)}{(11 - 7x)(7 - 4x)} = \dfrac{A}{11 - 7x} + \dfrac{B}{7 - 4x}$, $0 \leq x \leq 1$

$$2000(4 - 3x) = A(7 - 4x) + B(11 - 7x)$$

Let $x = \dfrac{11}{7}$: $-\dfrac{10{,}000}{7} = \dfrac{5}{7}A \implies A = -2000$

Let $x = \dfrac{7}{4}$: $-2500 = -\dfrac{5}{4}B \implies B = 2000$

$$\frac{2000(4 - 3x)}{(11 - 7x)(7 - 4x)} = \frac{-2000}{11 - 7x} + \frac{2000}{7 - 4x} = \frac{2000}{7 - 4x} - \frac{2000}{11 - 7x}$$

(b) $y_1 = \dfrac{2000}{7 - 4x}$

$y_2 = \dfrac{2000}{11 - 7x}$

Section 7.5 Systems of Inequalities

- You should be able to sketch the graph of an inequality in two variables:

   (a) Replace the inequality with an equal sign and graph the equation. Use a dashed line for < or >, a solid line for ≤ or ≥.

   (b) Test a point in each region formed by the graph. If the point satisfies the inequality, shade the whole region.

Solutions to Odd-Numbered Exercises

1. $x < 2$

Vertical boundary

Matches graph (g).

3. $2x + 3y \geq 6$

$\quad y \geq -\frac{2}{3}x + 2$

Line with negative slope

Matches (a).

5. $x^2 + y^2 < 9$

Circular boundary

Matches (e).

7. $xy > 1$ or $y > \dfrac{1}{x}$

Matches (f).

9. $x \geq 2$

Using a solid line, graph the vertical line $x = 2$ and shade to the right of this line.

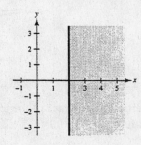

11. $y \geq -1$

Using a solid line, graph the horizontal line $y = -1$ and shade above this line.

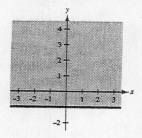

15. $2y - x \geq 4$

Using a solid line, graph $2y - x = 4$, and then shade above the line. (Use $(0, 0)$ as a test point.)

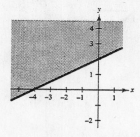

19. $y \leq \dfrac{1}{1 + x^2}$

Using a solid line, graph $y = \dfrac{1}{1 + x^2}$, and then shade below the curve. (Use $(0, 0)$ as a test point.)

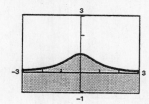

23. $x^2 + 5y - 10 \leq 0$

$$y \leq 2 - \frac{x^2}{5}$$

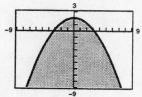

13. $y < 2 - x$

Using a dashed line, graph $y = 2 - x$, and then shade below the line. (Use $(0, 0)$ as a test point.)

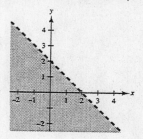

17. $y^2 - x < 0$

$$y^2 < x$$

Using a dashed line, graph the parabola $y^2 = x$, and then shade inside. (Use $(1, 0)$ as a test point.)

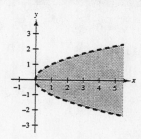

21. $y \geq \dfrac{2}{3}x - 1$

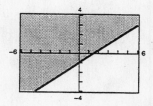

25. The line through $(-4, 0)$ and $(0, 2)$ is $y = (1/2)x + 2$. For the shaded region below the line, we have $y \leq (1/2)x + 2$.

27. The line through $(0, 2)$ and $(3, 0)$ is $y = -\frac{2}{3}x + 2$. For the shaded region above the line, we have:

$$y \geq -\frac{2}{3}x + 2$$
$$3y \geq -2x + 6$$
$$2x + 3y \geq 6$$
$$\frac{x}{3} + \frac{y}{2} \geq 1$$

29. The boundary is given by the function $y = 4 - |x|$. Hence, $y < 4 - |x|$.

31.
$$x + y \leq 1$$
$$-x + y \leq 1$$
$$y \geq 0$$

First, find the points of intersection of each pair of equations.

Vertex A	Vertex B	Vertex C
$x + y = 1$	$x + y = 1$	$-x + y = 1$
$-x + y = 1$	$y = 0$	$y = 0$
$(0, 1)$	$(1, 0)$	$(-1, 0)$

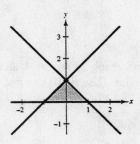

33.
$$x + y \leq 5$$
$$x \geq 2$$
$$y \geq 0$$

First, find the points of intersection of each pair of equations.

Vertex A	Vertex B	Vertex C
$x + y = 5$	$x + y = 5$	$x = 2$
$x = 2$	$y = 0$	$y = 0$
$(2, 3)$	$(5, 0)$	$(2, 0)$

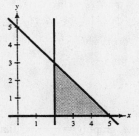

35.
$$-3x + 2y < 6$$
$$x - 4y > -2$$
$$2x + y < 3$$

First, find the points of intersection of each pair of equations.

Vertex A	Vertex B	Vertex C
$-3x + 2y = 6$	$-3x + 2y = 6$	$x - 4y = -2$
$x - 4y = -2$	$2x + y = 3$	$2x + y = 3$
$(-2, 0)$	$(0, 3)$	$\left(\frac{10}{9}, \frac{7}{9}\right)$

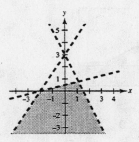

37. $2x + y > 2$

$6x + 3y < 2$

The lines are parallel. There are no points of intersection. There is no region in common to both inequalities.

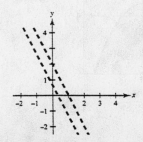

39.

$$x \geq 1$$
$$x - 2y \leq 3$$
$$3x + 2y \geq 9$$
$$x + y \leq 6$$

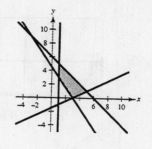

First, find the points of intersection of each pair of equations.

Vertex A	*Vertex B*	*Vertex C*
$x = 1$	$x = 1$	$x = 1$
$x - 2y = 3$	$3x + 2y = 9$	$x + y = 6$
$(1, -1)$	$(1, 3)$	$(1, 5)$

Vertex D	*Vertex E*	*Vertex F*
$x - 2y = 3$	$x - 2y = 3$	$3x + 2y = 9$
$3x + 2y = 9$	$x + y = 6$	$x + y = 6$
$(3, 0)$	$(5, 1)$	$(-3, 9)$

By shading each inequality, we find that the vertices of the region are $(1, 5)$, $(1, 3)$, $(3, 0)$, and $(5, 1)$.

41. $x^2 + y^2 \leq 9$

$x^2 + y^2 \geq 1$

There are no points of intersection. The region in common to both inequalities is the region between the circles.

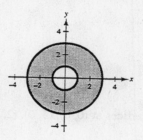

43. $x > y^2$

$x < y + 2$

Points of intersection:

$$y^2 = y + 2$$
$$y^2 - y - 2 = 0$$
$$(y + 1)(y - 2) = 0$$
$$y = -1, 2$$
$$(1, -1), (4, 2)$$

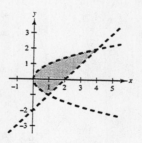

45. $y \leq \sqrt{3x} + 1$

$y \geq x^2 + 1$

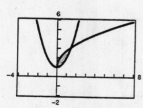

47. $y < x^3 - 2x + 1$

$y > -2x$

$x \leq 1$

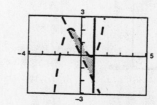

49. $x^2y \geq 1$

$0 < x \leq 4$

$y \leq 4$

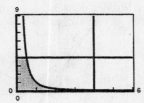

51. $y \leq -x + 4 \implies \dfrac{x}{4} + \dfrac{y}{4} \leq 1$

$x \geq 0 \qquad\qquad x \geq 0$

$y \geq 0 \qquad\qquad y \geq 0$

53. $(0, 4), (4, 0)$

Line: $y \geq 4 - x$

$(0, 2), (3, 1)$

Line: $y \geq 2 - \dfrac{1}{3}x$

$x \geq 0, \; y \geq 0$

55. $x^2 + y^2 \leq 16$

$x \geq 0$

$y \geq 0$

57. Rectangular region with vertices at $(2, 1), (5, 1), (5, 7),$ and $(2, 7)$

$x \geq 2$

$x \leq 5$

$y \geq 1$

$y \leq 7$

Thus, $2 \leq x \leq 5, \; 1 \leq y \leq 7.$

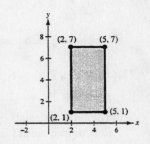

59. Triangle with vertices at $(0, 0), (5, 0), (2, 3)$

$(0, 0), (5, 0)$

Line: $y \geq 0$

$(0, 0), (2, 3)$

Line: $y \leq \dfrac{3}{2}x$

$(2, 3), (5, 0)$

Line: $y \leq -x + 5$

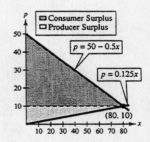

61. Account constraints:

$x \geq 5000$

$y \geq 5000$

$2x \leq y$

$x + y \leq 20,000$

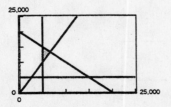

63. Assembly center constraint: $x + \dfrac{3}{2}y \leq 12$

Finishing center constraint: $\dfrac{4}{3}x + \dfrac{3}{2}y \leq 15$

Point of intersection: $(9, 2)$

Physical constraints: $x \geq 0$ and $y \geq 0$

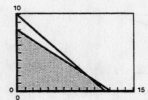

65. x = number of ounces of food X

y = number of ounces of food Y

Calcium: $20x + 10y \geq 280$

Iron: $15x + 10y \geq 160$

Vitamin B: $10x + 20y \geq 180$

$$x \geq 0$$

$$y \geq 0$$

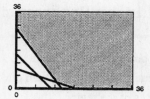

67. $xy \geq 500$ Body-building space

$2x + \pi y \geq 125$ Track (Two semi-circles and two lengths)

$x \geq 0$ Physical constraint

$y \geq 0$ Physical constraint

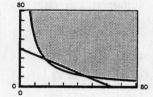

69.   Demand = Supply

$$50 - 0.5x = 0.125x$$

$$50 = 0.625x$$

$$80 = x$$

$$10 = p$$

Point of equilibrium: (80, 10)

The consumer surplus is the area of the triangle bounded by

$p \leq 50 - 0.5x$

$p \geq 10$

$x \geq 0.$

Consumer surplus = $\frac{1}{2}$(base)(height) = $\frac{1}{2}$(80)(40) = \$1600

The producer surplus is the area of the triangle bounded by

$p \geq 0.125x$

$p \leq 10$

$x \geq 0.$

Producer surplus = $\frac{1}{2}$(base)(height) = $\frac{1}{2}$(80)(10) = \$400

71. Demand = Supply

$$300 - x = 100 + x$$

$$x = 100$$

Consumer surplus = 5000

Producer surplus = 5000

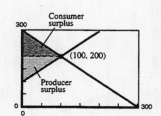

73. Test a point on either side of the boundary.

Section 7.6   Linear Programming

> ■ To solve a linear programming problem:
>   1. Sketch the solution set for the system of constraints.
>   2. Find the vertices of the region.
>   3. Test the objective function at each of the vertices.

Solutions to Odd-Numbered Exercises

1. $z = 4x + 5y$

At $(0, 6)$: $z = 4(0) + 5(6) = 30$

At $(0, 0)$: $z = 4(0) + 5(0) = 0$

At $(6, 0)$: $z = 4(6) + 5(0) = 24$

The minimum value is 0 at $(0, 0)$.
The maximum value is 30 at $(0, 6)$.

3. $z = 10x + 6y$

At $(0, 6)$: $z = 10(0) + 6(6) = 36$

At $(0, 0)$: $z = 10(0) + 6(0) = 0$

At $(6, 0)$: $z = 10(6) + 6(0) = 60$

The minimum value is 0 at $(0, 0)$.
The maximum value is 60 at $(6, 0)$.

5. $z = 3x + 2y$

At $(0, 5)$: $z = 3(0) + 2(5) = 10$

At $(4, 0)$: $z = 3(4) + 2(0) = 12$

At $(3, 4)$: $z = 3(3) + 2(4) = 17$

At $(0, 0)$: $z = 3(0) + 2(0) = 0$

The minimum value is 0 at $(0, 0)$.
The maximum value is 17 at $(3, 4)$.

7. $z = 5x + 0.5y$

At $(0, 5)$: $z = 5(0) + \frac{5}{2} = \frac{5}{2}$

At $(4, 0)$: $z = 5(4) + \frac{0}{2} = 20$

At $(3, 4)$: $z = 5(3) + \frac{4}{2} = 17$

At $(0, 0)$: $z = 5(0) + \frac{0}{2} = 0$

The minimum value is 0 at $(0, 0)$.
The maximum value is 20 at $(4, 0)$.

9. $z = 10x + 7y$

At $(0, 45)$: $z = 10(0) + 7(45) = 315$

At $(30, 45)$: $z = 10(30) + 7(45) = 615$

At $(60, 20)$: $z = 10(60) + 7(20) = 740$

At $(60, 0)$: $z = 10(60) + 7(0) = 600$

At $(0, 0)$: $z = 10(0) + 7(0) = 0$

The minimum value is 0 at $(0, 0)$.

The maximum value is 740 at $(60, 20)$.

11. $z = 25x + 30y$

At $(0, 45)$: $z = 25(0) + 30(45) = 1350$

At $(30, 45)$: $z = 25(30) + 30(45) = 2100$

At $(60, 20)$: $z = 25(60) + 30(20) = 2100$

At $(60, 0)$: $z = 25(60) + 30(0) = 1500$

At $(0, 0)$: $z = 25(0) + 30(0) = 0$

The minimum value is 0 at $(0, 0)$.
The maximum value is 2100 at any point along the
line segment connecting $(30, 45)$ and $(60, 20)$.

13. $z = 6x + 10y$

At $(0, 2)$: $z = 6(0) + 10(2) = 20$

At $(5, 0)$: $z = 6(5) + 10(0) = 30$

At $(0, 0)$: $z = 6(0) + 10(0) = 0$

The minimum value is 0 at $(0, 0)$.
The maximum value is 30 at $(5, 0)$.

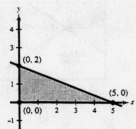

15. $z = 9z + 24y$

At $(0, 2)$: $z = 9(0) + 24(2) = 48$

At $(5, 0)$: $z = 9(5) + 24(0) = 45$

At $(0, 0)$: $z = 9(0) + 24(0) = 0$

The minimum value is 0 at $(0, 0)$.
The maximum value is 48 at $(0, 2)$.

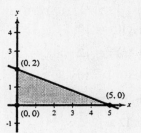

17. $z = 4x + 5y$

At $(10, 0)$: $z = 4(10) + 5(0) = 40$

At $(5, 3)$: $z = 4(5) + 5(3) = 35$

At $(0, 8)$: $z = 4(0) + 5(8) = 40$

The minimum value is 35 at $(5, 3)$.
C is unbounded. Therefore, there is no maximum.

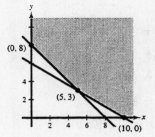

19. $z = 2x + 7y$

At $(10, 0)$: $z = 2(10) + 7(0) = 20$

At $(5, 3)$: $z = 2(5) + 7(3) = 31$

At $(0, 8)$: $z = 2(0) + 7(8) = 56$

The minimum value is 20 at $(10, 0)$.
C is unbounded. Therefore, there is no maximum.

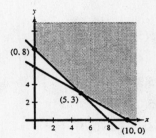

21. $z = 4x + y$

At $(36, 0)$: $z = 4(36) + 0 = 144$

At $(40, 0)$: $z = 4(40) + 0 = 160$

At $(24, 8)$: $z = 4(24) + 8 = 104$

The minimum value is 104 at $(24, 8)$.
The maximum value is 160 at $(40, 0)$.

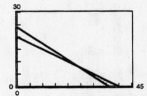

23. $z = x + 4y$

At $(36, 0)$: $z = 36 + 4(0) = 36$

At $(40, 0)$: $z = 40 + 4(0) = 40$

At $(24, 8)$: $z = 24 + 4(8) = 56$

The minimum value is 36 at $(36, 0)$.
The maximum value is 56 at $(24, 8)$.

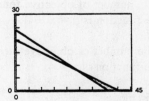

25. $z = 2x + 3y$

At $(36, 0)$: $z = 2(36) + 3(0) = 72$

At $(40, 0)$: $z = 2(40) + 3(0) = 80$

At $(24, 8)$: $z = 2(24) + 3(8) = 72$

Minimum at any point on the line segment
joining $(36, 0)$ and $(24, 8)$: 72.

Maximum at $(40, 0)$: 80.

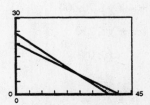

27. $z = 2x + y$

At $(0, 10)$: $z = 2(0) + (10) = 10$

At $(3, 6)$: $z = 2(3) + (6) = 12$

At $(5, 0)$: $z = 2(5) + (0) = 10$

At $(0, 0)$: $z = 2(0) + (0) = 0$

The maximum value is 12 at $(3, 6)$.

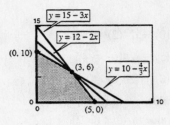

29. $z = x + y$

At $(0, 10)$: $z = (0) + (10) = 10$

At $(3, 6)$: $z = (3) + (6) = 9$

At $(5, 0)$: $z = (5) + (0) = 5$

At $(0, 0)$: $z = (0) + (0) = 0$

The maximum value is 10 at $(0, 10)$.

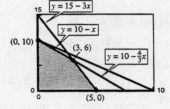

31. $z = x + 5y$

At $(0, 5)$: $z = 0 + 5(5) = 25$

At $(4, 4)$: $z = 4 + 5(4) = 24$

At $(5, 3)$: $z = 5 + 5(3) = 20$

At $(7, 0)$: $z = 7 + 5(0) = 7$

The maximum value is 25 at $(0, 5)$.

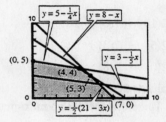

33. $z = 4x + 5y$

At $(0, 5)$: $z = 4(0) + 5(5) = 25$

At $(4, 4)$: $z = 4(4) + 5(4) = 36$

At $(5, 3)$: $z = 4(5) + 5(3) = 35$

At $(7, 0)$: $z = 4(7) + 5(0) = 28$

The maximum valve is 36 at $(4, 4)$.

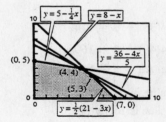

35. There are an infinite number of objective functions that would have a maximum at $(0, 4)$. One such objective function is $z = x + 5y$.

37. There are an infinite number of objective functions that would have a maximum at $(5, 0)$. One such objective function is $z = 4x + y$.

39. $x =$ number of \$250 models

$y =$ number of \$400 models

Constraints: $250x + 400y \le 70,000$

$x + y \le 250$

$x \ge 0$

$y \ge 0$

Objective function: $P = 45x + 50y$

Vertices: $(0, 175), (200, 50), (250, 0), (0, 0)$

—CONTINUED—

39. —CONTINUED—

At $(0, 175)$: $P = 45(0) + 50(175) = 8750$

At $(200, 50)$: $P = 45(200) + 50(50) = 11{,}500$

At $(250, 0)$: $P = 45(250) + 50(0) = 11{,}250$

At $(0, 0)$: $P = 45(0) + 50(0) = 0$

To maximize the profit, the merchant should stock 200 units of the model costing $250 and 50 units of the model costing $400. Then the maximum profit would be $11,500.

41. $x =$ fraction of type A

$y =$ fraction of type B

Constraints: $80x + 92y \le 90$

$x + y \le 1$

$x \le 0$

$y \le 0$

Objective function: $C = 1.13x + 1.28y$

Vertices: $\left(\frac{1}{6}, \frac{5}{6}\right)$

At $\left(\frac{1}{6}, \frac{5}{6}\right)$: $C = (1.13)\left(\frac{1}{6}\right) + (1.28)\left(\frac{5}{6}\right) = 1.255$

The minimum cost is $1.26 and occurs with a mixture that is $\frac{1}{6}$ A and $\frac{5}{6}$ B.

43. Objective function: $R = 1000x + 300y$

At $(0, 0)$: $R = 1000(0) + 300(0) = 0$

At $(0, 40)$: $R = 1000(0) + 300(40) = 12{,}000$

At $(8, 8)$: $R = 1000(8) + 300(8) = 10{,}400$

At $(9, 0)$: $R = 1000(9) + 300(0) = 9000$

The revenue will be maximum ($12,000) if the firm does 0 audits and 40 tax returns.

45. $x =$ fraction of Model A

$y =$ fraction of Model B

Constraints: $2.5x + 3y \le 4000$

$2x + y \le 2500$

$0.75x + 1.25y \le 1500$

$x \le 0$

$y \le 0$

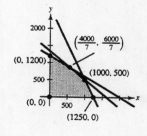

Objective function: $P = 50x + 52y$

Vertices: $(0, 0), (0, 1200), \left(\frac{4000}{7}, \frac{6000}{7}\right), (1000, 500), (1250, 0)$

At $(0, 0)$: $P = (50)(0) + 52(0) = 0$

At $(0, 1200)$: $P = 50(0) + 52(1200) = 62{,}400$

At $\left(\frac{4000}{7}, \frac{6000}{7}\right)$: $P = 50\left(\frac{4000}{7}\right) + 52\left(\frac{6000}{7}\right) \approx 73{,}142.86$

At $(1000, 500)$: $P = 50(1000) + 52(500) = 76{,}000$

At $(1250, 0)$: $P = 50(1250) + 52(0) = 62{,}500$

The maximum profit ($76,000) occurs when 1000 units of Model A and 500 units of Model B are produced.

47. Objective function: $z = 2.5x + y$

Constraints: $x \geq 0, y \geq 0, 3x + 5y \leq 15, 5x + 2y \leq 10$

At $(0, 0)$: $z = 0$

At $(2, 0)$: $z = 5$

At $\left(\frac{20}{19}, \frac{45}{19}\right)$: $z = \frac{95}{19} = 5$

At $(0, 3)$: $z = 3$

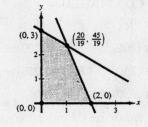

z is the maximum at any point on the line $5x + 2y = 10$ between the points $(2, 0)$ and $\left(\frac{20}{19}, \frac{45}{19}\right)$.

49. Objective function: $z = -x + 2y$

Constraints: $x \geq 0, y \geq 0, x \leq 10, x + y \leq 7$

At $(0, 0)$: $z = -0 + 2(0) = \ \ \ 0$

At $(0, 7)$: $z = -0 + 2(7) = \ \ 14$

At $(7, 0)$: $z = -7 + 2(0) = -7$

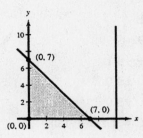

The constraint $x \leq 10$ is extraneous.
The maximum value of 14 occurs at $(0, 7)$.

51. Objective function: $z = 3x + 4y$

Constraints: $x \geq 0, y \geq 0, x + y \leq 1, 2x + y \leq 4$
The constraint $2x + y \leq 4$ is extraneous.
The maximum value of $z = 4$ occurs at $(0, 1)$.

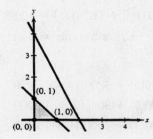

53. Constraints: $x \geq 0, y \geq 0, x + 3y \leq 15, 4x + y \leq 16$

Vertex	Value of $z = 3x + ty$
$(0, 0)$	$z = 0$
$(0, 5)$	$z = 5t$
$(3, 4)$	$z = 9 + 4t$
$(4, 0)$	$z = 12$

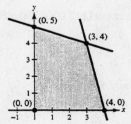

(a) For the maximum value to be at $(0, 5)$, $z = 5t$ must be greater than

$z = 9 + 4t$ and $z = 12$.

$5t > 9 + 4t$ and $5t > 12$

$\ \ t > 9$ $\qquad\qquad t > \frac{12}{5}$

Thus, $t > 9$.

(b) For the maximum value to be at $(3, 4)$, $z = 9 + 4t$ must be greater than $z = 5t$ and $z = 12$.

$9 + 4t > 5t$ and $9 + 4t > 12$

$\ \ \ 9 > t$ $\qquad\qquad\quad t > 3$

$\qquad\qquad\qquad\qquad t > \frac{3}{4}$

Thus, $\frac{3}{4} < t < 9$.

55. $\dfrac{\dfrac{9}{x}}{\left(\dfrac{6}{x}+2\right)} = \dfrac{\dfrac{9}{x}}{\dfrac{6+2x}{x}} = \dfrac{9}{x} \cdot \dfrac{x}{2(3+x)} = \dfrac{9}{2(3+x)} = \dfrac{9}{2(x+3)}, x \neq 0$

57. $\dfrac{\left(\dfrac{4}{x^2-9}+\dfrac{2}{x-2}\right)}{\left(\dfrac{1}{x+3}+\dfrac{1}{x-3}\right)} = \dfrac{\dfrac{4(x-2)+2(x^2-9)}{(x-2)(x^2-9)}}{\dfrac{(x-3)+(x+3)}{x^2-9}}$

$= \dfrac{2x^2+4x-26}{(x-2)(x^2-9)} \cdot \dfrac{x^2-9}{2x}$

$= \dfrac{2(x^2+2x-13)}{(x-2)(2x)}$

$= \dfrac{x^2+2x-13}{x(x-2)}, x \neq \pm 3$

❏ Review Exercises for Chapter 7

Solutions to Odd-Numbered Exercises

1. $x + y = 2 \implies \qquad\qquad y = 2 - x$
$ x - y = 0 \implies x - (2 - x) = 0$
$ 2x - 2 = 0$
$ x = 1$
$ y = 2 - 1 = 1$

Solution: $(1, 1)$

3. $x^2 - y^2 = 9$
$ x - y = 1 \implies \quad x = y + 1$
$ (y+1)^2 - y^2 = 9$
$ 2y + 1 = 9$
$ y = 4$
$ x = 5$

Solution: $(5, 4)$

5. $y = 2x^2$
$ y = x^4 - 2x^2 \implies 2x^2 = x^4 - 2x^2$
$ 0 = x^4 - 4x^2$
$ 0 = x^2(x^2 - 4)$
$ 0 = x^2(x+2)(x-2)$
$ x = 0, x = -2, x = 2$
$ y = 0, y = 8, y = 8$

Solutions: $(0, 0), (-2, 8), (2, 8)$

7. $y^2 - 2y + x = 0 \implies (y-1)^2 = 1 - x \implies y = 1 \pm \sqrt{1-x}$
$ x + y = 0 \implies \qquad\qquad y = -x$

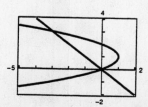

Points of intersection: $(0, 0)$ and $(-3, 3)$

9. $y = 2(6 - x)$

$y = 2^{x-2}$

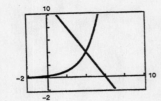

Point of intersection: $(4, 4)$

11. $\begin{array}{lll} 2x - y = 2 & \Longrightarrow & 16x - 8y = 16 \\ 6x + 8y = 39 & \Longrightarrow & 6x + 8y = 39 \\ \hline & & 22x \quad\quad = 55 \end{array}$

$x = \frac{55}{22} = \frac{5}{2}$

$y = 3$

Solution: $\left(\frac{5}{2}, 3\right)$

13. $\begin{array}{lll} 0.2x + 0.3y = 0.14 & \Longrightarrow & 20x + 30y = 14 & \Longrightarrow & 20x + 30y = \quad 14 \\ 0.4x + 0.5y = 0.20 & \Longrightarrow & 4x + 5y = 2 & \Longrightarrow & -20x - 25y = -10 \\ & & & & \hline \\ & & & & 5y = \quad 4 \end{array}$

$y = \frac{4}{5}$

$x = -\frac{1}{2}$

Solution: $\left(-\frac{1}{2}, \frac{4}{5}\right)$ or $(-0.5, 0.8)$

15. $\begin{array}{lll} 3x - 2y = 0 & \Longrightarrow & 3x - 2y = 0 \\ 3x + 2(y + 5) = 10 & \Longrightarrow & 3x + 2y = 0 \\ & & \hline \\ & & 6x \quad\quad = 0 \end{array}$

$x = 0$

$y = 0$

Solution: $(0, 0)$

17. $\begin{array}{lll} 1.25x - 2y = 3.5 & \Longrightarrow & 5x - 8y = \quad 14 \\ 5x - 8y = 14 & \Longrightarrow & -5x + 8y = -14 \\ & & \hline \\ & & 0 = \quad 0 \end{array}$

Infinite solutions

Let $y = a$, then $5x - 8a = 14 \Longrightarrow x = \frac{14}{5} + \frac{8}{5}a$.

Solution: $\left(\frac{14}{5} + \frac{8}{5}a, a\right)$

19. There are infinite linear systems with the solution $\left(\frac{4}{3}, 3\right)$. One possible solution is:

$3\left(\frac{4}{3}\right) + 3 = 7 \Longrightarrow 3x + y = 7$

$-6\left(\frac{4}{3}\right) + 3(3) = 1 \Longrightarrow -6x + 3y = 1$

21. Revenue $= 4.95x$

Cost $= 2.85x + 10,000$

Break even when Revenue $=$ Cost

$4.95x = 2.85x + 10,000$

$2.10x = 10,000$

$x \approx 4762$ units

23. Let $x =$ speed of the slower plane.

Let $y =$ speed of the faster plane.

Then, distance of first plane $+$ distance of second plane $= 275$ miles.

(rate of first plane)(time) $+$ (rate of second plane)(time) $= 275$ miles

$x\left(\frac{40}{60}\right) + y\left(\frac{40}{60}\right) = 275$

$y = x + 25$

$\frac{2}{3}x + \frac{2}{3}(x + 25) = 275$

$4x + 50 = 825$

$4x = 775$

$x = 193.75$ mph

$y = x + 25 = 218.75$ mph

25. Demand = Supply

$$37 - 0.0002x = 22 + 0.00001x$$

$$15 = 0.00021x$$

$$x = \frac{500,000}{7}$$

Point of equilibrium: $p = \frac{159}{7}$

$$\left(\frac{500,000}{7}, \frac{159}{7}\right)$$

27.
$$\begin{aligned} x + 3y - z &= 13 \\ 2x \qquad - 5z &= 23 \\ 4x - y - 2z &= 14 \end{aligned}$$

$$\begin{aligned} x + 3y - z &= 13 \\ - 6y - 3z &= -3 \\ - 13y + 2z &= -38 \end{aligned}$$

$$\begin{aligned} x + 3y - z &= 13 \\ - 6y - 3z &= -3 \\ \tfrac{17}{2}z &= -\tfrac{63}{2} \end{aligned}$$

$$\tfrac{17}{2}z = -\tfrac{63}{2} \implies z = -\tfrac{63}{17}$$

$$-6y - 3\left(-\tfrac{63}{17}\right) = -3 \implies y = \tfrac{40}{17}$$

$$x + 3\left(\tfrac{40}{17}\right) - \left(-\tfrac{63}{17}\right) = 13 \implies x = \tfrac{38}{17}$$

Solution: $\left(\tfrac{38}{17}, \tfrac{40}{17}, -\tfrac{63}{17}\right)$

29.
$$\begin{aligned} x - 2y + z &= -6 \\ 2x - 3y &= -7 \\ -x + 3y - 3z &= 11 \end{aligned}$$

$$\begin{aligned} x - 2y + z &= -6 \\ y - 2z &= 5 && -2\text{Eq.1} + \text{Eq.2} \\ y - 2z &= 5 && \text{Eq.1} + \text{Eq.3} \end{aligned}$$

$$\begin{aligned} x - 2y + z &= -6 \\ y - 2z &= 5 \\ 0 &= 0 && -\text{Eq.2} + \text{Eq.3} \end{aligned}$$

Let $z = a$, then $y = 2a + 5$.

$$x - 2(2a + 5) + a = -6$$

$$x - 3a - 10 = -6$$

$$x = 3a + 4$$

Solution: $(3a + 4, 2a + 5, a)$ where a is any real number.

31. There are an infinite number of linear systems with the solution $(4, -1, 3)$. One possible system is as follows:

$$2(4) + (-1) - 2(3) = 1 \implies 2x + y - 2z = 1$$
$$(4) + (-1) - (3) = 0 \implies x + y - z = 0$$
$$2(4) - 3(-1) - 2(3) = 5 \implies 2x - 3y - 2z = 5$$

33. $y = ax^2 + bx + c$ through $(0, -5)$, $(1, -2)$, and $(2, 5)$

$$\begin{aligned} (0, -5): -5 &= \qquad\quad + c \\ (1, -2): -2 &= a + b + c \implies a + b = 3 \\ (2, \ 5): \ 5 &= 4a + 2b + c \implies 2a + b = 5 \end{aligned}$$

$$\begin{aligned} 2a + b &= 5 \\ -a - b &= -3 \\ \hline a &= 2 \\ b &= 1 \end{aligned}$$

The equation of the parabola is $y = 2x^2 + x - 5$.

35. x = amount invested at 7%

y = amount invested at 9%

z = amount invested at 11%

$$\begin{array}{rcl} x + \quad y + \quad\quad z &=& 20{,}000 \\ 0.07x + 0.09y + 0.11z &=& 1780 \\ x - \quad y \quad\quad\quad &=& 3000 \\ x - \quad\quad\quad z &=& 1000 \end{array}$$

Solution: $x = \$8000,\ y = \$5000,\ z = \$7000$

37. $\dfrac{4 - x}{x^2 + 6x + 8} = \dfrac{A}{x + 2} + \dfrac{B}{x + 4}$

$4 - x = A(x + 4) + B(x + 2)$

Let $x = -2$: $6 = 2A \implies A = 3$

Let $x = -4$: $8 = -2B \implies B = -4$

$\dfrac{4 - x}{x^2 + 6x + 8} = \dfrac{3}{x + 2} - \dfrac{4}{x + 4}$

39. $\dfrac{x^2}{x^2 + 2x - 15} = 1 - \dfrac{2x - 15}{x^2 + 2x - 15} = 1 + \dfrac{A}{x + 5} + \dfrac{B}{x - 3}$

$-2x + 15 = A(x - 3) + B(x + 5)$

Let $x = -5$: $25 = -8A \implies A = -\dfrac{25}{8}$

Let $x = 3$: $9 = 8B \implies B = \dfrac{9}{8}$

$\dfrac{x^2}{x^2 + 2x - 15} = 1 + \dfrac{9}{8(x - 3)} - \dfrac{25}{8(x + 5)}$

41. $\dfrac{x^2 + 2x}{x^3 - x^2 + x - 1} = \dfrac{A}{x - 1} + \dfrac{Bx + C}{x^2 + 1}$

$x^2 + 2x = A(x^2 + 1) + (Bx + C)(x - 1)$

Let $x = 1$: $3 = 2A \implies A = \dfrac{3}{2}$

Let $x = 0$: $0 = A - C \implies C = \dfrac{3}{2}$

Let $x = 2$: $8 = 5A + 2B + C$

$$8 = \left(\dfrac{15}{2}\right) + 2B + \left(\dfrac{3}{2}\right) \implies B = -\dfrac{1}{2}.$$

$\dfrac{x^2 + 2x}{x^3 - x^2 + x - 1} = \dfrac{3/2}{x - 1} + \dfrac{-(1/2)x + 3/2}{x^2 + 1} = \dfrac{1}{2}\left(\dfrac{3}{x - 1} - \dfrac{x - 3}{x^2 + 1}\right)$

43. $x + 2y \le 160$

$3x + \ y \le 180$

$x \ge 0$

$y \ge 0$

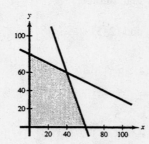

Vertex A	*Vertex B*	*Vertex C*	*Vertex D*	*Vertex E*	*Vertex F*
$x + 2y = 160$	$x + 2y = 160$	$3x + y = 180$	$x = 0$	$x + 2y = 160$	$3x + y = 180$
$3x + y = 180$	$x = 0$	$y = 0$	$y = 0$	$y = 0$	$x = 0$
$(40, 60)$	$(0, 80)$	$(60, 0)$	$(0, 0)$	$(160, 0)$	$(0, 180)$
				Outside the region	Outside the region

45. $3x + 2y \geq 24$

$\quad\quad x + 2y \geq 12$

$\quad\quad 2 \leq x \leq 15$

$\quad\quad\quad\quad y \leq 15$

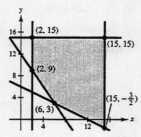

Vertex A	*Vertex B*	*Vertex C*	*Vertex D*	*Vertex E*
$3x + 2y = 24$	$3x + 2y = 24$	$3x + 2y = 24$	$3x + 2y = 24$	$x + 2y = 12$
$x + 2y = 12$	$x = 2$	$x = 15$	$y = 15$	$x = 2$
$(6, 3)$	$(2, 9)$	$\left(15, -\frac{21}{2}\right)$	$(-2, 15)$	$(2, 5)$
		Outside the region	Outside the region	Outside the region

Vertex F	*Vertex G*	*Vertex H*	*Vertex I*
$x + 2y = 12$	$x + 2y = 12$	$x = 2$	$x = 15$
$x = 15$	$y = 15$	$y = 15$	$y = 15$
$\left(15, -\frac{3}{2}\right)$	$(-18, 15)$	$(2, 15)$	$(15, 15)$
	Outside the region		

47. $y < x + 1$

$\quad y > x^2 - 1$

Vertices:

$x + 1 = x^2 - 1$

$\quad\quad 0 = x^2 - x - 2 = (x + 1)(x - 2)$

$x = -1 \quad$ or $\quad x = 2$

$y = 0 \quad\quad\quad\quad y = 3$

$(-1, 0) \quad\quad\quad (2, 3)$

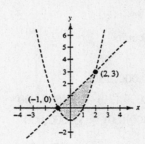

49. $2x - 3y \geq 0$

$\quad\quad 2x - y \leq 8$

$\quad\quad\quad\quad\quad y \geq 0$

Vertex A	*Vertex B*	*Vertex C*
$2x - 3y = 0$	$2x - 3y = 0$	$2x - y = 8$
$2x - y = 8$	$y = 0$	$y = 0$
$(6, 4)$	$(0, 0)$	$(4, 0)$

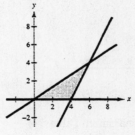

51

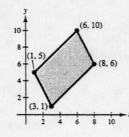

Line through $(1, 5)$, $(3, 1)$: $2x + y = 7$

Line through $(1, 5)$, $(6, 10)$: $-x + y = 4$

Line through $(6, 10)$, $(8, 6)$: $2x + y = 22$

Line through $(8, 6)$, $(3, 1)$: $-x + y = -2$

System of inequalities:

$$-x + y \leq 4$$
$$2x + y \leq 22$$
$$-x + y \geq -2$$
$$2x + y \geq 7$$

53. Let x = the number of bushels for Harrisburg, and y = the number of bushels for Philadelphia.

$$x \geq 400$$
$$y \geq 600$$
$$x + y \leq 1500$$

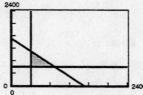

55.

$$\text{Demand} = \text{Supply}$$
$$160 - 0.001x = 70 + 0.002x$$
$$90 = 0.003x$$
$$x = 300{,}000 \text{ units}$$
$$p = \$130$$

Point of equilibrium: $(300{,}000, 130)$

Consumer surplus: $\frac{1}{2}(300{,}000)(30) = \$4{,}500{,}000$

Producer surplus: $\frac{1}{2}(300{,}000)(60) = \$9{,}000{,}000$

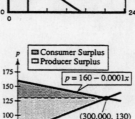

57. Maximize $z = 3x + 4y$ subject to the following constraints.

$$x \geq 0$$
$$y \geq 0$$
$$2x + 5y \leq 50$$
$$4x + y \leq 28$$

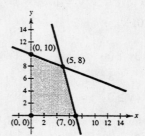

Vertex	Value of $z = 3x + 4y$
$(0, 0)$	$z = 0$
$(0, 10)$	$x = 40$
$(5, 8)$	$z = 47$, maximum value
$(7, 0)$	$z = 21$

59. Minimize $z = 1.75x + 2.25y$ subject to the following constraints.

$$2x + y \geq 25$$

$$3x + 2y \geq 45$$

$$x \geq 0$$

$$y \geq 0$$

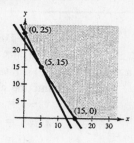

Vertex	Value of $z = 1.75x + 2.25y$
$(0, 25)$	$z = 56.25$
$(5, 15)$	$z = 42.5$
$(15, 0)$	$z = 26.25$, minimum value

61. Let $x =$ number of haircuts.

Let $y =$ number of perms.

Maximize $R = 17x + 60y$ subject to the following constraints.

$$x \geq 0$$

$$y \geq 0$$

$$\left(\tfrac{20}{60}\right)x + \left(\tfrac{70}{60}\right)y \leq 24 \implies 2x + 7y \leq 144$$

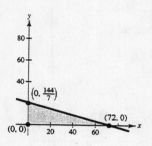

Vertex	Value of $R = 17x + 60y$
$(0, 0)$	$R = 0$
$(72, 0)$	$R = 1224$
$\left(0, \tfrac{144}{7}\right)$	$R \approx 1234.29$, maximum value

The revenue is maximum when $y = \tfrac{144}{7} \approx 20$ perms. (Round down since the student cannot work more than 24 hours. Note: Since we rounded down, the student would have enough time left to do 2 haircuts.)

63. Let $x =$ the number of bags of Brand X, and $y =$ the number of bags of Brand Y.

Objective function: Minimize $C = 15x + 30y$.

Constraints: $8x + 2y \geq 16$

$$x + y \geq 5$$

$$2x + 7y \geq 20$$

$$x \geq 0, y \geq 0$$

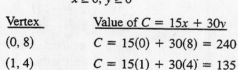

Vertex	Value of $C = 15x + 30y$
$(0, 8)$	$C = 15(0) + 30(8) = 240$
$(1, 4)$	$C = 15(1) + 30(4) = 135$
$(3, 2)$	$C = 15(3) + 30(2) = 105$, minimum value
$(10, 0)$	$C = 15(10) + 30(0) = 150$

To minimize cost, use three bags of Brand X and two bags of Brand Y. The cost per bag is $\tfrac{105}{5} = \$21$.

65. You want the equations to be multiples of each other. Multiply Equation 1 by $\tfrac{2}{3}$.

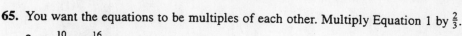

$$2x - \tfrac{10}{3}y = \tfrac{16}{3} \implies k_1 = -\tfrac{10}{3}, \; k_2 = \tfrac{16}{3}$$

$$2x + k_1 y = k_2$$

❏ Chapter Test for Chapter 7

1. $x - y = 4$
$3x + 2y = 2$

Solve the first equation for x.
$x = y + 4$
Substitute into the second equation.
$3(y + 4) + 2y = 2$
$3y + 12 + 2y = 2$
$\qquad\qquad 5y = -10$
$\qquad\qquad\ y = -2$
Substitute y back into rewritten first equation.
$x = -2 + 4 = 2$
Solution: $(2, -2)$

2. $y = x - 1$
$y = (x - 1)^3$
Substitute the first equation into the second equation.
$x - 1 = (x - 1)^3$
$x - 1 = x^3 - 3x^2 + 3x - 1$
$\qquad 0 = x^3 - 3x^2 + 2x$
$\qquad 0 = x(x - 2)(x - 1)$
$\qquad x = 0, 2, 1$
Substitute each of these values into the first equation.
$x = 0$: $y = 0 - 1 = -1$
$x = 2$: $y = 2 - 1 = 1$
$x = 1$: $y = 1 - 1 = 0$
Solutions: $(0, -1)$, $(2, 1)$,
$\qquad\qquad (1, 0)$

3. $2x - y^2 = 0$
$\quad x - y = 4$
Solve the second equation for x.
$x = y + 4$
Substitute into the first equation.
$2(y + 4) - y^2 = 0$
$-y^2 + 2y + 8 = 0$
$y^2 - 2y - 8 = 0$
$(y - 4)(y + 2) = 0$
$\qquad\qquad y = 4, -2$
Substitute each of these values into the rewritten second equation.
$y = \quad 4$: $x = 4 + 4 = 8$
$y = -2$: $x = -2 + 4 = 2$
Solutions: $(8, 4), (2, -2)$

4. $2x - 3y = 0$ $\quad\Rightarrow\quad$ $y = \frac{2}{3}x$
$2x + 3y = 12$ $\qquad\qquad y = -\frac{2}{3}x + 4$

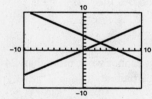

Solution: $(3, 2)$

5. $y = 9 - x^2$
$y = x + 3$

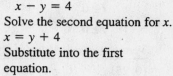

Solutions: $(-3, 0), (2, 5)$

6. $y = \log_3 x$
$y = -\frac{1}{3}x + 2$

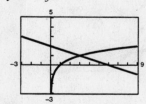

Solution: $(3, 1)$

7. $\quad 2x + \ 3y = \quad 17$
$\quad 5x - \ 4y = -15$

$\quad 8x + 12y = \quad 68$
$\underline{\quad 15x - 12y = -45}$
$\quad 23x \qquad\quad = \quad 23$
$\qquad\qquad x = \quad 1$
$2(1) + 3y = \quad 17$
$\qquad\quad 3y = \quad 15$
$\qquad\qquad y = \quad 5$
Solution: $(1, 5)$

8. $\ x - 2y + 3z = \quad 11$
$\ 2x \qquad - \ z = \quad 3$
$\qquad\quad 3y + \ z = -8$
Add -2 times the first equation to the second equation.
$x - 2y + 3z = \quad 11$
$\qquad 4y - 7z = -19$
$\qquad 3y + \ z = \ -8$
Add -3 times the second equation to 4 times the third equation.
$x - 2y + 3z = \quad 11$
$\qquad 4y - 7z = -19$
$\qquad\qquad 25z = \quad 25$
Solve for each variable.
$z = 1 \qquad 4y - 7(1) = -19$
$\qquad\qquad\qquad\qquad y = -3$
$x - 2(-3) + 3(1) = 11$
$\qquad\qquad\qquad x = 2$
Solution: $(2, -3, 1)$

9. Using the equations: $ax + by = c$
$\qquad\qquad\qquad\qquad\quad dx + ey = f$
choose values for the coefficients a, b, d, and e. Substitute $\frac{4}{3}$ for x and -5 for y and solve for c and f. Use the values for a, b, c, d, e, and f for the system of linear equations. Answers will vary.

10. $(0, 6)$: $6 = a(0)^2 + b(0) + c$

$\qquad\qquad 6 = c$

$(-2, 2)$: $2 = a(-2)^2 + b(-2) + c$

$\qquad\qquad 2 = 4a - 2b + c$

$\left(3, \frac{9}{2}\right)$: $\frac{9}{2} = a(3)^2 + b(3) + c$

$\qquad\qquad 9 = 18a + 6b + 2c$

$4a - 2b +\ c = 2$

$18a + 6b + 2c = 9$

$\qquad\qquad c = 6$

Add 9 times the first equation to -2 times the second equation.

$4a - 2b +\ c = 2$

$\quad\ -30b + 5c = 0$

$\qquad\qquad c = 6$

Solve for b and a.

$-30b + 5(6) = 0 \qquad 4a - 2(1) + 6 = 2$

$\qquad\qquad b = 1 \qquad\qquad\qquad a = -\frac{1}{2}$

Parabola: $y = -\frac{1}{2}x^2 + x + 6$

13. The line connecting the points $(0, 15)$ and $(9, 12)$:

$$y - 15 = \frac{15 - 12}{0 - 9}(x - 0)$$

$$y = -\frac{1}{3}x + 15$$

The line connecting the points $(9, 12)$ and $(12, 5)$:

$$y - 12 = \frac{12 - 5}{9 - 12}(x - 9)$$

$$y = -\frac{7}{3}x + 33$$

The line connecting the points $(12, 5)$ and $(12, 0)$:

$x = 12$

Set of inequalities to describe the region:

$y \leq -\frac{1}{3}x + 15$

$y \leq -\frac{7}{3}x + 33$

$y \geq 0$

$x \geq 0$

$x \leq 12$

11. $2x + y \leq 4 \qquad\quad y \leq -2x + 4$

$2x - y \geq 0 \implies y \leq 2x$

$\quad\ x \geq 0 \qquad\qquad x \geq 0$

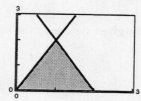

12. $y < -x^4 + x^2 + 4$

$\quad\ y > 4x$

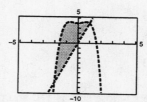

14. Begin by graphing the constraints.

$x \geq 0$

$y \geq 0$

$x + 4y \leq 3$

$3x + 2y \leq 36$

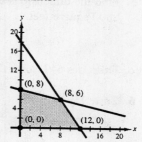

Vertex	Value of Objective Function
$(0, 0)$	$z = 20(0) + 12(0) = 0$
$(0, 8)$	$z = 20(0) + 12(8) = 96$
$(8, 6)$	$z = 20(8) + 12(6) = 232$
$(12, 0)$	$z = 20(12) + 12(0) = 240$

At $(12, 0)$ the maximum value of the objective function is 240.

15. Let x be the number of \$275 models. Let y be the number of \$400 models. The total monthly demand will not exceed 300 units: $x + y \leq 300$. The amount invested is no more than \$100,000 in inventory: $275x + 400y \leq 100,000$. x and y must be non-negative numbers: $x \geq 0$, $y \geq 0$. The objective function that will maximize profit is $z = 55x + 75y$.

Graph the constraints: $x + y \leq 300$

$\qquad\qquad\qquad\qquad 275x + 400y \leq 100,000$

$\qquad\qquad\qquad\qquad\qquad\quad x \geq 0$

$\qquad\qquad\qquad\qquad\qquad\quad y \geq 0$

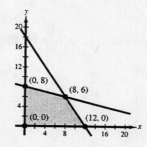

Vertex	Value of the Objective Function
$(0, 0)$	$z = 55(0) + 75(0) = 0$
$(0, 250)$	$z = 55(0) + 75(250) = 18,750$
$(160, 140)$	$z = 55(160) + 75(140) = 19,300$
$(300, 0)$	$z = 55(300) + 75(0) = 16,500$

A maximum profit is obtained when 160 units of the \$275 model and 140 units of the \$400 model are stocked.

❑ Practice Test for Chapter 7

For Exercises 1–3, solve the given system by the method of substitution.

1. $x + y = 1$
   $3x - y = 15$

2. $x - 3y = -3$
   $x^2 + 6y = 5$

3. $x + y + z = 6$
   $2x - y + 3z = 0$
   $5x + 2y - z = -3$

4. Find the two numbers whose sum is 110 and product is 2800.

5. Find the dimensions of a rectangle if its perimeter is 170 feet and its area is 2800 square feet.

For Exercises 6–7, solve the linear system by elimination.

6. $2x + 15y = 4$
   $x - 3y = 23$

7. $x + y = 2$
   $38x - 19y = 7$

8. Use a graphing utility to graph the two equations. Use the graph to approximate the solution of the system. Verify your answer analytically.

   $0.4x + 0.5y = 0.112$
   $0.3x - 0.7y = -0.131$

9. Herbert invests $17,000 in two funds that pay 11% and 13% simple interest, respectively. If he receives $2080 in yearly interest, how much is invested in each fund?

10. Find the least squares regression line for the points $(4, 3)$, $(1, 1)$, $(-1, -2)$, and $(-2, -1)$.

For Exercises 11–13, solve the system of equations.

11. $\begin{aligned} x + y &= -2 \\ 2x - y + z &= 11 \\ 4y - 3z &= -20 \end{aligned}$

12. $\begin{aligned} 4x - y + 5z &= 4 \\ 2x + y - z &= 0 \\ 2x + 4y + 8z &= 0 \end{aligned}$

13. $\begin{aligned} 3x + 2y - z &= 5 \\ 6x - y + 5z &= 2 \end{aligned}$

14. Find the equation of the parabola $y = ax^2 + bx + c$ passing through the points $(0, -1)$, $(1, 4)$ and $(2, 13)$.

15. Find the position equation $s = \frac{1}{2}at^2 + v_0 t + s_0$ given that $s = 12$ feet after 1 second, $s = 5$ feet after 2 seconds, and $s = 4$ after 3 seconds.

16. Graph $x^2 + y^2 \geq 9$.

17. Graph the solution of the system.

$$\begin{aligned} x + y &\leq 6 \\ x &\geq 2 \\ y &\geq 0 \end{aligned}$$

18. Derive a set of inequalities to describe the triangle with vertices $(0, 0)$, $(0, 7)$, and $(2, 3)$.

19. Find the maximum value of the objective function, $z = 30z + 26y$, subject to the following constraints.

$$\begin{aligned} x &\geq 0 \\ y &\geq 0 \\ 2x + 3y &\leq 21 \\ 5x + 3y &\leq 30 \end{aligned}$$

20. Graph the system of inequalities.

$$\begin{aligned} x^2 + y^2 &\leq 4 \\ (x - 2)^2 + y^2 &\geq 4 \end{aligned}$$

C H A P T E R   8
Matrices and Determinants

CHAPTER 8
Matrices and Determinants

Section 8.1   Matrices and Systems of Equations

■ You should be able to use elementary row operations to produce a row-echelon form (or reduced row-echelon form) of a matrix.

   1. Interchange two rows

   2. Multiply a row by a nonzero constant

   3. Add a multiple of one row to another row

■ You should be able to use either Gaussian elimination with back-substitution or Gauss-Jordan elimination to solve a system of linear equations.

Solutions to Odd-Numbered Exercises

1. Since the matrix has three rows and two columns, its order is 3×2.

3. Since the matrix has three rows and one column, its order is 3×1.

5. Since the matrix has two rows and two columns, its order is 2×2.

7. $4x - 3y = -5$
$-x + 3y = 12$

$$\begin{bmatrix} 4 & -3 & \vdots & -5 \\ -1 & 3 & \vdots & 12 \end{bmatrix}$$

9. $x + 10y - 2z = 2$
$5x - 3y + 4z = 0$
$2x + y = 6$

$$\begin{bmatrix} 1 & 10 & -2 & \vdots & 2 \\ 5 & -3 & 4 & \vdots & 0 \\ 2 & 1 & 0 & \vdots & 6 \end{bmatrix}$$

11. $$\begin{bmatrix} 1 & 2 & \vdots & 7 \\ 2 & -3 & \vdots & 4 \end{bmatrix}$$

$x + 2y = 7$
$2x - 3y = 4$

13. $$\begin{bmatrix} 2 & 0 & 5 & \vdots & -12 \\ 0 & 1 & -2 & \vdots & 7 \\ 6 & 3 & 0 & \vdots & 2 \end{bmatrix}$$

$2x + 5z = -12$
$y - 2z = 7$
$6x + 3y = 2$

15. $$\begin{bmatrix} 1 & 0 & 0 & 0 \\ 0 & 1 & 1 & 5 \\ 0 & 0 & 0 & 0 \end{bmatrix}$$

This matrix is in reduced row-echelon form.

17. $$\begin{bmatrix} 2 & 0 & 4 & 0 \\ 0 & -1 & 3 & 6 \\ 0 & 0 & 1 & 5 \end{bmatrix}$$

The first nonzero entries in rows one and two are not one. The matrix is not in row-echelon form.

19. $$\begin{bmatrix} 1 & 4 & 3 \\ 2 & 10 & 5 \end{bmatrix}$$

$$-2R_1 + R_2 \rightarrow \begin{bmatrix} 1 & 4 & 3 \\ 0 & \boxed{2} & -1 \end{bmatrix}$$

21.
$$\begin{bmatrix} 1 & 1 & 4 & -1 \\ 3 & 8 & 10 & 3 \\ -2 & 1 & 12 & 6 \end{bmatrix}$$

$$\begin{array}{c} -3R_1 + R_2 \rightarrow \\ 2R_1 + R_3 \rightarrow \end{array} \begin{bmatrix} 1 & 1 & 4 & -1 \\ 0 & 5 & \boxed{-2} & \boxed{6} \\ 0 & 3 & \boxed{20} & \boxed{4} \end{bmatrix}$$

$$\tfrac{1}{5}R_2 \rightarrow \begin{bmatrix} 1 & 1 & 4 & -1 \\ 0 & 1 & \boxed{-\tfrac{2}{5}} & \boxed{\tfrac{6}{5}} \\ 0 & 3 & \boxed{20} & \boxed{4} \end{bmatrix}$$

23. $\begin{bmatrix} 1 & 2 & 3 \\ 2 & -1 & -4 \\ 3 & 1 & -1 \end{bmatrix}$

(a) $\begin{bmatrix} 1 & 2 & 3 \\ 0 & -5 & -10 \\ 3 & 1 & -1 \end{bmatrix}$ (b) $\begin{bmatrix} 1 & 2 & 3 \\ 0 & -5 & -10 \\ 0 & -5 & -10 \end{bmatrix}$ (c) $\begin{bmatrix} 1 & 2 & 3 \\ 0 & -5 & -10 \\ 0 & 0 & 0 \end{bmatrix}$

(d) $\begin{bmatrix} 1 & 2 & 3 \\ 0 & 1 & 2 \\ 0 & 0 & 0 \end{bmatrix}$ (e) $\begin{bmatrix} 1 & 0 & -1 \\ 0 & 1 & 2 \\ 0 & 0 & 0 \end{bmatrix}$ This matrix is in reduced row-echelon form.

25.
$$\begin{bmatrix} 1 & 1 & 0 & 5 \\ -2 & -1 & 2 & -10 \\ 3 & 6 & 7 & 14 \end{bmatrix}$$

$$\begin{array}{c} 2R_1 + R_2 \rightarrow \\ 3R_1 + R_3 \rightarrow \end{array} \begin{bmatrix} 1 & 1 & 0 & 5 \\ 0 & 1 & 2 & 0 \\ 0 & 3 & 7 & -1 \end{bmatrix}$$

$$-3R_2 + R_3 \rightarrow \begin{bmatrix} 1 & 1 & 0 & 5 \\ 0 & 1 & 2 & 0 \\ 0 & 0 & 1 & -1 \end{bmatrix}$$

27.
$$\begin{bmatrix} 1 & -1 & -1 & 1 \\ 5 & -4 & 1 & 8 \\ -6 & 8 & 18 & 0 \end{bmatrix}$$

$$\begin{array}{c} -5R_1 + R_2 \rightarrow \\ 6R_1 + R_3 \rightarrow \end{array} \begin{bmatrix} 1 & -1 & -1 & 1 \\ 0 & 1 & 6 & 3 \\ 0 & 2 & 12 & 6 \end{bmatrix}$$

$$-2R_2 + R_3 \rightarrow \begin{bmatrix} 1 & -1 & -1 & 1 \\ 0 & 1 & 6 & 3 \\ 0 & 0 & 0 & 0 \end{bmatrix}$$

29.

$$\begin{bmatrix} 3 & 3 & 3 \\ -1 & 0 & -4 \\ 2 & 4 & -2 \end{bmatrix}$$

$$\tfrac{1}{3}R_1 \rightarrow \begin{bmatrix} 1 & 1 & 1 \\ -1 & 0 & -4 \\ 2 & 4 & -2 \end{bmatrix}$$

$$\begin{matrix} -R_2 + R_1 \rightarrow \\ -2R_1 + R_3 \rightarrow \end{matrix} \begin{bmatrix} 1 & 1 & 1 \\ 0 & 1 & -3 \\ 0 & 2 & -4 \end{bmatrix}$$

$$\begin{matrix} -R_2 + R_1 \rightarrow \\ \\ -2R_2 + R_3 \rightarrow \end{matrix} \begin{bmatrix} 1 & 0 & 4 \\ 0 & 1 & -3 \\ 0 & 0 & 2 \end{bmatrix}$$

$$\tfrac{1}{2}R_3 \rightarrow \begin{bmatrix} 1 & 0 & 4 \\ 0 & 1 & -3 \\ 0 & 0 & 1 \end{bmatrix}$$

$$\begin{matrix} -4R_3 + R_1 \rightarrow \\ 3R_3 + R_2 \rightarrow \\ \end{matrix} \begin{bmatrix} 1 & 0 & 0 \\ 0 & 1 & 0 \\ 0 & 0 & 1 \end{bmatrix}$$

31.

$$\begin{bmatrix} 1 & 2 & 3 & -5 \\ 1 & 2 & 4 & -9 \\ -2 & -4 & -4 & 3 \\ 4 & 8 & 11 & -14 \end{bmatrix}$$

$$\begin{matrix} -R_1 + R_2 \rightarrow \\ 2R_1 + R_3 \rightarrow \\ -4R_1 + R_4 \rightarrow \end{matrix} \begin{bmatrix} 1 & 2 & 3 & -5 \\ 0 & 0 & 1 & -4 \\ 0 & 0 & 2 & -7 \\ 0 & 0 & -1 & 6 \end{bmatrix}$$

$$\begin{matrix} -3R_2 + R_1 \rightarrow \\ \\ -2R_2 + R_3 \rightarrow \\ R_2 + R_4 \rightarrow \end{matrix} \begin{bmatrix} 1 & 2 & 0 & 7 \\ 0 & 0 & 1 & -4 \\ 0 & 0 & 0 & 1 \\ 0 & 0 & 0 & 2 \end{bmatrix}$$

$$\begin{matrix} -7R_3 + R_1 \rightarrow \\ 4R_3 + R_2 \rightarrow \\ \\ -2R_3 + R_4 \rightarrow \end{matrix} \begin{bmatrix} 1 & 2 & 0 & 0 \\ 0 & 0 & 1 & 0 \\ 0 & 0 & 0 & 1 \\ 0 & 0 & 0 & 0 \end{bmatrix}$$

33.
$$\begin{aligned} x - 2y &= 4 \\ y &= -3 \end{aligned}$$

$$\begin{aligned} x - 2(-3) &= 4 \\ x &= -2 \end{aligned}$$

Answer: $(-2, -3)$

35.
$$\begin{aligned} x - y + 2z &= 4 \\ y - z &= 2 \\ z &= -2 \end{aligned}$$

$$\begin{aligned} y - (-2) &= 2 \\ y &= 0 \end{aligned}$$

$$\begin{aligned} x - 0 + 2(-2) &= 4 \\ x &= 8 \end{aligned}$$

Answer: $(8, 0, -2)$

37.
$$\begin{bmatrix} 1 & 0 & \vdots & 7 \\ 0 & 1 & \vdots & -5 \end{bmatrix}$$

$$\begin{aligned} x &= 7 \\ y &= -5 \end{aligned}$$

Answer: $(7, -5)$

39.
$$\begin{bmatrix} 1 & 0 & 0 & \vdots & -4 \\ 0 & 1 & 0 & \vdots & -8 \\ 0 & 0 & 1 & \vdots & 2 \end{bmatrix}$$

$$\begin{aligned} x &= -4 \\ y &= -8 \\ z &= 2 \end{aligned}$$

Answer: $(-4, -8, 2)$

41. $x + 2y = 7$
$2x + y = 8$

$$\begin{bmatrix} 1 & 2 & \vdots & 7 \\ 2 & 1 & \vdots & 8 \end{bmatrix}$$

$-2R_1 + R_2 \rightarrow \begin{bmatrix} 1 & 2 & \vdots & 7 \\ 0 & -3 & \vdots & -6 \end{bmatrix}$

$-\frac{1}{3}R_2 \rightarrow \begin{bmatrix} 1 & 2 & \vdots & 7 \\ 0 & 1 & \vdots & 2 \end{bmatrix}$

$y = 2$

$x + 2(2) = 7 \implies x = 3$

Answer: $(3, 2)$

43. $-3x + 5y = -22$
$3x + 4y = 4$
$4x - 8y = 32$

$$\begin{bmatrix} -3 & 5 & \vdots & -22 \\ 3 & 4 & \vdots & 4 \\ 4 & -8 & \vdots & 32 \end{bmatrix}$$

$R_3 + R_1 \rightarrow \begin{bmatrix} 1 & -3 & \vdots & 10 \\ 3 & 4 & \vdots & 4 \\ 4 & -8 & \vdots & 32 \end{bmatrix}$

$\begin{matrix} \\ -3R_1 + R_2 \rightarrow \\ -4R_1 + R_3 \rightarrow \end{matrix} \begin{bmatrix} 1 & -3 & \vdots & 10 \\ 0 & 13 & \vdots & -26 \\ 0 & 4 & \vdots & -8 \end{bmatrix}$

$\begin{matrix} \\ \frac{1}{13}R_2 \rightarrow \\ -4R_2 + R_3 \rightarrow \end{matrix} \begin{bmatrix} 1 & -3 & \vdots & 10 \\ 0 & 1 & \vdots & -2 \\ 0 & 0 & \vdots & 0 \end{bmatrix}$

$y = -2$

$x - 3(-2) = 10 \implies x = 4$

Answer: $(4, -2)$

45. $8x - 4y = 7$
$5x + 2y = 1$

$$\begin{bmatrix} 8 & -4 & \vdots & 7 \\ 5 & 2 & \vdots & 1 \end{bmatrix}$$

$\begin{matrix} 3R_1 \rightarrow \\ 5R_2 \rightarrow \end{matrix} \begin{bmatrix} 24 & -12 & \vdots & 21 \\ 25 & 10 & \vdots & 5 \end{bmatrix}$

$-R_2 + R_1 \rightarrow \begin{bmatrix} -1 & -22 & \vdots & 16 \\ 25 & 10 & \vdots & 5 \end{bmatrix}$

$25R_1 + R_2 \rightarrow \begin{bmatrix} -1 & -22 & \vdots & 16 \\ 0 & -540 & \vdots & 405 \end{bmatrix}$

$\begin{matrix} -R_1 \rightarrow \\ -\frac{1}{540}R_2 \rightarrow \end{matrix} \begin{bmatrix} 1 & 22 & \vdots & -16 \\ 0 & 1 & \vdots & -\frac{3}{4} \end{bmatrix}$

$y = -\frac{3}{4}$

$x + 22\left(-\frac{3}{4}\right) = -16 \implies x = \frac{1}{2}$

Answer: $\left(\frac{1}{2}, -\frac{3}{4}\right)$

47. $-x + 2y = 1.5$
$2x - 4y = 3.0$

$$\begin{bmatrix} -1 & 2 & \vdots & 1.5 \\ 2 & -4 & \vdots & 3.0 \end{bmatrix}$$

$2R_1 + R_2 \rightarrow \begin{bmatrix} -1 & 2 & \vdots & 1.5 \\ 0 & 0 & \vdots & 6.0 \end{bmatrix}$

The system is inconsistent and there is no solution.

49.
$$x \quad\quad - 3z = -2$$
$$3x + y - 2z = 5$$
$$2x + 2y + z = 4$$

$$\begin{bmatrix} 1 & 0 & -3 & \vdots & -2 \\ 3 & 1 & -2 & \vdots & 5 \\ 2 & 2 & 1 & \vdots & 4 \end{bmatrix}$$

$$\begin{matrix} \\ -3R_1 + R_2 \rightarrow \\ -2R_1 + R_3 \rightarrow \end{matrix} \begin{bmatrix} 1 & 0 & -3 & \vdots & -2 \\ 0 & 1 & 7 & \vdots & 11 \\ 0 & 2 & 7 & \vdots & 8 \end{bmatrix}$$

$$\begin{matrix} \\ \\ -2R_2 + R_3 \rightarrow \end{matrix} \begin{bmatrix} 1 & 0 & -3 & \vdots & -2 \\ 0 & 1 & 7 & \vdots & 11 \\ 0 & 0 & -7 & \vdots & -14 \end{bmatrix}$$

$$\begin{matrix} \\ \\ -\frac{1}{7}R_3 \rightarrow \end{matrix} \begin{bmatrix} 1 & 0 & -3 & \vdots & -2 \\ 0 & 1 & 7 & \vdots & 11 \\ 0 & 0 & 1 & \vdots & 2 \end{bmatrix}$$

$$z = 2$$
$$y + 7(2) = 11 \Rightarrow y = -3$$
$$x - 3(2) = -2 \Rightarrow x = 4$$

Answer: $(4, -3, 2)$

51.
$$x + y - 5z = 3$$
$$x \quad\quad - 2z = 1$$
$$2x - y - z = 0$$

$$\begin{bmatrix} 1 & 1 & -5 & \vdots & 3 \\ 1 & 0 & -2 & \vdots & 1 \\ 2 & -1 & -1 & \vdots & 0 \end{bmatrix}$$

$$\begin{matrix} \\ -R_1 + R_2 \rightarrow \\ -2R_1 + R_3 \rightarrow \end{matrix} \begin{bmatrix} 1 & 1 & -5 & \vdots & 3 \\ 0 & -1 & 3 & \vdots & -2 \\ 0 & -3 & 9 & \vdots & -6 \end{bmatrix}$$

$$\begin{matrix} \\ \\ -3R_2 + R_3 \rightarrow \end{matrix} \begin{bmatrix} 1 & 1 & -5 & \vdots & 3 \\ 0 & -1 & 3 & \vdots & -2 \\ 0 & 0 & 0 & \vdots & 0 \end{bmatrix}$$

$$\begin{matrix} R_2 + R_1 \rightarrow \\ -R_2 \rightarrow \\ \end{matrix} \begin{bmatrix} 1 & 0 & -2 & \vdots & 1 \\ 0 & 1 & -3 & \vdots & 2 \\ 0 & 0 & 0 & \vdots & 0 \end{bmatrix}$$

$$z = a$$
$$y - 3a = 2 \Rightarrow y = 3a + 2$$
$$x - 2a = 1 \Rightarrow x = 2a + 1$$

Answer: $(2a + 1, 3a + 2, a)$

53. $x + 2y + z = 8$

$3x + 7y + 6z = 26$

$$\begin{bmatrix} 1 & 2 & 1 & \vdots & 8 \\ 3 & 7 & 6 & \vdots & 26 \end{bmatrix}$$

$-3R_1 + R_2 \rightarrow \begin{bmatrix} 1 & 2 & 1 & \vdots & 8 \\ 0 & 1 & 3 & \vdots & 2 \end{bmatrix}$

$-2R_2 + R_1 \rightarrow \begin{bmatrix} 1 & 0 & -5 & \vdots & 4 \\ 0 & 1 & 3 & \vdots & 2 \end{bmatrix}$

$z = a$

$y + 3a = 2 \implies y = -3a + 2$

$x - 5a = 4 \implies x = 5a + 4$

Answer: $(5a + 4, -3a + 2, a)$

55. $x + 2y = 0$

$-x - y = 0$

$$\begin{bmatrix} 1 & 2 & \vdots & 0 \\ -1 & -1 & \vdots & 0 \end{bmatrix}$$

$R_1 + R_2 \rightarrow \begin{bmatrix} 1 & 2 & \vdots & 0 \\ 0 & 1 & \vdots & 0 \end{bmatrix}$

$y = 0$

$x + 2(0) = 0 \implies$

$x = 0$

Answer: $(0, 0)$

57. $3x + 3y + 12z = 6$

$x + y + 4z = 2$

$2x + 5y + 20z = 10$

$-x + 2y + 8z = 4$

$z = a$

$y = -4a + 2$

$x = 0$

Answer: $(0, -4a + 2, a)$

$$\begin{bmatrix} 3 & 3 & 12 & \vdots & 6 \\ 1 & 1 & 4 & \vdots & 2 \\ 2 & 5 & 20 & \vdots & 10 \\ -1 & 2 & 8 & \vdots & 4 \end{bmatrix} \implies \begin{bmatrix} 1 & 0 & 0 & \vdots & 0 \\ 0 & 0 & 0 & \vdots & 0 \\ 0 & 1 & 4 & \vdots & 2 \\ 0 & 0 & 0 & \vdots & 0 \end{bmatrix}$$

59.
$$2x + y - z + 2w = -6$$
$$3x + 4y \quad\;\; + w = 1$$
$$x + 5y + 2z + 6w = -3$$
$$5x + 2y - z - w = 3$$

$$\begin{bmatrix} 2 & 1 & -1 & 2 & \vdots & -6 \\ 3 & 4 & 0 & 1 & \vdots & 1 \\ 1 & 5 & 2 & 6 & \vdots & -3 \\ 5 & 2 & -1 & -1 & \vdots & 3 \end{bmatrix} \Rightarrow \begin{bmatrix} 1 & 5 & 2 & 6 & \vdots & -3 \\ 0 & 1 & -1 & -3 & \vdots & 2 \\ 0 & 0 & 238 & 629 & \vdots & -306 \\ 0 & 0 & 0 & -71 & \vdots & 142 \end{bmatrix}$$

$$x \approx 1$$
$$y \approx 0$$
$$z \approx 4$$
$$w \approx -2$$

Answer: $(1, 0, 4\; -2)$

61.
$$x + y + z = 0$$
$$2x + 3y + z = 0$$
$$3x + 5y + z = 0$$

$$\begin{bmatrix} 1 & 1 & 1 & \vdots & 0 \\ 2 & 3 & 1 & \vdots & 0 \\ 3 & 5 & 1 & \vdots & 0 \end{bmatrix} \Rightarrow \begin{bmatrix} 1 & 0 & 2 & \vdots & 0 \\ 0 & 1 & -1 & \vdots & 0 \\ 0 & 0 & 0 & \vdots & 0 \end{bmatrix}$$

$$z = a$$
$$y = a$$
$$x = -2a$$

Answer: $(-2a, a, a)$

63.
$$z = a$$
$$y = -4a + 1$$
$$x = -3a - 2$$

One possible solution is:

$$x + y + 7z = (-3a - 2) + (-4a + 1) + 7a = -1$$
$$x + 2y + 11z = (-3a - 2) + 2(-4a + 1) + 11a = 0$$
$$2x + y + 10z = 2(-3a - 2) + (-4a + 1) + 10a = -3$$

65. $x =$ amount at 8%

$y =$ amount at 9%

$z =$ amount at 12%

$$x + y + z = 1{,}500{,}000$$
$$0.08x + 0.09y + 0.12z = 133{,}000$$
$$x \quad\quad - 4z = 0$$

$$\begin{bmatrix} 1 & 1 & 1 & \vdots & 1{,}500{,}000 \\ 0.08 & 0.09 & 0.12 & \vdots & 133{,}000 \\ 1 & 0 & -4 & \vdots & 0 \end{bmatrix}$$

$$\begin{matrix} -0.08R_1 + R_2 \rightarrow \\ -R_1 + R_3 \rightarrow \end{matrix} \begin{bmatrix} 1 & 1 & 1 & \vdots & 1{,}500{,}000 \\ 0 & 0.01 & 0.04 & \vdots & 13{,}000 \\ 0 & -1 & -5 & \vdots & -1{,}500{,}000 \end{bmatrix}$$

$$\begin{matrix} 100R_2 \\ R_2 + R_3 \end{matrix} \begin{bmatrix} 1 & 1 & 1 & \vdots & 1{,}500{,}000 \\ 0 & 1 & 4 & \vdots & 1{,}300{,}000 \\ 0 & 0 & -1 & \vdots & -200{,}000 \end{bmatrix}$$

$$-z = -200{,}000 \implies z = 200{,}000$$
$$y + 4(200{,}000) = 1{,}300{,}000 \implies y = 500{,}000$$
$$x + (500{,}000) + (200{,}000) = 1{,}500{,}000 \implies x = 800{,}000$$

Answer: $800{,}000 at 8%, $500{,}000 at 9%, $200{,}000 at 12%

67. $\dfrac{4x^2}{(x+1)^2(x-1)} = \dfrac{A}{x-1} + \dfrac{B}{x+1} + \dfrac{C}{(x+1)^2}$

$4x^2 = A(x+1)^2 + B(x+1)(x-1) + C(x-1)$

Let $x = 1$: $4 = 4A \implies A = 1$

Let $x = -1$: $4 = -2C \implies C = -2$

Let $x = 0$: $0 = A - B - C \implies 0 = 1 - B - (-2)$

$\implies B = 3$

Thus, $\dfrac{4x^2}{(x+1)^2(x-1)} = \dfrac{1}{x-1} + \dfrac{3}{x+1} - \dfrac{2}{(x+1)^2}.$

69. $f(x) = ax^2 + bx + c$

$f(1) = a + b + c = 8$

$f(2) = 4a + 2b + c = 13$

$f(3) = 9a + 3b + c = 20$

$$\begin{bmatrix} 1 & 1 & 1 & \vdots & 8 \\ 4 & 2 & 1 & \vdots & 13 \\ 9 & 3 & 1 & \vdots & 20 \end{bmatrix}$$

$\begin{matrix} \\ -4R_1 + R_2 \rightarrow \\ -9R_1 + R_3 \rightarrow \end{matrix} \begin{bmatrix} 1 & 1 & 1 & \vdots & 8 \\ 0 & -2 & -3 & \vdots & -19 \\ 0 & -6 & -8 & \vdots & -52 \end{bmatrix}$

$\begin{matrix} \\ -\frac{1}{2}R_2 \rightarrow \\ -3R_2 + R_3 \rightarrow \end{matrix} \begin{bmatrix} 1 & 1 & 1 & \vdots & 8 \\ 0 & 1 & \frac{3}{2} & \vdots & \frac{19}{2} \\ 0 & 0 & 1 & \vdots & 5 \end{bmatrix}$

$c = 5$

$b + \frac{3}{2}(5) = \frac{19}{2} \implies b = 2$

$a + 2 + 5 = 8 \implies a = 1$

Answer: $y = x^2 + 2x + 5$

73. $f(x) = ax^4 + bx^3 + cx^2 + dx + e$

$f(-2) = 16a - 8b + 4c - 2d + e = 0$

$f(-1) = a - b + c - d + e = 3$

$f(0) = e = 0$

$f(1) = a + b + c + d + e = 3$

$f(2) = 16a + 8b + 4c + 2d + e = 0$

Solving this system, you obtain

$a = -1,\ b = 0,\ c = 4,\ d = e = 0.$

Thus, $y = -x^4 + 4x^2.$

71. $f(x) = ax^3 + bx^2 + cx + d$

$f(-2) = -8a + 4b - 2c + d = 2$

$f(-1) = -a + b - c + d = -\frac{1}{4}$

$f(1) = a + b + c + d = -\frac{7}{4}$

$f(2) = 8a + 4b + 2c + d = 2$

Solving this system, you obtain

$a = \frac{1}{4},\ b = 1,\ c = -1$ and $d = -2.$

Thus, $y = \frac{1}{4}x^3 + x^2 - x - 2.$

75. (a) The points are $(0, 670),\ (1, 280),\ (2, 231).$

If $y = at^2 + bt + c$, then:

$670 = c$

$280 = a + b + c$

$231 = 4a + 2b + c$

Solving this system, you obtain

$a = 170.5,\ b = -560.5$ and $c = 670.$

Thus, $y = 170.5t^2 - 560.5t + 670.$

(b)

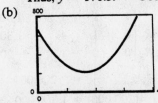

(c) For 1993, $t = 3$ and $y = 523.$

77. (a) $x_1 + x_3 = 600$

$\qquad x_1 = x_2 + x_4 \implies x_1 - x_2 - x_4 = 0$

$\qquad x_2 + x_5 = 500$

$\qquad x_3 + x_6 = 600$

$\qquad x_4 + x_7 = x_6 \implies x_4 - x_6 + x_7 = 0$

$\qquad x_5 + x_7 = 500$

$$\begin{bmatrix} 1 & 0 & 1 & 0 & 0 & 0 & 0 & \vdots & 600 \\ 1 & -1 & 0 & -1 & 0 & 0 & 0 & \vdots & 0 \\ 0 & 1 & 0 & 0 & 1 & 0 & 0 & \vdots & 500 \\ 0 & 0 & 1 & 0 & 0 & 1 & 0 & \vdots & 600 \\ 0 & 0 & 0 & 1 & 0 & -1 & 1 & \vdots & 0 \\ 0 & 0 & 0 & 0 & 1 & 0 & 1 & \vdots & 500 \end{bmatrix}$$

$$\begin{matrix} \\ -R_1 + R_2 \to \\ R_2 + R_3 \to \\ R_3 + R_4 \to \\ R_4 + R_5 \to \\ -R_5 + R_6 \to \end{matrix} \begin{bmatrix} 1 & 0 & 1 & 0 & 0 & 0 & 0 & \vdots & 600 \\ 0 & -1 & -1 & -1 & 0 & 0 & 0 & \vdots & -600 \\ 0 & 0 & -1 & -1 & 1 & 0 & 0 & \vdots & -100 \\ 0 & 0 & 0 & -1 & 1 & 1 & 0 & \vdots & 500 \\ 0 & 0 & 0 & 0 & 1 & 0 & 1 & \vdots & 500 \\ 0 & 0 & 0 & 0 & 0 & 0 & 0 & \vdots & 0 \end{bmatrix}$$

$$\begin{matrix} \\ -R_3 + R_2 \to \\ -R_4 + R_3 \to \\ -R_4 \to \\ \\ \\ \end{matrix} \begin{bmatrix} 1 & 0 & 1 & 0 & 0 & 0 & 0 & \vdots & 600 \\ 0 & -1 & 0 & 0 & -1 & 0 & 0 & \vdots & -500 \\ 0 & 0 & -1 & 0 & 0 & -1 & 0 & \vdots & -600 \\ 0 & 0 & 0 & 1 & -1 & -1 & 0 & \vdots & -500 \\ 0 & 0 & 0 & 0 & 1 & 0 & 1 & \vdots & 500 \\ 0 & 0 & 0 & 0 & 0 & 0 & 0 & \vdots & 0 \end{bmatrix}$$

Let $x_7 = t$ and $x_6 = s$, then:

$\qquad x_5 = 500 - t$

$\qquad x_4 = -500 + s + (500 - t) = s - t$

$\qquad x_3 = 600 - s$

$\qquad x_2 = 500 - (500 - t) = t$

$\qquad x_1 = 600 - (600 - s) = s$

(b) If $x_6 = x_7 = 0$, then $s = t = 0$, and

$\qquad x_1 = 0$

$\qquad x_2 = 0$

$\qquad x_3 = 600$

$\qquad x_4 = 0$

$\qquad x_5 = 500$

$\qquad x_6 = x_7 = 0$

(c) If $x_5 = 1000$ and $x_6 = 0$, then $s = 0$ and $t = -500$.

Thus, $x_1 = 0$

$\qquad x_2 = -500$

$\qquad x_3 = 600$

$\qquad x_4 = 500$

$\qquad x_5 = 1000$

$\qquad x_6 = 0$

$\qquad x_7 = -500$

79. (a) $200 + x_2 = x_1 \implies -x_1 - x_2 = 200$

$\qquad x_2 + 100 = x_4 \implies -x_2 + x_4 = 100$

$\qquad x_4 + 200 = x_3 \implies x_3 - x_4 = 200$

$\qquad x_1 + 100 = x_3 \implies x_1 - x_3 = -100$

Solving this system, you obtain:

$\qquad x_1 = 100 + t$

$\qquad x_2 = -100 + t$

$\qquad x_3 = 200 + t$

$\qquad x_4 = t$

(b) If $x_4 = 0$, then $t = 0$ and

$\qquad x_1 = 100, x_2 = -100, x_3 = 200.$

(c) f $x_4 = 100$, then $t = 100$ and $x_1 = 200, x_2 = 0,$

$\qquad x_3 = 300.$

81. For the year 2002, $t = 22$ and

$$s = 1.279 - 0.0049(20) = 1.1712 \text{ minutes (Men)}$$

$$s = 1.411 - 0.0078(20) = 1.2394 \text{ minutes (Women)}.$$

83. $f(x) = 2^{x-1}$

x	-1	0	1	2	3
y	$\frac{1}{4}$	$\frac{1}{2}$	1	2	4

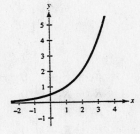

85. $h(x) = \log_2(x - 1) \implies 2^y = x - 1 \implies 2^y + 1 = x$

x	$\frac{3}{2}$	2	3	5	9
y	-1	0	1	2	3

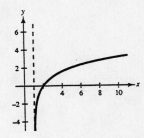

Section 8.2   Operations with Matrices

- $A = B$ if and only if they have the same order and $a_{ij} = b_{ij}$.
- You should be able to perform the operations of matrix addition, scalar multiplication, and matrix multiplication.
- Some properties of matrix addition and scalar multiplication are:
  (a) $A + B = B + A$
  (b) $A + (B + C) = (A + B) + C$
  (c) $(cd)A = c(dA)$
  (d) $1A = A$
  (e) $c(A + B) = cA + cB$
  (f) $(c + d)A = cA + dA$
- Some properties of matrix multiplication are:
  (a) $A(BC) = (AB)C$
  (b) $A(B + C) = AB + AC$
  (c) $(A + B)C = AC + BC$
  (d) $c(AB) = (cA)B = A(cB)$
- You should remember that $AB \neq BA$ in general.

Solutions to Odd-Numbered Exercises

1. $x = -4$, $y = 22$

3. $2x + 1 = 5$, $3y - 5 = 4$
$x = 2$, $y = 3$

5. (a) $A + B = \begin{bmatrix} 1 & -1 \\ 2 & -1 \end{bmatrix} + \begin{bmatrix} 2 & -1 \\ -1 & 8 \end{bmatrix} = \begin{bmatrix} 1+2 & -1-1 \\ 2-1 & -1+8 \end{bmatrix} = \begin{bmatrix} 3 & -2 \\ 1 & 7 \end{bmatrix}$

(b) $A - B = \begin{bmatrix} 1 & -1 \\ 2 & -1 \end{bmatrix} - \begin{bmatrix} 2 & -1 \\ -1 & 8 \end{bmatrix} = \begin{bmatrix} 1-2 & -1+1 \\ 2+1 & -1-8 \end{bmatrix} = \begin{bmatrix} -1 & 0 \\ 3 & -9 \end{bmatrix}$

(c) $3A = 3\begin{bmatrix} 1 & -1 \\ 2 & -1 \end{bmatrix} = \begin{bmatrix} 3(1) & 3(-1) \\ 3(2) & 3(-1) \end{bmatrix} = \begin{bmatrix} 3 & -3 \\ 6 & -3 \end{bmatrix}$

(d) $3A - 2B = \begin{bmatrix} 3 & -3 \\ 6 & -3 \end{bmatrix} - 2\begin{bmatrix} 2 & -1 \\ -1 & 8 \end{bmatrix} = \begin{bmatrix} 3 & -3 \\ 6 & -3 \end{bmatrix} + \begin{bmatrix} -4 & 2 \\ 2 & -16 \end{bmatrix} = \begin{bmatrix} -1 & -1 \\ 8 & -19 \end{bmatrix}$

7. $A = \begin{bmatrix} 6 & -1 \\ 2 & 4 \\ -3 & 5 \end{bmatrix}$, $B = \begin{bmatrix} 1 & 4 \\ -1 & 5 \\ 1 & 10 \end{bmatrix}$

(a) $A + B = \begin{bmatrix} 7 & 3 \\ 1 & 9 \\ -2 & 15 \end{bmatrix}$ (b) $A - B = \begin{bmatrix} 5 & -5 \\ 3 & -1 \\ -4 & -5 \end{bmatrix}$ (c) $3A = \begin{bmatrix} 18 & -3 \\ 6 & 12 \\ -9 & 15 \end{bmatrix}$

(d) $3A - 2B = \begin{bmatrix} 18 & -3 \\ 6 & 12 \\ -9 & 15 \end{bmatrix} - \begin{bmatrix} 2 & 8 \\ -2 & 10 \\ 2 & 20 \end{bmatrix} = \begin{bmatrix} 16 & -11 \\ 8 & 2 \\ -11 & -5 \end{bmatrix}$

9. $A = \begin{bmatrix} 2 & 2 & -1 & 0 & 1 \\ 1 & 1 & -2 & 0 & -1 \end{bmatrix}$, $B = \begin{bmatrix} 1 & 1 & -1 & 1 & 0 \\ -3 & 4 & 9 & -6 & -7 \end{bmatrix}$

(a) $A + B = \begin{bmatrix} 3 & 3 & -2 & 1 & 1 \\ -2 & 5 & 7 & -6 & -8 \end{bmatrix}$

(b) $A - B = \begin{bmatrix} 1 & 1 & 0 & -1 & 1 \\ 4 & -3 & -11 & 6 & 6 \end{bmatrix}$

(c) $3A = \begin{bmatrix} 6 & 6 & -3 & 0 & 3 \\ 3 & 3 & -6 & 0 & -3 \end{bmatrix}$

(d) $3A - 2B = \begin{bmatrix} 6 & 6 & -3 & 0 & 3 \\ 3 & 3 & -6 & 0 & -3 \end{bmatrix} - \begin{bmatrix} 2 & 2 & -2 & 2 & 0 \\ -6 & 8 & 18 & -12 & -14 \end{bmatrix} = \begin{bmatrix} 4 & 4 & -1 & -2 & 3 \\ 9 & -5 & -24 & 12 & 11 \end{bmatrix}$

11. $X = 3\begin{bmatrix} -2 & -1 \\ 1 & 0 \\ 3 & 4 \end{bmatrix} - 2\begin{bmatrix} 0 & 3 \\ 2 & 0 \\ -4 & -1 \end{bmatrix} = \begin{bmatrix} -6 & -3 \\ 3 & 0 \\ 9 & -12 \end{bmatrix} - \begin{bmatrix} 0 & 6 \\ 4 & 0 \\ -8 & -2 \end{bmatrix} = \begin{bmatrix} -6 & -9 \\ -1 & 0 \\ 17 & -10 \end{bmatrix}$

13. $X = -\frac{3}{2}A + \frac{1}{2}B = -\frac{3}{2}\begin{bmatrix} -2 & -1 \\ 1 & 0 \\ 3 & -4 \end{bmatrix} + \frac{1}{2}\begin{bmatrix} 0 & 3 \\ 2 & 0 \\ -4 & -1 \end{bmatrix} = \begin{bmatrix} 3 & 3 \\ -\frac{1}{2} & 0 \\ -\frac{13}{2} & \frac{11}{2} \end{bmatrix}$

15. (a) $AB = \begin{bmatrix} 1 & 2 \\ 4 & 2 \end{bmatrix}\begin{bmatrix} 2 & -1 \\ -1 & 8 \end{bmatrix} = \begin{bmatrix} 2-2 & -1+16 \\ 8-2 & -4+16 \end{bmatrix} = \begin{bmatrix} 0 & 15 \\ 6 & 12 \end{bmatrix}$

(b) $BA = \begin{bmatrix} 2 & -1 \\ -1 & 8 \end{bmatrix}\begin{bmatrix} 1 & 2 \\ 4 & 2 \end{bmatrix} = \begin{bmatrix} 2-4 & 4-2 \\ -1+32 & -2+16 \end{bmatrix} = \begin{bmatrix} -2 & 2 \\ 31 & 14 \end{bmatrix}$

(c) $A^2 = \begin{bmatrix} 1 & 2 \\ 4 & 2 \end{bmatrix}\begin{bmatrix} 1 & 2 \\ 4 & 2 \end{bmatrix} = \begin{bmatrix} 1+8 & 2+4 \\ 4+8 & 8+4 \end{bmatrix} = \begin{bmatrix} 9 & 6 \\ 12 & 12 \end{bmatrix}$

17. (a) $AB = \begin{bmatrix} 3 & -1 \\ 1 & 3 \end{bmatrix}\begin{bmatrix} 1 & -3 \\ 3 & 1 \end{bmatrix} = \begin{bmatrix} 3-3 & -9-1 \\ 1+9 & -3+3 \end{bmatrix} = \begin{bmatrix} 0 & -10 \\ 10 & 0 \end{bmatrix}$

(b) $BA = \begin{bmatrix} 1 & -3 \\ 3 & 1 \end{bmatrix}\begin{bmatrix} 3 & -1 \\ 1 & 3 \end{bmatrix} = \begin{bmatrix} 3-3 & -1-9 \\ 9+1 & -3+3 \end{bmatrix} = \begin{bmatrix} 0 & -10 \\ 10 & 0 \end{bmatrix}$

(c) $A^2 = \begin{bmatrix} 3 & -1 \\ 1 & 3 \end{bmatrix}\begin{bmatrix} 3 & -1 \\ 1 & 3 \end{bmatrix} = \begin{bmatrix} 9-1 & -3-3 \\ 3=3 & -1+9 \end{bmatrix} = \begin{bmatrix} 8 & -6 \\ 6 & 8 \end{bmatrix}$

19. (a) $AB = \begin{bmatrix} 1 & -1 & 7 \\ 2 & -1 & 8 \\ 3 & 1 & -1 \end{bmatrix}\begin{bmatrix} 1 & 1 & 2 \\ 2 & 1 & 1 \\ 1 & -3 & 2 \end{bmatrix} = \begin{bmatrix} 1-2+7 & 1-1-21 & 2-1+14 \\ 2-2+8 & 2-1-24 & 4-1+16 \\ 3+2-1 & 3+1+3 & 6+1-2 \end{bmatrix} = \begin{bmatrix} 6 & -21 & 15 \\ 8 & -23 & 19 \\ 4 & 7 & 5 \end{bmatrix}$

(b) $BA = \begin{bmatrix} 1 & 1 & 2 \\ 2 & 1 & 1 \\ 1 & -3 & 2 \end{bmatrix}\begin{bmatrix} 1 & -1 & 7 \\ 2 & -1 & 8 \\ 3 & 1 & -1 \end{bmatrix} = \begin{bmatrix} 1+2+6 & -1-1+2 & 7+8-2 \\ 2+2+3 & -2-1+1 & 14+8-1 \\ 1-6+6 & -1+3+2 & 7-24-2 \end{bmatrix} = \begin{bmatrix} 9 & 0 & 13 \\ 7 & -2 & 21 \\ 1 & 4 & -19 \end{bmatrix}$

(c) $A^2 = \begin{bmatrix} 1 & -1 & 7 \\ 2 & -1 & 8 \\ 3 & 1 & -1 \end{bmatrix}\begin{bmatrix} 1 & -1 & 7 \\ 2 & -1 & 8 \\ 3 & 1 & -1 \end{bmatrix} = \begin{bmatrix} 1-2+21 & -1+1+7 & 7-8-7 \\ 2-2+24 & -2+1+8 & 14-8-8 \\ 3+2-3 & -3-1-1 & 21+8+1 \end{bmatrix} = \begin{bmatrix} 20 & 7 & -8 \\ 24 & 7 & -2 \\ 2 & -5 & 30 \end{bmatrix}$

21. A is 3×2 and B is 3×3 $\implies$ AB is not defined.

23. A is 3×2, B is 2×2 $\implies$ AB is 3×2.

$AB = \begin{bmatrix} -1 & 3 \\ 4 & -5 \\ 0 & 2 \end{bmatrix}\begin{bmatrix} 1 & 2 \\ 0 & 7 \end{bmatrix} = \begin{bmatrix} -1 & 19 \\ 4 & -27 \\ 0 & 14 \end{bmatrix}$

25. A is 3×3, B is 3×3 $\implies$ AB is 3×3.

$AB = \begin{bmatrix} 5 & 0 & 0 \\ 0 & -8 & 0 \\ 0 & 0 & 7 \end{bmatrix}\begin{bmatrix} \frac{1}{5} & 0 & 0 \\ 0 & -\frac{1}{8} & 0 \\ 0 & 0 & \frac{1}{2} \end{bmatrix} = \begin{bmatrix} 1 & 0 & 0 \\ 0 & 1 & 0 \\ 0 & 0 & \frac{7}{2} \end{bmatrix}$

27. $\begin{bmatrix} 0 & 0 & 5 \\ 0 & 0 & -3 \\ 0 & 0 & 4 \end{bmatrix}\begin{bmatrix} 6 & -11 & 4 \\ 8 & 16 & 4 \\ 0 & 0 & 0 \end{bmatrix} = \begin{bmatrix} 0 & 0 & 0 \\ 0 & 0 & 0 \\ 0 & 0 & 0 \end{bmatrix}$

29. $\begin{bmatrix} 5 & 6 & -3 \\ -2 & 5 & 1 \\ 10 & -5 & 5 \end{bmatrix}\begin{bmatrix} 1 & -1 & 2 \\ 8 & 1 & 4 \\ 4 & -2 & 9 \end{bmatrix} = \begin{bmatrix} 41 & 7 & 7 \\ 42 & 5 & 25 \\ -10 & -25 & 45 \end{bmatrix}$

31. $\begin{bmatrix} -3 & 8 & -6 & 8 \\ -12 & 15 & 9 & 6 \\ 5 & -1 & 1 & 5 \end{bmatrix} \begin{bmatrix} 3 & 1 & 6 \\ 24 & 15 & 14 \\ 16 & 10 & 21 \\ 8 & -4 & 10 \end{bmatrix} = \begin{bmatrix} 151 & 25 & 48 \\ 516 & 279 & 387 \\ 47 & -20 & 87 \end{bmatrix}$

33. A is 2×4 and B is 2×4 $\implies$ AB is not defined.

35. $A = \begin{bmatrix} -1 & 1 \\ -2 & 1 \end{bmatrix}$, $X = \begin{bmatrix} x \\ y \end{bmatrix}$, $B = \begin{bmatrix} 4 \\ 0 \end{bmatrix}$

By Gauss-Jordan elimination on

$$\begin{bmatrix} -1 & 1 & \vdots & 4 \\ -2 & 1 & \vdots & 0 \end{bmatrix}$$

$$\begin{matrix} -R_1 \to \\ 2R_1 + R_2 \to \end{matrix} \begin{bmatrix} 1 & -1 & \vdots & -4 \\ 0 & -1 & \vdots & -8 \end{bmatrix}$$

$$\begin{matrix} R_2 + R_1 \to \\ -R_2 \to \end{matrix} \begin{bmatrix} 1 & 0 & \vdots & 4 \\ 0 & 1 & \vdots & 8 \end{bmatrix},$$

we have $x = 4$ and $y = 8$.

37. $A = \begin{bmatrix} 2 & 3 \\ 1 & 4 \end{bmatrix}$, $X = \begin{bmatrix} x \\ y \end{bmatrix}$, $B = \begin{bmatrix} 5 \\ 10 \end{bmatrix}$

By Gauss-Jordan elimination on

$$\begin{bmatrix} 1 & 4 & \vdots & 10 \\ 2 & 3 & \vdots & 5 \end{bmatrix}$$

$$-2R_1 + R_2 \to \begin{bmatrix} 1 & 4 & \vdots & 10 \\ 0 & -5 & \vdots & -15 \end{bmatrix}$$

$$\begin{matrix} -4R_2 + R_1 \to \\ -\frac{1}{5}R_2 \to \end{matrix} \begin{bmatrix} 1 & 0 & \vdots & -2 \\ 0 & 1 & \vdots & 3 \end{bmatrix},$$

we have $x = -2$ and $y = 3$.

39. $A = \begin{bmatrix} 2 & 0 \\ 4 & 5 \end{bmatrix}$

$f(A) = A^2 - 5A + 2 = \begin{bmatrix} 2 & 0 \\ 4 & 5 \end{bmatrix}\begin{bmatrix} 2 & 0 \\ 4 & 5 \end{bmatrix} - 5\begin{bmatrix} 2 & 4 \\ 0 & 5 \end{bmatrix} + 2\begin{bmatrix} 1 & 0 \\ 0 & 1 \end{bmatrix} = \begin{bmatrix} -4 & 0 \\ 8 & 2 \end{bmatrix}$

41. $A = \begin{bmatrix} 3 & 1 & 4 \\ 0 & 2 & 6 \\ 0 & 0 & 5 \end{bmatrix}$

$f(A) = \begin{bmatrix} 3 & 1 & 4 \\ 0 & 2 & 6 \\ 0 & 0 & 5 \end{bmatrix}^3 - 10\begin{bmatrix} 3 & 1 & 4 \\ 0 & 2 & 6 \\ 0 & 0 & 5 \end{bmatrix}^2 + 31\begin{bmatrix} 3 & 1 & 4 \\ 0 & 2 & 6 \\ 0 & 0 & 5 \end{bmatrix} - 30\begin{bmatrix} 1 & 0 & 0 \\ 0 & 1 & 0 \\ 0 & 0 & 1 \end{bmatrix} = \begin{bmatrix} 0 & 0 & 0 \\ 0 & 0 & 0 \\ 0 & 0 & 0 \end{bmatrix}$

43. $AC = \begin{bmatrix} 0 & 1 \\ 0 & 1 \end{bmatrix} \begin{bmatrix} 2 & 3 \\ 2 & 3 \end{bmatrix} = \begin{bmatrix} 2 & 3 \\ 2 & 3 \end{bmatrix}$

$BC = \begin{bmatrix} 1 & 0 \\ 1 & 0 \end{bmatrix} \begin{bmatrix} 2 & 3 \\ 2 & 3 \end{bmatrix} = \begin{bmatrix} 2 & 3 \\ 2 & 3 \end{bmatrix}$

$AC = BC$, but $A \neq B$.

For 45–53, A is of order 2×3, B is of order 2×3, C is of order 3×2 and D is of order 2×2.

45. $A + 2C$ is not possible. A and C are not of the same order.

47. AB is not possible. The number of columns of A does not equal the number of rows of B.

49. $BC - D$ is possible. The resulting order is 2×2.

51. (CA) is 3×3 so $(CA)D$ is not possible.

53. $D(A - 3B)$ is possible. The resulting order is 2×3.

55. $1.20 \begin{bmatrix} 60 & 40 & 20 \\ 30 & 90 & 60 \end{bmatrix} = \begin{bmatrix} 72 & 48 & 24 \\ 36 & 108 & 72 \end{bmatrix}$

57. $BA = \begin{bmatrix} 3.75 & 7.00 \end{bmatrix} \begin{bmatrix} 100 & 75 & 75 \\ 125 & 150 & 100 \end{bmatrix} = \begin{bmatrix} \$1250.00 & \$1331.25 & \$981.25 \end{bmatrix}$

The entries in the last matrix represent the profit for both crops at each of the three outlets.

59. $A^2 = \begin{bmatrix} i & 0 \\ 0 & i \end{bmatrix} \begin{bmatrix} i & 0 \\ 0 & i \end{bmatrix} = \begin{bmatrix} -1 & 0 \\ 0 & -1 \end{bmatrix}$ and $i^2 = -1$

$A^3 = A^2 A = \begin{bmatrix} -1 & 0 \\ 0 & -1 \end{bmatrix} \begin{bmatrix} i & 0 \\ 0 & i \end{bmatrix} = \begin{bmatrix} -i & 0 \\ .0 & -i \end{bmatrix}$ and $i^3 = -i$

$A^4 = A^3 A = \begin{bmatrix} -i & 0 \\ 0 & -i \end{bmatrix} \begin{bmatrix} i & 0 \\ 0 & i \end{bmatrix} = \begin{bmatrix} 1 & 0 \\ 0 & 1 \end{bmatrix}$ and $i^4 = 1$

61. $ST = \begin{bmatrix} 3 & 2 & 2 & 3 & 0 \\ 0 & 2 & 3 & 4 & 3 \\ 4 & 2 & 1 & 3 & 2 \end{bmatrix} \begin{bmatrix} 840 & 1100 \\ 1200 & 1350 \\ 1450 & 1650 \\ 2650 & 3000 \\ 3050 & 3200 \end{bmatrix} = \begin{bmatrix} \$15,770 & \$18,300 \\ \$26,500 & \$29,250 \\ \$21,260 & \$24,150 \end{bmatrix}$

The entries represent the wholesale and retail prices of the inventory at each outlet.

63. $P^2 = \begin{bmatrix} 0.6 & 0.1 & 0.1 \\ 0.2 & 0.7 & 0.1 \\ 0.2 & 0.2 & 0.8 \end{bmatrix} \begin{bmatrix} 0.6 & 0.1 & 0.1 \\ 0.2 & 0.7 & 0.1 \\ 0.2 & 0.2 & 0.8 \end{bmatrix} = \begin{bmatrix} 0.40 & 0.15 & 0.15 \\ 0.28 & 0.53 & 0.17 \\ 0.32 & 0.32 & 0.68 \end{bmatrix}$

This product represents the changes in party affiliation after *two* elections.

65. The product of two diagonal matrices of the same order is a diagonal matrix whose entries are the products of the corresponding diagonal entries of A and B.

67. $\log_4 \sqrt{32} = \log_4 2^{5/2} = \log_4 4^{5/4} = \frac{5}{4} \log_4 4 = \frac{5}{4}$

Section 8.3   The Inverse of a Square Matrix

- ■ You should be able to find the inverse, if it exists, of a square matrix.

  (a) Write the $n \times 2n$ matrix that consists of the given matrix A on the left and the $n \times n$ identity matrix I on the right to obtain $[A \ \vdots \ I]$. Note that we separate the matrices A and I by a dotted line. We call this process **adjoining** the matrices A and I.

  (b) If possible, row reduce A to I using elementary row operations of the *entire* matrix $[A \ \vdots \ I]$. The result will be the matrix $[I \ \vdots \ A]$. If this is not possible, then A is not invertible.

  (c) Check your work by multiplying to see that $AA^{-1} = I = A^{-1}A$.

- ■ You should be able to use inverse matrices to solve systems of equation.

- ■ You should be able to find inverses using a graphing utility.

Solutions to Odd-Numbered Exercises

1. $AB = \begin{bmatrix} 2 & 1 \\ 5 & 3 \end{bmatrix} \begin{bmatrix} 3 & -1 \\ -5 & 2 \end{bmatrix} = \begin{bmatrix} 2(3) + 1(-5) & 2(-1) + 1(2) \\ 5(3) + 3(-5) & 5(-1) + 3(2) \end{bmatrix} = \begin{bmatrix} 1 & 0 \\ 0 & 1 \end{bmatrix}$

$BA = \begin{bmatrix} 3 & -1 \\ -5 & 2 \end{bmatrix} \begin{bmatrix} 2 & 1 \\ 5 & 3 \end{bmatrix} = \begin{bmatrix} 3(2) + (-1)(5) & 3(1) + (-1)(3) \\ -5(2) + 2(5) & -5(1) + 2(3) \end{bmatrix} = \begin{bmatrix} 1 & 0 \\ 0 & 1 \end{bmatrix}$

3. $AB = \begin{bmatrix} 1 & 2 \\ 3 & 4 \end{bmatrix} \begin{bmatrix} -2 & 1 \\ \frac{3}{2} & -\frac{1}{2} \end{bmatrix} = \begin{bmatrix} -2+3 & 1-1 \\ -6+6 & 3-2 \end{bmatrix} = \begin{bmatrix} 1 & 0 \\ 0 & 1 \end{bmatrix}$

$BA = \begin{bmatrix} -2 & 1 \\ \frac{3}{2} & -\frac{1}{2} \end{bmatrix} \begin{bmatrix} 1 & 2 \\ 3 & 4 \end{bmatrix} = \begin{bmatrix} -2+3 & -4+4 \\ \frac{3}{2} - \frac{3}{2} & 3-2 \end{bmatrix} = \begin{bmatrix} 1 & 0 \\ 0 & 1 \end{bmatrix}$

5. $AB = \frac{1}{3} \begin{bmatrix} -2 & 2 & 3 \\ 1 & -1 & 0 \\ 0 & 1 & 4 \end{bmatrix} \begin{bmatrix} -4 & -5 & 3 \\ -4 & -8 & 3 \\ 1 & 2 & 0 \end{bmatrix} = \frac{1}{3} \begin{bmatrix} -8+8+3 & 10-16+6 & -6+6 \\ -4+4 & -5+8 & 3-3 \\ -4+4 & -8+8 & 3 \end{bmatrix}$

$= \frac{1}{3} \begin{bmatrix} 3 & 0 & 0 \\ 0 & 3 & 0 \\ 0 & 0 & 3 \end{bmatrix} = \begin{bmatrix} 1 & 0 & 0 \\ 0 & 1 & 0 \\ 0 & 0 & 1 \end{bmatrix}$

$BA = \frac{1}{3} \begin{bmatrix} -4 & -5 & 3 \\ -4 & -8 & 3 \\ 1 & 2 & 0 \end{bmatrix} \begin{bmatrix} -2 & 2 & 3 \\ 1 & -1 & 0 \\ 0 & 1 & 4 \end{bmatrix} = \frac{1}{3} \begin{bmatrix} 8-5 & -8+5+3 & -12+12 \\ 8-8 & -8+8+3 & -12+12 \\ -2+2 & 2-2 & 3 \end{bmatrix} = \begin{bmatrix} 1 & 0 & 0 \\ 0 & 1 & 0 \\ 0 & 0 & 1 \end{bmatrix}$

7. $AB = \begin{bmatrix} 2 & 0 & 1 & 1 \\ 3 & 0 & 0 & 1 \\ -1 & 1 & -2 & 1 \\ 4 & -1 & 1 & 0 \end{bmatrix} \begin{bmatrix} -1 & 2 & -1 & -1 \\ -4 & 9 & -5 & -6 \\ 0 & 1 & -1 & -1 \\ 3 & -5 & 3 & 3 \end{bmatrix} = \begin{bmatrix} 1 & 0 & 0 & 0 \\ 0 & 1 & 0 & 0 \\ 0 & 0 & 1 & 0 \\ 0 & 0 & 0 & 1 \end{bmatrix}$

$BA = \begin{bmatrix} -1 & 2 & -1 & -1 \\ -4 & 9 & -5 & -6 \\ 0 & 1 & -1 & -1 \\ 3 & -5 & 3 & 3 \end{bmatrix} \begin{bmatrix} 2 & 0 & 1 & 1 \\ 3 & 0 & 0 & 1 \\ -1 & 1 & -2 & 1 \\ 4 & -1 & 1 & 0 \end{bmatrix} = \begin{bmatrix} 1 & 0 & 0 & 0 \\ 0 & 1 & 0 & 0 \\ 0 & 0 & 1 & 0 \\ 0 & 0 & 0 & 1 \end{bmatrix}$

9. $[A \; \vdots \; I] = \begin{bmatrix} 2 & 0 & \vdots & 1 & 0 \\ 0 & 3 & \vdots & 0 & 1 \end{bmatrix}$

$\begin{matrix} \frac{1}{2}R_1 \rightarrow \\ \frac{1}{3}R_2 \rightarrow \end{matrix} \begin{bmatrix} 1 & 0 & \vdots & \frac{1}{2} & 0 \\ 0 & 1 & \vdots & 0 & \frac{1}{3} \end{bmatrix} = [I \; \vdots \; A^{-1}]$

$A^{-1} = \begin{bmatrix} \frac{1}{2} & 0 \\ 0 & \frac{1}{3} \end{bmatrix} = \frac{1}{6} \begin{bmatrix} 3 & 0 \\ 0 & 2 \end{bmatrix}$

11. $[A \; \vdots \; I] = \begin{bmatrix} 1 & -2 & \vdots & 1 & 0 \\ 2 & -3 & \vdots & 0 & 1 \end{bmatrix}$

$-2R_1 + R_2 \rightarrow \begin{bmatrix} 1 & -2 & \vdots & 1 & 0 \\ 0 & 1 & \vdots & -2 & 1 \end{bmatrix}$

$2R_2 + R_1 \rightarrow \begin{bmatrix} 1 & 0 & \vdots & -3 & 2 \\ 0 & 1 & \vdots & -2 & 1 \end{bmatrix} = [I \; \vdots \; A^{-1}]$

$A^{-1} = \begin{bmatrix} -3 & 2 \\ -2 & 1 \end{bmatrix}$

13. $[A \; \vdots \; I] = \begin{bmatrix} -1 & 1 & \vdots & 1 & 0 \\ -2 & 1 & \vdots & 0 & 1 \end{bmatrix}$

$-2R_1 + R_2 \rightarrow \begin{bmatrix} -1 & 1 & \vdots & 1 & 0 \\ 0 & -1 & \vdots & -2 & 1 \end{bmatrix}$

$R_2 + R_1 \rightarrow \begin{bmatrix} -1 & 0 & \vdots & -1 & 1 \\ 0 & -1 & \vdots & -2 & 1 \end{bmatrix}$

$\begin{matrix} -R_1 \rightarrow \\ -R_2 \rightarrow \end{matrix} \begin{bmatrix} 1 & 0 & \vdots & 1 & -1 \\ 0 & 1 & \vdots & 2 & -1 \end{bmatrix} = [I \; \vdots \; A^{-1}]$

$A^{-1} = \begin{bmatrix} 1 & -1 \\ 2 & -1 \end{bmatrix}$

15. $[A \; \vdots \; I] = \begin{bmatrix} 2 & 4 & \vdots & 1 & 0 \\ 4 & 8 & \vdots & 0 & 1 \end{bmatrix}$

$-2R_1 + R_2 \rightarrow \begin{bmatrix} 2 & 4 & \vdots & 1 & 0 \\ 0 & 0 & \vdots & -2 & 1 \end{bmatrix}$

The two zeros in the second row imply that the inverse does not exist.

17. $A = \begin{bmatrix} 2 & 7 & 1 \\ -3 & -9 & 2 \end{bmatrix}$

A has no inverse because it is not square.

19.

$$\begin{bmatrix} 1 & 1 & 1 & \vdots & 1 & 0 & 0 \\ 3 & 5 & 4 & \vdots & 0 & 1 & 0 \\ 3 & 6 & 5 & \vdots & 0 & 0 & 1 \end{bmatrix}$$

$$\begin{matrix} \\ -3R_1 + R_2 \rightarrow \\ -3R_1 + R_3 \rightarrow \end{matrix} \begin{bmatrix} 1 & 1 & 1 & \vdots & 1 & 0 & 0 \\ 0 & 2 & 1 & \vdots & -3 & 1 & 0 \\ 0 & 3 & 2 & \vdots & -3 & 0 & 1 \end{bmatrix}$$

$$\begin{matrix} -R_2 + R_1 \rightarrow \\ \tfrac{1}{2}R_2 \rightarrow \\ -3R_2 + R_3 \rightarrow \end{matrix} \begin{bmatrix} 1 & 0 & \tfrac{1}{2} & \vdots & \tfrac{5}{2} & -\tfrac{1}{2} & 0 \\ 0 & 1 & \tfrac{1}{2} & \vdots & -\tfrac{3}{2} & \tfrac{1}{2} & 0 \\ 0 & 0 & \tfrac{1}{2} & \vdots & \tfrac{3}{2} & -\tfrac{3}{2} & 1 \end{bmatrix}$$

$$\begin{matrix} -R_3 + R_1 \rightarrow \\ -R_3 + R_2 \rightarrow \\ 2R_3 \rightarrow \end{matrix} \begin{bmatrix} 1 & 0 & 0 & \vdots & 1 & 1 & -1 \\ 0 & 1 & 0 & \vdots & -3 & 2 & -1 \\ 0 & 0 & 1 & \vdots & 3 & -3 & 2 \end{bmatrix}$$

$$A^{-1} = \begin{bmatrix} 1 & 1 & -1 \\ -3 & 2 & -1 \\ 3 & -3 & 2 \end{bmatrix}$$

21.

$$[A \vdots I] = \begin{bmatrix} 1 & 0 & 0 & \vdots & 1 & 0 & 0 \\ 3 & 4 & 0 & \vdots & 0 & 1 & 0 \\ 2 & 5 & 5 & \vdots & 0 & 0 & 1 \end{bmatrix}$$

$$\begin{matrix} \\ -3R_1 + R_2 \rightarrow \\ -2R_1 + R_3 \rightarrow \end{matrix} \begin{bmatrix} 1 & 0 & 0 & \vdots & 1 & 0 & 0 \\ 0 & 4 & 0 & \vdots & -3 & 1 & 0 \\ 0 & 5 & 5 & \vdots & -2 & 0 & 1 \end{bmatrix}$$

$$\begin{matrix} \\ \\ -\tfrac{5}{4}R_2 + R_3 \rightarrow \end{matrix} \begin{bmatrix} 1 & 0 & 0 & \vdots & 1 & 0 & 0 \\ 0 & 4 & 0 & \vdots & -3 & 1 & 0 \\ 0 & 0 & 5 & \vdots & \tfrac{7}{4} & -\tfrac{5}{4} & 1 \end{bmatrix}$$

$$\begin{matrix} \\ \tfrac{1}{4}R_2 \rightarrow \\ \tfrac{1}{5}R_3 \rightarrow \end{matrix} \begin{bmatrix} 1 & 0 & 0 & \vdots & 1 & 0 & 0 \\ 0 & 1 & 0 & \vdots & -\tfrac{3}{4} & \tfrac{1}{4} & 0 \\ 0 & 0 & 1 & \vdots & \tfrac{7}{20} & -\tfrac{1}{4} & \tfrac{1}{5} \end{bmatrix}$$

$$= [I \vdots A^{-1}]$$

$$A^{-1} = \tfrac{1}{20}\begin{bmatrix} 20 & 0 & 0 \\ -15 & 5 & 0 \\ 7 & -5 & 4 \end{bmatrix} = \begin{bmatrix} 1 & 0 & 0 \\ -0.75 & 0.25 & 0 \\ 0.35 & -0.25 & 0.2 \end{bmatrix}$$

23. $[A \ \vdots \ I] = \begin{bmatrix} -8 & 0 & 0 & 0 & \vdots & 1 & 0 & 0 & 0 \\ 0 & 1 & 0 & 0 & \vdots & 0 & 1 & 0 & 0 \\ 0 & 0 & 4 & 0 & \vdots & 0 & 0 & 1 & 0 \\ 0 & 0 & 0 & -5 & \vdots & 0 & 0 & 0 & 1 \end{bmatrix}$

$\begin{array}{l} -\frac{1}{8}R_1 \rightarrow \\ \\ \frac{1}{4}R_3 \rightarrow \\ -\frac{1}{5}R_4 \rightarrow \end{array} \begin{bmatrix} 1 & 0 & 0 & 0 & \vdots & -\frac{1}{8} & 0 & 0 & 0 \\ 0 & 1 & 0 & 0 & \vdots & 0 & 1 & 0 & 0 \\ 0 & 0 & 1 & 0 & \vdots & 0 & 0 & \frac{1}{4} & 0 \\ 0 & 0 & 0 & 1 & \vdots & 0 & 0 & 0 & -\frac{1}{5} \end{bmatrix}$

$= [I \ \vdots \ A^{-1}]$

$A^{-1} = \begin{bmatrix} -\frac{1}{8} & 0 & 0 & 0 \\ 0 & 1 & 0 & 0 \\ 0 & 0 & \frac{1}{4} & 0 \\ 0 & 0 & 0 & -\frac{1}{5} \end{bmatrix}$

25. $A = \begin{bmatrix} 1 & 2 & -1 \\ 3 & 7 & -10 \\ -5 & -7 & -15 \end{bmatrix}$

$A^{-1} = \begin{bmatrix} -175 & 37 & -13 \\ 95 & -20 & 7 \\ 14 & -3 & 1 \end{bmatrix}$

27. $A = \begin{bmatrix} 1 & 1 & 2 \\ 3 & 1 & 0 \\ -2 & 0 & 3 \end{bmatrix}$

$A^{-1} = \frac{1}{2}\begin{bmatrix} -3 & 3 & 2 \\ 9 & -7 & -6 \\ -2 & 2 & 2 \end{bmatrix}$

29. $A = \begin{bmatrix} 0.1 & 0.2 & 0.3 \\ -0.3 & 0.2 & 0.2 \\ 0.5 & 0.4 & 0.4 \end{bmatrix}$

$A^{-1} = \frac{5}{11}\begin{bmatrix} 0 & -4 & 2 \\ -22 & 11 & 11 \\ 22 & -6 & -8 \end{bmatrix}$

31. $A = \begin{bmatrix} 1 & 0 & 3 & 0 \\ 0 & 2 & 0 & 4 \\ 1 & 0 & 3 & 0 \\ 0 & 2 & 0 & 4 \end{bmatrix}$

A^{-1} does not exist.

33. $A = \begin{bmatrix} 1 & -2 & -1 & -2 \\ 3 & -5 & -2 & -3 \\ 2 & -5 & -2 & -5 \\ -1 & 4 & 4 & 11 \end{bmatrix}$

$A^{-1} = \begin{bmatrix} -24 & 7 & 1 & -2 \\ -10 & 3 & 0 & -1 \\ -29 & 7 & 3 & -2 \\ 12 & -3 & -1 & 1 \end{bmatrix}$

35. $AA^{-1} = \begin{bmatrix} a & b \\ c & d \end{bmatrix} \left(\dfrac{1}{ad-bc} \right) \begin{bmatrix} d & -b \\ -c & a \end{bmatrix} = \dfrac{1}{ad-bc} \begin{bmatrix} a & b \\ c & d \end{bmatrix} \begin{bmatrix} d & -b \\ -c & a \end{bmatrix}$

$\qquad = \dfrac{1}{ad-bc} \begin{bmatrix} ad-bc & 0 \\ 0 & ad-bc \end{bmatrix} = \begin{bmatrix} 1 & 0 \\ 0 & 1 \end{bmatrix}$

$\quad A^{-1}A = \dfrac{1}{ad-bc} \begin{bmatrix} d & -b \\ -c & a \end{bmatrix} \begin{bmatrix} a & b \\ c & d \end{bmatrix} = \dfrac{1}{ad-bc} \begin{bmatrix} ad-bc & 0 \\ 0 & ad-bc \end{bmatrix} = \begin{bmatrix} 1 & 0 \\ 0 & 1 \end{bmatrix}$

37. $\begin{bmatrix} x \\ y \end{bmatrix} = \begin{bmatrix} -3 & 2 \\ -2 & 1 \end{bmatrix} \begin{bmatrix} 5 \\ 10 \end{bmatrix} = \begin{bmatrix} 5 \\ 0 \end{bmatrix}$

Answer: $(5, 0)$

39. $\begin{bmatrix} x \\ y \end{bmatrix} = \begin{bmatrix} -3 & 2 \\ -2 & 1 \end{bmatrix} = \begin{bmatrix} 4 \\ 2 \end{bmatrix} = \begin{bmatrix} -8 \\ -6 \end{bmatrix}$

Answer: $(-8, -6)$

41. $\begin{bmatrix} x \\ y \\ z \end{bmatrix} = \begin{bmatrix} 1 & 1 & -1 \\ -3 & 2 & -1 \\ 3 & -3 & 2 \end{bmatrix} \begin{bmatrix} 0 \\ 5 \\ 2 \end{bmatrix} = \begin{bmatrix} 3 \\ 8 \\ -11 \end{bmatrix}$

Answer: $(3, 8, -11)$

43. $\begin{bmatrix} x_1 \\ x_2 \\ x_3 \\ x_4 \end{bmatrix} = \begin{bmatrix} -24 & 7 & 1 & -2 \\ -10 & 3 & 0 & -1 \\ -29 & 7 & 3 & -2 \\ 12 & -3 & -1 & 1 \end{bmatrix} \begin{bmatrix} 0 \\ 1 \\ -1 \\ 2 \end{bmatrix} = \begin{bmatrix} 2 \\ 1 \\ 0 \\ 0 \end{bmatrix}$

Answer: $(2, 1, 0, 0)$

45. $A = \begin{bmatrix} 3 & 4 \\ 5 & 3 \end{bmatrix}$

$\quad A^{-1} = \dfrac{1}{9-20} \begin{bmatrix} 3 & -4 \\ -5 & 3 \end{bmatrix}$

$\quad \begin{bmatrix} x \\ y \end{bmatrix} = -\dfrac{1}{11} \begin{bmatrix} 3 & -4 \\ -5 & 3 \end{bmatrix} \begin{bmatrix} -2 \\ 4 \end{bmatrix} = -\dfrac{1}{11} \begin{bmatrix} -22 \\ 22 \end{bmatrix} = \begin{bmatrix} 2 \\ -2 \end{bmatrix}$

Answer: $(2, -2)$

47. $A = \begin{bmatrix} -0.4 & 0.8 \\ 2 & -4 \end{bmatrix}$

$\quad A^{-1} = \dfrac{1}{1.6-1.6} \begin{bmatrix} -4 & -0.8 \\ -2 & -0.4 \end{bmatrix} \Rightarrow A^{-1} \text{ does not exist.}$

No solution

49. $A = \begin{bmatrix} 3 & 6 \\ 6 & 14 \end{bmatrix}$

$\quad A^{-1} = \dfrac{1}{42-36} \begin{bmatrix} 14 & -6 \\ -6 & 3 \end{bmatrix}$

$\quad \begin{bmatrix} x \\ y \end{bmatrix} = \dfrac{1}{6} \begin{bmatrix} 14 & -6 \\ -6 & 3 \end{bmatrix} \begin{bmatrix} 6 \\ 11 \end{bmatrix} = \dfrac{1}{6} \begin{bmatrix} 18 \\ -3 \end{bmatrix} = \begin{bmatrix} 3 \\ -\frac{1}{2} \end{bmatrix}$

Answer: $\left(3, -\frac{1}{2}\right)$

51. $A = \begin{bmatrix} 4 & -1 & 1 \\ 2 & 2 & 3 \\ 5 & -2 & 6 \end{bmatrix}$

$\quad A^{-1} = \frac{1}{55} \begin{bmatrix} 18 & 4 & -5 \\ 3 & 19 & -10 \\ -14 & 3 & 10 \end{bmatrix}$

$\quad \begin{bmatrix} x \\ y \\ z \end{bmatrix} = \frac{1}{55} \begin{bmatrix} 18 & 4 & -5 & -5 \\ 3 & 19 & -10 & 10 \\ -14 & 3 & 10 & 1 \end{bmatrix} = \frac{1}{55} \begin{bmatrix} -55 \\ 165 \\ 110 \end{bmatrix} = \begin{bmatrix} -1 \\ 3 \\ 2 \end{bmatrix}$

Answer: $(-1, 3, 2)$

53. $A = \begin{bmatrix} 5 & -3 & 2 \\ 2 & 2 & -3 \\ -1 & 7 & -8 \end{bmatrix}$

$\quad A^{-1}$ does not exist.

No solution

55. $A = \begin{bmatrix} 7 & -3 & 0 & 2 \\ -2 & 1 & 0 & -1 \\ 4 & 0 & 1 & -2 \\ -1 & 1 & 0 & -1 \end{bmatrix}$

$A^{-1} = \begin{bmatrix} 0 & -1 & 0 & 1 \\ -1 & -5 & 0 & 3 \\ -2 & -4 & 1 & -2 \\ -1 & -4 & 0 & 1 \end{bmatrix}$

$\begin{bmatrix} x \\ y \\ z \\ w \end{bmatrix} = \begin{bmatrix} 0 & -1 & 0 & 1 \\ -1 & -5 & 0 & 3 \\ -2 & -4 & 1 & -2 \\ -1 & -4 & 0 & 1 \end{bmatrix} \begin{bmatrix} 41 \\ -13 \\ 12 \\ -8 \end{bmatrix} = \begin{bmatrix} 5 \\ 0 \\ -2 \\ 3 \end{bmatrix}$

Answer: $(5, 0, -2, 3)$

For 57–59 use $A = \begin{bmatrix} 1 & 1 & 1 \\ 0.065 & 0.07 & 0.09 \\ 0 & 2 & -1 \end{bmatrix}$. **Using the methods of this section, we have**

$A^{-1} = \frac{1}{11}\begin{bmatrix} 50 & -600 & -4 \\ -13 & 200 & 5 \\ -26 & 400 & -1 \end{bmatrix}$.

57. $X = A^{-1}B = \frac{1}{11}\begin{bmatrix} 50 & -600 & -4 \\ -13 & 200 & 5 \\ -26 & 400 & -1 \end{bmatrix} \begin{bmatrix} 25{,}000 \\ 1900 \\ 0 \end{bmatrix} = \begin{bmatrix} 10{,}000 \\ 5000 \\ 10{,}000 \end{bmatrix}$

Answer: $10,000 in AAA bonds, $5000 in A bonds, $10,000 in B bonds

59. $X = A^{-1}B = \frac{1}{11}\begin{bmatrix} 50 & -600 & -4 \\ -13 & 200 & 5 \\ -26 & 400 & -1 \end{bmatrix} \begin{bmatrix} 12{,}000 \\ 835 \\ 0 \end{bmatrix} = \begin{bmatrix} 9000 \\ 1000 \\ 2000 \end{bmatrix}$

Answer: $9000 in AAA bonds, $1000 in A bonds, $2000 in B bonds

61. The inverse matrix remained the same for each system.

63. $A = \begin{bmatrix} 2 & 0 & 4 \\ 0 & 1 & 4 \\ 1 & 1 & -1 \end{bmatrix}$

$A^{-1} = \frac{1}{14}\begin{bmatrix} 5 & -4 & 4 \\ -4 & 6 & 8 \\ 1 & 2 & -2 \end{bmatrix}$

$\begin{bmatrix} I_1 \\ I_2 \\ I_3 \end{bmatrix} = \frac{1}{14}\begin{bmatrix} 5 & -4 & 4 \\ -4 & 6 & 8 \\ 1 & 2 & -2 \end{bmatrix} \begin{bmatrix} 14 \\ 28 \\ 0 \end{bmatrix} = \begin{bmatrix} -3 \\ 8 \\ 5 \end{bmatrix}$

Answer: $I_1 = -3$ amps, $I_2 = 8$ amps, $I_3 = 5$ amps

65. (a) Given $A = \begin{bmatrix} a_{11} & 0 \\ 0 & a_{22} \end{bmatrix}$, $A^{-1} = \begin{bmatrix} \dfrac{1}{a_{11}} & 0 \\ 0 & \dfrac{1}{a_{22}} \end{bmatrix}$.

Given $A = \begin{bmatrix} a_{11} & 0 & 0 \\ 0 & a_{22} & 0 \\ 0 & 0 & a_{33} \end{bmatrix}$, $A^{-1} = \begin{bmatrix} \dfrac{1}{a_{11}} & 0 & 0 \\ 0 & \dfrac{1}{a_{22}} & 0 \\ 0 & 0 & \dfrac{1}{a_{33}} \end{bmatrix}$.

(b) In general, the inverse of the diagonal matrix A is

$$\begin{bmatrix} \dfrac{1}{a_{11}} & 0 & 0 & \cdots & 0 \\ 0 & \dfrac{1}{a_{22}} & 0 & \cdots & 0 \\ 0 & 0 & \dfrac{1}{a_{33}} & \cdots & 0 \\ \vdots & \vdots & \vdots & \vdots & \vdots \\ 0 & 0 & 0 & \cdots & \dfrac{1}{a_{nn}} \end{bmatrix} \quad (\text{assuming } a_{ii} \neq 0)$$

Section 8.4   The Determinant of a Square Matrix

- You should be able to determine the determinant of a matrix of order 2×2 by using the products of the diagonals.
- You should be able to use expansion by cofactors to find the determinant of a matrix of order 3 or greater.
- The determinant of a triangular matrix equals the product of the entries on the main diagonal.
- You should be able to calculate determinants using a graphing utility.

Solutions to Odd-Numbered Exercises

1. 5

3. $\begin{vmatrix} 2 & 1 \\ 3 & 4 \end{vmatrix} = 2(4) - 1(3) = 8 - 3 = 5$

5. $\begin{vmatrix} 5 & 2 \\ -6 & 3 \end{vmatrix} = 5(3) - 2(-6) = 15 + 12 = 27$

7. $\begin{vmatrix} -7 & 6 \\ \frac{1}{2} & 3 \end{vmatrix} = -7(3) - 6(\frac{1}{2}) = -21 - 3 = -24$

9. $\begin{vmatrix} 2 & 6 \\ 0 & 3 \end{vmatrix} = 2(3) - 6(0) = 6$

11. $\begin{vmatrix} 2 & -1 & 0 \\ 4 & 2 & 1 \\ 4 & 2 & 1 \end{vmatrix} = 2\begin{vmatrix} 2 & 1 \\ 2 & 1 \end{vmatrix} - 4\begin{vmatrix} -1 & 0 \\ 2 & 1 \end{vmatrix} + 4\begin{vmatrix} -1 & 0 \\ 2 & 1 \end{vmatrix} = 2(0) - 4(-1) + 4(-1) = 0$

13. $\begin{vmatrix} 6 & 3 & -7 \\ 0 & 0 & 0 \\ 4 & -6 & 3 \end{vmatrix} = 0\begin{vmatrix} 3 & -7 \\ -6 & 3 \end{vmatrix} - 0\begin{vmatrix} 6 & -7 \\ 4 & 3 \end{vmatrix} + 0\begin{vmatrix} 6 & 3 \\ 4 & -6 \end{vmatrix} = 0$

15. $\begin{vmatrix} -1 & 2 & 5 \\ 0 & 3 & 4 \\ 0 & 0 & 3 \end{vmatrix} = (-1)(3)(3) = -9$ (Upper Triangular)

17. $\begin{vmatrix} 0.3 & 0.2 & 0.2 \\ 0.2 & 0.2 & 0.2 \\ -0.4 & 0.4 & 0.3 \end{vmatrix} = -0.002$

19. $\begin{vmatrix} 1 & 4 & -2 \\ 3 & 6 & -6 \\ -2 & 1 & 4 \end{vmatrix} = 0$

21. $\begin{bmatrix} 3 & 4 \\ 2 & -5 \end{bmatrix}$

(a) $M_{11} = -5$

$M_{12} = 2$

$M_{21} = 4$

$M_{22} = 3$

(b) $C_{11} = M_{11} = -5$

$C_{12} = -M_{12} = -2$

$C_{21} = -M_{21} = -4$

$C_{22} = M_{22} = 3$

23. $\begin{bmatrix} 3 & -2 & 8 \\ 3 & 2 & -6 \\ -1 & 3 & 6 \end{bmatrix}$

(a) $M_{11} = \begin{vmatrix} 2 & -6 \\ 3 & 4 \end{vmatrix} = 12 + 18 = 30$

$\quad M_{12} = \begin{vmatrix} 3 & -6 \\ -1 & 6 \end{vmatrix} = 18 - 6 = 12$

$\quad M_{13} = \begin{vmatrix} 3 & 2 \\ -1 & 3 \end{vmatrix} = 9 + 2 = 11$

$\quad M_{21} = \begin{vmatrix} -2 & 8 \\ 3 & 6 \end{vmatrix} = -12 - 24 = -36$

$\quad M_{22} = \begin{vmatrix} 3 & 8 \\ -1 & 6 \end{vmatrix} = 18 + 8 = 26$

$\quad M_{23} = \begin{vmatrix} 3 & -2 \\ -1 & 3 \end{vmatrix} = 9 - 2 = 7$

$\quad M_{31} = \begin{vmatrix} -2 & 8 \\ 2 & -6 \end{vmatrix} = 12 - 16 = -4$

$\quad M_{32} = \begin{vmatrix} 3 & 8 \\ 3 & -6 \end{vmatrix} = -18 - 24 = -42$

$\quad M_{33} = \begin{vmatrix} 3 & -2 \\ 3 & 2 \end{vmatrix} = 6 + 6 = 12$

(b) $C_{11} = (-1)^2 M_{11} = 30$

$\quad C_{12} = (-1)^3 M_{12} = -12$

$\quad C_{13} = (-1)^4 M_{13} = 11$

$\quad C_{21} = (-1)^3 M_{21} = 36$

$\quad C_{22} = (-1)^4 M_{22} = 26$

$\quad C_{23} = (-1)^5 M_{23} = -7$

$\quad C_{31} = (-1)^4 M_{31} = -4$

$\quad C_{32} = (-1)^5 M_{32} = 42$

$\quad C_{33} = (-1)^6 M_{33} = 12$

25. (a) $\begin{vmatrix} -3 & 2 & 1 \\ 4 & 5 & 6 \\ 2 & -3 & 1 \end{vmatrix} = -3 \begin{vmatrix} 5 & 6 \\ -3 & 1 \end{vmatrix} - 2 \begin{vmatrix} 4 & 6 \\ 2 & 1 \end{vmatrix} + \begin{vmatrix} 4 & 5 \\ 2 & -3 \end{vmatrix} = -3(23) - 2(-8) - 22 = -75$

(b) $\begin{vmatrix} -3 & 2 & 1 \\ 4 & 5 & 6 \\ 2 & -3 & 1 \end{vmatrix} = -2 \begin{vmatrix} 4 & 6 \\ 2 & 1 \end{vmatrix} + 5 \begin{vmatrix} -3 & 1 \\ 2 & 1 \end{vmatrix} + 3 \begin{vmatrix} -3 & 1 \\ 4 & 6 \end{vmatrix} = -2(-8) + 5(-5) + 3(-22) = -75$

27. (a) $\begin{vmatrix} 5 & 0 & -3 \\ 0 & 12 & 4 \\ 1 & 6 & 3 \end{vmatrix} = 0 \begin{vmatrix} 0 & -3 \\ 6 & 3 \end{vmatrix} + 12 \begin{vmatrix} 5 & -3 \\ 1 & 3 \end{vmatrix} - 4 \begin{vmatrix} 5 & 0 \\ 1 & 6 \end{vmatrix} = 0(18) + 12(18) - 4(30) = 96$

(b) $\begin{vmatrix} 5 & 0 & -3 \\ 0 & 12 & 4 \\ 1 & 6 & 3 \end{vmatrix} = 0 \begin{vmatrix} 0 & 4 \\ 1 & 3 \end{vmatrix} + 12 \begin{vmatrix} 5 & -3 \\ 1 & 3 \end{vmatrix} - 6 \begin{vmatrix} 5 & -3 \\ 0 & 4 \end{vmatrix} = 0(-4) + 12(18) - 6(20) = 96$

29. (a) $\begin{vmatrix} 6 & 0 & -3 & 5 \\ 4 & 13 & 6 & -8 \\ -1 & 0 & 7 & 4 \\ 8 & 6 & 0 & 2 \end{vmatrix} = -4\begin{vmatrix} 0 & -3 & 5 \\ 0 & 7 & 4 \\ 6 & 0 & 2 \end{vmatrix} + 13\begin{vmatrix} 6 & -3 & 5 \\ -1 & 7 & 4 \\ 8 & 0 & 2 \end{vmatrix} - 6\begin{vmatrix} 6 & 0 & 5 \\ -1 & 0 & 4 \\ 8 & 6 & 2 \end{vmatrix} - 8\begin{vmatrix} 6 & 0 & -3 \\ -1 & 0 & 7 \\ 8 & 6 & 0 \end{vmatrix}$

$$= -4(-282) + 13(-298) - 6(-174) - 8(-234) = 170$$

(b) $\begin{vmatrix} 6 & 0 & -3 & 5 \\ 4 & 13 & 6 & -8 \\ -1 & 0 & 7 & 4 \\ 8 & 6 & 0 & 2 \end{vmatrix} = 0\begin{vmatrix} 4 & 6 & -8 \\ -1 & 7 & 4 \\ 8 & 0 & 2 \end{vmatrix} + 13\begin{vmatrix} 6 & -3 & 5 \\ -1 & 7 & 4 \\ 8 & 0 & 2 \end{vmatrix} + 0\begin{vmatrix} 6 & -3 & 5 \\ 4 & 6 & -8 \\ 8 & 0 & 2 \end{vmatrix} + 6\begin{vmatrix} 6 & -3 & 5 \\ 4 & 6 & -8 \\ -1 & 7 & 4 \end{vmatrix}$

$$= 0 + 13(-298) + 0 + 6(674) = 170$$

31. Expand by Column 3.

$$\begin{vmatrix} 1 & 4 & -2 \\ 3 & 2 & 0 \\ -1 & 4 & 3 \end{vmatrix} = -2\begin{vmatrix} 3 & 2 \\ -1 & 4 \end{vmatrix} + 3\begin{vmatrix} 1 & 4 \\ 3 & 2 \end{vmatrix} = -2(14) + 3(-10) = -58$$

33. $\begin{vmatrix} 2 & 4 & 6 \\ 0 & 3 & 1 \\ 0 & 0 & -5 \end{vmatrix} = (2)(3)(-5) = -30$ (Upper Triangular)

35. Expand by Column 3.

$$\begin{vmatrix} 2 & 6 & 6 & 2 \\ 2 & 7 & 3 & 6 \\ 1 & 5 & 0 & 1 \\ 3 & 7 & 0 & 7 \end{vmatrix} = 6\begin{vmatrix} 2 & 7 & 6 \\ 1 & 5 & 1 \\ 3 & 7 & 7 \end{vmatrix} - 3\begin{vmatrix} 2 & 6 & 2 \\ 1 & 5 & 1 \\ 3 & 7 & 7 \end{vmatrix} = 6(-20) - 3(16) = -168$$

37. Expand by Column 1.

$$\begin{vmatrix} 5 & 3 & 0 & 6 \\ 4 & 6 & 4 & 12 \\ 0 & 2 & -3 & 4 \\ 0 & 1 & -2 & 2 \end{vmatrix} = 5\begin{vmatrix} 6 & 4 & 12 \\ 2 & -3 & 4 \\ 1 & -2 & 2 \end{vmatrix} - 4\begin{vmatrix} 3 & 0 & 6 \\ 2 & -3 & 4 \\ 1 & -2 & 2 \end{vmatrix} = 5(0) - 4(0) = 0$$

39. Expand by Column 2.

$$\begin{vmatrix} 3 & 2 & 4 & -1 & 5 \\ -2 & 0 & 1 & 3 & 2 \\ 1 & 0 & 0 & 4 & 0 \\ 6 & 0 & 2 & -1 & 0 \\ 3 & 0 & 5 & 1 & 0 \end{vmatrix} = -2\begin{vmatrix} -2 & 1 & 3 & 2 \\ 1 & 0 & 4 & 0 \\ 6 & 2 & -1 & 0 \\ 3 & 5 & 1 & 0 \end{vmatrix} = (-2)(-2)\begin{vmatrix} 1 & 0 & 4 \\ 6 & 2 & -1 \\ 3 & 5 & 1 \end{vmatrix} = 4(103) = 412$$

41. $\begin{vmatrix} 3 & 8 & -7 \\ 0 & -5 & 4 \\ 8 & 1 & 6 \end{vmatrix} = -126$

43. $\begin{vmatrix} 7 & 0 & -14 \\ -2 & 5 & 4 \\ -6 & 2 & 12 \end{vmatrix} = 0$

45. $\begin{vmatrix} 1 & -1 & 8 & 4 \\ 2 & 6 & 0 & -4 \\ 2 & 0 & 2 & 6 \\ 0 & 2 & 8 & 0 \end{vmatrix} = -336$

47. $\begin{vmatrix} 3 & -2 & 4 & 3 & 1 \\ -1 & 0 & 2 & 1 & 0 \\ 5 & -1 & 0 & 3 & 2 \\ 4 & 7 & -8 & 0 & 0 \\ 1 & 2 & 3 & 0 & 2 \end{vmatrix} = 410$

49. $\begin{vmatrix} w & x \\ y & z \end{vmatrix} = wz - xy$

$-\begin{vmatrix} y & z \\ w & x \end{vmatrix} = -(xy - wz) = wz - xy$

Thus, $\begin{vmatrix} w & x \\ y & z \end{vmatrix} = -\begin{vmatrix} y & z \\ w & x \end{vmatrix}.$

51. $\begin{vmatrix} w & x \\ y & z \end{vmatrix} = wz - xy$

$\begin{vmatrix} w & x + cw \\ y & z + cy \end{vmatrix} = w(z + cy) - y(x + cw) = wz - xy$

Thus, $\begin{vmatrix} w & x \\ y & z \end{vmatrix} = \begin{vmatrix} w & x + cw \\ y & z + cy \end{vmatrix}.$

53. $\begin{vmatrix} 1 & x & x^2 \\ 1 & y & y^2 \\ 1 & z & z^2 \end{vmatrix} = \begin{vmatrix} y & y^2 \\ z & z^2 \end{vmatrix} - \begin{vmatrix} x & x^2 \\ z & z^2 \end{vmatrix} + \begin{vmatrix} x & x^2 \\ y & y^2 \end{vmatrix}$

$= (yz^2 - y^2z) - (xz^2 - x^2z) + (xy^2 - x^2y)$

$= yz^2 - xz^2 - y^2z + x^2z + xy(y - x)$

$= z^2(y - x) - z(y^2 - x^2) + xy(y - x)$

$= z^2(y - x) - z(y - x)(y + x) + xy(y - x)$

$= (y - x)[z^2 - z(y + x) + xy]$

$= (y - x)[z^2 - zy - zx + xy]$

$= (y - x)[z^2 - zx - zy + xy]$

$= (y - x)[z(z - x) - y(z - x)]$

$= (y - x)(z - x)(z - y)$

55. $\begin{vmatrix} x - 1 & 2 \\ 3 & x - 2 \end{vmatrix} = 0$

$(x - 1)(x - 2) - 6 = 0$

$x^2 - 3x - 4 = 0$

$(x + 1)(x - 4) = 0$

$x = -1 \text{ or } x = 4$

57. $\begin{vmatrix} 4u & -1 \\ -1 & 2v \end{vmatrix} = 8uv - 1$

59. $\begin{vmatrix} e^{2x} & e^{3x} \\ 2e^{2x} & 3e^{3x} \end{vmatrix} = 3e^{5x} - 2e^{5x} = e^{5x}$

61. $\begin{vmatrix} x & \ln x \\ 1 & \dfrac{1}{x} \end{vmatrix} = 1 - \ln x$

63. (a) $\begin{vmatrix} -1 & 0 \\ 0 & 3 \end{vmatrix} = -3$ **(b)** $\begin{vmatrix} 2 & 0 \\ 0 & -1 \end{vmatrix} = -2$

(c) $\begin{bmatrix} -1 & 0 \\ 0 & 3 \end{bmatrix} \begin{bmatrix} 2 & 0 \\ 0 & -1 \end{bmatrix} = \begin{bmatrix} -2 & 0 \\ 0 & -3 \end{bmatrix}$ **(d)** $\begin{vmatrix} -2 & 0 \\ 0 & -3 \end{vmatrix} = 6$

65. (a) $\begin{vmatrix} -1 & 2 & 1 \\ 1 & 0 & 1 \\ 0 & 1 & 0 \end{vmatrix} = 2$ **(b)** $\begin{vmatrix} -1 & 0 & 0 \\ 0 & 2 & 0 \\ 0 & 0 & 3 \end{vmatrix} = -6$

(c) $\begin{bmatrix} -1 & 2 & 1 \\ 1 & 0 & 1 \\ 0 & 1 & 0 \end{bmatrix} \begin{bmatrix} -1 & 0 & 0 \\ 0 & 2 & 0 \\ 0 & 0 & 3 \end{bmatrix} = \begin{bmatrix} 1 & 4 & 3 \\ -1 & 0 & 3 \\ 0 & 2 & 0 \end{bmatrix}$

(d) $\begin{vmatrix} 1 & 4 & 3 \\ -1 & 0 & 3 \\ 0 & 2 & 0 \end{vmatrix} = -12$

67. Let $A = \begin{bmatrix} 1 & 3 \\ -2 & 4 \end{bmatrix}$ and $B = \begin{bmatrix} -4 & 0 \\ 3 & 5 \end{bmatrix}$

$|A| = \begin{vmatrix} 1 & 3 \\ -2 & 4 \end{vmatrix} = 10$, $|B| = \begin{vmatrix} -4 & 0 \\ 3 & 5 \end{vmatrix} = -20$, $|A| + |B| = -10$

$A + B = \begin{bmatrix} -3 & 3 \\ 1 & 9 \end{bmatrix}$, $|A + B| = \begin{vmatrix} -3 & 3 \\ 1 & 9 \end{vmatrix} = -30$

Thus, $|A + B| \neq |A| + |B|$. Your answer may differ, depending on how you choose A and B.

69. A square matrix is a square array of numbers. The determinant of a square matrix is a real number.

71. (a) Columns 2 and 3 are interchanged.

(b) Rows 1 and 3 are interchanged.

73. (a) 5 is factored out of the first row of A.

(b) 4 and 3 are factored out of columns 2 and 3.

Section 8.5   Applications of Matrices and Determinants

■ You should be able to find the area of a triangle with vertices (x_1, y_1), (x_2, y_2), and (x_3, y_3).

$$\text{Area} = \pm \frac{1}{2} \begin{vmatrix} x_1 & y_1 & 1 \\ x_2 & y_2 & 1 \\ x_3 & y_3 & 1 \end{vmatrix}$$

The $\pm$ symbol indicates that the appropriate sign should be chosen so that the area is positive.

■ You should be able to test to see if three points, (x_1, y_1), (x_2, y_2), and (x_3, y_3), are collinear.

$$\begin{vmatrix} x_1 & y_1 & 1 \\ x_2 & y_2 & 1 \\ x_3 & y_3 & 1 \end{vmatrix} = 0, \text{ if and only if they are collinear.}$$

■ You should be able to find the equation of the line through (x_1, y_1) and (x_2, y_2) by evaluating

$$\begin{vmatrix} x & y & 1 \\ x_1 & y_1 & 1 \\ x_2 & y_2 & 1 \end{vmatrix} = 0.$$

■ You should be able to use Cramer's Rule to solve a system of linear equations.

■ Now you should be able to solve a system of linear equations by substitution, elimination, elementary row operations on an augmented matrix, using the inverse matrix, or Cramer's Rule.

■ You should be able to encode and decode messages by using an invertible $n \times n$ matrix.

Solutions to Odd-Numbered Exercises

1. Vertices: $(-2, -3)$, $(2, -3)$, $(0, 4)$

$$\text{Area} = \frac{1}{2} \begin{vmatrix} -2 & -3 & 1 \\ 2 & -3 & 1 \\ 0 & 4 & 1 \end{vmatrix} = \frac{1}{2} \left(-2 \begin{vmatrix} -3 & 1 \\ 4 & 1 \end{vmatrix} - 2 \begin{vmatrix} -3 & 1 \\ 4 & 1 \end{vmatrix} \right) = \frac{1}{2}(14 + 14) = 14 \text{ square units}$$

3. Vertices: $(0, 0)$, $(3, 1)$, $(1, 5)$

$$\text{Area} = \frac{1}{2} \begin{vmatrix} 0 & 0 & 1 \\ 3 & 1 & 1 \\ 1 & 5 & 1 \end{vmatrix} = \frac{1}{2} \begin{vmatrix} 3 & 1 \\ 1 & 5 \end{vmatrix} = 7 \text{ square units}$$

5. Vertices: $\left(0, \frac{1}{2}\right)$, $\left(\frac{5}{2}, 0\right)$, $(4, 3)$

$$\text{Area} = \frac{1}{2} \begin{vmatrix} 0 & \frac{1}{2} & 1 \\ \frac{5}{2} & 0 & 1 \\ 4 & 3 & 1 \end{vmatrix} = \frac{1}{2}\left(2 + \frac{15}{2} - \frac{5}{4}\right) = \frac{33}{8} \text{ square units}$$

7. Vertices: $(4, 5)$, $(6, 1)$, $(7, 9)$

$$\text{Area} = \frac{1}{2}\begin{vmatrix} 4 & 5 & 1 \\ 6 & 1 & 1 \\ 7 & 9 & 1 \end{vmatrix} = \frac{1}{2}(20) = 10 \text{ square units}$$

9. Vertices: $(-3, 5)$, $(2, 6)$, $(3, -5)$

$$\text{Area} = -\frac{1}{2}\begin{vmatrix} -3 & 5 & 1 \\ 2 & 6 & 1 \\ 3 & -5 & 1 \end{vmatrix} = -\frac{1}{2}\begin{vmatrix} -3 & 5 & 1 \\ 5 & 1 & 0 \\ 6 & -10 & 0 \end{vmatrix} = -\frac{1}{2}\begin{vmatrix} 5 & 1 \\ 6 & -10 \end{vmatrix} = 28 \text{ square units}$$

11. $4 = \pm\dfrac{1}{2}\begin{vmatrix} -5 & 1 & 1 \\ 0 & 2 & 1 \\ -2 & x & 1 \end{vmatrix}$

$\pm 8 = -5\begin{vmatrix} 2 & 1 \\ x & 1 \end{vmatrix} - 2\begin{vmatrix} 1 & 1 \\ 2 & 1 \end{vmatrix}$

$\pm 8 = -5(2 - x) - 2(-1)$

$\pm 8 = 5x - 8$

$x = \dfrac{8 \pm 8}{5}$

$x = \dfrac{16}{5}$ OR $x = 0$

13. $3x + 4y = -2$

$5x + 3y = 4$

$x = \dfrac{\begin{vmatrix} -2 & 4 \\ 4 & 3 \end{vmatrix}}{\begin{vmatrix} 3 & 4 \\ 5 & 3 \end{vmatrix}} = \dfrac{-22}{-11} = 2$

$y = \dfrac{\begin{vmatrix} 3 & -2 \\ 5 & 4 \end{vmatrix}}{\begin{vmatrix} 3 & 4 \\ 5 & 3 \end{vmatrix}} = \dfrac{22}{-11} = -2$

Answer: $(2, -2)$

15. $4x - y + z = -5$

$2x + 2y + 3z = 10$ $\qquad D = \begin{vmatrix} 4 & -1 & 1 \\ 2 & 2 & 3 \\ 5 & -2 & 6 \end{vmatrix} = 55$

$5x - 2y + 6z = 1$

$x = \dfrac{\begin{vmatrix} -5 & -1 & 1 \\ 10 & 2 & 3 \\ 1 & -2 & 6 \end{vmatrix}}{55} = \dfrac{-55}{55} = -1$

$y = \dfrac{\begin{vmatrix} 4 & -5 & 1 \\ 2 & 10 & 3 \\ 5 & 1 & 6 \end{vmatrix}}{55} = \dfrac{165}{55} = 3$

$z = \dfrac{\begin{vmatrix} 4 & -1 & -5 \\ 2 & 2 & 10 \\ 5 & -2 & 1 \end{vmatrix}}{55} = \dfrac{110}{55} = 2$

Answer: $(-1, 3, 2)$

17. $3x + 3y + 5z = 1$

$3x + 5y + 9z = 2 \qquad D = \begin{vmatrix} 3 & 3 & 5 \\ 3 & 5 & 9 \\ 5 & 9 & 17 \end{vmatrix} = 4$

$5x + 9y + 17z = 4$

$$x = \frac{\begin{vmatrix} 1 & 3 & 5 \\ 2 & 5 & 9 \\ 4 & 9 & 17 \end{vmatrix}}{4} = 0, \; y = \frac{\begin{vmatrix} 3 & 1 & 5 \\ 3 & 2 & 9 \\ 5 & 4 & 17 \end{vmatrix}}{4} = -\frac{1}{2}, \; z = \frac{\begin{vmatrix} 3 & 3 & 1 \\ 3 & 5 & 2 \\ 5 & 9 & 4 \end{vmatrix}}{4} = \frac{1}{2}$$

Answer: $\left(0, -\frac{1}{2}, \frac{1}{2}\right)$

19. Vertices: $\overset{A}{(0, 25)}, \; \overset{B}{(10, 0)}, \; \overset{C}{(28, 5)}$

Area $= \frac{1}{2} \begin{vmatrix} 0 & 25 & 1 \\ 10 & 0 & 1 \\ 28 & 5 & 1 \end{vmatrix} = 250$ square miles

21. Points: $(3, -1), \; (0, -3), \; (12, 5)$

$\begin{vmatrix} 3 & -1 & 1 \\ 0 & -3 & 1 \\ 12 & 5 & 1 \end{vmatrix} = \begin{vmatrix} 3 & -1 & 1 \\ 0 & -3 & 1 \\ 0 & 9 & -3 \end{vmatrix} = 3 \begin{vmatrix} -3 & 1 \\ 9 & -3 \end{vmatrix} = 0$

The points are collinear.

23. Points: $\left(2, -\frac{1}{2}\right), \; (-4, 4), \; (6, -3)$

$\begin{vmatrix} 2 & -\frac{1}{2} & 1 \\ -4 & 4 & 1 \\ 6 & -3 & 1 \end{vmatrix} = \begin{vmatrix} -4 & \frac{5}{2} & 0 \\ -10 & 7 & 0 \\ 6 & -3 & 1 \end{vmatrix} = 3 \begin{vmatrix} -4 & \frac{5}{2} \\ -10 & 7 \end{vmatrix} = -3 \neq 0$

The points are not collinear.

25. Points: $(0, 2), \; (1, 2.4), \; (-1, 1.6)$

$\begin{vmatrix} 0 & 2 & 1 \\ 1 & 2.4 & 1 \\ -1 & 1.6 & 1 \end{vmatrix} = \begin{vmatrix} 0 & 2 & 1 \\ 1 & 2.4 & 1 \\ 0 & 4.0 & 2 \end{vmatrix} = - \begin{vmatrix} 2 & 1 \\ 4 & 2 \end{vmatrix} = 0$

The points are collinear.

27. Points: $(0, 0), \; (5, 3)$

Equation: $\begin{vmatrix} x & y & 1 \\ 0 & 0 & 1 \\ 5 & 3 & 1 \end{vmatrix} = 5y - 3x = 0 \quad$ OR $\quad 3x - 5y = 0$

29. Points: $(-4, 3), \; (2, 1)$

Equation: $\begin{vmatrix} x & y & 1 \\ -4 & 3 & 1 \\ 2 & 1 & 1 \end{vmatrix} = 3x + 2y - 4 - 6 + 4y - x = 0 \quad$ OR $\quad x + 3y - 5 = 0$

31. Points: $\left(-\frac{1}{2}, 3\right)$, $\left(\frac{5}{2}, 1\right)$

Equation: $\begin{vmatrix} x & y & 1 \\ -\frac{1}{2} & 3 & 1 \\ \frac{5}{2} & 1 & 1 \end{vmatrix} = 3x + \frac{5}{2}y - \frac{1}{2} - \frac{15}{2} + \frac{1}{2}y - x = 0$ OR $2x + 3y - 8 = 0$

33. $\begin{vmatrix} 2 & -5 & 1 \\ 4 & x & 1 \\ 5 & -2 & 1 \end{vmatrix} = 0$

$2\begin{vmatrix} x & 1 \\ -2 & 1 \end{vmatrix} + 5\begin{vmatrix} 4 & 1 \\ 5 & 1 \end{vmatrix} + \begin{vmatrix} 4 & x \\ 5 & -2 \end{vmatrix} = 0$

$2(x + 2) + 5(-1) + (-8 - 5x) = 0$

$-3x - 9 = 0$

$x = -3$

35. The uncoded row matrices are the rows of the 7×3 matrix on the left.

$\begin{array}{ccc} \text{T} & \text{R} & 0 \\ \text{U} & \text{B} & \text{L} \\ \text{E} & & \text{I} \\ \text{N} & & \text{R} \\ \text{I} & \text{V} & \text{E} \\ \text{R} & & \text{C} \\ \text{I} & \text{T} & \text{Y} \end{array} \begin{bmatrix} 20 & 18 & 15 \\ 21 & 2 & 12 \\ 5 & 0 & 9 \\ 14 & 0 & 18 \\ 9 & 22 & 5 \\ 18 & 0 & 3 \\ 9 & 20 & 25 \end{bmatrix} \begin{bmatrix} 1 & -1 & 0 \\ 1 & 0 & -1 \\ -6 & 2 & 3 \end{bmatrix} = \begin{bmatrix} -52 & 10 & 27 \\ -49 & 3 & 34 \\ -49 & 13 & 27 \\ -94 & 22 & 54 \\ 1 & 1 & -7 \\ 0 & -12 & 9 \\ -121 & 41 & 55 \end{bmatrix}$

Answer: $[-52, 10, 27], [-49, 3, 34], [-49, 13, 27], [-94, 22, 54], [1, 1, -7], [0, -12, 9], [-121, 41, 55]$

In Exercises 37–39, use the matrix $A = \begin{bmatrix} 1 & 2 & 2 \\ 3 & 7 & 9 \\ -1 & -4 & -7 \end{bmatrix}$.

37. L A N   D I N G _ S U C C   E S S   F U L

$[12 \quad 1 \quad 14]$ $[4 \quad 9 \quad 14]$ $[7 \quad 0 \quad 19]$ $[21 \quad 3 \quad 3]$ $[5 \quad 19 \quad 19]$ $[6 \quad 21 \quad 12]$

$[12 \quad 1 \quad 14]A = [\quad 1 \quad -25 \quad -65]$

$[4 \quad 9 \quad 14]A = [\quad 17 \quad 15 \quad -9]$

$[7 \quad 0 \quad 19]A = [-12 \quad -62 \quad -119]$

$[21 \quad 3 \quad 3]A = [\quad 27 \quad 51 \quad 48]$

$[5 \quad 19 \quad 19]A = [\quad 43 \quad 67 \quad 48]$

$[6 \quad 21 \quad 12]A = [\quad 57 \quad 111 \quad 117]$

Cryptogram: $1 \;-25 \;-65 \; 17 \; 15 \;-9 \;-12 \;-62 \;-119 \; 27 \; 51 \; 48 \; 43 \; 67 \; 48 \; 57 \; 111 \; 117$

39. H   A   P   P   Y   _   B   I   R   T   H   D   A   Y   _

[8   1   16]   [16   25   0]   [2   9   18]   [20   8   4]   [1   25   0]

$[\,8 \quad 1 \quad 16\,]A = [\,5 \quad -41 \quad -87\,]$

$[\,16 \quad 25 \quad 0\,]A = [\,91 \quad 207 \quad 257\,]$

$[\,2 \quad 9 \quad 18\,]A = [\,11 \quad -5 \quad -41\,]$

$[\,20 \quad 8 \quad 4\,]A = [\,40 \quad 80 \quad 84\,]$

$[\,1 \quad 25 \quad 0\,]A = [\,76 \quad 177 \quad 227\,]$

Cryptogram: $-5 \quad -41 \quad -87 \quad 91 \quad 207 \quad 257 \quad 11 \quad -5 \quad -41 \quad 40 \quad 80 \quad 84 \quad 76 \quad 177 \quad 227$

41. $A^{-1} = \begin{bmatrix} 1 & 2 \\ 3 & 5 \end{bmatrix}^{-1} = \begin{bmatrix} -5 & 2 \\ 3 & -1 \end{bmatrix}$

$\begin{bmatrix} 11 & 21 \\ 64 & 112 \\ 25 & 50 \\ 29 & 53 \\ 23 & 46 \\ 40 & 75 \\ 55 & 92 \end{bmatrix} \begin{bmatrix} -5 & 2 \\ 3 & -1 \end{bmatrix} = \begin{bmatrix} 8 & 1 \\ 16 & 16 \\ 25 & 0 \\ 14 & 5 \\ 23 & 0 \\ 25 & 5 \\ 1 & 18 \end{bmatrix}$
H A
P P
Y
N E
W
Y E
A R
     Message: HAPPY NEW YEAR

43. $A^{-1} = \begin{bmatrix} 1 & 2 & 2 \\ 3 & 7 & 9 \\ -1 & -4 & -7 \end{bmatrix}^{-1} = \begin{bmatrix} -13 & 6 & 4 \\ 12 & -5 & -3 \\ -5 & 2 & 1 \end{bmatrix}$

$\begin{bmatrix} 20 & 17 & -15 \\ -12 & -56 & -104 \\ 1 & -25 & -65 \\ 62 & 143 & 181 \end{bmatrix} \begin{bmatrix} -13 & 6 & 4 \\ 12 & -5 & -3 \\ -5 & 2 & 1 \end{bmatrix} = \begin{bmatrix} 19 & 5 & 14 \\ 4 & 0 & 16 \\ 12 & 1 & 14 \\ 5 & 19 & 0 \end{bmatrix}$
S E N
D   P
L A N
E S
     Message: SEND PLANES

45. Let *A* be the 2 × 2 matrix needed to decode the message.

$\begin{bmatrix} -18 & -18 \\ 1 & 16 \end{bmatrix} A = \begin{bmatrix} 0 & 18 \\ 15 & 14 \end{bmatrix}$
R
O N

$A = \begin{bmatrix} -18 & -18 \\ 1 & 16 \end{bmatrix}^{-1} \begin{bmatrix} 0 & 18 \\ 15 & 14 \end{bmatrix} = \begin{bmatrix} -\frac{8}{135} & -\frac{1}{15} \\ \frac{1}{270} & \frac{1}{15} \end{bmatrix} \begin{bmatrix} 0 & 18 \\ 15 & 14 \end{bmatrix} = \begin{bmatrix} -1 & -2 \\ 1 & 1 \end{bmatrix}$

$\begin{bmatrix} 8 & 21 \\ -15 & -10 \\ -13 & -13 \\ 5 & 10 \\ 5 & 25 \\ 5 & 19 \\ -1 & 6 \\ 20 & 40 \\ -18 & -18 \\ 1 & 16 \end{bmatrix} \begin{bmatrix} -1 & -2 \\ 1 & 1 \end{bmatrix} = \begin{bmatrix} 13 & 5 \\ 5 & 20 \\ 0 & 13 \\ 5 & 0 \\ 20 & 15 \\ 14 & 9 \\ 7 & 8 \\ 20 & 0 \\ 0 & 18 \\ 15 & 14 \end{bmatrix}$
M E
E T
  M
E
T O
N I
G H
T
  R
O N
     Message: MEET ME TONIGHT RON

❑ **Review Exercises for Chapter 8**

Solutions to Odd-Numbered Exercises

1. $\begin{bmatrix} 3 & -10 & \vdots & 15 \\ 5 & 4 & \vdots & 22 \end{bmatrix}$

3. $\begin{bmatrix} 5 & 1 & 7 & \vdots & -9 \\ 4 & 2 & 0 & \vdots & 10 \\ 9 & 4 & 2 & \vdots & 3 \end{bmatrix}$ $\begin{aligned} 5x + y + 7z &= -9 \\ 4x + 2y \quad &= 10 \\ 9x + 4y + 2z &= 3 \end{aligned}$

5. $\begin{bmatrix} 0 & 1 & 1 \\ 1 & 2 & 3 \\ 2 & 2 & 2 \end{bmatrix}$

$\begin{array}{l} R_1 + R_2 \rightarrow \\ -R_1 + R_2 \rightarrow \\ -2R_1 + R_3 \rightarrow \end{array} \begin{bmatrix} 1 & 3 & 4 \\ 0 & -1 & -1 \\ 0 & -4 & -6 \end{bmatrix}$

$\begin{array}{l} 3R_2 + R_1 \rightarrow \\ -R_2 \rightarrow \\ -4R_2 + R_3 \rightarrow \end{array} \begin{bmatrix} 1 & 0 & 1 \\ 0 & 1 & 1 \\ 0 & 0 & -2 \end{bmatrix}$

$\begin{array}{l} -R_3 + R_1 \rightarrow \\ -R_3 + R_2 \rightarrow \\ -\frac{1}{2}R_3 \rightarrow \end{array} \begin{bmatrix} 1 & 0 & 0 \\ 0 & 1 & 0 \\ 0 & 0 & 1 \end{bmatrix}$

7. $\begin{bmatrix} 3 & -2 & 1 & 0 \\ 4 & -3 & 0 & 1 \end{bmatrix} \Rightarrow \begin{bmatrix} 1 & 0 & 3 & -2 \\ 0 & 1 & 4 & -3 \end{bmatrix}$

9. $\begin{bmatrix} 1 & 3 & 4 \\ 0 & 1 & 1 \\ 2 & 4 & 6 \end{bmatrix} \Rightarrow \begin{bmatrix} 1 & 0 & 1 \\ 0 & 1 & 1 \\ 0 & 0 & 0 \end{bmatrix}$

11. $\begin{bmatrix} 5 & 4 & \vdots & 2 \\ -1 & 1 & \vdots & -22 \end{bmatrix}$

$\begin{array}{l} 4R_2 + R_1 \rightarrow \\ R_1 + R_2 \rightarrow \end{array} \begin{bmatrix} 1 & 8 & \vdots & -86 \\ 0 & 9 & \vdots & -108 \end{bmatrix}$

$\begin{array}{l} -8R_2 + R_1 \rightarrow \\ \frac{1}{9}R_2 \rightarrow \end{array} \begin{bmatrix} 1 & 0 & \vdots & 10 \\ 0 & 1 & \vdots & -12 \end{bmatrix}$

$x = 10, \ y = -12$

Answer: $(10, -12)$

13. $\begin{bmatrix} 2 & 1 & \vdots & 0.3 \\ 3 & -1 & \vdots & -1.3 \end{bmatrix}$

$-R_1 + R_2 \rightarrow \begin{bmatrix} 2 & 1 & \vdots & 0.3 \\ 1 & -2 & \vdots & -1.6 \end{bmatrix}$

$\begin{bmatrix} 1 & -2 & \vdots & -1.6 \\ 2 & 1 & \vdots & 0.3 \end{bmatrix}$

$-2R_1 + R_2 \rightarrow \begin{bmatrix} 1 & -2 & \vdots & -1.6 \\ 0 & 5 & \vdots & 3.5 \end{bmatrix}$

$\begin{array}{l} 2R_2 + R_1 \rightarrow \\ \frac{1}{5}R_2 \rightarrow \end{array} \begin{bmatrix} 1 & 0 & \vdots & -0.2 \\ 0 & 1 & \vdots & 0.7 \end{bmatrix}$

$x = -0.2, \ y = 0.7$

15.
$$\begin{bmatrix} 2 & 1 & 2 & \vdots & 4 \\ 2 & 2 & 0 & \vdots & 5 \\ 2 & -1 & 6 & \vdots & 2 \end{bmatrix}$$

$$\begin{matrix} \\ -R_1 + R_2 \rightarrow \\ -R_1 + R_3 \rightarrow \end{matrix} \begin{bmatrix} 2 & 1 & 2 & \vdots & 4 \\ 0 & 1 & -2 & \vdots & 1 \\ 0 & -2 & 4 & \vdots & -2 \end{bmatrix}$$

$$\begin{matrix} -R_1 + R_2 \rightarrow \\ \\ -R_1 + R_3 \rightarrow \end{matrix} \begin{bmatrix} 2 & 1 & 2 & \vdots & 4 \\ 0 & 1 & -2 & \vdots & 1 \\ 0 & -2 & 4 & \vdots & -2 \end{bmatrix}$$

Let $z = a$, then $y - 2a = 1 \implies y = 2a + 1$, and $2x + 4a = 3 \implies x = -2a + \frac{3}{2}$.

Answer: $\left(-2a + \frac{3}{2}, 2a + 1, a\right)$

17.
$$\begin{bmatrix} 4 & 4 & 4 & \vdots & 5 \\ 4 & -2 & -8 & \vdots & 1 \\ 5 & 3 & 8 & \vdots & 6 \end{bmatrix}$$

$$\begin{matrix} R_3 - R_1 \rightarrow \\ -4R_1 + R_2 \rightarrow \\ -5R_1 + R_3 \rightarrow \end{matrix} \begin{bmatrix} 1 & -1 & 4 & \vdots & 1 \\ 0 & 2 & -24 & \vdots & -3 \\ 0 & 8 & -12 & \vdots & 1 \end{bmatrix}$$

$$\begin{matrix} R_2 + R_1 \rightarrow \\ \frac{1}{2}R_2 \rightarrow \\ -8R_2 + R_3 \rightarrow \end{matrix} \begin{bmatrix} 1 & 0 & -8 & \vdots & -\frac{1}{2} \\ 0 & 1 & -12 & \vdots & -\frac{3}{2} \\ 0 & 0 & 84 & \vdots & 13 \end{bmatrix}$$

$$\begin{matrix} 8R_3 + R_1 \rightarrow \\ 12R_3 + R_2 \rightarrow \\ \frac{1}{84}R_3 \rightarrow \end{matrix} \begin{bmatrix} 1 & 0 & 0 & \vdots & \frac{31}{42} \\ 0 & 1 & 0 & \vdots & \frac{5}{14} \\ 0 & 0 & 1 & \vdots & \frac{13}{84} \end{bmatrix}$$

$x = \frac{31}{42}, y = \frac{5}{14}, z = \frac{13}{84}$

Answer: $\left(\frac{31}{42}, \frac{5}{14}, \frac{13}{84}\right)$

19.
$$\begin{bmatrix} -1 & 1 & 2 & \vdots & 1 \\ 2 & 3 & 1 & \vdots & -2 \\ 5 & 4 & 2 & \vdots & 4 \end{bmatrix}$$

$$\begin{matrix} -R_1 \rightarrow \\ 2R_1 + R_2 \rightarrow \\ 5R_1 + R_3 \rightarrow \end{matrix} \begin{bmatrix} 1 & -1 & -2 & \vdots & -1 \\ 0 & 5 & 5 & \vdots & 0 \\ 0 & 9 & 12 & \vdots & 9 \end{bmatrix}$$

$$\begin{matrix} R_2 + R_1 \rightarrow \\ \frac{1}{5}R_2 \rightarrow \\ -9R_2 + R_3 \rightarrow \end{matrix} \begin{bmatrix} 1 & 0 & -1 & \vdots & -1 \\ 0 & 1 & 1 & \vdots & 0 \\ 0 & 0 & 3 & \vdots & 9 \end{bmatrix}$$

$$\begin{matrix} R_3 + R_1 \rightarrow \\ -R_3 + R_2 \rightarrow \\ \frac{1}{3}R_3 \rightarrow \end{matrix} \begin{bmatrix} 1 & 0 & 0 & \vdots & 2 \\ 0 & 1 & 0 & \vdots & -3 \\ 0 & 0 & 1 & \vdots & 3 \end{bmatrix}$$

$x = 2, y = -3, z = 3$

Answer: $(2, -3, 3)$

21.
$$\begin{bmatrix} 1 & 2 & 6 & \vdots & 1 \\ 2 & 5 & 15 & \vdots & 4 \\ 3 & 1 & 3 & \vdots & -6 \end{bmatrix}$$

$$\begin{matrix} \\ -2R_1 + R_2 \rightarrow \\ -3R_1 + R_3 \rightarrow \end{matrix} \begin{bmatrix} 1 & 2 & 6 & \vdots & 1 \\ 0 & 1 & 3 & \vdots & 2 \\ 0 & -5 & -15 & \vdots & -9 \end{bmatrix}$$

$$\begin{matrix} -2R_2 + R_1 \rightarrow \\ \\ 5R_2 + R_3 \rightarrow \end{matrix} \begin{bmatrix} 1 & 0 & 0 & \vdots & -3 \\ 0 & 1 & 3 & \vdots & 2 \\ 0 & 0 & 0 & \vdots & 1 \end{bmatrix}$$

$x = -3$

$y + 3z = 2$

$0 = 1$

Inconsistent, no solution

23. If a system of linear equations has a unique solution, the augmented matrix reduces to a form in which the number of rows with nonzero entries on the coefficient side of the matrix equals the number of variables.

25. $\begin{bmatrix} 1 & -3 & \vdots & -2 \\ 1 & 1 & \vdots & 2 \end{bmatrix} \Rightarrow \begin{bmatrix} 1 & 0 & \vdots & 1 \\ 0 & 1 & \vdots & 1 \end{bmatrix}$

$x = 1, \ y = 1$

27. $\begin{bmatrix} 1 & 2 & -1 & \vdots & 7 \\ 0 & -1 & -1 & \vdots & 4 \\ 4 & 0 & -1 & \vdots & 16 \end{bmatrix} \Rightarrow \begin{bmatrix} 1 & 0 & 0 & \vdots & 3 \\ 0 & 1 & 0 & \vdots & 0 \\ 0 & 0 & 1 & \vdots & -4 \end{bmatrix}$

$x = 3, \ y = 0, \ z = -4$

29. $\begin{bmatrix} 2 & 1 & 0 \\ 0 & 5 & -4 \end{bmatrix} - 3\begin{bmatrix} 5 & 3 & -6 \\ 0 & -2 & 5 \end{bmatrix} = \begin{bmatrix} 2 & 1 & 0 \\ 0 & 5 & -4 \end{bmatrix} - \begin{bmatrix} 15 & 9 & -18 \\ 0 & -6 & 15 \end{bmatrix}$

$$= \begin{bmatrix} -13 & -8 & 18 \\ 0 & 11 & -19 \end{bmatrix}$$

31. $\begin{bmatrix} 1 & 2 \\ 5 & -4 \\ 6 & 0 \end{bmatrix}\begin{bmatrix} 6 & -2 & 8 \\ 4 & 0 & 0 \end{bmatrix} = \begin{bmatrix} 1(6) + 2(4) & 1(-2) + 2(0) & 1(8) + 2(0) \\ 5(6) + (-4)(4) & 5(-2) + (-4)(0) & 5(8) + (-4)(0) \\ 6(6) + (0)(4) & 6(-2) + (0)(0) & 6(8) + (0)(0) \end{bmatrix}$

$$= \begin{bmatrix} 14 & -2 & 8 \\ 14 & -10 & 40 \\ 36 & -12 & 48 \end{bmatrix}$$

33. $\begin{bmatrix} 1 & 5 & 6 \\ 2 & -4 & 0 \end{bmatrix}\begin{bmatrix} 6 & 4 \\ -2 & 0 \\ 8 & 0 \end{bmatrix} = \begin{bmatrix} 1(6) + 5(-2) + 6(8) & 1(4) + 5(0) + 6(0) \\ 2(6) - 4(-2) + 0(8) & 2(4) - 4(0) + 0(0) \end{bmatrix}$

$$= \begin{bmatrix} 44 & 4 \\ 20 & 8 \end{bmatrix}$$

35. $\begin{bmatrix} 1 & 3 & 2 \\ 0 & 2 & -4 \\ 0 & 0 & 3 \end{bmatrix}\begin{bmatrix} 4 & -3 & 2 \\ 0 & 3 & -1 \\ 0 & 0 & 2 \end{bmatrix} = \begin{bmatrix} 1(4) & 1(-3) + 3(3) & 1(2) + 3(-1) + 2(2) \\ 0 & 2(3) & 2(-1) + (-4)(2) \\ 0 & 0 & 3(2) \end{bmatrix}$

$$= \begin{bmatrix} 4 & 6 & 3 \\ 0 & 6 & -10 \\ 0 & 0 & 6 \end{bmatrix}$$

37. $3\begin{bmatrix} 8 & -2 & 5 \\ 1 & 3 & -1 \end{bmatrix} + 6\begin{bmatrix} 4 & -2 & -3 \\ 2 & 7 & 6 \end{bmatrix} = \begin{bmatrix} 48 & -18 & -3 \\ 15 & 51 & 33 \end{bmatrix}$

39. $\begin{bmatrix} 4 & 1 \\ 11 & -7 \\ 12 & 3 \end{bmatrix}\begin{bmatrix} 3 & -5 & 6 \\ 2 & -2 & -2 \end{bmatrix} = \begin{bmatrix} 14 & -22 & 22 \\ 19 & -41 & 80 \\ 42 & -66 & 66 \end{bmatrix}$

41. $X = 3A - 2B = 3\begin{bmatrix} -4 & 0 \\ 1 & -5 \\ -3 & 2 \end{bmatrix} - 2\begin{bmatrix} 1 & 2 \\ -2 & 1 \\ 4 & 4 \end{bmatrix} = \begin{bmatrix} -14 & -4 \\ 7 & -17 \\ -17 & -2 \end{bmatrix}$

43. $X = \frac{1}{3}[B - 2A] = \frac{1}{3}\left(\begin{bmatrix} 1 & 2 \\ -2 & 1 \\ 4 & 4 \end{bmatrix} - 2\begin{bmatrix} -4 & 0 \\ 1 & -5 \\ -3 & 2 \end{bmatrix}\right) = \frac{1}{3}\begin{bmatrix} 9 & 2 \\ -4 & 11 \\ 10 & 0 \end{bmatrix}$

45. $\begin{bmatrix} 5 & 4 \\ -1 & 1 \end{bmatrix}\begin{bmatrix} x \\ y \end{bmatrix} = \begin{bmatrix} 2 \\ -22 \end{bmatrix} \Rightarrow \begin{bmatrix} 5x + 4y \\ -x + y \end{bmatrix} = \begin{bmatrix} 2 \\ -22 \end{bmatrix} \Rightarrow \begin{array}{r} 5x + 4y = 2 \\ -x + y = -22 \end{array}$

47. $\begin{bmatrix} 2 & 6 \\ 3 & -6 \end{bmatrix}^{-1} = \begin{bmatrix} \frac{1}{5} & \frac{1}{5} \\ \frac{1}{10} & -\frac{1}{15} \end{bmatrix}$

49. $\begin{bmatrix} 2 & 0 & 3 \\ -1 & 1 & 1 \\ 2 & -2 & 1 \end{bmatrix}^{-1} = \begin{bmatrix} \frac{1}{2} & -1 & -\frac{1}{2} \\ \frac{1}{2} & -\frac{2}{3} & -\frac{5}{6} \\ 0 & \frac{2}{3} & \frac{1}{3} \end{bmatrix}$

51. $\begin{vmatrix} 50 & -30 \\ 10 & 5 \end{vmatrix} = 50(5) - (-30)(10) = 550$

53. $\begin{vmatrix} 10 & 8 \\ -6 & -4 \end{vmatrix} = 10(-4) - (-6)(8) = -40 + 48 = 8$

55. $\begin{vmatrix} 1 & 0 & -2 \\ 0 & 1 & 0 \\ -2 & 0 & 1 \end{vmatrix} = 1\begin{vmatrix} 1 & -2 \\ -2 & 1 \end{vmatrix} = 1(1) - (-2)(-2) = 1 - 4 = -3$

57. $\begin{vmatrix} 3 & 0 & -4 & 0 \\ 0 & 8 & 1 & 2 \\ 6 & 1 & 8 & 2 \\ 0 & 3 & -4 & 1 \end{vmatrix} = 3\begin{vmatrix} 8 & 1 & 2 \\ 1 & 8 & 2 \\ 3 & -4 & 1 \end{vmatrix} + (-4)\begin{vmatrix} 0 & 8 & 2 \\ 6 & 1 & 2 \\ 0 & 3 & 1 \end{vmatrix}$ (Expansion along Row 1)

$= 3[3(8 - (-8)) - 1(1 - 6) + 2(-4 - 24)] - 4[0 - 6(8 - 6) + 0]$

$= 3[128 + 5 - 56] - 4[-12]$

$= 279$

59. $x + 2y = -1$

$3x + 4y = -5$

$\begin{bmatrix} 1 & 2 \\ 3 & 4 \end{bmatrix}^{-1} = \begin{bmatrix} -2 & 1 \\ \frac{3}{2} & -\frac{1}{2} \end{bmatrix} \Rightarrow \begin{bmatrix} x \\ y \end{bmatrix} = \begin{bmatrix} -2 & 1 \\ \frac{3}{2} & -\frac{1}{2} \end{bmatrix}\begin{bmatrix} -1 \\ -5 \end{bmatrix} = \begin{bmatrix} -3 \\ 1 \end{bmatrix}$

$x = -3, y = 1$

Answer: $(-3, 1)$

61. $-3x - 3y - 4z = 2$

$\qquad\quad y + z = -1$

$\quad 4x + 3y + 4z = -1$

$$\begin{bmatrix} -3 & -3 & -4 \\ 0 & 1 & 1 \\ 4 & 3 & 4 \end{bmatrix}^{-1} = \begin{bmatrix} 1 & 0 & 1 \\ 4 & 4 & 3 \\ 4 & -3 & -3 \end{bmatrix} \Rightarrow \begin{bmatrix} x \\ y \\ z \end{bmatrix} = \begin{bmatrix} 1 & 0 & 1 \\ 4 & 4 & 3 \\ -4 & -3 & -3 \end{bmatrix}\begin{bmatrix} 2 \\ -1 \\ -1 \end{bmatrix} = \begin{bmatrix} 1 \\ 1 \\ -2 \end{bmatrix}$$

$x = 1, y = 1, z = -2$

Answer: $(1, 1, -2)$

63. $\quad x + 3y + 2z = 2$

$\quad -2x - 5y - z = 10$

$\quad 2x + 4y \quad\;\; = -12$

$$\begin{bmatrix} 1 & 3 & 2 \\ -2 & -5 & -1 \\ 2 & 4 & 0 \end{bmatrix}^{-1} = \begin{matrix} 2 & 4 & \frac{7}{2} \\ -1 & -2 & -\frac{3}{2} \\ 1 & 1 & \frac{1}{2} \end{matrix} \Rightarrow \begin{bmatrix} x \\ y \\ z \end{bmatrix} = \begin{matrix} 2 & 4 & \frac{7}{2} \\ -1 & -2 & -\frac{3}{2} \\ 1 & 1 & \frac{1}{2} \end{matrix}\begin{bmatrix} 2 \\ 10 \\ -12 \end{bmatrix} = \begin{bmatrix} 2 \\ -4 \\ 6 \end{bmatrix}$$

$x = 2, y = -4, z = 6$

Answer: $(2, -4, 6)$

65. $-x + y + z = 6$

$\quad 4x - 3y + z = 20$

$\quad 2x - y + 3z = 8$

$$\begin{bmatrix} -1 & 1 & 1 \\ 4 & -3 & 1 \\ 2 & -1 & 3 \end{bmatrix}^{-1} \quad \text{does not exist.}$$

The system is inconsistent and has no solution.

67. $x = \dfrac{\begin{vmatrix} 5 & 2 \\ 1 & 1 \end{vmatrix}}{\begin{vmatrix} 1 & 2 \\ -1 & 1 \end{vmatrix}} = \dfrac{3}{3} = 1$

$y = \dfrac{\begin{vmatrix} 1 & 5 \\ -1 & 1 \end{vmatrix}}{\begin{vmatrix} 1 & 2 \\ -1 & 1 \end{vmatrix}} = \dfrac{6}{3} = 2$

69. $x = \dfrac{\begin{vmatrix} 11 & 8 \\ 21 & -24 \end{vmatrix}}{\begin{vmatrix} 20 & 8 \\ 12 & -24 \end{vmatrix}} = \dfrac{-432}{-576} = \dfrac{3}{4}$

$y = \dfrac{\begin{vmatrix} 20 & 11 \\ 12 & 21 \end{vmatrix}}{\begin{vmatrix} 20 & 8 \\ 12 & -24 \end{vmatrix}} = \dfrac{288}{-576} = -\dfrac{1}{2}$

71. $x = \dfrac{\begin{vmatrix} 5 & 6 \\ 11 & 14 \end{vmatrix}}{\begin{vmatrix} 3 & 6 \\ 6 & 14 \end{vmatrix}} = \dfrac{4}{6} = \dfrac{2}{3}$

$y = \dfrac{\begin{vmatrix} 3 & 5 \\ 6 & 11 \end{vmatrix}}{\begin{vmatrix} 3 & 6 \\ 6 & 14 \end{vmatrix}} = \dfrac{3}{6} = \dfrac{1}{2}$

73. Cramer's Rule does not apply because the determinant of the coefficient matrix is zero.

75. $x = $ number of carnations

$y = $ number of roses

$$x + \quad y = 12$$
$$0.75x + 1.50y = 12.00$$

$$\begin{bmatrix} 1 & 1 & \vdots & 12 \\ 0.75 & 1.50 & \vdots & 12 \end{bmatrix} -0.75R_1 + R_2 \begin{bmatrix} 1 & 1 & \vdots & 12 \\ 0 & 0.75 & \vdots & 3 \end{bmatrix}$$

$0.75y = 3 \implies y = 4$

$x + (4) = 12 \implies x = 8$

Answer: 8 carnations, 4 roses

77. $(-1, 2), \ (0, 3), \ (1, 6)$

$f(x) = ax^2 + bx + c$

$$\left.\begin{aligned} f(-1) &= a - b + c = 2 \implies a - b = -1 \\ f(0) &= c = 3 \\ f(1) &= a + b + c = 6 \implies a + b = \quad 3 \end{aligned}\right\} a = 1, b = 2$$

Thus, $y = x^2 + 2x + 3.$

79. $13a + \quad 91b = 1107$

$91a + 819b = 8404.7$

(a) $a \approx 59.9, \ b \approx 3.6$

$y = 59.9 + 3.6t$

(b)

(c) The median price of one-family homes sold in the United States has been increasing by an average of 3.6 thousand dollars ($3600) each year.

(d) $y(15) = 59.9 + 3.6(15) = 113.9$ which corresponds to a price of $113,900.

81. $(1, 0), \ (5, 0), \ (5, 8)$

$$\text{Area} = \frac{1}{2} \begin{vmatrix} 1 & 0 & 1 \\ 5 & 0 & 1 \\ 5 & 8 & 1 \end{vmatrix} = \frac{1}{2}(32)$$

$= 16$ square units

83. $(1, 2), \ (4, -5), \ (3, 2)$

$$\text{Area} = \frac{1}{2} \begin{vmatrix} 1 & 2 & 1 \\ 4 & -5 & 1 \\ 3 & 1 & 1 \end{vmatrix} = \frac{1}{2}(14)$$

$= 7$ square units

85. $(-4, 0), \ (4, 4)$

$$\begin{vmatrix} x & y & 1 \\ -4 & 0 & 1 \\ 4 & 4 & 1 \end{vmatrix} = 0$$

$-4x + 8y - 16 = 0$

$x - 2y + 4 = 0$

87. $\left(-\frac{5}{2}, 3\right), \ \left(\frac{7}{2}, 1\right)$

$$\begin{vmatrix} x & y & 1 \\ -\frac{5}{2} & 3 & 1 \\ \frac{7}{2} & 1 & 1 \end{vmatrix} = 0$$

$-2x - 6y + 13 = 0$

89. Expansion by Row 3

$$\begin{vmatrix} a_{11} & a_{12} & a_{13} \\ a_{21} & a_{22} & a_{23} \\ a_{31} + c_1 & a_{32} + c_2 & a_{33} + c_3 \end{vmatrix} = (a_{31} + c_1) \begin{vmatrix} a_{12} & a_{13} \\ a_{22} & a_{23} \end{vmatrix} - (a_{32} + c_2) \begin{vmatrix} a_{11} & a_{13} \\ a_{21} & a_{23} \end{vmatrix} + (a_{33} + c_3) \begin{vmatrix} a_{11} & a_{12} \\ a_{21} & a_{22} \end{vmatrix}$$

$$= a_{31} \begin{vmatrix} a_{12} & a_{13} \\ a_{22} & a_{23} \end{vmatrix}$$

$$- a_{32} \begin{vmatrix} a_{11} & a_{13} \\ a_{21} & a_{23} \end{vmatrix} + a_{33} \begin{vmatrix} a_{11} & a_{12} \\ a_{21} & a_{22} \end{vmatrix}$$

$$+ c_1 \begin{vmatrix} a_{12} & a_{13} \\ a_{22} & a_{23} \end{vmatrix} - c_2 \begin{vmatrix} a_{11} & a_{13} \\ a_{21} & a_{23} \end{vmatrix} + c_3 \begin{vmatrix} a_{11} & a_{12} \\ a_{21} & a_{22} \end{vmatrix}$$

$$= \begin{vmatrix} a_{11} & a_{12} & a_{13} \\ a_{21} & a_{22} & a_{23} \\ a_{31} & a_{32} & a_{33} \end{vmatrix} + \begin{vmatrix} a_{11} & a_{12} & a_{13} \\ a_{21} & a_{22} & a_{23} \\ c_1 & c_2 & c_3 \end{vmatrix}$$

Note: Expand each of these matrices by Row 3 to see the previous step.

91. Since each of the three rows is multiplied by 4, $|4A| = 4^3 |A| = 4^3(2) = 128$.

❑ Chapter Test for Chapter 8

1. $\begin{bmatrix} 1 & -1 & 5 \\ 6 & 2 & 3 \\ 5 & 3 & 3 \end{bmatrix} \Rightarrow \begin{bmatrix} 1 & -1 & 5 \\ 0 & 8 & -27 \\ 0 & 8 & -28 \end{bmatrix}$

$\Rightarrow \begin{bmatrix} 1 & -1 & 5 \\ 0 & 8 & -27 \\ 0 & 0 & -1 \end{bmatrix}$

$\Rightarrow \begin{bmatrix} 1 & -1 & 0 \\ 0 & 8 & 0 \\ 0 & 0 & 1 \end{bmatrix}$

$\Rightarrow \begin{bmatrix} 1 & 0 & 0 \\ 0 & 1 & 0 \\ 0 & 0 & 1 \end{bmatrix}$

2. $\begin{bmatrix} 1 & 0 & -1 & 2 \\ -1 & 1 & 1 & -3 \\ 1 & 1 & -1 & 1 \\ 3 & 2 & -3 & 4 \end{bmatrix} \Rightarrow \begin{bmatrix} 1 & 0 & -1 & 2 \\ 0 & 1 & 0 & -1 \\ 0 & 1 & 0 & -1 \\ 0 & 2 & 0 & -2 \end{bmatrix}$

$\Rightarrow \begin{bmatrix} 1 & 0 & -1 & 2 \\ 0 & 1 & 0 & -1 \\ 0 & 0 & 0 & 0 \\ 0 & 0 & 0 & 0 \end{bmatrix}$

3. $\begin{bmatrix} 4 & 3 & -2 & \vdots & 14 \\ -1 & -1 & 2 & \vdots & -5 \\ 3 & 1 & -4 & \vdots & 8 \end{bmatrix} \Rightarrow \begin{bmatrix} 1 & 0 & 0 & \vdots & 1 \\ 0 & 1 & 0 & \vdots & 3 \\ 0 & 0 & 1 & \vdots & -\frac{1}{2} \end{bmatrix} \Rightarrow (x, y, z) = \left(1, 3, -\frac{1}{2}\right)$

4. $-2 = a(-2)^2 + b(-2) + c$

$\quad 2 = a(2)^2 + b(2) + c$

$-2 = a(4)^2 + b(4) + c$

$4a - 2b + c = -2$

$4a + 2b + c = 2$

$16a + 4b + c = -2$

Row-reducing the augmented matrix yields

$\begin{bmatrix} 4 & -2 & 1 & \vdots & -2 \\ 4 & 2 & 1 & \vdots & 2 \\ 16 & 4 & 1 & \vdots & -2 \end{bmatrix} \Rightarrow \begin{bmatrix} 1 & 0 & 0 & \vdots & -\frac{1}{2} \\ 0 & 1 & 0 & \vdots & 1 \\ 0 & 0 & 1 & \vdots & 2 \end{bmatrix}$.

Thus, $y = -\frac{1}{2}x^2 + x + 2$.

5. (a) $A - B = \begin{bmatrix} 5 & 4 & 4 \\ -4 & -4 & 0 \end{bmatrix} - \begin{bmatrix} 4 & -1 & 6 \\ -4 & 0 & -3 \end{bmatrix} = \begin{bmatrix} 1 & 5 & -2 \\ 0 & -4 & 3 \end{bmatrix}$

(b) $3A = 3\begin{bmatrix} 5 & 4 & 4 \\ -4 & -4 & 0 \end{bmatrix} = \begin{bmatrix} 15 & 12 & 12 \\ -12 & -12 & 0 \end{bmatrix}$

(c) $3A - 2B = \begin{bmatrix} 15 & 12 & 12 \\ -12 & -12 & 0 \end{bmatrix} - 2\begin{bmatrix} 4 & -1 & 6 \\ -4 & 0 & -3 \end{bmatrix} = \begin{bmatrix} 7 & 14 & 0 \\ -4 & -12 & 6 \end{bmatrix}$

6. $AB = \begin{bmatrix} 2 & -2 & 6 \\ 3 & -1 & 7 \\ 2 & 0 & -2 \end{bmatrix}\begin{bmatrix} 4 & 4 \\ 3 & 2 \\ 1 & -2 \end{bmatrix} = \begin{bmatrix} 8 & -8 \\ 16 & -4 \\ 6 & 12 \end{bmatrix}$

7. $A^{-1} = \dfrac{1}{ad - bc}\begin{bmatrix} d & -b \\ -c & a \end{bmatrix} = \dfrac{1}{30 - 40}\begin{bmatrix} -5 & -4 \\ -10 & -6 \end{bmatrix} = \dfrac{1}{10}\begin{bmatrix} 5 & 4 \\ 10 & 6 \end{bmatrix} = \begin{bmatrix} \frac{1}{2} & \frac{2}{5} \\ 1 & \frac{3}{5} \end{bmatrix}$

8. $X = A^{-1}B = \begin{bmatrix} \frac{1}{2} & \frac{2}{5} \\ 1 & \frac{3}{5} \end{bmatrix}\begin{bmatrix} 10 \\ 20 \end{bmatrix} = \begin{bmatrix} 13 \\ 22 \end{bmatrix} \implies (x, y) = (13, 22)$

9. $\det(A) = \begin{vmatrix} 4 & 0 & 3 \\ 1 & -8 & 2 \\ 3 & 2 & 2 \end{vmatrix} = 4(-16 - 4) - 0 + 3(2 + 24)$

$= -80 + 78 = -2$

10. $\begin{vmatrix} -5 & 0 & 1 \\ 3 & 2 & 1 \\ 4 & 4 & 1 \end{vmatrix} = -5(2 - 4) - 0 + 1(12 - 8) = 10 + 4 = 14$

Area $= \frac{1}{2}(14) = 7$

❏ Practice Test for Chapter 8

1. Write the matrix in reduced row-echelon form.

$$\begin{bmatrix} 1 & -2 & 4 \\ 3 & -5 & 9 \end{bmatrix}$$

For Exercises 2–4, use matrices to solve the system of equations.

2. $3x + 5y = 3$
$2x - y = -11$

3. $2x + 3y = -3$
$3x + 2y = 8$
$x + y = 1$

4. $x + 3z = -5$
$2x + y = 0$
$3x + y - z = 3$

5. Multiply $\begin{bmatrix} 1 & 4 & 5 \\ 2 & 0 & -3 \end{bmatrix} \begin{bmatrix} 1 & 6 \\ 0 & -7 \\ -1 & 2 \end{bmatrix}$

6. Given $A = \begin{bmatrix} 9 & 1 \\ -4 & 8 \end{bmatrix}$ and $B = \begin{bmatrix} 6 & -2 \\ 3 & 5 \end{bmatrix}$, find $3A - 5B$.

7. Find $f(A)$:

$$f(x) = x^2 - 7x + 8, \quad A = \begin{bmatrix} 3 & 0 \\ 7 & 1 \end{bmatrix}$$

8. True or false:

$(A + B)(A + 3B) = A^2 + 4AB + 3B^2$ where A and B are matrices.

(Assume that A^2, AB, and B^2 exist.)

For Exercises 9–10, find the inverse of the matrix, if it exists.

9. $\begin{bmatrix} 1 & 2 \\ 3 & 5 \end{bmatrix}$

10. $\begin{bmatrix} 1 & 1 & 1 \\ 3 & 6 & 5 \\ 6 & 10 & 8 \end{bmatrix}$

11. Use an inverse matrix to solve the systems.

(a) $x + 2y = 4$
$3x + 5y = 1$

(b) $x + 2y = 3$
$3x + 5y = -2$

For Exercises 12–13, find the determinant of the matrix.

12. $\begin{bmatrix} 6 & -1 \\ 3 & 4 \end{bmatrix}$

13. $\begin{bmatrix} 1 & 3 & -1 \\ 5 & 9 & 0 \\ 6 & 2 & -5 \end{bmatrix}$

14. Use a graphing utility to find the determinant of the matrix.

$$\begin{bmatrix} 1 & 4 & 2 & 3 \\ 0 & 1 & -2 & 0 \\ 3 & 5 & -1 & 1 \\ 2 & 0 & 6 & 1 \end{bmatrix}$$

15. Evaluate $\begin{vmatrix} 6 & 4 & 3 & 0 & 6 \\ 0 & 5 & 1 & 4 & 8 \\ 0 & 0 & 2 & 7 & 3 \\ 0 & 0 & 0 & 9 & 2 \\ 0 & 0 & 0 & 0 & 1 \end{vmatrix}$.

16. Use a determinant to find the area of the triangle with vertices $(0, 7)$, $(5, 0)$, and $(3, 9)$.

17. Use a determinant to find the equation of the line through $(2, 7)$ and $(-1, 4)$.

For Exercises 18–20, use Cramer's Rule to find the indicated value.

18. Find x.

$$6x - 7y = 4$$
$$2x + 5y = 11$$

19. Find z.

$$3x \quad + z = 1$$
$$y + 4z = 3$$
$$x - y \quad = 2$$

20. Find y.

$$721.4x - 29.1y = 33.77$$
$$45.9x + 105.6y = 19.85$$

CHAPTER 9
Sequences, Probability, and Statistics

CHAPTER 9
Sequences, Probability, and Statistics

Section 9.1   Sequences and Summation Notation

- Given the general nth term in a sequence, you should be able to find, or list, some of the terms.
- You should be able to find an expression for the nth term of a sequence.
- You should be able to use and evaluate factorials.
- You should be able to use sigma notation for a sum.

Solutions to Odd-Numbered Exercises

1. $a_n = 2n + 1$

$a_1 = 2(1) + 1 = 3$

$a_2 = 2(2) + 1 = 5$

$a_3 = 2(3) + 1 = 7$

$a_4 = 2(4) + 1 = 9$

$a_5 = 2(5) + 1 = 11$

3. $a_n = 2^n$

$a_1 = 2^1 = 2$

$a_2 = 2^2 = 4$

$a_3 = 2^3 = 8$

$a_4 = 2^4 = 16$

$a_5 = 2^5 = 32$

5. $a_n = (-2)^n$

$a_1 = (-2)^1 = -2$

$a_3 = (-2)^2 = 4$

$a_3 = (-2)^3 = -8$

$a_4 = (-2)^4 = 16$

$a_5 = (-2)^5 = -32$

7. $a_n = \dfrac{n + 1}{n}$

$a_1 = \dfrac{1 + 1}{1} = 2$

$a_2 = \dfrac{3}{2}$

$a_3 = \dfrac{4}{3}$

$a_4 = \dfrac{5}{4}$

$a_5 = \dfrac{6}{5}$

9. $a_n = \dfrac{6n}{3n^2 - 1}$

$a_1 = \dfrac{6(1)}{3(1)^2 - 1} = 3$

$a_2 = \dfrac{6(2)}{3(2)^2 - 1} = \dfrac{12}{11}$

$a_3 = \dfrac{6(3)}{3(3)^2 - 1} = \dfrac{9}{13}$

$a_4 = \dfrac{6(4)}{3(4)^2 - 1} = \dfrac{24}{47}$

$a_5 = \dfrac{6(5)}{3(5)^2 - 1} = \dfrac{15}{37}$

11. $a_n = \dfrac{1 + (-1)^n}{n}$

$a_1 = 0$

$a_2 = \dfrac{2}{2} = 1$

$a_3 = 0$

$a_4 = \dfrac{2}{4} = \dfrac{1}{2}$

$a_5 = 0$

13. $a_n = 3 - \dfrac{1}{2^n}$

$a_1 = 3 - \dfrac{1}{2} = \dfrac{5}{2}$

$a_2 = 3 - \dfrac{1}{4} = \dfrac{11}{4}$

$a_3 = 3 - \dfrac{1}{8} = \dfrac{23}{8}$

$a_4 = 3 - \dfrac{1}{16} = \dfrac{47}{16}$

$a_5 = 3 - \dfrac{1}{32} = \dfrac{95}{32}$

15. $a_n = \dfrac{1}{n^{3/2}}$

$a_1 = \dfrac{1}{1} = 1$

$a_2 = \dfrac{1}{2^{3/2}}$

$a_3 = \dfrac{1}{3^{3/2}}$

$a_4 = \dfrac{1}{4^{3/2}} = \dfrac{1}{8}$

$a_5 = \dfrac{1}{5^{3/2}}$

17. $a_n = \dfrac{3^n}{n!}$

$a_1 = \dfrac{3^1}{1!} = \dfrac{3}{1} = 3$

$a_2 = \dfrac{3^2}{2!} = \dfrac{9}{2}$

$a_3 = \dfrac{27}{6} = \dfrac{9}{2}$

$a_4 = \dfrac{81}{24} = \dfrac{27}{8}$

$a_5 = \dfrac{243}{120} = \dfrac{81}{40}$

19. $a_n = \dfrac{(-1)^n}{n^2}$

$a_1 = \dfrac{-1}{1} = -1$

$a_2 = \dfrac{1}{4}$

$a_3 = \dfrac{-1}{9}$

$a_4 = \dfrac{1}{16}$

$a_5 = \dfrac{-1}{25}$

21. $a_n = \dfrac{2}{3}$

$a_1 = \dfrac{2}{3}$

$a_2 = \dfrac{2}{3}$

$a_3 = \dfrac{2}{3}$

$a_4 = \dfrac{2}{3}$

$a_5 = \dfrac{2}{3}$

23. $a_{25} = (-1)^{25}[3(25) - 2] = -73$

25. $a_1 = 28$ and $a_{k+1} = a_k - 4$

$a_1 = 28$

$a_2 = a_1 - 4 = 28 - 4 = 24$

$a_3 = a_2 - 4 = 24 - 4 = 20$

$a_4 = a_3 - 4 = 20 - 4 = 16$

$a_5 = a_4 - 4 = 16 - 4 = 12$

27. $a_1 = 3$ and $a_{k+1} = 2(a_k - 1)$

$a_1 = 3$

$a_2 = 2(a_1 - 1) = 2(3 - 1) = 4$

$a_3 = 2(a_2 - 1) = 2(4 - 1) = 6$

$a_4 = 2(a_3 - 1) = 2(6 - 1) = 10$

$a_5 = 2(a_4 - 1) = 2(10 - 1) = 18$

29. $a_n = \dfrac{2}{3} n$

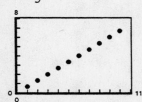

31. $a_n = 16(-0.5)^{n-1}$

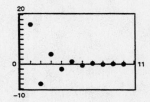

33. $a_n = \dfrac{2n}{n + 1}$

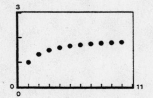

35. $a_n = \dfrac{8}{n+1}$

$a_n \to 0$ as $n \to \infty$

$a_1 = 4, \ a_{10} = \dfrac{8}{11}$

Matches graph (c).

37. $a_n = 4(0.5)^{n-1}$

$a_n \to 0$ as $n \to \infty$

$a_1 = 4, \ a_{10} \approx 0.008$

Matches graph (d).

39. $\dfrac{4!}{6!} = \dfrac{4!}{6 \cdot 5 \cdot 4!} = \dfrac{1}{30}$

41. $\dfrac{10!}{8!} = \dfrac{10 \cdot 9 \cdot 8!}{8!} = 90$

43. $\dfrac{(n+1)!}{n!} = \dfrac{(n+1)n!}{n!} = n+1$

45. $\dfrac{(2n-1)!}{(2n+1)!} = \dfrac{(2n-1)!}{(2n+1)(2n)(2n-1)!}$

$\qquad = \dfrac{1}{2n(2n+1)}$

47. $1, 4, 7, 10, 13, \ldots$

$a_n = 1 + (n-1)3 = 3n - 2$

49. $0, 3, 8, 15, 24, \ldots$

$a_n = n^2 - 1$

51. $\dfrac{2}{3}, \dfrac{3}{4}, \dfrac{4}{5}, \dfrac{5}{6}, \dfrac{6}{7}, \ldots$

$a_n = \dfrac{n+1}{n+2}$

53. $\dfrac{1}{2}, \dfrac{-1}{4}, \dfrac{1}{8}, \dfrac{-1}{16}, \ldots$

$a_n = \dfrac{(-1)^{n+1}}{2^n}$

55. $1 + \dfrac{1}{1}, 1 + \dfrac{1}{2}, 1 + \dfrac{1}{3}, 1 + \dfrac{1}{4}, 1 + \dfrac{1}{5}, \ldots$

$a_n = 1 + \dfrac{1}{n}$

57. $1, \dfrac{1}{2}, \dfrac{1}{6}, \dfrac{1}{24}, \dfrac{1}{120}, \ldots$

$a_n = \dfrac{1}{n!}$

59. $1, -1, 1, -1, 1, \ldots$

$a_n = (-1)^{n+1}$

61. $a_1 = 6$ and $a_{k+1} = a_k + 2$

$a_1 = 6$

$a_2 = a_1 + 2 = 6 + 2 = 8$

$a_3 = a_2 + 2 = 8 + 2 = 10$

$a_4 = a_3 + 2 = 10 + 2 = 12$

$a_5 = a_4 + 2 = 12 + 2 = 14$

In general, $a_n = 2n + 4$.

63. $a_1 = 81$ and $a_{k+1} = \dfrac{1}{3} a_k$

$a_1 = 81$

$a_2 = \dfrac{1}{3} a_1 = \dfrac{1}{3}(81) = 27$

$a_3 = \dfrac{1}{3} a_2 = \dfrac{1}{3}(27) = 9$

$a_4 = \dfrac{1}{3} a_3 = \dfrac{1}{3}(9) = 3$

$a_5 = \dfrac{1}{3} a_4 = \dfrac{1}{3}(3) = 1$

In general, $a_n = 81\left(\dfrac{1}{3}\right)^{n-1} = 81(3)\left(\dfrac{1}{3}\right)^n = \dfrac{253}{3^n}$.

65. $\displaystyle\sum_{i=1}^{5} (2i+1) = (2+1) + (4+1) + (6+1) + (8+1) + (10+1) = 35$

67. $\displaystyle\sum_{k=1}^{4} 10 = 10 + 10 + 10 + 10 = 40$

69. $\displaystyle\sum_{i=0}^{4} i^2 = 0^2 + 1^2 + 2^2 + 3^2 + 4^2 = 30$

71. $\displaystyle\sum_{k=0}^{3}\frac{1}{k^2+1} = \frac{1}{1} + \frac{1}{1+1} + \frac{1}{1+4} + \frac{1}{9+1} = \frac{9}{5}$

73. $\displaystyle\sum_{i=1}^{4}[(i-1)^2 + (i+1)^3] = [(0)^2 + (2)^3] + [(1)^2 + (3)^3] + [(2)^2 + (4)^3] + [(3)^2 + (5)^3] = 238$

75. $\displaystyle\sum_{i=1}^{4}2^i = 2^1 + 2^2 + 2^3 + 2^4 = 30$ **77.** $\displaystyle\sum_{j=1}^{6}(24 - 3j) = 81$

79. $\displaystyle\sum_{k=0}^{4}\frac{(-1)^k}{k+1} = \frac{47}{60}$

81. $\displaystyle\frac{1}{3(1)} + \frac{1}{3(2)} + \frac{1}{3(3)} + \cdots + \frac{1}{3(9)} = \sum_{i=1}^{9}\frac{1}{3i}$

83. $\displaystyle\left[2\left(\frac{1}{8}\right) + 3\right] + \left[2\left(\frac{2}{8}\right) + 3\right] + \left[2\left(\frac{3}{8}\right) + 3\right] + \cdots + \left[2\left(\frac{8}{8}\right) + 3\right] = \sum_{i=1}^{8}\left[2\left(\frac{i}{8}\right) + 3\right]$

85. $\displaystyle 3 - 9 + 27 - 81 + 243 - 729 = \sum_{i=1}^{6}(-1)^{i+1}3^i$

87. $\displaystyle\frac{1}{1^2} - \frac{1}{2^2} + \frac{1}{3^2} - \frac{1}{4^2} + \cdots + -\frac{1}{20^2} = \sum_{i=1}^{20}\frac{(-1)^{i+1}}{i^2}$

89. $\displaystyle\frac{1}{4} + \frac{3}{8} + \frac{7}{16} + \frac{15}{32} + \frac{31}{64} = \sum_{i=1}^{5}\frac{2i-1}{2^{i+1}}$

91. $A_n = 5000\left(1 + \dfrac{0.08}{4}\right)^n$, $n = 1, 2, 3, \ldots$

(a) $A_1 = \$5100.00$
$A_2 = \$5202.00$
$A_3 = \$5306.04$
$A_4 = \$5412.16$
$A_5 = \$5520.40$
$A_6 = \$5630.81$
$A_7 = \$5743.43$
$A_8 = \$5858.30$

(b) $A_{40} = \$11,040.20$

93. $a_n = 510.13 + 16.37n + 3.23n^2$, $n = 1, \ldots, 11$

$a_1 = 529.73$
$a_2 = 555.79$
$a_3 = 588.31$
$a_4 = 627.29$
$a_5 = 672.73$
$a_6 = 724.63$
$a_7 = 782.99$
$a_8 = 847.81$
$a_9 = 919.09$
$a_{10} = 996.83$
$a_{11} = 1081.03$

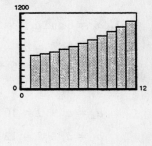

95. $\displaystyle\sum_{n=5}^{14}(129.9 + 0.9n^3) = \$11,131.5$ **million**

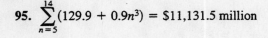

97. $a_1 = 1,\ a_2 = 1,\ a_{k+2} = a_{k+1} + a_k$

$a_1 = 1$ $\qquad\qquad$ $b_1 = \frac{1}{1} = 1$

$a_2 = 1$ $\qquad\qquad$ $b_2 = \frac{2}{1} = 2$

$a_3 = 1 + 1 = 2$ $\qquad$ $b_3 = \frac{3}{2}$

$a_4 = 2 + 1 = 3$ $\qquad$ $b_4 = \frac{5}{3}$

$a_5 = 3 + 2 = 5$ $\qquad$ $b_5 = \frac{8}{5}$

$a_6 = 5 + 3 = 8$ $\qquad$ $b_6 = \frac{13}{8}$

$a_7 = 8 + 5 = 13$ $\qquad$ $b_7 = \frac{21}{13}$

$a_8 = 13 + 8 = 21$ $\qquad$ $b_8 = \frac{34}{21}$

$a_9 = 21 + 13 = 34$ $\qquad$ $b_9 = \frac{55}{34}$

$a_{10} = 34 + 21 = 55$ $\qquad$ $b_{10} = \frac{89}{55}$

$a_{11} = 55 + 34 = 89$

$a_{12} = 89 + 55 = 144$

Section 9.2 Arithmetic Sequences

- You should be able to recognize an arithmetic sequence, find its common difference, and find its nth term.
- You should be able to find the nth partial sum of an arithmetic sequence with common difference d using the formula

$$S_n = \frac{n}{2}(a_1 + a_n).$$

Solutions to Odd-Numbered Exercises

1. $10, 8, 6, 4, 2, \ldots$

Arithmetic sequence, $d = -2$

3. $1, 2, 4, 8, 16, 32, \ldots$

Not an arithmetic sequence

5. $\frac{9}{4}, 2, \frac{7}{4}, \frac{3}{2}, \frac{5}{4}, \ldots$

Arithmetic sequence, $d = -\frac{1}{4}$

7. $-12, -8, -4, 0, 4, \ldots$

Arithmetic sequence, $d = 4$

9. $5.3, 5.7, 6.1, 6.5, 6.9, \ldots$

Arithmetic sequence, $d = 0.4$

11. $a_n = 5 + 3n$

$8, 11, 14, 17, 20$

Arithmetic sequence, $d = 3$

13. $a_n = \dfrac{1}{n + 1}$

$\dfrac{1}{2}, \dfrac{1}{3}, \dfrac{1}{4}, \dfrac{1}{5}, \dfrac{1}{6}$

Not an arithmetic sequence

15. $a_n = 100 - 3n$

$97, 94, 91, 88, 85$

Arithmetic sequence, $d = -3$

17. $a_n = 3 + \dfrac{(-1)^n 2}{n}$

$1, 4, \dfrac{7}{3}, \dfrac{7}{2}, \dfrac{13}{5}$

Not an arithmetic sequence

19. $a_1 = 15$, $a_{k+1} = a_k + 4$
$a_2 = 15 + 4 = 19$
$a_3 = 19 + 4 = 23$
$a_4 = 23 + 4 = 27$
$a_5 = 27 + 4 = 31$
$a_n = 11 + 4n$

21. $a_1 = 200$, $a_{k+1} = a_k - 10$
$a_2 = 200 - 10 = 190$
$a_3 = 190 - 10 = 180$
$a_4 = 180 - 10 = 170$
$a_5 = 170 - 10 = 160$
$a_n = 210 - 10n$

23. $a_1 = \frac{3}{2}$, $a_{k+1} = a_k - \frac{1}{4}$
$a_2 = \frac{3}{2} - \frac{1}{4} = \frac{5}{4}$
$a_3 = \frac{5}{4} - \frac{1}{4} = 1$
$a_4 = 1 - \frac{1}{4} = \frac{3}{4}$
$a_5 = \frac{3}{4} - \frac{1}{4} = \frac{1}{2}$
$a_n = \frac{7}{4} - \frac{1}{4}n$

25. $a_1 = 5$, $d = 6$
$a_1 = 5$
$a_2 = 5 + 6 = 11$
$a_3 = 11 + 6 = 17$
$a_4 = 17 + 6 = 23$
$a_5 = 23 + 6 = 29$

27. $a_1 = -2.6$, $d = -0.4$
$a_1 = -2.6$
$a_2 = -2.6 + (-0.4) = -3.0$
$a_3 = -3.0 + (-0.4) = -3.4$
$a_4 = -3.4 + (-0.4) = -3.8$
$a_5 = -3.8 + (-0.4) = -4.2$

29. $a_1 = 2$, $a_{12} = 46$
$46 = 2 + (12 - 1)d$
$44 = 11d$
$4 = d$
$a_1 = 2$
$a_2 = 2 + 4 = 6$
$a_3 = 6 + 4 = 10$
$a_4 = 10 + 4 = 14$
$a_5 = 14 + 4 = 18$

31. $a_8 = 26$, $a_{12} = 42$
$26 = a_8 = a_1 + (n - 1)d = a_1 + 7d$
$42 = a_{12} = a_1 + (n - 1)d = a_1 + 11d$
Answer: $d = 4$, $a_1 = -2$
$a_1 = -2$
$a_2 = -2 + 4 = 2$
$a_3 = 2 + 4 = 6$
$a_4 = 6 + 4 = 10$
$a_5 = 10 + 4 = 14$

33. $a_1 = 1$, $d = 3$
$a_n = a_1 + (n - 1)d = 1 + (n - 1)(3)$

35. $a_1 = 100$, $d = -8$
$a_n = a_1 + (n - 1)d = 100 + (n - 1)(-8)$

37. $a_1 = x$, $d = 2x$
$a_n = a_1 + (n - 1)d = x + (n - 1)(2x)$

39. $4, \frac{3}{2}, -1, -\frac{7}{2}, \ldots$
$d = -\frac{5}{2}$
$a_n = a_1 + (n - 1)d = 4 + (n - 1)\left(-\frac{5}{2}\right)$

41. $a_1 = 5$, $a_4 = 15$
$a_4 = a_1 + 3d \implies 15 = 5 + 3d \implies d = \frac{10}{3}$
$a_n = a_1 + (n - 1)d = 5 + (n - 1)\left(\frac{10}{3}\right)$

43. $a_3 = 94$, $a_6 = 85$
$a_6 = a_3 + 3d \implies 85 = 94 + 3d \implies d = -3$
$a_1 = a_3 - 2d \implies a_1 = 94 - 2(-3) = 100$
$a_n = a_1 + (n - 1)d = 100 + (n - 1)(-3)$

45. $a_n = -\frac{2}{3}n + 6$
$d = -\frac{2}{3}$ so the sequence is decreasing, and $a_1 = 5\frac{1}{3}$.
Matches (b).

47. $a_n = 2 + \frac{3}{4}n$
$d = \frac{3}{4}$ so the sequence is increasing, and $a_1 = 2\frac{3}{4}$.
Matches (c).

49. $a_n = 15 - \frac{3}{2}n$

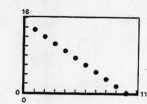

51. $a_n = 0.2n + 3$

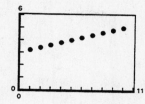

53. Since $a_n = dn + c$, its geometric pattern is linear.

55. $8, 20, 32, 44, \ldots$

$a_1 = 8, \ d = 12, \ n = 10$

$a_{10} = 8 + 9(12) = 116$

$S_{10} = \frac{10}{2}(8 + 116) = 620$

57. $-6, -2, 2, 6, \ldots$

$a_1 = -6, \ d = 4, \ n = 50$

$a_{50} = -6 + 49(4) = 190$

$S_{50} = \frac{50}{2}(-6 + 190) = 4600$

59. $40, 37, 34, 31, \ldots$

$a_1 = 40, d = -3, n = 10$

$a_{10} = 40 + 9(-3) = 13$

$S_{10} = \frac{10}{2}(40 + 13) = 265$

61. $a_1 = 100, \ a_{25} = 220, \ n = 25$

$S_n = \frac{n}{2}[a_1 + a_n]$

$S_{25} = \frac{25}{2}(100 + 220) = 4000$

63. $a_1 = 1, \ a_{50} = 50, \ n = 50$

$\sum_{n=1}^{50} n = \frac{50}{2}(1 + 50) = 1275$

65. $a_1 = 5, \ a_{100} = 500, n = 100$

$\sum_{n=1}^{100} 5n = \frac{100}{2}(5 + 500) = 25,250$

67. $\sum_{n=11}^{30} n - \sum_{n=1}^{10} n = \frac{20}{2}(11 + 30) - \frac{10}{2}(1 + 10) = 355$

69. $a_1 = 4, \ a_{500} = 503, n = 500$

$\sum_{n=1}^{500} (n + 3) = \frac{500}{2}(4 + 503) = 126,750$

71. $a_1 = 7, a_{20} = 45, \ n = 20$

$\sum_{n=1}^{20} (2n + 5) = \frac{20}{2}(7 + 45) = 520$

73. $a_0 = 1000, \ a_{50} = 750, \ n = 51$

$\sum_{n=0}^{50} (100 - 5n) = \frac{51}{2}(1000 + 750) = 44,625$

75. $a_1 = \frac{742}{3}, \ a_{60} = 90, \ n = 60$

$\sum_{i=1}^{60} \left(250 - \frac{8}{3}i\right) = \frac{60}{2}\left(\frac{742}{3} + 90\right) = 10,120$

77. $a_1 = 1, \ a_{100} = 199, \ n = 100$

$\sum_{n=1}^{100} (2n - 1) = \frac{100}{2}(1 + 199) = 10,000$

79. (a) $a_1 = 32,500, \ d = 1500$

$a_6 = a_1 + 5d = 32,500 + 5(1500) = \$40,000$

(b) $S_6 = \frac{6}{2}[32,500 + 40,000] = \$217,500$

81. $a_1 = 20, \ d = 4, \ n = 30$

$a_{30} = 20 + 29(4) = 136$

$S_{30} = \frac{30}{2}(20 + 36) = 2340$ seats

83. $a_1 = 14, a_{18} = 31$

$S_{18} = \frac{18}{2}(14 + 31) = 405$ bricks

85. $a_1 = 25, \ a_2 = 25 + 2 = 27$, etc. $\implies d = 2$ and $n = 15$.

$a_{15} = 2(15) + 23 = 53$

$S_{15} = \frac{15}{2}(25 + 53) = \frac{15}{2} \cdot 78 = 585$ seats

87. $(1 + 2 + \cdots + 12) + (1 + 2 + \cdots + 12) = \frac{12}{2}(1 + 12) \times 2 = 12 \cdot 13 = 156$ times

89. (a) $1 + 3 = 4$

$1 + 3 + 5 = 9$

$1 + 3 + 5 + 7 = 16$

$1 + 3 + 5 + 7 + 9 = 25$

$1 + 3 + 5 + 7 + 9 + 11 = 36$

(b) $S_n = n^2$

$S_7 = 1 + 3 + 5 + 7 + 9 + 11 + 13 = 49 = 7^2$

(c) $S_n = \frac{n}{2}[1 + (2n - 1)] = \frac{n}{2}(2n) = n^2$

91. $S_{20} = \frac{20}{2}\{a_1 + [a_1 + (20 - 1)(3)]\} = 650$

$10(2a_1 + 57) = 650$

$2a_1 + 57 = 65$

$2a_1 = 8$

$a_1 = 4$

Section 9.3   Geometric Sequences

- You should be able to identify a geometric sequence, find its common ratio, and find the nth term.

- You should be able to find the nth partial sum of a geometric sequence with common ratio r using the formula.

$$S_n = a_1\left(\frac{1 - r^n}{1 - r}\right)$$

- You should know that if $|r| < 1$, then

$$\sum_{n=1}^{\infty} a_1 r^{n-1} = \frac{a_1}{1 - r}.$$

Solutions to Odd-Numbered Exercises

1. $5, 15, 45, 135, \ldots$

Geometric sequence, $r = 3$

3. $3, 12, 21, 30, \ldots$

Not a geometric sequence

Note: It is an arithmetic sequence
with $d = 9$.

5. $1, -\frac{1}{2}, \frac{1}{4}, -\frac{1}{8}, \ldots$

Geometric sequence, $r = -\frac{1}{2}$

7. $\frac{1}{2}, \frac{2}{3}, \frac{3}{4}, \frac{4}{5}, \ldots$

Not a geometric sequence

9. $1, \frac{1}{2}, \frac{1}{3}, \frac{1}{4}, \ldots$

Not a geometric sequence

11. $a_1 = 2, \; r = 3$

$a_1 = 2$

$a_2 = 2(3) = 6$

$a_3 = 6(3) = 18$

$a_4 = 18(3) = 54$

$a_5 = 54(3) = 162$

13. $a_1 = 1, \; r = \frac{1}{2}$

$a_1 = 1$

$a_2 = 1\left(\frac{1}{2}\right) = \frac{1}{2}$

$a_3 = \frac{1}{2}\left(\frac{1}{2}\right) = \frac{1}{4}$

$a_4 = \frac{1}{4}\left(\frac{1}{2}\right) = \frac{1}{8}$

$a_5 = \frac{1}{8}\left(\frac{1}{2}\right) = \frac{1}{16}$

15. $a_1 = 5, \; r = -\frac{1}{10}$

$a_1 = 5$

$a_2 = 5\left(-\frac{1}{10}\right) = -\frac{1}{2}$

$a_3 = \left(-\frac{1}{2}\right)\left(-\frac{1}{10}\right) = \frac{1}{20}$

$a_4 = \frac{1}{20}\left(-\frac{1}{10}\right) = -\frac{1}{200}$

$a_5 = \left(-\frac{1}{200}\right)\left(-\frac{1}{10}\right) = \frac{1}{2000}$

17. $a_1 = 1, \; r = e$

$a_1 = 1$

$a_2 = 1(e) = e$

$a_3 = (e)(e) = e^2$

$a_4 = (e^2)(e) = e^3$

$a_5 = (e^3)(e) = e^4$

19. $a_1 = 3, \; r = \frac{x}{2}$

$a_1 = 3$

$a_2 = 3\left(\frac{x}{2}\right) = \frac{3x}{2}$

$a_3 = \left(\frac{3x}{2}\right)\left(\frac{x}{2}\right) = \frac{3x^2}{4}$

$a_4 = \left(\frac{3x^3}{4}\right)\left(\frac{x}{2}\right) = \frac{3x^3}{8}$

$a_5 = \left(\frac{3x^3}{8}\right)\left(\frac{x}{2}\right) = \frac{3x^4}{16}$

21. $a_1 = 64, \; a_{k+1} = \frac{1}{2}a_k$

$a_1 = 64$

$a_2 = \frac{1}{2}(64) = 32$

$a_3 = \frac{1}{2}(32) = 16$

$a_4 = \frac{1}{2}(16) = 8$

$a_5 = \frac{1}{2}(8) = 4$

$a_n = 64\left(\frac{1}{2}\right)^{n-1} = 128\left(\frac{1}{2}\right)^n$

23. $a_1 = 4, \; a_{k+1} = 3a_k$

$a_1 = 4$

$a_2 = 3(4) = 12$

$a_3 = 3(12) = 36$

$a_4 = 3(36) = 108$

$a_5 = 3(108) = 324$

$a_n = 4(3)^{n-1} = \frac{4}{3}(3)^n$

25. $a_k = 6, \; a_{k+1} = -\frac{3}{2}a_k$

$a_1 = 6$

$a_2 = -\frac{3}{2}(6) = -9$

$a_3 = -\frac{3}{2}(-9) = \frac{27}{2}$

$a_4 = -\frac{3}{2}\left(\frac{27}{2}\right) = -\frac{81}{4}$

$a_5 = -\frac{3}{2}\left(-\frac{81}{4}\right) = \frac{243}{8}$

$a_n = 6\left(-\frac{3}{2}\right)^{n-1}$

27. $a_1 = 4, \; r = \frac{1}{2}, \; n = 10$

$a_n = a_1 r^{n-1}$

$a_{10} = 4\left(\frac{1}{2}\right)^9 = \left(\frac{1}{2}\right)^7 = \frac{1}{128}$

29. $a_1 = 6, \; r = -\frac{1}{3}, \; n = 12$

$a_n = a_1 r^{n-1}$

$a_{12} = 6\left(-\frac{1}{3}\right)^{11} = \frac{-2}{3^{10}}$

31. $a_1 = 100, \; r = e^x, \; n = 9$

$a_n = a_1 r^{n-1}$

$a_9 = 100(e^x)^8 = 100e^{8x}$

33. $a_1 = 500, \; r = 1.02, \; n = 40$

$a_n = a_1 r^{n-1}$

$a_{40} = 500(1.02)^{39} \approx 1082.37$

35. $a_1 = 16, \; a_4 = \frac{27}{4}, \; n = 3$

$\frac{27}{4} = 16r^3 \;\Rightarrow\; r = \frac{3}{4}$

$a_n = a_1 r^{n-1}$

$a_3 = 16\left(\frac{3}{4}\right)^2 = 9$

37. $a_2 = a_1 r = -18 \;\Rightarrow\; a_1 = \frac{-18}{r}$

$a_5 = a_1 r^4 = (a_1 r)r^3 = -18r^3 = \frac{2}{3} \;\Rightarrow\; r = -\frac{1}{3}$

$a_1 = \frac{-18}{r} = \frac{-18}{-1/3} = 54$

$a_6 = a_1 r^5 = 54\left(\frac{-1}{3}\right)^5 = \frac{54}{243} = -\frac{2}{9}$

39. $a_n = 18\left(\frac{2}{3}\right)^{n-1}$

$r = \frac{2}{3} < 1$, so the sequence is decreasing.

Matches (a).

41. $a_n = 18\left(\frac{3}{2}\right)^{n-1}$

$r = \frac{3}{2} > 1$, so the sequence is increasing.

Matches (b).

43. $a_n = 12(-0.75)^{n-1}$

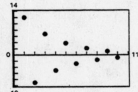

45. $a_n = 2(1.3)^{n-1}$

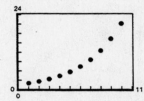

47. Given real numbers r between -1 and 1, as the exponent increases, r^n approaches zero.

49. $A = P\left(1 + \dfrac{r}{n}\right)^{nt} = 1000\left(1 + \dfrac{0.10}{n}\right)^{n(10)}$

   (a) $n = 1$, $A = 1000(1 + 0.10)^{10} \approx \$2593.74$

   (b) $n = 2$, $A = 1000\left(1 + \dfrac{0.10}{2}\right)^{2(10)} \approx \$2653.30$

   (c) $n = 4$, $A = 1000\left(1 + \dfrac{0.10}{4}\right)^{4(10)} \approx \$2685.06$

   (d) $n = 12$, $A = 1000\left(1 + \dfrac{0.10}{12}\right)^{12(10)} \approx \$2707.04$

   (e) $n = 365$, $A = 1000\left(1 + \dfrac{0.10}{365}\right)^{365(10)} \approx \$2717.91$

51. $V_5 = 135,000(0.70)^5 = \$22,689.45$

53. $8, -4, 2, -1, \frac{1}{2}$

$S_1 = 8$

$S_2 = 8 + (-4) = 4$

$S_3 = 8 + (-4) + 2 = 6$

$S_4 = 8 + (-4) + 2 + (-1) = 5$

55. $\displaystyle\sum_{n=1}^{9} 2^{n-1} \implies a_1 = 1, r = 2$

$S_9 = \dfrac{1(1 - 2^9)}{1 - 2} = 511$

57. $\displaystyle\sum_{i=1}^{7} 64\left(-\frac{1}{2}\right)^{i-1} \implies a_1 = 64, r = -\frac{1}{2}$

$S_7 = 64\left[\dfrac{1 - (-1/2)^7}{1 - (-1/2)}\right] = \dfrac{128}{3}\left[1 - \left(-\frac{1}{2}\right)^7\right] = 43$

59. $\displaystyle\sum_{n=0}^{20} 3\left(\frac{3}{2}\right)^{n} = \sum_{n=1}^{21} 3\left(\frac{3}{2}\right)^{n-1} \implies a_1 = 3, r = \frac{3}{2}$

$S_{21} = 3\left[\dfrac{1 - (3/2)^{21}}{1 - (3/2)}\right] = -6\left[1 - \left(\frac{3}{2}\right)^{21}\right] \approx 29,921.31$

61. $\displaystyle\sum_{i=1}^{10} 8\left(-\frac{1}{4}\right)^{i-1} \implies a_1 = 8,\ r = -\frac{1}{4}$

$\displaystyle S_{10} = 8\left[\frac{1 - (-1/4)^{10}}{1 - (-1/4)}\right] = \frac{32}{5}\left[1 - \left(-\frac{1}{4}\right)^{10}\right] \approx 6.4$

63. $\displaystyle\sum_{n=0}^{5} 300(1.06)^n = \sum_{n=1}^{6} 300(1.06)^{n-1} \implies a_1 = 300,\ r = 1.06$

$\displaystyle S_6 = 300\left[\frac{1 - (1.06)^6}{1 - 1.06}\right] \approx 2092.60$

65. $5 + 15 + 45 + \cdots + 3645$

$r = 3$ and $3645 = 5(3)^{n-1} \implies n = 7$

Thus, the sum can be written as $\displaystyle\sum_{n=1}^{7} 5(3)^{n-1}$.

67. $\displaystyle A = \sum_{n=1}^{60} 100\left(1 + \frac{0.10}{12}\right)^n = 100\left(1 + \frac{0.10}{12}\right) \cdot \frac{\left[1 - \left(1 + \frac{0.10}{12}\right)^{60}\right]}{\left[1 - \left(1 + \frac{0.10}{12}\right)\right]} \approx \$7808.24$

69. Let $N = 12t$ be the total number of deposits.

$\displaystyle A = P\left(1 + \frac{r}{12}\right) + P\left(1 + \frac{r}{12}\right)^2 + \cdots + P\left(1 + \frac{r}{12}\right)^N$

$\displaystyle = \left(1 + \frac{r}{12}\right)\left[P + P\left(r + \frac{r}{12}\right) + \cdots + P\left(1 + \frac{r}{12}\right)^{N-1}\right]$

$\displaystyle = P\left(1 + \frac{r}{12}\right)\sum_{n=1}^{N}\left(1 + \frac{r}{12}\right)^{n-1}$

$\displaystyle = P\left(1 + \frac{r}{12}\right)\frac{1 - \left(1 + \frac{r}{12}\right)^N}{1 - \left(1 + \frac{r}{12}\right)}$

$\displaystyle = P\left(1 + \frac{r}{12}\right)\left(-\frac{12}{r}\right)\left[1 - \left(1 + \frac{r}{12}\right)^N\right]$

$\displaystyle = P\left(\frac{12}{r} + 1\right)\left[-1 + \left(1 + \frac{r}{12}\right)^N\right]$

$\displaystyle = P\left[\left(1 + \frac{r}{12}\right)^N - 1\right]\left(1 + \frac{12}{r}\right)$

$\displaystyle = P\left[\left(1 + \frac{r}{12}\right)^{12t} - 1\right]\left(1 + \frac{12}{r}\right)$

71. $P = \$50$, $r = 7\%$, $t = 20$ years

(a) Compounded monthly: $A = 50\left[\left(1 + \dfrac{0.07}{12}\right)^{12(20)} - 1\right]\left(1 + \dfrac{12}{0.07}\right) \approx \$26{,}198.27$

(b) Compounded continuously: $A = \dfrac{50e^{0.07/12}(e^{0.07(20)} - 1)}{e^{0.07/12} - 1} \approx \$26{,}263.88$

73. $P = \$100$, $r = 10\%$, $t = 40$ years

(a) Compounded monthly: $A = 100\left[\left(1 + \dfrac{0.10}{12}\right)^{12(40)} - 1\right]\left(1 + \dfrac{12}{0.10}\right) \approx \$637{,}678.02$

(b) Compounded continuously: $A = \dfrac{100e^{0.10/12}(e^{(0.10)(40)} - 1)}{e^{0.10/12} - 1} \approx \$645{,}861.43$

75. $P = w\displaystyle\sum_{n=1}^{12t}\left[\left(1 + \dfrac{r}{12}\right)^{-1}\right]^{n}$

$= w\left(1 + \dfrac{r}{12}\right)^{-1}\left[\dfrac{1 - \left(1 + \dfrac{r}{12}\right)^{-12t}}{1 - \left(1 - \dfrac{r}{12}\right)^{-1}}\right]$

$= w\left(\dfrac{1}{1 + \dfrac{r}{12}}\right)\dfrac{\left[1 - \left(1 + \dfrac{r}{12}\right)^{-12t}\right]}{1 - \dfrac{1}{\left(1 + \dfrac{r}{12}\right)}}$

$= w\dfrac{\left[1 - \left(1 + \dfrac{r}{12}\right)^{-12t}\right]}{\left(1 + \dfrac{r}{12}\right) - 1}$

$= w\left(\dfrac{12}{r}\right)\left[1 - \left(1 - \dfrac{r}{12}\right)^{-12t}\right]$

77. $\displaystyle\sum_{n=0}^{5}\dfrac{16^2}{4}\left(\dfrac{1}{2}\right)^{n} \approx 126$

Total area of shaded region is approximately 126 square inches

79. $S_n = \displaystyle\sum_{i=1}^{n}0.01(2)^{i-1}$

$S_{29} = \$5{,}368{,}709.11$

$S_{30} = \$10{,}737{,}418.23$

$S_{31} = \$21{,}474{,}836.47$

81. $a_1 = 1$, $r = \dfrac{1}{2}$

$\displaystyle\sum_{n=0}^{\infty}\left(\dfrac{1}{2}\right)^{n} = \dfrac{a_1}{1 - r} = \dfrac{1}{1 - (1/2)} = 2$

83. $a_1 = 1$, $r = -\dfrac{1}{2}$

$\displaystyle\sum_{n=0}^{\infty}\left(-\dfrac{1}{2}\right)^{n} = \sum_{n=1}^{\infty}\left(-\dfrac{1}{2}\right)^{n-1} = \dfrac{a_1}{1 - r} = \dfrac{1}{1 - (-1/2)} = \dfrac{2}{3}$

85. $a_1 = 4$, $r = \dfrac{1}{4}$

$\displaystyle\sum_{n=0}^{\infty}4\left(\dfrac{1}{4}\right)^{n} = \dfrac{a_1}{1 - r} = \dfrac{4}{1 - (1/4)} = \dfrac{16}{3}$

87. $8 + 6 + \dfrac{9}{2} + \dfrac{27}{8} + \cdots = \displaystyle\sum_{n=0}^{\infty} 8\left(\dfrac{3}{4}\right)^n = \dfrac{8}{1 - 3/4} = 32$

89. $0.\overline{36} = \displaystyle\sum_{n=0}^{\infty} 0.36(0.01)^n = \dfrac{0.36}{1 - 0.01} = \dfrac{0.36}{0.99} = \dfrac{36}{99} = \dfrac{4}{11}$

91. $0.3\overline{18} = 0.3 + \displaystyle\sum_{n=0}^{\infty} 0.018(0.01)^n = \dfrac{3}{10} + \dfrac{0.018}{1 - 0.01}$

$\qquad = \dfrac{3}{10} + \dfrac{0.018}{0.99} = \dfrac{3}{10} + \dfrac{18}{990} = \dfrac{3}{10} + \dfrac{2}{110}$

$\qquad = \dfrac{35}{110} = \dfrac{7}{22}$

93. $f(x) = 6\left[\dfrac{1 - (0.5)^x}{1 - (0.5)}\right]$, $\displaystyle\sum_{n=0}^{\infty} 6\left(\dfrac{1}{2}\right)^n = \dfrac{6}{1 - 1/2} = 12$

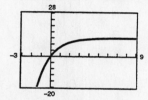

The horizontal asymptote of $f(x)$ is $y = 12$. This corresponds to the sum of the series.

95. (a) Total distance $= \left[\displaystyle\sum_{n=0}^{\infty} 32(0.81)^n\right] - 16 = \dfrac{32}{1 - 0.81} - 16 \approx 152.42$ feet

$\qquad$ (b) $t = 1 + 2\displaystyle\sum_{n=1}^{\infty} (0.9)^n = 1 + 2\left[\dfrac{0.9}{1 - 0.9}\right] = 19$ seconds

97. Time $= \dfrac{\text{Distance}}{\text{Speed}} = \dfrac{200}{50} + \dfrac{200}{42} = 200\left[\dfrac{92}{2100}\right]$ hours

$\qquad$ Speed $= \dfrac{\text{Distance}}{\text{Time}} = \dfrac{400}{200\left[\frac{92}{2100}\right]} = \dfrac{2(2100)}{92} \approx 45.65$ mph

99. Your friend mows at the rate of $\frac{1}{4}$ lawns/hour, and your rate is $\frac{1}{6}$ lawns/hour. Together, the time would be

$\dfrac{1}{\frac{1}{4} + \frac{1}{6}} = \dfrac{1}{\frac{10}{24}} = \dfrac{24}{10} = 2.4$ hours.

Section 9.4   Mathematical Induction

- You should be sure that you understand the principle of mathematical induction. If P_n is a statement involving the positive integer n, where P_1 is true and the truth of P_k implies the truth of P_{k+1}, then P_n is true for all positive integers n.

- You should be able to verify (by induction) the formulas for the sums of powers of integers and be able to use these formulas.

- You should be able to work with finite differences.

Solutions to Odd-Numbered Exercises

1. $P_k = \dfrac{5}{k(k+1)}$

$P_{k+1} = \dfrac{5}{(k+1)[(k+1)+1]} = \dfrac{5}{(k+1)(k+2)}$

3. $P_k = \dfrac{k^2(k+1)^2}{4}$

$P_{k+1} = \dfrac{(k+1)^2[(k+1)+1]^2}{4} = \dfrac{(k+1)^2(k+2)^2}{4}$

5. 1. When $n = 1$, $S_1 = 2 = 1(1+1)$.

    2. Assume that

$$S_k = 2 + 4 + 6 + 8 + \cdots + 2k = k(k+1).$$

Then,

$$S_{k+1} = 2 + 4 + 6 + 8 + \cdots + 2k + 2(k+1)$$
$$= S_k + 2(k+1) = k(k+1) + 2(k+1) = (k+1)(k+2).$$

We conclude by mathematical induction that the formula is valid for all positive integer values of n.

7. 1. When $n = 1$, $S_1 = 2 = \dfrac{1}{2}[5(1) - 1]$.

    2. Assume that

$$S_k = 2 + 7 + 12 + 17 + \cdots + (5k - 3) = \frac{k}{2}(5k - 1).$$

Then,

$$S_{k+1} = 2 + 7 + 12 + 17 + \cdots + (5k - 3) + [5(k+1) - 3]$$

$$= S_k + (5k + 5 - 3) = \frac{k}{2}(5k - 1) + 5k + 2$$

$$= \frac{5k^2 - k + 10k + 4}{2} = \frac{5k^2 + 9k + 4}{2}$$

$$= \frac{(k+1)(5k+4)}{2} = \frac{(k+1)}{2}[5(k+1) - 1].$$

We conclude by mathematical induction that the formula is valid for all positive integer values of n.

9. 1. When $n = 1$, $S_1 = 1 = 2^1 - 1$.

2. Assume that
$$S_k = 1 + 2 + 2^2 + 2^3 + \cdots + 2^{k-1} = 2^k - 1.$$
Then,
$$S_{k+1} = 1 + 2 + 2^2 + 2^3 + \cdots + 2^{k-1} + 2^k$$
$$= S_k + 2^k = 2^k - 1 + 2^k = 2(2^k) - 1 = 2^{k+1} - 1.$$

Therefore, by mathematical induction, the formula is valid for all positive integer values of n.

11. 1. When $n = 1$, $S_1 = 1 = \dfrac{1(1 + 1)}{2}$.

2. Assume that
$$S_k = 1 + 2 + 3 + 4 + \cdots + k = \frac{k(k + 1)}{2}.$$
Then,
$$S_{k+1} = 1 + 2 + 3 + 4 + \cdots + k + (k + 1)$$
$$= S_k + (k + 1) = \frac{k(k + 1)}{2} + \frac{2(k + 1)}{2} = \frac{(k + 1)(k + 2)}{2}.$$

Therefore, we conclude that this formula holds for all positive integer values of n.

13. 1. When $n = 1$, $S_1 = 1^3 = 1 = \dfrac{1(1 + 1)^2}{4}$.

2. Assume that
$$S_k = 1^3 + 2^3 + 3^3 + 4^3 + \cdots + k^3 = \frac{k^2(k + 1)^2}{4}.$$
Then,
$$S_{k+1} = 1^3 + 2^3 + 3^3 + 4^3 + \cdots + k^3 + (k + 1)^3$$
$$= S_k + (k + 1)^3 = \frac{k^2(k + 1)^2}{4} + (k + 1)^3 = \frac{k^2(k + 1)^2 + 4(k + 1)^3}{4}$$
$$= \frac{(k + 1)^2[k^2 + 4(k + 1)]}{4} = \frac{(k + 1)^2(k^2 + 4k + 4)}{4} = \frac{(k + 1)^2(k + 2)^2}{4}.$$

Therefore, we conclude that this formula holds for all positive integer values of n.

15. 1. When $n = 1$, $S_1 = \dfrac{(1)^2(1+1)^2(2(1)^2 + 2(1) - 1)}{12} = 1$.

2. Assume that

$$S_k = \sum_{i=1}^{k} i^5 = \frac{k^2(k+1)^2(2k^2 + 2k - 1)}{12}.$$

Then,

$$S_{k+1} = \sum_{i=1}^{k+1} i^5 = \sum_{i=1}^{k} i^5 + (k+1)^5$$

$$= \frac{k^2(k+1)^2(2k^2 + 2k - 1)}{12} + \frac{12(k+1)^5}{12}$$

$$= \frac{(k+1)^2[k^2(2k^2 + 2k - 1) + 12(k+1)^3]}{12}$$

$$= \frac{(k+1)^2[2k^4 + 2k^3 - k^2 + 12(k^3 + 3k^2 + 3k + 1)]}{12}$$

$$= \frac{(k+1)^2[2k^4 + 14k^3 + 35k^2 + 36k + 12]}{12}$$

$$= \frac{(k+1)^2(k^2 + 4k + 4)(2k^2 + 6k + 3)}{12}$$

$$= \frac{(k+1)^2(k+2)^2[2(k+1)^2 + 2(k+1) - 1]}{12}.$$

Therefore, we conclude that this formula holds for all positive integer values of n.

17. 1. When $n = 1$, $S_1 = 2 = \dfrac{1(2)(3)}{3}$.

2. Assume that

$$S_k = 1(2) + 2(3) + 3(4) + \cdots + k(k+1) = \frac{k(k+1)(k+2)}{3}.$$

Then,

$$S_{k+1} = 1(2) + 2(3) + 3(4) + \cdots + k(k+1) + (k+1)(k+2)$$

$$= S_k + (k+1)(k+2) = \frac{k(k+1)(k+2)}{3} + \frac{3(k+1)(k+2)}{3}$$

$$= \frac{(k+1)(k+2)(k+3)}{3}.$$

Thus, this formula is valid for all positive integer values of n.

19. $\displaystyle\sum_{n=1}^{20} n = \frac{20(20+1)}{2} = 210$

21. $\displaystyle\sum_{n=1}^{6} n^2 = \frac{6(6+1)[2(6)+1]}{6} = 91$

23. $\displaystyle\sum_{n=1}^{5} n^4 = \frac{5(5+1)(2(5)+1)(3(5)^2 + 3(5) - 1)}{30} = 979$

25. $\displaystyle\sum_{n=1}^{6} (n^2 - n) = \sum_{n=1}^{6} n^2 - \sum_{n=1}^{6} n = \frac{6(6+1)[2(6)+1]}{6} - \frac{6(6+1)}{2} = 91 - 21 = 70$

27. $\sum_{i=1}^{6}(6i - 8i^3) = 6\sum_{i=1}^{6}i - 8\sum_{i=1}^{6}i^3$

$$= 6\left[\frac{9(6+1)}{2}\right] - 8\left[\frac{(6)^2(6+1)^2}{4}\right] = 6(21) - 8(441) = -3402$$

29. $1 + 5 + 9 + 13 + \cdots = \sum_{n=0}^{\infty}(1 + 4n)$ **31.** $1 + \frac{9}{10} + \frac{81}{100} + \frac{729}{1000} + \cdots = \sum_{n=1}^{\infty}\left(\frac{9}{10}\right)^{n-1}$

33. $\frac{1}{4} + \frac{1}{12} + \frac{1}{24} + \frac{1}{40} + \cdots + \frac{1}{2n(n-1)} + \cdots = \sum_{n=2}^{\infty}\frac{1}{2n(n-1)}$

35. 1. When $n = 4$, $4! = 24$ and $2^4 = 16$, thus $4! > 2^4$.

2. Assume $k! > 2^k$, $k > 4$. Then, $(k+1)! = k!(k+1) > 2^k(2)$ since $k + 1 > 2$. Thus, $(k+1)! > 2^{k+1}$.

Therefore, by mathematical induction, the formula is valid for all integers n such that $n \geq 4$.

37. 1. When $n = 2$, $\frac{1}{\sqrt{1}} + \frac{1}{\sqrt{2}} \approx 1.707$ and $\sqrt{2} \approx 1.414$, thus $\frac{1}{\sqrt{1}} + \frac{1}{\sqrt{2}} > \sqrt{2}$.

2. Assume

$$\frac{1}{\sqrt{1}} + \frac{1}{\sqrt{2}} + \frac{1}{\sqrt{3}} + \cdots + \frac{1}{\sqrt{k}} > \sqrt{k}, k > 2.$$

Then,

$$\frac{1}{\sqrt{1}} + \frac{1}{\sqrt{2}} + \frac{1}{\sqrt{3}} + \cdots + \frac{1}{\sqrt{k}} + \frac{1}{\sqrt{k+1}} > \sqrt{k} + \frac{1}{\sqrt{k+1}}.$$

Now we need to show that

$$\sqrt{k} + \frac{1}{\sqrt{k+1}} > \sqrt{k+1}, k > 2.$$

This is true since

$$\sqrt{k(k+1)} > k$$

$$\sqrt{k(k+1)} + 1 > k + 1$$

$$\frac{\sqrt{k(k+1)} + 1}{\sqrt{k+1}} > \frac{k+1}{\sqrt{k+1}}$$

$$\sqrt{k} + \frac{1}{\sqrt{k+1}} > \sqrt{k+1}.$$

Therefore,

$$\frac{1}{\sqrt{1}} + \frac{1}{\sqrt{2}} + \frac{1}{\sqrt{3}} + \ldots + \frac{1}{\sqrt{k}} + \frac{1}{\sqrt{k+1}} > \sqrt{k+1}.$$

Therefore, by mathematical induction, the formula is valid for all integers n such that $n \geq 2$.

39. 1. When $n = 1$, $(ab)^1 = a^1b^1 = ab$.

2. Assume that $(ab)^k = a^kb^k$.

Then, $(ab)^{k+1} = (ab)^k(ab)$

$= a^kb^kab$

$= a^{k+1}b^{k+1}$.

Thus, $(ab)^n = a^nb^n$.

41. 1. When $n = 1$, $(x_1)^{-1} = x_1^{-1}$.

2. Assume that

$$(x_1x_2x_3\cdots x_k)^{-1} = x_1^{-1}x_2^{-1}x_3^{-1}\cdots x_k^{-1}.$$

Then,

$$(x_1x_2x_3\cdots x_kx_{k+1})^{-1} = [(x_1x_2x_3\cdots x_k)x_{k+1}]^{-1}$$

$$= (x_1x_2x_3\ldots x_k)^{-1}x_{k+1}^{-1}$$

$$= x_1^{-1}x_2^{-1}x_3^{-1}\cdots x_k^{-1}x_{k+1}^{-1}.$$

Thus, the formula is valid.

43. 1. When $n = 1, x(y_1) = xy_1$.

2. Assume that

$$x(y_1 + y_2 + \cdots + y_k) = xy_1 + xy_2 + \cdots + xy_k.$$

Then,

$$xy_1 + xy_2 + \cdots + xy_k + xy_{k+1} = x(y_1 + y_2 + \cdots + y_k) + xy_{k+1}$$
$$= x[(y_1 + y_2 + \cdots + y_k) + y_{k+1}]$$
$$= x(y_1 + y_2 + \cdots + y_k + y_{k+1}).$$

Hence, the formula holds.

45. 1. When $n = 1, \sin(x + \pi) = \sin x \cdot \cos \pi + \cos x \cdot \sin \pi$
$$= -\sin x = (-1)^1 \sin x.$$

2. Assume $\sin(x + k\pi) = (-1)^k \sin x$.

Then,

$$\sin(x + (k + 1)\pi) = \sin[(x + k\pi) + \pi]$$
$$= \sin(x + k\pi) \cos \pi + \cos(x + k\pi) \sin \pi$$
$$= \sin(x + k\pi)(-1) + 0$$
$$= (-1)(-1)^k \sin x$$
$$= (-1)^{k+1} \sin x.$$

Thus, $\sin(x + n\pi) = (-1)^n \sin x$ for all positive integers n.

47. 1. When $n = 1, [1^3 + 3(1)^2 + 2(1)] = 6$ and 3 is a factor.

2. Assume that 3 is a factor of $(k^3 + 3k^2 + 2k)$. Then,

$$[(k + 1)^3 + 3(k + 1)^2 + 2(k + 1)] = k^3 + 3k^2 + 3k + 1 + 3k^2 + 6k + 3 + 2k + 2$$
$$= (k^3 + 3k^2 + 2k) + (3k^2 + 9k + 6)$$
$$= (k^3 + 3k^2 + 2k) + 3(k^2 + 3k + 2).$$

Since 3 is a factor of $(k^3 + 3k^2 + 2k)$ by our assumption, and 3 is a factor of $3(k^2 + 3k + 2)$ then 3 is a factor of the whole sum.

Thus, 3 is a factor of $(n^3 + 3n^2 + 2n)$ for every positive integer n.

49. See the domino illustration and Figure 7.4.

51. $a_0 = 1, a_n = a_{n-1} + 2$

$a_0 = 1$

$a_1 = a_0 + 2 = 1 + 2 = 3$

$a_2 = a_1 + 2 = 3 + 2 = 5$

$a_3 = a_2 + 2 = 5 + 2 = 7$

$a_4 = a_3 + 2 = 7 + 2 = 9$

53. $a_0 = 4, a_1 = 2, a_n = a_{n-1} - a_{n-2}$

$a_0 = 4$

$a_1 = 2$

$a_2 = a_1 - a_0 = 2 - 4 = -2$

$a_3 = a_2 - a_1 = -2 - 2 = -4$

$a_4 = a_3 - a_2 = -4(-2) = -2$

55. $f(1) = 0, a_n = a_{n-1} + 3$

$a_1 = f(1) = 0$

$a_2 = a_1 + 3 = 0 + 3 = 3$

$a_3 = a_2 + 3 = 3 + 3 = 6$

$a_4 = a_3 + 3 = 6 + 3 = 9$

$a_5 = a_4 + 3 = 9 + 3 = 12$

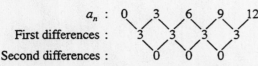

a_n : 0   3   6   9   12

First differences :   3   3   3   3

Second differences :   0   0   0

Since the first differences are equal, the sequence has a linear model.

57. $f(1) = 3, a_n = a_{n-1} - n$

$a_1 = f(1) = 3$

$a_2 = a_1 - 2 = 3 - 2 = 1$

$a_3 = a_2 - 3 = 1 - 3 = -2$

$a_4 = a_3 - 4 = -2 - 4 = -6$

$a_5 = a_4 - 5 = -6 - 5 = -11$

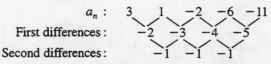

a_n : 3   1   -2   -6   -11

First differences :   -2   -3   -4   -5

Second differences :   -1   -1   -1

Since the second differences are all the same, the sequence has a quadratic model.

59. $a_0 = 0, a_n = a_{n-1} + n$

$a_0 = 0$

$a_1 = a_0 + 1 = 0 + 1 = 1$

$a_2 = a_1 + 2 = 1 + 2 = 3$

$a_3 = a_2 + 3 = 3 + 3 = 6$

$a_4 = a_3 + 4 = 6 + 4 = 10$

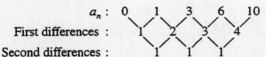

a_n : 0   1   3   6   10

First differences :   1   2   3   4

Second differences :   1   1   1

Since the second differences are equal, the sequence has a quadratic model.

61. $f(1) = 2, a_n = a_{n-1} + 2$

$a_1 = f(1) = 2$

$a_2 = a_1 + 2 = 2 + 2 = 4$

$a_3 = a_2 + 2 = 4 + 2 = 6$

$a_4 = a_3 + 2 = 6 + 2 = 8$

$a_5 = a_4 + 2 = 8 + 2 = 10$

a_n : 2   4   6   8   10

First differences :   2   2   2   2

Second differences :   0   0   0

Since the first differences are equal, the sequence has a linear model.

63. $a_0 = 1, a_n = a_{n-1} + n^2$

$a_0 = 1$

$a_1 = 1 + 1^2 = 2$

$a_2 = 2 + 2^2 = 6$

$a_3 = 6 + 3^2 = 15$

$a_4 = 15 + 4^2 = 31$

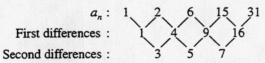

a_n : 1   2   6   15   31

First differences :   1   4   9   16

Second differences :   3   5   7

Since neither the first differences, nor the second differences are equal, the sequence does not have a linear or a quadratic model.

65. $a_0 = 3, a_1 = 3, a_4 = 15$

Let $a_n = an^2 + bn + c$. Thus

$a_0 = a(0)^2 + b(0) + c = 3 \implies c = 3$

$a_1 = a(1)^2 + b(1) + c = 3 \implies a + b + c = 3$

$$a + b \qquad = 0$$

$a_4 = a(4)^2 + b(4) + c = 15 \implies 16a + 4b + c = 15$

$$16a + 4b \qquad = 12$$

$$4a + b \qquad = 3$$

By elimination: $-a - b = 0$

$$\underline{4a + b = 3}$$

$$3a \qquad = 3$$

$$a = 1 \implies b = -1$$

Thus, $a_n = n^2 - n + 3$.

67. $a_0 = -3, a_2 = 1, a_4 = 9$

Let $a_n = an^2 + bn + c$. Then

$a_0 = a(0)^2 + b(0) + c = -3 \implies c = -3$

$a_2 = a(2)^2 + b(2) + c = 1 \implies 4a + 2b + c = 1$

$$4a + 2b \qquad = 4$$

$$2a + b \qquad = 2$$

$a_4 = a(4)^2 + b(4) + c = 9 \implies 16a + 4b + c = 9$

$$16a + 4b \qquad = 12$$

$$4a + b \qquad = 3$$

By elimination: $-2a - b = -2$

$$\underline{4a + b = 3}$$

$$2a \qquad = -1$$

$$a = \tfrac{1}{2} \implies b = 1$$

Thus, $a_n = \tfrac{1}{2}n^2 + n - 3$.

69. $\qquad y = x^2$

$$-3x + 2y = 2 \implies -3x + 2x^2 = 2$$

$$2x^2 - 3x - 2 = 0$$

$$(2x + 1)(x - 2) = 0$$

$$x = -\tfrac{1}{2} \text{ or } x = 2$$

$$x = -\tfrac{1}{4} \text{ or } x = 4$$

Points of intersection: $\left(-\tfrac{1}{2}, \tfrac{1}{4}\right), (2, 4)$

71. $x - y \qquad = -1$

$\quad x + 2y - 2z = 3$

$\quad 3x - y + 2z = 3$

Using an augmented matrix, we have

$$\begin{bmatrix} 1 & -1 & 0 & \vdots & -1 \\ 1 & 2 & -2 & \vdots & 3 \\ 3 & -1 & 2 & \vdots & 3 \end{bmatrix}$$

$$\begin{matrix} \\ -R_1 + R_2 \rightarrow \\ -3R_1 + R_3 \rightarrow \end{matrix} \begin{bmatrix} 1 & -1 & 0 & \vdots & -1 \\ 0 & 3 & -2 & \vdots & 4 \\ 0 & 2 & 2 & \vdots & 6 \end{bmatrix}$$

$$\begin{matrix} \\ -R_3 + R_2 \rightarrow \\ \tfrac{1}{2}R_3 \rightarrow \end{matrix} \begin{bmatrix} 1 & -1 & 0 & \vdots & -1 \\ 0 & 1 & -4 & \vdots & -2 \\ 0 & 1 & 1 & \vdots & 3 \end{bmatrix}$$

$$\begin{matrix} R_2 + R_1 \rightarrow \\ \\ -R_2 + R_3 \rightarrow \end{matrix} \begin{bmatrix} 1 & 0 & -4 & \vdots & -3 \\ 0 & 1 & -4 & \vdots & -2 \\ 0 & 0 & 5 & \vdots & 5 \end{bmatrix}$$

$$\begin{matrix} 4R_3 + R_1 \rightarrow \\ 4R_3 + R_2 \rightarrow \\ \tfrac{1}{5}R_3 \rightarrow \end{matrix} \begin{bmatrix} 1 & 0 & 0 & \vdots & 1 \\ 0 & 1 & 0 & \vdots & 2 \\ 0 & 0 & 1 & \vdots & 1 \end{bmatrix}$$

Thus, $x = 1, y = 2, z = 1$.

Answer: $(1, 2, 1)$

Section 9.5   The Binomial Theorem

■   You should be able to use the Binomial Theorem

$$(x + y)^n = x^n + nx^{n-1}y + \frac{n(n-1)}{2!}x^{n-2}y^2 + \cdots + {}_nC_r x^{n-r}y^r + \cdots + y^n$$

where ${}_nC_r = \dfrac{n!}{(n-r)!r!}$, to expand $(x + y)^n$.

■   You should be able to use Pascal's Triangle.

Solutions to Odd-Numbered Exercises

1. ${}_5C_3 = \dfrac{5!}{3!2!} = \dfrac{5 \cdot 4}{2 \cdot 1} = 10$

3. ${}_{12}C_0 = \dfrac{12!}{0!12!} = 1$

5. ${}_{20}C_{15} = \dfrac{20!}{15!5!} = \dfrac{20 \cdot 19 \cdot 18 \cdot 17 \cdot 16}{5 \cdot 4 \cdot 3 \cdot 2 \cdot 1} = 15,504$

7. $_{100}C_{98} = \dfrac{100!}{98!2!} = \dfrac{100 \cdot 99}{2 \cdot 1} = 4950$

9. $_{100}C_2 = \dfrac{100!}{2!98!} = \dfrac{100 \cdot 99}{2 \cdot 1} = 4950$

11. The first and last number in each row is 1. Each number is found by adding the two numbers immediately above it.

13.
```
              1
            1   1
          1   2   1
        1   3   3   1
      1   4   6   4   1
    1   5  10  10   5   1
  1   6  15  20  15   6   1
1   7  21  35 (35) 21   7   1
```

$_7C_4 = 35$, the 5th entry in the 8th row.

15.
```
                1
              1   1
            1   2   1
          1   3   3   1
        1   4   6   4   1
      1   5  10  10   5   1
    1   6  15  20  15   6   1
  1   7  21  35  35  21   7   1
1   8  28  56  70 (56) 28   8   1
```

$_8C_5 = 56$, the 6th entry in the 9th row.

17. $(x + 1)^4 = {}_4C_0 x^4 + {}_4C_1 x^3(1) + {}_4C_2 x(1)^2 + {}_4C_3 x(1)^3 + {}_4C_4(1)^4$

$\qquad\qquad = x^4 + 4x^3 + 6x^2 + 4x + 1$

19. $(a + 2)^3 = {}_3C_0 a^3 + {}_3C_1 a^2(2) + {}_3C_2 a(2)^2 + {}_3C_3(2)^3$

$\qquad\qquad = a^3 + 3a^2(2) + 3a(2)^2 + (2)^3$

$\qquad\qquad = a^3 + 6a^2 + 12a + 8$

21. $(y - 2)^4 = {}_4C_0 y^4 - {}_4C_1 y^3(2) + {}_4C_2 y^2(2)^2 - {}_4C_3 y(2)^3 + {}_4C_4(2)^4$

$\qquad\qquad = y^4 - 4y^3(2) + 6y^2(4) - 4y(8) + 16$

$\qquad\qquad = y^4 - 8y^3 + 24y^2 - 32y + 16$

23. $(x + y)^5 = {}_5C_0 x^5 + {}_5C_1 x^4 y + {}_5C_2 x^3 y^2 + {}_5C_3 x^2 y^3 + {}_5C_4 xy^4 + {}_5C_5 y^5$

$\qquad\qquad = x^5 + 5x^4 y + 10x^3 y^2 + 10x^2 y^3 + 5xy^4 + y^5$

25. $(r + 3s)^6 = {}_6C_0 r^6 + {}_6C_1 r^5(3s) + {}_6C_2 r^4(3s)^2 + {}_6C_3 r^3(3s)^3 + {}_6C_4 r^2(3s)^4$

$\qquad\qquad\quad + {}_6C_5 r(3s)^5 + {}_6C_6(3s)^6$

$\qquad\qquad = r^6 + 18r^5 s + 135r^4 s^2 + 540r^3 s^3 + 1215r^2 s^4 + 1458rs^5 + 729s^6$

27. $(x - y)^5 = {}_5C_0 x^5 - {}_5C_1 x^4 y + {}_5C_2 x^3 y^2 - {}_5C_3 x^2 y^3 - {}_5C_4 xy^4 - {}_5C_5 y^5$

$\qquad\qquad = x^5 - 5x^4 y + 10x^3 y^2 - 10x^2 y^3 + 5xy^4 - y^5$

29. $(1 - 2x)^3 = {}_3C_0 1^3 - {}_3C_1 1^2(2x) + {}_3C_2 1(2x)^2 - {}_3C_3(2x)^3$

$\qquad\qquad = 1 - 3(2x) + 3(2x)^2 - (2x)^3$

$\qquad\qquad = 1 - 6x + 12x^2 - 8x^3$

31. $(x^2 + 5)^4 = {}_4C_0(x^2)^4 + {}_4C_1(x^2)^3(5) + {}_4C_2(x^2)^2(5)^2 + {}_4C_3(x^2)(5)^3 + {}_4C_4(5)^4$

$\qquad\qquad = x^8 + 4x^6(5) + 6x^4(25) + 4x^2(125) + 625$

$\qquad\qquad = x^8 + 20x^6 + 150x^4 + 500x^2 + 625$

33. $\left(\dfrac{1}{x} + y\right)^5 = {}_5C_0\left(\dfrac{1}{x}\right)^5 + {}_5C_1\left(\dfrac{1}{x}\right)^4 y + {}_5C_2\left(\dfrac{1}{x}\right)^3 y^2 + {}_5C_3\left(\dfrac{1}{x}\right)^2 y^3 + {}_5C_4\left(\dfrac{1}{x}\right)y^4 + {}_5C_5 y^5$

$$= \dfrac{1}{x^5} + \dfrac{5y}{x^4} + \dfrac{10y^2}{x^3} + \dfrac{10y^3}{x^2} + \dfrac{5y^4}{x} + y^5$$

35. $2(x-3)^4 + 5(x-3)^2 = 2[x^4 - 4(x^3)(3) + 6(x^2)(3^2) - 4(x)(3^3) + 3^4] + 5[x^2 - 2(x)(3) + 3^2]$

$$= 2(x^4 - 12x^3 + 54x^2 - 108x + 81) + 5(x^2 - 6x + 9)$$

$$= 2x^4 - 24x^3 + 113x^2 - 246x + 207$$

37. 5th Row of Pascal's Triangle: 1   5   10   10   5   1

$(2t - s)^5 = 1(2t)^5 + 5(2t)^4(-s) + 10(2t)^3(-s)^2 + 10(2t)^2(-s)^3 + 5(2t)(-s)^4 + 1(-s)^5$

$$= 32t^5 - 80t^4s + 80t^3s^2 - 40t^2s^3 + 10ts^4 - s^5$$

39. 4th Row of Pascal's Triangle: 1   4   6   4   1

$(3 - 2z)^4 = 3^4 - 4(3)^3(2z) + 6(3)^2(2z)^2 - 4(3)(2z)^3 + (2z)^4$

$$= 81 - 216z + 216z^2 - 96z^3 + 16z^4$$

41. The term involving x^5 in the expansion of $(x + 3)^{12}$ is

$${}_{12}C_7 x^5(3)^7 = \dfrac{12!}{7!5!} \cdot 3^7 x^5 = 1{,}732{,}104x^5.$$ The coefficient is 1,732,104.

43. The term involving $x^8 y^2$ in the expansion of $(x - 2y)^{10}$ is

$${}_{10}C_2 x^8(-2y)^2 = \dfrac{10!}{2!8!} \cdot 4x^8 y^2 = 180x^8 y^2.$$ The coefficient is 180.

45. The coefficient of $x^4 y^5$ in the expansion of $(3x - 2y)^9$ is

$${}_9C_5(3)^4(-2)^5 = \dfrac{9!}{5!4!}(81)(-32) = -326{,}592.$$

47. The coefficient of $x^8 y^6 = (x^2)^4 y^6$ in the expansion of $(x^2 + y)^{10}$ is ${}_{10}C_6 = 210$.

49. There are $n + 1$ terms in the expansion of $(x + y)^n$.

51. $\left(\sqrt{x} + 3\right)^4 = \left(\sqrt{x}\right)^4 + 4\left(\sqrt{x}\right)^3(3) + 6\left(\sqrt{x}\right)^2(3)^2 + 4\left(\sqrt{x}\right)(3)^3 + (3)^4$

$$= x^2 + 12x\sqrt{x} + 54x + 108\sqrt{x} + 81$$

$$= x^2 + 12x^{3/2} + 54x + 108x^{1/2} + 81$$

53. $(x^{2/3} - y^{1/3})^3 = (x^{2/3})^3 - 3(x^{2/3})^2(y^{1/3}) + 3(x^{2/3})(y^{1/3})^2 - (y^{1/3})^3$

$$= x^2 - 3x^{4/3}y^{1/3} + 3x^{2/3}y^{2/3} - y$$

55. $\dfrac{f(x+h) - f(x)}{h} = \dfrac{(x+h)^3 - x^3}{h}$

$$= \dfrac{x^3 + 3x^2h + 3xh^2 + h^3 - x^3}{h}$$

$$= \dfrac{h(3x^2 + 3xh + h^2)}{h}$$

$$= 3x^2 + 3xh + h^2, \ h \neq 0$$

57. $\dfrac{f(x+h) - f(x)}{h} = \dfrac{\sqrt{x+h} - \sqrt{x}}{h}$

$$= \dfrac{\sqrt{x+h} - \sqrt{x}}{h}$$

59. $(1 + i)^4 = {}_4C_0 1^4 + {}_4C_1(1)^3 i + {}_4C_2(1)^2 i^2 + {}_4C_3 1 \cdot i^3 + {}_4C_4 i^4$

$= 1 + 4i - 6 - 4i + 1$

$= -4$

61. $(2 - 3i)^6 = {}_6C_0 2^6 - {}_6C_1 2^5(3i) + {}_6C_2 2^4(3i)^2 - {}_6C_3 2^3(3i)^3 + {}_6C_4 2^2(3i)^4 - {}_6C_5 2(3i)^5 + {}_6C_6(3i)^6$

$= 64 - 576i - 2160 + 4320i + 4860 - 2916i - 729$

$= 2035 + 828i$

63. $\left(-\dfrac{1}{2} + \dfrac{\sqrt{3}}{2}i\right)^3 = \dfrac{1}{8}(-1 + \sqrt{3}i)^3$

$= \dfrac{1}{8}\left[(-1)^3 + 3(-1)^2(\sqrt{3}i) + 3(-1)(\sqrt{3}i)^2 + (\sqrt{3}i)^3\right]$

$= \dfrac{1}{8}\left[-1 + 3\sqrt{3}i + 9 - 3\sqrt{3}i\right]$

$= 1$

65. ${}_7C_4\left(\dfrac{1}{2}\right)^4\left(\dfrac{1}{2}\right)^3 = 35\left(\dfrac{1}{16}\right)\left(\dfrac{1}{8}\right) \approx 0.273$

67. ${}_8C_4\left(\dfrac{1}{3}\right)^4\left(\dfrac{2}{3}\right)^4 = 70\left(\dfrac{1}{81}\right)\left(\dfrac{16}{81}\right) \approx 0.171$

69. $(1.02)^8 = (1 + 0.02)^8 = 1 + 8(0.02) + 28(0.02)^2 + 56(0.02)^3 + 70(0.02)^4 + 56(0.02)^5$

$+ 28(0.02)^6 + 8(0.02)^7 + (0.02)^8$

$= 1 + 0.16 + 0.0112 + 0.000448 + \cdots \approx 1.172$

71. $(2.99)^{12} = (3 - 0.01)^{12}$

$= 3^{12} - 12(3)^{11}(0.01) + 66(3)^{10}(0.01)^2 - 220(3)^9(0.01)^3 + 495(3)^8(0.01)^4$

$- 792(3)^7(0.01)^5 + 924(3)^6(0.01)^6 - 792(3)^5(0.01)^7 + 495(3)^4(0.01)^8$

$- 220(3)^3(0.01)^9 + 66(3)^2(0.01)^{10} - 12(3)(0.01)^{11} + (0.01)^{12}$

$\approx 510{,}568.785$

73. $f(x) = x^3 - 4x$

$g(x) = f(x + 4)$

$= (x + 4)^3 - 4(x + 4)$

$= x^3 + 3x^2(4) + 3x(4)^2 + (4)^3 - 4x - 16$

$= x^3 + 12x^2 + 48x + 64 - 4x - 16$

$= x^3 + 12x^2 + 44x + 48$

The graph of g is the same as the graph of f shifted 4 units to the left.

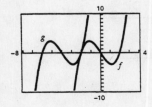

75. $f(x) = -x^2 + 3x + 2$

$g(x) = f(x - 2)$

$= -(x - 2)^2 + 3(x - 2) + 2$

$= -x^2 + 4x - 4 + 3x - 6 + 2$

$= -x^2 + 7x - 8$

The graph of g is the same as the graph of f shifted 2 units to the right

77. $_nC_{n-r} = \dfrac{n!}{[n - (n - r)]!(n - r)!}$

$= \dfrac{n!}{r!(n - r)!}$

$= \dfrac{n!}{(n - r)!r!}$

$= \,_nC_r$

79. $_nC_r + \,_nC_{r-1} = \dfrac{n!}{(n - r)!r!} + \dfrac{n!}{(n - r + 1)!(r - 1)!}$

$= \dfrac{n!(n - r + 1)}{(n - r)!r!(n - r + 1)} + \dfrac{n!}{(n - r + 1)!(r - 1)!}\cdot\dfrac{r}{r}$

$= \dfrac{n!(n - r + 1)}{(n - r + 1)!r!} + \dfrac{n!r}{(n - r + 1)!r!}$

$= \dfrac{n!(n - r + 1 + r)}{(n - r + 1)!r!}$

$= \dfrac{n!(n + 1)}{(n - r + 1)!r!}$

$= \dfrac{(n + 1)!}{(n + 1 - r)!r!} = \,_{n+1}C_r$

81. (a) $_{12}C_5 = 792$

(b) $(_6C_5)^2 = 36$

(c) $_{11}C_5 + \,_{11}C_4 = 792$

(d) $_6C_5 + \,_6C_5 = 12$

(a) and (c) are equal.

83. $f(x) = (1 - x)^3$

$g(x) = 1 - 3x$

$h(x) = 1 - 3x + 3x^2$

$p(x) = 1 - 3x + 3x^2 - x^3$

Since $p(x)$ is the expansion of $f(x)$, they have the same graph.

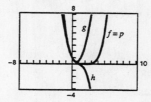

85. $f(t) = 0.1506t^2 + 0.7361t + 21.1374, \, 0 \leq t \leq 22$

(a) $g(t) = f(t + 10)$

$= 0.1506(t + 10)^2 + 0.7361(t + 10) + 21.1374$

$= 0.1506(t^2 + 20t + 100) + 0.7361t + 7.361 + 21.1374$

$= 0.1506t^2 + 3.7481t + 43.5584$

(b)

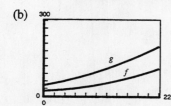

87. $g(x) = f(x) + 8$

$g(x)$ is shifted 8 units up from $f(x)$.

89. $g(x) = f(-x)$

$g(x)$ is the reflection of $f(x)$ in the y-axis.

Section 9.6 Counting Principles

- You should know The Fundamental Counting Principle.

- $_nP_r = \dfrac{n!}{(n-r)!}$ is the number of permutations of n elements taken r at a time.

- Given a set of n objects that has n_1 of one kind, n_2 of a second kind, and so on, the number of distinguishable permutations is

$$\frac{n!}{n_1!n_2!\cdots n_k!}.$$

- $_nC_r = \dfrac{n!}{(n-r)!r!}$ is the number of combinations of n elements taken r at a time.

Solutions to Odd-Numbered Exercises

1. Odd integers: 1, 3, 5, 7, 9, 11

6 ways

3. Prime integers: 2, 3, 5, 7, 11

5 ways

5. Divisible by 4: 4, 8, 12

3 ways

7. Sum is 8: $1+7$, $2+6$, $3+5$, $4+4$, $5+3$, $6+2$, $7+1$

7 ways

9. Amplifiers: 2 choices

Compact disc players: 4 choices

Speakers: 6 choices

Total: $2 \cdot 4 \cdot 6 = 48$ ways

11. Chemist: 3 choices

Statistician: 4 choices

Total: $3 \cdot 4 = 12$ ways

13. $2^6 = 64$

15. 1st Position: 2 choices

2nd Position: 3 choices

3rd Position: 2 choices

4th Position: 1 choice

Total: $2 \cdot 3 \cdot 2 \cdot 1 = 12$ ways

Label the four people A, B, C, and D and suppose that A and B are willing to take the first position. The twelve combinations are as follows.

ABCD	BACD
ABDC	BADC
ACBD	BCAD
ACDB	BCDA
ADBC	BDAC
ADCB	BDCA

17. $26 \cdot 26 \cdot 10 \cdot 10 \cdot 10 \cdot 10 = 6,760,000$

19. (a) $9 \cdot 10 \cdot 10 = 900$

(b) $9 \cdot 9 \cdot 8 = 648$

(c) $9 \cdot 10 \cdot 2 = 180$

(d) $10 \cdot 10 \cdot 10 - 400 = 600$

21. $40^3 = 64,000$

23. (a) $6 \cdot 5 \cdot 4 \cdot 3 \cdot 2 \cdot 1 = 720$

(b) $6 \cdot 1 \cdot 4 \cdot 1 \cdot 2 \cdot 1 = 48$

25. $_nP_r = \dfrac{n!}{(n-r)!}$

So, $_4P_4 = \dfrac{4!}{0!} = 4! = 24$.

27. $_8P_3 = \dfrac{8!}{5!} = 8 \cdot 7 \cdot 6 = 336$

29. $_5P_4 = \dfrac{5!}{1!} = 120$

31. $\qquad 14 \cdot {}_nP_3 = {}_{n+2}P_4 \qquad$ Note $n \geq 3$ for this to be defined.

$$14\left[\frac{n!}{(n-3)!}\right] = \frac{(n+2)!}{(n-2)!}$$

$14n(n-1)(n-2) = (n+2)(n+1)n(n-1) \qquad$ (We can divide here by $n(n-1)$ since $n \neq 0, n \neq 1$.)

$$14n - 28 = n^2 + 3n + 2$$
$$0 = n^2 - 11n + 30$$
$$0 = (n-5)(n-6)$$
$$n = 5 \quad \text{or} \quad n = 6$$

33. $_{20}P_5 = 1,860,480$

35. $_{100}P_3 = 970,200$

37. $_{20}C_5 = 15,504$

39. $_{100}P_{80} \approx 3.836 \times 10^{139}$

This number is too large for some calculators to evaluate.

41.

ABCD	BACD	CABD	DABC
ABDC	BADC	CADB	DACB
ACBD	BCAD	CBAD	DBAC
ACDB	BCDA	CBDA	DBCA
ADBC	BDAC	CDAB	DCAB
ADCB	BDCA	CDBA	DCBA

43. $5! = 120$ ways

45. $_{12}P_4 = \dfrac{12!}{8!} = 12 \cdot 11 \cdot 10 \cdot 9 = 11,880$ ways

47. $\dfrac{7!}{2!1!3!1!} = \dfrac{7!}{2!3!} = 420$

49. $\dfrac{7!}{2!1!1!1!1!1!} = \dfrac{7!}{2!} = 7 \cdot 6 \cdot 5 \cdot 4 \cdot 3 = 2520$

51. $_6C_2 = 15$

The 15 ways are listed below.

AB, AC, AD, AE, AF,

BC, BD, BE, BF, CD,

CE, CF, DE, DF, EF

53. $_{20}C_4 = 4845$ groups

55. $_{40}C_6 = 3,838,380$ ways

57. $_{100}C_4 = 3,921,225$ subsets

59. $_7C_2 = 21$ lines

61. (a) $_8C_4 = \dfrac{8!}{(8-4)!4!} = \dfrac{8!}{4!4!} = \dfrac{8 \cdot 7 \cdot 6 \cdot 5}{4 \cdot 3 \cdot 2} = 70$ ways

(b) $_3C_2 \cdot {}_5C_2 = \dfrac{3!}{(3-2)!2!} \cdot \dfrac{5!}{(5-2)!2!} = 3 \cdot 10 = 30$ ways

63. (a) $_8C_4 = \dfrac{8!}{4!4!} = 70$ ways

(b) There are $2^4 = 16$ ways that a group of four can be formed without any couples in the group. Therefore, if at least one couple is to be in the group, there are $70 - 60 = 54$ ways that could occur.

(c) $2 \cdot 2 \cdot 2 \cdot 2 = 16$ ways

65. $_5C_2 - 5 = 10 - 5 = 5$ diagonals   **67.** $_8C_2 - 8 = 28 - 8 = 20$ diagonals

69. $_nP_{n-1} = \dfrac{n!}{(n-(n-1))!} = \dfrac{n!}{1!} = \dfrac{n!}{0!} = {}_nP_n$

71. $_nC_{n-1} = \dfrac{n!}{[n-(n-1)]!(n-1)!} = \dfrac{n!}{(1)!(n-1)!} = \dfrac{n!}{(n-1)!1!} = {}_nC_1$

73. From the graph of $y = \sqrt{x-3} - x + 6$, you see that there is one zero, $x \approx 8.303$. Analytically,

$$\sqrt{x-3} = x - 6$$
$$x - 3 = x^2 - 12x + 36$$
$$0 = x^2 - 13x + 39.$$

By the Quadratic Formula, $x = \dfrac{13 \pm \sqrt{(-13)^2 - 4(39)}}{2} = \dfrac{13 \pm \sqrt{13}}{2}$.

Selecting the larger solution, $x = \dfrac{13 + \sqrt{13}}{2} \approx 8.303$.

75. $\log_2(x-3) = 5$
$$2^5 = x - 3$$
$$2^5 + 3 = x$$
$$x = 35$$

Section 9.7   Probability

> You should know the following basic principles of probability.
> - If an event E has $n(E)$ equally likely outcomes and its sample space has $n(S)$ equally likely outcomes, then the probability of event E is
>
> $$P(E) = \frac{n(E)}{n(S)}, \text{ where } 0 \le P(E) \le 1.$$
>
> - If A and B are mutually exclusive events, then $P(A \cup B) = P(A) + P(B)$.
>   If A and B are not mutually exclusive events, then $P(A \cup B) = P(A) + P(B) - P(A \cap B)$.
> - If A and B are independent events, then the probability that both A and B will occur is $P(A)P(B)$.
> - The probability of the complement of an event A is $P(A') = 1 - P(A)$.

Solutions to Odd-Numbered Exercises

1. $\{(h, 1), (h, 2), (h, 3), (h, 4), (h, 5), (h, 6),$
$(t, 1), (t, 2), (t, 3), (t, 4), (t, 5), (t, 6)\}$

3. $\{ABC, ACB, BAC, BCA, CAB, CBA\}$

5. $\{(A, B), (A, C), (A, D), (A, E), (B, C), (B, D), (B, E), (C, D), (C, E), (D, E)\}$

7. $E = \{HHT, HTH, THH\}$

$$P(E) = \frac{n(E)}{n(S)} = \frac{3}{8}$$

9. $E = \{HHH, HHT, HTH, HTT, THH, THT, TTH\}$

$$P(E) = \frac{n(E)}{n(S)} = \frac{7}{8}$$

11. $E = \{K, K, K, K, Q, Q, Q, Q, J, J, J, J\}$

$$P(E) = \frac{n(E)}{n(S)} = \frac{12}{52} = \frac{3}{13}$$

13. $E = \{K, K, Q, Q, J, J\}$

$$P(E) = \frac{n(E)}{n(S)} = \frac{6}{52} = \frac{3}{26}$$

15. $E = \{(1, 3), (2, 2), (3, 1)\}$

$$P(E) = \frac{n(E)}{n(S)} = \frac{3}{36} = \frac{1}{12}$$

17. not $E = \{(5, 6), (6, 5), (6, 6)\}$

$$n(E) = n(S) - n(\text{not } E) = 36 - 3 = 33$$

$$P(E) = \frac{n(E)}{n(S)} = \frac{33}{36} = \frac{11}{12}$$

19. $E_3 = \{(1, 2), (2, 1)\}, \; n(E_3) = 2$
$E_5 = \{(1, 4), (2, 3), (3, 2), (4, 1)\}, \; n(E_5) = 4$
$E_7 = \{(1, 6), (2, 5), (3, 4), (4, 3), (5, 2), (6, 1)\}, \; n(E_7) = 6$
$E = E_3 \cup E_5 \cup E_7$
$n(E) = 2 + 4 + 6 = 12$

$$P(E) = \frac{n(E)}{n(s)} = \frac{12}{36} = \frac{1}{3}$$

21. $P(E) = \dfrac{{}_3C_2}{{}_6C_2} = \dfrac{3}{15} = \dfrac{1}{5}$

23. $P(E) = \dfrac{{}_4C_2}{{}_6C_2} = \dfrac{6}{15} = \dfrac{2}{5}$

25. $1 - P(E) = 1 - 0.7 = 0.3$

27. $1 - p = 1 - 0.15 = 0.85$

29. (a) $0.37(2.5) = 0.925$ million $= 925,000$

(b) 0.18

31. (a) $\frac{290}{500} = 0.58$

(b) $\frac{478}{500} = 0.956$

(c) $\frac{2}{500} = 0.004$

33. (a) $\frac{672}{1254}$

(b) $\frac{582}{1254}$

(c) $\frac{672 - 124}{1254} = \frac{548}{1254}$

35. $p + p + 2p = 1$

$p = 0.25$

Taylor: $0.50 = \frac{1}{2}$

Moore: $0.25 = \frac{1}{4}$

Jenkins: $0.25 = \frac{1}{4}$

37. (a) $\frac{_{15}C_{10}}{_{20}C_{10}} = \frac{3003}{184,756} = \frac{21}{1292} \approx 0.016$

(b) $\frac{_{15}C_8 \cdot _5C_2}{_{20}C_{10}} = \frac{64,350}{184,756} = \frac{225}{646} \approx 0.348$

(c) $\frac{_{15}C_9 \cdot _5C_1}{_{20}C_{10}} + \frac{_{15}C_{10}}{_{20}C_{10}} = \frac{25,025 + 3003}{184,756} = \frac{28,028}{184,756} = \frac{49}{323} \approx 0.152$

39. Total ways to insert letters: $4! = 24$ ways

4 correct: 1 way

3 correct: not possible

2 correct: 6 ways

1 correct: 8 ways

0 correct: 9 ways

(a) $\frac{8}{24} = \frac{1}{3}$

(b) $\frac{8 + 6 + 1}{24} = \frac{15}{24} = \frac{5}{8}$

41. (a) $\frac{1}{_5P_5} = \frac{1}{120}$

(b) $\frac{1}{_4P_4} = \frac{1}{24}$

43. (a) $\frac{4}{52} \cdot \frac{4}{52} = \frac{1}{169}$

(b) $\frac{4}{52} \cdot \frac{3}{51} = \frac{1}{221}$

45. (a) $\frac{_9C_4}{_{12}C_4} = \frac{126}{495} = \frac{14}{55}$ (4 good units)

(b) $\frac{(_9C_2)(_3C_2)}{_{12}C_4} = \frac{108}{495} = \frac{12}{55}$ (2 good units)

(c) $\frac{(_9C_3)(_3C_1)}{_{12}C_4} = \frac{252}{495} = \frac{28}{55}$ (3 good units)

At least 2 good units: $\frac{12}{55} + \frac{28}{55} + \frac{14}{55} = \frac{54}{55}$

47. (a) $P(EE) = \frac{15}{30} \cdot \frac{15}{30} = \frac{1}{4}$

(b) $P(E)$ or $P(OE) = 2\left(\frac{15}{30}\right)\left(\frac{15}{30}\right) = \frac{1}{2}$

(c) $P(N_1 < 10, N_2 < 10) = \frac{9}{30} \cdot \frac{9}{30} = \frac{9}{100}$

(d) $P(N_1 N_1) = \frac{30}{30} \cdot \frac{1}{30} = \frac{1}{30}$

49. (a) $P(SS) = (0.985)^2 \approx 0.9702$

(b) $P(S) = 1 - P(FF) = 1 - (0.015)^2 \approx 0.9998$

(c) $P(FF) = (0.015)^2 \approx 0.0002$

51. (a) $\left(\frac{1}{4}\right)^5 = \frac{1}{1024}$

(b) $\left(\frac{3}{4}\right)^5 = \frac{243}{1024}$

(c) $1 - \frac{243}{1024} = \frac{781}{1024}$

53. $(0.32)^2 = 0.1024$

55. $1 - \frac{(45)^2}{(60)^2} = 1 - \left(\frac{45}{60}\right)^2 = 1 - \left(\frac{3}{4}\right)^2 = 1 - \frac{9}{16} = \frac{7}{16}$

57. (a) As you consider successive people with distinct birthdays, the probabilities must decrease to take into account the birth dates already used. Since the birth dates of people are independent events, multiply the respective probabilities of distinct birthdays.

(b) $\frac{365}{365} \cdot \frac{364}{365} \cdot \frac{363}{365} \cdot \frac{362}{365}$

(c) $P_1 = \frac{365}{365} = 1$

$P_2 = \frac{365}{365} \cdot \frac{364}{365} = \frac{364}{365} P_1 = \frac{365 - (2-1)}{365} P_1$

$P_3 = \frac{365}{365} \cdot \frac{364}{365} \cdot \frac{363}{365} = \frac{363}{365} P_2 = \frac{365 - (3-1)}{365} P_2$

$P_n = \frac{365}{365} \cdot \frac{364}{365} \cdot \frac{363}{365} \cdot \ldots \cdot \frac{365 - (n-1)}{365} = \frac{365 - (n-1)}{365} P_{n-1}$

(d) Q_n is the probability that the birthdays are *not* distinct which is equivalent to at least 2 people having the same birthday.

(e)

n	10	15	20	23	30	40	50
P_n	0.88	0.75	0.59	0.49	0.29	0.11	0.03
Q_n	0.12	0.25	0.41	(0.51)	0.71	0.89	0.97

(f) 23,   See the chart above.

59. $1 - 0.546 = 0.454$

Section 9.8   Exploring Data: Measures of Central Tendency

Solutions to Odd-Numbered Exercises

1. Mean: $\dfrac{5 + 7 + 7 + 8 + 9 + 12 + 14}{7} = \dfrac{62}{7} \approx 8.86$

Median: 8

Mode: 7

3. Mean: $\dfrac{5 + 7 + 7 + 8 + 9 + 12 + 24}{7} = \dfrac{72}{7} \approx 10.29$

Median: 8

Mode: 7

5. Mean: $\dfrac{5 + 7 + 7 + 9 + 12 + 14}{6} = \dfrac{54}{6} = 9$

Median: $\dfrac{7 + 9}{2} = 8$

Mode: 7

7. The mean is sensitive to extreme values, because the magnitude of the extreme value is extended into the algebraic formula used to compute the mean.

9. Mean: 26.2

Median: 25.5

Mode: 23

11. Mean: 30.36

Median: 30

Mode: none

13. Mean: $\dfrac{62.00 + 62.50 + 67.00 + 67.99 + 69.84 + 75.35 + 75.35 + 77.92 + 84.98 + 91.76 + 93.98 + 97.82}{12}$

$= \dfrac{926.49}{12} \approx \$77.21$

Median: $75.35

15. (a) $1 + 24 + 45 + 54 + 50 + 19 + 7 = 200$ families

(b) Mean: $\dfrac{1(0) + 24(1) + 45(2) + 54(3) + 50(4) + 19(5) + 7(6)}{200} = \dfrac{613}{200} = 3.065 \approx 3.07$

Median: 3

Mode: 3

17. Mean: $\dfrac{8 + 9.5 + 10 + 10.5 + 10.5 + 10.5 + 11 + 12}{8} = \dfrac{82}{8} \approx 10.25$

Median: 10.t

Mode: 10.5

The median and mode give the most representative description.

19. (a)

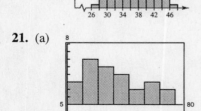

(b) The average tread life is 36,540 miles.

21. (a)

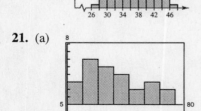

(b)

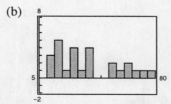

(c)

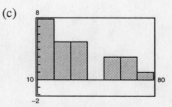

(d) The median is 27.

(e) The mean is 34.12.

The width of the intervals are changed.

The endpoints of the intervals are changed.

Section 9.9   Exploring Data: Measures of Dispersion

Solutions to Even-Numbered Exercises

1. $\bar{x} = \dfrac{2 + 4 + 8 + 10}{4} = \dfrac{24}{4} = 6$

   $v = \dfrac{(2-6)^2 + (4-6)^2 + (8-6)^2 + (10-6)^2}{4} = \dfrac{40}{4} = 10$

   $\sigma = \sqrt{v} = \sqrt{10} \approx 3.16$

3. $\bar{x} = \dfrac{0 + 1 + 1 + 2 + 2 + 2 + 3 + 3 + 4}{9} = \dfrac{18}{9} = 2$

   $v = \dfrac{(0-2)^2 + (1-2)^2 + (1-2)^2 + (2-2)^2 + (2-2)^2 + (2-2)^2 + (3-2)^2 + (3-2)^2 + (4-2)^2}{9}$

   $= \dfrac{12}{9} = \dfrac{4}{3}$

   $\sigma = \sqrt{v} = \sqrt{\dfrac{4}{3}} \approx 1.15$

5. $\bar{x} = \dfrac{1 + 2 + 3 + 4 + 5 + 6 + 7}{7} = 4$

   $v = \dfrac{(1-4)^2 + (2-4)^2 + (3-4)^2 + (4-4)^2 + (5-4)^2 + (6-4)^2 + (7-4)^2}{7} = \dfrac{28}{7} = 4$

   $\sigma = \sqrt{v} = \sqrt{4} = 2$

7. The mean is $\bar{x} = 6$. Then,

   $\sigma = \sqrt{\dfrac{2^2 + 4^2 + 6^2 + 6^2 + 13^2 + 5^2}{6} - 6^2} = \sqrt{\dfrac{286}{6} - 36} = \sqrt{\dfrac{35}{3}} \approx 3.42$

9. $\bar{x} = \dfrac{73}{7}$, $n = 7$

   $\sigma = \sqrt{\dfrac{18^2 + 17^2 + 4^2 + 10^2 + 8^2 + 14^2 + 2^2}{7} - \left(\dfrac{73}{7}\right)^2}$

   $= \sqrt{\dfrac{993}{7} - \left(\dfrac{73}{7}\right)^2}$

   $\approx \sqrt{33.102} \approx 5.75$

11. $\bar{x} = 47$, $v = 226$, $\sigma = 15.03$

13. $\bar{x} = 1.1$, $v = 0.38$, $\sigma = 0.62$

15. $\bar{x} = 5.8$, $v = 2.71$, $\sigma = 1.65$

17. (a) $\bar{x} = 12$, $\sigma = 2.83$

    (b) $\bar{x} = 20$, $\sigma = 2.83$

    (c) $\bar{x} = 12$, $\sigma = 1.41$

    (d) $\bar{x} = 9$, $\sigma = 1.41$

19. It will increase the mean by 5, but the standard deviation will not change.

21. Set A: $\bar{x} = \dfrac{1(4) + 2(4) + 3(4) + 4(4)}{16} = \dfrac{5}{2}$

$$\sigma = \sqrt{\dfrac{4(1)^2 + 4(2)^2 + 4(3)^2 + 4(4)^2}{16} - \left(\dfrac{5}{2}\right)^2} = 1.118$$

Set B: $\bar{x} = \dfrac{1(8) + 2(1) + 3(2) + 4(4)}{15} = \dfrac{32}{15}$

$$\sigma = \sqrt{\dfrac{8(1)^2 + 1(2)^2 + 2(3)^2 + 4(4)^2}{15} - \left(\dfrac{32}{15}\right)^2} = 1.310$$

Set C: $\bar{x} = \dfrac{1(0) + 2(6) + 3(8) + 4(2)}{16} = \dfrac{11}{4}$

$$\sigma = \sqrt{\dfrac{0(1)^2 + 6(2)^2 + 8(3)^2 + 2(4)^2}{16} - \left(\dfrac{11}{4}\right)^2} = 0.661$$

Set D: $\bar{x} = \dfrac{1(2) + 2(6) + 3(6) + 4(2)}{16} = \dfrac{5}{2}$

$$\sigma = \sqrt{\dfrac{2(1)^2 + 6(2)^2 + 6(3)^2 + 2(4)^2}{16} - \left(\dfrac{5}{2}\right)^2} = 0.866$$

The smallest to largest standard deviation: C, D, A, B

23. (a) With $\bar{x} = 235$ and $\sigma = 28$ the intervals containing at least $\frac{3}{4}$ of the scores must lie within two standard deviations of the mean. $235 \pm 2(28) \implies [179, 291]$

With $\bar{x} = 235$ and $\sigma = 28$ the intervals con-taining at least $\frac{8}{9}$ of the scores must lie within two standard deviations of the mean. $235 \pm 3(28) \implies [151, 319]$

(b) With $\bar{x} = 235$ and $\sigma = 16$ the intervals con-taining at least $\frac{3}{4}$ of the scores must lie within two standard deviations of the mean. $235 \pm 2(16) \implies [203, 267]$

With $\bar{x} = 235$ and $\sigma = 16$ the intervals con-taining at least $\frac{8}{9}$ of the scores must lie within two standard deviations of the mean. $235 \pm 3(16) \implies [187, 283]$

(c) With $\bar{x} = 225$ and $\sigma = 28$ the intervals containing at least $\frac{3}{4}$ of the scores must lie within two standard deviations of the mean. $225 \pm 2(28) \implies [169, 281]$

With $\bar{x} = 225$ and $\sigma = 28$ the intervals containing at least $\frac{8}{9}$ of the scores must lie within two standard deviations of the mean. $225 \pm 3(28) \implies [141, 309]$

25. (a) $\bar{x} = 197.32, \quad \sigma = 14.81$

    (b) 96% of the data lie within two standard deviations of the mean.

27. (a) $\bar{x} = 2.63, \quad \sigma = 0.12$

    (b) In Hawaii and Utah

29.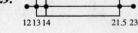

12 13 14   21.5 23

31.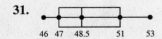

46 47   48.5   51   53

33.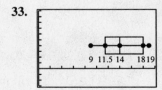

9 11.5 14   18 19

35.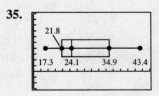

21.8

17.3   24.1   34.9   43.4

37. From the plots, you can see that the lifetimes of the units in the sample made by the new design are greater than the lifetimes of the units in the sample made by the original design. The median increased by more than 1 year.

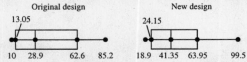

Original design   New design

13.05   24.15

10   28.9   62.6   85.2   18.9   41.35   63.95   99.5

❑ Review Exercises for Chapter 9

Solutions to Odd-Numbered Exercises

1. $a_n = 2 + \dfrac{6}{n}$

$a_1 = 2 + \dfrac{6}{1} = 8$

$a_2 = 2 + \dfrac{6}{2} = 5$

$a_3 = 2 + \dfrac{6}{3} = 4$

$a_4 = 2 + \dfrac{6}{4} = \dfrac{7}{2}$

$a_5 = 2 + \dfrac{6}{5} = \dfrac{16}{5}$

3. $a_n = \dfrac{72}{n!}$

$a_1 = \dfrac{72}{1!} = 72$

$a_2 = \dfrac{72}{2!} = 36$

$a_3 = \dfrac{72}{3!} = 12$

$a_4 = \dfrac{72}{4!} = 3$

$a_5 = \dfrac{72}{5!} = \dfrac{3}{5}$

5. $a_n = \dfrac{3}{2}n$

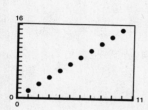

7. $a_n = \dfrac{3n}{n + 2}$

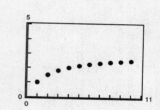

9. $\dfrac{1}{2(1)} + \dfrac{1}{2(2)} + \dfrac{1}{2(3)} + \cdots + \dfrac{1}{2(20)} = \displaystyle\sum_{k=1}^{20} \dfrac{1}{2k}$

11. $\dfrac{1}{2} + \dfrac{2}{3} + \dfrac{3}{4} + \cdots + \dfrac{9}{10} = \displaystyle\sum_{k=1}^{9} \dfrac{k}{k + 1}$

13. $\displaystyle\sum_{i=1}^{6} 5 = 6(5) = 30$

15. $\displaystyle\sum_{j=1}^{4} \dfrac{6}{j^2} = \dfrac{6}{1^2} + \dfrac{6}{2^2} + \dfrac{6}{3^2} + \dfrac{6}{4^2} + \ = 6 + \dfrac{3}{2} + \dfrac{2}{3} + \dfrac{3}{8} = \dfrac{205}{24}$

17. $\displaystyle\sum_{k=1}^{10} 2k^3 = 2(1)^3 + 2(2)^3 + 2(3)^3 + \cdots + 2(10)^3 = 6050$

19. $\displaystyle\sum_{n=0}^{10} (n^2 + 3) = \displaystyle\sum_{n=0}^{10} n^2 + \displaystyle\sum_{n=0}^{10} 3 = \dfrac{10(11)(21)}{6} + 11(3) = 418$

21. $a_1 = 3, d = 4$

$a_1 = 3$

$a_2 = 3 + 4 = 7$

$a_3 = 7 + 4 = 11$

$a_4 = 11 + 4 = 15$

$a_5 = 15 + 4 = 19$

23. $a_4 = 10 \quad a_{10} = 28$

$a_{10} = a_4 + 6d$

$28 = 10 + 6d$

$18 = 6d$

$3 = d$

$a_1 = a_4 - 3d$

$a_1 = 10 - 3(3)$

$a_1 = 1$

$a_2 = 1 + 3 = 4$

$a_3 = 4 + 3 = 7$

$a_4 = 7 + 3 = 10$

$a_5 = 10 + 3 = 13$

25. $a_1 = 35, a_{k+1} = a_k - 3$

$a_1 = 35$

$a_2 = a_1 - 3 = 35 - 3 = 32$

$a_3 = a_2 - 3 = 32 - 3 = 29$

$a_4 = a_3 - 3 = 29 - 3 = 26$

$a_5 = a_4 - 3 = 26 - 3 = 23$

$a_n = 35 + (n - 1)(-3) = 38 - 3n$

27. $a_1 = 9, a_{k+1} = a_k + 7$

$a_1 = 9$

$a_2 = a_1 + 7 = 9 + 7 = 16$

$a_3 = a_2 + 7 = 16 + 7 = 23$

$a_4 = a_3 + 7 = 23 + 7 = 30$

$a_5 = a_4 + 7 = 30 + 7 = 37$

$a_n = 9 + (n - 1)(7) = 2 + 7n$

29. $a_n = 100 + (n - 1)(-3) = 103 - 3n$

$$\sum_{n=1}^{20}(103 - 3n) = \sum_{n=1}^{20}103 - 3\sum_{n=1}^{20}n = 20(103) - 3\left[\frac{(20)(21)}{2}\right] = 1430$$

31. $\displaystyle\sum_{j=1}^{10}(2j - 3) = 2\sum_{j=1}^{10}j - \sum_{j=1}^{10}3 = 2\left[\frac{10(11)}{2}\right] - 10(3) = 80$

33. $\displaystyle\sum_{k=1}^{11}\left(\frac{2}{3}k + 4\right) = \frac{2}{3}\sum_{k=1}^{11}k + \sum_{k=1}^{11}4 = \frac{2}{3} \cdot \frac{(11)(12)}{2} + 11(4) = 88$

35. $\displaystyle\sum_{k=1}^{100}5k = 5\left[\frac{(100)(101)}{2}\right] = 25{,}250$

37. (a) $34{,}000 + 4(2250) = \$43{,}000$

(b) $\displaystyle\sum_{k=1}^{5}[34{,}000 + (k - 1)(2250)] = \sum_{k=1}^{5}(31{,}750 + 2250k) = \$192{,}500$

39. $a_1 = 4, \ r = -\frac{1}{4}$

$a_1 = 4$

$a_2 = 4\left(-\frac{1}{4}\right) = -1$

$a_3 = -1\left(-\frac{1}{4}\right) = \frac{1}{4}$

$a_4 = \frac{1}{4}\left(-\frac{1}{4}\right) = -\frac{1}{16}$

$a_5 = -\frac{1}{16}\left(-\frac{1}{4}\right) = \frac{1}{64}$

41. $a_1 = 9$, $a_3 = 4$

$a_3 = a_1 r^2$

$4 = 9r^2$

$\frac{4}{9} = r^2 \implies r = \pm\frac{2}{3}$

$a_1 = 9$ $a_1 = 9$

$a_2 = 9\left(\frac{2}{3}\right) = 6$ $a_2 = 9\left(-\frac{2}{3}\right) = -6$

$a_3 = 6\left(\frac{2}{3}\right) = 4$ **OR** $a_3 = -6\left(-\frac{2}{3}\right) = 4$

$a_4 = 4\left(\frac{2}{3}\right) = \frac{8}{3}$ $a_4 = 4\left(-\frac{2}{3}\right) = -\frac{8}{3}$

$a_5 = \frac{8}{3}\left(\frac{2}{3}\right) = \frac{16}{9}$ $a_5 = -\frac{8}{3}\left(-\frac{2}{3}\right) = \frac{16}{9}$

43. $a_1 = 120$, $a_{k+1} = \frac{1}{3}a_k$

$a_1 = 120$

$a_2 = \frac{1}{3}(120) = 40$

$a_3 = \frac{1}{3}(40) = \frac{40}{3}$

$a_4 = \frac{1}{3}\left(\frac{40}{3}\right) = \frac{40}{9}$

$a_5 = \frac{1}{3}\left(\frac{40}{9}\right) = \frac{40}{27}$

$a_n = 120\left(\frac{1}{3}\right)^{n-1}$

45. $a_1 = 25$, $a_{k+1} = -\frac{3}{5}a_k$

$a_1 = 25$

$a_2 = -\frac{3}{5}(25) = -15$

$a_3 = -\frac{3}{5}(-15) = 9$

$a_4 = -\frac{3}{5}(9) = -\frac{27}{5}$

$a_5 = -\frac{3}{5}\left(-\frac{27}{5}\right) = \frac{81}{25}$

$a_n = 25\left(-\frac{3}{5}\right)^{n-1}$

47. $a_2 = a_1 r$

$-8 = 16r$

$-\frac{1}{2} = r$

$a_n = 16\left(-\frac{1}{2}\right)^{n-1}$

$\displaystyle\sum_{n=1}^{20} 16\left(-\frac{1}{2}\right)^{n-1} = 16\left[\frac{1 - (-1/2)^{20}}{1 - (-1/2)}\right] \approx 10.67$

49. $\displaystyle\sum_{i=1}^{7} 2^{i-1} = \frac{1 - 2^7}{1 - 2} = 127$

51. $\displaystyle\sum_{i=1}^{\infty} \left(\frac{7}{8}\right)^{i-1} = \frac{1}{1 - 7/8} = 8$

53. $\displaystyle\sum_{k=1}^{\infty} 4\left(\frac{2}{3}\right)^{k-1} = \frac{4}{1 - 2/3} = 12$

55. $\displaystyle\sum_{i=1}^{10} 10\left(\frac{3}{5}\right)^{i-1} \approx 24.849$

57. (a) $a_t = 120{,}000(0.7)^t$

(b) $a_5 = 120{,}000(0.7)^5 = \$20{,}168.40$

59. $A = \displaystyle\sum_{i=1}^{24} 200\left(1 + \frac{0.06}{12}\right)^t \approx \$5111.82$

61. 1. When $n = 1$, $1 = \frac{1}{2}(3(1) - 1)$.

2. Assume that

$S_k = 1 + 4 + \cdots + (3k - 2) = \frac{k}{2}(3k - 1)$.

Then,

$S_{k+1} = 1 + 4 + \cdots + (3k - 2) + (3(k + 1) - 2) = S_k + (3k + 1)$

$= \frac{k}{2}(3k - 1) + (3k + 1) = \frac{k(3k - 1) + 2(3k + 1)}{2}$

$= \frac{3k^2 + 5k + 2}{2} = \frac{(k + 1)(3k + 2)}{2} = \frac{(k + 1)}{2}[3(k + 1) - 1]$.

Therefore, by mathematical induction, the formula is valid for all positive integer values of n.

63. 1. When $n = 1$, $a = a\left(\dfrac{1 - r}{1 - r}\right)$.

2. Assume that

$$S_k = \sum_{i=0}^{k-1} ar^i = \frac{a(1 - r^k)}{1 - r}.$$

Then,

$$S_{k+1} = \sum_{i=0}^{k} ar^i = \sum_{i=0}^{k-1} ar^i + ar^k = \frac{a(1 - r^k)}{1 - r} + ar^k$$

$$= \frac{a(1 - r^k + r^k - r^{k+1})}{1 - r} = \frac{a(1 - r^{k+1})}{1 - r}.$$

Therefore, by mathematical induction, the formula is valid for all positive integer values of n.

65. $_6C_4 = \dfrac{6!}{2!4!} = 15$

67. $_8P_5 = \dfrac{8!}{3!} = 6720$

69. $\left(\dfrac{x}{2} + y\right)^4 = \left(\dfrac{x}{2}\right)^4 + 4\left(\dfrac{x}{2}\right)^3 y + 6\left(\dfrac{x}{2}\right)^2 y^2 + 4\left(\dfrac{x}{2}\right)y^3 + y^4$

$$= \frac{x^4}{16} + \frac{x^3y}{2} + \frac{3x^2y^2}{2} + 2xy^3 + y^4$$

71. $\left(\dfrac{2}{x} - 3x\right)^6 = \left(\dfrac{2}{x}\right)^6 + 6\left(\dfrac{2}{x}\right)^5(-3x) + 15\left(\dfrac{2}{x}\right)^4(-3x)^2 + 20\left(\dfrac{2}{x}\right)^3(-3x)^3$

$$+ 15\left(\dfrac{2}{x}\right)^2(-3x)^4 + 6\left(\dfrac{2}{x}\right)(-3x)^5 + (-3x)^6$$

$$= \frac{64}{x^6} - \frac{576}{x^4} + \frac{2160}{x^2} - 4320 + 4860x^2 - 2916x^4 + 729x^6$$

73. $(5 + 2i)^4 = (5)^4 + 4(5)^3(2i) + 6(5)^2(2i)^2 + 4(5)(2i)^3 + (2i)^4$

$$= 625 + 1000i + 600i^2 + 160i^3 + 16i^4$$

$$= 625 + 1000i - 600 - 160i + 16 = 41 + 840i$$

75. $(26)\,(26)\,(10)\,(26)\,(26)\,(26) = 118{,}813{,}760$

77.

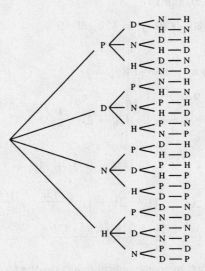

There are 24 possible arrangements.

79. $\frac{10}{10} \cdot \frac{1}{9} = \frac{1}{9}$

81. Chance of rolling a 3 with one die is $\frac{1}{6}$. With two dice $E = \{(1, 5), (2, 4), (3, 3), (4, 2), (5, 1)\}$ and $P(E) = \frac{5}{36}$. The probability of rolling a 3 with one die is higher.

83. $1 - P(HHHHH) = 1 - \left(\frac{1}{2}\right)^5 = \frac{31}{32}$

85. $P(2 \text{ pairs}) = \dfrac{(_{13}C_2)(_4C_2)(_4C_2)(_{44}C_1)}{(_{52}C_5)} = 0.0475$

87. (a) $\bar{x} = 4, \quad \sigma = 1.79$

(b) $\bar{x} = 10, \quad \sigma = 1.79$

(c) $\bar{x} = 8, \quad \sigma = 0.89$

(d) $\bar{x} = 20, \quad \sigma = 0.89$

89. Mean: 10.7

Median: 11.5, $\sigma = 4.65$

91. Mean: 166.28

Median: 174.75, $\sigma = 34.16$

93. Mean: 50.36

Median: 51, $\sigma = 7.39$

1 battery had a lifetime more than 2 standard deviations from the mean.

95. (a) Mean: 30.47

Median: 30.5

Mode: 30, 32

(b) $\sigma = 2.79$

(c) 33% have lengths that differ from the mean by more than 1 standard deviation.

❏ Chapter Test for Chapter 9

1. $a_n = \left(-\frac{2}{3}\right)^{n-1}$ $\quad$ $a_1 = \left(-\frac{2}{3}\right)^{1-1} = \left(-\frac{2}{3}\right)^0 = 1$ $\qquad$ **2.** $a_1 = 12,\ a_{k+1} = a_k + 4$

$\qquad\qquad\qquad\qquad\quad a_2 = -\frac{2}{3}$ $\qquad\qquad\qquad\qquad\qquad\qquad\qquad a_2 = 12 + 4 = 16$

$\qquad\qquad\qquad\qquad\quad a_3 = \left(-\frac{2}{3}\right)^2 = \frac{4}{9}$ $\qquad\qquad\qquad\qquad\qquad\qquad a_3 = 16 + 4 = 20$

$\qquad\qquad\qquad\qquad\quad a_4 = \left(-\frac{2}{3}\right)^3 = -\frac{8}{27}$ $\qquad\qquad\qquad\qquad\qquad a_4 = 20 + 4 = 24$

$\qquad\qquad\qquad\qquad\quad a_5 = \left(-\frac{2}{3}\right)^4 = \frac{16}{81}$ $\qquad\qquad\qquad\qquad\qquad\quad a_5 = 24 + 4 = 28$

3. $a_n = dn + c,\ c = a_1 - d = 5000 - (-100) = 5100$

$\quad \Rightarrow\ a_n = -100n + 5100 = 5000 - 100(n - 1)$

4. $a_n = a_1 r^{n-1},\ a_1 = 4,\ r = \dfrac{1}{2}\ \Rightarrow\ a_n = 4\left(\dfrac{1}{2}\right)^{n-1}$ $\qquad$ **5.** $\displaystyle\sum_{n=1}^{12} \frac{2}{3n+1}$

6. $3 + 6 + \cdots + 150 = 3(1 + 2 + \cdots + 50) = 3 \cdot \dfrac{50(51)}{2} = 3825$

7. $2(r) = -3\ \Rightarrow\ r = -\dfrac{3}{2}$

8. $r = \dfrac{a_2}{a_1} = 2\ \Rightarrow\ S_8 = a_1\left(\dfrac{1 - r^8}{1 - r}\right) = 3\left(\dfrac{1 - 2^8}{1 - 2}\right) = 765$

9. $\quad A_1 = 50\left(1 + \dfrac{0.08}{12}\right)^1 = 50(1.00667)$

$\qquad A_{300} = 50\left(1 + \dfrac{0.08}{12}\right)^{300} 50(1.00667)^{300}$

$\qquad S_n = 50(1.00667)\left[\dfrac{1 - (1.00667)^{300}}{1 - 1.00667}\right] \approx \$47{,}868.33$

10. $_{20}C_3 = \dfrac{20!}{(17!)(3!)} = \dfrac{20 \cdot 19 \cdot 18 \cdot 17!}{17!\ 3!} = \dfrac{20 \cdot 19 \cdot 18}{6} = 1140$

11. $_8C_3 = \dfrac{8!}{5!\ 3!} = \dfrac{8 \cdot 7 \cdot 6}{3 \cdot 2} = 56$ $\qquad\qquad\qquad$ **12.** $26 \cdot 10 \cdot 10 \cdot 10 = 26{,}000$ ways

13. $_{25}C_4 = \dfrac{25!}{21!\ 4!} = \dfrac{25 \cdot 24 \cdot 23 \cdot 22}{24} = 12{,}650$ ways

14. There are $_2C_4 = 6$ ways to select 2 spark plugs. Only one way corresponds to both being selected. Probability $= \frac{1}{6}$.

15. $\bar{x} = \dfrac{2(8) + 6(9) + 5(10) + 3(11) + 2(12) + 13}{19} = \dfrac{190}{19} = 10$

$\qquad \sigma = \sqrt{\dfrac{2(8^2) + 6(9^2) + 5(10)^2 + 3(11^2) + 2(12^2) + 13^2}{19} - 10^2}$

$\qquad\quad = \sqrt{\dfrac{1934}{19} - 100} \approx \sqrt{1.78947} \approx 1.34$

16.

24 26 28 $\qquad$ 43 46

❏ Practice Test for Chapter 9

1. Write out the first five terms of the sequence $a_n = \dfrac{2n}{(n+2)!}$.

2. Write an expression for the nth term of the sequence $\left\{ \dfrac{4}{3}, \dfrac{5}{9}, \dfrac{6}{27}, \dfrac{7}{81}, \dfrac{8}{243}, \dots \right\}$.

3. Find the sum $\displaystyle\sum_{i=1}^{6} (2i - 1)$.

4. Write out the first five terms of the arithmetic sequence where $a_1 = 23$ and $d = -2$.

5. Find a_n for the arithmetic sequence with $a_1 = 12$, $d = 3$, and $n = 50$.

6. Find the sum of the first 200 positive integers.

7. Write out the first five terms of the geometric sequence with $a_1 = 7$ and $r = 2$.

8. Evaluate $\displaystyle\sum_{n=0}^{9} 6\left(\dfrac{2}{3}\right)^n$.

9. Evaluate $\displaystyle\sum_{n=0}^{\infty} (0.03)^n$.

10. Use mathematical induction to prove that $1 + 2 + 3 + 4 + \cdots + n = \dfrac{n(n+1)}{2}$.

11. Use mathematical induction to prove that $n! > 2^n$, $n \geq 4$.

12. Evaluate $_{13}C_4$. Verify with a graphing utility.

13. Find the term involving x^7 in $(x - 2)^{12}$.

14. Evaluate $_{30}P_4$.

15. How many ways can six people sit at a table with six chairs?

16. Twelve cars run in a race. How many different ways can they come in first, second, and third place? (Assume that there are no ties.)

17. Two six-sided dice are tossed. Find the probability that the total of the two dice is less than 5.

18. Two cards are selected at random form a deck of 52 playing cards without replacement. Find the probability that the first card is a King and the second card is a black ten.

19. A manufacturer has determined that for every 1000 units it produces, 3 will be faulty. What is the probability that an order of 50 units will have one or more faulty units?

20. Find the mean, variance, and standard deviation of the data 3, 5, 8, 10, 16, 16.

CHAPTER 10
Topics in Analytic Geometry

CHAPTER 10
Topics in Analytic Geometry

Section 10.1   Introduction to Conics: Parabolas

- A **parabola** is the set of all points (x, y) that are equidistant from a fixed line (**directrix**) and a fixed point (**focus**) not on the line.

- The standard equation of a parabola with vertex (h, k) and:
  - (a) Vertical axis $x = h$ and directrix $y = k - p$ is:
    $(x - h)^2 = 4p(y - k), p \neq 0$
  - (b) Horizontal axis $y = k$ and directrix $x = h - p$ is:
    $(y - k)^2 = 4p(x - h), p \neq 0$

- The tangent line to a parabola at a point P makes **equal angles** with:
  - (a) the line through P and the focus
  - (b) the axis of the parabola

Solutions to Odd-Numbered Exercises

1. $y^2 = -4x$
Vertex: $(0, 0)$
Opens to the left since p is negative.
Matches graph (e).

3. $x^2 = -8y$
Vertex: $(0, 0)$
Opens downward since p is negative.
Matches graph (d).

5. $(y - 1)^2 = 4(x - 3)$
Vertex: $(3, 1)$
Opens to the right since p is positive.
Matches graph (a).

7. $y = \frac{1}{2}x^2$
$x^2 = 2y$
$x^2 = 4\left(\frac{1}{2}\right)y \implies h = 0, k = 0, p = \frac{1}{2}$
Vertex: $(0, 0)$
Focus: $\left(\frac{1}{2}, 0\right)$
Directrix: $y = -\frac{1}{2}$

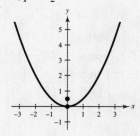

9. $y^2 = -6x$
$y^2 = 4\left(-\frac{3}{2}\right)x \implies h = 0, k = 0, p = -\frac{3}{2}$
Vertex: $(0, 0)$
Focus: $\left(-\frac{3}{2}, 0\right)$
Directrix: $x = \frac{3}{2}$

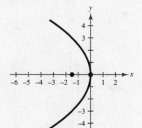

11. $x^2 + 8y = 0$
$x^2 = 4(-2)y \implies h = 0, k = 0, p = -2$
Vertex: $(0, 0)$
Focus: $(0, -2)$
Directrix: $y = 2$

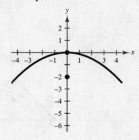

13. $(x - 1)^2 + 8(y + 2) = 0$
$(x - 1)^2 = 4(-2)(y + 2)$
$h = 1, k = -2, p = -2$
Vertex: $(1, -2)$
Focus: $(1, -4)$
Directrix: $y = 0$

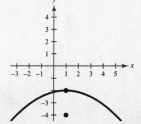

433

15. $\left(y + \frac{1}{2}\right)^2 = 2(x - 5)$

$\qquad = 4\left(\frac{1}{2}\right)(x - 5)$

$h = 5, k = -\frac{1}{2}, p = \frac{1}{2}$

Vertex: $\left(5, -\frac{1}{2}\right)$

Focus: $\left(\frac{11}{2}, -\frac{1}{2}\right)$

Directrix: $x = \frac{9}{2}$

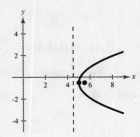

17. $\qquad y = \frac{1}{4}(x^2 - 2x + 5)$

$\qquad 4y - 4 = (x - 1)^2$

$\qquad (x - 1)^2 = 4(1)(y - 1)$

$h = 1, k = 1, p = 1$

Vertex: $(1, 1)$

Focus: $(1, 2)$

Directrix: $y = 0$

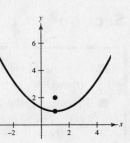

19. $y^2 + 6y + 8x + 25 = 0$

$\qquad y^2 + 6y + 9 = -8x - 25 + 9$

$\qquad\qquad (y + 3)^2 = 4(-2)(x + 2)$

$h = -2, k = -3, p = -2$

Vertex: $(-2, -3)$

Focus: $(-4, -3)$

Directrix: $x = 0$

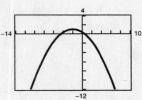

21. $\qquad y = -\frac{1}{6}(x^2 + 4x - 2)$

$\qquad\qquad -6y = x^2 + 4x - 2$

$\qquad -6y + 2 + 4 = x^2 + 4x + 4$

$\qquad\qquad -6y + 6 = (x + 2)^2$

$\qquad\qquad (x + 2)^2 = -6(y - 1)$

$\qquad\qquad (x + 2)^2 = 4\left(-\frac{3}{2}\right)(y - 1)$

$h = -2, k = 1, p = -\frac{3}{2}$

Vertex: $(-2, 1)$

Focus: $\left(-2, -\frac{1}{2}\right)$

Directrix: $y = \frac{5}{2}$

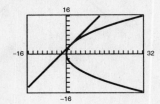

23. $y^2 + x + y = 0$

$\qquad y^2 + y + \frac{1}{4} = -x + \frac{1}{4}$

$\qquad \left(y + \frac{1}{2}\right)^2 = 4\left(-\frac{1}{4}\right)\left(x - \frac{1}{4}\right)$

$h = \frac{1}{4}, k = -\frac{1}{2}, p = -\frac{1}{4}$

Vertex: $\left(\frac{1}{4}, -\frac{1}{2}\right)$

Focus: $\left(0, -\frac{1}{2}\right)$

Directrix: $x = \frac{1}{2}$

To use a graphing calculator, enter:

$y_1 = -\frac{1}{2} + \sqrt{\frac{1}{4} - x}$

$y_2 = -\frac{1}{2} - \sqrt{\frac{1}{4} - x}$

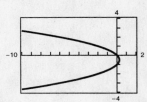

25. $y^2 - 8x = 0 \implies y = \pm\sqrt{8x}$

$\qquad x - y + 2 = 0 \implies y = x + 2$

The point of tangency is $(2, 4)$.

27. $y = -\sqrt{-6x}$

29. $x^2 = 4py \implies y = \dfrac{x^2}{4p}$

(a)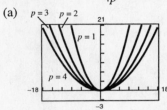

As p increases, the parabola opens wider.

(c)

p-value	Length of chord
1	4
2	8
3	12
4	16

In general, the chord through the focus parallel to the directrix has a length of $4p$.

(b) $p = 1$: focus $(0, 1)$
$p = 2$: focus $(0, 2)$
$p = 3$: focus $(0, 3)$
$p = 4$: focus $(0, 4)$

(d) Once the focus is located, move $2|p|$ units in both directions from the focus parallel to the directrix. This yields two points on the graph of the parabola, as shown in the following figures.

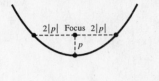 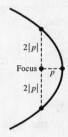

31. Vertex: $(0, 0) \implies h = 0, k = 0$

Graph opens upward.

$x^2 = 4py$

Point on graph: $(3, 6)$

$3^2 = 4p(6)$

$9 = 24p$

$\dfrac{3}{8} = p$

Thus, $x^2 = 4\left(\dfrac{3}{8}\right)y \implies y = \dfrac{2}{3}x^2$.

33. Vertex: $(0, 0) \implies h = 0, k = 0$

Focus: $\left(0, -\dfrac{3}{2}\right) \implies p = -\dfrac{3}{2}$

$(x - h)^2 = 4p(y - k)$

$\quad x^2 = 4\left(-\dfrac{3}{2}\right)y$

$\quad x^2 = -6y$

35. Vertex: $(0, 0) \implies h = 0, k = 0$

Focus: $(-2, 0) \implies p = -2$

$(y - k)^2 = 4p(x - h)$

$\qquad y^2 = 4(-2)x$

$\qquad y^2 = -8x$

37. Vertex: $(0, 0) \implies h = 0, k = 0$

Directrix: $y = -1 \implies p = 1$

$(x - h)^2 = 4p(y - k)$

$(x - 0)^2 = 4(1)(y - 0)$

$\qquad x^2 = 4y$ or $y = \dfrac{1}{4}x^2$

39. Vertex: $(0, 0) \implies h = 0, k = 0$

Directrix: $y = 2 \implies p = -2$

$(x - h)^2 = 4p(y - k)$

$(x - 0)^2 = 4(-2)(y - 0)$

$\qquad x^2 = -8y$ or $y = -\dfrac{1}{8}x^2$

41. Vertex: $(0, 0) \implies h = 0, k = 0$

Horizontal axis and passes through the point $(4, 6)$

$(y - k)^2 = 4p(x - h)$

$(y - 0)^2 = 4p(x - 0)$

$\qquad y^2 = 4px$

$\qquad 6^2 = 4p(4)$

$\qquad 36 = 16p \implies p = \dfrac{9}{4}$

$\qquad y^2 = 4\left(\dfrac{9}{4}\right)x$

$\qquad y^2 = 9x$

43. Vertex: (3, 1) and opens downward. Passes
through (2, 0) and (4, 0).

$$y = -(x - 2)(x - 4)$$
$$= -x^2 + 6x - 8$$
$$= -(x - 3)^2 + 1$$
$$(x - 3)^2 = -(y - 1)$$

45. Vertex: $(-2, 0)$ and opens to the right.
Passes through (0, 2).

$$(y - 0)^2 = 4p(x + 2)$$
$$2^2 = 4p(0 + 2)$$
$$\tfrac{1}{2} = p$$
$$y^2 = 4\left(\tfrac{1}{2}\right)(x + 2)$$
$$y^2 = 2(x + 2)$$

47. Vertex: (3, 2)
Focus: (1, 2)
Horizontal axis
$$p = 1 - 3 = -2$$
$$(y - 2)^2 = 4(-2)(x - 3)$$
$$(y - 2)^2 = -8(x - 3)$$

49. Vertex: (0, 4)
Directrix: $y = 2$
Vertical axis
$$p = 4 - 2 = 2$$
$$(x - 0)^2 = 4(2)(y - 4)$$
$$x^2 = 8(y - 4)$$

51. Focus: (2, 2)
Directrix: $x = -2$
Horizontal axis
Vertex: (0, 2)
$$p = 2 - 0 = 2$$
$$(y - 2)^2 = 4(2)(x - 0)$$
$$(y - 2)^2 = 8x$$

53. $(y - 3)^2 = 6(x + 1)$

For the upper half of the parabola:
$$y - 3 = +\sqrt{6(x + 1)}$$
$$y = \sqrt{6(x + 1)} + 3$$

55. Vertex: $(0, 0) \Rightarrow h = 0, k = 0$
Focus: $(0, 3.5) \Rightarrow p = 3.5$
$$(x - h)^2 = 4p(y - k)$$
$$(x - 0)^2 = 4(3.5)(y - 0)$$
$$x^2 = 14y \text{ or } y = \tfrac{1}{14}x^2$$

57. (a) Converting 16 meters to 1600 centimeters, and superimposing the coordinate plane over the parabola so that its
vertex is (0, 0), shows us that the points (± 800, 3) are on the parabola.

$$(x - 0)^2 = 4p(y - 0)$$
$$x^2 = 4py$$

At (± 800, 3) we have:

$$640{,}000 = 12p$$
$$p = \frac{640{,}000}{12}$$
$$x^2 = 4\left(\frac{640{,}000}{12}\right)y$$
$$y = \frac{3x^2}{640{,}000}$$

(b) Let $y = 1$, then:

$$1 = \frac{3x^2}{640{,}000}$$
$$x^2 = \frac{640{,}000}{3}$$
$$x = \frac{800\sqrt{3}}{3} \approx 462 \text{ centimeters}$$

59. Vertex: (0, 0)

$$y^2 = 4px$$

Point: (1000, 800)

$$800^2 = 4p(1000) \implies p = 160$$
$$y^2 = 4(160)x$$
$$y^2 = 640x$$

61. $y = 30{,}000 - \dfrac{x^2}{39{,}204} = 0$ for $x = 34{,}294.6$ feet.

Thus, the bomb should be dropped

$$\frac{34{,}294.6}{792} \approx 43.3$$

seconds prior to being over the target.

63. The slope of the line $y - y_1 = \dfrac{x_1}{2p}(x - x_1)$ is $m = \dfrac{x_1}{2p}$.

65. $x^2 = 2y \Rightarrow p = \dfrac{1}{2}$

Point: $(x_1, y_1) = (4, 8)$

Use: $y - y_1 = \dfrac{x_1}{2p}(x - x_1)$

$$y - 8 = \dfrac{4}{2(1/2)}(x - 4)$$

$$y - 8 = 4x - 16$$

$$y = 4x - 8 \Rightarrow 0 = 4x - y - 8$$

x-intercept: $(2, 0)$

67. $y = -2x^2 \Rightarrow x^2 = -\dfrac{1}{2}y \Rightarrow p = -\dfrac{1}{8}$

Point: $(x_1, y_1) = (-1, -2)$

Use: $y - y_1 = \dfrac{x_1}{2p}(x - x_1)$

$$y + 2 = \dfrac{-1}{2(-1/8)}(x + 1)$$

$$y + 2 = 4(x + 1)$$

$$y = 4x + 2 \Rightarrow 0 = 4x - y + 2$$

x-intercept: $\left(-\dfrac{1}{2}, 0\right)$

69. (a) $y = \dfrac{-16}{v^2}x^2 + s$

$$= \dfrac{-16}{32^2}x^2 + 75 = -\dfrac{1}{64}x^2 + 75$$

(b) $y = 0 = -\dfrac{1}{64}x^2 + 75 \Rightarrow x^2 = (75)(64) \Rightarrow x \approx 69.28$ feet

71. $f(x) = (x - 3)[x - (2 + i)][x - (2 - i)]$

$$= (x - 3)(x - 2 - i)(x - 2 + i)$$

$$= (x - 3)(x^2 - 4x + 5)$$

$$= x^3 - 7x^2 + 17x - 15$$

73. $g(x) = 6x^4 + 7x^3 - 29x^2 - 28x + 20$

Possible rational roots: $\pm 1, \pm 2, \pm 4, \pm 5, \pm 10, \pm 20,$
$\pm\frac{1}{2}, \pm\frac{5}{2}, \pm\frac{1}{3}, \pm\frac{2}{3}, \pm\frac{4}{3}, \pm\frac{5}{3}, \pm\frac{10}{3}, \pm\frac{20}{3}, \pm\frac{1}{6}, \pm\frac{5}{6}$

$x = \pm 2$ are both solutions.

2	6	7	−29	−28	20
		12	38	18	−20
−2	6	19	9	−10	0
		−12	−14	10	
	6	7	−5	0	

$g(x) = (x - 2)(x + 2)(6x^2 + 7x - 5)$

$$= (x - 2)(x + 2)(2x - 1)(3x + 5)$$

The zeros of $g(x)$ are $x = \pm 2$, $x = \frac{1}{2}$, $x = -\frac{5}{3}$.

Section 10.2 Ellipses

- An **ellipse** is the set of all points (x, y) the sum of whose distances from two distinct fixed points (**foci**) is constant.
- The standard equation of an ellipse with center (h, k) and major and minor axes of lengths $2a$ and $2b$ is:

  (a) $\dfrac{(x - h)^2}{a^2} + \dfrac{(y - k)^2}{b^2} = 1$ if the major axis is horizontal.

  (b) $\dfrac{(x - h)^2}{b^2} + \dfrac{(y - k)^2}{a^2} = 1$ if the major axis is vertical.

- $c^2 = a^2 - b^2$ where c is the distance from the center to a focus.
- The eccentricity of an ellipse is $e = \dfrac{c}{a}$

Solutions to Odd-Numbered Exercises

1. $\dfrac{x^2}{4} + \dfrac{y^2}{9} = 1$

Center: $(0, 0)$

$a = 3, b = 2$

Vertical major axis

Matches graph (b).

3. $\dfrac{x^2}{4} + \dfrac{y^2}{25} = 1$

Center: $(0, 0)$

$a = 5, b = 2$

Vertical major axis

Matches graph (d).

5. $\dfrac{(x - 2)^2}{16} + (y + 1)^2 = 1$

Center: $(2, -1)$

$a = 4, b = 1$

Horizontal major axis

Matches graph (a).

7. $\dfrac{x^2}{25} + \dfrac{y^2}{16} = 1$

Center: $(0, 0)$

$a = 5, b = 4, c = 3$

Foci: $(\pm 3, 0)$

Vertices: $(\pm 5, 0)$

$e = \dfrac{3}{5}$

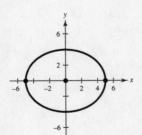

9. $\dfrac{x^2}{16} + \dfrac{y^2}{25} = 1$

$a = 5, b = 4, c = 3$

Center: $(0, 0)$

Foci: $(0, \pm 3)$

Vertices: $(0, \pm 5)$

$e = \dfrac{3}{5}$

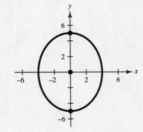

11. $\dfrac{x^2}{9} + \dfrac{y^2}{5} = 1$

Center: $(0, 0)$

$a = 3, b = \sqrt{5}, c = 2$

Foci: $(\pm 2, 0)$

Vertices: $(\pm 3, 0)$

$e = \dfrac{2}{3}$

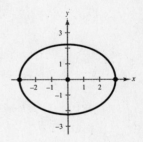

13. $\dfrac{(x - 1)^2}{9} + \dfrac{(y - 5)^2}{25} = 1$

$a = 5, b = 3, c = 4$

Center: $(1, 5)$

Foci: $(1, 9), (1, 1)$

Vertices: $(1, 10), (1, 0)$

$e = \dfrac{4}{5}$

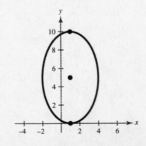

15. $9x^2 + 4y^2 + 36x - 24y + 36 = 0$

$9(x^2 + 4x + 4) + 4(y^2 - 6y + 9) = -36 + 36 + 36$

$$\frac{(x+2)^2}{4} + \frac{(y-3)^2}{9} = 1$$

$a = 3, b = 2, c = \sqrt{5}$

Center: $(-2, 3)$

Foci: $\left(-2, 3 \pm \sqrt{5}\right)$

Vertices: $(-2, 6), (-2, 0)$

$e = \dfrac{\sqrt{5}}{3}$

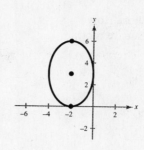

17. $16x^2 + 25y^2 - 32x + 50y + 16 = 0$

$16(x^2 - 2x + 1) + 25(y^2 + 2y + 1) = -16 + 16 + 25$

$$\frac{(x-1)^2}{25/16} + (y+1)^2 = 1$$

$a = \dfrac{5}{4}, b = 1, c = \dfrac{3}{4}$

Center: $(1, -1)$

Foci: $\left(\dfrac{7}{4}, -1\right), \left(\dfrac{1}{4}, -1\right)$

Vertices: $\left(\dfrac{9}{4}, -1\right), \left(-\dfrac{1}{4}, -1\right)$

$e = \dfrac{3}{5}$

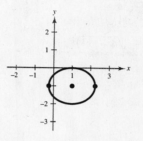

19. $5x^2 + 3y^2 = 15$

$$\frac{x^2}{3} + \frac{y^2}{5} = 1$$

Center: $(0, 0)$

$a = \sqrt{5}, b = \sqrt{3}, c = \sqrt{2}$

Foci: $\left(0, \pm\sqrt{2}\right)$

Vertices: $\left(0, \pm\sqrt{5}\right)$

To graph, solve for y.

$$y^2 = \frac{15 - 5x^2}{3}$$

$$y_1 = \sqrt{\frac{15 - 5x^2}{3}}$$

$$y_2 = -\sqrt{\frac{15 - 5x^2}{3}}$$

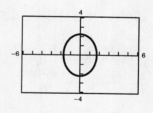

21. $12x^2 + 20y^2 - 12x + 40y - 37 = 0$

$12\left(x^2 - x + \dfrac{1}{4}\right) + 20(y^2 + 2y + 1) = 37 + 3 + 20$

$$\frac{[x - (1/2)]^2}{5} + \frac{(y+1)^2}{3} = 1$$

$a = \sqrt{5}, b = \sqrt{3}, c = \sqrt{2}$

Center: $\left(\dfrac{1}{2}, -1\right)$

Foci: $\left(\dfrac{1}{2} \pm \sqrt{2}, -1\right)$

Vertices: $\left(\dfrac{1}{2} \pm \sqrt{5}, -1\right)$

$e = \dfrac{\sqrt{10}}{5}$

To graph, solve for y.

$$(y+1)^2 = 3\left[1 - \frac{(x - 0.5)^2}{5}\right]$$

$$y_1 = -1 + \sqrt{3\left[1 - \frac{(x - 0.5)^2}{5}\right]}$$

$$y_2 = -1 - \sqrt{3\left[1 - \frac{(x - 0.5)^2}{5}\right]}$$

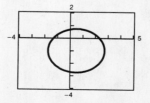

23. For the right half of the ellipse, solve for x and use the positive square root.

$$\frac{(x-3)^2}{9} + \frac{y^2}{4} = 1$$

$$4(x-3)^2 + 9y^2 = 36$$

$$4(x-3)^2 = 36 - 9y^2$$

$$(x-3)^2 = \frac{9(4-y^2)}{4}$$

$$x - 3 = \frac{2}{3}\sqrt{4 - y^2}$$

$$x = 3 + \frac{2}{3}\sqrt{4 - y^2}$$

$$= \frac{3}{2}\left(2 + \sqrt{4 - y^2}\right)$$

25. Vertices: $(\pm 5, 0)$

$a = 5, c = 2 \Rightarrow b = \sqrt{21}$

Foci: $(\pm 2, 0)$

Horizontal major axis

Center: $(0, 0)$

$$\frac{(x-h)^2}{a^2} + \frac{(y-k)^2}{b^2} = 1$$

$$\frac{x^2}{25} + \frac{y^2}{21} = 1$$

27. Center: $(0, 0)$

$a = 2, b = 1$

Vertical major axis

$$\frac{(x-h)^2}{b^2} + \frac{(y-k)^2}{a^2} = 1$$

$$\frac{x^2}{1} + \frac{y^2}{4} = 1$$

29. Foci: $(\pm 5, 0) \Rightarrow c = 5$

Center: $(0, 0)$

Horizontal major axis

Major axis of length 12 $\Rightarrow 2a = 12$

$$a = 6$$

$6^2 - b^2 = 5^2 \Rightarrow b^2 = 11$

$$\frac{(x-h)^2}{a^2} + \frac{(y-k)^2}{b^2} = 1$$

$$\frac{x^2}{36} + \frac{y^2}{11} = 1$$

31. Vertices: $(0, \pm 5) \Rightarrow a = 5$

Center: $(0, 0)$

Vertical major axis

$$\frac{(x-h)^2}{b^2} + \frac{(y-k)^2}{a^2} = 1$$

$$\frac{x^2}{b^2} + \frac{y^2}{25} = 1$$

Point: $(4, 2)$

$$\frac{4^2}{b^2} + \frac{2^2}{25} = 1$$

$$\frac{16}{b^2} = 1 - \frac{4}{25} = \frac{21}{25}$$

$$400 = 21b^2$$

$$\frac{400}{21} = b^2$$

$$\frac{x^2}{400/21} + \frac{y^2}{25} = 1$$

$$\frac{21x^2}{400} + \frac{y^2}{25} = 1$$

33. Center: $(2, 3)$

$a = 3, \quad b = 1$

Vertical major axis

$$\frac{(x-h)^2}{b^2} + \frac{(y-k)^2}{a^2} = 1$$

$$\frac{(x-2)^2}{1} + \frac{(y-3)^2}{9} = 1$$

35. Center: $(2, 2)$

$a = 3, \quad b = 2$

Horizontal major axis

$$\frac{(x-h)^2}{a^2} + \frac{(y-k)^2}{b^2} = 1$$

$$\frac{(x-2)^2}{9} + \frac{(y-2)^2}{4} = 1$$

37. Vertices: $(0, 2), (4, 2) \Rightarrow a = 2$

Minor axis of length $2 \Rightarrow b = 1$

Center: $(2, 2) = (h, k)$

$\dfrac{(x - h)^2}{a^2} + \dfrac{(y - k)^2}{b^2} = 1$

$\dfrac{(x - 2)^2}{4} + \dfrac{(y - 2)^2}{1} = 1$

39. Foci: $(0, 0), (0, 8) \Rightarrow c = 4$

Major axis of length $16 \Rightarrow a = 8$

$b^2 = a^2 - c^2 = 64 - 16 = 48$

Center: $(0, 4) = (h, k)$

$\dfrac{(x - h)^2}{b^2} + \dfrac{(y - k)^2}{a^2} = 1$

$\dfrac{x^2}{48} + \dfrac{(y - 4)^2}{64} = 1$

41. Vertices: $(3, 1), (3, 9) \Rightarrow a = 4$

Center: $(3, 5)$

Minor axis of length $6 \Rightarrow b = 3$

Vertical major axis

$\dfrac{(x - h)^2}{b^2} + \dfrac{(y - k)^2}{a^2} = 1$

$\dfrac{(x - 3)^2}{9} + \dfrac{(y - 5)^2}{16} = 1$

43. Center: $(0, 4)$

Vertices: $(-4, 4), (4, 4) \Rightarrow a = 4$

$a = 2c \Rightarrow 4 = 2c \Rightarrow c = 2$

$2^2 = 4^2 - b^2 \Rightarrow b^2 = 12$

Horizontal major axis

$\dfrac{(x - h)^2}{a^2} + \dfrac{(y - k)^2}{b^2} = 1$

$\dfrac{x^2}{16} + \dfrac{(y - 4)^2}{12} = 1$

45. (a) The length of the string is $2a$.

(b) The path is an ellipse because the sum of the distances from the two thumbtacks is always the length of the string, that is, it is constant.

47. $\dfrac{x^2}{a^2} + \dfrac{y^2}{b^2} = 1$

(a) $a + b = 20 \Rightarrow b = 20 - a$

$A = \pi ab = \pi a(20 - a)$

(b) $264 = \pi a(20 - a)$

$0 = -\pi a^2 + 20\pi a - 264$

$0 = \pi a^2 - 20\pi a + 264$

$a = 14$ or $a = 6$. The equation of an ellipse with an area of 264 is $\dfrac{x^2}{196} + \dfrac{y^2}{36} = 1$.

(c)

a	8	9	10	11	12	13
A	301.6	311.0	314.2	311.0	301.6	285.9

The area is maximum when $a = 10$ and the ellipse is a circle.

(d)

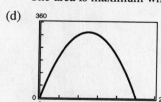

The area is maximum (314.16) when $a = b = 10$ and the ellipse is a circle.

49. Vertices: $(\pm 5, 0) \Rightarrow a = 5$

Eccentricity: $\dfrac{3}{5} \Rightarrow c = \dfrac{3}{5}a = 3$

$b^2 = a^2 - c^2 = 25 - 9 = 16$

Center: $(0, 0) = (h, k)$

$\dfrac{(x - h)^2}{a^2} + \dfrac{(y - k)^2}{b^2} = 1$

$\dfrac{x^2}{25} + \dfrac{y^2}{16} = 1$

51. $2a = 36.23 \Rightarrow a = 18.115$

$e = \dfrac{c}{a} = 0.97 \Rightarrow c = (18.115)(0.97) = 17.57155$

$b^2 = a^2 - c^2 = 18.115^2 - 17.57155^2 \approx 19.39$

The equation of the ellipse is:

$\dfrac{x^2}{(18.115)^2} + \dfrac{y^2}{19.39} = 1$

$\dfrac{x^2}{328.15} + \dfrac{y^2}{19.39} = 1$

53. (a) $b^2 = a^2 - c^2 = a^2 - \left(\dfrac{c^2}{a^2}\right)a^2 = a^2(1 - e^2)$

$\dfrac{(x - h)^2}{a^2} + \dfrac{(y - k)^2}{b^2} = 1$

$\dfrac{(x - h)^2}{a^2} + \dfrac{(y - k)^2}{a^2(1 - e^2)} = 1$

(b)

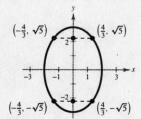

(c) As e approaches 0, the ellipse approaches a circle.

55. $\dfrac{x^2}{4} + \dfrac{y^2}{1} = 1$

$a = 2, b = 1, c = \sqrt{3}$

Points on the ellipse: $(\pm 2, 0), (0, \pm 1)$

Length of latus recta: $\dfrac{2b^2}{a} = \dfrac{2(1)^2}{2} = 1$

Additional points: $\left(-\sqrt{3}, \pm\dfrac{1}{2}\right), \left(\sqrt{3}, \pm\dfrac{1}{2}\right)$

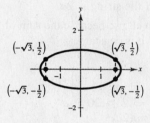

57. $9x^2 + 4y^2 = 36$

$\dfrac{x^2}{4} + \dfrac{y^2}{9} = 1$

$a = 3, b = 2, c = \sqrt{5}$

Points on the ellipse: $(\pm 2, 0), (0, \pm 3)$

Length of latus recta: $\dfrac{2b^2}{a} = \dfrac{2 \cdot 2^2}{3} = \dfrac{8}{3}$

Additional points: $\left(\pm\dfrac{4}{3}, -\sqrt{5}\right), \left(\pm\dfrac{4}{3}, \sqrt{5}\right)$

59. False. This equation is not second-degree in y.

Section 10.3   Hyperbolas

- A **hyperbola** is the set of all points (x, y) the difference of whose distances from two distinct fixed points **(foci)** is constant.

- The standard equation of a hyperbola with center (h, k) and transverse and conjugate axes of lengths $2a$ and $2b$ is:

  (a) $\dfrac{(x - h)^2}{a^2} - \dfrac{(y - k)^2}{b^2} = 1$ if the traverse axis is horizontal.

  (b) $\dfrac{(y - k)^2}{a^2} - \dfrac{(x - h)^2}{b^2} = 1$ if the traverse axis is vertical.

- $c^2 = a^2 + b^2$ where c is the distance from the center to a focus.

- The asymptotes of a hyperbola are:

  (a) $y = k \pm \dfrac{b}{a}(x - h)$ if the transverse axis is horizontal.

  (b) $y = k \pm \dfrac{a}{b}(x - h)$ the transverse axis is vertical.

- The eccentricity of a hyperbola is $e = \dfrac{c}{a}$.

- To classify a nondegenerate conic from its general equation $Ax^2 + Cy^2 + Dx + Ey + F = 0$:
  (a) If $A = C$ $(A \neq 0, C \neq 0)$, then it is a circle.
  (b) If $AC = 0$ $(A = 0$ or $C = 0$, but not both$)$, then it is a parabola.
  (c) If $AC > 0$, then it is an ellipse.
  (d) If $AC < 0$, then it is a hyperbola.

Solutions to Odd-Numbered Exercises

1. $\dfrac{x^2}{16} - \dfrac{y^2}{4} = 1$

   Center: $(0, 0)$

   $a = 4, b = 2$

   Horizontal transverse axis

   Matches graph (a).

3. $\dfrac{y^2}{9} - \dfrac{x^2}{16} = 1$

   Center: $(0, 0)$

   $a = 3, b = 4$

   Vertical transverse axis

   Matches graph (d).

5. $x^2 - y^2 = 1$

   $a = 1, b = 1, c = \sqrt{2}$

   Center: $(0, 0)$

   Vertices: $(\pm 1, 0)$

   Foci: $\left(\pm \sqrt{2}, 0\right)$

   Asymptotes: $y = \pm x$

   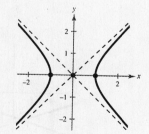

7. $\dfrac{y^2}{1} - \dfrac{x^2}{4} = 1$

   $a = 1, b = 2, c = \sqrt{5}$

   Center: $(0, 0)$

   Vertices: $(0, \pm 1)$

   Foci: $\left(0, \pm \sqrt{5}\right)$

   Asymptotes: $y = \pm \dfrac{1}{2}x$

   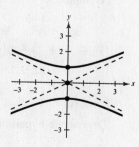

9. $\dfrac{y^2}{25} - \dfrac{x^2}{144} = 1$

$a = 5, b = 12, c = 13$

Center: $(0, 0)$

Vertices: $(0, \pm 5)$

Foci: $(0, \pm 13)$

Asymptotes: $y = \pm\dfrac{5}{12}x$

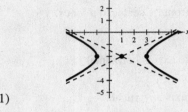

11. $\dfrac{(x-1)^2}{4} - \dfrac{(y+2)^2}{1} = 1$

$a = 2, b = 1, c = \sqrt{5}$

Center: $(1, -2)$

Vertices: $(-1, -2), (3, -2)$

Foci: $\left(1 \pm \sqrt{5}, -2\right)$

Asymptotes: $y = -2 \pm \dfrac{1}{2}(x-1)$

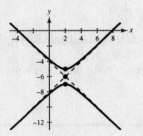

13. $(y+6)^2 - (x-2)^2 = 1$

$a = 1, b = 1, c = \sqrt{2}$

Center: $(2, -6)$

Vertices: $(2, -5), (2, -7)$

Foci: $\left(2, -6 \pm \sqrt{2}\right)$

Asymptotes: $y = -6 \pm (x-2)$

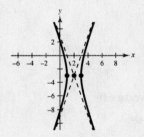

15. $\qquad 9x^2 - y^2 - 36x - 6y + 18 = 0$

$9\left(x^2 - 4x + 4\right) - \left(y^2 + 6y + 9\right) = -18 + 36 - 9$

$$\dfrac{(x-2)^2}{1} - \dfrac{(y+3)^2}{9} = 1$$

$a = 1, b = 3, c = \sqrt{10}$

Center: $(2, -3)$

Vertices: $(1, -3), (3, -3)$

Foci: $\left(2 \pm \sqrt{10}, -3\right)$

Asymptotes: $y = -3 \pm 3(x-2)$

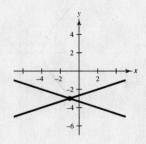

17. $\qquad x^2 - 9y^2 + 2x - 54y - 80 = 0$

$\left(x^2 + 2x + 1\right) - 9\left(y^2 + 6y + 9\right) = 80 + 1 - 81$

$\qquad\qquad (x+1)^2 - 9(y+3)^2 = 0$

$$y + 3 = \pm\dfrac{1}{3}(x+1)$$

Degenerate hyperbola is two lines intersecting at $(-1, -3)$.

19. $2x^2 - 3y^2 = 6$

$$\frac{x^2}{3} - \frac{y^2}{2} = 1$$

$a = \sqrt{3}, b = \sqrt{2}, c = \sqrt{5}$

Center: $(0, 0)$

Vertices: $(\pm\sqrt{3}, 0)$

Foci: $(\pm\sqrt{5}, 0)$

Asymptotes: $y = \pm\sqrt{\frac{2}{3}}\,x$

To use a graphing calculator, solve first for y.

$$y^2 = \frac{2x^2 - 6}{3}$$

$\left. \begin{array}{l} y_1 = \sqrt{\dfrac{2x^2 - 6}{3}} \\[2ex] y_2 = -\sqrt{\dfrac{2x^2 - 6}{3}} \end{array} \right\}$ Hyperbola

$\left. \begin{array}{l} y_3 = \sqrt{\dfrac{2}{3}}\,x \\[2ex] y_4 = -\sqrt{\dfrac{2}{3}}\,x \end{array} \right\}$ Asymptotes

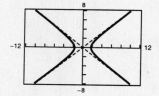

21. $9y^2 - x^2 + 2x + 54y + 62 = 0$

$9(y^2 + 6y + 9) - (x^2 - 2x + 1) = -62 - 1 + 81$

$$\frac{(y + 3)^2}{2} - \frac{(x - 1)^2}{18} = 1$$

$a = \sqrt{2}, b = 3\sqrt{2}, c = 2\sqrt{5}$

Center: $(1, -3)$

Vertices: $(1, -3 \pm \sqrt{2})$

Foci: $(1, -3 \pm 2\sqrt{5})$

Asymptotes: $y = -3 \pm \frac{1}{3}(x - 1)$

To use a graphing calculator, solve for y first.

$9(y + 3)^2 = 18 + (x - 1)^2$

$$y = -3 \pm \sqrt{\frac{18 + (x - 1)^2}{9}}$$

$\left. \begin{array}{l} y_1 = -3 + \dfrac{1}{3}\sqrt{18 + (x - 1)^2} \\[2ex] y_2 = -3 - \dfrac{1}{3}\sqrt{18 + (x - 1)^2} \end{array} \right\}$ Hyperbola

$\left. \begin{array}{l} y_3 = -3 + \dfrac{1}{3}(x - 1) \\[2ex] y_4 = -3 - \dfrac{1}{3}(x - 1) \end{array} \right\}$ Asymptotes

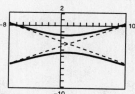

23. Vertices: $(0, \pm 2) \Rightarrow a = 2$

Foci: $(0, \pm 4) \Rightarrow c = 4$

$b^2 = c^2 - a^2 = 16 - 4 = 12$

Center: $(0, 0) = (h, k)$

$$\frac{(y - k)^2}{a^2} - \frac{(x - h)^2}{b^2} = 1$$

$$\frac{y^2}{4} - \frac{x^2}{12} = 1$$

25. Vertices: $(\pm 1, 0) \Rightarrow a = 1$

Asymptotes: $y = \pm 3x \Rightarrow \dfrac{b}{a} = 3, b = 3$

Center: $(0, 0) = (h, k)$

$$\frac{(x - h)^2}{a^2} - \frac{(y - k)^2}{b^2} = 1$$

$$\frac{x^2}{1} - \frac{y^2}{9} = 1$$

27. Foci: $(0, \pm 8) \Rightarrow c = 8$

Asymptotes: $y = \pm 4x \Rightarrow \dfrac{a}{b} = 4 \Rightarrow a = 4b$

Center: $(0, 0) = (h, k)$

$c^2 = a^2 + b^2 \Rightarrow 64 = 16b^2 + b^2$

$$\frac{64}{17} = b^2 \Rightarrow a^2 = \frac{1024}{17}$$

$$\frac{(y - k)^2}{a^2} - \frac{(x - h)^2}{b^2} = 1$$

$$\frac{y^2}{1024/17} - \frac{x^2}{64/17} = 1$$

$$\frac{17y^2}{1024} - \frac{17x^2}{65} = 1$$

29. Vertices: $(2, 0), (6, 0) \Rightarrow a = 2$

Foci: $(0, 0), (8, 0) \Rightarrow c = 4$

$b^2 = c^2 - a^2 = 16 - 4 = 12$

Center: $(4, 0) = (h, k)$

$$\frac{(x - h)^2}{a^2} - \frac{(y - k)^2}{b^2} = 1$$

$$\frac{(x - 4)^2}{4} - \frac{y^2}{12} = 1$$

31. Vertices: $(4, 1), (4, 9) \Rightarrow a = 4$

Foci: $(4, 0), (4, 10) \Rightarrow c = 5$

$b^2 = c^2 - a^2 = 25 - 16 = 9$

Center: $(4, 5) = (h, k)$

$$\frac{(y - k)^2}{a^2} - \frac{(x - h)^2}{b^2} = 1$$

$$\frac{(y - 4)^2}{16} - \frac{(x - 4)^2}{9} = 1$$

33. Vertices: $(2, 3), (2, -3) \Rightarrow a = 3$

Solution point: $(0, 5)$

Center: $(2, 0) = (h, k)$

$$\frac{(y - k)^2}{a^2} - \frac{(x - h)^2}{b^2} = 1$$

$$\frac{y^2}{9} - \frac{(x - 2)^2}{b^2} = 1 \Rightarrow b^2 = \frac{9(x - 2)^2}{y^2 - 9} = \frac{9(-2)^2}{25 - 9} = \frac{36}{16} = \frac{9}{4}$$

$$\frac{y^2}{9} - \frac{(x - 2)^2}{9/4} = 1$$

35. Vertices: $(0, 2), (6, 2) \Rightarrow a = 3$

Asymptotes: $y = \dfrac{2}{3}x, \, y = 4 - \dfrac{2}{3}x$

$\dfrac{b}{a} = \dfrac{2}{3} \Rightarrow b = 2$

Center: $(3, 2) = (h, k)$

$$\frac{(x - h)^2}{a^2} - \frac{(y - k)^2}{b^2} = 1$$

$$\frac{(x - 3)^2}{9} - \frac{(y - 2)^2}{4} = 1$$

37. $x = 3 - \frac{2}{3}\sqrt{9 + (y - 1)^2}$ represents the left branch of the hyperbola.

39. Foci: $(\pm150, 0) \implies c = 150$

Center: $(0, 0) = (h, k)$

$$\frac{d_2}{186,000} - \frac{d_1}{186,000} = 0.001 \implies 2a = 186, a = 93$$

$$b^2 = c^2 - a^2 = 150^2 - 93^2 = 13,851$$

$$\frac{x^2}{93^2} - \frac{y^2}{13,851} = 1$$

$$x^2 = 93^2\left(1 + \frac{75^2}{13,851}\right) \approx 12,161$$

$$x \approx 110.3 \text{ miles}$$

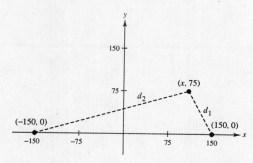

41. Center: $(0, 0) = (h, k)$

Focus: $(24, 0) \implies c = 24$

Solution point: $(24, 24)$

$$24^2 = a^2 + b^2 \implies b^2 = 24^2 - a^2$$

$$\frac{(x - h)^2}{a^2} - \frac{(y - k)^2}{b^2} = 1$$

$$\frac{x^2}{a^2} - \frac{y^2}{24^2 - a^2} = 1 \implies \frac{24^2}{a^2} - \frac{24^2}{24^2 - a^2} = 1$$

Solving yields $a^2 = \dfrac{(3 - \sqrt{5})24^2}{2} \approx 220.0124$ and $b^2 \approx 355.9876$.

Thus, we have $\dfrac{x^2}{220.0124} - \dfrac{y^2}{355.9876} = 1$.

The right vertex is at $(a, 0) \approx (14.83, 0)$.

43. $x^2 + y^2 - 6x + 4y + 9 = 0$

$A = 1, C = 1$

$A = C \implies$ Circle

45. $4x^2 - y^2 - 4x - 3 = 0$

$A = 4, C = -1$

$AC = 4(-1)$

$\quad = -4 < 0 \implies$ Hyperbola

47. $4x^2 + 3y^2 + 8x - 24y + 51 = 0$

$A = 4, C = 3$

$AC = 4(3) = 12 > 0 \implies$ Ellipse

49. $25x^2 - 10x - 200y - 119 = 0$

$A = 25, C = 0$

$AC = 25(0) = 0 \implies$ Parabola

Section 10.4   Rotation and Systems of Quadratic Equations

> ■ The general second-degree equation $Ax^2 + Bxy + Cy^2 + Dx + Ey + F = 0$ can be rewritten as
> $A'(x')^2 + C'(y')^2 + D'x' + E'y' + F' = 0$ by rotating the coordinate axes through the angle
> θ, where $\cot 2\theta = (A - C)/B$.
>
> ■ $x = x' \cos \theta - y' \sin \theta$
> $y = x' \sin \theta + y' \cos \theta$
>
> ■ The graph of the nondegenerate equation $Ax^2 + Bxy + Cy^2 + Dx + Ey + F = 0$ is:
>
>   (a) An ellipse or circle if $B^2 - 4AC < 0$.
>
>   (b) A parabola if $B^2 - 4AC = 0$.
>
>   (c) A hyperbola if $B^2 - 4AC > 0$.

Solutions to Odd-Numbered Exercises

1. $\theta = 90°$; Point: (0, 3)

$x' = x \cos \theta - y \sin \theta = 0(\cos 90°) - 3(\sin 90°) = -3$

$y' = x \sin \theta + y \cos \theta = 0(\sin 90°) + 3(\cos 90°) = 0$

Thus, $(x', y') = (-3, 0)$.

3. $\theta = 30°$; Point: (1, 4)

$x' = x \cos \theta - y \sin \theta = 1(\cos 30°) - 4(\sin 30°) = \dfrac{\sqrt{3}}{2} - \dfrac{4}{2} = \dfrac{1}{2}(\sqrt{3} - 4)$

$y' = x \sin \theta + y \cos \theta = 1(\sin 30°) + 4(\cos 30°) = \dfrac{1}{2} + \dfrac{4\sqrt{3}}{2} = \dfrac{1}{2}(1 + 4\sqrt{3})$

Thus, $(x', y') = \left(\dfrac{1}{2}(\sqrt{3} - 4), \dfrac{1}{2}(1 + 4\sqrt{3})\right)$.

5. $xy + 1 = 0$

$A = 0, B = 1, C = 0$

$\cot 2\theta = \dfrac{A - C}{B} = 0 \Rightarrow 2\theta = \dfrac{\pi}{2} \Rightarrow \theta = \dfrac{\pi}{4}$

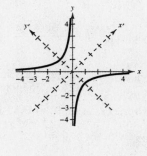

$x = x' \cos \dfrac{\pi}{4} - y' \sin \dfrac{\pi}{4}$

$\qquad = x'\left(\dfrac{\sqrt{2}}{2}\right) - y'\left(\dfrac{\sqrt{2}}{2}\right)$

$\qquad = \dfrac{x' - y'}{\sqrt{2}}$

$y = x' \sin \dfrac{\pi}{4} + y' \cos \dfrac{\pi}{4}$

$\qquad = x'\left(\dfrac{\sqrt{2}}{2}\right) + y'\left(\dfrac{\sqrt{2}}{2}\right)$

$\qquad = \dfrac{x' + y'}{\sqrt{2}}$

$xy + 1 = 0$

$\left(\dfrac{x' - y'}{\sqrt{2}}\right)\left(\dfrac{x' + y'}{\sqrt{2}}\right) + 1 = 0$

$\dfrac{(y')^2}{2} - \dfrac{(x')^2}{2} = 1$

7. $x^2 - 10xy + y^2 + 1 = 0$

    $A = 1, B = -10, C = 1$

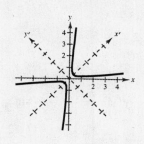

$$\cot 2\theta = \frac{A - C}{B} = 0 \Rightarrow 2\theta = \frac{\pi}{2} \Rightarrow \theta = \frac{\pi}{4}$$

$$x = x'\cos\frac{\pi}{4} - y'\sin\frac{\pi}{4} \qquad\qquad y = x'\sin\frac{\pi}{4} + y'\cos\frac{\pi}{4}$$

$$= x'\left(\frac{\sqrt{2}}{2}\right) - y'\left(\frac{\sqrt{2}}{2}\right) \qquad\qquad = x'\left(\frac{\sqrt{2}}{2}\right) + y'\left(\frac{\sqrt{2}}{2}\right)$$

$$= \frac{x' - y'}{\sqrt{2}} \qquad\qquad\qquad\qquad = \frac{x' + y'}{\sqrt{2}}$$

$x^2 - 10xy + y^2 + 1 = 0$

$$\left(\frac{x' - y'}{\sqrt{2}}\right)^2 - 10\left(\frac{x' - y'}{\sqrt{2}}\right)\left(\frac{x' + y'}{\sqrt{2}}\right) + \left(\frac{x' + y'}{\sqrt{2}}\right)^2 + 1 = 0$$

$$\frac{(x')^2}{2} - x'y' + \frac{(y')^2}{2} - 5(x')^2 + 5(y')^2 + \frac{(x')^2}{2} + x'y' + \frac{(y')^2}{2} + 1 = 0$$

$$-4(x')^2 + 6(y')^2 = -1$$

$$\frac{(x')^2}{1/4} - \frac{(y')^2}{1/6} = 1$$

9. $xy - 2y - 4x = 0$

    $A = 0, B = 1, C = 0$

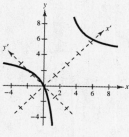

$$\cot 2\theta = \frac{A - C}{B} = 0 \Rightarrow 2\theta = \frac{\pi}{2} \Rightarrow \theta = \frac{\pi}{4}$$

$$x = x'\cos\frac{\pi}{4} - y'\sin\frac{\pi}{4} \qquad\qquad y = x'\sin\frac{\pi}{4} + y'\cos\frac{\pi}{4}$$

$$= x'\left(\frac{\sqrt{2}}{2}\right) - y'\left(\frac{\sqrt{2}}{2}\right) \qquad\qquad = x'\left(\frac{\sqrt{2}}{2}\right) + y'\left(\frac{\sqrt{2}}{2}\right)$$

$$= \frac{x' - y'}{\sqrt{2}} \qquad\qquad\qquad\qquad = \frac{x' + y'}{\sqrt{2}}$$

$xy - 2y - 4x = 0$

$$\left(\frac{x' - y'}{\sqrt{2}}\right)\left(\frac{x' + y'}{\sqrt{2}}\right) - 2\left(\frac{x' + y'}{\sqrt{2}}\right) - 4\left(\frac{x' - y'}{\sqrt{2}}\right) = 0$$

$$\frac{(x')^2}{2} - \frac{(y')^2}{2} - \sqrt{2}x' - \sqrt{2}y' - 2\sqrt{2}x' + 2\sqrt{2}y' = 0$$

$$\left[(x')^2 - 6\sqrt{2}x' + (3\sqrt{2})^2\right] - \left[(y')^2 - 2\sqrt{2}y' + (\sqrt{2})^2\right] = 0 + (3\sqrt{2})^2 - (\sqrt{2})^2$$

$$(x' - 3\sqrt{2})^2 - (y' - \sqrt{2})^2 = 16$$

$$\frac{(x' - 3\sqrt{2})^2}{16} - \frac{(y' - \sqrt{2})^2}{16} = 1$$

11. $5x^2 - 2xy + 5y^2 - 12 = 0$

$A = 5, B = -2, C = 5$

$\cot 2\theta = \dfrac{A - C}{B} = 0 \Rightarrow 2\theta = \dfrac{\pi}{2} \Rightarrow \theta = \dfrac{\pi}{4}$

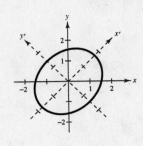

$x = x'\cos\dfrac{\pi}{4} - y'\sin\dfrac{\pi}{4} \qquad y = x'\sin\dfrac{\pi}{4} + y'\cos\dfrac{\pi}{4}$

$\quad = x'\left(\dfrac{\sqrt{2}}{2}\right) - y'\left(\dfrac{\sqrt{2}}{2}\right) \qquad = x'\left(\dfrac{\sqrt{2}}{2}\right) + y'\left(\dfrac{\sqrt{2}}{2}\right)$

$\quad = \dfrac{x' - y'}{\sqrt{2}} \qquad\qquad\quad = \dfrac{x' + y'}{\sqrt{2}}$

$$5x^2 - 2xy + 5y^2 - 12 = 0$$

$$5\left(\dfrac{x' - y'}{\sqrt{2}}\right)^2 - 2\left(\dfrac{x' - y'}{\sqrt{2}}\right)\left(\dfrac{x' + y'}{\sqrt{2}}\right) + 5\left(\dfrac{x' + y'}{\sqrt{2}}\right)^2 - 12 = 0$$

$$\dfrac{5(x')^2}{2} - 5x'y' + \dfrac{5(y')^2}{2} - (x')^2 + (y')^2 + \dfrac{5(x')^2}{2} + 5x'y' + \dfrac{5(y')^2}{2} - 12 = 0$$

$$4(x')^2 + 6(y')^2 = 12$$

$$\dfrac{(x')^2}{3} + \dfrac{(y')^2}{2} = 1$$

13. $3x^2 - 2\sqrt{3}xy + y^2 + 2x + 2\sqrt{3}y = 0$

$A = 3, B = -2\sqrt{3}, C = 1$

$\cot 2\theta = \dfrac{A - C}{B} = -\dfrac{1}{\sqrt{3}} \Rightarrow \theta = 60°$

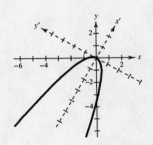

$x = x'\cos 60° - y'\sin 60°$

$\quad = x'\left(\dfrac{1}{2}\right) - y'\left(\dfrac{\sqrt{3}}{2}\right) = \dfrac{x' - \sqrt{3}y'}{2}$

$y = x'\sin 60° + y'\cos 60°$

$\quad = x'\left(\dfrac{\sqrt{3}}{2}\right) + y'\left(\dfrac{1}{2}\right) = \dfrac{\sqrt{3}x' - y'}{2}$

$$3x^2 - 2\sqrt{3}xy + y^2 + 2x + 2\sqrt{3}y = 0$$

$$3\left(\dfrac{x' - \sqrt{3}y'}{2}\right)^2 - 2\sqrt{3}\left(\dfrac{x' - \sqrt{3}y'}{2}\right)\left(\dfrac{\sqrt{3}x' + y'}{2}\right) + \left(\dfrac{\sqrt{3}x' + y'}{2}\right)^2 + 2\left(\dfrac{x' - \sqrt{3}y'}{2}\right) + 2\sqrt{3}\left(\dfrac{\sqrt{3}x' + y'}{2}\right) = 0$$

$$\dfrac{3(x')^2}{4} - \dfrac{6\sqrt{3}x'y'}{4} + \dfrac{9(y')^2}{4} - \dfrac{6(x')^2}{4} + \dfrac{4\sqrt{3}x'y'}{4} + \dfrac{6(y')^2}{4} + \dfrac{3(x')^2}{4} + \dfrac{2\sqrt{3}x'y'}{4} + \dfrac{(y')^2}{4}$$

$$+ x' - \sqrt{3}y' + 3x' + \sqrt{3}y' = 0$$

$$4(y')^2 + 4x' = 0$$

$$x' = -(y')^2$$

15. $9x^2 + 24xy + 16y^2 + 90x - 130y = 0$

$A = 9, B = 24, C = 16$

$$\cot 2\theta = \frac{A - C}{B} = -\frac{7}{24} \Rightarrow \theta \approx 53.13°$$

$$\cos 2\theta = -\frac{7}{25}$$

$$\sin \theta = \sqrt{\frac{1 - \cos \theta}{2}} = \sqrt{\frac{1 - (-7/25)}{2}} = \frac{4}{5}$$

$$\cos \theta = \sqrt{\frac{1 + \cos 2\theta}{2}} = \sqrt{\frac{1 + (-7/25)}{2}} = \frac{3}{5}$$

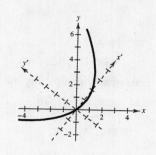

$x = x'\cos\theta - y'\sin\theta$

$\qquad = x'\left(\dfrac{3}{5}\right) - y'\left(\dfrac{4}{5}\right) = \dfrac{3x' - 4y'}{5}$

$y = x'\sin\theta + y'\cos\theta$

$\qquad = x'\left(\dfrac{4}{5}\right) + y'\left(\dfrac{3}{5}\right)$

$\qquad = \dfrac{4x' + 3y'}{5}$

$$9x^2 + 24xy + 16y^2 + 90x - 130y = 0$$

$$9\left(\frac{3x' - 4y'}{5}\right)^2 + 24\left(\frac{3x' - 4y'}{5}\right)\left(\frac{4x' + 3y'}{5}\right) + 16\left(\frac{3x' - 4y'}{5}\right)^2 + 90\left(\frac{3x' - 4y'}{5}\right) - 130\left(\frac{4x' + 3y'}{5}\right) = 0$$

$$\frac{81(x')^2}{25} - \frac{216x'y'}{25} + \frac{144(y')^2}{25} + \frac{288(x')^2}{25} - \frac{168x'y'}{25} - \frac{288(y')^2}{25} + \frac{256(x')^2}{25} + \frac{384x'y'}{25}$$

$$+ \frac{144(y')^2}{25} + 54x' - 72y' - 104x' - 78y' = 0$$

$$25(x')^2 - 50x' - 150y' = 0$$

$$(x')^2 - 2x' + 1 = 6y' + 1$$

$$y' = \frac{(x')^2}{6} - \frac{x'}{3}$$

17. $x^2 + xy + y^2 = 10$

$$\cot 2\theta = \frac{A - C}{B} = \frac{1 - 1}{0} = 0 \Rightarrow \theta = \frac{\pi}{4} \text{ or } 45°$$

To graph the conic using a graphing calculator, we need to solve for y in terms of x.

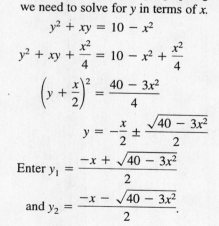

$$y^2 + xy = 10 - x^2$$

$$y^2 + xy + \frac{x^2}{4} = 10 - x^2 + \frac{x^2}{4}$$

$$\left(y + \frac{x}{2}\right)^2 = \frac{40 - 3x^2}{4}$$

$$y = -\frac{x}{2} \pm \frac{\sqrt{40 - 3x^2}}{2}$$

Enter $y_1 = \dfrac{-x + \sqrt{40 - 3x^2}}{2}$

and $y_2 = \dfrac{-x - \sqrt{40 - 3x^2}}{2}$.

19. $17x^2 + 32xy - 7y^2 = 75$

$\cot 2\theta = \dfrac{A - C}{B} = \dfrac{17 + 7}{32} = \dfrac{24}{32} = \dfrac{3}{4} \Rightarrow \theta \approx 26.57°$

Solve for y in terms of x by completing the square.

$$-7y^2 + 32xy = -17x^2 + 75$$

$$y^2 - \frac{32}{7}xy = \frac{17}{7}x^2 - \frac{75}{7}$$

$$y^2 - \frac{32}{7}xy + \frac{256}{49}x^2 = \frac{119}{49}x^2 - \frac{525}{49} + \frac{256}{49}x^2$$

$$\left(y - \frac{16}{7}x\right)^2 = \frac{375x^2 - 525}{49}$$

$$y = \frac{16}{7}x \pm \sqrt{\frac{375x^2 - 525}{49}}$$

$$y = \frac{16x \pm 5\sqrt{15x^2 - 21}}{7}$$

Use $y_1 = \dfrac{16x + 5\sqrt{15x^2 - 21}}{7}$

and $y_2 = \dfrac{16x - 5\sqrt{15x^2 - 21}}{7}$.

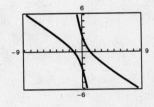

21. $32x^2 + 50xy + 7y^2 = 52$

$\cot 2\theta = \dfrac{A - C}{B} = \dfrac{32 - 7}{50} = \dfrac{1}{2} \Rightarrow \theta \approx 31.72°$

Solve for y in terms of x by completing the square.

$$7y^2 + 50xy = 52 - 32x^2$$

$$y^2 + \frac{50}{7}xy = \frac{52 - 32x^2}{7}$$

$$y^2 + \frac{50}{7}xy + \frac{625}{49}x^2 = \frac{52 - 32x^2}{7} + \frac{625x^2}{49}$$

$$\left(y + \frac{25}{7}x\right)^2 = \frac{364 + 401x^2}{49}$$

$$y = -\frac{25x}{7} \pm \frac{\sqrt{364 + 401x^2}}{7}$$

Enter $y_1 = \dfrac{-25x + \sqrt{364 + 401x^2}}{7}$

and $y_2 = \dfrac{-25x - \sqrt{364 + 401x^2}}{7}$.

23. $xy + 3 = 0$

$B^2 - 4AC = 1 \Rightarrow$ The graph is a hyperbola.

$\cot 2\theta = \dfrac{A - C}{B} = 0 \Rightarrow \theta = 45°$

Matches graph (e).

25. $-2x^2 + 3xy + 2y^2 + 3 = 0$

$B^2 - 4AC = (3)^2 - 4(-2)(2) = 25 \Rightarrow$ The graph is a hyperbola.

$\cot 2\theta = \dfrac{A - C}{B} = -\dfrac{4}{3} \Rightarrow \theta \approx -18.43°$

Matches graph (b).

27. $3x^2 + 2xy + y^2 - 10 = 0$

$B^2 - 4AC = (2)^2 - 4(3)(1) = -8 \Rightarrow$ The graph is an ellipse or circle.

$\cot 2\theta = \dfrac{A - C}{B} = 1 \Rightarrow \theta = 22.5°$

Matches graph (d).

29. $16x^2 - 24xy + 9y^2 - 30x - 40y = 0$

$B^2 - 4AC = (-24)^2 - 4(16)(9) = 0$

Parabola

31. $13x^2 - 8xy + 7y^2 - 45 = 0$

$B^2 - 4AC = (-8)^2 - 4(13)(7) = -300$

Ellipse or circle

33. $x^2 - 6xy - 5y^2 + 4x - 22 = 0$

$B^2 - 4AC = (-6)^2 - 4(1)(-5) = 56$

Hyperbola

35. $x^2 + 4xy + 4y^2 - 5x - y - 3 = 0$

$B^2 - 4AC = (4)^2 - 4(1)(4) = 0$

Parabola

37. $y^2 - 4x^2 = 0$

$\qquad y^2 = 4x^2$

$\qquad y = \pm 2x$

Two intersecting lines

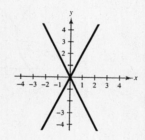

39. $x^2 + 2xy + y^2 - 1 = 0$

$\qquad (x + y)^2 - 1 = 0$

$\qquad (x + y)^2 = 1$

$\qquad x + y = \pm 1$

$\qquad y = -x \pm 1$

Two parallel lines

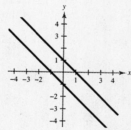

41.

$\qquad -x^2 + y^2 + 4x - 6y + 4 = 0 \Rightarrow (y - 3)^2 - (x - 2)^2 = 1$

$\qquad \underline{x^2 + y^2 - 4x - 6y + 12 = 0 \Rightarrow (x - 2)^2 + (y - 3)^2 = 1}$

$\qquad\qquad\qquad 2y^2 - 12y + 16 = 0$

$\qquad\qquad\qquad 2(y - 2)(y - 4) = 0$

$\qquad\qquad\qquad y = 2 \text{ or } y = 4$

For $y = 2$: $x^2 + 2^2 - 4x - 6(2) + 12 = 0$

$\qquad\qquad\qquad x^2 - 4x + 4 = 0$

$\qquad\qquad\qquad (x - 2)^2 = 0$

$\qquad\qquad\qquad x = 2$

For $y = 4$: $x^2 + 4^2 - 4x - 6(4) + 12 = 0$

$\qquad\qquad\qquad x^2 - 4x + 4 = 0$

$\qquad\qquad\qquad (x - 2)^2 = 0$

$\qquad\qquad\qquad x = 2$

The points of intersection are $(2, 2)$ and $(2, 4)$.

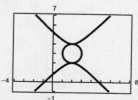

43. $-4x^2 - y^2 - 32x + 24y - 64 = 0$
$\underline{4x^2 + y^2 + 56x - 24y + 304 = 0}$
$24x + 240 = 0$
$x = -10$

When $x = -10$: $4(-10)^2 + y^2 + 56(-10) - 24y + 304 = 0$
$y^2 - 24y + 144 = 0$
$(y - 12)^2 = 0$
$y = 12$

The point of intersection is $(-10, 12)$.
In standard form the equations are:

$$\frac{(x + 4)^2}{36} + \frac{(y - 12)^2}{144} = 1$$

$$\frac{(x + y)^2}{9} + \frac{(y - 12)^2}{36} = 1$$

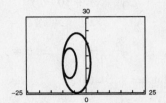

45. $x^2 - y^2 - 12x + 12y - 36 = 0$
$\underline{x^2 + y^2 - 12x - 12y + 36 = 0}$
$2x^2 - 24x = 0$
$2x(x - 12) = 0$
$x = 0 \text{ or } x = 12$

When $x = 0$: $y^2 - 12y + 36 = 0$
$(y - 6)^2 = 0$
$y = 6$

When $x = 12$: $12^2 + y^2 - 12(12) - 12y + 36 = 0$
$y^2 - 12y + 36 = 0$
$(y - 6)^2 = 0$
$y = 6$

The points of intersection are $(0, 6)$ and $(12, 6)$.
In standard form the equations are:

$$\frac{(x - 6)^2}{36} + \frac{(y - 6)^2}{36} = 1$$

$$(x - 6)^2 + (y - 6)^2 = 36$$

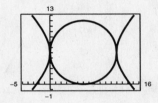

47. $-16x^2 - y^2 + 24y - 80 = 0$
$\underline{16x^2 + 25y^2 - 400 = 0}$
$24y^2 + 24y - 480 = 0$
$24(y + 5)(y - 4) = 0$
$y = -5 \text{ or } y = 4$

When $y = -5$: $16x^2 + 25(-5)^2 - 400 = 0$
$16x^2 = -225$
No real solution

When $y = 4$: $16x^2 + 25(4)^2 - 400 = 0$
$16x^2 = 0$
$x = 0$

The point of intersection is $(0, 4)$.
In standard form the equations are:

$$\frac{x^2}{4} + \frac{(y - 12)^2}{64} = 1$$

$$\frac{x^2}{25} + \frac{y^2}{16} = 1$$

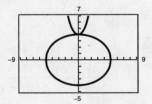

49. $x^2 + y^2 - 25 = 0 \Rightarrow y^2 = 25 - x^2$

$$9x - 4y^2 = 0 \Rightarrow 9x - 4(25 - x^2) = 0$$
$$4x^2 + 9x - 100 = 0$$
$$(4x + 25)(x - 4) = 0$$
$$x = -\tfrac{25}{4} \text{ or } x = 4$$

When $x = -\tfrac{25}{4}$: $y^2 = 25 - \left(-\tfrac{25}{4}\right)^2$

$$y^2 = -\tfrac{225}{16}$$

No real solution

When $x = 4$: $y^2 = 25 - 4^2$

$$y^2 = 9$$
$$y = \pm 3$$

The points of intersection are $(4, 3)$, $(4, -3)$.
In standard form the equations are:

$$x^2 + y^2 = 25$$
$$y^2 = \tfrac{9}{4}x$$

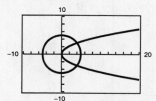

51.

$$x^2 + 2y^2 - 4x + 6y - 5 = 0$$
$$x + y + 5 = 0 \Rightarrow y = -x - 5$$
$$x^2 + 2(-x - 5)^2 - 4x + 6(-x - 5) - 5 = 0$$
$$x^2 + 2x^2 + 20x + 50 - 4x - 6x - 30 - 5 = 0$$
$$3x^2 + 10x + 15 = 0$$

No real solution

No points of intersection
In standard form we have:

$$\frac{(x - 2)^2}{27/2} + \frac{(y + 3/2)^2}{27/4} = 1$$
$$x + y = -5$$

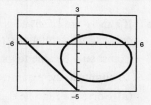

53.

$$xy + x - 2y + 3 = 0 \Rightarrow y = \frac{-x - 3}{x - 2}$$
$$x^2 + 4y^2 - 9 = 0$$
$$x^2 + 4\left(\frac{-x - 3}{x - 2}\right)^2 = 9$$
$$x^2(x - 2)^2 + 4(-x - 3)^2 = 9(x - 2)^2$$
$$x^2(x^2 - 4x + 4) + 4(x^2 + 6x + 9) = 9(x^2 - 4x + 4)$$
$$x^4 - 4x^3 + 4x^2 + 4x^2 + 24x + 36 = 9x^2 - 36x + 36$$
$$x^4 - 4x^3 - x^2 + 60x = 0$$
$$x(x + 3)(x^2 - 7x + 20) = 0$$
$$x = 0 \text{ or } x = -3$$

Note: $x^2 - 7x + 20 = 0$ has no real solution.

When $x = 0$: $\quad y = \dfrac{-0 - 3}{0 - 2} = \dfrac{3}{2}$

When $x = -3$: $y = \dfrac{-(-3) - 3}{-3 - 2} = 0$

The points of intersection are $\left(0, \dfrac{3}{2}\right)$, $(-3, 0)$.

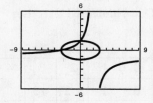

55. $(x')^2 + (y')^2 = (x \cos \theta + y \sin \theta)^2 + (y \cos \theta - x \sin \theta)^2$

$$= x^2 \cos^2 \theta + 2xy \cos \theta \sin \theta + y^2 \sin^2 \theta + y^2 \cos^2 \theta - 2xy \cos \theta \sin \theta + x^2 \sin^2 \theta$$
$$= x^2(\cos^2 \theta + \sin^2 \theta) + y^2(\sin^2 \theta + \cos^2 \theta) = x^2 + y^2 = r^2$$

57. $g(x) = \dfrac{2}{2 - x}$

y-intercept: $(0, 1)$
Vertical asymptote: $x = 2$
Horizontal asymptote: $y = 0$

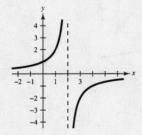

59. $h(t) = \dfrac{t^2}{2 - t} = -t - 2 + \dfrac{4}{2 - t}$

Intercept: $(0, 0)$
Vertical asymptote: $t = 2$
Slant asymptote: $y = -t - 2$

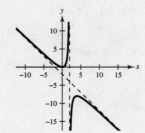

Section 10.5   Parametric Equations

- If f and g are continuous functions of t on an interval I, then the set of ordered pairs $(f(t), g(t))$ is a *plane curve C*. The equations $x = f(t)$ and $y = g(t)$ are *parametric equations* for C and t is the *parameter.*
- You should be able to graph plane curves with your graphing utility.
- To eliminate the parameter:
  (a) Solve for t in one equation and substitute into the second equation.
  (b) Use trigonometric identities.
- You should be able to find the parametric equations for a graph.

Solutions to Odd-Numbered Exercises

1. $x = \sqrt{t}, y = 1 - t$

(a)

t	0	1	2	3	4
x	0	1	$\sqrt{2}$	$\sqrt{3}$	2
y	1	0	-1	-2	-3

(b)

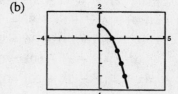

(c) $x = \sqrt{t} \implies t = x^2$. Hence, $y = 1 - t = 1 - x^2$.
The graph is the entire parabola rather than just the right half.

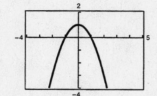

3. $x = t, y = -2t$

$y = -2x$

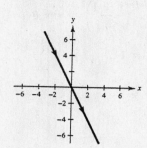

5. $x = 3t - 1, y = 2t + 1$

$y = 2\left(\dfrac{x+1}{3}\right) + 1$

$2x - 3y + 5 = 0$

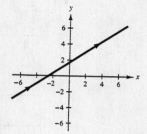

7. $x = \dfrac{1}{4}t, y = t^2$

$y = (4x)^2$

$y = 16x^2$

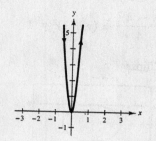

9. $x = t + 1, y = t^2$

$y = (x - 1)^2$

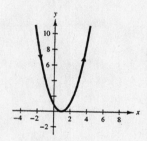

11. $x = t^3, y = \dfrac{t^2}{2}$

$y = \dfrac{1}{2}\left(\sqrt[3]{x}\right)^2$

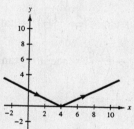

13. $x = 2t \implies t = \dfrac{x}{2}$

$y = |t - 2|$

$y = \left|\dfrac{x}{2} - 2\right| = \dfrac{1}{2}|x - 4|$

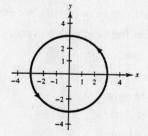

15. $x = 3\cos\theta \implies \left(\dfrac{x}{3}\right)^2 = \cos^2\theta$

$y = 3\sin\theta \implies \left(\dfrac{y}{3}\right)^2 = \sin^2\theta$

$\left(\dfrac{x}{3}\right)^2 + \left(\dfrac{y}{3}\right)^2 = 1$

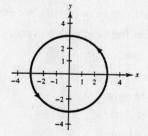

17. $x = 4\sin 2\theta \implies \left(\dfrac{x}{4}\right)^2 = \sin^2 2\theta$

$y = 2\cos 2\theta \implies \left(\dfrac{y}{2}\right)^2 = \cos^2 2\theta$

$\dfrac{x^2}{16} + \dfrac{y^2}{4} = \sin^2 2\theta + \cos^2 2\theta$

$\dfrac{x^2}{16} + \dfrac{y^2}{4} = 1$

Ellipse

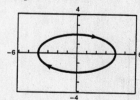

19. $x = 4 + 2\cos\theta \implies \left(\dfrac{x-4}{2}\right)^2 = \cos^2\theta$

$y = -1 + \sin\theta \qquad (y+1)^2 = \sin^2\theta$

$\left(\dfrac{x-4}{2}\right)^2 + (y+1)^2 = \cos^2\theta + \sin^2\theta$

$\dfrac{(x-4)^2}{4} + (y+1)^2 = 1$

Ellipse

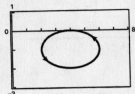

21. $x = 4 + 2\cos\theta \implies \left(\dfrac{x-4}{2}\right)^2 = \cos^2\theta$

$y = -1 + 4\sin\theta \implies \left(\dfrac{y+1}{4}\right)^2 = \sin^2\theta$

$\left(\dfrac{x-4}{2}\right)^2 + \left(\dfrac{y+1}{4}\right)^2 = 1$

$\dfrac{(x-4)^2}{4} + \dfrac{(y+1)^2}{16} = 1$

Ellipse

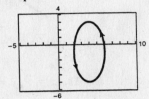

23. $x = 4\sec\theta \implies \left(\dfrac{x}{4}\right)^2 = \sec^2\theta$

$y = 3\tan\theta \implies \left(\dfrac{y}{3}\right)^2 = \tan^2\theta$

$\left(\dfrac{x}{4}\right)^2 - \left(\dfrac{y}{3}\right)^2 = \sec^2\theta - \tan^2\theta$

$\dfrac{x^2}{16} - \dfrac{y^2}{9} = 1$

Hyperbola

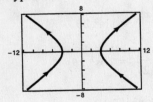

25. $x = e^{-t} \implies \dfrac{1}{x} = e^t$

$y = e^{3t} \implies y = (e^t)^3$

$y = \left(\dfrac{1}{x}\right)^3$

$y = \dfrac{1}{x^3}, \; x > 0, \, y > 0$

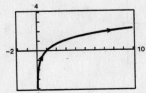

27. $x = t^3 \quad \implies x^{1/3} = t$

$y = 3\ln t \implies y = \ln t^3$

$y = \ln(x^{1/3})^3$

$y = \ln x$

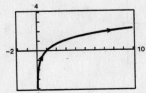

29. By eliminating the parameters in (a)–(d), we get $y = 2x + 1$. They differ from each other in restricted domain and in orientation.

(a) Domain: $-\infty < x < \infty$

   Orientation: Left to right

(b) Domain: $-1 \le x \le 1$

   Orientation: Depends on θ

(c) Domain: $0 < x < \infty$

   Orientation: Right to left

(d) Domain: $0 < x < \infty$

   Orientation: Left to right

31. $x = x_1 + t(x_2 - x_1)$

$y = y_1 + t(y_2 - y_1)$

$\dfrac{x - x_1}{x_2 - x_1} = t$

$y = y_1 + \left(\dfrac{x - x_1}{x_2 - x_1}\right)(y_2 - y_1)$

$y - y_1 = \dfrac{y_2 - y_1}{x_2 - x_1}(x - x_1)$

Notice that this is the point-slope form of the line,

$y - y_1 = m(x - x_1)$.

33. $x = h + a \cos \theta$

$y = k + b \sin \theta$

$\dfrac{x - h}{a} = \cos \theta, \quad \dfrac{y - k}{b} = \sin \theta$

$\dfrac{(x - h)^2}{a^2} + \dfrac{(y - k)^2}{b^2} = 1$

35. From Exercise 31:

$x = 5t$

$y = -2t$

Solution not unique

37. From Exercise 32:

$x = 2 + 4 \cos \theta$

$y = 1 + 4 \sin \theta$

Solution not unique

39. From Exercise 33:

$a = 5, c = 4$, and hence, $b = 3$.

$x = 5 \cos \theta$

$y = 3 \sin \theta$

Center: $(0, 0)$

Solution not unique

41. $y = 3x - 2$

Examples:

$x = t \qquad\qquad x = 2t$

$y = 3t - 2 \qquad y = 6t - 2$

43. $x = 2(\theta - \sin \theta)$

$y = 2(1 - \cos \theta)$

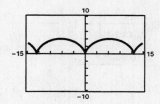

45. $x = 3 \cos^3 \theta, \; y = 3 \sin^3 \theta$

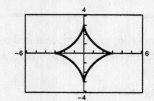

47. $x = 2 \cot \theta, \; y = 2 \sin^2 \theta$

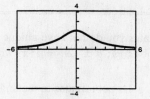

49. Matches graph (b).

51. Matches graph (d).

53. $x = (v_0 \cos \theta)t, \; y = h + (v_0 \sin \theta)t - 16t^2$

(a) $100 \text{ miles/hour} = \dfrac{100 \text{ mi/hr} \cdot 5280 \text{ ft/mi}}{3600 \text{ sec/hr}}$

$\qquad\qquad\qquad = 146.67 \text{ ft/sec}$

$\quad x = (146.67 \cos \theta)t$

$\quad y = 3 + (146.67 \sin \theta)t - 16t^2$

(b) $\theta = 15°$

$\quad x = (146.67 \cos 15°)t = 141.7t$

$\quad y = 3 + (146.67 \sin 15°)t - 16t^2$

$\qquad = 3 + 38.0t - 16t^2$

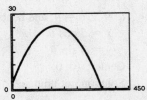

It is not a home run because $y < 10$ when $x = 400$.

(c) $\theta = 23°$

$\quad x = (146.67 \cos 23°)t = 135.0t$

$\quad y = 3 + (146.67 \sin 23°)t - 16t^2$

$\qquad = 3 + 57.3t - 16t^2$

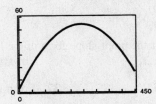

Yes, it is a home run because $y > 10$ when $x = 400$.

(d) $\theta = 19.4°$ is the minimum angle.

55. $y = h + (v_0 \sin \theta)t - 16t^2$ and $t = \dfrac{x}{v_0 \cos \theta}$

$\quad = h + (v_0 \sin \theta)\left(\dfrac{x}{v_0 \cos \theta}\right) - 16\left(\dfrac{x}{v_0 \cos \theta}\right)^2$

$\quad = h + (\tan \theta)x - \dfrac{16 \sec^2 \theta}{v_0{}^2}x^2$

57. False, the graph of $x = t^2$, $y = t^2$ is the portion of the line $y = x$ for $x \geq 0$.

Section 10.6   Polar Coordinates

■ In polar coordinates you do not have unique representation of points. The point (r, θ) can be represented by $(r, \theta \pm 2n\pi)$ or by $(-r, \theta \pm (2n + 1)\pi)$ where n is any integer. The pole is represented by $(0, \theta)$ where θ is any angle.

■ To convert from polar coordinates to rectangular coordinates, use the following relationships.

$\quad x = r \cos \theta$

$\quad y = r \sin \theta$

■ To convert from rectangular coordinates to polar coordinates, use the following relationships.

$\quad r = \pm\sqrt{x^2 + y^2}$

$\quad \tan \theta = y/x$

If θ is in the same quadrant as the point (x, y), then r is positive. If θ is in the opposite quadrant as the point (x, y), then r is negative.

■ You should be able to convert rectangular equations to polar form and vice versa.

Solutions to Odd-Numbered Exercises

1. Polar coordinates: $\left(4, \dfrac{\pi}{2}\right)$

$$x = 4\cos\left(\dfrac{\pi}{2}\right) = 0$$

$$y = 4\sin\left(\dfrac{\pi}{2}\right) = 4$$

Rectangular coordinates: $(0, 4)$

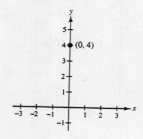

3. Polar coordinates: $\left(-1, \dfrac{5\pi}{4}\right)$

$$x = -1\cos\left(\dfrac{5\pi}{4}\right) = \dfrac{\sqrt{2}}{2}$$

$$y = -1\sin\left(\dfrac{5\pi}{4}\right) = \dfrac{\sqrt{2}}{2}$$

Rectangular coordinates: $\left(\dfrac{\sqrt{2}}{2}, \dfrac{\sqrt{2}}{2}\right)$

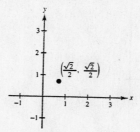

5. Polar coordinates: $\left(4, -\dfrac{\pi}{3}\right)$

$$x = 4\cos\left(-\dfrac{\pi}{3}\right) = 2$$

$$y = 4\sin\left(-\dfrac{\pi}{3}\right) = -2\sqrt{3}$$

Rectangular coordinates: $\left(2, -2\sqrt{3}\right)$

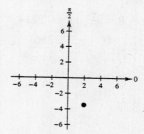

7. Polar coordinates: $\left(0, -\dfrac{7\pi}{6}\right)$

$$x = 0\cos\left(-\dfrac{7\pi}{6}\right) = 0$$

$$y = 0\sin\left(-\dfrac{7\pi}{6}\right) = 0$$

Rectangular coordinates: $(0, 0)$

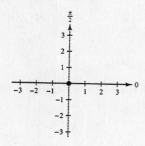

9. Polar coordinates: $\left(\sqrt{2}, 2.36\right)$

$$x = \sqrt{2}\cos(2.36) \approx -1.004$$

$$y = \sqrt{2}\sin(2.36) \approx 0.996$$

Rectangular coordinates: $(-1.004, 0.996)$

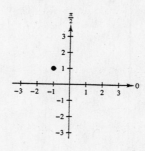

11. $(r, \theta) = \left(2, \dfrac{3\pi}{4}\right) \implies (x, y) = (-1.414, 1.414) = \left(-\sqrt{2}, \sqrt{2}\right)$

13. $(r, \theta) = (-4.5, 1.3) \implies (x, y) = (-1.204, -4.336)$

15. Rectangular coordinates: $(1, 1)$

$r = \sqrt{2}$, $\tan \theta = 1$, $\theta = \dfrac{\pi}{4}$

Polar coordinates: $\left(\sqrt{2}, \dfrac{\pi}{4}\right)$, $\left(-\sqrt{2}, \dfrac{5\pi}{4}\right)$

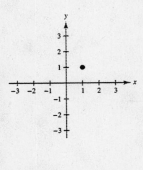

17. Rectangular coordinates: $(-6, 0)$

$r = 6$, $\tan \theta = 0$, $\theta = 0$

Polar coordinates: $(6, \pi)$, $(-6, 0)$

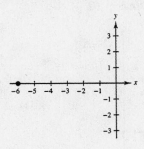

19. Rectangular coordinates: $(-3, 4)$

$r = \sqrt{9 + 16} = 5$, $\tan \theta = -\dfrac{4}{3}$, $\theta \approx 2.214$

Polar coordinates: $(5, 2.214)$, $(-5, 5.356)$

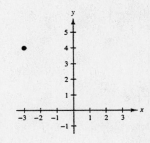

21. Rectangular coordinates: $\left(-\sqrt{3}, -\sqrt{3}\right)$

$r = \sqrt{3 + 3} = \sqrt{6}$, $\tan \theta = 1$, $\theta = \dfrac{\pi}{4}$

Polar coordinates: $\left(\sqrt{6}, \dfrac{5\pi}{4}\right)$, $\left(-\sqrt{6}, \dfrac{\pi}{4}\right)$

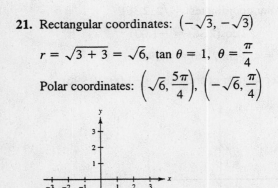

23. Rectangular coordinates: $(4, 6)$

$r = \sqrt{16 + 36} = 2\sqrt{13}$, $\tan \theta = \dfrac{3}{2}$, $\theta \approx 0.983$

Polar coordinates: $(2\sqrt{13}, 0.983)$, $(-2\sqrt{13}, 4.124)$

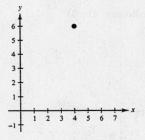

25. $(x, y) = (3, -2) \implies r = \sqrt{3^2 + (-2)^2} = \sqrt{13}$

$\theta = \arctan\left(-\dfrac{2}{3}\right) \approx -0.588$

$(r, \theta) \approx (\sqrt{13}, -0.588)$

27. $(x, y) = (\sqrt{3}, 2) \implies r = \sqrt{3 + 2^2} = \sqrt{7}$

$$\theta = \arctan\left(\frac{2}{\sqrt{3}}\right) \approx 0.857$$

$(r, \theta) \approx (\sqrt{7}, 0.857)$

29. $(x, y) = \left(\frac{5}{2}, \frac{4}{3}\right) \implies r = \sqrt{\left(\frac{5}{2}\right)^2 + \left(\frac{4}{3}\right)^2} = \frac{17}{6}$

$$\theta = \arctan\left(\frac{4/3}{5/2}\right) \approx 0.490$$

$(r, \theta) \approx \left(\frac{17}{6}, 0.490\right)$

31. True, the distances from the origin are the same.

33. $x^2 + y^2 = 9$
$r = 3$

35. $x^2 + y^2 - 2ax = 0$
$r^2 - 2ar\cos\theta = 0$
$r(r - 2a\cos\theta) = 0$
$r = 2a\cos\theta$

37. $y = 4$
$r\sin\theta = 4$
$r = 4\csc\theta$

39. $x = 10$
$r\cos\theta = 10$
$r = 10\sec\theta$

41. $3x - y + 2 = 0$
$3r\cos\theta - r\sin\theta + 2 = 0$
$r(3\cos\theta - \sin\theta) = -2$
$$r = \frac{-2}{3\cos\theta - \sin\theta}$$

43. $xy = 4$
$(r\cos\theta)(r\sin\theta) = 4$
$r^2 = 4\ \sec\theta\csc\theta = 8\csc 2\theta$

45. $(x^2 + y^2)^2 - 9(x^2 - y^2) = 0$
$(r^2)^2 - 9(r^2\cos^2\theta - r^2\sin^2\theta) = 0$
$r^2[r^2 - 9(\cos 2\theta)] = 0$
$r^2 = 9\cos 2\theta$

47. $r = 4\sin\theta$
$r^2 = 4r\sin\theta$
$x^2 + y^2 = 4y$
$x^2 + y^2 - 4y = 0$

49. $\theta = \frac{\pi}{6}$
$\tan\theta = \frac{\sqrt{3}}{3}$
$\frac{y}{x} = \frac{\sqrt{3}}{3}$
$y = \frac{\sqrt{3}}{3}x$
$\sqrt{3}x - 3y = 0$

51. $r = 2\csc\theta$
$r\sin\theta = 2$
$y = 2$

53. $r = 2\sin 3\theta$
$r = 2(3\sin\theta - 4\sin^3\theta)$
$r^4 = 6r^3\sin\theta - 8r^3\sin^3\theta$
$(x^2 + y^2)^2 = 6(x^2 + y^2)y - 8y^3$
$(x^2 + y^2)^2 = 6x^2y - 2y^3$

55. $r = \dfrac{6}{2 - 3\sin\theta}$
$r(2 - 3\sin\theta) = 6$
$2r = 6 + 3r\sin\theta$
$2(\pm\sqrt{x^2 + y^2}) = 6 + 3y$
$4(x^2 + y^2) = (6 + 3y)^2$
$4x^2 + 4y^2 = 36 + 36y + 9y^2$
$4x^2 - 5y^2 - 36y - 36 = 0$

57.
$$r = 3$$
$$r^2 = 9$$
$$x^2 + y^2 = 9$$

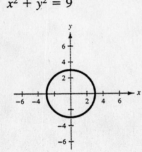

59.
$$\theta = \frac{\pi}{4}$$
$$\tan \theta = \tan \frac{\pi}{4}$$
$$\frac{y}{x} = 1$$
$$y = x$$
$$x - y = 0$$

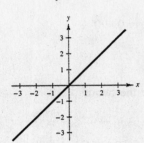

61.
$$r = 3 \sec \theta$$
$$r \cos \theta = 3$$
$$x = 3$$
$$x - 3 = 0$$

63.
$$r = 2(h \cos \theta + k \sin \theta)$$
$$r = 2\left(h\left(\frac{x}{r}\right) + k\left(\frac{y}{r}\right)\right)$$
$$r = \frac{2hx + 2ky}{r}$$
$$r^2 = 2hx + 2ky$$
$$x^2 + y^2 = 2hx + 2ky$$
$$x^2 - 2hx + y^2 - 2ky = 0$$
$$(x^2 - 2hx + h^2) + (y^2 - 2ky + k^2) = h^2 + k^2$$
$$(x - h)^2 + (y - k)^2 = h^2 + k^2$$

Center: (h, k)

Radius: $\sqrt{h^2 + k^2}$

65. (a) $(r_1, \theta_1) = (x_1, y_1)$ where $x_1 = r_1 \cos \theta_1$ and $y_1 = r_1 \sin \theta_1$.

$(r_2, \theta_2) = (x_2, y_2)$ where $x_2 = r_2 \cos \theta_2$ and $y_2 = r_2 \sin \theta_2$.

Then $x_1^2 + y_1^2 = r_1^2 \cos^2 \theta_1 + r_1^2 \sin^2 \theta_1 = r_1^2$ and $x_2^2 + y_2^2 = r_2^2$. Thus,

$$
\begin{aligned}
d &= \sqrt{(x_1 - x_2)^2 + (y_1 - y_2)^2} \\
&= \sqrt{x_1^2 - 2x_1x_2 + x_2^2 + y_1^2 - 2y_1y_2 + y_2^2} \\
&= \sqrt{(x_1^2 + y_1^2) + (x_2^2 + y_2^2) - 2(x_1x_2 + y_1y_2)} \\
&= \sqrt{r_1^2 + r_2^2 - 2(r_1r_2 \cos \theta_1 \cos \theta_2 + r_1r_2 \sin \theta_1 \sin \theta_2)} \\
&= \sqrt{r_1^2 + r_2^2 - 2r_1r_2 \cos(\theta_1 - \theta_2)}.
\end{aligned}
$$

(b) If $\theta_1 = \theta_2$, the points are on the same line through the origin. In this case,
$$d = \sqrt{r_1^2 + r_2^2 - 2r_1r_2 \cos(0)} = \sqrt{(r_1 - r_2)^2} = |r_1 - r_2|.$$

(c) If $\theta_1 - \theta_2 = 90°, d = \sqrt{r_1^2 + r_2^2}$, the Pythagorean Theorem.

(d) For instance, $\left(3, \frac{\pi}{6}\right), \left(4, \frac{\pi}{3}\right)$ gives $d \approx 2.053$ and $\left(-3, \frac{7\pi}{6}\right), \left(-4, \frac{4\pi}{3}\right)$ gives $d \approx 2.053$. (same!)

67. $D = \begin{vmatrix} 5 & -7 \\ -3 & 1 \end{vmatrix} = 5 - 21 = -16$

$D_x = \begin{vmatrix} -11 & -7 \\ -3 & 1 \end{vmatrix} = -11 - 21 = -32$

$D_y = \begin{vmatrix} 5 & -11 \\ -3 & -3 \end{vmatrix} = -15 - 33 = -48$

$x = \dfrac{D_x}{D} = \dfrac{-32}{-16} = 2$

$y = \dfrac{D_y}{D} = \dfrac{-48}{-16} = 3$

69. $D = \begin{vmatrix} 3 & -2 & 1 \\ 2 & 1 & -3 \\ 1 & -3 & 9 \end{vmatrix} = 35$

$D_x = \begin{vmatrix} 0 & -2 & 1 \\ 0 & 1 & -3 \\ 8 & -3 & 9 \end{vmatrix} = 40$

$D_y = \begin{vmatrix} 3 & 0 & 1 \\ 2 & 0 & -3 \\ 1 & 8 & 9 \end{vmatrix} = 88$

$D_z = \begin{vmatrix} 3 & -2 & 0 \\ 2 & 1 & 0 \\ 1 & -3 & 8 \end{vmatrix} = 56$

$x = \dfrac{D_x}{D} = \dfrac{40}{35} = \dfrac{8}{7}$

$y = \dfrac{D_y}{D} = \dfrac{88}{35}$

$z = \dfrac{D_z}{D} = \dfrac{56}{35} = \dfrac{8}{5}$

71. $D = \begin{vmatrix} 1 & 1 & 1 & -3 \\ 3 & -1 & -2 & 1 \\ -1 & 1 & -1 & 2 \\ 0 & 1 & 0 & 1 \end{vmatrix} = 20$

$D_x = \begin{vmatrix} -8 & 1 & 1 & -3 \\ 7 & -1 & -2 & 1 \\ -2 & 1 & -1 & 2 \\ -6 & 1 & 0 & 1 \end{vmatrix} = 20$

$D_y = \begin{vmatrix} 1 & -8 & 1 & -3 \\ 3 & 7 & -2 & 1 \\ -1 & -2 & -1 & 2 \\ 0 & -6 & 0 & 1 \end{vmatrix} = -80$

$D_z = \begin{vmatrix} 1 & 1 & -8 & -3 \\ 3 & -1 & 7 & 1 \\ -1 & 1 & -2 & 2 \\ 0 & 1 & -6 & 1 \end{vmatrix} = 20$

$D_w = \begin{vmatrix} 1 & 1 & 1 & -8 \\ 3 & -1 & -2 & 7 \\ -1 & 1 & -1 & -2 \\ 0 & 1 & 0 & -6 \end{vmatrix} = 40$

$x = \dfrac{D_x}{D} = \dfrac{20}{20} = 1$

$y = \dfrac{D_y}{D} = -\dfrac{80}{20} = -4$

$z = \dfrac{D_z}{D} = \dfrac{20}{20} = 1$

$w = \dfrac{D_w}{D} = \dfrac{40}{20} = 2$

Section 10.7 Graphs of Polar Equations

■ When graphing polar equations:

1. Test for symmetry

    (a) $\theta = \pi/2$: Replace (r, θ) by $(r, \pi - \theta)$ or $(-r, -\theta)$.

    (b) Polar axis: Replace (r, θ) by $(r, -\theta)$ or $(-r, \pi - \theta)$.

    (c) Pole: Replace (r, θ) by $(r, \pi + \theta)$ or $(-r, \theta)$.

    (d) $r = f(\sin \theta)$ is symmetric with respect to the line $\theta = \pi/2$.

    (e) $r = f(\cos \theta)$ is symmetric with respect to the polar axis.

2. Find the θ values for which $|r|$ is maximum.

3. Find the θ values for which $r = 0$.

4. Know the different types of polar graphs.

    (a) Limaçons

        $r = a \pm b \cos \theta$

        $r = a \pm b \sin \theta$

    (c) Circles

        $r = a \cos \theta$

        $r = a \sin \theta$

        $r = a$

    (b) Rose Curves, $n \geq 2$

        $r = a \cos n\theta$

        $r = a \sin n\theta$

    (d) Lemniscates

        $r^2 = a^2 \cos 2\theta$

        $r^2 = a^2 \sin 2\theta$

■ You should be able to graph polar equations of the form $r = f(\theta)$ with your graphing utility. If your utility does not have a polar mode, use

$$x = f(t) \cos t$$
$$y = f(t) \sin t$$

in parametric mode.

Solutions to Odd-Numbered Exercises

1. $r = 3 \cos 2\theta$ is a rose curve.   **3.** $r = 2 - \cos \theta$ is a limaçon.   **5.** $r = 6 \sin 2\theta$ is a rose curve.

7. $r = 10 + 6 \cos \theta$

$\theta = \dfrac{\pi}{2}$: $-r = 10 + 6 \cos(-\theta)$

           $-r = 10 + 6 \cos \theta$

        Not an equivalent equation

Polar axis: $r = 10 + 6 \cos(-\theta)$

           $r = 10 + 6 \cos \theta$

        Equivalent equation

Pole: $-r = 10 + 6 \cos \theta$

        Not an equivalent equation

Answer: Symmetric with respect to polar axis

9. $r = \dfrac{2}{1 + \sin \theta}$

$\theta = \dfrac{\pi}{2}$: $r = \dfrac{2}{1 + \sin(\pi - \theta)}$

$r = \dfrac{2}{1 + \sin \pi \cos \theta - \cos \pi \sin \theta}$

$r = \dfrac{2}{1 + \sin \theta}$

Equivalent equation

Polar axis: $r = \dfrac{2}{1 + \sin(-\theta)}$

$r = \dfrac{2}{1 - \sin \theta}$

Not an equivalent equation

Pole: $-r = \dfrac{2}{1 + \sin \theta}$

Not an equivalent equation

Answer: Symmetric with respect to $\theta = \pi/2$

11. $r = 4 \sec \theta \csc \theta$

$\theta = \dfrac{\pi}{2}$: $-r = 4 \sec(-\theta) \csc(-\theta)$

$-r = -4 \sec \theta \csc \theta$

$r = 4 \sec \theta \csc \theta$

Equivalent equation

Polar axis: $-r = 4 \sec(\pi - \theta) \csc(\pi - \theta)$

$-r = 4(-\sec \theta) \csc \theta$

$r = 4 \sec \theta \csc \theta$

Equivalent equation

Pole: $r = 4 \sec(\pi + \theta) \csc(\pi + \theta)$

$r = 4(-\sec \theta)(-\csc \theta)$

$r = 4 \sec \theta \csc \theta$

Equivalent equation

Answer: Symmetric with respect to $\theta = \pi/2$, pole axis, and pole

13. $|r| = |10(1 - \sin \theta)|$

$= 10|1 - \sin \theta| \le 10(2) = 20$

$|1 - \sin \theta| = 2$

$1 - \sin \theta = 2$ or $1 - \sin \theta = -2$

$\sin \theta = -1 \qquad \sin \theta = 3$

$\theta = \dfrac{3\pi}{2} \qquad$ Not possible

Maximum: $|r| = 20$ when $\theta = \dfrac{3\pi}{2}$.

$r = 0$ when $1 - \sin \theta = 0$

$\sin \theta = 1$

$\theta = \dfrac{\pi}{2}$.

15. $|r| = |4 \cos 3\theta| = 4 |\cos 3\theta| \le 4$

$|\cos 3\theta| = 1$

$\cos 3\theta = \pm 1$

$\theta = 0, \dfrac{\pi}{3}, \dfrac{2\pi}{3}$

Maximum: $|r| = 4$ when $\theta = 0, \dfrac{\pi}{3}, \dfrac{2\pi}{3}$.

$r = 0$ when $\cos 3\theta = 0$

$\theta = \dfrac{\pi}{6}, \dfrac{\pi}{2}, \dfrac{5\pi}{6}$.

17. Circle: $r = 5$

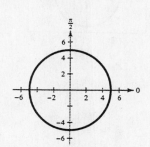

19. $r = \dfrac{\pi}{6}$ is a circle.

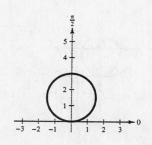

21. $r = 3 \sin \theta$

Symmetric with respect to $\theta = \dfrac{\pi}{2}$

Circle with radius of $\dfrac{3}{2}$

23. $r = 3(1 - \cos\theta)$
Cardioid

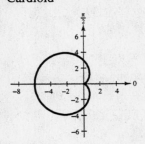

25. $r = 4 + 4\sin\theta$
Symmetric with respect to

$\theta = \dfrac{\pi}{2}$

$\dfrac{a}{b} = \dfrac{4}{4} = 1 \implies$ Cardioid

$|r| = 8$ when $\theta = \dfrac{\pi}{2}$.

$r = 0$ when $\theta = \dfrac{3\pi}{2}$.

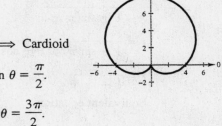

27. $r = 3 - 2\cos\theta$
Dimpled limaçon

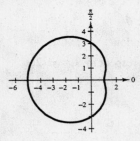

29. $r = 2 + \sin\theta$
Convex limaçon

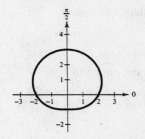

31. $r = 2\cos 3\theta$
Rose curve

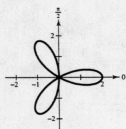

33. $r = 3\sin 2\theta$
Symmetric with respect
to $\theta = \dfrac{\pi}{2}$, polar axis, and pole

Rose curve ($n = 2$) with
4 petals

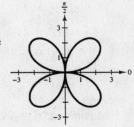

35. $r = \dfrac{\theta}{2}$

Symmetric with respect
to $\theta = \dfrac{\pi}{2}$ and polar axis
Spiral

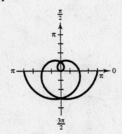

37.

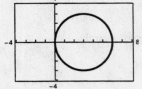

39.

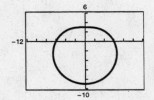

41.

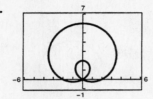

43.

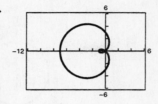

45.

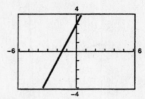

47.

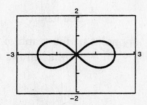

49.

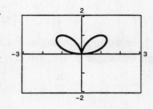

51.

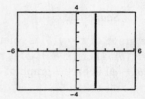

53.

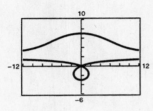

55. $r = 3 - 4 \cos \theta,\ 0 \le \theta < 2\pi$

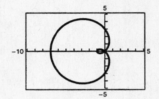

57. $r = 2 + \sin \theta,\ 0 \le \theta < 2\pi$

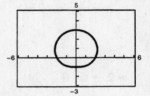

59. $r = 2 \cos\left(\dfrac{3\theta}{2}\right),\ 0 \le \theta < 4\pi$

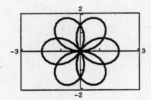

61. $r^2 = 4 \sin 2\theta,\ 0 \le \theta < \dfrac{\pi}{2}$

$\left(\text{Use } r_1 = \sqrt{4 \sin 2\theta} \text{ and } r_2 = -\sqrt{4 \sin 2\theta}.\right)$

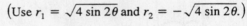

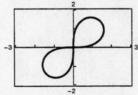

63. $r = 2 - \sec \theta$

$x = -1$ is an asymptote.

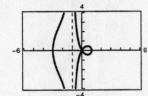

65. $r = \dfrac{2}{\theta}$

$y = 2$ is an asymptote.

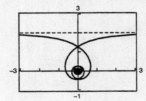

67.

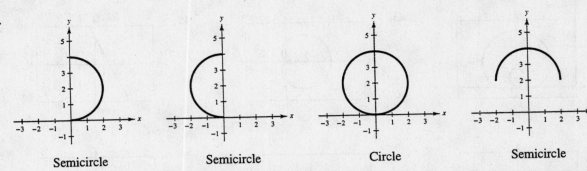

Semicircle   Semicircle   Circle   Semicircle

69. The graph of $r = f(\theta)$ is rotated about the pole through an angle ϕ. Let (r, θ) be any point on the graph of $r = f(\theta)$. Then $(r, \theta + \phi)$ is rotated through the angle ϕ, and since $r = f((\theta + \phi) - \phi) = f(\theta)$, it follows that $(r, \theta + \phi)$ is on the graph of $r = f(\theta - \phi)$.

71. (a) $r = 2 - \sin\left(\theta - \dfrac{\pi}{4}\right)$

$= 2 - \dfrac{\sqrt{2}}{2}(\sin\theta - \cos\theta)$

(b) $r = 2 - \sin\left(\theta - \dfrac{\pi}{2}\right)$

$= 2 + \cos\theta$

(c) $r = 2 - \sin(\theta - \pi)$

$= 2 + \sin\theta$

(d) $r = 2 - \sin\left(\theta - \dfrac{3\pi}{2}\right)$

$= 2 - \cos\theta$

73. (a) $r = 1 - \sin\theta$

(b) $r = 1 - \sin\left(\theta - \dfrac{\pi}{4}\right)$

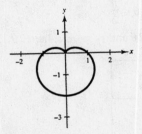

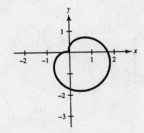

75. $r = 2 + k\cos\theta$

$k = 0$

Circle

$k = 1$

Convex limaçon

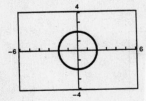

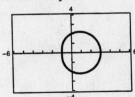

$k = 2$

Cardioid

$k = 3$

Limaçon with inner loop

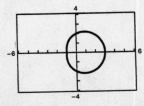

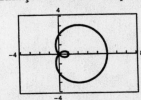

Section 10.8   Polar Equations of Conics

■ The graph of a polar equation of the form

$$r = \frac{ep}{1 \pm e \cos \theta} \quad \text{or} \quad r = \frac{ep}{1 \pm e \sin \theta}$$

is a conic, where $e > 0$ is the eccentricity and $|p|$ is the distance between the focus (pole) and the directrix.

(a) If $e < 1$, the graph is an ellipse.

(b) If $e = 1$, the graph is a parabola.

(c) If $e > 1$, the graph is a hyperbola.

■ Guidelines for finding polar equations of conics:

(a) Horizontal directrix above the pole: $r = \dfrac{ep}{1 + e \sin \theta}$

(b) Horizontal directrix below the pole: $r = \dfrac{ep}{1 - e \sin \theta}$

(c) Vertical directrix to the right of the pole: $r = \dfrac{ep}{1 + e \cos \theta}$

(d) Vertical directrix to the left of the pole: $r = \dfrac{ep}{1 - e \cos \theta}$

Solutions to Odd-Numbered Exercises

1.

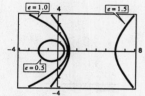

3.

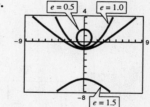

5. Matches (b).

   (Parabola $e = 1$)

7. Matches (d).

   (Hyperbola $e = 2$)

9. $r = \dfrac{2}{1 - \cos \theta}$

$e = 1$ so the graph is a parabola.

Vertex: $(r, \theta) = (1, \pi)$

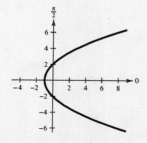

11. $r = \dfrac{5}{1 + \sin \theta}$

$e = 1$ so the graph is a parabola.

Vertex: $(r, \theta) = \left(\dfrac{5}{2}, \dfrac{\pi}{2}\right)$

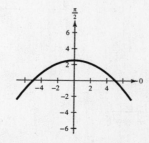

13. $r = \dfrac{2}{2 - \cos \theta} = \dfrac{1}{1 - (1/2) \cos \theta}$

$e = \dfrac{1}{2} < 1$, the graph is an ellipse.

Vertices: $(r, \theta) = (2, 0), \left(\dfrac{2}{3}, \pi\right)$

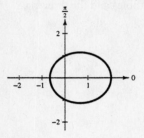

15. $r = \dfrac{4}{2 + \sin \theta} = \dfrac{2}{1 + (1/2) \sin \theta}$

$e = \dfrac{1}{2} < 1$, the graph is an ellipse.

Vertices: $(r, \theta) = \left(\dfrac{4}{3}, \dfrac{\pi}{2}\right), \left(4, \dfrac{3\pi}{2}\right)$

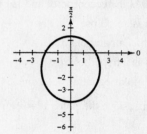

17. $r = \dfrac{3}{2 + 4 \sin \theta} = \dfrac{3/2}{1 + 2 \sin \theta}$

$e = 2 > 1$, the graph is a hyperbola.

Vertices: $(r, \theta) = \left(\dfrac{1}{2}, \dfrac{\pi}{2}\right), \left(-\dfrac{3}{2}, \dfrac{3\pi}{2}\right)$

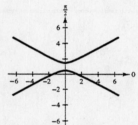

19. $r = \dfrac{3}{2 - 6 \cos \theta} = \dfrac{3/2}{1 - 3 \cos \theta}$

$e = 3 > 1$, the graph is a hyperbola.

Vertices: $(r, \theta) = \left(-\dfrac{3}{4}, 0\right), \left(\dfrac{3}{8}, \pi\right)$

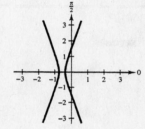

21.

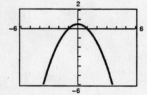

23.

25.

27.

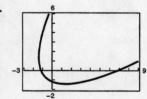

29.

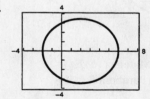

31. $e = 1, x = -1, p = 1$

Vertical directrix to the left of the pole

$r = \dfrac{1(1)}{1 - 1 \cos \theta} = \dfrac{1}{1 - \cos \theta}$

33. $e = \dfrac{1}{2}, y = 1, p = 1$

Horizontal directrix above the pole

$r = \dfrac{(1/2)(1)}{1 + (1/2) \sin \theta} = \dfrac{1}{2 + \sin \theta}$

35. $e = 2, x = 1, p = 1$

Vertical directrix to the right of the pole

$r = \dfrac{2(1)}{1 + 2 \cos \theta} = \dfrac{2}{1 + 2 \cos \theta}$

37. Vertex: $\left(1, -\dfrac{\pi}{2}\right) \implies e = 1, p = 2$

Horizontal directrix below the pole

$$r = \frac{1(2)}{1 - 1 \sin \theta} = \frac{2}{1 - \sin \theta}$$

41. Center: $(3, \pi)$; $c = 3, a = 5, e = \dfrac{3}{5}$

Vertical directrix to the right of the pole

$$r = \frac{(3/5)p}{1 + (3/5) \cos \theta} = \frac{3p}{5 + 3 \cos \theta}$$

$$2 = \frac{3p}{5 + 3 \cos 0}$$

$$p = \frac{16}{3}$$

$$r = \frac{3(16/3)}{5 + 3 \cos \theta} = \frac{16}{5 + 3 \cos \theta}$$

45. Center: $\left(5, \dfrac{3\pi}{2}\right)$; $c = 5, a = 4, e = \dfrac{5}{4}$

Horizontal directrix below the pole

$$r = \frac{(5/4)p}{1 - (5/4) \sin \theta} = \frac{5p}{4 - 5 \sin \theta}$$

$$1 = \frac{5p}{4 - 5 \sin(3\pi/2)}$$

$$p = \frac{9}{5}$$

$$r = \frac{5(9/5)}{4 - 5 \sin \theta} = \frac{9}{4 - 5 \sin \theta}$$

49. $\dfrac{x^2}{169} + \dfrac{y^2}{144} = 1$

$a = 13, b = 12, c = 5, e = \dfrac{5}{13}$

$$r^2 = \frac{144}{1 - (25/169) \cos^2 \theta} = \frac{24{,}336}{169 - 25 \cos^2 \theta}$$

39. Vertex: $(5, \pi) \implies e = 1, p = 10$

Vertical directrix to the left of the pole

$$r = \frac{1(10)}{1 - 1 \cos \theta} = \frac{10}{1 - \cos \theta}$$

43. Center: $(8, 0)$; $c = 8, a = 12, e = \dfrac{2}{3}$

Vertical directrix to the left of the pole

$$r = \frac{(2/3)p}{1 - (2/3) \cos \theta} = \frac{2p}{3 - 2 \cos \theta}$$

$$20 = \frac{2p}{3 - 2 \cos 0}$$

$$p = 10$$

$$r = \frac{2(10)}{3 - 2 \cos \theta} = \frac{20}{3 - 2 \cos \theta}$$

47.

$$\frac{x^2}{a^2} + \frac{y^2}{b^2} = 1$$

$$\frac{r^2 \cos^2 \theta}{a^2} + \frac{r^2 \sin^2 \theta}{b^2} = 1$$

$$\frac{r^2 \cos^2 \theta}{a^2} + \frac{r^2(1 - \cos^2 \theta)}{b^2} = 1$$

$$r^2 b^2 \cos^2 \theta + r^2 a^2 - r^2 a^2 \cos^2 \theta = a^2 b^2$$

$$r^2(b^2 - a^2) \cos^2 \theta + r^2 a^2 = a^2 b^2$$

For an ellipse, $b^2 - a^2 = -c^2$. Hence,

$$-r^2 c^2 \cos^2 \theta + r^2 a^2 = a^2 b^2$$

$$-r^2 \left(\frac{c}{a}\right)^2 \cos^2 \theta + r^2 = b^2, \quad e = \frac{c}{a}$$

$$-r^2 e^2 \cos^2 \theta + r^2 = b^2$$

$$r^2(1 - e^2 \cos^2 \theta) = b^2$$

$$r^2 = \frac{b^2}{1 - e^2 \cos^2 \theta}.$$

51. $\dfrac{x^2}{9} - \dfrac{y^2}{16} = 1$

$a = 3, b = 4, c = 5, e = \dfrac{5}{3}$

$$r^2 = \frac{-16}{1 - (25/9) \cos^2 \theta} = \frac{144}{25 \cos^2 \theta - 9}$$

53. Hyperbola

One focus: $\left(5, \dfrac{\pi}{2}\right)$

Vertices: $\left(4, \dfrac{\pi}{2}\right), \left(4, -\dfrac{\pi}{2}\right)$

$a = 4, c = 5, b = 3, e = \dfrac{5}{4}$

$r^2 = \dfrac{-3^2}{1 - (5/4)^2 \sin^2 \theta} = \dfrac{-144}{16 - 25 \sin^2 \theta}$

$\quad = \dfrac{144}{25 \sin^2 \theta - 16}$

55. When $\theta = 0, r = c + a = ea + a = a(1 + e)$.

Therefore,

$$a(1 + e) = \frac{ep}{1 - e \cos 0}$$

$$a(1 + e)(1 - e) = ep$$

$$a(1 - e^2) = ep.$$

Thus, $r = \dfrac{ep}{1 - e \cos \theta} = \dfrac{(1 - e^2)a}{1 - e \cos \theta}.$

57. $r = \dfrac{[1 - (0.0167)^2](92.957 \times 10^6)}{1 - 0.0167 \cos \theta} \approx \dfrac{9.2931 \times 10^7}{1 - 0.0167 \cos \theta}$

Perihelion distance: $r = 92.957 \times 10^6 (1 - 0.0167) \approx 9.1405 \times 10^7$

Aphelion distance: $r = 92.957 \times 10^6 (1 + 0.0167) \approx 9.4509 \times 10^7$

59. Radius of earth ≈ 4000 miles. Choose $r = \dfrac{ep}{1 - e \cos \theta}.$

Vertices: $(126{,}000, 0)$ and $(4119, \pi)$

$a = \dfrac{126{,}000 + 4119}{2} = 65{,}059.5$

$c = 65{,}059.5 - 4119 = 60{,}940.5$

$e = \dfrac{c}{a} = \dfrac{60{,}940.5}{65{,}059.5} \approx 0.93669$

$2a = \dfrac{ep}{1 - e \cos 0} + \dfrac{ep}{1 - e \cos(\pi)} = \dfrac{ep}{1 - e} + \dfrac{ep}{1 + e} = \dfrac{2ep}{1 - e^2}$

Thus, $p = \dfrac{a(1 - e^2)}{e} \approx 8516.4.$

Thus, $r = \dfrac{ep}{1 - e \cos \theta} = \dfrac{(0.93669)(8516.4)}{1 - (0.93669) \cos \theta} \approx \dfrac{7977.22}{1 - 0.9367 \cos \theta}.$

When $\theta = 60°$, $r \approx 15{,}004.5$ and the distance from the surface of the earth to the satellite is $15{,}004.5 - 4000 = 11{,}004.5$ miles.

Review Exercises for Chapter 10

Solutions to Odd-Numbered Exercises

1. $4x^2 + y^2 = 4$

$$\frac{x^2}{1} + \frac{y^2}{4} = 1$$

Ellipse with center $(0, 0)$ and a vertical major axis
Matches graph (d).

3. $y^2 = -4x$

$$x = -\frac{y^2}{4}$$

Parabola with vertex $(0, 0)$ and opening to left.
Matches graph (a).

5. $4x - y^2 = 0$

$y^2 = 4x$

The graph is a parabola.
Vertex: $(0, 0)$

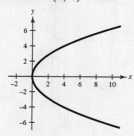

7. $x^2 - 6x + 2y + 9 = 0$

$$(x - 3)^2 = -2y$$

The graph is a parabola.
Vertex: $(3, 0)$

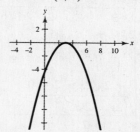

9. $x^2 + y^2 - 2x - 4y + 5 = 0$

$$(x - 1)^2 + (y - 2)^2 = 0$$

The graph is a degenerate circle. $(1, 2)$ is the only
point that satisfies this equation.

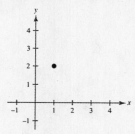

11. $4x^2 + y^2 = 16$

$$\frac{x^2}{4} + \frac{y^2}{16} = 1$$

The graph is an ellipse.
Center: $(0, 0)$
Vertices: $(0, \pm 4)$

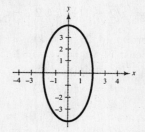

13. $x^2 + 9y^2 + 10x - 18y + 25 = 0$

$$(x + 5)^2 + 9(y - 1)^2 = 9$$

$$\frac{(x + 5)^2}{9} + \frac{(y - 1)^2}{1} = 1$$

The graph is an ellipse.
Center: $(-5, 1)$
Vertices: $(-8, 1)\,(-2, 1)$

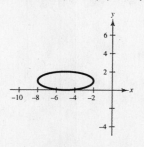

15. $5y^2 - 4x^2 = 20$

$$\frac{y^2}{4} - \frac{x^2}{5} = 1$$

The graph is a hyperbola.
Center: $(0, 0)$
Vertices: $(\pm 2, 0)$

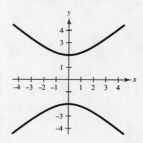

17. $3x^2 + 2y^2 - 12x + 12y + 29 = 0$

$3(x^2 - 4x + 4) + 2(y^2 + 6y + 9) = -29 + 12 + 18$

$3(x - 2)^2 + 2(y + 3)^2 = 1$

Ellipse: $2(y + 3)^2 = 1 - 3(x - 2)^2$

$(y + 3) = \pm\sqrt{\frac{1}{2} - \frac{3}{2}(x - 2)^2}$

$y = -3 \pm \sqrt{\frac{1}{2} - \frac{3}{2}(x - 2)^2}$

Graph: $y_1 = -3 + \sqrt{\frac{1}{2} - \frac{3}{2}(x - 2)^2}$

$y_2 = -3 - \sqrt{\frac{1}{2} - \frac{3}{2}(x - 2)^2}$

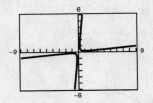

19. $x^2 - 10xy + y^2 + 1 = 0$

Since $B^2 - 4AC = (-10)^2 - 4(1)(1) > 0$, the graph is a hyperbola.
To use a graphing calculator, we need to solve for y in terms of x.

$(y^2 - 10xy + 25x^2) = -x^2 - 1 + 25x^2$

$(y - 5x)^2 = 24x^2 - 1$

$y = 5x \pm \sqrt{24x^2 - 1}$

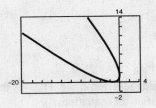

21. $y^2 + (2x - 2\sqrt{2})y + x^2 + 2\sqrt{2}x + 2 = 0$

$y = \dfrac{-(2x - 2\sqrt{2}) \pm \sqrt{(2x - 2\sqrt{2})^2 - 4(x^2 + 2\sqrt{2}x + 2)}}{2}$

$= \dfrac{2(\sqrt{2} - x) \pm \sqrt{(4x^2 - 8x\sqrt{2} + 8) - 4x^2 - 8\sqrt{2}x - 8}}{2}$

$= \dfrac{2(\sqrt{2} - x) \pm \sqrt{-16x\sqrt{2}}}{2}$

$= (\sqrt{2} - x) \pm 2\sqrt[4]{2}\sqrt{-x}$ Parabola

23. Vertex: $(4, 2) = (h, k)$
Focus: $(4, 0) \Rightarrow p = -2$
$(x - h)^2 = 4p(y - k)$
$(x - 4)^2 = -8(y - 2)$

25. Vertex: $(0, 2) = (h, k)$
Directrix: $x = -3 \Rightarrow p = 3$
$(y - k^2) = 4p(x - h)$
$(y - 2)^2 = 12x$

27. Vertices: $(-3, 0), (7, 0) \Rightarrow a = 5, (h, k) = (2, 0)$
Foci: $(0, 0), (4, 0) \Rightarrow c = 2$
$b^2 = a^2 - c^2 = 25 - 4 = 21$

$\dfrac{(x - h)^2}{a^2} + \dfrac{(y - k)^2}{b^2} = 1$

$\dfrac{(x - 2)^2}{25} + \dfrac{y^2}{21} = 1$

29. Vertices: $(0, \pm 6) \Rightarrow a = 6, (h, k) = (0, 0)$
Passes through $(2, 2)$

$\dfrac{(x - h)^2}{b^2} + \dfrac{(y - k)^2}{a^2} = 1$

$\dfrac{x^2}{b^2} + \dfrac{y^2}{36} = 1 \Rightarrow b^2 = \dfrac{36(4)}{36 - 4} = \dfrac{36x^2}{36 - y^2} = \dfrac{9}{2}$

$\dfrac{x^2}{9/2} + \dfrac{y^2}{36} = 1$

$\dfrac{2x^2}{9} + \dfrac{y^2}{36} = 1$

31. Vertices: $(0, \pm 1) \Rightarrow a = 1, (h, k) = (0, 0)$
Foci: $(0, \pm 3) \Rightarrow c = 3$
$$b^2 = c^2 - a^2 = 9 - 1 = 8$$
$$\frac{(y - k)^2}{a^2} - \frac{(x - h)^2}{b^2} = 1$$
$$y^2 - \frac{x^2}{8} = 1$$

33. Foci: $(0, 0), (8, 0) \Rightarrow c = 4, (h, k) = (4, 0)$
Asymptotes: $y = \pm 2(x - 4) \Rightarrow \dfrac{b}{a} = 2, b = 2a$
$$b^2 = c^2 - a^2 \Rightarrow 4a^2 = 16 - a^2 \Rightarrow a^2 = \frac{16}{5}, b^2 = \frac{64}{5}$$
$$\frac{(x - h)^2}{a^2} - \frac{(y - k)^2}{b^2} = 1$$
$$\frac{(x - 4)^2}{16/5} - \frac{y^2}{64/5} = 1$$
$$\frac{5(x - 4)^2}{16} - \frac{5y^2}{64} = 1$$

35. $x = 1 + 4t, y = 2 - 3t$
$$t = \frac{x - 1}{4}$$
$$y = 2 - 3\left(\frac{x - 1}{4}\right)$$
$$3x + 4y = 11$$

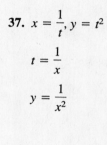

37. $x = \dfrac{1}{t}, y = t^2$
$$t = \frac{1}{x}$$
$$y = \frac{1}{x^2}$$

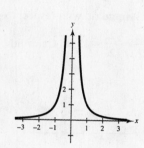

39. $x = 6\cos\theta, y = 6\sin\theta$
$$\cos\theta = \frac{x}{6}, \sin\theta = \frac{y}{6}$$
$$\frac{x^2}{36} + \frac{y^2}{36} = 1$$
$$x^2 + y^2 = 36$$

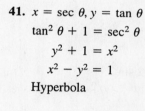

41. $x = \sec\theta, y = \tan\theta$
$$\tan^2\theta + 1 = \sec^2\theta$$
$$y^2 + 1 = x^2$$
$$x^2 - y^2 = 1$$
Hyperbola

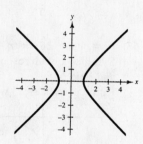

43. False. The y^4-term is not degree 2.

45. $x = 3\tan\theta$
$y = 4\sec\theta$

47. $(x^2 + y^2)^2 = ax^2y$
$$(r^2)^2 = ar^2\cos^2\theta\, r\sin\theta$$
$$r = a\cos^2\theta\sin\theta$$

49. $$r = 3\cos\theta$$
$$r^2 = 3r\cos\theta$$
$$x^2 + y^2 = 3x$$

51. $r = \dfrac{2}{1 + \sin\theta}$
$$r + r\sin\theta = 2$$
$$\sqrt{x^2 + y^2} + y = 2$$
$$\sqrt{x^2 + y^2} = 2 - y$$
$$x^2 + y^2 = 4 - 4y + y^2$$
$$x^2 + 4y - 4 = 0$$

53. $$r^2 = \cos 2\theta$$
$$r^2 = 1 - 2\sin^2\theta$$
$$r^4 = r^2 - 2r^2\sin^2\theta$$
$$(x^2 + y^2)^2 = x^2 + y^2 - 2y^2$$
$$(x^2 + y^2)^2 - x^2 + y^2 = 0$$

55. $r = 4$

Circle of radius 4 centered at the pole

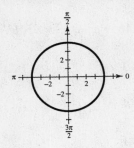

57. $r = 4 \sin 2\theta$

Symmetric with respect to $\theta = \pi/2$
Rose curve ($n = 2$) with 4 petals

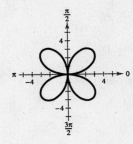

59. $r = -2 - 2 \cos \theta$

Symmetric with respect to polar axis

$\dfrac{a}{b} = \dfrac{2}{2} = 1 \Rightarrow$ Cardioid

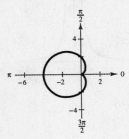

61. $r = \dfrac{2}{1 - \sin \theta}, e = 1$

Parabola symmetric with $\theta = \pi/2$ and the vertex at $(1, 3\pi/2)$

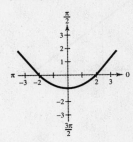

63. $r^2 = 4 \sin^2 2\theta \Rightarrow r = \pm 2 \sin 2\theta$

Symmetric with respect to $\theta = \pi/2$, polar axis, and pole
Rose curve ($n = 2$) with 4 petals

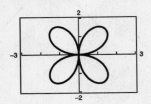

65. $r = \dfrac{3}{\cos(\theta - \pi/4)}$

The graph is a line.

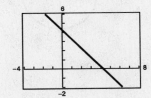

67. Center: $(5, \pi/2)$

Solution point: $(0, 0) \Rightarrow$ Radius $= 5 \Rightarrow a = 10$

$r = a \sin \theta$

$r = 10 \sin \theta$

69. Parabola: $r = \dfrac{ep}{1 - e \cos \theta}, e = 1$

Vertex: $(2, \pi)$

Focus: $(0, 0) \Rightarrow p = 4$

$r = \dfrac{4}{1 - \cos \theta}$

71. Ellipse: $r = \dfrac{ep}{1 - e \cos \theta}$;   Vertices: $(5, 0), (1, \pi) \Rightarrow a = 3$;   One focus: $(0, 0) \Rightarrow c = 2$

$e = \dfrac{c}{a} = \dfrac{2}{3}, p = \dfrac{5}{2}$

$r = \dfrac{(2/3)(5/2)}{1 - (2/3) \cos \theta} = \dfrac{5/3}{1 - (2/3) \cos \theta} = \dfrac{5}{3 - 2 \cos \theta}$

❏ Cumulative Test for Chapters 7–10

1.
$$2x - y^2 = 0$$
$$x - y = 4 \implies x = y + 4$$
$$2(y + 4) - y^2 = 0$$
$$y^2 - 2y - 8 = 0$$
$$(y - 4)(y + 2) = 0$$
$$y = 4 \implies x = 8 \quad (8, 4)$$
$$y = -2 \implies x = 2 \quad (2, -2)$$

2. $y = \log_3 x = \dfrac{\log_{10} x}{\log_{10} 3}$

$$y = -\frac{1}{3}x + 2$$

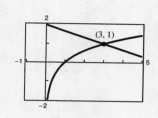

3. $\begin{bmatrix} 1 & -2 & 3 & \vdots & 11 \\ 2 & 0 & -1 & \vdots & 3 \\ 0 & 3 & 1 & \vdots & -8 \end{bmatrix} \implies \begin{bmatrix} 1 & 0 & 0 & \vdots & 2 \\ 0 & 1 & 0 & \vdots & -3 \\ 0 & 0 & 1 & \vdots & 1 \end{bmatrix}$

$x = 2, y = -3, z = 1$

$(2, -3, 1)$

4. Let $x =$ number of \$275 models. Let $y =$ number of \$400 models.

Constraints: $275 + 400y \le 100,000$
$$x + y \le 300$$
$$x \ge 0$$
$$y \ge 0$$

The intersection of the two lines is:

$275x + 400y = 275(300 - y) + 400y = 100,000$
$$125y = 17,500$$
$$y = 140$$
$$x = 160$$

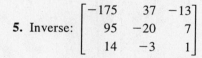

At $(0, 0)$: $\quad P = 275(0) + 400(0) = 0$

At $(300, 0)$: $\quad P = 275(300) + 400(0) = 82,500$

At $(160, 140)$: $P = 275(160 + 400(140) = 100,000$

At $(0, 250)$: $\quad P = 275(0) + 400(250) = 100,000$

Sell 160 of \$275 units and 140 of \$400 units. (In fact, you could sell 250 of \$400 units.)

5. Inverse: $\begin{bmatrix} -175 & 37 & -13 \\ 95 & -20 & 7 \\ 14 & -3 & 1 \end{bmatrix}$

6. $\begin{vmatrix} 0 & 0 & 1 \\ 6 & 2 & 1 \\ 8 & 10 & 1 \end{vmatrix} = 60 - 16 = 44$

Area $= \frac{1}{2}(44) = 22$

7. $\displaystyle\sum_{i=0}^{\infty} 3\left(\frac{1}{2}\right)^i = \frac{3}{1 - (1/2)} = 6$

8. $(z - 3)^4 = z^4 - 12z^3 + 54z^2 - 108z + 81$

9. There are 2 ways to select the first digit (4 or 5), and 2 ways for the second digit. Hence, $p = \frac{1}{4}$.

10. Mean: 24.17, Median: 24, $\sigma = 3.67$

11. (a)

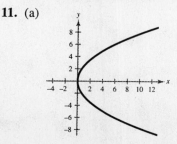

(b)

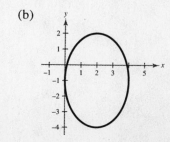

12. Center: $(0, 2)$

Vertical axis

$\dfrac{a}{b} = \dfrac{1}{2}$

$c = 2, \ b^2 = c^2 - a^2 = c^2 - \left(\dfrac{b}{2}\right)^2 \ \Rightarrow \ \dfrac{5}{4}b^2 = 4 \ \Rightarrow \ b^2 = \dfrac{16}{5}$

$a^2 = \dfrac{b^2}{4} = \dfrac{4}{5}$

Thus, $\dfrac{(y - 2)^2}{4/5} - \dfrac{x^2}{16/5} = 1.$

13. $x = 2 + (6 - 2)t = 2 + 4t, \ y = -3 + (4 - (-3))t = -3 + 7t$

14. $x^2 + y^2 - 6y = 0$

$r^2 - 6r \sin \theta = 0$

$r = 6 \sin \theta$

15. (a)

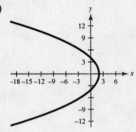

(b)

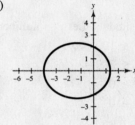

❑ Practice Test for Chapter 10

1. Find the vertex, focus and directrix of the parabola $x^2 - 6x - 4y + 1 = 0$.

2. Find an equation of the parabola with its vertex at $(2, -5)$ and focus at $(2, -6)$.

3. Find the center, foci, vertices, and eccentricity of the ellipse $x^2 + 4y^2 - 2x + 32y + 61 = 0$.

4. Find an equation of the ellipse with vertices $(0, \pm 6)$ and eccentricity $e = \frac{1}{2}$.

5. Find the center, vertices, foci, and asymptotes of the hyperbola $16y^2 - x^2 - 6x - 128y + 231 = 0$.

6. Find an equation of the hyperbola with vertices at $(\pm 3, 2)$ and foci at $(\pm 5, 2)$.

7. Rotate the axes to eliminate the xy-term. Sketch the graph of the resulting equation, showing both sets of axes.

   $5x^2 + 2xy + 5y^2 - 10 = 0$

8. Use the discriminant to determine whether the graph of the equation is a parabola, ellipse, or hyperbola.

   (a) $6x^2 - 2xy + y^2 = 0$ (b) $x^2 + 4xy + 4y^2 - x - y + 17 = 0$

For Exercises 9 and 10, eliminate the parameter and write the corresponding rectangular equation.

9. $x = 3 - 2 \sin \theta, y = 1 + 5 \cos \theta$ 10. $x = e^{2t}, y = e^{4t}$

11. Convert the polar point $\left(\sqrt{2}, (3\pi)/4\right)$ to rectangular coordinates.

12. Convert the rectangular point $\left(\sqrt{3}, -1\right)$ to polar coordinates.

13. Convert the rectangular equation $4x - 3y = 12$ to polar form.

14. Convert the polar equation $r = 5 \cos \theta$ to rectangular form.

15. Sketch the graph of $r = 1 - \cos \theta$.

16. Sketch the graph of $r = 5 \sin 2\theta$.

17. Sketch the graph of $r = \dfrac{3}{6 - \cos \theta}$.

18. Find a polar equation of the parabola with its vertex at $\left(6, \pi/2\right)$ and focus at $(0, 0)$.

C H A P T E R 1 1
Analytic Geometry in Three Dimensions

CHAPTER 11
Analytic Geometry in Three Dimensions

Section 11.1 The Three-Dimensional Coordinate System

> ■ You should be able to plot points in the three-dimensional coordinate system.
>
> ■ The distance between the points (x_1, y_1, z_1) and (x_2, y_2, z_2) is
> $$d = \sqrt{(x_2 - x_1)^2 + (y_2 - y_1)^2 + (z_2 - z_1)^2}.$$
>
> ■ The midpoint of the line segment joining the points (x_1, y_1, z_1) and (x_2, y_2, z_2) is
> $$\left(\frac{x_1 + x_2}{2}, \frac{y_1 + y_2}{2}, \frac{z_1 + z_2}{2}\right).$$
>
> ■ The equation of the sphere with center (h, k, j) and radius r is
> $$(x - h)^2 + (y - k)^2 + (z - j)^2 = r^2.$$
>
> ■ You should be able to find the trace of a surface in space.

Solutions to Odd-Numbered Exercises

1. A: $(2, 1, 3)$
   B: $(1, -2, 2)$

3. A: $(-1, -3, 2)$
   B: $(3, -2, -4)$

5.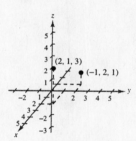

7.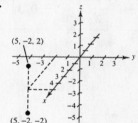

9. $x = -3, y = 4, z = 5$: $(-3, 4, 5)$

11. $y = z = 0, x = 10$: $(10, 0, 0)$

13. In the xy-plane, the z-coordinate is 0.
    In the xz-plane, the y-coordinate is 0.
    In the yz-plane, the x-coordinate is 0.

15. Octant IV

17. Octants I, II, III, IV (above the xy-plane)

19. Octants II, IV, VI, VIII

21. $d_1 = \sqrt{(3 - 0)^2 + (3 - 0)^2 + (2 - 0)^2} = \sqrt{9 + 9 + 4} = \sqrt{22}$
    $d_2 = \sqrt{(3 - 3)^2 + (3 - 3)^2 + (0 - 2)^2} = \sqrt{4} = 2$
    $d_3 = \sqrt{(3 - 0)^2 + (3 - 0)^2 + (0 - 0)^2} = \sqrt{9 + 9} = \sqrt{18}$
    $d_1^2 = d_2^2 + d_3^2 = 22$

23. $d_1 = \sqrt{(-2 - 0)^2 + (5 - 0)^2 + (2 - 2)^2} = \sqrt{4 + 25} = \sqrt{29}$
    $d_2 = \sqrt{(0 - 0)^2 + (4 - 0)^2 + (0 - 2)^2} = \sqrt{16 + 4} = \sqrt{20}$
    $d_3 = \sqrt{(0 + 2)^2 + (4 - 5)^2 + (0 - 2)^2} = \sqrt{4 + 1 + 4} = \sqrt{9} = 3$
    $d_1^2 = d_2^2 + d_3^2 = 29$

25. $d_1 = \sqrt{(2-0)^2 + (2-0)^2 + (1-0)^2} = \sqrt{4+4+1} = \sqrt{9} = 3$

$d_2 = \sqrt{(2-2)^2 + (-4-2)^2 + (4-1)^2} = \sqrt{36+9} = \sqrt{45}$

$d_3 = \sqrt{(2-0)^2 + (-4-0)^2 + (4-0)^2} = \sqrt{4+16+16} = \sqrt{36} = 6$

$d_2^2 = d_1^2 + d_3^2 = 45$ Right triangle

27. $d_1 = \sqrt{(5-1)^2 + (-1+3)^2 + (2+2)^2} = \sqrt{16+4+16} = \sqrt{36} = 6$

$d_2 = \sqrt{(5+1)^2 + (-1-1)^2 + (2-2)^2} = \sqrt{36+4} = \sqrt{40}$

$d_3 = \sqrt{(-1-1)^2 + (1+3)^2 + (2+2)^2} = \sqrt{4+16+16} = \sqrt{36} = 6$

$d_1 = d_3$ Isosceles triangle

29. Midpoint: $\left(\dfrac{3-3}{2}, \dfrac{-6+2}{2}, \dfrac{10+2}{2}\right) = (0, -2, 6)$

31. Midpoint: $\left(\dfrac{4-4}{2}, \dfrac{-2+2}{2}, \dfrac{5+8}{2}\right) = \left(0, 0, \dfrac{13}{2}\right)$

33. $x_m = \dfrac{x_2 + x_1}{2} \implies x_2 = 2x_m - x_1$. Similarly for y_2 and z_2.

$(2x_m - x_1, 2y_m - y_1, 2z_m - z_1)$

35. $(x-3)^2 + (y-2)^2 + (z-4)^2 = 16$

37. $(x-0)^2 + (y-4)^2 + (z-3)^2 = 4^2$

$x^2 + (y-4)^2 + (z-3)^2 = 16$

39. Radius $= \dfrac{\text{diameter}}{2} = 5$: $(x+3)^2 + (y-7)^2 + (z-5)^2 = 5^2 = 25$

41. Center: $\left(\dfrac{3+0}{2}, \dfrac{0+0}{2}, \dfrac{0+6}{2}\right) = \left(\dfrac{3}{2}, 0, 3\right)$

Radius: $\sqrt{\left(3 - \dfrac{3}{2}\right)^2 + (0-0)^2 + (0-3)^2} = \sqrt{\dfrac{9}{4} + 9} = \sqrt{\dfrac{45}{4}}$

Sphere: $\left(x - \dfrac{3}{2}\right)^2 + (y-0)^2 + (z-3)^2 = \dfrac{45}{4}$

43. $(x^2 - 4x + 4) + (y^2 + 2y + 1) + (z^2 - 6z + 9) = -10 + 4 + 1 + 9$

$(x+2)^2 + (y+1)^2 + (z-3)^2 = 4$

Center: $(2, -1, 3)$

Radius: 2

45. $(x^2 + 4x + 4) + y^2 + (z^2 - 8z + 16) = -19 + 4 + 16$

$(x+2)^2 + y^2 + (z-4)^2 = 1$

Center: $(-2, 0, 4)$

Radius: 1

47. $\qquad\qquad x^2 + y^2 + z^2 - 2x - \dfrac{2}{3}y - 8z = -\dfrac{73}{9}$

$(x^2 - 2x + 1) + \left(y^2 - \dfrac{2}{3}y + \dfrac{1}{9}\right) + (z^2 - 8z + 16) = -\dfrac{73}{9} + 1 + \dfrac{1}{9} + 16$

$(x-1)^2 + \left(y - \dfrac{1}{3}\right)^2 + (z-4)^2 = 9$

Center: $\left(1, \dfrac{1}{3}, 4\right)$

49. (a)

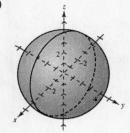

(b)

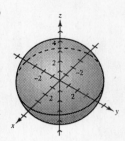

51. $z^2 = 16 - x^2 - y^2 \implies \begin{cases} z_1 = \sqrt{16 - x^2 - y^2} \\ z_2 = -\sqrt{16 - x^2 - y^2} \end{cases}$

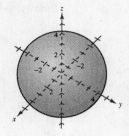

53. $(z - 5)^2 = 4 - (x - 3)^2 - (y - 4)^2 \implies \begin{cases} z_1 = 5 + \sqrt{4 - (x - 3)^2 - (y - 4)^2} \\ z_2 = 5 - \sqrt{4 - (x - 3)^2 - (y - 4)^2} \end{cases}$

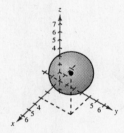

55. Distance between aircraft $= d_2 - d_1$

$$= (0.16t^2 - 3.93t + 20) - (0.07t^2 - 2.43t + 12)$$

$$= 0.09t^2 - 1.5t + 8$$

Section 11.2 Vectors in Space

■ Vectors in space $\mathbf{v} = \langle v_1, v_2, v_3 \rangle$ have many of the same properties as vectors in the plane.

■ The dot product of two vectors $\mathbf{u} = \langle u_1, u_2, u_3 \rangle$ and $\mathbf{v} = \langle v_1, v_2, v_3 \rangle$ in space is $\mathbf{u} \cdot \mathbf{v} = u_1v_1 + u_2v_2 + u_3v_3$.

■ Two nonzero vectors $\mathbf{u}$ and $\mathbf{v}$ are said to be parallel if there is some scalar c such that $\mathbf{u} - c\mathbf{v}$.

■ You should be able to use vectors to solve real life problems.

Solutions to Even-Numbered Exercises

1. $\mathbf{v} = \langle 0 - 2, 3 - 0, 2 - 1 \rangle = \langle -2, 3, 1 \rangle$

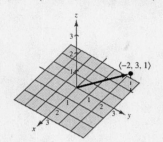

3. $\mathbf{v} = \langle 3 - (-1), 2 - (-2), 5 - 1 \rangle = \langle 4, 4, 4 \rangle$
any vector of form $c\langle 1, 1, 1 \rangle, c > 0$, is parallel to
$\mathbf{v}$ any vector of form $c\langle 1, 1, 1 \rangle, c < 0$, is opposite
direction.

5. $\mathbf{v} = \langle q_1, q_2, q_3 \rangle$. Since $\mathbf{v}$ lies in the yz-plane, $q_1 = 0$. Since $\mathbf{v}$ makes an angle of $45°$, $q_2 = q_3$. Finally, $\|\mathbf{v}\| = 4$
implies that $q_2^2 + q_3^2 = 16$. Thus, $q_2 = q_3 = 2\sqrt{2}$ and $\mathbf{v} = \langle 0, 2\sqrt{2}, 2\sqrt{2} \rangle$.

7. (a)

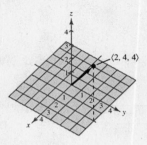

(b)

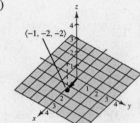

(c)

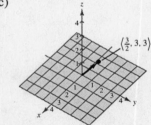

(d)

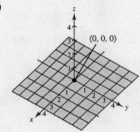

9. $\mathbf{z} = \mathbf{u} - 2\mathbf{v} = \langle -1, 3, 2 \rangle - 2\langle 1, -2, -2 \rangle = \langle -3, 7, 6 \rangle$

11. $2\mathbf{z} - 4\mathbf{u} = \mathbf{w} \implies \mathbf{z} = \frac{1}{2}(4\mathbf{u} + \mathbf{w}) = \frac{1}{2}(4\langle -1, 3, 2 \rangle + \langle 5, 0, -5 \rangle)$
$$= \langle \tfrac{1}{2}, 6, \tfrac{3}{2} \rangle$$

13. $\|\mathbf{v}\| = \sqrt{4^2 + 1^2 + 4^2} = \sqrt{33}$

15. $\mathbf{v} = \langle 1 - 1, 0 - (-3), -1 - 4 \rangle = \langle 0, 3, -5 \rangle$
$\|\mathbf{v}\| = \sqrt{0 + 3^2 + (-5)^2} = \sqrt{34}$

17. (a) $\dfrac{\mathbf{u}}{\|\mathbf{u}\|} = \dfrac{\langle 8, 3, -1 \rangle}{\sqrt{74}} = \dfrac{1}{\sqrt{74}}(8\mathbf{i} + 3\mathbf{j} - \mathbf{k})$

(b) $-\dfrac{1}{\sqrt{74}}(8\mathbf{i} + 3\mathbf{j} - \mathbf{k})$

19. $6\mathbf{u} - 4\mathbf{v} = 6\langle -1, 3, 4 \rangle - 4\langle 5, 4.5, -6 \rangle$
$= \langle -6, 18, 24 \rangle + \langle -20, -18, 24 \rangle$
$= \langle -26, 0, 48 \rangle$

21. $\mathbf{u} + \mathbf{v} = \langle -1, 3, 4 \rangle + \langle 5, 4.5, -6 \rangle = \langle 4, 7.5, -2 \rangle$
$\|\mathbf{u} + \mathbf{v}\| = \sqrt{4^2 + 7.5^2 + (-2)^2} = \frac{1}{2}\sqrt{305} \approx 34.93$

23. $\mathbf{u} \cdot \mathbf{v} = \langle 4, 4, -1 \rangle \cdot \langle 2, -5, -8 \rangle$
$= 8 - 20 + 8 = -4$

25. $\cos \theta = \dfrac{\mathbf{u} \cdot \mathbf{v}}{\|\mathbf{u}\| \|\mathbf{v}\|} = \dfrac{-8}{\sqrt{8}\sqrt{25}} \implies \theta = 124.45°$

27. $\cos \theta = \dfrac{\mathbf{u} \cdot \mathbf{v}}{\|\mathbf{u}\| \|\mathbf{v}\|} = \dfrac{-270}{\sqrt{405}\sqrt{180}} = -1 \implies$ parallel

29. $\mathbf{v} = \langle 7 - 5, 3 - 4, -1 - 1 \rangle = \langle 2, -1, -2 \rangle$
$\mathbf{u} = \langle 4 - 7, 5 - 3, 3 - (-1) \rangle = \langle -3, 2, 4 \rangle$
Since $\mathbf{u}$ and $\mathbf{v}$ are not parallel, the points are not
collinear.

31. $\mathbf{v} = \langle -1 - 1, 2 - 3, 5 - 2 \rangle = \langle -2, -1, 3 \rangle$
$\mathbf{u} = \langle 3 - (-1), 4 - 2, -1 - 5 \rangle = \langle 4, 2, -6 \rangle$
Since $\mathbf{u} = -2\mathbf{v}$, the points are collinear.

33. Let the points be P_1, P_2, P_3, P_4.

$\overrightarrow{P_1P_2} = \langle 1, 2, 3 \rangle$, $\overrightarrow{P_3P_4} = \langle 1, 2, 3 \rangle$ Parallel and equal length

$\overrightarrow{P_1P_3} = \langle -2, 1, 1 \rangle$, $\overrightarrow{P_2P_4} = \langle -2, 1, 1 \rangle$ Parallel and equal length

35. $\mathbf{v} = \langle 2, -4, 7 \rangle = \langle q_1 - 1, q_2 - 5, q_3 - 0 \rangle \implies$

$\left. \begin{array}{l} 2 = q_1 - 1 \\ -4 = q_2 - 5 \\ 7 = q_3 \end{array} \right\} \implies \left. \begin{array}{l} q_1 = 3 \\ q_2 = 1 \\ q_3 = 7 \end{array} \right\} \implies$ Terminal point is $\langle 3, 1, 7 \rangle$.

37. $c\mathbf{u} = c\mathbf{i} + 2c\mathbf{j} + 3c\mathbf{k}$

$\|c\mathbf{u}\| = \sqrt{c^2 + 4c^2 + 9c^2} = |c|\sqrt{14} = 3 \implies c = \pm\dfrac{3}{\sqrt{14}}$

39. Sphere: $(x - x_1)^2 + (y - y_1)^2 + (z - z_1)^2 = 16$

Section 11.3   The Cross Product of Two Vectors

- The cross product of two vectors $\mathbf{u} = u_1\mathbf{i} + u_2\mathbf{j} + u_3\mathbf{k}$ and $\mathbf{v} = v_1\mathbf{i} + v_2\mathbf{j} + v_3\mathbf{k}$ is given by

$$\mathbf{u} \times \mathbf{v} = (u_2v_3 - u_3v_2)\mathbf{i} - (u_1v_3 - u_3v_1)\mathbf{j} + (u_1v_2 - u_2v_1)\mathbf{k}$$

$$= \begin{vmatrix} \mathbf{i} & \mathbf{j} & \mathbf{k} \\ u_1 & u_2 & u_3 \\ v_1 & v_2 & v_3 \end{vmatrix}.$$

- The cross product satisfies the following algebraic properties.

  (a) $\mathbf{u} \times \mathbf{v} = -(\mathbf{v} \times \mathbf{u})$

  (b) $\mathbf{u} \times (\mathbf{v} + \mathbf{w}) = (\mathbf{u} \times \mathbf{v}) + (\mathbf{u} \times \mathbf{w})$

  (c) $c(\mathbf{u} \times \mathbf{v}) = (c\mathbf{u}) \times \mathbf{v} = \mathbf{u} \times (c\mathbf{v})$

  (d) $\mathbf{u} \times \mathbf{0} = \mathbf{0} \times \mathbf{u} = \mathbf{0}$

  (e) $\mathbf{u} \times \mathbf{u} = \mathbf{0}$

  (f) $\mathbf{u} \cdot (\mathbf{v} \times \mathbf{w}) = (\mathbf{u} \times \mathbf{v}) \cdot \mathbf{w}$

- The following geometric properties of the cross product are valid, where θ is the angle between the vectors $\mathbf{u}$ and $\mathbf{v}$:

  (a) $\mathbf{u} \times \mathbf{v}$ is orthogonal to both $\mathbf{u}$ and $\mathbf{v}$.

  (b) $\|\mathbf{u} \times \mathbf{v}\| = \|\mathbf{u}\| \|\mathbf{v}\| \sin \theta$

  (c) $\mathbf{u} \times \mathbf{v} = \mathbf{0}$ if and only if $\mathbf{u}$ and $\mathbf{v}$ are scalar multiples.

  (d) $\|\mathbf{u} \times \mathbf{v}\|$ is the area of the parallelogram having $\mathbf{u}$ and $\mathbf{v}$ as sides.

- The triple scalar product is the volume of the parallelepiped having $\mathbf{u}, \mathbf{v}$ and $\mathbf{w}$ as sides.

$$\mathbf{u} \cdot (\mathbf{v} \times \mathbf{w}) = \begin{vmatrix} u_1 & u_2 & u_3 \\ v_1 & v_2 & v_3 \\ w_1 & w_2 & w_3 \end{vmatrix}$$

Solutions to Odd-Numbered Exercises

1. $\mathbf{j} \times \mathbf{i} = \begin{vmatrix} \mathbf{i} & \mathbf{j} & \mathbf{k} \\ 0 & 1 & 0 \\ 1 & 0 & 0 \end{vmatrix} = -\mathbf{k}$

3. $\mathbf{i} \times \mathbf{k} = \begin{vmatrix} \mathbf{i} & \mathbf{j} & \mathbf{k} \\ 1 & 0 & 0 \\ 0 & 0 & 1 \end{vmatrix} = -\mathbf{j}$

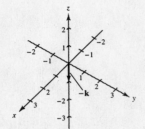

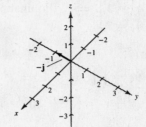

5. $\mathbf{u} \times \mathbf{v} = \begin{vmatrix} \mathbf{i} & \mathbf{j} & \mathbf{k} \\ 1 & -4 & 0 \\ 2 & 6 & 0 \end{vmatrix} = 14\mathbf{k} = \langle 0, 0, 14 \rangle$

7. $\mathbf{u} \times \mathbf{v} = \begin{vmatrix} \mathbf{i} & \mathbf{j} & \mathbf{k} \\ 7 & -5 & 2 \\ -1 & 4 & -1 \end{vmatrix} = \langle -3, 5, 23 \rangle$

9. $\mathbf{u} \times \mathbf{v} = \begin{vmatrix} \mathbf{i} & \mathbf{j} & \mathbf{k} \\ 6 & 2 & 1 \\ 1 & 3 & -2 \end{vmatrix} = \langle -7, 13, 16 \rangle$

$= -7\mathbf{i} + 13\mathbf{j} + 16\mathbf{k}$

11. $\mathbf{u} \times \mathbf{v} = \begin{vmatrix} \mathbf{i} & \mathbf{j} & \mathbf{k} \\ 1 & \frac{3}{2} & \frac{-5}{2} \\ \frac{1}{2} & \frac{-3}{4} & \frac{1}{4} \end{vmatrix} = \langle \frac{-3}{2}, -\frac{3}{2}, -\frac{3}{2} \rangle$

$= \frac{-3}{2}\mathbf{i} - \frac{3}{2}\mathbf{j} - \frac{3}{2}\mathbf{k}$

13. $\mathbf{u} \times \mathbf{v} = \begin{vmatrix} \mathbf{i} & \mathbf{j} & \mathbf{k} \\ 2 & 4 & 3 \\ 0 & -2 & 1 \end{vmatrix} = \langle 10, -2, -4 \rangle$

15. $\mathbf{u} \times \mathbf{v} = \begin{vmatrix} \mathbf{i} & \mathbf{j} & \mathbf{k} \\ 1 & -5 & 1 \\ \frac{1}{3} & -\frac{1}{3} & \frac{2}{3} \end{vmatrix} = \langle -3, -\frac{11}{3}, -\frac{1}{3} \rangle$

$= -3\mathbf{i} - \frac{11}{3}\mathbf{j} - \frac{1}{3}\mathbf{k}$

17. $\mathbf{u} \times \mathbf{v} = \begin{vmatrix} \mathbf{i} & \mathbf{j} & \mathbf{k} \\ 3 & 1 & 0 \\ 0 & 1 & 1 \end{vmatrix} = \mathbf{i} - 3\mathbf{j} + 3\mathbf{k}$

$\|\mathbf{u} \times \mathbf{v}\| = \sqrt{19}$

Unit vector $= \dfrac{\mathbf{u} \times \mathbf{v}}{\|\mathbf{u} \times \mathbf{v}\|} = \dfrac{1}{\sqrt{19}}(\mathbf{i} - 3\mathbf{j} + 3\mathbf{k})$

19. $\mathbf{u} \times \mathbf{v} = \begin{vmatrix} \mathbf{i} & \mathbf{j} & \mathbf{k} \\ -2 & 1 & 3 \\ 1 & 4 & 6 \end{vmatrix} = -6\mathbf{i} + 15\mathbf{j} - 9\mathbf{k}$

$\|\mathbf{u} \times \mathbf{v}\| = \sqrt{342}$

Unit vector $= \dfrac{\mathbf{u} \times \mathbf{v}}{\|\mathbf{u} \times \mathbf{v}\|} = \dfrac{1}{\sqrt{342}}(-6\mathbf{i} + 15\mathbf{j} - 9\mathbf{k})$

21. $\mathbf{u} \times \mathbf{v} = \begin{vmatrix} \mathbf{i} & \mathbf{j} & \mathbf{k} \\ 1 & 1 & -1 \\ 1 & 1 & 1 \end{vmatrix} = 2\mathbf{i} - 2\mathbf{j}$

$\|\mathbf{u} \times \mathbf{v}\| = 2\sqrt{2}$

Unit vector $= \dfrac{\mathbf{u} \times \mathbf{v}}{\|\mathbf{u} \times \mathbf{v}\|} = \dfrac{1}{2\sqrt{2}}(2\mathbf{i} - 2\mathbf{j})$

$= \dfrac{1}{\sqrt{2}}\mathbf{i} - \dfrac{1}{\sqrt{2}}\mathbf{j}$

23. $\mathbf{u} \times \mathbf{v} = \begin{vmatrix} \mathbf{i} & \mathbf{j} & \mathbf{k} \\ 0 & 0 & 1 \\ 1 & 0 & 1 \end{vmatrix} = \mathbf{j}$

Area $= \|\mathbf{u} \times \mathbf{v}\| = \|\mathbf{j}\| = 1$

25. $\mathbf{u} \times \mathbf{v} = \begin{vmatrix} \mathbf{i} & \mathbf{j} & \mathbf{k} \\ 3 & 4 & 6 \\ 2 & -1 & 5 \end{vmatrix} = 26\mathbf{i} + 3\mathbf{j} + 11\mathbf{k}$

Area $= \|\mathbf{u} \times \mathbf{v}\| = \sqrt{26^2 + 3^2 + (-11)^2} = \sqrt{806}$

27. $\mathbf{u} \times \mathbf{v} = \begin{vmatrix} \mathbf{i} & \mathbf{j} & \mathbf{k} \\ 2 & 2 & -3 \\ 0 & 2 & 3 \end{vmatrix} = \langle 12, -6, 4 \rangle$

Area $= \|\mathbf{u} \times \mathbf{v}\| = \sqrt{12^2 + (-6)^2 + 4^2} = 14$

29. $\overrightarrow{AB} = \langle 3 - 2, 1 - (-1), 2 - 4 \rangle = \langle 1, 2, -2 \rangle$ is parallel to $\overrightarrow{DC} = \langle 0 - (-1), 5 - 3, 6 - 8 \rangle = \langle 1, 2, -2 \rangle$.

$\overrightarrow{AD} = \langle -3, 4, 4 \rangle$ is parallel to $\overrightarrow{BC} = \langle -3, 4, 4 \rangle$.

$$\overrightarrow{AB} \times \overrightarrow{AD} = \begin{vmatrix} \mathbf{i} & \mathbf{j} & \mathbf{k} \\ 1 & 2 & -2 \\ -3 & 4 & 4 \end{vmatrix} = \langle 16, 2, 10 \rangle$$

Area $= \|\overrightarrow{AB} \times \overrightarrow{AD}\| = \sqrt{16^2 + 2^2 + 10^2} = \sqrt{360} = 6\sqrt{10}$

31. $\mathbf{u} = \langle 4, -2, 6 \rangle, \quad \mathbf{v} = \langle -4, 0, 3 \rangle$

$$\mathbf{u} \times \mathbf{v} = \begin{vmatrix} \mathbf{i} & \mathbf{j} & \mathbf{k} \\ 4 & -2 & 6 \\ -4 & 0 & 3 \end{vmatrix} = \langle -6, -36, -8 \rangle$$

Area $= \frac{1}{2}\|\mathbf{u} \times \mathbf{v}\| = \frac{1}{2}\sqrt{(-6)^2 + (-36)^2 + (-8)^2}$
$\qquad = \frac{1}{2}\sqrt{1396} = \sqrt{349}$

33. $\mathbf{u} = \langle -2 - 2, -2 - 3, 0 - (-5) \rangle = \langle -4, -5, 5 \rangle$

$\mathbf{v} = \langle 3 - 2, 0 - 3, 6 - (-5) \rangle = \langle 1, -3, 11 \rangle$

$$\mathbf{u} \times \mathbf{v} = \begin{vmatrix} \mathbf{i} & \mathbf{j} & \mathbf{k} \\ -4 & -5 & 5 \\ 1 & -3 & 11 \end{vmatrix} = \langle -40, 49, 17 \rangle$$

Area $= \frac{1}{2}\|\mathbf{u} \times \mathbf{v}\| = \frac{1}{2}\sqrt{(-40)^2 + 49^2 + 17^2}$
$\qquad = \frac{1}{2}\sqrt{4290}$

35. $\mathbf{u} \cdot (\mathbf{v} \times \mathbf{w}) = \begin{vmatrix} 2 & 3 & 3 \\ 4 & 4 & 0 \\ 0 & 0 & 4 \end{vmatrix} = 2(16) - 3(16) + 3(0) = -16$

37. $\mathbf{u} \cdot (\mathbf{v} \times \mathbf{w}) = \begin{vmatrix} 1 & 1 & 0 \\ 0 & 1 & 1 \\ 1 & 0 & 1 \end{vmatrix} = 1 + 1 = 2$

Volume $= |\mathbf{u} \cdot (\mathbf{v} \times \mathbf{w})| = 2$

39. $\mathbf{u} = \langle 4, 0, 0 \rangle, \quad \mathbf{v} = \langle 0, -2, 3 \rangle, \quad \mathbf{w} = \langle 0, 5, 3 \rangle$

$$\mathbf{u} \cdot (\mathbf{v} \times \mathbf{w}) = \begin{vmatrix} 4 & 0 & 0 \\ 0 & -2 & 3 \\ 0 & 5 & 3 \end{vmatrix} = 4(-21) = -84$$

Volume $= |-84| = 84$

41. $\mathbf{V} \times \mathbf{F} = \begin{vmatrix} \mathbf{i} & \mathbf{j} & \mathbf{k} \\ 0 & \frac{1}{2}\cos 40° & \frac{1}{2}\sin 40° \\ 0 & 0 & -20 \end{vmatrix} = -10 \cos 40° \, \mathbf{i}$

$\|\mathbf{V} \times \mathbf{F}\| = 10 \cos 40° \approx 7.44$ ft/lbs

43. $\mathbf{u} \times \mathbf{u} = \begin{vmatrix} \mathbf{i} & \mathbf{j} & \mathbf{k} \\ u_1 & u_2 & u_3 \\ u_1 & u_2 & u_3 \end{vmatrix} = (u_2 u_3 - u_2 u_3)\mathbf{i} - (u_1 u_3 - u_1 u_3)\mathbf{j} - (u_1 u_2 - u_1 u_2)\mathbf{k} = \mathbf{0}$

45. $\mathbf{u} \times \mathbf{v} = \begin{vmatrix} \mathbf{i} & \mathbf{j} & \mathbf{k} \\ \cos \beta & \sin \beta & 0 \\ \cos \alpha & \sin \alpha & 0 \end{vmatrix} = (\cos \beta \sin \alpha - \cos \alpha \sin \beta)\mathbf{k}$

The area of the triangle is $\frac{1}{2}bh = \frac{1}{2}(1) \sin(\alpha - \beta)$. Moreover, the area $\frac{1}{2}\|\mathbf{u} \times \mathbf{v}\| = \frac{1}{2}(\cos \beta \sin \alpha - \cos \alpha \sin \beta)$. Hence, $\sin(\alpha - \beta) = \cos \beta \sin \alpha - \cos \alpha \sin \beta$.

Section 11.4 Lines and Planes in Space

- The parametric equations of the line in space parallel to the vector $\langle a, b, c \rangle$ and passing through the point (x_1, y_2, z_3) are

$$x = x_1 + at, \quad y = y_1 + bt, \quad z = z_1 + ct.$$

- The standard equation of the plane in space containing the point (x_1, y_1, z_1) and having normal vector (a, b, c) is

$$a(x - x_1) + b(y - y_1) + (z - z_1) = 0.$$

- You should be able to find the angle between two planes by calculating the angle between their normal vectors.

- You should be able to sketch a plane in space.

- The distance between a point Q and a plane having normal $\mathbf{n}$ is

$$D = \|\text{proj}_\mathbf{n}\overline{PQ}\| = \frac{|\overline{PQ} \cdot \mathbf{n}|}{\|\mathbf{n}\|}$$

where P is a point in the plane.

Solutions to Odd-Numbered Exercises

1. $x = 1 + 3t, \ y = 2 - t, \ z = 2 + 5t$

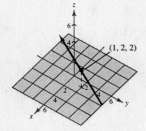

3. $y = 0 \Rightarrow t = 2 \Rightarrow x = 7$ and $z = 12$: $(7, 0, 12)$

5. $x = x_1 + at = 0 - 2t = -2t, \ y = y_1 + bt = 0 + 4t = 4t, \ z = z_1 + ct = 0 + t = t$

(a) Parametric equations: $x = -2t, y = 4t, z = t$

(b) Symmetric equations: $\dfrac{-x}{2} = \dfrac{y}{4} = z$

7. $x = x_1 + at = -4 + \dfrac{1}{2}t, \ y = y_1 + bt = 1 + \dfrac{4}{3}t, \ z = z_1 + ct = 0 - t$

(a) Parametric equations: $x = -4 + \dfrac{1}{2}t, y = 1 + \dfrac{4}{3}t, z = -t$

Equivalently: $x = -4 + 3t, y = 1 + 8t, z = -6t$

(b) Symmetric equations: $\dfrac{x + 4}{3} = \dfrac{y - 1}{8} = \dfrac{z}{-6}$

9. $x = x_1 + at = 2 + 2t, \ y = y_1 + bt = -3 - 3t, \ z = z_1 + ct = 5 + t$

(a) Parametric equations: $x = 2 + 2t, y = -3 - 3t, z = 5 + t$

(b) Symmetric equations: $\dfrac{x - 2}{2} = \dfrac{y + 3}{-3} = z - 5$

11. (a) $\vec{\mathbf{v}} = \left\langle 3 - \left(-\dfrac{3}{2}\right), -5 - \dfrac{3}{2}, -4 - 2 \right\rangle = \left\langle \dfrac{9}{2}, -\dfrac{13}{2}, -6 \right\rangle$

Use $\langle 9, -13, -12 \rangle$: $x = 3 + 9t, y = -5 - 13t, z = -4 - 12t$

(b) $\dfrac{x - 3}{9} = \dfrac{y + 5}{-13} = \dfrac{z + 4}{-12}$

13. The line is $x = -4 + 3t, y = -1 - t, z = 7$, or $(x + 4)/3 = (y + 1)/-1, z = 7$.

Only (b) and (c) satisfy the equation.

15.

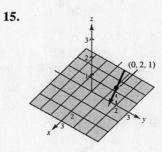

17. $a(x - x_1) + b(y - y_1) + c(z - z_1) = 0$
$0(x - 3) + 1(y - 4) + 0(z + 2) = 0$
$y - 4 = 0$

19. $-2(x - 5) + 1(y - 6) - 2(z - 3) = 0$
$-2x + y - 2z + 10 = 0$

21. $\mathbf{n} = \langle -1, -2, 1 \rangle \implies -1(x - 2) - 2(y - 0) + 1(z - 0) = 0$
$-x - 2y + z + 2 = 0$

23. $\mathbf{u} = \langle 2 - 0, 1 - 0, 3 - 0 \rangle = \langle 2, 1, 3 \rangle$

$\mathbf{v} = \langle -2 - 0, 1 - 0, 3 - 0 \rangle = \langle -2, 1, 3 \rangle$

$\overline{\mathbf{n}} = \mathbf{u} \times \mathbf{v} = \begin{vmatrix} \mathbf{i} & \mathbf{j} & \mathbf{k} \\ 2 & 1 & 3 \\ -2 & 1 & 3 \end{vmatrix} = -12\mathbf{j} + 4\mathbf{k}$

$0(x - 0) - 12(y - 0) + 4(z - 0) = 0$
$-12y + 4z = 0$
$-3y + z = 0$

25. $\mathbf{n} = \mathbf{j}$: $0(x - 2) + 1(y - 5) + 0(z - 3) = 0$
$y - 5 = 0$

27. $\mathbf{n} = \mathbf{j} - \mathbf{k}$:
Point: $(0, 0, 0)$
$0(x - 0) + 1(y - 0) - 1(z - 0) = 0$
$y - z = 0$

29. $\mathbf{n}_1 = \langle 3, 1, -4 \rangle, \mathbf{n}_2 = \langle -9, -3, 12 \rangle = -3\mathbf{n}_1$
Parallel

31. $\mathbf{n}_1 = \langle 2, 0, -1 \rangle, \mathbf{n}_2 = \langle 4, 1, 8 \rangle$
$\mathbf{n}_1 \cdot \mathbf{n}_2 = 8 - 8 = 0$ Orthogonal

33.

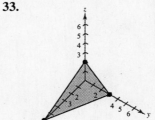

35.

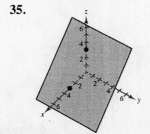

37.

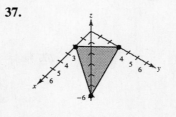

39.

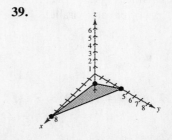

41. $D = \dfrac{|\overrightarrow{PQ} \cdot \mathbf{n}|}{\|\mathbf{n}\|}$

$P = (4, 0, 0)$ on plane, $Q = (0, 0, 0)$, $\mathbf{n} = \langle 3, 2, 1 \rangle$

$D = \dfrac{|\langle -4, 0, 0 \rangle \cdot \langle 3, 2, 1 \rangle|}{\sqrt{14}} = \dfrac{12}{\sqrt{14}}$

43. (a) Sphere: $(x - 4)^2 + (y + 1)^2 + (z - 1)^2 = 4$

   (b) Two planes parallel to given plane. Let $Q = (x, y, z)$ be a point on one of these planes, and pick $P = (0, 0, 10)$ on the given plane. By the distance formula,

$$\pm 2 = \dfrac{|\overrightarrow{PQ} \cdot \mathbf{n}|}{\|\mathbf{n}\|} = \dfrac{\langle x, y, z - 10 \rangle \cdot \langle 4, -3, 1 \rangle}{\sqrt{26}}$$

$$\pm 2\sqrt{26} = 4x - 3y + z - 10$$

$$4x - 3y + z = 10 \pm 2\sqrt{26}$$

❑ Review Exercises for Chapter 11

Solutions to Even-Numbered Exercises

1. $A(2, -3, -2)$, $B(-2, 4, 4)$ **3.** $(-5, 4, 0)$ **5.** $xy > 0$ in Octants I, II, VII or VIII.

7. $d_1 = \sqrt{(3 - 0)^2 + (-2 - 3)^2 + (0 - 2)^2} = \sqrt{9 + 25 + 4} = \sqrt{38}$

$d_2 = \sqrt{(0 - 0)^2 + (5 - 3)^2 + (-3 - 2)^2} = \sqrt{4 + 25} = \sqrt{29}$

$d_3 = \sqrt{(0 - 3)^2 + (5 - (-2))^2 + (-3 - 0)^2} = \sqrt{9 + 49 + 9} = \sqrt{67}$

$d_1^2 + d_2^2 = 38 + 29 = 67 = d_3^2$

9. Midpoint: $\left(\dfrac{8 + 5}{2}, \dfrac{-2 + 6}{2}, \dfrac{3 + 7}{2} \right) = \left(\dfrac{13}{2}, 2, 5 \right)$ **11.** $(x - 2)^2 + (y - 3)^2 + (z - 5)^2 = 1$

13. Radius: $\dfrac{15}{2}$

$(x - 3)^2 + (y + 2)^2 + (z - 6)^2 = \dfrac{225}{4}$

15. $(x^2 - 4x + 4) + (y^2 - 6y + 9) + z^2 = -4 + 4 + 9$

$(x - 2)^2 + (y - 3)^2 + z^2 = 9$

Center: $(2, 3, 0)$

Radius: 3

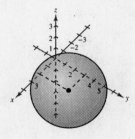

17. $\overrightarrow{PQ} = \langle 3 - 2, 3 - (-1), 0 - 4 \rangle = \langle 1, 4, -4 \rangle$ **19.** $\overrightarrow{PQ} = \langle -3 - 7, 2 - (-4), 10 - 3 \rangle = \langle -10, 6, 7 \rangle$

21. Since $-\dfrac{2}{3} \langle 39, -12, 21 \rangle = \langle -26, 8, -14 \rangle$, the vectors are parallel.   **23.** Since $\mathbf{u} \cdot \mathbf{v} = 0$, the angle is 90°.

25 $\mathbf{u} \times \mathbf{v} = \begin{vmatrix} \mathbf{i} & \mathbf{j} & \mathbf{k} \\ -2 & 8 & 2 \\ 1 & 1 & -1 \end{vmatrix} = \langle -10, 0, -10 \rangle$ **27.** $\|\mathbf{u}\| = \sqrt{2^2 + (-3)^2 + 3^2} = \sqrt{4 + 9 + 9} = \sqrt{22}$

29. $\mathbf{u} \cdot \mathbf{v} = \langle 2, -3, 3 \rangle \cdot \langle 1, 4, -2 \rangle$ **31.** $\mathbf{u} \cdot \mathbf{u} = \langle 2, -3, 3 \rangle \cdot \langle 2, -3, 3 \rangle = 4 + 9 + 9$

$= 2 - 12 - 6 = -16$ $= 22 = \|\mathbf{u}\|^2$

33. $\mathbf{u} \times \mathbf{v} = \begin{vmatrix} \mathbf{i} & \mathbf{j} & \mathbf{k} \\ 2 & -3 & 3 \\ 1 & 4 & -2 \end{vmatrix} = -6\mathbf{i} + 7\mathbf{j} + 11\mathbf{k}$

$\mathbf{v} \times \mathbf{u} = \begin{vmatrix} \mathbf{i} & \mathbf{j} & \mathbf{k} \\ 1 & 4 & -2 \\ 2 & -3 & 3 \end{vmatrix} = 6\mathbf{i} - 7\mathbf{j} - 11\mathbf{k}$

35. $\mathbf{u}(\mathbf{v} + \mathbf{w}) = \mathbf{u} \times \langle -1, 5, 0 \rangle = \begin{vmatrix} \mathbf{i} & \mathbf{j} & \mathbf{k} \\ 2 & -3 & 3 \\ 1 & 5 & 0 \end{vmatrix} = -15\mathbf{i} - 3\mathbf{j} + 7\mathbf{k}$

$(\mathbf{u} \times \mathbf{v}) + (\mathbf{u} \times \mathbf{w}) = (-6\mathbf{i} + 7\mathbf{j} + 11\mathbf{k}) + (-9\mathbf{i} - 10\mathbf{j} - 4\mathbf{k}) = -15\mathbf{i} - 3\mathbf{j} + 7\mathbf{k}$

37. The parallelepiped is determined by the vectors with initial points at $(3, 2, -1)$.

$\mathbf{u} = \langle 1, -3, 7 \rangle, \quad \mathbf{v} = \langle -1, 3, 2 \rangle, \quad \mathbf{w} = \langle -8, 2, 3 \rangle$

$\text{Volume} = |\mathbf{u} \cdot (\mathbf{v} \times \mathbf{w})| = \begin{vmatrix} 1 & -3 & 7 \\ -1 & 3 & 2 \\ -8 & 2 & 3 \end{vmatrix} = 1(5) + 3(13) + 7(22) = 198$

39. $\mathbf{v}_1 = \langle -1, 3, 5 \rangle, \quad \mathbf{v}_2 = \langle 7, 1, 2 \rangle$

$\mathbf{n} = \mathbf{v}_1 \times \mathbf{v}_2 = \begin{vmatrix} \mathbf{i} & \mathbf{j} & \mathbf{k} \\ -1 & 3 & 5 \\ 7 & 1 & 2 \end{vmatrix} = \mathbf{i} + 37\mathbf{j} - 22\mathbf{k}$

$\mathbf{u} = \frac{1}{\sqrt{1854}} \langle 1, 37, -22 \rangle$

41. $\mathbf{v} = \langle 5, 20, -3 \rangle$

$x = 5t, \quad y = -10 + 20t, \quad z = 3 - 3t$

43. $\mathbf{v} = \langle 1, 1, 1 \rangle$

$x = 3 + t, \quad y = 1 + t, \quad z = 2 + t$

45. $2x - y - 2z = 4$

$x - y - z = 8$

Subtracting, $x - z = -4$ or $x = z - 4$. Then let $z = t$ and hence, $x = t - 4$ and $y = x - z - 8 = -12$.

$x = t - 4, \quad y = -12, \quad z = t$

47. $\mathbf{u} = \langle 5, -5, -2 \rangle, \quad \mathbf{v} = \langle 3, 5, 2 \rangle$

$\mathbf{n} = \mathbf{u} \times \mathbf{v} = \begin{vmatrix} \mathbf{i} & \mathbf{j} & \mathbf{k} \\ 5 & -5 & -2 \\ 3 & 5 & 2 \end{vmatrix} = \langle 0, -16, 40 \rangle$

Plane: $0(x + 1) - 16(y - 3) + 40(z - 4) = 0$

$-2(y - 3) + 5(z - 4) = 0$

$-2y + 5z - 14 = 0$

49. $\mathbf{n} = \langle 1, 1, 1 \rangle$ normal vector

Plane: $1(x - 3) + 1(y - 1) + 1(z - 2) = 0$

$x + y + z - 6 = 0$

51. $D = \dfrac{|\vec{PQ} \cdot \mathbf{n}|}{\|\mathbf{n}\|}$

$\mathbf{n} = \langle 1, -10, 37 \rangle$

$Q = (0, 0, 0)$

$P = (2, 0, 0)$ in plane

$D = \dfrac{|\langle -2, 0, 0 \rangle \cdot \langle 1, -10, 3 \rangle|}{\sqrt{1 + 100 + 9}}$

$= \dfrac{2}{\sqrt{110}} \approx 0.191$

53. Let $Q = (0, 0, 1)$ be a point in first plane.

$P = \left(0, 0, -\frac{1}{2} \right)$ in plane

$\mathbf{n} = \langle 6, -4, -2 \rangle$

$D = \dfrac{\vec{PQ} \cdot \mathbf{n}}{\|\mathbf{n}\|}$

$= \dfrac{\left| \langle 0, 0, \frac{3}{2} \rangle \cdot \langle 6, -4, -2 \rangle \right|}{\sqrt{36 + 16 + 4}}$

$= \dfrac{3}{\sqrt{56}} = \dfrac{3}{2\sqrt{14}}$

❑ Chapter Test Solutions for Chapter 11

1. $x = 3, y = -9, z = 0$: $(3, -9, 0)$

2. Octants I or III

3. Midpoint $= \left(\dfrac{8 + 6}{2}, \dfrac{-2 + 4}{2}, \dfrac{5 - 1}{2} \right) = (7, 1, 2)$

4. Diameter $= \sqrt{(8 - 6)^2 + (-2 - 4)^2 + (5 + 1)^2} = \sqrt{4 + 36 + 36} = \sqrt{76}$

Radius $= \sqrt{19}$

$(x - 7)^2 + (y - 1)^2 + (z - 2)^2 = 19$

5. $AB = \sqrt{76}$ (Problem 4 above)

$AC = \sqrt{(8 + 4)^2 + (-2 - 3)^2 + (5 - 0)^2} = \sqrt{144 + 25 + 25} = \sqrt{194}$

$BC = \sqrt{(6 + 4)^2 + (4 - 3)^2 + (-1 - 0)^2} = \sqrt{100 + 1 + 1} = \sqrt{102}$

No. $\left(\sqrt{76} \right)^2 + \left(\sqrt{102} \right)^2 \neq \left(\sqrt{194} \right)^2$

6. $\mathbf{u} = \langle -2, 6, -6 \rangle$, $\mathbf{v} = \langle -12, 5, -5 \rangle$

(a) $\mathbf{u} \cdot \mathbf{v} = 24 + 30 + 30 = 84$

(c) $\cos \theta = \dfrac{\mathbf{u} \cdot \mathbf{v}}{\|\mathbf{u}\| \, \|\mathbf{v}\|} = \dfrac{84}{\sqrt{76}\sqrt{194}} \approx 0.6918$

$\Rightarrow \theta \approx 46.23°$ or 0.8068 radians

(e) $x = 8 - 2t, y = -2 + 6t, z = 5 - 6t$

$\dfrac{x - 8}{-2} = \dfrac{y + 2}{6} = \dfrac{z - 5}{-6}$

(b) $\mathbf{u} \times \mathbf{v} = \begin{vmatrix} \mathbf{i} & \mathbf{j} & \mathbf{k} \\ -2 & 6 & -6 \\ -12 & 5 & -5 \end{vmatrix} = \langle 0, 62, 62 \rangle$

(d) Area $= \dfrac{1}{2}\|\mathbf{u} \times \mathbf{v}\| = \dfrac{1}{2}\sqrt{62^2 + 62^2} = 31\sqrt{2}$

(f) Normal vector: $\mathbf{n} = \mathbf{u} \times \mathbf{v} = \langle 0, 62, 62 \rangle$ or $\langle 0, 1, 1 \rangle$.

$0(x - 8) + 1(y + 2) + 1(z - 5) = 0$

$y + z - 3 = 0$

7. $\mathbf{v} = \langle 4, 10, 5 \rangle$

$\dfrac{\mathbf{v} \cdot \mathbf{i}}{\|\mathbf{v}\| \, \|\mathbf{i}\|} = \dfrac{4}{\sqrt{141}} \Rightarrow \theta \approx 70.3°$

$\dfrac{\mathbf{v} \cdot \mathbf{j}}{\|\mathbf{v}\| \, \|\mathbf{j}\|} = \dfrac{10}{\sqrt{141}} \Rightarrow \theta \approx 32.6°$

$\dfrac{\mathbf{v} \cdot \mathbf{k}}{\|\mathbf{v}\| \, \|\mathbf{k}\|} = \dfrac{5}{\sqrt{141}} \Rightarrow \theta \approx 65.1°$

8. $\mathbf{n} = \langle 3, -2, 1 \rangle$, $Q = (4, 3, 8)$, $P = (0, 0, 6)$ in plane $\overrightarrow{PQ} = \langle 4, 3, 2 \rangle$.

$D = \dfrac{|\overrightarrow{PQ} \cdot \mathbf{n}|}{\|\mathbf{n}\|} = \dfrac{8}{\sqrt{14}} = \dfrac{4\sqrt{14}}{7}$

9.

❏ Practice Test for Chapter 11

1. Find the lengths of the sides of the triangle with vertices $(0, 0, 0)$, $(1, 2, -4)$, and $(0, -2, -1)$. Show that the triangle is a right triangle.

2. Find the standard form of the equation of a sphere having center $(0, 4, 1)$ and radius 5.

3. Find the center and radius of the sphere $x^2 + y^2 + z^2 + 2x - 4z - 11 = 0$.

4. Find the vector $\mathbf{u} - 3\mathbf{v}$ given $\mathbf{u} = \langle 1, 0, -1 \rangle$ and $\mathbf{v} = \langle 4, 3, -6 \rangle$.

5. Find the length of $\frac{1}{2}\mathbf{v}$ if $\mathbf{v} = \langle 2, 4, -6 \rangle$.

6. Find the dot product of $\mathbf{u} = \langle 2, 1, -3 \rangle$ and $\mathbf{v} = \langle 1, 1, -2 \rangle$.

7. Determine whether $\mathbf{u} = \langle 1, 1, -1 \rangle$ and $\mathbf{v} = \langle -3, -3, 3 \rangle$ are orthogonal, parallel, or neither.

8. Find the cross product of $\mathbf{u} = \langle -1, 0, 2 \rangle$ and $\mathbf{v} = \langle 1, -1, 3 \rangle$. What is $\mathbf{v} \times \mathbf{u}$?

9. Use the triple scalar product to find the volume of the parallelepiped having adjacent edges $\mathbf{u} = \langle 1, 1, 1 \rangle$, $\mathbf{v} = \langle 0, -1, 1 \rangle$, and $\mathbf{w} = \langle 1, 0, 4 \rangle$.

10. Find a set of parametric equations for the line through the points $(0, -3, 3)$ and $(2, -3, 4)$.

11. Find an equation of the plane passing through $(1, 2, 3)$ and perpendicular to the vector $\mathbf{n} = \langle 1, -1, 0 \rangle$.

12. Find an equation of the plane passing through the three points $A = (0, 0, 0)$, $B = (1, 1, 1)$, and $C = (1, 2, 3)$.

13. Determine whether the planes $x + y - z = 12$ and $3x - 4y - z = 9$ are parallel, orthogonal or neither.

14. Find the distance between the point $(1, 1, 1)$ and the plane $x + 2y + z = 6$.

CHAPTER 12
Limits and an Introduction to Calculus

CHAPTER 12
Limits and an Introduction to Calculus

Section 12.1 Introduction to Limits

■ If $f(x)$ becomes arbitrarily close to a unique number L as x approaches c from either side, then the limit of $f(x)$ as x approaches c is L:

$$\lim_{x \to c} f(x) = L.$$

■ You should be able to use a calculator to find a limit.

■ You should be able to use a graph to find a limit.

■ You should understand how limits can fail to exist:

(a) $f(x)$ approaches a different number from the right of c than it approaches from the left of c.

(b) $f(x)$ increases or decreases without bound as x approaches c.

(c) $f(x)$ oscillates between two fixed values as x approaches c.

■ You should know and be able to use the elementary properties of limits.

Solutions to Odd-Numbered Exercises

1. (a)

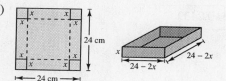

(b) $V = (\text{base})\text{height} = (24 - 2x)^2 x = 4x(12 - x)^2$

(c) $\lim_{x \to 4} V = 1024$

x	3	3.5	3.9	4	4.1	4.5	5
V	972.0	1011.5	1023.5	1024.0	1023.5	1012.5	980.0

(d)

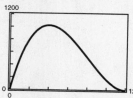

3. $\lim_{x \to 3}(3 - 2x) = -3$

x	2.9	2.99	2.999	3.0	3.001	3.01	3.1
$f(x)$	-2.8	-2.98	-2.998	-3	-3.002	-3.02	-3.2

The limit is reached.

5. $\lim_{x \to 2} \dfrac{x - 2}{x^2 - 4} = \dfrac{1}{4}$

x	1.9	1.99	1.999	2.0	2.001	2.01	2.1
$f(x)$	0.2564	0.2506	0.2501	?	0.2499	0.2494	0.2439

The limit is not reached.

7. (a)

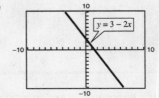 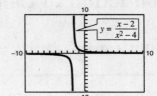

No, in each case x is approaching an integer and the utility may not be evaluating the function at the integer values. It depends on the viewing rectangle.

(b)

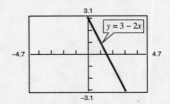 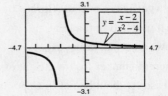

Yes, in each case x is approaching an integer and the utility is evaluating the function at integer (decimal) values.

9. $\lim\limits_{x \to 1} \dfrac{x-1}{x^2 + 2x - 3} = \dfrac{1}{4}$

x	0.9	0.99	0.999	1.0	1.001	1.01	1.1
$f(x)$	0.2564	0.2506	0.2501	?	0.2499	0.2464	0.2439

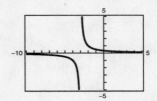

11. $\lim\limits_{x \to 0} \dfrac{\sqrt{x+3} - \sqrt{3}}{x} \approx 0.2887$ $\left(\text{Actual limit is } \dfrac{1}{2\sqrt{3}}. \right)$

x	0.9	0.99	0.999	1.0	1.001	1.01	1.1
$f(x)$	0.2911	0.2889	0.2887	?	0.2887	0.2884	0.2863

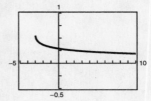

13. $\lim\limits_{x \to -4} \dfrac{[x/(x+2)] - 2}{x + 4} = \dfrac{1}{2}$

x	-4.1	-4.01	-4.001	-4.0	-3.999	-3.99	-3.9
$f(x)$	0.4762	0.4975	0.4998	?	0.5003	0.5025	0.5263

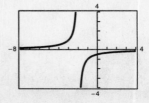

15. Make sure your calculator is set in radian mode.

$$\lim_{x \to 0} \frac{\sin x}{x} = 1$$

x	-0.1	-0.01	-0.001	0	0.001	0.01	0.1
$f(x)$	0.9983	0.99998	0.9999998	?	0.9999998	0.99998	0.9983

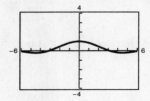

17. $\lim\limits_{x \to 2} (3 - x) = 1$

19. $\lim\limits_{x \to -1} \sin \dfrac{\pi x}{2} = -1$

21. The limit does not exist because $f(x)$ approaches different values from the left of $x = -2$ and the right of $x = -2$.

23. The limit does not exist because $f(x)$ oscillates between 2 and -2.

25.

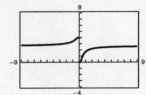

The limit does not exist.

27.

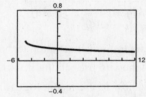

$\lim\limits_{x \to 4} f(x) = \frac{1}{6}$

29.

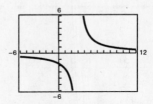

The limit does not exist.

31. No. The limit may or may not exist. And if it exists, it may not be equal to 4.

33. $\lim\limits_{x \to 2} (x^2 + 3x - 4) = 2^2 + 3(2) - 4 = 6$

35. $\lim\limits_{x \to 3} \dfrac{12}{x} = \dfrac{12}{3} = 4$

37. $\lim\limits_{x \to -1} \dfrac{x^2 - 1}{x} = \dfrac{(-1)^2 - 1}{-1} = 0$

39. $\lim\limits_{x \to 2} e^x = e^2 \approx 7.389$

41. $\lim\limits_{x \to \pi} \sin 2x = \sin 2\pi = 0$

43. $\lim\limits_{x \to 1/2} \arcsin x = \arcsin \dfrac{1}{2} = \dfrac{\pi}{6} \approx 0.5236$

45. The limit does not exist. As x approaches 2 from the left, $f(x)$ approaches 5. As x approaches 2 from the right, $f(x)$ approaches 6.

47. (a) $\lim\limits_{x \to c} [-2g(x)] = -2(5) = -10$

(b) $\lim\limits_{x \to c} [f(x) + g(x)] = 4 + 5 = 9$

(c) $\lim\limits_{x \to c} \dfrac{f(x)}{g(x)} = \dfrac{4}{5}$

(d) $\lim\limits_{x \to c} \sqrt{f(x)} = \sqrt{4} = 2$

49.

$$\lim_{x \to 9} f(x) = 6$$

Domain: $x \ge 0,\ x \ne 9$ or $[0, 9),\ (9, \infty)$

It is difficult to determine the domain solely by the graph because it is not obvious that the function is undefined at $x = 9$.

Section 12.2   Techniques for Evaluating Limits

- You can use direct substitution to find the limit of a polynomial function $p(x)$:
$$\lim_{x \to c} p(x) = p(c).$$

- You can use direct substitution to find the limt of a rational function $r(x) = \dfrac{p(x)}{q(x)}$, as long as $q(c) \ne 0$:
$$\lim_{x \to c} r(x) = r(c) = \frac{p(c)}{q(c)},\ q(c) \ne 0.$$

- You should be able to use cancellation techniques to find a limit.
- You should know how to use rationalization techniques to find a limit.
- You should know how to use technology to find a limit.
- You should be able to calculate one-sided limits.

Solutions to Odd-Numbered Exercises

1. $\lim\limits_{x \to 5} (12 - x^2) = 12 - 5^2 = 12 - 25 = -13$

3. $\lim\limits_{x \to -3} \dfrac{2x}{x^2 + 1} = \dfrac{2(-3)}{(-3)^2 + 1} = \dfrac{-6}{10} = \dfrac{-3}{5}$

5. $\lim\limits_{x \to -3} \ln e^x = \ln e^{-3} = -3$

7. $\lim\limits_{\theta \to 1} \sin \dfrac{2\pi\theta}{3} = \sin \dfrac{\pi(1)}{3} = \dfrac{\sqrt{3}}{2} \approx 0.8660$

9. $g(x) = \dfrac{-2x^2 + x}{x}$

$g_2(x) = -2x + 1$

(a) $\lim\limits_{x \to 0} g(x) = 1$

(b) $\lim\limits_{x \to -1} g(x) = 3$

(c) $\lim\limits_{x \to -2} g(x) = 5$

11. $g(x) = \dfrac{x^3 - x}{x - 1}$

$g_2(x) = x^2 + x = x(x + 1)$

(a) $\lim\limits_{x \to 1} g(x) = 2$

(b) $\lim\limits_{x \to -1} g(x) = 0$

(c) $\lim\limits_{x \to 0} g(x) = 0$

13. $\lim\limits_{x \to 7} \dfrac{x - 7}{x^2 - 49} = \lim\limits_{x \to 7} \dfrac{x - 7}{(x - 7)(x + 7)} = \lim\limits_{x \to 7} \dfrac{1}{x + 7} = \dfrac{1}{14}$

15. $\lim\limits_{x \to -1} \dfrac{1 - 2x - 3x^2}{1 + x} = \lim\limits_{x \to -1} \dfrac{(1 + x)(1 - 3x)}{1 + x} = \lim\limits_{x \to -1} (1 - 3x) = 4$

17. $\displaystyle\lim_{y\to 0} \frac{\sqrt{3+y}-\sqrt{3}}{y} = \lim_{y\to 0} \frac{\sqrt{3+y}-\sqrt{3}}{y} \cdot \frac{\sqrt{3+y}+\sqrt{3}}{\sqrt{3+y}+\sqrt{3}}$

$\displaystyle\qquad = \lim_{y\to 0} \frac{(3+y)-3}{y(\sqrt{3+y}+\sqrt{3})}$

$\displaystyle\qquad = \lim_{y\to 0} \frac{1}{\sqrt{3+y}+\sqrt{3}} = \frac{1}{2\sqrt{3}}$

19. $\displaystyle\lim_{x\to -3} \frac{\sqrt{x+7}-2}{x+3} = \lim_{x\to -3} \frac{(\sqrt{x+7}-2)}{x+3} \cdot \frac{\sqrt{x+7}+2}{\sqrt{x+7}+2}$

$\displaystyle\qquad = \lim_{x\to -3} \frac{(x+7)-4}{(x+3)(\sqrt{x+7}+2)}$

$\displaystyle\qquad = \lim_{x\to -3} \frac{1}{\sqrt{x+7}+2} = \frac{1}{4}$

21. $\displaystyle\lim_{x\to 0} \frac{\dfrac{1}{1+x}-1}{x} = \lim_{x\to 0} \frac{1-(1+x)}{(1+x)x} = \lim_{x\to 0} \frac{-1}{1+x} = -1$

23. $\displaystyle\lim_{x\to 0} \frac{\sec x}{\tan x} = \lim_{x\to 0} \frac{1}{\cos x} \cdot \frac{\cos x}{\sin x} = \lim_{x\to 0} \frac{1}{\sin x}$, does not exist.

25.

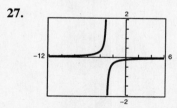

x	-0.1	-0.01	-0.001	0	0.001	0.01	0.1
$f(x)$	-0.4881	-0.4988	-0.5000	?	-0.5001	-0.5013	-0.5132

$\displaystyle\lim_{x\to 0} \frac{\sqrt{1-x}-1}{x} = \lim_{x\to 0} \frac{\sqrt{1-x}-1}{x} \cdot \frac{\sqrt{1-x}+1}{\sqrt{1-x}+1}$

$\displaystyle\qquad = \lim_{x\to 0} \frac{(1-x)-1}{x(\sqrt{1-x}+1)} = \lim_{x\to 0} \frac{-1}{\sqrt{1-x}+1} = \frac{-1}{2}$

27.

x	-0.1	-0.01	-0.001	0	0.001	0.01	0.1
$f(x)$	-0.1149	-0.1115	-0.1111	?	-0.1111	-0.1107	-0.1075

$\displaystyle\lim_{x\to 0} \frac{\dfrac{1}{3+x}-\dfrac{1}{3}}{x} = \lim_{x\to 0} \frac{3-(3+x)}{3(3+x)x} = \lim_{x\to 0} \frac{-x}{3(3+x)x}$

$\displaystyle\qquad = \lim_{x\to 0} \frac{-1}{3(3+x)} = -\frac{1}{9} \approx -0.1111$

29.

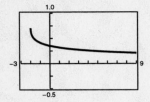

$$\lim_{x \to 0} \frac{\sqrt{x+2} - \sqrt{2}}{x} \approx -0.3536 \quad \left(\text{exact is } \frac{1}{2\sqrt{2}}\right)$$

31.

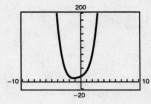

$$\lim_{x \to 2} \frac{x^5 - 32}{x - 2} = 80$$

33.

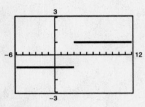

$$\lim_{x \to 3^+} \frac{|x - 3|}{x - 3} = 1$$

$$\lim_{x \to 3^-} \frac{|x - 3|}{x - 3} = -1$$

Limit does not exist.

35.

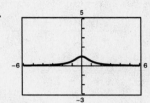

$$\lim_{x \to 1-} \frac{1}{x^2 + 1} = \lim_{x \to 1+} \frac{1}{x^2 + 1} = \lim_{x \to 1} \frac{1}{x^2 + 1} = \frac{1}{2}$$

37. $\lim_{x \to 2^-} f(x) = 2 - 1 = 1$

$\lim_{x \to 2^+} f(x) = 2(2) - 3 = 1$

$\lim_{x \to 2} f(x) = 1$

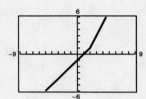

39. $\lim_{h \to 0} \dfrac{f(x+h) - f(x)}{h} = \lim_{h \to 0} \dfrac{3(x+h) - 1 - (3x - 1)}{h}$

$$= \lim_{h \to 0} \frac{3x + 3h - 1 - 3x + 1}{h}$$

$$= \lim_{h \to 0} \frac{3h}{h} = 3$$

41. $\lim_{h \to 0} \dfrac{f(x+h) - f(x)}{h} = \lim_{h \to 0} \dfrac{((x+h)^2 - 3(x+h)) - (x^2 - 3x)}{h}$

$$= \lim_{h \to 0} \frac{x^2 + 2xh + h^2 - 3x - 3h - x^2 + 3x}{h}$$

$$= \lim_{h \to 0} \frac{2xh + h^2 - 3h}{h} = \lim_{h \to 0} (2x + h - 3) = 2x - 3$$

43. $\lim_{x \to 0^+} x \ln x = 0$

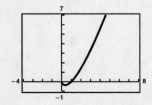

45. $\lim_{x \to 0} \dfrac{\sin 2x}{x} = 2$

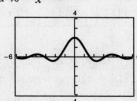

47. $\lim\limits_{x \to 0} \dfrac{\tan x}{x} = 1$

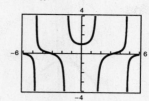

49. $\lim\limits_{x \to 1} \dfrac{1 - \sqrt[3]{x}}{1 - x} = \dfrac{1}{3} \approx 0.333$

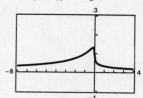

51.

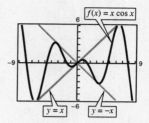

$\lim\limits_{x \to 0} f(x) = 0$

53.

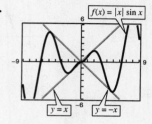

$\lim\limits_{x \to 0} f(x) = 0$

55.

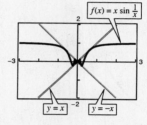

$\lim\limits_{x \to 0} f(x) = 0$

57. (a) Can be evaluated by direct substitution:

$$\lim\limits_{x \to 0} x^2 \sin x^2 = 0^2 \sin 0^2 = 0$$

(b) Cannot be evaluated by direct substitution:

$$\lim\limits_{x \to 0} \dfrac{\sin x^2}{x^2} = 1$$

59. $\lim\limits_{t \to 1} \dfrac{(-16(1) + 128) - (-16t^2 + 128)}{1 - t} = \lim\limits_{t \to 1} \dfrac{16t^2 - 16}{1 - t} = \lim\limits_{t \to 1} \dfrac{16(t - 1)(t + 1)}{1 - t}$

$$= \lim\limits_{t \to 1} -16(t + 1) = -32 \dfrac{\text{ft}}{\text{sec}}$$

Section 12.3   The Tangent Line Problem

■ You should be able to visually approximate the slope of a graph.

■ The slope m of the graph of f at the point $(x, f(x))$ is given by

$$m = \lim\limits_{h \to 0} \dfrac{f(x + h) - f(x)}{h}$$

provided this limit exists.

■ You should be able to use the limit definition to find the slope of a graph.

■ The derivative of f at x is given by

$$f'(x) = \lim\limits_{h \to 0} \dfrac{f(x + h) - f(x)}{h}$$

provided this limit exists. Notice that this is the same limit as that for the tangent line slope.

■ You should be able to use the limit definition to find the derivative of a function.

Solutions to Odd-Numbered Exercises

1. Slope is 0 at (x, y).

3. Slope is $\frac{1}{2}$ at (x, y).

5. (a)

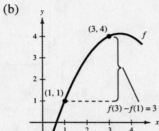

(b)

(c)

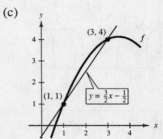

7.

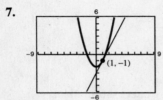

slope ≈ 2

9.

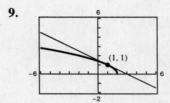

slope $\approx -\frac{1}{2}$

11. (a) $y = 0.301t^3 - 5.380t^2 - 14.495t + 1767.199$

(b)

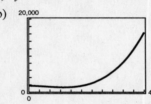

(c)

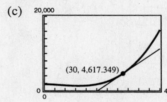

The slope at $t = 30$ is about 475. This means that the per capita debt is increasing at the rate of \$475 per year in 1980.

13. $m_{sec} = \dfrac{g(1 + h) - g(1)}{h} = \dfrac{4 - 3(1 + h) - 1}{h} = \dfrac{-3h}{h}$

$m = \lim\limits_{h \to 0} \dfrac{-3h}{h} = -3$

15. $m_{sec} = \dfrac{f(3 + h) - f(3)}{h} = \dfrac{(3 + h)^2 - 2(3 + h) - (9 - 6)}{h} = \dfrac{9 + 6h + h^2 - 6 - 2h - 3}{h}$

$= \dfrac{h^2 + 4h}{h} = h + 4, \ h \neq 0$

$m = \lim\limits_{h \to 0} (h + 4) = 4$

17. $m_{sec} = \dfrac{g(2 + h) - g(2)}{h} = \dfrac{\dfrac{4}{2 + h} - 2}{h} = \dfrac{4 - 2(2 + h)}{(2 + h)h} = \dfrac{-2}{2 + h}, \ h \neq 0$

$m = \lim\limits_{h \to 0} \left(\dfrac{-2}{2 + h} \right) = -1$

19. $m_{\sec} = \dfrac{h(9 + h) - h(9)}{h} = \dfrac{\sqrt{9 + h} - 3}{h} \cdot \dfrac{\sqrt{9 + h} + 3}{\sqrt{9 + h} + 3}$

$\qquad = \dfrac{(9 + h) - 9}{h[\sqrt{9 + h} + 3]} = \dfrac{1}{\sqrt{9 + h} + 3}, \ h \neq 0$

$\qquad m = \lim\limits_{h \to 0} \dfrac{1}{\sqrt{9 + h} + 3} = \dfrac{1}{6}$

21. $m_{\sec} = \dfrac{g(x + h) - g(x)}{h} = \dfrac{4 - (x + h)^2 - (4 - x^2)}{h}$

$\qquad = \dfrac{-2xh - h^2}{h} = -2x - h, \ h \neq 0$

$\qquad m = \lim\limits_{h \to 0} (-2x - h) = -2x$

At $(0, 4)$, $m = -2(0) = 0$.

At $(-1, 3)$, $m = -2(-1) = 2$.

23. $m_{\sec} = \dfrac{g(x + h) - g(x)}{h} = \dfrac{\dfrac{1}{x + h + 4} - \dfrac{1}{x + 4}}{h} = \dfrac{(x + 4) - (x + 4 + h)}{(x + h + 4)(x + 4)(h)}$

$\qquad = \dfrac{-h}{(x + h + 4)(x + 4)h} = \dfrac{-1}{(x + h + 4)(x + 4)}, \ h \neq 0$

$\qquad m = \lim\limits_{h \to 0} \dfrac{-1}{(x + h + 4)(x + 4)} = \dfrac{-1}{(x + 4)^2}$

At $\left(0, \frac{1}{4}\right)$, $m = \dfrac{-1}{(0 + 4)^2} = \dfrac{-1}{16}$.

At $\left(-2, \frac{1}{2}\right)$, $m = \dfrac{-1}{(-2 + 4)^2} = \dfrac{-1}{4}$.

25. $f'(x) = \lim\limits_{h \to 0} \dfrac{f(x + h) - f(x)}{h} = \lim\limits_{h \to 0} \dfrac{5 - 5}{h} = 0$

27. $g'(x) = \lim\limits_{h \to 0} \dfrac{g(x + h) - g(x)}{h} = \lim\limits_{h \to 0} \dfrac{\left(6 - \frac{2}{3}(x + h)\right) - \left(6 - \frac{2}{3}x\right)}{h}$

$\qquad = \lim\limits_{h \to 0} \dfrac{-\frac{2}{3}h}{h} = \dfrac{-2}{3}$

29. $f'(x) = \lim\limits_{h \to 0} \dfrac{f(x + h) - f(x)}{h} = \lim\limits_{h \to 0} \dfrac{\dfrac{1}{(x + h)^2} - \dfrac{1}{x^2}}{h} = \lim\limits_{h \to 0} \dfrac{x^2 - (x^2 + 2xh + h^2)}{(x + h)^2 x^2 h} = \lim\limits_{h \to 0} -\dfrac{2x - h}{(x + h)^2 x^2}$

$\qquad = -\dfrac{2x}{x^4} = -\dfrac{2}{x^3}$

31. $m_{\sec} = \dfrac{f(2 + h) - f(2)}{h} = \dfrac{(2 + h)^2 - 1 - 3}{h} = \dfrac{4h + h^2}{h} = 4 + h, \ h \neq 0$

$\qquad m = \lim\limits_{h \to 0} (4 + h) = 4$

Tangent line: $y - 3 = 4(x - 2)$

$\qquad\qquad\qquad y = 4x - 5$

33. $m_{\text{sec}} = \dfrac{f(3 + h) - f(3)}{h} = \dfrac{\sqrt{3 + h + 1} - 2}{h} \cdot \dfrac{\sqrt{4 + h} + 2}{\sqrt{4 + h} + 2} = \dfrac{(4 + h) - 4}{h[\sqrt{4 + h} + 2]} = \dfrac{1}{\sqrt{4 + h} + 2}$

$m = \lim\limits_{h \to 0} \dfrac{1}{\sqrt{4 + h} + 2} = \dfrac{1}{4}$

Tangent line: $y - 2 = \dfrac{1}{4}(x_3)$

$4y = x + 5$

35.

x	-2	-1.5	-1	-0.5	0	0.5	1	1.5	2
$f(x)$	2	1.125	0.5	0.125	0	0.125	0.5	0.125	2
$f'(x)$	-2	-1.5	-1	-0.5	0	0.5	1	1.5	2

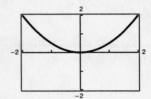

37.

x	-2	-1.5	-1	-0.5	0	0.5	1	1.5	2
$f(x)$	-2	$-.844$	$-.25$	$-.031$	0	$.031$	0.25	$.844$	2
$f'(x)$	3	1.688	$.75$	$.188$	0	$.188$	$.75$	1.688	3

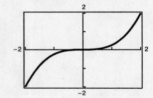

39. $f'(x) = \lim\limits_{h \to 0} \dfrac{f(x + h) - f(x)}{h} = \lim\limits_{h \to 0} \dfrac{[(x + h)^2 - 4(x + h) + 3] - [x^2 - 4x + 3]}{h}$

$= \lim\limits_{h \to 0} \dfrac{(x^2 + 2xh + h^2 - 4x - 4h + 3) - (x^2 - 4x + 3)}{h}$

$= \lim\limits_{h \to 0} \dfrac{2xh + h^2 - 4h}{h} = \lim\limits_{h \to 0} 2x + h - 4 = 2x - 4$

$f'(x) = 0 = 2x - 4 \implies x = 2$

f has a horizontal tangent at $(2, -1)$.

41. $f'(x) = \lim\limits_{h \to 0} \dfrac{f(x + h) - f(x)}{h} = \lim\limits_{h \to 0} \dfrac{3(x + h)^3 - 9(x + h) - (3x^3 - 9x)}{h}$

$= \lim\limits_{h \to 0} \dfrac{9x^2h + 9xh^2 + 3h^3 - 9h}{h} = 9x^2 - 9$

$f'(x) = 0 = 9x^2 - 9 \implies x = \pm 1$

f has horizontal tangents at $(1, -6)$ and $(-1, 6)$.

43. Matches (b).

45. Matches (d).

(derivative is -1 for $x < 0$, 1 for $x > 0$.)

Section 12.4   Limits at Infinity and Limits of Sequences

- ■ The limit at infinity
$$\lim_{x \to \infty} f(x) = L$$
  means that $f(x)$ get arbitrarily close to L as x increases without bound.

- ■ Similarly, the limit at infinity
$$\lim_{x \to -\infty} f(x) = L$$
  means that $F(x)$ get arbitrarily close to L as x decreases without bound.

- ■ You should be able to calculate limits at infinity, especially those arising from rational functions.

- ■ Limits of functions can be used to evaluate limits of sequences. If f is a function such that $\lim_{x \to \infty} f(x) = L$ and if a_n is a sequence such that $f(n) = a_n$, then $\lim_{n \to \infty} a_n = L$.

Solutions to Odd-Numbered Exercises

1. Horizontal asymptote $y = 0$.
Not defined at $x = 0$.
Matches (b).

3. Horizontal asymptote $y = 4$.
Matches (e).

5. Horizontal asymptote $y = 4$.
Not defined at $x = 0$.
Matches (f).

7. $\displaystyle\lim_{x \to \infty} \frac{2}{x^2} = 0$

9. $\displaystyle\lim_{x \to \infty} \frac{2 + x}{2 - x} = -1$

11. $\displaystyle\lim_{x \to -\infty} \frac{4x - 3}{2x + 1} = 2$

13. $\displaystyle\lim_{t \to \infty} \frac{t^2}{t + 2}$ does not exist

15. $\displaystyle\lim_{x \to -\infty} \left(-2 + \frac{2}{x} \right) = -2 + 0 = -2$

17. $\displaystyle\lim_{x \to -\infty} \frac{x}{(x + 1)^2} = \lim_{x \to -\infty} \frac{x}{x^2 + 2x + 1} = 0$

19. $\displaystyle\lim_{t \to \infty} \left(\frac{1}{3t^2} - \frac{5t}{t + 2} \right) = 0 - 5 = -5$

21.
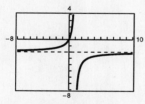
Horizontal asymptote: $y = -2$

23.
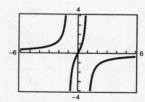
Horizontal asymptote: $y = 0$

25.
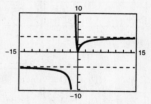
Horizontal asymptotes:
$y = 4, \ y = -4$

27.

x	10^0	10^1	10^2	10^3	10^4	10^5	10^6
$f(x)$	-0.7321	-0.0995	-0.00999	-0.001	-1×10^{-4}	-1×10^{-5}	-1×10^{-6}

$\displaystyle\lim_{x \to \infty} \left(x - \sqrt{x^2 + 2} \right) = 0$

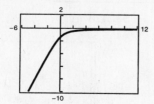

29.

x	10^0	10^1	10^2	10^3	10^4	10^5	10^6
$f(x)$	$-.7082$	$-.7454$	$-.7495$	$-.74995$	$-.749995$	$-.75$	$-.75$

$$\lim_{x \to \infty} 3\left(2x - \sqrt{4x^2 + x}\right) = -\frac{3}{4}$$

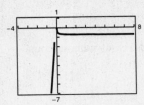

31. (a) Average cost $= \dfrac{c}{x} = \dfrac{1.35x + 4570}{x}$

          when $x = 100$, average cost $= \$47.50$
          when $x = 1000$, average cost $= \$5.92$

(b) $\displaystyle\lim_{x \to \infty} \frac{c}{x} = \lim_{x \to \infty} \frac{1.35x + 4570}{x} = \$1.35$

33. $3, \dfrac{4}{3}, 1, \dfrac{6}{7}, \dfrac{7}{9} \cdot \displaystyle\lim_{n \to \infty} \frac{n + 2}{2n - 1} = \frac{1}{2}$

35. $1, \dfrac{3}{5}, \dfrac{2}{5}, \dfrac{5}{17}, \dfrac{3}{13} \cdot \displaystyle\lim_{n \to \infty} \frac{n + 2}{n^2 + 1} = 0$

37. $\dfrac{1}{7}, \dfrac{1}{3}, \dfrac{9}{17}, \dfrac{8}{11}, \dfrac{25}{27} \cdot \displaystyle\lim_{n \to \infty} \frac{n^2}{5n + 2}$ does not exist

39. $2, 3, 4, 5, 6 \cdot \displaystyle\lim_{n \to \infty} \frac{(n + 1)!}{n!} = \lim_{n \to \infty} (n + 1)$ does not exist

41. $10, 12, 26, 64, 138 \cdot \displaystyle\lim_{n \to \infty} 2[5 + (n - 1)^3]$ does not exist

43. $\dfrac{1}{3}, -\dfrac{2}{5}, \dfrac{3}{7}, -\dfrac{4}{9}, \dfrac{5}{11} \cdot \displaystyle\lim_{n \to \infty} (-1)^{n-1} \frac{n}{2n + 1}$ does not exist

45. $\displaystyle\lim_{n \to \infty} a_n = \frac{3}{2}$

n	10^0	10^1	10^2	10^3	10^4	10^5	10^6
a_n	2	1.55	1.505	1.5005	1.50005	1.500005	1.5000005

47. $\displaystyle\lim_{n \to \infty} a_n = 4$

n	10^0	10^1	10^2	10^3	10^4	10^5	10^6
a_n	2	3.8	3.98	3.998	3.9998	3.99998	3.999998

49. $\displaystyle\lim_{n \to \infty} a_n = \frac{16}{3}$

n	10^0	10^1	10^2	10^3	10^4	10^5	10^6
a_n	16	6.16	5.4136	5.341336	5.3341	5.33341	5.333341

51. Converges to 0

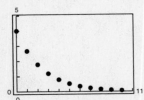

53. Diverges

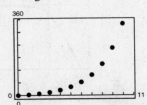

Section 12.5   The Area Problem

- You should know the following summation formulas and properties.

  (a) $\sum_{i=1}^{n} c = cn$

  (b) $\sum_{i=1}^{n} i = \frac{n(n+1)}{2}$

  (c) $\sum_{i=1}^{n} i^2 = \frac{n(n+1)(2n+1)}{6}$

  (d) $\sum_{i=1}^{n} i^3 = \frac{n^2(n+1)^2}{4}$

  (e) $\sum_{i=1}^{n} (a_i \pm b_i) = \sum_{i=1}^{n} a_i \pm \sum_{i=1}^{n} b_i$

  (f) $\sum_{i=1}^{n} ka_i = k\sum_{i=1}^{n} a_i$

- You should be able to evaluate a limit of a summation, $\lim_{n \to \infty} S(n)$.

- You should be able to approximate the area of a region using rectangles. By increasing the number of rectangles, the approximation improves.

- The area of a plane region bounded by f between $x = a$ and $x = b$ is the limit of the sum of the approximating rectangles:

$$A = \lim_{n \to \infty} \sum_{i=1}^{n} f\left(a + \frac{(b-a)i}{n}\right)\left(\frac{b-a}{n}\right)$$

- You should be able to use the limit definition of area to find the area bounded by simple functions in the plane.

Solutions to Odd-Numbered Exercises

1. $\sum_{i=1}^{60} i = \frac{n(n+1)}{2} = \frac{60(61)}{2} = 1830$

2. $\sum_{k=1}^{20} k^3 = \frac{n^2(n+1)^2}{4} = \frac{20^2(21)^2}{4} = 44,100$

5. $\sum_{j=1}^{25} (j^2 + j) = \frac{25(26)(51)}{6} + \frac{25(26)}{2} = 5850$

7. $S(n) = \sum_{i=1}^{n} \frac{4i^2}{n^3} = \frac{4}{n^3} \cdot \frac{n(n+1)(2n+1)}{6} = \frac{4n^3 + 6n^2 + 2n}{3n^3}$

n	10^0	10^1	10^2	10^3
$S(n)$	4	1.54	1.353	1.335

$\lim_{n \to \infty} S(n) = \frac{4}{3}$

9. $S(n) = \sum_{i=1}^{n} \frac{3}{n^3}(1 + i^2) = \frac{3}{n^3}\left[n + \frac{n(n+1)(2n+1)}{6}\right] = \frac{3}{n^2} + \frac{6n^2 + 9n + 3}{6n^2}$

n	10^0	10^1	10^2	10^3
$S(n)$	6	1.185	1.0154	1.0015

$\lim_{n \to \infty} S(n) = 1$

11. $S(n) = \sum_{i=1}^{n} \frac{2i+3}{n^2} = \frac{1}{n^2}\left(2\left(\frac{n(n+1)}{2}\right) + 3n\right)\frac{n+1}{N} + \frac{3}{n}$

n	10^0	10^1	10^2	10^3
$S(n)$	5	1.4	1.04	1.004

$\lim_{n \to \infty} S(n) = 1$

13. $S(n) = \sum_{i=1}^{n} \left(\frac{i^2}{n^3} + \frac{2}{n}\right)\left(\frac{1}{n}\right) = \frac{1}{n}\left[\frac{n(n+1)(2n+1)}{6n^3} + \frac{2n}{n}\right] = \frac{1}{6n^3}(2n^2 + 3n + 1) + \frac{2}{n}$

n	10^0	10^1	10^2	10^3
$S(n)$	3	0.2385	0.02338	0.00233

$\lim_{n \to \infty} S(n) = 0$

15. $S(n) = \sum_{i=1}^{n} \left[1 - \left(\frac{i}{n}\right)^2 \right]\left(\frac{1}{n}\right) = \frac{1}{n}\left[n - \frac{1}{n^2}\left(\frac{n(n+1)(2n+1)}{6}\right) \right] = 1 - \frac{2n^2 + 3n + 1}{6n^2}$

n	10^0	10^1	10^2	10^3
$S(n)$	0	0.615	0.66165	0.66617

$\lim_{n\to\infty} S(n) = \frac{2}{3}$

17. Width of each rectangle is $\frac{1}{2}$. The height is obtained by evaluating f at the right hand endpoint of each interval.

$A \approx \sum_{i=1}^{4} f\left(\frac{i}{2}\right)\left(\frac{1}{2}\right) = \sum_{i=1}^{4} \left(\frac{i}{2} + 1\right)\frac{1}{2} = 9\left(\frac{1}{2}\right) = 4.5$

19. The width of each rectangle is $\frac{1}{4}$. The height is obtained by evaluating f at the right hand endpoint of each interval.

$A \approx \sum_{i=1}^{8} f\left(\frac{i}{4}\right)\left(\frac{1}{4}\right) = \sum_{i=1}^{8} \frac{1}{4}\left(\frac{i}{4}\right)^3 \left(\frac{1}{4}\right) = 1.265625$

21. Width of each rectangle is $8/n$. The height is

$f\left(\frac{8i}{n}\right) = -\frac{1}{2}\left(\frac{8i}{n}\right) + 4.$

$A \approx \sum_{i=1}^{n} \left[-\frac{1}{2}\left(\frac{8i}{n}\right) + 4 \right]\left(\frac{8}{n}\right)$ (Note: exact area is 16)

n	4	8	20	50
Approximate area	12	14	15.2	15.68

23. The width of each rectangle is $3/n$. The height is

$\frac{1}{9}\left(\frac{3i}{n}\right)^3.$

$A \approx \sum_{i=1}^{n} \frac{1}{9}\left(\frac{3i}{n}\right)^3\left(\frac{3}{n}\right)$

n	4	8	20	50
Approximate area	3.52	2.85	2.45	2.34

25. $A \approx \sum_{i=1}^{n} f\left(\frac{i}{n}\right)\left(\frac{1}{n}\right)$

$= \sum_{i=1}^{n} 3\left(\frac{i}{n}\right)\left(\frac{1}{n}\right)$

$= \frac{3}{n^2} \sum_{i=1}^{n} i$

$= \frac{3}{n^2} \frac{n(n+1)}{2}$

$= \frac{3n^2 + 3n}{2n^2}$

$A = \lim_{n\to\infty} \frac{3n^2 + 3n}{2n^2} = \frac{3}{2}$

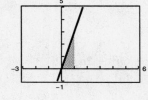

27. $A \approx \sum_{i=1}^{n} f\left(-1 + \frac{2i}{n}\right)\left(\frac{2}{n}\right)$

$= \sum_{i=1}^{n} \left[3 - \left(-1 + \frac{2i}{n}\right)^2 \right]\frac{2}{n}$

$= \sum_{i=1}^{n} \left[3 - 1 + \frac{4i}{n} - \frac{4i^2}{n^2} \right]\left(\frac{2}{n}\right)$

$= \frac{4}{n} \sum_{i=1}^{n} 1 + \frac{8}{n^2} \sum_{i=1}^{n} i - \frac{8}{n^3} \sum_{i=1}^{n} i^2$

$= \frac{4}{n}(n) + \frac{8}{n^2}\frac{n(n+1)}{2} - \frac{8}{n^3}\frac{n(n+1)(2n+1)}{6}$

$A = \lim_{n\to\infty} \left[4 + 4\frac{n(n+1)}{n^2} - \frac{4}{3}\frac{n(n+1)(2n+1)}{n^3} \right] = 4 + 4 - \frac{8}{3} = \frac{16}{3}$

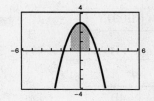

29. $A \approx \sum_{i=1}^{n} g\left(1 + \frac{i}{n}\right)\left(\frac{1}{n}\right)$

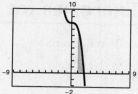

$= \sum_{i=1}^{n}\left[8 - \left(1 + \frac{i}{n}\right)^3\right]\frac{1}{n}$

$= \sum_{i=1}^{n}\left[7 - \frac{3i}{n} + \frac{3i^2}{n^2} + \frac{i^3}{n^3}\right]\frac{1}{n}$

$= \frac{7}{n}\sum_{i=1}^{n} 1 - \frac{3}{n^2}\sum_{i=1}^{n} i - \frac{3}{n^3}\sum_{i=1}^{n} i^2 - \frac{1}{n^4}\sum_{i=1}^{n} i^3$

$= \frac{7}{n}(n) - \frac{3}{n^2}\frac{n(n+1)}{2} - \frac{3}{n^3}\frac{n(n+1)(2n+1)}{6} - \frac{1}{n^4}\frac{n^2(n+1)^2}{4}$

$A = \lim_{n\to\infty}\left[7 - \frac{3}{2}\frac{n(n+1)}{n^2} - \frac{1}{2n^3}n(n+1)(2n+1) - \frac{1}{n^4}\frac{n^2(n+1)^2}{4}\right]$

$= 7 - \frac{3}{2} - 1 - \frac{1}{4} = \frac{17}{4}$

31. $A \approx \sum_{i=1}^{n} f\left(1 + \frac{3i}{n}\right)\left(\frac{3}{n}\right)$

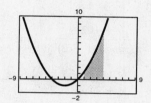

$= \sum_{i=1}^{n}\left[\frac{1}{4}\left(1 + \frac{3i}{n}\right)^2 + \left(1 + \frac{3i}{n}\right)\right]\left(\frac{3}{n}\right)$

$= \sum_{i=1}^{4}\left(\frac{1}{4} + \frac{3}{2}\frac{i}{n} + \frac{9}{4}\frac{i^2}{n^2} + 1 + \frac{3i}{n}\right)\left(\frac{3}{n}\right)$

$= \frac{15}{4n}\sum_{i=1}^{n} 1 + \frac{27}{2n^2}\sum_{i=1}^{n} i + \frac{27}{4n^3}\frac{n(n+1)(2n+1)}{6}$

$= \frac{15}{4n}(n) + \frac{27}{2n^2}\left(\frac{n(n+1)}{2}\right) + \frac{27}{4n^3}\frac{n(n+1)(2n+1)}{6}$

$A = \lim_{n\to\infty}\left[\frac{15}{4} + \frac{27}{4}\frac{n(n+1)}{n^2} + \frac{9}{8n^3}n(n+1)(2n+1)\right] = \frac{15}{4} + \frac{27}{4} + \frac{9}{4} = \frac{51}{4}$

33. The area is less than 8 and more than 4.
Matches (b).

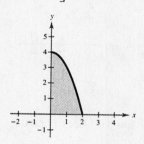

35. Area $\approx 0.167 \left(= \frac{1}{6}\right)$

37. Area ≈ 1.0

39. $y = (-3.0 \cdot 10^{-6})x^3 + 0.002x^2 - 1.05x + 400$

Note that $y = 0$ when $x = 500$.

Area $\approx 105{,}208.33$ square feet

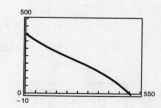

❑ Review Exercises for Chapter 12

Solutions to Odd-Numbered Exercises

1. $\lim\limits_{x\to 2}\dfrac{x-2}{3x^2-4x-4}=\dfrac{1}{8}$

x	1.9	1.99	1.999	2	2.001	2.01	2.1
$f(x)$	.1299	.1255	.1250	?	.1250	.1245	.1205

3. $\lim\limits_{x\to 3}(5x-4)=5(3)-4=11$

5. $\lim\limits_{x\to 2}(5x-3)(3x+5)=(5(2)-3)(3(2)+5)=(7)(11)=77$

7. $\lim\limits_{t\to 3}\dfrac{t^2+1}{t}=\dfrac{9+1}{3}=\dfrac{10}{3}$

9. $\lim\limits_{t\to -2}\dfrac{t+2}{t^2-4}=\lim\limits_{t\to -2}\dfrac{t+2}{(t+2)(t-2)}=\lim\limits_{t\to -2}\dfrac{1}{t-2}=-\dfrac{1}{4}$

11. $\lim\limits_{x\to -2}\dfrac{x^2-4}{x^3+8}=\lim\limits_{x\to -2}\dfrac{(x+2)(x-2)}{(x+2)(x^2-2x+4)}=\lim\limits_{x\to -2}\dfrac{x-2}{x^2-2x+4}$

$$=\dfrac{-4}{12}=\dfrac{-1}{3}$$

13. $\lim\limits_{u\to 0}\dfrac{\sqrt{4+u}-2}{u}=\lim\limits_{u\to 0}\dfrac{\sqrt{4+u}-2}{u}\cdot\dfrac{\sqrt{4+u}+2}{\sqrt{4+u}+2}$

$$=\lim\limits_{u\to 0}\dfrac{(4+u)-4}{u(\sqrt{4+u}+2)}=\lim\limits_{u\to 0}\dfrac{1}{\sqrt{4+u}+2}=\dfrac{1}{4}$$

15. $\lim\limits_{x\to -1}\dfrac{\dfrac{1}{x+2}-1}{x+1}=\lim\limits_{x\to -1}\dfrac{1-(x-2)}{(x+2)(x+1)}=\lim\limits_{x\to -1}\dfrac{-(x+1)}{(x+2)(x+1)}$

$$=\lim\limits_{x\to -1}\dfrac{-1}{(x+2)}=-1$$

17. $\lim\limits_{x\to 6^-}\dfrac{|x-6|}{x-6}=-1$

19. $\lim\limits_{x\to 1}f(x)$ does not exist

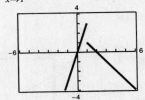

21. $\lim\limits_{x\to 0}g(x)=2$

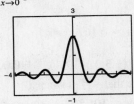

23. (a)

x	1.1	1.01	1.001	1.0001
$f(x)$	0.5680	0.5764	0.5772	0.5773

$\lim\limits_{x\to 1^+}f(x)\approx 0.577$

(b) $\lim\limits_{x\to 1^+}\dfrac{\sqrt{2x+1}-\sqrt{3}}{x-1}\cdot\dfrac{\sqrt{2x+1}+\sqrt{3}}{\sqrt{2x+1}+\sqrt{3}}$

$$=\lim\limits_{x\to 1^+}\dfrac{(2x+1)-3}{(x-1)(\sqrt{2x+1}+\sqrt{3})}$$

$$=\lim\limits_{x\to 1^+}\dfrac{2(x-1)}{(x-1)(\sqrt{2x+1}+\sqrt{3})}$$

$$\lim\limits_{x\to 1^+}\dfrac{2}{\sqrt{2x+1}+\sqrt{3}}=\dfrac{2}{2\sqrt{3}}=\dfrac{1}{\sqrt{3}}=\dfrac{\sqrt{3}}{3}$$

25. (a) $\lim\limits_{x \to c} [f(x)]^3 = 4^3 = 64$

(b) $\lim\limits_{x \to c} [3f(x) - g(x)] = 3(4) - 5 = 7$

(c) $\lim\limits_{x \to c} [f(x)\, g(x)] = (4)(5) = 20$

(d) $\lim\limits_{x \to c} \dfrac{f(x)}{g(x)} = \dfrac{4}{5}$

27. $\lim\limits_{h \to 0} \dfrac{f(x + h) - f(x)}{h} = \lim\limits_{h \to 0} \dfrac{3(x + h) - (x + h)^2 - (3x - x^2)}{h}$

$\qquad = \lim\limits_{h \to 0} \dfrac{3x + 3h - x^2 - 2xh - h^2 - 3x + x^2}{h} = \lim\limits_{h \to 0} \dfrac{3h - 2xh - h^2}{h}$

$\qquad = \lim\limits_{h \to 0} (3 - 2x - h) = 3 - 2x$

29.

Slope at $(2, f(2))$ is approximately 2.

31. $m = \lim\limits_{h \to 0} \dfrac{g(x + h) - g(x)}{h} = \lim\limits_{h \to 0} \dfrac{(x + h)^2 - 4(x + h) - (x^2 - 4x)}{h}$

$\qquad = \lim\limits_{h \to 0} \dfrac{x^2 + 2xh + h^2 - 4x - 4h - x^2 - 4x}{h}$

$\qquad = \lim\limits_{h \to 0} \dfrac{2xh + h^2 - 4h}{h} = \lim\limits_{h \to 0} (2x + h - 4) = 2x - 4$

(a) At $(0, 0)$, $m = 2(0) - 4 = -4$.

(b) At $(5, 5)$, $m = 2(5) - 4 = 6$.

33. $m = \lim\limits_{h \to 0} \dfrac{f(x + h) - f(x)}{h} = \lim\limits_{h \to 0} \dfrac{\dfrac{4}{x + h - 6} - \dfrac{4}{x - 6}}{h}$

$\qquad = \lim\limits_{h \to 0} \dfrac{4(x - 6) - 4(x + h - 6)}{(x + h - 6)(x - 6)h} = \lim\limits_{h \to 0} \dfrac{-4h}{(x + h - 6)(x - 6)h}$

$\qquad = \lim\limits_{h \to 0} \dfrac{-4}{(x + h - 6)(x - 6)} = \dfrac{-4}{(x - 6)^2}$

(a) At $(7, 4)$, $m = \dfrac{-4}{(7 - 6)^2} = -4$

(b) At $(8, 2)$, $m = \dfrac{-4}{(8 - 6)^2} = -1$.

35. $g'(x) = \lim\limits_{h \to 0} \dfrac{g(x + h) - g(x)}{h} = \lim\limits_{h \to 0} \dfrac{-4 - (-4)}{h} = 0$

37. $h'(x) = \lim\limits_{h \to 0} \dfrac{h(x + h) - h(x)}{h} = \lim\limits_{h \to 0} \dfrac{\left[5 - \frac{1}{2}(x + h)\right] - \left[5 - \frac{1}{2}x\right]}{h}$

$\qquad = \lim\limits_{h \to 0} \dfrac{-\frac{1}{2}h}{h} = -\dfrac{1}{2}$

39. $f'(t) = \lim\limits_{h \to 0} \dfrac{f(t + h) - f(t)}{h} = \lim\limits_{h \to 0} \dfrac{\sqrt{t + h + 5} - \sqrt{t + 5}}{h} \cdot \dfrac{\sqrt{t + h + 5} + \sqrt{t + 5}}{\sqrt{t + h + 5} + \sqrt{t + 5}}$

$\qquad = \lim\limits_{h \to 0} \dfrac{(t + h + 5) - (t - 5)}{h\left(\sqrt{t + h + 5} + \sqrt{t + 5}\right)} = \lim\limits_{h \to 0} \dfrac{1}{\sqrt{t + h + 5} + \sqrt{t + 5}}$

$\qquad = \dfrac{1}{2\sqrt{t + 5}}$

41. $y = 8.73t^2 - 6.23t + 0.54$

43. $f(x) = 2x$ matches (d) (slope 2).

45. $f(x) = |x - 2|$ matches (b) (slope 1 for $x > 2$, -1 for $x < 2$).

47. $\displaystyle\lim_{x \to \infty} \frac{4x}{2x - 3} = \frac{4}{2} = 2$

49. $\displaystyle\lim_{x \to \infty} \frac{2x}{x^2 - 25} = 0$

51. $\displaystyle\lim_{x \to \infty} \left(4 - \frac{7}{x^3}\right) = 4 - 0 = 4$

53. $\displaystyle\lim_{n \to \infty} \frac{1}{2n^2}[3 - 2n(n + 1)] = \lim_{n \to \infty} \left[\frac{3}{2n^2} - \frac{2(n + 1)}{2n}\right] = -1$

55. $\displaystyle\sum_{i=1}^{n} \left(\frac{4i^2}{n^2} - \frac{i}{n}\right)\frac{1}{n} = \frac{4}{n^3} \sum_{i=1}^{n} i^2 - \frac{1}{n^2} \sum_{i=1}^{n} i$

$$= \frac{4}{n^3} \frac{n(n + 1)(2n + 1)}{6} - \frac{1}{n^2} \frac{n(n + 1)}{2}$$

$$= \frac{4n(n + 1)(2n + 1) - 3n^2(n + 1)}{6n^3}$$

$$= \frac{n(n + 1)(8n + 4 - 3n)}{6n^3} = \frac{(n + 1)(5n + 4)}{6n^2}$$

n	10^0	10^1	10^2	10^3
$S(n)$	3	0.99	0.8484	0.8348

$\displaystyle\lim_{n \to \infty} S(n) = \frac{5}{6}$

57. Area $\approx \frac{1}{2}\left[\frac{7}{2} + 3 + \frac{5}{2} + 2 + \frac{3}{2} + 1\right] = \frac{1}{2}\frac{27}{2} = \frac{27}{4} = 6.75$

59.

n	4	8	20	50
Approx. Area	6.7266	5.9707	5.5341	5.3629

$\left(\text{Exact area is } \frac{16}{3}\right)$

61. $A \approx \displaystyle\sum_{i=1}^{5} (10 - (2i))(2) = 40$ approximate area

$A = \displaystyle\lim_{n \to \infty} \sum_{i=1}^{n} \left(10 - \frac{10i}{n}\right)\left(\frac{10}{n}\right)$

$= \displaystyle\lim_{n \to \infty} \left[\frac{100}{n} \sum_{i=1}^{n} 1 - \frac{100}{n^2} \sum_{i=1}^{n} i\right]$

$= \displaystyle\lim_{n \to \infty} \left[\frac{100}{n}(n) - \frac{100}{n^2}\left(\frac{n(n + 1)}{2}\right)\right]$

$= \displaystyle\lim_{n \to \infty} \left[100 - 50 \frac{n(n + 1)}{n^2}\right] = 100 - 50 = 50$ exact area

63. $A \approx \displaystyle\sum_{i=1}^{6} \left[\left(-1 + \frac{i}{2} \right)^2 + 4 \right] \left(\frac{1}{2} \right)$

$= \displaystyle\sum_{i=1}^{6} \left[5 - i + \frac{i^2}{4} \right] \frac{1}{2}$

$= \dfrac{5}{2} \displaystyle\sum_{i=1}^{6} 1 = \dfrac{1}{2} \displaystyle\sum_{i=1}^{6} i + \dfrac{1}{8} \displaystyle\sum_{i=1}^{6} i^2$

$= \dfrac{5}{2}(6) - \dfrac{1}{2}(21) + \dfrac{1}{8}(91) = \dfrac{127}{8} = 15.875$ approximate area

$A = \displaystyle\lim_{n \to \infty} \sum_{i=1}^{n} \left[\left(-1 + \frac{3i}{n} \right)^2 + 4 \right] \left(\frac{3}{n} \right) = \lim_{n \to \infty} \sum_{i=1}^{n} \left[5 - \frac{6i}{n} + \frac{9i^2}{n^2} \right] \frac{3}{n}$

$= \displaystyle\lim_{n \to \infty} \left[\frac{15}{n} \sum_{i=1}^{n} 1 - \frac{18}{n^2} \sum_{i=1}^{n} i + \frac{27}{n^3} \sum_{i=1}^{n} i^2 \right]$

$= \displaystyle\lim_{n \to \infty} \left[\frac{15}{n}(n) - \frac{18}{n^2} \frac{n(n+1)}{2} + \frac{27}{n^3} \frac{n(n+1)(2n+1)}{6} \right]$

$= 15 - 9 + 9 = 15$ exact area

65. $A \approx \displaystyle\sum_{i=1}^{4} 2 \left[\left(-1 + \frac{i}{2} \right)^2 - \left(-1 + \frac{i}{3} \right)^3 \right] \left(\frac{1}{2} \right) = \sum_{i=1}^{4} \left[\left(-1 + \frac{i}{2} \right)^2 - \left(-1 + \frac{i}{3} \right)^3 \right] = \frac{1}{2}$ approximate area

$A = \displaystyle\lim_{n \to \infty} \sum_{i=1}^{n} 2 \left[\left(-1 + \frac{2i}{n} \right)^2 - \left(-1 + \frac{2i}{n} \right)^3 \right] \left(\frac{2}{n} \right)$

$= \displaystyle\lim_{n \to \infty} \sum_{i=1}^{n} \frac{4}{n} \left(1 - \frac{4i}{n} + \frac{4i^2}{n^2} - \left(-1 + \frac{6i}{n} - \frac{12i^2}{n^2} + \frac{8i^3}{n^3} \right) \right)$

$= \displaystyle\lim_{n \to \infty} \sum_{i=1}^{n} \frac{4}{n} \left(2 - \frac{10i}{n} + \frac{16i^2}{n^2} - \frac{8i^3}{n^3} \right)$

$= \displaystyle\lim_{n \to \infty} \left[\frac{8}{n} \sum_{i=1}^{n} 1 - \frac{40}{n^2} \sum_{i=1}^{n} i + \frac{64}{n^3} \sum_{i=1}^{n} i^2 - \frac{32}{n^4} \sum_{i=1}^{n} i^3 \right]$

$= \displaystyle\lim_{n \to \infty} \left[\frac{8}{n}(n) - \frac{40}{n^2} \frac{n(n+1)}{2} + \frac{64}{n^3} \frac{n(n+1)(2n+1)}{6} - \frac{32}{n^4} \frac{n^2(n+1)^2}{4} \right]$

$= 8 - 20 + \dfrac{64}{3} - 8 = \dfrac{4}{3}$ exact area

67. Area $\approx 100[125 + 125 + 120 + 112 + 90 + 90 + 95 + 88 + 75 + 35]$

$= 95,500$ sq. ft. (using left hand endpoints)

Area $= 100[125 + 120 + 112 + 90 + 90 + 95 + 88 + 75 + 35 + 0]$

$= 83,000$ sq. ft. (using right hand endpoint)

❑ Cumulative Test for Chapters 11–12

1. Center $= (2, 2, 4)$

Radius $= \sqrt{2^2 + 2^2 + 4^2} = \sqrt{24}$

$(x - 2)^2 + (y - 2)^2 + (z - 4)^2 = 24$

2. $\langle -3, 4, 1 \rangle \cdot \langle 5, 0, 2 \rangle = -15 + 2 = -13$

$$\mathbf{u} \times \mathbf{v} = \begin{vmatrix} \mathbf{i} & \mathbf{j} & \mathbf{k} \\ -3 & 4 & 1 \\ 5 & 0 & 2 \end{vmatrix} = \langle 8, 11, -20 \rangle$$

3. Vector is $\langle 5 + 2, 8 - 3, 25 - 0 \rangle = \langle 7, 5, 25 \rangle$

$x = -2 + 7t, y = 3 + 5t, z = 25t$

4. $\mathbf{u} = \langle -2, 3, 0 \rangle, \mathbf{v} = \langle 5, 8, 25 \rangle$

$$\mathbf{u} \times \mathbf{v} = \begin{vmatrix} \mathbf{i} & \mathbf{j} & \mathbf{k} \\ -2 & 3 & 0 \\ 5 & 8 & 25 \end{vmatrix} = \langle 75, 50, -31 \rangle$$

Normal to plane.

Plane: $75x + 50y - 31z = 0$

5. $\mathbf{u} = \langle 3, 6, 4 \rangle, \mathbf{v} = \langle 0, 10, 6 \rangle$

$$\mathbf{u} \times \mathbf{v} = \begin{vmatrix} \mathbf{i} & \mathbf{j} & \mathbf{k} \\ 3 & 6 & 4 \\ 0 & 10 & 6 \end{vmatrix} = \langle -4, -18, 30 \rangle$$

Area $= \frac{1}{2} \| \mathbf{u} \times \mathbf{v} \| = \frac{1}{2} \sqrt{16 + 324 + 900} = \sqrt{310} \approx 17.61$

6. $\mathbf{n} = \langle 2, -5, 1 \rangle, Q = (0, 0, 25), P = (0, 0, 10)$ in plane $\overrightarrow{PQ} = \langle 0, 0, 15 \rangle$.

$$D = \frac{|\overrightarrow{PQ} \cdot \mathbf{n}|}{\| \mathbf{n} \|} = \frac{15}{\sqrt{30}} = \frac{\sqrt{30}}{2} \approx 2.74$$

7. $\lim\limits_{x \to 4} (5x - x^2) = 5(4) - 4^2 = 4$

8. $\lim\limits_{x \to -2} \dfrac{x + 2}{(x + 2)(x - 1)} = \lim\limits_{x \to -2} \dfrac{1}{x - 1} = -\dfrac{1}{3}$

9. $\lim\limits_{x \to 0} \dfrac{\sqrt{x + 4} - 2}{x} \cdot \dfrac{\sqrt{x + 4} + 2}{\sqrt{x + 4} + 2} = \lim\limits_{x \to 0} \dfrac{(x + 4) - 4}{x(\sqrt{x + 4} + 2)} = \lim\limits_{x \to 0} \dfrac{1}{\sqrt{x + 4} + 2} = \dfrac{1}{2 + 2} = \dfrac{1}{4}$

10. $\lim\limits_{x \to 4} \dfrac{|x - 4|}{x - 4}$ Does not exist.

11. $\lim\limits_{x \to \infty} \dfrac{6x}{3x + 5} = \dfrac{6}{3} = 2$

12. $\lim\limits_{x \to \infty} \dfrac{2x}{x^2 + 3x - 2} = 0$

13. $m_{\text{sec}} = \dfrac{h(1 + h) - h(1)}{h} = \dfrac{3 - (1 + h)^2 - 2}{h} = \dfrac{3 - (1 + 2h + h^2) - 2}{h} = \dfrac{-2h - h^2}{h}$

Slope $= \lim\limits_{h \to 0} \dfrac{-h^2 - 2h}{h} = \lim\limits_{h \to 0} (-h - 2) = -2$

14. $m_{\text{sec}} = \dfrac{f(-2 + h) - f(-2)}{h} = \dfrac{\sqrt{-2 + h + 3} - 1}{h} = \dfrac{\sqrt{h + 1} - 1}{h}$

$= \dfrac{\sqrt{h + 1} - 1}{h} \cdot \dfrac{\sqrt{h + 1} + 1}{\sqrt{h + 1} + 1} = \dfrac{(h + 1) - 1}{h(\sqrt{h + 1} + 1)} = \dfrac{h}{h(\sqrt{h + 1} + 1)}$

Slope $= \lim\limits_{h \to 0} \dfrac{h}{h(\sqrt{h + 1} + 1)} = \dfrac{1}{2}$

15. $\sum\limits_{k=1}^{20} 3k^2 - 2k = 3 \cdot \dfrac{20(21)(41)}{6} - 2\dfrac{20(21)}{2} = 8610 - 420 = 8190$

16.

n	4	8	20	50
Approximate Area	8.1250	8.4063	8.5650	8.6264

17. Width: $\dfrac{1}{n}$

Height: $f\left(\dfrac{i}{n}\right) = 1 - \left(\dfrac{i}{n}\right)^3$

$A \approx \displaystyle\sum_{i=1}^{n} \left(1 - \left(\dfrac{i}{n}\right)^3\right)\left(\dfrac{1}{n}\right) = \dfrac{1}{n}\sum_{i=1}^{n} - \dfrac{1}{n^4}\sum_{i=1}^{n} i^3$

$= \dfrac{1}{n}(n) - \dfrac{1}{n^4}\left[\dfrac{n^2(n+1)^2}{4}\right]$

$A = \lim_{n\to\infty}\left[n - \dfrac{1}{n^4}\left[\dfrac{n^2(n+1)^2}{4}\right]\right] = 1 - \dfrac{1}{4} = \dfrac{3}{4}$

❑ **Practice Test for Chapter 12**

1. Use a graphing utility to complete the table and use the result to estimate the limit

$$\lim_{x \to 3} \frac{x - 3}{x^2 - 9}.$$

x	2.9	2.99	3	3.01	3.1
$f(x)$			?		

2. Graph the function $f(x) = \dfrac{\sqrt{x + 4} - 2}{x}$ and estimate the limit

$$\lim_{x \to 0} \frac{\sqrt{x + 4} - 2}{x}.$$

3. Find the limit $\lim_{x \to 2} e^{x - 2}$ by direct substitution.

4. Find the limit $\lim_{x \to 1} \dfrac{x^3 - 1}{x - 1}$ analytically.

5. Use a graphing utility to estimate the limit

$$\lim_{x \to 0} \frac{\sin 5x}{2x}.$$

6. Find the limit $\lim_{x \to -2} \dfrac{|x + 2|}{x + 2}$.

7. Use the limit process to find the slope of the graph of $f(x) = \sqrt{x}$ at the point $(4, 2)$.

8. Find the derivative of the function $f(x) = 3x - 1$.

9. Find the limits.

(a) $\lim_{x \to \infty} \dfrac{3}{x^4}$

(b) $\lim_{x \to -\infty} \dfrac{x^2}{x^2 + 3}$

(c) $\lim_{x \to \infty} \dfrac{|x|}{1 - x}$

10. Write the first four terms of the sequence $a_n = \dfrac{1 - n^2}{2n^2 + 1}$, and find the limit of the sequence.

11. Find the sum $\displaystyle\sum_{i=1}^{25} (i^2 + i)$

12. Write the sum $\displaystyle\sum_{i=1}^{n} \frac{i^2}{n^3}$ as a rational function $s(n)$, and find $\lim_{n \to \infty} s(n)$.

13. Find the area of the region bounded by $f(x) = 1 - x^2$ over the interval $0 \le x \le 1$.

Practice Test Solutions
❏ Chapter 1 Practice Test Solutions

1.

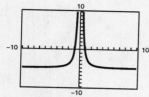

x-intercepts: ± 0.894

2.

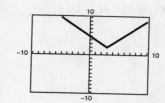

No x-intercepts

3. $3x - 5y = 15$

Line

x-intercept: $(5, 0)$

y-intercept: $(0, -3)$

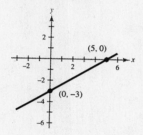

4. $y = \sqrt{9 - x}$

Domain: $(-\infty, 9]$

x-intercept: $(9, 0)$

y-intercept: $(0, 3)$

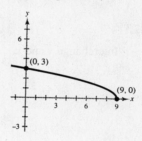

5. $5x + 4 = 7x - 8$

$4 + 8 = 7x - 5x$

$12 = 2x$

$x = 6$

6. $\dfrac{x}{3} - 5 = \dfrac{x}{5} + 1$

$15\left(\dfrac{x}{3} - 5\right) = 15\left(\dfrac{x}{5} + 1\right)$

$5x - 75 = 3x + 15$

$2x = 90$

$x = 45$

7. No, y is not a function of x. For example, $(0, 2)$ and $(0, -2)$ both satisfy the equation.

8. $f(0) = \dfrac{|0 - 2|}{(0 - 2)} = \dfrac{2}{-2} = -1$

$f(2)$ is not defined.

$f(4) = \dfrac{|4 - 2|}{(4 - 2)} = \dfrac{2}{2} = 1$

9. The domain of

$f(x) = \dfrac{5}{x^2 - 16}$

is all $x \neq \pm 4$.

10. The domain of $g(t) = \sqrt{4 - t}$

consists of all t satisfying

$4 - t \geq 0$ or $t \leq 4$.

11.

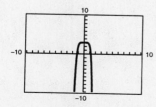

$f(x) = 3 - x^6$

is even.

12.

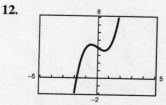

Relative minimum:
$(0.577, 3.615)$

Relative maximum:
$(-0.577, 4.385)$

13. $f(x) = x^3 - 3$ is a vertical shift of 3 units downward of $y = x^3$.

14. $f(x) = \sqrt{x - 6}$ is a horizontal shift 6 units to the right of $y = \sqrt{x}$.

15. $(g \circ f)(x) = g(f(x)) = g(\sqrt{x}) = (\sqrt{x})^2 - 2 = x - 2$

Domain: $x \geq 0$

16. $\left(\dfrac{f}{g}\right)(x) = \dfrac{f(x)}{g(x)} = \dfrac{3x^2}{16 - x^4}$

The domain is all $x \neq \pm 2$.

17. $(f \circ g)(x) = f\left(\dfrac{x - 1}{3}\right) = 3\left(\dfrac{x - 1}{3}\right) + 1 = (x - 1) + 1 = x$

$(g \circ f)(x) = g(3x + 1) = \dfrac{(3x + 1) - 1}{3} = \dfrac{3x}{3} = x$

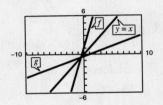

18. $y = \sqrt{9 - x^2}, \ 0 \leq x \leq 3$

$x = \sqrt{9 - y^2}$ Interchange x and y.

$x^2 = 9 - y^2$

$y^2 = 9 - x^2$

$y = \sqrt{9 - x^2}, \ 0 \leq x \leq 3$

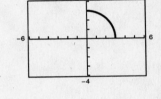

19. $y = 0.882 + 0.912x$

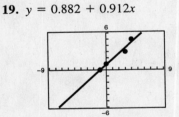

20.

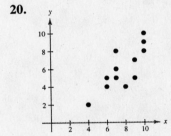

❏ Chapter 2 Practice Test Solutions

1. *x*-intercepts: $(1, 0)$, $(5, 0)$
y-intercept: $(0, 5)$
Vertex: $(3, -4)$

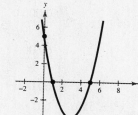

2. $a = 0.01, b = -90$

$$\frac{-b}{2a} = \frac{90}{2(.01)} = 4500 \text{ units}$$

3. Vertex: $(1, 7)$
Opening downward through $(2, 5)$

$y = a(x - 1)^2 + 7$ Standard form

$5 = a(2 - 1)^2 + 7$

$5 = a + 7$

$a = -2$

$y = -2(x - 1)^2 + 7$

$\quad = -2(x^2 - 2x + 1) + 7$

$\quad = -2x^2 + 4x + 5$

4. $y = \pm a(x - 2)(3x - 4)$ where a is any nonzero real number.

$$y = \pm(3x^2 - 10x + 8)$$

5. Leading coefficient: -3
Degree: 5
Moves down to the right and up to the left.

6. $0 = x^5 - 5x^3 + 4x$

$\quad = x(x^4 - 5x^2 + 4)$

$\quad = x(x^2 - 1)(x^2 - 4)$

$\quad = x(x + 1)(x - 1)(x + 2)(x - 2)$

$x = 0, x = \pm 1, x = \pm 2$

7.

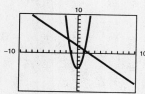

$(1.257, \ 0.743)$

$(-1.591, \ 3.591)$

8.

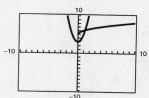

$(1.248, \ 6.117)$

9. $\dfrac{2}{1 + i} = \dfrac{2}{1 + i} \cdot \dfrac{1 - i}{1 - i} = \dfrac{2 - 2i}{1 + 1} = 1 - i$

10. $\dfrac{3 + i}{2} - \dfrac{i + 1}{4} = \dfrac{6 + 2i - i - 1}{4} = \dfrac{5}{4} + \dfrac{i}{4}$

11. $28 + 5x - 3x^2 = 0$

$(4 - x)(7 + 3x) = 0$

$4 - x = 0 \Longrightarrow x = 4$

$7 + 3x = 0 \Longrightarrow x = -\dfrac{7}{3}$

12. $(x - 2)^2 = 24$

$x - 2 = \pm\sqrt{24}$

$x - 2 = \pm 2\sqrt{6}$

$x = 2 \pm 2\sqrt{6}$

13. $x^2 - 4x - 9 = 0$

$x^2 - 4x + 2^2 = 9 + 2^2$

$(x - 2)^2 = 13$

$x - 2 = \pm\sqrt{13}$

$x = 2 \pm \sqrt{13}$

14. $x^2 + 5x - 1 = 0$

$a = 1, \ b = 5, \ c = -1$

$x = \dfrac{-5 \pm \sqrt{(5)^2 - 4(1)(-1)}}{2(1)}$

$\quad = \dfrac{-5 \pm \sqrt{25 + 4}}{2} = \dfrac{-5 \pm \sqrt{29}}{2}$

15. $3x^2 - 2x + 4 = 0$

$a = 3, \ b = -2, \ c = 4$

$$x = \frac{-(-2) \pm \sqrt{(-2)^2 - 4(3)(4)}}{2(3)}$$

$$= \frac{2 \pm \sqrt{4 - 48}}{6}$$

$$= \frac{2 \pm \sqrt{-44}}{6}$$

$$= \frac{2 \pm 2i\sqrt{11}}{6}$$

$$= \frac{1 \pm i\sqrt{11}}{3} = \frac{1}{3} \pm \frac{\sqrt{11}}{3}i$$

16.

$$60,000 = xy$$

$$y = \frac{60,000}{x}$$

$$2x + 2y = 1100$$

$$2x + 2\left(\frac{60,000}{x}\right) = 1100$$

$$x + \frac{60,000}{x} = 550$$

$$x^2 + 60,000 = 550x$$

$$x^2 - 550x + 60,000 = 0$$

$$(x - 150)(x - 400) = 0$$

$x = 150 \quad \text{or} \quad x = 400$

$y = 400 \qquad y = 150$

Length: 400 feet

Width: 150 feet

17. $\qquad x(x + 2) = 624$

$x^2 + 2x - 624 = 0$

$(x - 24)(x + 26) = 0$

$\qquad x = 24 \quad \text{or} \quad x = -26, \text{ (extraneous solution)}$

$\qquad x + 2 = 26$

18. $x^3 - 10x^2 + 24x = 0$

$x(x^2 - 10x + 24) = 0$

$x(x - 4)(x - 6) = 0$

$\qquad x = 0, \ x = 4, \ x = 6$

19. $\sqrt[3]{6 - x} = 4 \implies 6 - x = 64 \implies -x = 58$

$\qquad x = -58$

20. $(x^2 - 8)^{2/5} = 4$

$x^2 - 8 = \pm 4^{5/2}$

$x^2 - 8 = 32 \quad \text{or} \quad x^2 - 8 = -32$

$x^2 = 40 \qquad\qquad x^2 = -24$

$x = \pm\sqrt{40} \qquad\quad x = \pm\sqrt{-24}$

$x = \pm 2\sqrt{10} \qquad\quad x = \pm 2\sqrt{6}\,i$

21. $\quad x^4 - x^2 - 12 = 0(x^2 - 4)(x^2 + 3) = 0$

$x^2 = 4 \quad \text{or} \quad x^2 = -3$

$x^2 = \pm 2 \qquad x = \pm\sqrt{3}\,i$

22. $f(x) = x(x - 3)(x + 2)$

$= x(x^2 - x - 6)$

$= x^3 - x^2 - 6x$

Any nonzero multiple will also be correct.

23.

$$x - 3\overline{)3x^4 + 0x^3 - 7x^2 + 2x - 10}$$

quotient: $3x^3 + 9x^2 + 20x + 62 + \dfrac{176}{x - 3}$

$\underline{3x^4 - 9x^3}$

$9x^3 - 7x^2$

$\underline{9x^3 - 27x^2}$

$20x^2 + 2x$

$\underline{20x^2 - 60x}$

$62x - 10$

$\underline{62x - 186}$

176

24.

$$
\begin{array}{r|rrrrrr}
-5 & 3 & 13 & 0 & 0 & 12 & -1 \\
 & & -15 & 10 & -50 & 250 & -1310 \\
\hline
 & 3 & -2 & 10 & -50 & 262 & -1311
\end{array}
$$

$$\frac{3x^5 + 13x^4 + 12x - 1}{x + 5} = 3x^4 - 2x^3 + 10x^2 - 50x + 262 - \frac{1311}{x + 5}$$

25. $0 = 6x^3 - 5x^2 + 4x - 15$

Possible rational roots: $\pm 1, \ \pm 3, \ \pm 5, \ \pm 15, \ \pm \frac{1}{2}, \ \pm \frac{3}{2}, \ \pm \frac{5}{2}, \ \pm \frac{15}{2}, \ \pm \frac{1}{3}, \ \pm \frac{5}{3}, \ \pm \frac{1}{6}, \ \pm \frac{5}{6}$

26. $\begin{aligned} f(x) &= (x - 2)[x - (3 + i)][x - (3 - i)] \\ &= (x - 2)[x^2 - x(3 - i) - x(3 + i) + (3 + i)(3 - i)] \\ &= (x - 2)[x^2 - 6x + 10] \\ &= x^3 - 8x^2 + 22x - 20 \end{aligned}$

27. Vertical asymptote: $x = 0$

Horizontal asymptote: $y = \frac{1}{2}$

x-intercept: $(1, 0)$

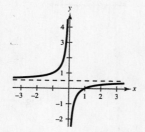

28. Vertical asymptote: $x = 0$

Slant asymptote: $y = 3x$

x-intercept: $\left(\pm \dfrac{2}{\sqrt{3}}, 0 \right)$

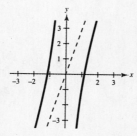

29. $y = 8$ is a horizontal asymptote since the degree on the numerator equals the degree of the denominator. There are no vertical asymptotes.

30. $x = 1$ is a vertical asymptote.

$$\frac{4x^2 - 2x + 7}{x - 1} = 4x + 2 + \frac{9}{x - 1} \text{ so } y = 4x + 2 \text{ is a slant asymptote.}$$

❏ Chapter 3 Practice Test Solutions

1. $x^{3/5} = 8$

$\quad x = 8^{5/3} = \left(\sqrt[3]{8} \right)^5 = 2^5 = 32$

2. $3^{x-1} = \frac{1}{81}$

$\quad 3^{x-1} = 3^{-4}$

$\quad x - 1 = -4$

$\quad x = -3$

3. $f(x) = 2^{-x} = \left(\frac{1}{2}\right)^x$

x	-2	-1	0	1	2
$f(x)$	4	2	1	$\frac{1}{2}$	$\frac{1}{4}$

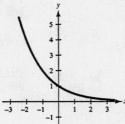

4. $g(x) = e^x + 1$

x	-2	-1	0	1	2
$g(x)$	1.14	1.37	2	3.72	8.39

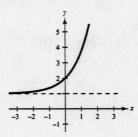

5. $A = P\left(1 + \dfrac{r}{n}\right)^{nt}$

(a) $A = 5000\left(1 + \dfrac{0.09}{12}\right)^{12(3)} \approx \$6543.23$

(b) $A = 5000\left(1 + \dfrac{0.09}{4}\right)^{4(3)} \approx \$6530.25$

(c) $A = 5000e^{(0.09)(3)} \approx \$6549.82$

6. $7^{-2} = \dfrac{1}{49}$

$\log_7 \dfrac{1}{49} = -2$

7. $x = 4 = \log_2 \frac{1}{64}$

$2^{x-4} = \frac{1}{64}$

$2^{x-4} = 2^{-6}$

$x - 4 = -6$

$x = -2$

8. $\log \sqrt[4]{\frac{8}{25}} = \frac{1}{4} \log_b \frac{8}{25}$

$\qquad = \frac{1}{4}[\log_b 8 - \log_b 25]$

$\qquad = \frac{1}{4}[\log_b 2^3 - \log_b 5^2]$

$\qquad = \frac{1}{4}[3 \log_b 2 - 2 \log_b 5]$

$\qquad = \frac{1}{4}[3(0.3562) - 2(0.8271)]$

$\qquad = -0.1464$

9. $5 \ln x - \dfrac{1}{2} \ln y + 6 \ln z = \ln x^5 - \ln \sqrt{y} + \ln z^6 = \ln\left(\dfrac{x^5 z^6}{\sqrt{y}}\right)$

10. $\log_9 28 = \dfrac{\log 28}{\log 9} \approx 1.5166$

11. $\log N = 0.6646$

$N = 10^{0.6646} \approx 4.62$

12.

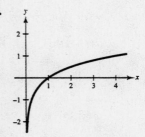

13. Domain:

$x^2 - 9 > 0$

$(x + 3)(x - 3) > 0$

$x < -3 \ \text{ or } \ x > 3$

14.

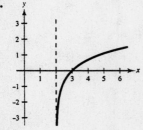

15. $\dfrac{\ln x}{\ln y} \neq \ln(x - y)$ since $\dfrac{\ln x}{\ln y} = \log_y x.$

16. $5^3 = 41$

$$x = \log_5 41 = \frac{\ln 41}{\ln 5} \approx 2.3074$$

17. $x - x^2 = \log_5 \frac{1}{25}$

$5^{x - x^2} = \frac{1}{25}$

$5^{x - x^2} = 5^{-2}$

$x - x^2 = -2$

$0 = x^2 - x - 2$

$0 = (x + 1)(x - 2)$

$x = -1$ or $x = 2$

18. $\log_2 x + \log_2(x - 3) = 2$

$\log_2[x(x - 3)] = 2$

$x(x - 3) = 2^2$

$x^2 - 3x = 4$

$x^2 - 3x - 4 = 0$

$(x + 1)(x - 4) = 0$

$x = 4$

$x = -1$

No solution (extraneous solution)

19. $\dfrac{e^x + e^{-x}}{3} = 4$

$e^x(e^x + e^{-x}) = 12e^x$

$e^{2x} + 1 = 12e^x$

$e^{2x} - 12e^x + 1 = 0$

$$e^x = \frac{12 \pm \sqrt{144 - 4}}{2}$$

$e^x = 11.9161$ or $e^x = 0.0839$

$x = \ln 11.9161$ $x = \ln 0.0839$

$x \approx 2.4779$ $x \approx -2.4779$

20. $A = Pe^{et}$

$12{,}000 = 6000e^{0.13t}$

$2 = e^{0.13t}$

$0.13t = \ln 2$

$$t = \frac{\ln 2}{0.13}$$

$t \approx 5.3319$ yr or 5 yr 4 mo

21. There are 2 points of intersection:

(0.0169, −2.983),

(1.731, 1.647)

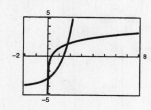

22. $y = 1.0597x^{1.9792}$

❑ Chapter 4 Practice Test Solutions

1. $350° = 350\left(\dfrac{\pi}{180}\right) = \dfrac{35\pi}{18}$

2. $\dfrac{5\pi}{9} = \dfrac{5\pi}{9} \cdot \dfrac{180}{\pi} = 100°$

3. $135° \, 14' \, 12'' = \left(135 + \dfrac{14}{60} + \dfrac{12}{3600}\right)^{\circ}$

$\approx 135.2367°$

4. $-22.569° = -(22° + 0.569(60)')$

$= -22° \, 34.14'$

$= -(22° \, 34' + 0.14(60)'')$

$\approx -22° \, 34' \, 8''$

5. $\cos \theta = \dfrac{2}{3}$

$x = 2, \; r = 3, \; y = \pm\sqrt{9-4} = \pm\sqrt{5}$

$\tan \theta = \dfrac{y}{x} = \pm\dfrac{\sqrt{5}}{2}$

6. $\sin \theta = 0.9063$

$\theta = \arcsin(0.9063)$

$\theta = 65°$ or $\dfrac{13\pi}{36}$

7. $\tan 20° = \dfrac{35}{x}$

$x = \dfrac{35}{\tan 20°} \approx 96.1617$

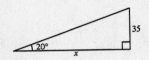

8. $\theta = \dfrac{6\pi}{5}$, θ is in Quadrant III.

Reference angle: $\dfrac{6\pi}{5} - \pi = \dfrac{\pi}{5}$ or $36°$

9. $\csc 3.92 = \dfrac{1}{\sin 3.92} \approx -1.4242$

10. $\tan \theta = 6 = \dfrac{6}{1}$, θ lies in Quadrant III.

$y = -6, \; x = -1, \; r = \sqrt{36+1} = \sqrt{37},$

so $\sec \theta = \dfrac{\sqrt{37}}{-1} \approx -6.0828.$

11. Period: 4π

Amplitude: 3

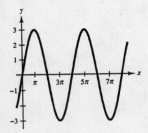

12. Period: 2π

Amplitude: 2

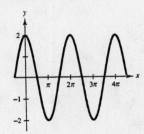

13. Period: $\dfrac{\pi}{2}$

14. Period: 2π

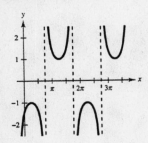

15.

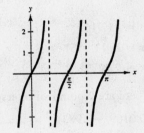

16.

17. $\theta = \arcsin 1$

$\sin \theta = 1$

$\theta = \dfrac{\pi}{2}$

18. $\theta = \arctan(-3)$

$\tan \theta = -3$

$\theta \approx -1.249$ or $-71.565°$

19. $\sin\left(\arccos\dfrac{4}{\sqrt{35}}\right)$

$\sin\theta = \dfrac{\sqrt{19}}{\sqrt{35}} \approx 0.7368$

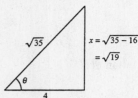

$x = \sqrt{35-16}$
$= \sqrt{19}$

20. $\cos\left(\arcsin\dfrac{x}{4}\right)$

$\cos\theta = \dfrac{\sqrt{16-x^2}}{4}$

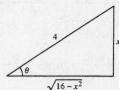

21. Given $A = 40°$, $c = 12$

$B = 90° - 40° = 50°$

$\sin 40° = \dfrac{a}{12}$

$a = 12\sin 40° \approx 7.713$

$\cos 40° = \dfrac{b}{12}$

$b = 12\cos 40° \approx 9.193$

22. Given $B = 6.84°$, $a = 21.3$

$A = 90° - 6.84° = 83.16°$

$\sin 83.16° = \dfrac{21.3}{c}$

$c = \dfrac{21.3}{\sin 83.16°} \approx 21.453$

$\tan 83.16° = \dfrac{21.3}{b}$

$b = \dfrac{21.3}{\tan 83.16°} \approx 2.555$

23. Given $a = 5$, $b = 9$

$c = \sqrt{25 + 81} = \sqrt{106}$

≈ 10.296

$\tan A = \dfrac{5}{9}$

$A = \arctan\dfrac{5}{9} \approx 29.055°$

$B = 90° - 29.055° = 60.945°$

24. $\sin 67° = \dfrac{x}{20}$

$x = 20\sin 67° \approx 18.41$ feet

25. $\tan 5° = \dfrac{250}{x}$

$x = \dfrac{250}{\tan 5°}$

≈ 2857.513 feet

≈ 0.541 mi

❑ Chapter 5 Practice Test Solutions

1. $\tan x = \dfrac{4}{11}$, $\sec x < 0$ $\implies$ x is in Quadrant III.

$y = -4$, $x = -11$, $r = \sqrt{16 + 121} = \sqrt{137}$

$\sin x = -\dfrac{4}{\sqrt{137}} = -\dfrac{4\sqrt{137}}{137}$ $\csc x = -\dfrac{\sqrt{137}}{4}$

$\cos x = -\dfrac{11}{\sqrt{137}} = -\dfrac{11\sqrt{137}}{137}$ $\sec x = -\dfrac{\sqrt{137}}{11}$

$\tan x = \dfrac{4}{11}$ $\cot x = \dfrac{11}{4}$

2. $\dfrac{\sec^2 x + \csc^2 x}{\csc^2 x(1 + \tan^2 x)} = \dfrac{\sec^2 x + \csc^2 x}{\csc^2 x + (\csc^2 x)\tan^2 x} = \dfrac{\sec^2 x + \csc^2 x}{\csc^2 x + \dfrac{1}{\sin^2 x} \cdot \dfrac{\sin^2 x}{\cos^2 x}}$

$\qquad = \dfrac{\sec^2 x + \csc^2 x}{\csc^2 x + \dfrac{1}{\cos^2 x}} = \dfrac{\sec^2 x + \csc^2 x}{\csc^2 x + \sec^2 x} = 1$

3. $\ln|\tan \theta| - \ln|\cot \theta| = \ln\left|\dfrac{\tan \theta}{\cot \theta}\right| = \ln\left|\dfrac{\sin \theta/\cos \theta}{\cos \theta/\sin \theta}\right| = \ln\left|\dfrac{\sin^2 \theta}{\cos^2 \theta}\right| = \ln|\tan^2 \theta| = 2\ln|\tan \theta|$

4. $\cos\left(\dfrac{\pi}{2} - x\right) = \dfrac{1}{\csc x}$ is true since $\cos\left(\dfrac{\pi}{2} - x\right) = \sin x = \dfrac{1}{\csc x}$.

5. $\sin^4 x + (\sin^2 x)\cos^2 x = \sin^2 x(\sin^2 x + \cos^2 x) = \sin^2 x(1) = \sin^2 x$

6. $(\csc x + 1)(\csc x - 1) = \csc^2 x - 1 = \cot^2 x$

7. $\dfrac{\cos^2 x}{1 - \sin x} \cdot \dfrac{1 + \sin x}{1 + \sin x} = \dfrac{\cos^2 x(1 + \sin x)}{1 - \sin^2 x} = \dfrac{\cos^2 x(1 + \sin x)}{\cos^2 x} = 1 + \sin x$

8. $\dfrac{1 + \cos \theta}{\sin \theta} + \dfrac{\sin \theta}{1 + \cos \theta} = \dfrac{(1 + \cos \theta)^2 + \sin^2 \theta}{\sin \theta(1 + \cos \theta)}$

$\qquad = \dfrac{1 + 2\cos \theta + \cos^2 \theta + \sin^2 \theta}{\sin \theta(1 + \cos \theta)} = \dfrac{2 + 2\cos \theta}{\sin \theta(1 + \cos \theta)} = \dfrac{2}{\sin \theta} = 2\csc \theta$

9. $\tan^4 x + 2\tan^2 x + 1 = (\tan^2 x + 1)^2 = (\sec^2 x)^2 = \sec^2 x) = \sec^{4} {}^{x}$

10. (a) $\sin 105° = \sin(60° + 45°) = \sin 60° \cos 45° + \cos 60° \sin 45°$

$\qquad = \dfrac{\sqrt{3}}{2} \cdot \dfrac{\sqrt{2}}{2} + \dfrac{1}{2} \cdot \dfrac{\sqrt{2}}{2} = \dfrac{\sqrt{2}}{4}\left(\sqrt{3} + 1\right)$

(b) $\tan 15° = \tan(60° - 45°) = \dfrac{\tan 60° - \tan 45°}{1 + \tan 60° \tan 45°}$

$\qquad = \dfrac{\sqrt{3} - 1}{1 + \sqrt{3}} \cdot \dfrac{1 - \sqrt{3}}{1 - \sqrt{3}} = \dfrac{2\sqrt{3} - 1 - 3}{1 - 3} = \dfrac{2\sqrt{3} - 4}{-2} = 2 - \sqrt{3}$

11. $(\sin 42°)\cos 38° - (\cos 42°)\sin 38° = \sin(42° - 38°) = \sin 4°$

12. $\tan\left(\theta + \dfrac{\pi}{4}\right) = \dfrac{\tan \theta + \tan(\pi/4)}{1 - (\tan \theta)\tan(\pi/4)} = \dfrac{\tan \theta + 1}{1 - \tan \theta(1)} = \dfrac{1 + \tan \theta}{1 - \tan \theta}$

13. $\sin(\arcsin x - \arccos x) = \sin(\arcsin x)\cos(\arccos x) - \cos(\arcsin x)\sin(\arccos x)$

$\qquad = (x)(x) - \left(\sqrt{1 - x^2}\right)\left(\sqrt{1 - x^2}\right) = x^2 - (1 - x^2) = 2x^2 - 1$

14. (a) $\cos(120°) = \cos[2(60°)] = 2\cos^2 60° - 1 = 2\left(\dfrac{1}{2}\right)^2 - 1 = -\dfrac{1}{2}$

(b) $\tan(300°) = \tan[2(150°)] = \dfrac{2\tan 150°}{1 - \tan^2 150°} = \dfrac{-2\sqrt{3}/3}{1 - (1/3)} = -\sqrt{3}$

15. (a) $\sin 22.5° = \sin \dfrac{45°}{2} = \sqrt{\dfrac{1 - \cos 45°}{2}} = \sqrt{\dfrac{1 - \sqrt{2}/2}{2}} = \dfrac{\sqrt{2 - \sqrt{2}}}{2}$

(b) $\tan \dfrac{\pi}{12} = \tan \dfrac{\pi/6}{2} = \dfrac{\sin(\pi/6)}{1 + \cos(\pi/6)} = \dfrac{1/2}{1 + \sqrt{3}/2} = \dfrac{1}{2 + \sqrt{3}} = 2 - \sqrt{3}$

16. $\sin \theta = \dfrac{4}{5}$, θ lies in Quadrant II $\Rightarrow \cos \theta = -\dfrac{3}{5}$.

$\cos \dfrac{\theta}{2} = \sqrt{\dfrac{1 + \cos \theta}{2}} = \sqrt{\dfrac{1 - 3/5}{2}} = \sqrt{\dfrac{2}{10}} = \dfrac{1}{\sqrt{5}} = \dfrac{\sqrt{5}}{5}$

17. $(\sin^2 x) \cos^2 x = \dfrac{1 - \cos^2 x}{2} \cdot \dfrac{1 + \cos^2 x}{2} = \dfrac{1}{4}[1 - \cos^2 2x] = \dfrac{1}{4}\left[1 - \dfrac{1 + \cos 4x}{2}\right]$

$= \dfrac{1}{8}[2 - (1 + \cos 4x)] = \dfrac{1}{8}[1 - \cos 4x]$

18. $6(\sin 5\theta) \cos 2\theta = 6\left\{\dfrac{1}{2}[\sin(5\theta + 2\theta) + \sin(5\theta - 2\theta)]\right\} = 3[\sin 7\theta + \sin 3\theta]$

19. $\sin(x + \pi) + \sin(x - \pi) = 2\left(\sin \dfrac{[(x + \pi) + (x - \pi)]}{2}\right) \cos \dfrac{[(x + \pi) - (x - \pi)]}{2}$

20. $\dfrac{\sin 9x + \sin 5x}{\cos 9x - \cos 5x} = \dfrac{2 \sin 7x \cos 2x}{-2 \sin 7x \sin 2x} = -\dfrac{\cos 2x}{\sin 2x} = -\cot 2x$

21. $\dfrac{1}{2}[\sin(u + v) - \sin(u - v)] = \dfrac{1}{2}\{(\sin u) \cos v + (\cos u) \sin v - [(\sin u) \cos v - (\cos u) \sin v]\}$

$= \dfrac{1}{2}[2(\cos u) \sin v] = (\cos u) \sin v$

22. $4 \sin^2 x = 1$

$\sin^2 x = \dfrac{1}{4}$

$\sin x = \pm\dfrac{1}{2}$

$\sin x = \dfrac{1}{2} \qquad \text{or} \quad \sin x = -\dfrac{1}{2}$

$x = \dfrac{\pi}{6} \text{ or } \dfrac{5\pi}{6} \qquad x = \dfrac{7\pi}{6} \text{ or } \dfrac{11\pi}{6}$

23. $\tan^2 \theta + \left(\sqrt{3} - 1\right) \tan \theta - \sqrt{3} = 0$

$(\tan \theta - 1)\left(\tan \theta + \sqrt{3}\right) = 0$

$\tan \theta = 1 \qquad \text{or} \quad \tan \theta = -\sqrt{3}$

$\theta = \dfrac{\pi}{4} \text{ or } \dfrac{5\pi}{4} \qquad \theta = \dfrac{2\pi}{3} \text{ or } \dfrac{5\pi}{3}$

24. $\sin 2x = \cos x$

$2(\sin x) \cos x - \cos x = 0$

$\cos x(2 \sin x - 1) = 0$

$\cos x = 0 \qquad \text{or} \quad \sin x = \dfrac{1}{2}$

$x = \dfrac{\pi}{2} \text{ or } \dfrac{3\pi}{2} \qquad x = \dfrac{\pi}{6} \text{ or } \dfrac{5\pi}{6}$

25. $\tan^2 x - 6 \tan x + 4 = 0$

$$\tan x = \frac{-(-6) \pm \sqrt{(-6)^2 - 4(1)(4)}}{2(1)}$$

$$\tan x = \frac{6 \pm \sqrt{20}}{2} = 3 \pm \sqrt{5}$$

$\tan x = 3 + \sqrt{5}$ or $\tan x = 3 - \sqrt{5}$

$x \approx 1.3821$ or 4.5237 $x = 0.6524$ or 3.7940

❑ Chapter 6   Practice Test Solutions

1. $C = 180° - (40° + 12°) = 128°$

$a = \sin 40° \left(\dfrac{100}{\sin 12°} \right) \approx 309.164$

$c = \sin 128° \left(\dfrac{100}{\sin 12°} \right) \approx 379.012$

2. $\sin A = 5 \left(\dfrac{\sin 150°}{20} \right) = 0.125$

$A \approx 7.181°$

$B \approx 180° - (150° + 7.181°) = 22.819°$

$b = \sin 22.819° \left(\dfrac{20}{\sin 150°} \right) \approx 15.513$

3. Area $= \frac{1}{2} ab \sin C$

$\qquad = \frac{1}{2}(3)(5) \sin 130°$

$\qquad \approx 5.745$ square units

4. $h = b \sin A$

$\quad = 35 \sin 22.5°$

$\quad \approx 13.394$

$a = 10$

Since $a < h$ and A is acute, the triangle has no solution.

5. $\cos A = \dfrac{(53)^2 + (38)^2 - (49)^2}{2(53)(38)} \approx 0.4598$

$A \approx 62.627°$

$\cos B = \dfrac{(49)^2 + (38)^2 - (53)^2}{2(49)(38)} \approx 0.2782$

$B \approx 73.847°$

$C \approx 180° - (42.627° + 73.847°)$

$\quad = 43.526°$

6. $c^2 = (100)^2 + (300)^2 - 2(100)(300) \cos 29°$

$\qquad \approx 47,522.8176$

$c \approx 218$

$\cos A = \dfrac{(300)^2 + (218)^2 - (100)^2}{2(300)(218)} \approx 0.97495$

$A \approx 12.85°$

$B \approx 180° - (12.85° + 29°) = 138.15°$

7. $s = \dfrac{a + b + c}{2} = \dfrac{4.1 + 6.8 + 5.5}{2} = 8.2$

Area $= \sqrt{s(s - a)(s - b)(s - c)}$

$\qquad = \sqrt{8.2(8.2 - 4.1)(8.2 - 6.8)(8.2 - 5.5)}$

$\qquad = 11.273$ square units

8. $x^2 = (40)^2 + (70)^2 - 2(40)(70) \cos 168°$

$\qquad \approx 11,977.6266$

$x \approx 190.442$ miles

9. $\mathbf{w} = 4(3\mathbf{i} + \mathbf{j}) - 7(-\mathbf{i} + 2\mathbf{j})$

$\quad = 19\mathbf{i} - 10\mathbf{j}$

10. $\dfrac{\mathbf{v}}{\|\mathbf{v}\|} = \dfrac{5\mathbf{i} + 3\mathbf{j}}{\sqrt{25 + 9}} = \dfrac{5}{\sqrt{34}}\mathbf{i} - \dfrac{3}{\sqrt{34}}\mathbf{j}$

$\qquad = \dfrac{5\sqrt{34}}{34}\mathbf{i} - \dfrac{3\sqrt{34}}{34}\mathbf{j}$

11. $\mathbf{u} = 6\mathbf{i} + 5\mathbf{j}$ $\mathbf{v} = 2\mathbf{i} - 3\mathbf{j}$

$\mathbf{u} \cdot \mathbf{v} = 6(2) + 5(-3) = -3$

$\|\mathbf{u}\| = \sqrt{61}$ $\|\mathbf{v}\| = \sqrt{13}$

$\cos\theta = \dfrac{-3}{\sqrt{61}\sqrt{13}}$

$\theta \approx 96.116°$

12. $4(\mathbf{i}\cos 30° + \mathbf{j}\sin 30°) = 4\left(\dfrac{\sqrt{3}}{2}\mathbf{i} + \dfrac{1}{2}\mathbf{j}\right)$
$$= \langle 4\sqrt{3}, 2\rangle$$

13. $\text{proj}_\mathbf{v}\mathbf{u} = \left(\dfrac{\mathbf{u}\cdot\mathbf{v}}{\|\mathbf{v}\|^2}\right)\mathbf{v} = \dfrac{-10}{20}\langle -2, 4\rangle = \langle 1, -2\rangle$

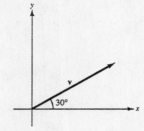

14. $r = \sqrt{25 + 25} = \sqrt{50} = 5\sqrt{2}$

$\tan\theta = \dfrac{-5}{5} = -1$

Since z is in Quadrant IV,

$\theta = 315°$

$z = 5\sqrt{2}(\cos 315° + i\sin 315°).$

15. $\cos 225° = -\dfrac{\sqrt{2}}{2}$ $\sin 225° = -\dfrac{\sqrt{2}}{2}$

$z = 6\left(-\dfrac{\sqrt{2}}{2} - i\dfrac{\sqrt{2}}{2}\right)$

$= -3\sqrt{2} - 3\sqrt{2}i$

16. $[7(\cos 23° + i\sin 23°)][4(\cos 7° + i\sin 7°)] = 7(4)[\cos(23° + 7°) + i\sin(23° + 7°)]$
$$= 28(\cos 30° + i\sin 30°)$$

17. $\dfrac{9\left(\cos\dfrac{5\pi}{4} + i\sin\dfrac{5\pi}{4}\right)}{3(\cos\pi + i\sin\pi)} = \dfrac{9}{3}\left[\cos\left(\dfrac{5\pi}{4} - \pi\right) + i\sin\left(\dfrac{5\pi}{4} - \pi\right)\right] = 3\left(\cos\dfrac{\pi}{4} + i\sin\dfrac{\pi}{4}\right)$

18. $(2 + 2i)^8 = [2\sqrt{2}(\cos 45° + i\sin 45°)]^8 = \left(2\sqrt{2}\right)^8[\cos(8)(45°) + i\sin(8)(45°)]$
$$= 4096[\cos 360° + i\sin 360°] = 4096$$

19. $z = 8\left(\cos\dfrac{\pi}{3} + i\sin\dfrac{\pi}{3}\right)$, $n = 3$

The cube roots of z are:

For $k = 0$, $\sqrt[3]{8}\left[\cos\dfrac{\pi/3}{3} + i\sin\dfrac{\pi/3}{3}\right] = 2\left(\cos\dfrac{\pi}{9} + i\sin\dfrac{\pi}{9}\right).$

For $k = 1$, $\sqrt[3]{8}\left[\cos\dfrac{\pi/3 + 2\pi}{3} + i\sin\dfrac{\pi/3 + 2\pi}{3}\right] = 2\left(\cos\dfrac{7\pi}{9} + i\sin\dfrac{7\pi}{9}\right).$

For $k = 2$, $\sqrt[3]{8}\left[\cos\dfrac{\pi/3 + 4\pi}{3} + i\sin\dfrac{\pi/3 + 4\pi}{3}\right] = 2\left(\cos\dfrac{13\pi}{9} + i\sin\dfrac{13\pi}{9}\right).$

20. $x^4 = -i = 1\left(\cos\dfrac{3\pi}{2} + i\sin\dfrac{3\pi}{2}\right)$

For $k = 0$, $\cos\dfrac{3\pi/2}{4} + i\sin\dfrac{3\pi/2}{4} = \cos\dfrac{3\pi}{8} + i\sin\dfrac{3\pi}{8}$.

For $k = 1$, $\cos\dfrac{3\pi/2 + 2\pi}{4} + i\sin\dfrac{3\pi/2 + 2\pi}{4} = \cos\dfrac{7\pi}{8} + i\sin\dfrac{7\pi}{8}$.

For $k = 2$, $\cos\dfrac{3\pi/2 + 4\pi}{4} + i\sin\dfrac{3\pi/2 + 4\pi}{4} = \cos\dfrac{11\pi}{8} + i\sin\dfrac{11\pi}{8}$.

For $k = 3$, $\cos\dfrac{3\pi/2 + 6\pi}{4} + i\sin\dfrac{3\pi/2 + 6\pi}{4} = \cos\dfrac{15\pi}{8} + i\sin\dfrac{15\pi}{8}$.

❑ Chapter 7 Practice Test Solutions

1.
$$x + y = 1$$
$$3x - y = 15 \implies y = 3x - 15$$
$$x + (3x - 15) = 1$$
$$4x = 16$$
$$x = 4$$
$$y = -3$$

2.
$$x - 3y = -3 \implies x = 3y - 3$$
$$x^2 + 5y = 5$$
$$(3y - 3)^2 + 6y = 5$$
$$9y^2 - 18y + 9 + 6y = 5$$
$$9y^2 - 12y + 4 = 0$$
$$(3y - 2)^2 = 0$$
$$y = \tfrac{2}{3}$$
$$x = -1$$

3.
$$x + y + z = 6 \implies z = 6 - x - y$$
$$2x - y + 3z = 0 \qquad 2x - y + 3(6 - x - y) = 0 \implies -x - 4y = -18$$
$$5x + 2y - z = -3 \qquad 5x + 2y - (6 - x - y) = -3 \implies 6x + 3y = 3$$
$$x = 18 - 4y$$
$$6(18 - 4y) + 3y = 3$$
$$-21y = -105$$
$$y = 5$$
$$x = 18 - 4y = -2$$
$$z = 6 - x - y = 3$$

4.
$$x + y = 110 \implies y = 110 - x$$
$$xy = 2800$$
$$x(110 - x) = 2800$$
$$0 = x^2 - 110x + 2800$$
$$0 = (x - 40)(x - 70)$$
$$x = 40 \ \text{ or } \ x = 70$$
$$y = 70 \qquad y = 40$$

5.
$$2x + 2y = 170 \implies y = \frac{170 - 2x}{2} = 85 - x$$
$$xy = 2800$$
$$x(85 - x) = 2800$$
$$0 = x^2 - 85x + 2800$$
$$0 = (x - 25)(x - 60)$$
$$x = 25 \ \text{ or } \ x = 60$$
$$y = 60 \qquad y = 25$$
Dimensions: 60' × 25'

6.
$$2x + 15y = 4 \implies 2x + 15y = 4$$
$$x - 3y = 23 \implies \underline{5x - 15y = 115}$$
$$7x = 119$$
$$x = 17$$
$$y = \frac{x - 23}{3} = -2$$

7.
$$x + y = 2 \implies 19x + 19y = 38$$
$$38x - 19y = 7 \implies \underline{38x - 19y = 7}$$
$$57x = 45$$
$$x = \frac{45}{57} = \frac{15}{19}$$
$$y = 2 - x = \frac{38}{19} - \frac{15}{19} = \frac{23}{19}$$

8. $y_1 = 2(0.112 - 0.4x)$

$y_2 = \dfrac{(0.13 + 0.3x)}{0.7}$

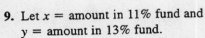

$$0.4x + 0.5y = 0.112 \implies 0.28x + 0.35y = 0.0784$$
$$0.3x - 0.7y = -0.131 \implies \underline{0.15x - 0.35y = -0.0655}$$
$$\overline{0.43x = 0.0129}$$

$$x = \frac{0.0129}{0.43} = 0.03$$

$$y = \frac{0.112 - 0.4x}{0.5} = 0.20$$

9. Let $x =$ amount in 11% fund and
$y =$ amount in 13% fund.

$$x + y = 17000 \implies y = 17000 - x$$
$$0.11x + 0.13y = 2080$$
$$0.11x + 0.13(17000 - x) = 2080$$
$$-0.02x = -130$$
$$x = \$6500$$
$$y = \$10500$$

10. Using a graphing utility, you obtain
$y = 0.7857x - 0.1429$. Analytically, $(4, 3)$, $(1, 1)$, $(-1, -2)$, $(-2, -1)$.

$$n = 4, \sum_{i=1}^{4} x_i = 2, \sum_{i=1}^{4} y_i = 1, \sum_{i=1}^{4} x_i^2 = 22, \sum_{i=1}^{4} x_i y_i = 17$$

$$4b + 2a = 1 \implies 4b + 2a = 1$$
$$2b + 22a = 17 \implies \underline{-4b - 44a = -34}$$
$$-42a = -33$$

$$a = \frac{33}{42} = \frac{11}{14}$$

$$b = \frac{1}{4}\left(1 - 2\left(\frac{33}{42}\right)\right) = -\frac{1}{7}$$

$$y = ax + b = \frac{11}{14}x - \frac{1}{7}$$

11.
$$x + y = -2 \implies -2x - 2y = 4 \qquad -9y + 3z = 45$$
$$2x - y + z = 11 \implies 2x - \underline{y + z = 11} \qquad \underline{4y - 3z = -20}$$
$$4y - 3z = -20 \qquad\qquad -3y + z = 15 \qquad -5y = 25$$
$$y = -5$$
$$x = 3$$
$$z = 0$$

12.

$$
\begin{aligned}
4x - y + 5z &= 4 &\implies&& 4x - y + 5z &= 4 \\
2x + y - z &= 0 &\implies&& -4x - 2y + 2z &= 0 \\
2x + 4y + 8z &= 0 &&& -3y + 7z &= 4
\end{aligned}
$$

$$
\begin{aligned}
2x + 4y + 8z &= 0 \\
-2x - y + z &= 0 \\
\hline
3y + 9z &= 0 \\
-3y + 7z &= 4 \\
\hline
16z &= 4 \\
z &= \tfrac{1}{4} \\
y &= -\tfrac{3}{4} \\
x &= \tfrac{1}{2}
\end{aligned}
$$

13.

$$
\begin{aligned}
3x + 2y - z &= 5 &\implies&& 6x + 4y - 2z &= 10 \\
6x - y + 5z &= 2 &\implies&& -6x + y - 5z &= -2 \\
&&& \hline && 5y - 7z &= 8 \\
&&&& y &= \frac{8 + 7z}{5}
\end{aligned}
$$

$$
\begin{aligned}
3x + 2y - z &= 5 \\
12x - 2y + 10z &= 4 \\
\hline
15x \quad + 9z &= 9 \\
x &= \frac{9 - 9z}{15} = \frac{3 - 3z}{5}
\end{aligned}
$$

Let $z = a$, then $x = \dfrac{3 - 3a}{5}$ and $y = \dfrac{8 + 7a}{5}$.

14. $y = ax^2 + bx + c$ passes through $(0, -1)$, $(1, 4)$, and $(2, 13)$.

At $(0, -1)$: $-1 = a(0)^2 + b(0) + c \implies c = -1$

At $(1, 4)$: $4 = a(1)^2 + b(1) - 1 \implies 5 = a + b \implies 5 = a + b$

At $(2, 13)$: $13 = a(2)^2 + b(2) - 1 \implies 14a + 2b \implies \dfrac{-7 = -2a - b}{-2 = -a}$

$$
\begin{aligned}
a &= 2 \\
b &= 3
\end{aligned}
$$

Thus, $y = 2x^2 + 3x - 1$.

15. $s = \frac{1}{2}at^2 + v_0t + s_0$ passes through $(1, 12)$, $(2, 5)$, and $(3, 4)$.

At $(1, 12)$: $12 = \frac{1}{2}a + v_0 + s_0 \implies \quad 24 = \quad a + 2v_0 + 2s_0$

At $(2, 5)$: $\quad 5 = 2a + 2v_0 + s_0 \implies \quad -5 = -2a - 2v_0 - \quad s_0$

At $(3, 4)$: $\quad 4 = \frac{9}{2}a + 3v_0 + s_0 \implies \quad \overline{19 = \quad -a \qquad + \quad s_0}$

$$15 = \quad 6a + 6v_0 + 3s_0$$
$$-8 = -9a - 6v_0 - 2s_0$$
$$\overline{7 = -3a \qquad + \quad s_0}$$
$$-19 = \quad a \qquad - \quad s_0$$
$$\overline{-12 = -2a}$$
$$a = 6$$
$$s_0 = 25$$
$$v_0 = -16$$

Thus, $s = \frac{1}{2}(6)t^2 - 16t + 25 = 3t^2 - 16t + 25$.

16. $x^2 + y^2 \geq 9$

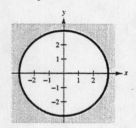

17. $x + y \leq 6$

$\qquad x \geq 2$

$\qquad y \geq 0$

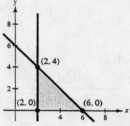

18. Line through $(0, 0)$ and $(0, 7)$: $x = 0$

Line through $(0, 0)$ and $(2, 3)$: $y = \frac{3}{2}x$ or $3x - 2y = 0$

Line through $(0, 7)$ and $(2, 3)$: $y = -2x + 7$ or $2x + y = 7$

Inequalities: $\qquad x \geq 0$

$$3x - 2y \leq 0$$
$$2x + y \leq 7$$

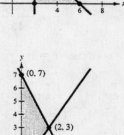

19. Vertices: $(0, 0)$, $(0, 7)$, $(6, 0)$, $(3, 5)$

$z = 30x + 26y$

At $(0, 0)$: $z = 0$

At $(0, 7)$: $z = 182$

At $(6, 0)$: $z = 180$

At $(3, 5)$: $z = 220$

The maximum value is z is 220.

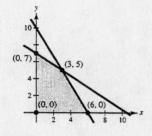

20. $x^2 + y^2 \le 4$

$(x - 2)^2 + y^2 \ge 4$

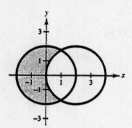

❑ Chapter 8 Practice Test Solutions

1.
$$\begin{bmatrix} 1 & -2 & 4 \\ 3 & -5 & 9 \end{bmatrix}$$

$-3R_1 + R_2 \rightarrow \begin{bmatrix} 1 & -2 & 4 \\ 0 & 1 & -3 \end{bmatrix}$

$2R_2 + R_1 \rightarrow \begin{bmatrix} 1 & 0 & -2 \\ 0 & 1 & -3 \end{bmatrix}$

2. $3x + 5y = \ \ \ 3$

$\ \ 2x - \ \ y = -11$

$$\begin{bmatrix} 3 & 5 & \vdots & 3 \\ 2 & -1 & \vdots & -11 \end{bmatrix}$$

$-R_2 + R_1 \rightarrow \begin{bmatrix} 1 & 6 & \vdots & 14 \\ 2 & -1 & \vdots & -11 \end{bmatrix}$

$-2R_1 + R_2 \rightarrow \begin{bmatrix} 1 & 6 & \vdots & 14 \\ 0 & -13 & \vdots & -39 \end{bmatrix}$

$-\frac{1}{13}R_2 \rightarrow \begin{bmatrix} 1 & 6 & \vdots & 14 \\ 0 & 1 & \vdots & 3 \end{bmatrix}$

$-6R_2 + R_1 \rightarrow \begin{bmatrix} 1 & 0 & \vdots & -4 \\ 0 & 1 & \vdots & 3 \end{bmatrix}$

Answer: $x = -4, y = 3$

3. $2x + 3y = -3$

$\ \ 3x - 2y = \ \ \ 8$

$\ \ \ x + \ \ y = \ \ \ 1$

$$\begin{bmatrix} 2 & 3 & \vdots & -3 \\ 3 & 2 & \vdots & 8 \\ 1 & 1 & \vdots & 1 \end{bmatrix}$$

$\begin{matrix} R_1 \rightarrow \\ \\ R_3 \rightarrow \end{matrix} \begin{bmatrix} 1 & 1 & \vdots & 1 \\ 3 & 2 & \vdots & 8 \\ 2 & 3 & \vdots & -3 \end{bmatrix}$

$\begin{matrix} -3R_1 + R_2 \rightarrow \\ -2R_1 + R_3 \rightarrow \end{matrix} \begin{bmatrix} 1 & 1 & \vdots & 1 \\ 0 & -1 & \vdots & 5 \\ 0 & 1 & \vdots & -5 \end{bmatrix}$

$\begin{matrix} R_2 + R_1 \rightarrow \\ -R_2 \rightarrow \\ -R_2 + R_3 \rightarrow \end{matrix} \begin{bmatrix} 1 & 0 & \vdots & 6 \\ 0 & 1 & \vdots & -5 \\ 0 & 0 & \vdots & 0 \end{bmatrix}$

Answer: $x = 6, y = -5$

4. $x \qquad + 3z = -5$
$\qquad 2x + y \qquad = \quad 0$
$\qquad 3x + y - \ z = -3$

$$\begin{bmatrix} 1 & 0 & 3 & \vdots & -5 \\ 2 & 1 & 0 & \vdots & 0 \\ 3 & 1 & -1 & \vdots & 3 \end{bmatrix}$$

$\begin{array}{c} -2R_1 + R_2 \rightarrow \\ -3R_1 + R_3 \rightarrow \end{array} \begin{bmatrix} 1 & 0 & 3 & \vdots & -5 \\ 0 & 1 & -6 & \vdots & 10 \\ 0 & 1 & -10 & \vdots & 18 \end{bmatrix}$

$\begin{array}{c} \\ \\ -R_2 + R_3 \rightarrow \end{array} \begin{bmatrix} 1 & 0 & 3 & \vdots & -5 \\ 0 & 1 & -6 & \vdots & 10 \\ 0 & 0 & -4 & \vdots & 8 \end{bmatrix}$

$\begin{array}{c} -3R_3 + R_1 \rightarrow \\ 6R_3 + R_2 \rightarrow \\ -\frac{1}{4}R_4 \rightarrow \end{array} \begin{bmatrix} 1 & 0 & 0 & \vdots & 1 \\ 0 & 1 & 0 & \vdots & -2 \\ 0 & 0 & 1 & \vdots & -2 \end{bmatrix}$

Answer: $x = 1, y = -2, z = -2$

5. $\begin{bmatrix} 1 & 4 & 5 \\ 2 & 0 & -3 \end{bmatrix} \begin{bmatrix} 1 & 6 \\ 0 & -7 \\ -1 & 2 \end{bmatrix} = \begin{bmatrix} -4 & -12 \\ 5 & 6 \end{bmatrix}$

6. $3A - 5B = 3\begin{bmatrix} 9 & 1 \\ -4 & 8 \end{bmatrix} - 5\begin{bmatrix} 6 & -2 \\ 3 & 5 \end{bmatrix}$

$\qquad = \begin{bmatrix} 27 & 3 \\ -12 & 24 \end{bmatrix} - \begin{bmatrix} 30 & -10 \\ 15 & 25 \end{bmatrix}$

$\qquad = \begin{bmatrix} -3 & 13 \\ -27 & -1 \end{bmatrix}$

7. $f(A) = \begin{bmatrix} 3 & 0 \\ 7 & 1 \end{bmatrix}^2 - 7\begin{bmatrix} 3 & 0 \\ 7 & 1 \end{bmatrix} + 8\begin{bmatrix} 1 & 0 \\ 0 & 1 \end{bmatrix}$

$\qquad = \begin{bmatrix} 3 & 0 \\ 7 & 1 \end{bmatrix}\begin{bmatrix} 3 & 0 \\ 7 & 1 \end{bmatrix} - \begin{bmatrix} 21 & 0 \\ 49 & 7 \end{bmatrix} + \begin{bmatrix} 8 & 0 \\ 0 & 8 \end{bmatrix}$

$\qquad = \begin{bmatrix} 9 & 0 \\ 28 & 1 \end{bmatrix} - \begin{bmatrix} 21 & 0 \\ 49 & 7 \end{bmatrix} + \begin{bmatrix} 8 & 0 \\ 0 & 8 \end{bmatrix}$

$\qquad = \begin{bmatrix} -4 & 0 \\ -21 & 2 \end{bmatrix}$

8. False since

$$(A + B)(A + 3B) = A(A + 3B) + B(A + 3B)$$
$$= A^2 + 3AB + BA + 3B^2.$$

9.
$$\begin{bmatrix} 1 & 2 & \vdots & 1 & 0 \\ 3 & 5 & \vdots & 0 & 1 \end{bmatrix}$$

$$-3R_1 + R_2 \rightarrow \begin{bmatrix} 1 & 2 & \vdots & 1 & 0 \\ 0 & -1 & \vdots & -3 & 1 \end{bmatrix}$$

$$\begin{matrix} 2R_2 + R_1 \rightarrow \\ -R_2 \rightarrow \end{matrix} \begin{bmatrix} 1 & 0 & \vdots & -5 & 2 \\ 0 & 1 & \vdots & 3 & -1 \end{bmatrix}$$

$$A^{-1} = \begin{bmatrix} -5 & 2 \\ 3 & -1 \end{bmatrix}$$

10.
$$\begin{bmatrix} 1 & 1 & 1 & \vdots & 1 & 0 & 0 \\ 3 & 6 & 5 & \vdots & 0 & 1 & 0 \\ 6 & 10 & 8 & \vdots & 0 & 0 & 1 \end{bmatrix}$$

$$\begin{matrix} -3R_1 + R_2 \rightarrow \\ -6R_1 + R_3 \rightarrow \end{matrix} \begin{bmatrix} 1 & 1 & 1 & \vdots & 1 & 0 & 0 \\ 0 & 3 & 2 & \vdots & -3 & 1 & 0 \\ 0 & 4 & 2 & \vdots & -6 & 0 & 1 \end{bmatrix}$$

$$\begin{matrix} -R_2 + R_1 \rightarrow \\ \frac{1}{3}R_2 \rightarrow \\ -4R_2 + R_3 \rightarrow \end{matrix} \begin{bmatrix} 1 & 0 & \frac{1}{3} & \vdots & 2 & -\frac{1}{3} & 0 \\ 0 & 1 & \frac{2}{3} & \vdots & -1 & \frac{1}{3} & 0 \\ 0 & 0 & -\frac{2}{3} & \vdots & -2 & -\frac{4}{3} & 1 \end{bmatrix}$$

$$\begin{matrix} \frac{1}{2}R_3 + R_1 \rightarrow \\ R_3 + R_2 \rightarrow \\ -\frac{3}{2}R_3 \rightarrow \end{matrix} \begin{bmatrix} 1 & 0 & 0 & \vdots & 1 & -1 & \frac{1}{2} \\ 0 & 1 & 0 & \vdots & -3 & -1 & 1 \\ 0 & 0 & 1 & \vdots & 3 & 2 & -\frac{3}{2} \end{bmatrix}$$

$$A^{-1} = \begin{bmatrix} 1 & -1 & \frac{1}{2} \\ -3 & -1 & 1 \\ 3 & 2 & -\frac{3}{2} \end{bmatrix}$$

11. (a) $x + 2y = 4$
$3x + 5y = 1$

$$\begin{bmatrix} 1 & 2 & \vdots & 1 & 0 \\ 3 & 5 & \vdots & 0 & 1 \end{bmatrix}$$

$$-3R_1 + R_2 \rightarrow \begin{bmatrix} 1 & 2 & \vdots & 1 & 0 \\ 0 & -1 & \vdots & -3 & 1 \end{bmatrix}$$

$$\begin{matrix} -2R_2 + R_1 \rightarrow \\ -R_2 \rightarrow \end{matrix} \begin{bmatrix} 1 & 0 & \vdots & -5 & 2 \\ 0 & 1 & \vdots & 3 & -1 \end{bmatrix}$$

$$X = A^{-1}B = \begin{bmatrix} -5 & 2 \\ 3 & -1 \end{bmatrix} \begin{bmatrix} 4 \\ 1 \end{bmatrix} = \begin{bmatrix} -18 \\ 11 \end{bmatrix}$$

$$x = -18, y = 11$$

(b) $x + 2y = 3$
$3x + 5y = -2$

$$X = A^{-1}B - \begin{bmatrix} -5 & 2 \\ 3 & -1 \end{bmatrix} \begin{bmatrix} 3 \\ -2 \end{bmatrix} = \begin{bmatrix} -19 \\ 11 \end{bmatrix}$$

$$x = -19, y = 11$$

12. $\begin{vmatrix} 6 & -1 \\ 3 & 4 \end{vmatrix} = 24 - (-3) = 27$

13. $\begin{vmatrix} 1 & 3 & -1 \\ 5 & 9 & 0 \\ 6 & 2 & -5 \end{vmatrix} = 1(-45) + 3(25) + (-1)(-44) = 74$

14. $\begin{vmatrix} 1 & 4 & 2 & 3 \\ 0 & 1 & -2 & 0 \\ 3 & 5 & -2 & 1 \\ 2 & 0 & 6 & 1 \end{vmatrix} = -7$

15. $\begin{vmatrix} 6 & 4 & 3 & 0 & 6 \\ 0 & 5 & 1 & 4 & 8 \\ 0 & 0 & 2 & 7 & 3 \\ 0 & 0 & 0 & 9 & 2 \\ 0 & 0 & 0 & 0 & 1 \end{vmatrix} = 6(5)(2)(9)(1) = 540$

16. $\text{Area} = \dfrac{1}{2} \begin{vmatrix} 0 & 7 & 1 \\ 5 & 0 & 1 \\ 3 & 9 & 1 \end{vmatrix} = \dfrac{1}{2}(31)$

$= 15.5$ square units

17. $\begin{vmatrix} x & y & 1 \\ 2 & 7 & 1 \\ -1 & 4 & 1 \end{vmatrix} = 3x - 3y + 15 = 0$

$\text{OR} = x - y + 5 = 0$

18. $x = \dfrac{\begin{vmatrix} 4 & -7 \\ 11 & 5 \end{vmatrix}}{\begin{vmatrix} 6 & -7 \\ 2 & 5 \end{vmatrix}} = \dfrac{97}{44}$

19. $z = \dfrac{\begin{vmatrix} 3 & 0 & 1 \\ 0 & 1 & 3 \\ 1 & -1 & 2 \end{vmatrix}}{\begin{vmatrix} 3 & 0 & 1 \\ 0 & 1 & 4 \\ 1 & -1 & 0 \end{vmatrix}} = \dfrac{14}{11}$

20. $y = \dfrac{\begin{vmatrix} 721.4 & 33.77 \\ 45.9 & 19.85 \end{vmatrix}}{\begin{vmatrix} 721.4 & -29.1 \\ 45.9 & 105.6 \end{vmatrix}} = \dfrac{12{,}769.747}{77{,}515.530} \approx 0.1647$

❏ Chapter 9 Practice Test Solutions

1. $a_n = \dfrac{2n}{(n+2)!}$

$a_1 = \dfrac{2(1)}{3!} = \dfrac{2}{6} = \dfrac{1}{3}$

$a_2 = \dfrac{2(2)}{4!} = \dfrac{4}{24} = \dfrac{1}{6}$

$a_3 = \dfrac{2(3)}{5!} = \dfrac{6}{120} = \dfrac{1}{20}$

$a_4 = \dfrac{2(4)}{6!} = \dfrac{8}{720} = \dfrac{1}{90}$

$a_5 = \dfrac{2(5)}{7!} = \dfrac{10}{5040} = \dfrac{1}{504}$

Terms: $\dfrac{1}{3}, \dfrac{1}{6}, \dfrac{1}{20}, \dfrac{1}{90}, \dfrac{1}{504}$

2. $a_n = \dfrac{n+3}{3^n}$

3. $\displaystyle\sum_{i=1}^{6}(2i-1) = 1 + 3 + 5 + 7 + 9 + 11 = 36$

4. $a_1 = 23, d = -2$

$a_2 = a_1 + d = 21$

$a_3 = a_2 + d = 19$

$a_4 = a_3 + d = 17$

$a_5 = a_4 + d = 15$

Terms: 23, 21, 19, 17, 15

5. $a_1 = 12, d = 3, n = 50$

$a_n = a_1 + (n-1)d$

$a_{50} = 12 + (50-1)3 = 159$

6. $a_1 = 1$

$a_{200} = 200$

$S_n = \dfrac{n}{2}(a_1 + a_n)$

$S_{200} = \dfrac{200}{2}(1 + 200) = 20{,}100$

7. $a_1 = 7, r = 2$

$a_2 = a_1 r = 14$

$a_3 = a_1 r^2 = 28$

$a_4 = a_1 r^3 = 56$

$a_5 = a_1 r^4 = 112$

Terms: 7, 14, 28, 56, 112

8. $\displaystyle\sum_{n=0}^{9}6\left(\dfrac{2}{3}\right)^n$

$a_1 = 6, r = \dfrac{2}{3}, n = 10$

$S_n = \dfrac{a_1(1 - r^n)}{1 - r}$

$= \dfrac{6\left(1 - (2/3)^{10}\right)}{1 - (2/3)} \approx 17.6879$

9. $\displaystyle\sum_{n=0}^{\infty}(0.03)^n$

$a_1 = 1, r = 0.03$

$S = \dfrac{a_1}{1-r} = \dfrac{1}{1 - 0.03}$

$= \dfrac{1}{0.97} = \dfrac{100}{97} \approx 1.0309$

10. For $n = 1$, $1 = \dfrac{1(1+1)}{2}$.

Assume that $1 + 2 + 3 = 4 + \cdots + k = \dfrac{k(k+1)}{2}$.

Now for $n = k + 1$,

$1 + 2 + 3 + 4 + \cdots + k + (k+1) = \dfrac{k(k+1)}{2} + k + 1$

$= \dfrac{k(k+1)}{2} + \dfrac{2(k+1)}{2}$

$= \dfrac{(k+1)(k+2)}{2}.$

Thus, $1 + 2 + 3 + 4 + \cdots + n = \dfrac{n(n+1)}{2}$ for all integers $n \geq 1$.

11. For $n = 4$, $4! > 2^4$. Assume that $k! > 2^k$. Then $(k + 1)! = (k + 1)(k!) > (k + 1)2^k > 2 \cdot 2^k = 2^{k+1}$.
Thus, $n! > 2^n$ for all integers $n \geq 4$.

12. $_{13}C_4 = \dfrac{13!}{(13 - 4)!4!} = 715$

13. $_{12}C_5 x^7(-2)^5 = -25{,}344x^7$

14. $_{30}P_4 = \dfrac{30!}{(30 - 4)!} = 657{,}720$

15. $6! = 720$ ways

16. $_{12}P_3 = 1320$

17. $P(2) + P(3) + P(4) = \frac{1}{36} + \frac{2}{36} + \frac{3}{36}$
$\qquad\qquad\qquad\qquad = \frac{6}{36} = \frac{1}{6}$

18. $P(K, B10) = \frac{4}{52} \cdot \frac{2}{51} = \frac{2}{663}$

19. Let A = probability of no faulty units.
$P(A) = \left(\frac{997}{1000}\right)^{50} \approx 0.8605$
$P(A') = 1 - P(A) \approx 0.1395$

20. Mean $= \dfrac{3 + 5 + 8 + 10 + 16 + 16}{6} = \dfrac{58}{6} = 9.67$

Variance $= \dfrac{\left[\left(3 - \frac{58}{6}\right)^2 + \left(5 - \frac{58}{6}\right)^2 + \left(8 - \frac{58}{6}\right)^2 + \left(10 - \frac{58}{6}\right)^2 + \left(16 + \frac{58}{6}\right)^2 + \left(16 - \frac{58}{6}\right)^2\right]}{6}$

$= \dfrac{149.33}{6} = 24.89$

Standard deviation $= \sqrt{24.89} = 4.99$

❑ Chapter 10 Practice Test Solutions

1. $x^2 - 6x - 4y + 1 = 0$

$$x^2 - 6x + 9 = 4y - 1 + 9$$
$$(x - 3)^2 = 4y + 8$$
$$(x - 3)^2 = 4(1)(y + 2) \Rightarrow p = 1$$

Vertex: $(3, -2)$
Focus: $(3, -1)$
Directrix: $y = -3$

2. Vertex: $(2, -5)$
Focus: $(2, -6)$
Vertical axis; opens downward with $p = -1$

$$(x - h)^2 = 4p(y - k)$$
$$(x - 2)^2 = 4(-1)(y + 5)$$
$$x^2 - 4x + 4 = -4y - 20$$
$$x^2 - 4x + 4y + 24 = 0$$

3. $x^2 + 4y^2 - 2x + 32y + 61 = 0$

$$(x^2 - 2x + 1) + 4(y^2 + 8y + 16) = -61 + 1 + 64$$
$$(x - 1)^2 + 4(y + 4)^2 = 4$$
$$\frac{(x - 1)^2}{4} + \frac{(y + 4)^2}{1} = 1$$

$a = 2, b = 1, c = \sqrt{3}$
Horizontal major axis
Center: $(1, -4)$
Foci: $\left(1 \pm \sqrt{3}, -4\right)$
Vertices: $(3, -4), (-1, -4)$

Eccentricity: $e = \dfrac{\sqrt{3}}{2}$

4. Vertices: $(0, \pm 6)$

Eccentricity: $e = \dfrac{1}{2}$

Center: $(0, 0)$
Vertical major axis

$a = 6, e = \dfrac{c}{a} = \dfrac{c}{6} = \dfrac{1}{2} \Rightarrow c = 3$

$b^2 = (6)^2 - (3)^2 = 27$

$\dfrac{x^2}{27} + \dfrac{y^2}{36} = 1$

5. $16y^2 - x^2 - 6x - 128y + 231 = 0$

$$16(y^2 - 8y + 16) - (x^2 + 6x + 9) = -231 + 256 - 9$$
$$16(y - 4)^2 - (x + 3)^2 = 16$$
$$\frac{(y - 4)^2}{1} - \frac{(x + 3)^2}{16} = 1$$

$a = 1, b = 4, c = \sqrt{17}$
Center: $(-3, 4)$
Vertical transverse axis
Vertices: $(-3, 5), (-3, 3)$
Foci: $\left(-3, 4 \pm \sqrt{17}\right)$

Asymptotes: $y = 4 \pm \dfrac{1}{4}(x + 3)$

6. Vertices: $(\pm 3, 2)$
Foci: $(\pm 5, 2)$
Center: $(0, 2)$
Horizontal transverse axis

$a = 3, c = 5, b = 4$

$$\frac{(x - 0)^2}{9} - \frac{(y - 2)^2}{16} = 1$$

$$\frac{x^2}{9} - \frac{(y - 2)^2}{16} = 1$$

7. $5x^2 + 2xy + 5y^2 - 10 = 0$

$A = 5, B = 2, C = 5$

$\cot 2\theta = \dfrac{5-5}{2} = 0$

$2\theta = \dfrac{\pi}{2} \Rightarrow \theta = \dfrac{\pi}{4}$

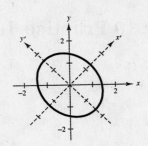

$x = x' \cos \dfrac{\pi}{4} - y' \sin \dfrac{\pi}{4}$　　　　$x = x' \cos \dfrac{\pi}{4} + y' \sin \dfrac{\pi}{4}$

$= \dfrac{x' - y'}{\sqrt{2}}$　　　　　　　　　　　$= \dfrac{x' + y'}{\sqrt{2}}$

$$5\left(\dfrac{x'-y'}{\sqrt{2}}\right)^2 + 2\left(\dfrac{x'-y'}{\sqrt{2}}\right)\left(\dfrac{x'+y'}{\sqrt{2}}\right) + 5\left(\dfrac{x'+y'}{\sqrt{2}}\right)^2 - 10 = 0$$

$$\dfrac{5(x')^2}{2} - \dfrac{10x'y'}{2} + \dfrac{5(y')^2}{2} + (x')^2 - (y')^2 + \dfrac{5(x')^2}{2} + \dfrac{10x'y'}{2} + \dfrac{5(y')^2}{2} - 10 = 0$$

$$6(x')^2 + 4(y')^2 - 10 = 0$$

$$\dfrac{3(x')^2}{5} + \dfrac{2(y')^2}{5} = 1$$

$$\dfrac{(x')^2}{5/3} + \dfrac{(y')^2}{5/2} = 1$$

Ellipse centered at the origin

8. (a) $6x^2 - 2xy + y^2 = 0$

$A = 6, B = -2, C = 1$

$B^2 - 4AC = (-2)^2 - 4(6)(1) = -20 < 0$

Ellipse

(b) $x^2 + 4xy + 4y^2 - x - y + 17 = 0$

$A = 1, B = 4, C = 4$

$B^2 - 4AC = (4)^2 - 4(1)(4) = 0$

Parabola

9. $x = 3 - 2 \sin \theta, y = 1 + 5 \cos \theta$

$\dfrac{x-3}{-2} = \sin \theta, \dfrac{y-1}{5} = \cos \theta$

$\left(\dfrac{x-3}{-2}\right)^2 + \left(\dfrac{y-1}{5}\right)^2 = 1$

$\dfrac{(x-3)^2}{4} + \dfrac{(y-1)^2}{25} = 1$

10. $x = e^{2t}, y = e^{4t}$

$x > 0, y > 0$

$y = (e^{2t})^2 = (x)^2 = x^2, x < 0, y > 0$

11. Polar: $\left(\sqrt{2}, \dfrac{3\pi}{4}\right)$

$x = \sqrt{2} \cos \dfrac{3\pi}{4} = \sqrt{2}\left(-\dfrac{1}{\sqrt{2}}\right) = -1$

$y = \sqrt{2} \sin \dfrac{3\pi}{4} = \sqrt{2}\left(\dfrac{1}{\sqrt{2}}\right) = 1$

Rectangular: $(-1, 1)$

12. Rectangular: $\left(\sqrt{3}, -1\right)$

$r = \pm\sqrt{(\sqrt{3})^2 + (-1)^2} = \pm 2$

$\tan \theta = \dfrac{\sqrt{3}}{-1} = -\sqrt{3}$

$\theta = \dfrac{2\pi}{3}$ or $\theta = \dfrac{5\pi}{3}$

Polar: $\left(-2, \dfrac{2\pi}{3}\right)$ or $\left(2, \dfrac{5\pi}{3}\right)$

13. Rectangular: $4x - 3y = 12$

Polar: $4r \cos \theta - 3r \sin \theta = 12$

$r(4 \cos \theta - 3 \sin \theta) = 12$

$r = \dfrac{12}{4 \cos \theta - 3 \sin \theta}$

14. Polar: $r = 5 \cos \theta$

$r^2 = 5r \cos \theta$

Rectangular:　　$x^2 + y^2 = 5x$

$x^2 + y^2 - 5x = 0$

15. $r = 1 - \cos\theta$

Cardioid

Symmetry: Polar axis

Maximum value of $|r|$: $r = 2$ when $\theta = \pi$.

Zero of r: $r = 0$ when $\theta = 0$

θ	0	$\dfrac{\pi}{2}$	π	$\dfrac{3\pi}{2}$
r	0	1	2	1

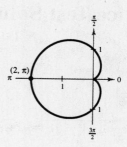

16. $r = 5\sin 2\theta$

Rose curve with four petals

Symmetry: Polar axis, $\theta = \dfrac{\pi}{2}$, and pole

Maximum value of $|r|$: $|r| = 5$ when $\theta = \dfrac{\pi}{4}, \dfrac{3\pi}{4}, \dfrac{5\pi}{4}, \dfrac{7\pi}{4}$

Zeros of r: $r = 0$ when $\theta = 0, \dfrac{\pi}{2}, \pi, \dfrac{3\pi}{2}$

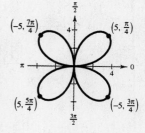

17. $r = \dfrac{3}{6 - \cos\theta}$

$r = \dfrac{\frac{1}{2}}{1 - \frac{1}{6}\cos\theta}$

$e = \dfrac{1}{6} < 1$, so the graph is an ellipse.

θ	0	$\dfrac{\pi}{2}$	π	$\dfrac{3\pi}{2}$
r	$\dfrac{3}{5}$	$\dfrac{1}{2}$	$\dfrac{3}{7}$	$\dfrac{1}{2}$

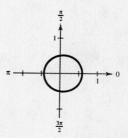

18. Parabola

Vertex: $\left(6, \dfrac{\pi}{2}\right)$

Focus: $(0, 0)$

$e = 1$

$r = \dfrac{ep}{1 + e\sin\theta}$

$r = \dfrac{p}{1 + \sin\theta}$

$6 = \dfrac{p}{1 + \sin(\pi/2)}$

$6 = \dfrac{p}{2}$

$12 = p$

$r = \dfrac{12}{1 + \sin\theta}$

❑ Chapter 11 Practice Test Solutions

1. Let $A = (0, 0, 0)$, $B = (1, 2, -4)$, $C = (0, -2, -1)$

Side AB: $\sqrt{1^2 + 2^2 + 4^2} = \sqrt{21}$

Side AC: $\sqrt{0^2 + 2^2 + 1^2} = \sqrt{5}$

Side BC: $\sqrt{(-1)^2 + (-2 - 2)^2 + (-1 + 4)^2} = \sqrt{1 + 16 + 9} = \sqrt{26}$

$BC^2 = AB^2 + AC^2$

$26 = 21 + 5$

2. $(x - 0)^2 + (y - 4)^2 + (z - 1)^2 = 5^2$

$\quad x^2 + (y - 4)^2 + (z - 1)^2 = 25$

3. $(x^2 + 2x + 1) + y^2 + (z^2 - 4z + 4) = 1 + 4 + 11$

$\quad\quad (x + 1)^2 + y^2 + (z - 2)^2 = 16$

Center: $(-1, 0, 2)$

Radius: 4

4. $\mathbf{u} - 3\mathbf{v} = \langle 1, 0, -1 \rangle - 3\langle 4, 3, -6 \rangle = \langle 1, 0, -1 \rangle - \langle 12, 9, -18 \rangle$

$\quad\quad\quad\quad = \langle -11, -9, 17 \rangle$

5. $\frac{1}{2}\mathbf{v} = \frac{1}{2}\langle 2, 4, -6 \rangle = \langle 1, 2, -3 \rangle$

$\left\| \frac{1}{2}\mathbf{v} \right\| = \sqrt{1^2 + 2^2 + (-3)^2} = \sqrt{14}$

6. $\mathbf{u} \cdot \mathbf{v} = \langle 2, 1, -3 \rangle \cdot \langle 1, 1, -2 \rangle$

$\quad\quad = 2 + 1 + 6 = 9$

7. Because $\mathbf{v} = \langle -3, -3, 3 \rangle = -3\langle 1, 1, -1 \rangle = -3\mathbf{u}$, $\mathbf{u}$ and $\mathbf{v}$ are parallel.

8. $\mathbf{u} \times \mathbf{v} = \begin{vmatrix} \mathbf{i} & \mathbf{j} & \mathbf{k} \\ -1 & 0 & 2 \\ 1 & -1 & 3 \end{vmatrix} = \langle 2, 5, 1 \rangle$

$\mathbf{v} \times \mathbf{u} = -(\mathbf{u} \times \mathbf{v}) = \langle -2, -5, -1 \rangle$

9. $\mathbf{u} \cdot (\mathbf{v} \times \mathbf{w}) = \begin{vmatrix} 1 & 1 & 1 \\ 0 & -1 & 1 \\ 1 & 0 & 4 \end{vmatrix}$

$\quad\quad\quad = 1(-4) - 1(-1) + 1(1)$

$\quad\quad\quad = -4 + 1 + 1 = -2$

Volume $= |\mathbf{u} \cdot (\mathbf{v} \times \mathbf{w})| = |-2| = 2$

10. $\mathbf{v} = \langle (2 - 0), -3 - (-3), 4 - 3 \rangle = \langle 2, 0, 1 \rangle$

$x = 2 + 2t, y = -3, z = 4 + t$

11. $1(x - 1) - 1(y - 2) + 0(z - 3) = 0$

$\quad\quad\quad x - 1 - y + 2 = 0$

$\quad\quad\quad\quad x - y + 1 = 0$

12. $\overrightarrow{AB} = \langle 1, 1, 1 \rangle$, $\overrightarrow{AC} = \langle 1, 2, 3 \rangle$

$\mathbf{n} = \overrightarrow{AB} \times \overrightarrow{AC} = \begin{vmatrix} \mathbf{i} & \mathbf{j} & \mathbf{k} \\ 1 & 1 & 1 \\ 1 & 2 & 3 \end{vmatrix} = \langle 1, -2, 1 \rangle$

Plane: $1(x - 0) - 2(y - 0) + (z - 0) = 0$

$\quad\quad\quad x - 2y + z = 0$

13. $\mathbf{n}_1 = \langle 1, 1, -1 \rangle$, $\mathbf{n}_2 = \langle 3, -4, -1 \rangle$

$\mathbf{n}_1 \cdot \mathbf{n}_2 = 3 - 4 + 1 = 0 \implies$ Orthogonal planes

14. $\mathbf{n} = \langle 1, 2, 1 \rangle$, $Q = (1, 1, 1)$, $P = (0, 0, 6)$ on plane.

$\overrightarrow{PQ} = \langle 1, 1, -5 \rangle$

$D = \dfrac{|\overrightarrow{PQ} \cdot \mathbf{n}|}{\|\mathbf{n}\|} = \dfrac{|1 + 2 - 5|}{\sqrt{1 + 4 + 1}} = \dfrac{2}{\sqrt{6}} = \dfrac{\sqrt{6}}{3}$

❑ Chapter 12 Practice Test Solutions

1.

x	2.9	2.99	3	3.01	3.1
$f(x)$	0.1695	0.1669	?	0.1664	0.1639

$$\lim_{x \to 3} \frac{x - 3}{x^2 - 9} \approx 0.1667$$

2. $\lim\limits_{x \to 0} \dfrac{\sqrt{x + 4} - 2}{x} \approx \dfrac{1}{4}$

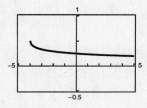

3. $\lim\limits_{x \to 2} e^{x-2} = e^{2-2} = e^0 = 1$

4. $\lim\limits_{x \to 1} \dfrac{x^3 - 1}{x - 1} = \lim\limits_{x \to 1} \dfrac{(x - 1)(x^2 + x + 1)}{x - 1}$

$\qquad = \lim\limits_{x \to 1} (x^2 + x + 1) = 3$

5. $\lim\limits_{x \to 0} \dfrac{\sin 5x}{2x} \approx 2.5$

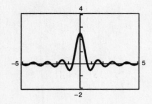

6. The limit does not exist. If $f(x) = \dfrac{|x + 2|}{x + 2}$, then $f(x) = 1$ for $x > -2$, and $f(x) = -1$ for $x < -2$.

7. $m_{\text{sec}} = \dfrac{f(4 + h) - f(4)}{h}$

$\qquad = \dfrac{\sqrt{4 + h} - 2}{h}$

$\qquad = \dfrac{\sqrt{4 + h} - 2}{h} \cdot \dfrac{\sqrt{4 + h} + 2}{\sqrt{4 + h} + 2}$

$\qquad = \dfrac{(4 + h) - 4}{h[\sqrt{4 + h} + 2]}$

$\qquad = \dfrac{h}{h[\sqrt{4 + h} + 2]}$

$\qquad = \dfrac{1}{\sqrt{4 + h} + 2}, \ h \neq 0$

$m = \lim\limits_{h \to 0} \dfrac{1}{\sqrt{4 + h} + 2} = \dfrac{1}{\sqrt{4} + 2} = \dfrac{1}{4}$

8. $f^1(x) = \lim\limits_{h \to 0} \dfrac{f(x + h) - f(x)}{h}$

$\qquad = \lim\limits_{h \to 0} \dfrac{[3(x + h) - 1] - [3x - 1]}{h}$

$\qquad = \lim\limits_{h \to 0} \dfrac{3x + 3h - 1 - 3x + 1}{h}$

$\qquad = \lim\limits_{h \to 0} \dfrac{3h}{h} = \lim\limits_{h \to 0} 3 = 3$

9. (a) $\lim\limits_{x \to \infty} \dfrac{3}{x^4} = 0$

(b) $\lim\limits_{x \to -\infty} \dfrac{x^2}{x^2 + 3} = 1$

(c) $\lim\limits_{x \to \infty} \dfrac{|x|}{1 - x} = -1$

10. $a_1 = 0, \ a_2 = \dfrac{1 - 4}{8 + 1} = -\dfrac{1}{3}, \ a_3 = \dfrac{1 - 9}{18 + 1} = -\dfrac{8}{19}$

$a_4 = \dfrac{1 - 16}{33} = -\dfrac{15}{33}$

$\lim\limits_{n \to \infty} a_n = \lim\limits_{n \to \infty} \dfrac{1 - n^2}{2n^2 + 1} = -\dfrac{1}{2}$

11. $\displaystyle\sum_{i=1}^{25} i^2 + \sum_{i=1}^{25} i = \frac{25(26)(51)}{6} + \frac{25(26)}{2} = \frac{25(26)}{6}[51 + 3] = \frac{25(26)(54)}{6} = 5850$

12. $\displaystyle\sum_{i=1}^{n} \frac{i^2}{n^3} = \frac{1}{n^3} \sum_{i=1}^{n} i^2 = \frac{1}{n^3}\left[\frac{n(n+1)(2n+1)}{6}\right] = \frac{2n^2 + 3n + 1}{6n^2} = s(n)$

$\displaystyle\lim_{n\to\infty} s(n) = \frac{1}{3}$

13. Width of rectangles: $\dfrac{b-a}{n} = \dfrac{1}{n}$

Height: $f\left(a + \dfrac{(b-a)i}{n}\right) = f\left(\dfrac{i}{n}\right) = 1 - \left(\dfrac{i}{n}\right)^2$

$a_n \approx \displaystyle\sum_{i=1}^{n}\left[1 - \frac{i^2}{n^2}\right]\frac{1}{n} = \sum_{i=1}^{n}\frac{1}{n} - \sum_{i=1}^{n}\frac{i^2}{n^3} = 1 - \frac{1}{n^2}\frac{n(n+1)(2n+1)}{6}$

$A = \displaystyle\lim_{n\to\infty} A_n = 1 - \frac{1}{3} = \frac{2}{3}$